Historic Documents of 2007

Historic Documents of 2007

Includes Cumulative Index, 2003–2007

CQ Press
A Division of SAGE
Washington, D.C.

CQ Press
2300 N Street, NW, Suite 800
Washington, DC 20037

Phone: 202-729-1900; toll-free, 1-866-4CQ-PRESS (1-866-427-7737)

Web: www.cqpress.com

Cover design: McGaughy Design, Centerville, Virginia
Cover photos: AP Images
Composition: Judy Myers, Graphic Design, Alpine, California
Indexer: Julia Petrakis

♾ The paper used in this publication exceeds the requirements of the American National Standard for Information Sciences—Permanence of Paper for Printed Library Materials, ANSI Z39.48-1992.

Printed and bound in the United States of America

12 11 10 09 08 1 2 3 4 5

The Library of Congress cataloged the first issue of this title as follows:
Historic documents. 1972–
Washington. Congressional Quarterly Inc.

1. United States—Politics and government—1945– —Yearbooks.
2. World politics—1945– —Yearbooks. I. Congressional Quarterly Inc.
E839.5H57 917.3'03'9205 72-97888

ISSN 0892-080X
ISBN 978-0-87289-907-0

Contents

Thematic Table of Contents xvii

List of Document Sources xxi

Preface xxvii

How to Use This Book xxviii

Overview of 2007 xxix

JANUARY

President Bush on the Death of Gerald R. Ford 3
The eulogy for former president Gerald R. Ford, delivered by President George W. Bush on January 2, 2007, at state funeral services held at the National Cathedral in Washington, D.C.

President Bush on the "Surge" of U.S. Troops to Iraq 9
A nationally televised speech made by President George W. Bush on January 10, 2007, describing his order of some 20,000 additional U.S. troops to Iraq.

State of the Union Address and Democratic Response 21
The State of the Union address delivered by President George W. Bush to a joint session of Congress on January 23, 2007, and the Democratic response, delivered shortly afterward, by Sen. Jim Webb of Virginia.

FEBRUARY

Palestinian Peace Efforts and the Hamas Takeover of Gaza 39
First, the text of the Mecca Agreement, issued on February 8, 2007, calling for an end to Fatah and Hamas factional conflict and the establishment of a unity government. Second, an unofficial translation of Palestinian president Mahmoud Abbas's speech to the PLO Central Council Meeting in Ramallah on June 20, 2007, addressing the Fatah takeover of Gaza. Third, the Israeli Security Cabinet's declaration designating Gaza as hostile territory and describing Hamas as a "terrorist organization," issued on September 19, 2007, by Prime Minister Ehud Olmert and the Israel Ministry of Foreign Affairs.

FEDERAL RESPONSE TO IMPORTED FOOD SAFETY CONCERNS 51
First, testimony by David M. Walker, comptroller general of the United States, before the House Subcommittee on Agriculture, Rural Development, FDA, and Related Agencies on February 8, 2007, in which he described the federal government's oversight of food safety as being at high risk of failure. Second, a press release from the U.S. Department of Health and Human Services, dated December 11, 2007, announcing a memorandum of agreement between the United States and China intended to improve the quality and safety of certain Chinese food exports to the United States.

RUSSIAN PRESIDENT PUTIN ON WORLD AFFAIRS AND RUSSIAN POLITICS 62
First, a speech made by Russian president Vladimir Putin on February 10, 2007, to the Munich Conference on Security Policy in Munich, Germany, in which he discussed world politics and many aspects of U.S. foreign policy. Second, Putin's concluding remarks to the annual congress of the United Russia Party on October 1, 2007.

INTERNATIONAL EFFORTS TO DENUCLEARIZE NORTH KOREA 77
First, "Initial Actions for the Implementation of the Joint Statement," issued on February 13, 2007, at the conclusion of a round of Six-Party Talks in Beijing, China, among representatives from China, Japan, North Korea, Russia, South Korea, and the United States. Second, "Second-Phase Actions for the Implementation of the September 2005 Joint Statement," issued by the Chinese foreign minister on October 3, 2007, following another round of Six-Party Talks.

MARCH

ACTING SURGEON GENERAL MORITSUGU ON THE DANGERS OF UNDERAGE DRINKING 91
"The Surgeon General's Call to Action to Prevent and Reduce Underage Drinking," released on March 6, 2007, by acting surgeon general Kenneth P. Moritsugu.

FEDERAL APPEALS COURT ON D.C. HANDGUN BAN 101
The majority and minority opinions in the case of *Parker v. District of Columbia,* in which a three-judge panel of the U.S. Court of Appeals for the District of Columbia Circuit ruled on March 9, 2007, that the D.C. handgun ban was an unconstitutional infringement on the Second Amendment.

UN RESOLUTION AND U.S. INTELLIGENCE ESTIMATE ON IRAN'S NUCLEAR THREAT 112
First, United Nations Security Council Resolution 1747, adopted on March 24, 2007, demanding Iranian compliance with UN requirements to provide information about its nuclear programs and imposing sanctions against Iran if it did not comply within sixty days. Second,

the unclassified version of "Iran: Nuclear Intentions and Capabilities," a National Intelligence Estimate drafted in November 2007 by the National Intelligence Council and made public on December 3, 2007.

World Health Organization on Male Circumcision and HIV 130
A World Health Organization press release on March 28, 2007, officially recommending the use of male circumcision in some situations to reduce the risk of HIV infection.

APRIL

Supreme Court on EPA Regulation of Greenhouse Gases 139
The majority opinion, written by Justice John Paul Stevens, in the case of *Massachusetts v. Environmental Protection Agency,* in which the Supreme Court ruled on April 2, 2007, that the Environmental Protection Agency has the authority under the Clean Air Act of 1970 to regulate greenhouse gases.

Catholic Bishops on Human Rights Abuses in Zimbabwe 149
"God Hears the Cry of the Oppressed," a pastoral letter by the Zimbabwe Catholic Bishops' Conference posted on the Internet on March 30, 2007, distributed in Zimbabwe's Catholic Churches during the week of Easter, and officially dated April 5, 2007. The letter was signed by Robert C. Ndlovu, archbishop of Harare; Pius Alec M. Ncube, archbishop of Bulawayo; Alexio Churu Muchabaiwa, bishop of Mutare; Michael D. Bhasera, bishop of Masvingo; Angel Floro, bishop of Gokwe; Martin Munyanyi, bishop of Gweru; Dieter B. Scholz SJ, bishop of Chinhoyi; Albert Serrano, bishop of Hwange; and Patrick M. Mutume, auxiliary bishop of Mutare.

President Bush on Border Security and Immigration Reform 158
President George W. Bush's remarks on April 9, 2007, on border security and immigration reform.

Remarks and Findings on the Shootings at Virginia Tech 167
First, President George W. Bush's remarks on April 17, 2007, at the Virginia Tech Memorial Convocation for the victims of a mass shooting on the campus the day before. Second, the "Summary of Key Findings" issued on August 30, 2007, by the Virginia Tech Review Panel, an independent commission appointed by Virginia governor Tim Kaine to investigate the shootings.

Supreme Court on the "Partial-Birth" Abortion Ban 176
The majority and minority opinions in the cases of *Gonzales v. Carhart* and *Gonzales v. Planned Parenthood Federation of America,* in which the Supreme Court on April 18, 2007, upheld the constitutionality of a federal act barring a medical procedure opponents call "partial-birth" abortion.

Turkey's General Staff and the European Union on Turkish Politics 190
First, a statement posted on April 27, 2007, on the Web site of the Turkish General Staff, discussing the country's political situation. Second, "Turkey 2007 Progress Report," adopted by the European Commission on November 6, 2007.

Report on Israel's Handling of Its 2006 War with Hezbollah 200
The Interim Report of the Winograd Commission, which examined Israel's handling of its July–August 2006 war with the Hezbollah militia in Lebanon. This unclassified summary of the report was made public on April 30, 2007.

MAY

Leaders of Northern Ireland on Their New Government 213
Speeches delivered on May 8, 2007, by Rev. Ian Paisley and Martin McGuinness, following their installation as first minister and deputy first minister, respectively, of the provincial government of Northern Ireland.

Outgoing and Incoming Prime Ministers on British Politics 221
First, a speech by British prime minister Tony Blair, on May 10, 2007, announcing his intention to resign his office. Second, a statement by new prime minister Gordon Brown on June 27, 2007.

Chrysler Executives on Takeover, Export Deal, and Job Reductions 232
Three press releases issued by Chrysler announcing the purchase of Chrysler by Cerberus Capital Management (May 14, 2007), the cooperative agreement between Chrysler and China's Chery Automobile Co. (July 3, 2007), and a new round of job reductions and shift eliminations (November 1, 2007).

Sarkozy on His Inauguration as President of France 240
A speech delivered by Nicolas Sarkozy in Paris on May 16, 2007, following his inauguration as president of France.

World Bank Statements on the Resignation of President Wolfowitz 247
Two statements released by the World Bank on May 17, 2007, regarding the resignation of Paul Wolfowitz as president of the bank. First, a statement by the bank's board of executive directors. Second, a statement by Wolfowitz.

Ambassador Crocker on U.S.-Iranian Diplomacy 257
A statement by Ryan C. Crocker, U.S. ambassador to Iraq, on May 28, 2007, following negotiations among U.S., Iranian, and Iraqi diplomats in Baghdad.

Yar'Adua on His Inauguration as President of Nigeria 266
The address by Umaru Yar'Adua upon his inauguration as president of Nigeria on May 29, 2007.

JUNE

Council of Europe Investigation into the CIA's "Secret Prisons" in Europe 275
Selections from "Secret Detentions and Illegal Transfers of Detainees Involving Council of Europe Member States: Second Report," a report submitted on June 7, 2007, to the Committee on Legal Affairs and Human Rights of the Parliamentary Assembly of the Council of Europe, by Dick Marty, the committee's special rapporteur.

Chinese President, Rights Activists on Preparation for the Olympic Games 284
First, a June 25, 2007, speech by Chinese president Hu Jintao to leaders of the Communist Party assembled at the Central Party School in Beijing. Second, "One World, One Dream and Universal Human Rights" an open letter addressed to "Chinese and World Leaders," posted on the Internet on August 8, 2007, and signed by thirty-seven Chinese human rights activists, intellectuals, lawyers, and others.

Supreme Court on Regulation of Issue Ads 296
The majority opinion, written by Chief Justice John G. Roberts, in the case of *Federal Election Commission v. Wisconsin Right to Life, Inc.*, in which the Court ruled on June 25, 2007, that provisions of the Bipartisan Campaign Reform Act of 2002 restricting issue ads were unconstitutional as applied to ads produced by the Wisconsin organization.

Supreme Court on Voluntary School Desegregation 306
The majority and minority opinions in the case of *Parents Involved in Community Schools v. Seattle School District No.1,* in which the Supreme Court on June 28, 2007, ruled to restrict the use of race-based plans for assigning students to elementary and secondary schools to achieve racial diversity.

JULY

President Bush on Clemency for Former Aide in the CIA Leak Case 329
A statement issued by President George W. Bush on July 2, 2007, commuting the prison sentence given to I. Lewis "Scooter" Libby for perjury and obstruction of justice in connection with a CIA leak case.

U.S. National Intelligence Estimate on Terrorist Threats 335
The unclassified summary of "The Terrorist Threat to the U.S. Homeland," a National Intelligence Estimate released by the director of national intelligence on July 17, 2007.

United Nations on the Political Situation and Violence in Somalia 343
First, a report, dated July 18, 2007, to the United Nations Security Council from the monitoring group established in 1992 to oversee an international arms embargo against all factions in Somalia. Second, a quarterly report to the Security Council by Secretary-General Ban Ki-moon on the situation in Somalia.

Executive Order and Statement on CIA Interrogation Techniques 354
First, the executive order signed by President George W. Bush on July 20, 2007, laying out legal standards for the detention and interrogation of suspected terrorists by the Central Intelligence Agency. Second, a December 6, 2007, statement by Gen. Mike Hayden, director of the Central Intelligence Agency, to CIA employees, explaining that the agency had videotaped the interrogations of some terrorism suspects in 2002 and 2003 but later destroyed the tapes.

President's Commission on Caring for America's Wounded Soldiers 364
The final report of the President's Commission on Care for America's Returning Wounded Warriors, released on July 25, 2007.

United Nations on the Violence in the Democratic Republic of the Congo 374
First, a statement given at a news conference in Kinshasa on July 27, 2007, by Yakin Ertürk, special rapporteur on violence against women for the United Nations Human Rights Council, detailing her observations during a July 16–27 visit to the Democratic Republic of the Congo. Second, a November 14, 2007, report by Secretary-General Ban Ki-moon to the United Nations Security Council on the situation in the Congo.

United Nations and Human Rights Coalition on the Violence in Darfur 385
First, Resolution 1769, adopted by the United Nations Security Council on July 31, 2007, authorizing deployment of a 26,000-person peacekeeping force in the Darfur region of Sudan. Second, "UNAMID Deployment on the Brink: The Road to Security in Darfur Blocked by Government Obstructions," a report released on December 19, 2007, by a coalition of thirty-five international human rights advocacy groups led by Human Rights Watch.

AUGUST

U.S. Intelligence Agencies and General Petraeus on Security in Iraq 407
First, "Prospects for Iraq's Stability: Some Security Progress but Political Reconciliation Elusive," a National Intelligence Estimate issued by the National Intelligence Council on August 2, 2007. Second, "Report to Congress on the Situation in Iraq," testimony by Gen. David H. Petraeus, commander of U.S. forces in Iraq, on September 10–11, 2007.

President Bush on the Passage of Electronic Surveillance Legislation 429
Remarks by President George W. Bush on August 5, 2007, as he signed legislation that temporarily expanded executive branch authority to conduct warrantless surveillance of foreign targets.

UN Secretary-General on the Postwar Situation in Liberia 435
"Fifteenth Progress Report of the Secretary-General on the United Nations Mission to Liberia," submitted by Secretary-General Ban Ki-moon to the Security Council on August 8, 2007.

Afghan and Pakistani Presidents Sign a "Peace Jirga" Communiqué 442
The "declaration" signed by Afghan president Hamid Karzai and Pakistani president Pervez Musharraf at the end of a "peace jirga" (assembly) of some 700 representatives from Afghanistan and Pakistan, held in Kabul on August 9–12, 2007.

Federal Reserve Board on the State of the U.S. Economy 449
Four statements from the Federal Reserve's Federal Open Market Committee, on August 17, September 18, October 31, and December 11, 2007, announcing its decisions on key interest rate levels.

President Bush and Attorney General Gonzales on His Resignation 457
Statements dated August 27, 2007, from Alberto R. Gonzales, announcing his resignation as U.S. attorney general, and from President George W. Bush, accepting the resignation.

Senator Craig on His Arrest for Solicitation and Possible Resignation 466
Three press releases issued on August 28, September 1, and October 4, 2007, by the office of Sen. Larry Craig, R-Idaho, in connection with his arrest for alleged solicitation of sex in a men's restroom at the Minneapolis-St. Paul Airport.

SEPTEMBER

Government Accountability Office on Iraqi Actions to Meet "Benchmarks" 477
"Securing, Stabilizing, and Rebuilding Iraq: Iraqi Government Has Not Met Most Legislative, Security, and Economic Benchmarks," a report presented to congressional committees on September 4, 2007, by the U.S. Government Accountability Office.

Congressional Testimony on the Minneapolis Bridge Collapse 489
Statements from Transportation Secretary Mary E. Peters and the department's inspector general, Calvin E. Scovel III, delivered September 5, 2007, to the House Committee on Transportation and Infrastructure concerning the federal government's response to the August 1 collapse of the I-35W bridge in Minneapolis, Minnesota.

Japanese Prime Ministers Abe and Fukuda on Government Changes 504
First, a statement by Japanese prime minister Shinzo Abe at a September 12, 2007, news conference during which he announced his decision to resign. Second, a speech by the new prime minister, Yasuo Fukuda, to the Japanese Diet (parliament) on October 1, 2007, announcing his government's legislative program.

CBO Director and President Bush on the Housing Crisis 514
First, testimony on the housing crisis, given by Peter R. Orszag, director of the Congressional Budget Office, on September 19, 2007, before the Joint Economic Committee. Second, remarks on December 6, 2007, by President George W. Bush, outlining a voluntary industry plan for helping certain homeowners avoid foreclosure.

Reports on Violence and Unrest in Myanmar 528
First, a statement made by Buddhist monks during a demonstration in Yangon on September 19, 2007, calling on all monks to boycott the government. Second, a Memorandum on the Situation of Human Rights in the Union of Myanmar, sent to the secretary-general of the United Nations on November 7, 2007, by Kyaw Tint Swe, the ambassador to the UN from Myanmar. Third, a report to the UN Human Rights Council, dated December 7, 2007, from Paulo Sérgio Pinheiro, the UN special rapporteur on human rights in Myanmar.

UN Secretary-General on the Situation in Afghanistan 547
"The Situation in Afghanistan and Its Implications for International Peace and Security," a report released to the United Nations Security Council on September 21, 2007, by Secretary-General Ban Ki-moon.

FBI Report on Crime in the United States 560
The Federal Bureau of Investigation's annual report, "Crime in the United States, 2006," released on September 24, 2007.

Supreme Court on Constitutionality of Lethal Injections 569
The petition for a writ of certiorari (review) filed with the Supreme Court in the case of *Baze v. Rees,* asking the Court to "provide guidance to the lower counts on the applicable legal standard for method of execution cases." The Court on September 25, 2007, agreed to hear arguments on the first three questions set out in the petition.

Prime Minister Surayud on a "Sustainable" Democracy in Thailand 575
A September 26, 2007, address by retired general Surayud Chulanont, then prime minister of Thailand, to a meeting of the Asia Society in New York City.

OCTOBER

PRESIDENT BUSH ON HIS VETOES OF CHILD HEALTH INSURANCE LEGISLATION 585
Two messages from President George W. Bush, dated October 3 and December 12, 2007, vetoing legislation that would have significantly expanded funding for the State Children's Health Insurance Program (SCHIP).

WORLD BANK ON ITS AGRICULTURE PROGRAMS IN AFRICA 592
"World Bank Assistance to Agriculture in Sub-Saharan Africa: An IEG Review," a report prepared by the Independent Evaluation Group of the World Bank and made public on October 12, 2007.

INTERNATIONAL MONETARY FUND ON THE WORLD ECONOMIC OUTLOOK 603
"World Economic Outlook," a report published by the International Monetary Fund on October 17, 2007.

PRESIDENT BUSH AND CUBAN FOREIGN MINISTER ON U.S. POLICY TOWARD CUBA 613
First, President George W. Bush's speech at the State Department on October 24, 2007, restating U.S. policy toward Cuba. Second, Foreign Minister Felipe Pérez Roque's speech to the United Nations General Assembly on October 30, 2007, during the debate on a resolution (subsequently adopted) calling for an end to the U.S. economic embargo against Cuba.

UNITED NATIONS ON THE POLITICAL SITUATION IN LEBANON 624
First, a report to the UN Security Council by Secretary-General Ban Ki-moon, dated October 24, 2007, on the implementation of Security Council Resolution 1559 (adopted in 2004), which required the withdrawal of all foreign military forces from Lebanon. Second, a statement by the Security Council on December 11, 2007, calling for political reconciliation in Lebanon.

NOVEMBER

GOVERNMENT ACCOUNTABILITY OFFICE ON THE SAFETY OF IMPORTED DRUGS 635
Testimony by Marcia Crosse, director of health care for the Government Accountability Office, to the House Energy and Commerce Subcommittee on Oversight and Inspections on November 1, 2007.

PAKISTANI LEADERS ON THE POLITICAL SITUATION IN PAKISTAN 647
First, the "Proclamation of Emergency," issued on November 3, 2007, by Pakistani president Pervez Musharraf, acting in his capacity as chief of staff of the army. Second, the November 10, 2007, address by Benazir Bhutto, leader of the Pakistan People's Party, to foreign diplomats at a meeting in Islamabad.

Intergovernmental Panel of Scientists on Climate Change 657
"Summary for Policymakers," a report prepared by the Intergovernmental Panel on Climate Change and released on November 17, 2007.

UN Refugee Agency on the Status of Iraqi Refugees 674
A statement attributed to United Nations High Commissioner for Refugees spokesperson Jennifer Pagonis at a November 23, 2007, news briefing in Geneva, Switzerland, responding to reports that Iraqi refugees were returning home.

President Bush on Middle East Peace Talks 680
Remarks by President George W. Bush at the Annapolis Conference on the Middle East, on November 27, 2007, in which he announced a joint understanding between Israeli prime minister Ehud Olmert and Palestinian president Mahmoud Abbas on the convening of negotiations toward a peace treaty.

DECEMBER

Gore on Accepting the Nobel Peace Prize for Work on Climate Change 691
The address delivered by former U.S. vice president Al Gore as he accepted the 2007 Nobel Peace Prize in Oslo, Norway, on December 10, 2007.

International "Troika" on Independence Options for Kosovo 699
Report of the diplomatic group known as the "Troika" named by the European Union, Russia, and the United States to negotiate an agreement between the Republic of Serbia and the government of Kosovo. Members of the Troika were Wolfgang Ischinger, the German ambassador to Britain; Aleksandr Botsan-Kharchenko, head of the Russian Foreign Ministry's Balkans department; and Frank Wisner, a retired U.S. ambassador. The report, which is dated December 4, 2007, was submitted on December 7 to UN secretary-general Ban Ki-Moon and published officially on December 10, 2007.

New York Attorney General on the Student Loan Settlement 709
A press release issued on December 11, 2007, by New York attorney general Andrew M. Cuomo, announcing a settlement with a student loan company over deceptive advertising practices.

Mitchell Report on Investigation into Steroid Use in Baseball 715
"Report to the Commissioner of Baseball of an Independent Investigation into the Illegal Use of Steroids and Other Performance-Enhancing Substances by Players in Major League Baseball," made public on December 13, 2007, by former U.S. senator George J. Mitchell.

European Union Leaders on the Treaty of Lisbon 721
"Treaty at a Glance," a document produced by the European Union as a summary of the Treaty of Lisbon, which was signed by EU leaders in Lisbon on December 13, 2007.

U.S. Comptroller General on the Nation's Long-Term Deficit 728
A speech delivered by Comptroller General David M. Walker at the National Press Club in Washington, D.C., on December 17, 2007, discussing the government's long-term deficit problems and outlining a plan to help bring the deficits under control.

Kenyan President Kibaki on His Reelection to a Second Term 738
A speech delivered by Mwai Kibaki on December 30, 2007, shortly after he had been sworn in for a second term as president of Kenya, following disputed elections.

Credits 745

Index 747

Thematic Table of Contents

HEALTH AND SOCIAL SERVICES

INTERNATIONAL AFFAIRS
Africa

INTERNATIONAL AFFAIRS
Asia

INTERNATIONAL AFFAIRS
Europe

INTERNATIONAL AFFAIRS
Latin America and the Caribbean

INTERNATIONAL AFFAIRS
Middle East

INTERNATIONAL AFFAIRS
Russia and Former Soviet Republics

INTERNATIONAL AFFAIRS

Global Issues

NATIONAL SECURITY AND TERRORISM

RIGHTS, RESPONSIBILITIES, AND JUSTICE

List of Document Sources

CONGRESS

U.S. Congress. Congressional Budget Office. "Turbulence in Mortgage Markets: Implications for the Economy and Policy Options." September 19, 2007. 518

U.S. Congress. Government Accountability Office. "A Call for Stewardship." December 17, 2007. 731

U.S. Congress. Government Accountability Office. "Drug Safety: Preliminary Findings Suggest Weaknesses in FDA's Program for Inspecting Foreign Drug Manufacturers." November 1, 2007. 639

U.S. Congress. Government Accountability Office. "Federal Oversight of Food Safety: High-Risk Designation Can Bring Needed Attention to Fragmented System." February 8, 2007. 55

U.S. Congress. Government Accountability Office. "Securing, Stabilizing, and Rebuilding Iraq: Iraqi Government Has Not Met Most Legislative, Security, and Economic Benchmarks." September 4, 2007. 482

U.S. Senate. Office of Senator Larry Craig. "Craig Reaction to Court Ruling." October 4, 2007. 472

U.S. Senate. Office of Senator Larry Craig. "Senator Craig Announces Intent to Resign from the Senate." September 1, 2007. 471

U.S. Senate. Office of Senator Larry Craig. "Statement of Senator Craig." August 28, 2007. 470

EXECUTIVE DEPARTMENTS AND AGENCIES

Central Intelligence Agency. "Statement to Employees by Director of the Central Intelligence Agency, General Mike Hayden on the Taping of Early Detainee Interrogations." December 6, 2007. 362

U.S. Department of Defense. "Report to Congress on the Situation in Iraq." September 10, 2007. 421

U.S. Department of Health and Human Services. "New Agreement Will Enhance the Safety of Food and Feed Imported from the People's Republic of China." December 11, 2007. 60

U.S. Department of Health and Human Services. Office of the Surgeon General. "The Surgeon General's Call to Action to Prevent and Reduce Underage Drinking." March 6, 2007. 94

U.S. Department of Justice. Federal Bureau of Investigation. *Crime in the United States, 2006.* September 24, 2007. 563

U.S. Department of Justice. "Remarks of Attorney General Alberto R. Gonzales Announcing His Resignation." August 27, 2007. 463

U.S. Department of State. Embassy of the United States, Baghdad, Iraq. "Remarks by Ambassador Ryan C. Crocker at the Press Availability after Meeting with Iranian Officials." May 28, 2007. 262

U.S. Department of State. Under Secretary for Public Diplomacy and Public Affairs. Bureau of Public Affairs: Press Office. Office of the Spokesman. "North Korea—Denuclearization Action Plan." February 13, 2007. 83

U.S. Department of State. Under Secretary for Public Diplomacy and Public Affairs. Bureau of Public Affairs: Press Office. Office of the Spokesman. "Six-Party Talks—Second-Phase Actions for the Implementation of the September 2005 Joint Statement." October 3, 2007. 85

U.S. Department of Transportation. Office of Inspector General. "Federal Highway Administration's Oversight of Structurally Deficient Bridges." September 5, 2007. 497

U.S. Department of Transportation. "Statement of the Hon. Mary E. Peters, Sec. of Transportation, before the Committee on Transportation and Infrastructure, U.S. House of Representatives." September 5, 2007. 493

U.S. Director of National Intelligence. National Intelligence Council. "Iran: Nuclear Intentions and Capabilities." December 3, 2007. 125

U.S. Director of National Intelligence. National Intelligence Council. "Prospects for Iraq's Stability: Some Security Progress but Political Reconciliation Elusive." August 2, 2007. 417

U.S. Director of National Intelligence. National Intelligence Council. "The Terrorist Threat to the US Homeland." July 17, 2007. 340

U.S. Federal Reserve System. "Federal Open Market Committee Statement." August 17, 2007. 453

U.S. Federal Reserve System. "Federal Open Market Committee Statement and Board Approval of Discount Rate Requests of the Federal Reserve Banks of Boston, New York, Cleveland, St. Louis, Minneapolis, Kansas City, and San Francisco." September 18, 2007. 453

U.S. Federal Reserve System. "Federal Open Market Committee Statement and Board Approval of Discount Rate Requests of the Federal Reserve Banks of New York, Richmond, Atlanta, Chicago, St. Louis, and San Francisco." October 31, 2007. 454

U.S. Federal Reserve System. "Federal Open Market Committee Statement and Board Approval of Discount Rate Requests of the Federal Reserve Banks of New York, Philadelphia, Cleveland, Richmond, Atlanta, Chicago, and St. Louis." December 11, 2007. 455

INTERNATIONAL NONGOVERNMENTAL ORGANIZATIONS

Asian Human Rights Commission. "Burma: The Alms Bowl and the Duty to Defy." September 19, 2007. 533

China Elections & Governance. "Hu Jintao's Speech at Party School, full text." June 25, 2007. 288

Council of Europe. Parliamentary Assembly. Committee on Legal Affairs and Human Rights. "Secret Detentions and Illegal Transfers of Detainees Involving Council of Europe Member States: Second Report." June 7, 2007. 279

European Union. Commission of the European Communities. "Turkey 2007 Progress Report." November 6, 2007. 196

European Union. "The Treaty at a Glance." December 13, 2007. 724

Human Rights Watch. "UNAMID Deployment on the Brink: The Road to Security in Darfur Blocked by Government Obstructions." December 19, 2007. 396

International Monetary Fund. "World Economic and Financial Surveys, World Economic Outlook: Globalization and Inequality." October 17, 2007. 607

Jerusalem Media & Communications Centre. "Speech of President Abbas before the PLO Central Council Meeting, Ramallah." June 20, 2007. 47

Jerusalem Media & Communications Centre. "The Text of the Mecca Agreement." February 8, 2007. 46

The Nobel Foundation. "Nobel Lecture by Al Gore." Oslo, Norway. December 10, 2007. 694

Pakistan Peoples Party. "Mohtarma Benazir Bhutto's Address to Diplomats at PPP Foreign Liaison Committee Reception." November 10, 2007. 653

United Kingdom. Labour Party. "Tony Blair's Statement on 10th May." May 10, 2007. 226

United Nations. General Assembly. "Annex to the Letter Dated 5 November 2007 from the Permanent Representative of Myanmar to the United Nations Addressed to the Secretary-General. Memorandum on the Situation of Human Rights in the Union of Myanmar." November 7, 2007. 534

United Nations. General Assembly. Human Rights Council. "Report of the Special Rapporteur on the Situation of Human Rights in Myanmar, Paulo Sérgio Pinheiro." December 7, 2007. 538

United Nations. General Assembly and Security Council. "The Situation in Afghanistan and Its Implications for International Peace and Security: Report of the Secretary-General." September 21, 2007. 554

United Nations. MONUC: UN Mission in DR Congo. "South Kivu: 4,500 Sexual Violence Cases in the First Six Months of This Year Alone." July 27, 2007. 378

United Nations. Office of the High Commissioner for Refugees. "Iraq: UNHCR Cautious About Returns." November 23, 2007. 677

United Nations. Security Council. "Fifteenth Progress Report of the Secretary-General on the United Nations Mission in Liberia." August 8, 2007. 438

United Nations. Security Council. "Letter Dated 10 December 2007 from the Secretary-General to the President of the Security Council. Enclosure: Report of the European Union/United States/Russian Federation Troika on Kosovo." December 10, 2007. 703

United Nations. Security Council. "Report of the Monitoring Group on Somalia Pursuant to Security Council Resolution 1724 (2006)." July 18, 2007. 348

United Nations. Security Council. "Report of the Secretary-General on the Situation in Somalia." November 7, 2007. 350

United Nations. Security Council. "Resolution 1747 (2007)" [concerning Iran's nuclear programs]. March 24, 2007. 118

United Nations. Security Council. "Resolution 1769 (2007)" [concerning the situation in Darfur]. July 31, 2007. 391

United Nations. Security Council. "Sixth Semi-Annual Report of the Secretary-General on the Implementation of Security Council Resolution 1559 (2004)" [concerning Lebanon]. October 24, 2007. 628

United Nations. Security Council. "Statement by the President of the Security Council" [on the political situation in Lebanon]. December 11, 2007. 631

United Nations. Security Council. "Twenty-Fourth Report of the Secretary-General on the United Nations Organization Mission in the Democratic Republic of the Congo." November 14, 2007. 381

United Nations. World Health Organization. "WHO and UNAIDS Announce Recommendations from Expert Consultation on Male Circumcision for HIV Prevention." March 28, 2007. 133

United Nations. World Meteorological Organization and United Nations Environment Program. Intergovernmental Panel on Climate Change. "Climate Change 2007: Synthesis Report. Summary for Policymakers." November 17, 2007. 661

World Bank. Independent Evaluation Group. "World Bank Assistance to Agriculture in Sub-Saharan Africa." October 12, 2007. 595

World Bank. "Statements of Executive Directors and President Wolfowitz." May 17, 2007. 252

Zimbabwe Catholic Bishops' Conference. "God Hears the Cry of the Oppressed." Released March 30, 2007. 154

JUDICIARY

State of New York. Office of the Attorney General. "Cuomo Announces Settlement with Student Loan Company Tied to NCAA Division I Schools: Lender to End Kickbacks and Co-Branding Agreements." December 11, 2007. 712

Supreme Court of the United States. *Baze v. Rees* [order granting petition for a writ of certiorari]. September 25, 2007. 572

Supreme Court of the United States. *Federal Election Commission v. Wisconsin Right to Life, Inc.* June 25, 2007. 300

Supreme Court of the United States. *Gonzales v. Carhart.* April 18, 2007. 180

Supreme Court of the United States. *Massachusetts vs. EPA.* April 2, 2007. 144

Supreme Court of the United States. *Parents Involved in Community Schools v. Seattle School District No. 1.* June 28, 2007. 311

U.S. Court of Appeals for the District of Columbia Circuit. *Shelly Parker v. District of Columbia.* March 9, 2007. 104

NON-U.S. GOVERNMENTS

Federal Republic of Nigeria. Federal Ministry of Information and Communications. "Inaugural Address of Umaru Musa Yar'Adua, President of the Federal Republic of Nigeria and Commander-in-Chief of the Armed Forces." May 29, 2007. 270

French Republic. Embassy of France in the United States. "Speech by Nicolas Sarkozy, President of the Republic, at the Investiture Ceremony." May 16, 2007. 244

High Commission for Pakistan in United Kingdom. "Pakistan and Afghanistan Hold the First Ever Joint Peace Jirga." August 12, 2007. 447

Islamic Republic of Pakistan. Associated Press of Pakistan. "Text of 'Proclamation of Emergency.'" November 3, 2007. 652

Japan. Prime Minister of Japan and His Cabinet. "Press Conference by Prime Minister Shinzo Abe." September 12, 2007. 508
Japan. Prime Minister of Japan and His Cabinet. "Policy Speech by Prime Minister Yasuo Fukuda to the 168th Session of the Diet." October 1, 2007. 509
Kingdom of Thailand. Royal Thai Government. "'Thailand: Moving Toward a More Sustainable and Democratic Future.' Keynote Address by His Excellency General Surayud Chulanont (Ret.) Prime Minister of the Kingdom of Thailand at the Asia Society." September 26, 2007. 578
Republic of Cuba. Permanent Mission to the United Nations. "Statement by Felipe Pérez Roque, Minister of Foreign Affairs of the Republic of Cuba, Under Agenda Item 'Necessity of Ending the Economic, Commercial and Financial Embargo Imposed by the United States of America Against Cuba.'" October 30, 2007. 620
Republic of Kenya. State House. "Acceptance Speech by His Excellency Hon. Mwai Kibaki, C.G.H., M.P., President and Commander-in-Chief of the Armed Forces of the Republic of Kenya Following His Re-Election to Serve a Second Term." December 30, 2007. 742
Republic of Turkey. Office of the Chief of the General Staff. "On Reactionary Activities, Army's Duty." April 27, 2007. 194
Russian Federation. President of Russia. "Concluding Remarks at the VIIIth United Russia Party Congress." October 1, 2007. 75
Russian Federation. President of Russia. "Speech and the Following Discussion at the Munich Conference on Security Policy." February 10, 2007. 69
State of Israel. Ministry of Foreign Affairs. "Security Cabinet Declares Gaza Hostile Territory." September 19, 2007. 50
State of Israel. Ministry of Foreign Affairs. "Winograd Commission Submits Interim Report." April 30, 2007. 204
United Kingdom. Northern Ireland Executive. Office of the First Minister and Deputy First Minister. "Speech Delivered by First Minister Reverend Ian Paisley Today in Parliament Buildings, Belfast." May 8, 2007. 217
United Kingdom. Northern Ireland Executive. Office of the First Minister and Deputy First Minister. "Speech Delivered by Deputy First Minister Martin McGuinness Today in Parliament Buildings, Belfast." May 8, 2007. 219
United Kingdom. 10 Downing Street. "Statement at Downing Street." June 27, 2007. 230

U.S. NONGOVERNMENTAL ORGANIZATIONS

Chrysler LLC. "Chery and Chrysler Group Finalize Cooperative Agreement." July 3, 2007. 236
Chrysler LLC. "Chrysler Announces Product and Plant Changes." November 1, 2007. 238
Chrysler LLC. "DaimlerChrysler Definitive Agreement to Sell Chrysler Group (including Chrysler Financial Corporation) to Cerberus Capital Management, L.P." May 14, 2007. 236
The Democratic Party. "The Democratic Response to the State of the Union." January 23, 2007. 33
Human Rights Watch. "Letter by Liu Xiaobo, Ding Zilin, Bao Tong and Other Activists One Year Before the Beijing Games." August 8, 2007. 292

Major League Baseball. "Report to the Commissioner of Baseball of an Independent Investigation into the Illegal Use of Steroids and Other Performance Enhancing Substances by Players in Major League Baseball." December 13, 2007. 719

U.S. STATE AND LOCAL GOVERNMENTS

Commonwealth of Virginia. Official Site of the Governor of Virginia. "Report of the Virginia Tech Review Panel." August 30, 2007. 172

WHITE HOUSE AND THE PRESIDENT

U.S. Executive Office of the President. "Address Before a Joint Session of the Congress on the State of the Union." January 23, 2007. 25

U.S. Executive Office of the President. "Address to the Nation on the War on Terror in Iraq." January 10, 2007. 15

U.S. Executive Office of the President. "Eulogy at the National Funeral Service for Former President Gerald R. Ford." January 2, 2007. 7

U.S. Executive Office of the President. "Executive Order 13440—Interpretation of the Geneva Conventions Common Article 3 as Applied to a Program of Detention and Interrogation Operated by the Central Intelligence Agency." July 20, 2007. 359

U.S. Executive Office of the President. "Message to the House of Representatives on the President's Second Veto of Child Health Insurance Legislation." December 12, 2007. 590

U.S. Executive Office of the President. "Message to the House of Representatives Returning Without Approval the 'Children's Health Insurance Program Reauthorization Act of 2007.'" October 3, 2007. 589

U.S. Executive Office of the President. "President Bush Offers Condolences at Virginia Tech Memorial Convocation." April 17, 2007. 171

U.S. Executive Office of the President. "Remarks at the Annapolis Conference in Annapolis, Maryland." November 27, 2007. 684

U.S. Executive Office of the President. "Remarks at the Department of State." October 24, 2007. 616

U.S. Executive Office of the President. "Remarks Following a Meeting with Secretary of the Treasury Henry M. Paulson, Jr., and Secretary of Housing and Urban Development Alphonso R. Jackson." December 6, 2007. 524

U.S. Executive Office of the President. "Remarks on Border Security and Immigration Reform in Yuma." April 9, 2007. 162

U.S. Executive Office of the President. "Remarks on the Resignation of Attorney General Alberto R. Gonzales in Waco, Texas." August 27, 2007. 464

U.S. Executive Office of the President. "Statement on Congressional Passage of Intelligence Reform Legislation." August 5, 2007. 433

U.S. Executive Office of the President. "Statement on Granting Executive Clemency to I. Lewis Libby." July 2, 2007. 332

U.S. President's Commission on Care for America's Returning Wounded Warriors. "Serve, Support, Simplify." July 25, 2007. 368

Preface

Serious concerns over the state of the U.S economy, political turmoil in Pakistan, fierce debate over the funding of children's health insurance, and an improved security situation in Iraq are just a few of the topics of national and international importance chosen for discussion in *Historic Documents of 2007*. This edition marks the thirty-sixth volume of a CQ Press project that began with *Historic Documents of 1972*. This series allows students, librarians, journalists, scholars, and others to research and understand the most important domestic and foreign issues and events of the year through primary source documents. To aid research, many of the lengthy documents that were written for specialized audiences have been excerpted to highlight the most important sections. The official statements, news conferences, speeches, special studies, and court decisions presented here should be of lasting public and academic interest.

Historic Documents of 2007 opens with an "Overview of 2007," a sweeping narrative of the key events and issues of the year that provides context for the documents that follow. The balance of the book is organized chronologically; each article comprises an introduction entitled "Document in Context" and one or more related documents on a specific event, issue, or topic. The introductions provide context and an account of further developments during the year. At the front of the volume, a thematic table of contents (page xvii) and a list of documents organized by source (page xxi) follow the standard table of contents and help readers locate events and documents.

As events, issues, and consequences become more complex and far-reaching, these introductions and documents yield important information and deepen understanding about the world's increasing interconnectedness. As memories of current events fade, these selections will continue to further understanding of the events and issues that have shaped the lives of people around the world.

How to Use This Book

Each of the sixty-nine articles in this edition consists of two parts: a comprehensive introduction followed by one or more primary source documents. The articles are arranged in chronological order by month. Articles with multiple documents are placed according to the date of the first document. There are several ways to find events and documents of interest:

By date: If the approximate date of an event or document is known, browse through the titles for that month in the table of contents. Alternatively, browse the monthly tables of contents that appear at the beginning of each month's articles.

By theme: To find a particular topic or subject area, browse the thematic table of contents.

By document type or source: To find a particular type of document or document source, such as the White House or Congress, review the list of document sources.

By index: The five-year index allows researchers to locate references to specific events or documents as well as entries on the same or related subjects. The index in this volume covers the years 2003–2007. A separate volume, *Historic Documents Index, 1972–2005,* may also be useful.

Each article begins with a section entitled "Document in Context." This feature provides historical and intellectual contexts for the documents that follow. Documents are reproduced with the spelling, capitalization, and punctuation of the original or official copy. Ellipsis points indicate textual omissions, and brackets are used for editorial insertions within documents for text clarification. Full citations appear at the end of each document. If a document is not available on the Internet, this too is noted. For further reading on a particular topic, consult the "Other Historic Documents of Interest" section at the end of each article. These features provide cross-references for related articles in this edition of *Historic Documents* as well as in previous editions. References to articles from past volumes include the year and page number for easy retrieval.

Overview of 2007

A meltdown of the international financial system and a temporary improvement in the security situation in Iraq were among the most important developments of 2007. Both came late in the year and suggested that 2008—an election year in the United States—could be especially stressful, possibly featuring a recession in the United States as part of a broader economic slowdown in much of the world.

The U.S. economy began to overtake the ongoing conflict in Iraq as the top concern of the many Americans who were struggling to cope with a collapsing housing market, tightening credit conditions, and rising prices for gasoline, food, and health care. The U.S. housing boom that began in 1995 finally went bust in 2007. A rash of delinquencies on subprime mortgages—those given to the riskiest borrowers—exacerbated a natural correction in the housing market, displacing many families from their homes in the process. By some estimates, as many as half a million families were expected to face foreclosure in 2008. Slowing home sales also cost many construction workers their jobs.

The White House and Congress were slow to respond to the housing crisis. In December, President George W. Bush announced a voluntary plan that would provide some relief to some distressed homeowners. Congress considered various options for helping stranded homeowners but did not take final action on any major plan by the end of the year.

The U.S. housing crisis, which was distressing on its own merits, also touched off an unanticipated crisis in worldwide financial markets. For several years, banks and investment houses had been bundling subprime mortgage payments into complicated securities packages that were then sold to investors on Wall Street and elsewhere. When large numbers of homeowners began to default on their mortgage payments in late 2006 and early 2007, the value of mortgage-backed securities began to plummet. By early August, some of the world's largest banks had been forced to write down billions of dollars in losses, and the availability of credit not only for mortgages but also for other types of loans had contracted. Central banks around the world injected cash into the system to reassure jittery investors. In the United States, the Federal Reserve took additional steps to loosen credit by lowering key interest rates. Although markets seemed to stabilize toward the end of the year, dismal earnings reports from major banks, concerns that the subprime problem might have even wider repercussions in coming months, and suggestions that the "real" economy was slowing heightened concerns that the United States could be heading into a recession.

The slowing economy and high gasoline prices (which hovered around three dollars per gallon for much of the year) continued to take a toll on the American automobile industry. Successful labor negotiations with the United Auto Workers helped the "Big Three" U.S. automakers pare back wages and benefits, making them more competitive with their nonunion foreign rivals. But the three companies' continued reliance on sales of light trucks and sports utility vehicles offset some of those gains as customers increasingly turned to small cars with better gas mileage.

Rising gas prices and widespread concern over climate change helped persuade automakers to drop their long-standing opposition to any increase in federal fuel economy standards. That eased the way for Congress to clear legislation in December that raised the standard to an average of 35 miles per gallon across a manufacturer's entire fleet by the 2020 model year. In a move that many environmental advocates criticized as a political payoff to the auto industry, however, the Environmental Protection Agency (EPA) denied California and several other states permission to set greenhouse gas emission standards that would have forced automakers to meet the higher fuel economy standard by 2016.

Seeking to give lower-income Americans some relief from soaring health care costs, Democrats—who had won control of both chambers of Congress in the November 2006 elections—tried to expand a popular federal program providing health insurance to low- and middle-income children. President Bush opposed the legislation and vetoed it twice, saying it went in "the wrong direction" because it would encourage millions of families to give up their private health insurance and sign up for the government program instead. Democrats were unable to override the vetoes and settled for a short-term extension of the existing program.

Democrats vs. Republicans

The fight over children's health insurance was emblematic of the year in Congress. Again and again, the slim Democratic majorities tried to push an ambitious legislative program through Congress, sometimes with the help of some powerful Republican allies. But time after time, they were blocked by determined Republican minorities in both chambers who used procedural rules to defeat or force changes in Democratic measures and by a president who made effective use of vetoes and veto threats. Nowhere was this more evident than in the Democrats' plans to hasten the withdrawal of U.S. troops from Iraq. The pattern was set in May, when Bush vetoed the first piece of Iraq-related legislation Congress sent him, which set a deadline for starting to bring the troops home, and the House failed to override. Subsequent efforts by Democrats to legislate change in U.S. policy in Iraq never cleared Congress. Democrats were also stymied in their initial efforts to place meaningful limits on the president's intelligence-gathering authority, including a surveillance program that eavesdropped on U.S. citizens. Legislation seeking to limit that authority was pending at the end of the year.

The Democrats did carry the day on a few of their top priorities, including legislation lowering interest rates on federal student loans and a measure raising the minimum wage for the first time in a decade. Democrats also worked with President Bush to enact comprehensive immigration reform. As in 2006, however, their efforts were blocked by Republicans and some Democrats; opponents argued that provisions in the bill that would allow some illegal aliens to earn citizenship were tantamount to amnesty for people who had broken the law.

Democrats also exercised congressional oversight of executive branch actions to a degree not seen in some years. One target was the Food and Drug Administration, which had come under increasing criticism after some imported pet food, seafood, and drugs were found to carry dangerous contaminants. Other targets were the Pentagon and Veterans Affairs Department for the way they treated disabled soldiers and veterans. Investigations by a broad range of commissions found widespread evidence of mismanagement, personnel shortages, and other problems that were preventing disabled soldiers and veterans from receiving adequate medical attention and disability payments. By the end of the year, several steps had been taken to improve the process, but much more work still needed to be done.

A congressional investigation into whether the Justice Department had fired several of the government's top prosecutors for political, partisan reasons contributed to the resignation of Attorney General Alberto R. Gonzales in August. Gonzales, who had served as White House counsel during Bush's first term, was a central figure in ongoing controversies over the legal definition of torture and the limits of secret surveillance of U.S. citizens. His changing and sometimes conflicting explanations of his personal role in and the Justice Department's position on the firings and the secret surveillance programs undermined his credibility among congressional Republicans as well as Democrats. In the conclusion of another embarrassment for the administration, I. Lewis "Scooter" Libby, the former top aide to Vice President Dick Cheney, in March was found guilty of lying about his role in revealing the name of a covert CIA operative to the media. In July, Bush commuted Libby's prison sentence but did not formally pardon him.

Supreme Court Opinions

The Supreme Court issued several notable decisions in 2007. In its first ruling on any issue related to global warming, the Court on April 2 ruled that the Environmental Protection Agency had authority under the Clean Air Act to regulate carbon dioxide and other heat-trapping gases in vehicle tailpipe emissions—the greenhouse gases that many scientists said contributed to global warming. The Bush administration had refused to regulate the emissions, arguing that they were not air pollutants within the meaning of the Clean Air law.

The Bush administration maintained its opposition to mandatory cuts in greenhouse gas emissions at an international conference in Bali, Indonesia, in December, where delegates were debating what should replace the Kyoto Protocol on climate change when that treaty expired. The administration did agree, however, to participate in United Nations negotiations aimed at developing a new climate control treaty by 2009, which meant that many of the important decisions about a new treaty would be made after the election of a new president in November 2008. All leading Democratic candidates and a leading Republican candidate, Sen. John McCain, had taken positions on climate change that were sharply different from those of the Bush administration.

In a 5–4 decision that heartened abortion opponents, the Supreme Court upheld a federal ban on so-called partial-birth abortions, a procedure sometimes used to terminate pregnancy after the third month. The decision represented the first time since the Court recognized a women's constitutional right to abortion in 1973 that it had upheld a ban on a specific procedure. It also was the first time the Court had approved a restriction on abortion that did not include an exemption to protect the health of the mother. In another 5–4 ruling, the Court overturned key provisions of the Bipartisan Campaign Reform Act of 2002, which were aimed at curbing controversial issue ads. These ads often praised or scorned a candidate's position on a specific issue without directly instructing the viewer to vote for or against a candidate. The Court majority held that the curbs as applied in the specific case before it were an unconstitutional restriction on the advertiser's free speech rights.

The Court, also by a 5–4 vote, ruled that race could not be the determining factor in assigning students to public schools. Four of the justices in the majority would have gone even further to state that schools could never consider race. The four dissenting justices said the ruling was a "cruel distortion" of the meaning of the historic decision in *Brown v. Board of Education* (1954), in which a unanimous Court called for an end to segregated schools.

The Supreme Court also agreed to hear arguments on two other controversial matters in its 2007–2008 term. One case challenged the constitutionality of lethal injection, the most common method used to execute death row inmates. The question in the second case was whether the Second Amendment confers the right to keep and bear arms on individuals under certain conditions as well as on organized militias such as the National Guard; the case involved a District of Columbia gun control law considered to be one of the toughest in the nation.

Whether gun control laws were an effective tool in controlling crime had long been a matter of some controversy. The question arose again in April after a deeply disturbed student shot and killed thirty-two students at Virginia Tech University in Blacksburg, Virginia, before taking his own life. It was the deadliest campus shooting in U.S. history. The shooter used two handguns he had purchased even though his mental illness should have placed his name on a federal background check list that would have prohibited the sales.

The Troop "Surge" in Iraq

The conflict in Iraq entered its fifth year in the spring of 2007, making it one of the longest conflicts fought by the U.S. military. President Bush decided at the beginning of the year to try something new, or at least to take a somewhat bolder approach to combating the insurgents who had launched a campaign against the U.S. occupation of Iraq in mid-2003. The most visible component of this new approach was a "surge" of approximately 30,000 additional troops into Iraq, putting the U.S. presence there at about 160,000 soldiers and marines. The additional troops, along with Iraqi military units that were still unable to operate entirely on their own, were deployed in a new counterinsurgency strategy. Rather than operating from big, protected military bases, thousands of soldiers were stationed in dozens of small outposts in Baghdad and other areas, enabling them to interact with the locals and, presumably, to be more effective in fighting the insurgents hiding among the civilian population.

Another central component of the new U.S. approach to Iraq involved extensive cooperation with, and encouragement of, a rebellion by Iraq's Sunni Muslim minority against the insurgency, which was based in the Sunni community. This "Sunni awakening" stemmed from ordinary citizens' frustration with the violence that had disrupted their communities and killed or maimed thousands of Iraqis. By the middle of 2007, the United States was supporting tens of thousands of Sunni fighters who pledged to battle the insurgents; some of these fighters almost certainly had been insurgents themselves before they switched sides. The United States also benefited from a cease-fire called by Moqtada al-Sadr, the leader of one of Iraq's biggest Shiite Muslim militias. Sadr halted fighting by his Mahdi Army in August for tactical reasons—not as a favor to Washington—but U.S. policy benefited nonetheless.

The combination of the troop surge, the Sunni awakening, Sadr's cease-fire, and several other developments led to a sharp reduction in violence in Iraq during the last four months of the year. By nearly every measure, Iraq was a safer place in December than it had been in January or even August. Iraq was still one of the world's most violent countries, but the relative calm created an opening for Iraq's political leaders to attempt to resolve their differences that had contributed to the violence. Some politicians made some efforts to reach out to competing factions, and the parliament began responding to U.S. pressure to enact legislation intended to reassure the Sunni and Kurdish minorities of their rights. The overall result, however, was that the political reconciliation that U.S. officials hoped to promote, through the surge, was still more a dream than a reality.

Elsewhere in the Middle East

The Bush administration's other dreams for the Middle East also were frustrated during the year. This was particularly the case in the seemingly endless conflict between the Israelis and the Palestinians. In 2006, Hamas, the radical Islamist faction, had shocked the world and embarrassed Washington by winning Palestinian elections that the United States had promoted as part of President Bush's vision for democracy in the Middle East. In 2007, Hamas and the more moderate, U.S.-favored Fatah faction tried to settle their differences but failed, leading to an intense period of violence in June that gave Hamas full control over the impoverished Gaza Strip. The result was that the two Palestinian territories were controlled by two separate factions: Hamas in Gaza and Fatah in the West Bank, posing new dangers to Israel and to hopes of negotiating a peaceful settlement to the Israeli-Palestinian conflict.

President Bush chose this unpromising circumstance to launch a new effort to negotiate such a settlement. He began with an international conference in Annapolis, Maryland, in November. There, in front of world leaders, Israeli prime minister Ehud Olmert and Palestinian president Mahmoud Abbas pledged to try to reach a major peace agreement before President Bush left office in January 2009. The odds against them were long at the outset and grew longer still during their first follow-up negotiating sessions in December, which stalled on procedural matters and underlying policy disputes.

Looming over nearly all of the year's developments in the Middle East was the Iranian government's continuing drive for greater influence in the region. The Bush administration accused Tehran of fomenting violence in Iraq by supplying weapons and training to anti-U.S. factions there; Washington's concerns were so great, in fact, that President Bush authorized the first direct diplomacy between the United States and Iran in a generation. A series of meetings between the U.S. and Iranian ambassadors in Baghdad produced only limited results but did establish a precedent for more diplomacy in the future. Another subject of dispute was Iran's work to develop the technology to enrich uranium—work that Israelis and many experts in the West believed was intended to build nuclear weapons but that Iranian officials insisted was for peaceful purposes only. Unexpected affirmation of the Iranian position came from U.S. intelligence agencies, which issued a National Intelligence Estimate in December suggesting that Tehran wanted to build nuclear weapons but had stopped work on its weapons program in 2003. This estimate, which was harshly criticized by many conservatives and by Israelis, undercut the administration's push at the United Nations for tough new sanctions against Iran and appeared to end, at least for the moment, any prospect for a preemptive U.S. military strike to destroy Iran's nuclear facilities.

Pakistan: On the Brink?

Very few places in the world worried U.S. foreign policy makers more than Pakistan. With its nuclear weapons, an unstable political system, a long history of antagonism with neighboring India (also a nuclear power), the likely presence of Osama bin Laden and other leaders of the al Qaeda network in the mountainous border area adjacent to Afghanistan, and a vocal minority of radical Islamists, Pakistan had all the hallmarks of a strategically located country that could descend into anarchy and take much of South Asia with it. Additionally, Pakistan's president, Gen. Pervez Musharraf, who had supported Washington in its "war on terror" against al Qaeda, had become increasingly unpopular and faced elections in 2007.

Pakistan underwent a series of political crises during the year, including dissension over Musharraf's reelection by the parliament in October—something that was accomplished only after the president promised to resign from his other post as head of the army once he won the vote. The drama over Musharraf's election paled in comparison to what happened next. Later in October, throngs of Pakistanis welcomed home former prime minister Benazir Bhutto, who had been living in exile and had tried to negotiate a deal with Musharraf to allow her to return. Hours after she arrived in Karachi on October 18, a bomb exploded next to her motorcade, killing an estimated 140 people and wounding some 400 others. Bhutto was unhurt, and she blamed the government for failing to act against those who sought to kill her.

Two months later, on December 27, as she was campaigning in Rawalpindi for parliamentary elections scheduled for the following month, Bhutto was assassinated in another attack. It was unclear, by year's end, whether she was killed by gunfire or by a bomb that exploded near her car. In either case, the results were the same: thousands of angry supporters blamed the government, which in turn blamed Islamist extremists. The assassination deprived Bhutto's Pakistan People's Party of her charismatic leadership just before crucial elections, which were postponed to mid-February 2008. The one outcome that appeared certain was that the year's events had sapped much of Musharraf's political clout; whether the country's long-bickering civilian politicians could step in to fill the void was far from certain.

North Korea's Promise on Nuclear Weapons

Potentially significant progress was achieved during the year in negotiations concerning North Korea's nuclear weapons program. Since 2003, China had led "Six-Party Talks" intended to convince North Korea to give up its nuclear weapons program in return for economic aid and assurances that it would not be attacked by the United States. In addition to China and North Korea, the other parties participating were Japan, Russia, South Korea, and the United States.

North Korea agreed, in principle, in 2005 to abandon "all nuclear weapons and existing nuclear weapons programs," but one year later tested a nuclear weapon. The test appeared to be a failure, but it unnerved Asian neighbors and brought a swift denunciation from the UN Security Council.

The Bush administration, which had been reluctant to negotiate directly with North Korea, finally took that approach early in 2007. The result was another significant agreement by North Korea, announced on February 13: an action plan to implement the promises Pyongyang had made in 2005. Under the agreement, North Korea agreed to "shut down and seal" a plutonium-producing nuclear plant in Yongbyon; in return, other countries in the Six-Party Talks agreed to supply North Korea with large quantities of fuel oil to keep its fragile economy running. More negotiations were necessary before North Korea would agree to additional steps to implement these agreements, resulting in yet another accord on October 3. A key part of that agreement was a pledge by North Korea to give the five other countries a "complete and correct declaration" of all of its nuclear facilities, something U.S. officials said was necessary to determine whether Pyongyang had conducted a program separate from the Yongbyon reactor—using enriched uranium—to develop nuclear weapons. By year's end, North Korea had dismantled much of the Yongbyon reactor but appeared to be stalling on its pledge to reveal details of its other nuclear activities, thus raising new suspicions about its true intentions.

January

President Bush on the Death of Gerald R. Ford

JANUARY 2, 2007

Gerald R. Ford, the thirty-eighth president of the United States, died on December 26, 2006, at age ninety-three. Ford would be remembered as the president who calmed a country in turmoil following the resignation of Richard M. Nixon in August 1974, only to stir it up again a month later by pardoning Nixon for his role in the Watergate scandal. Although Ford's presidency was brief—he lost a bid for election in his own right to Jimmy Carter in 1976—he grappled with the end of the Vietnam War and the worst recession since the Great Depression of the 1930s.

Through it all, Ford was widely regarded more as a common man than a celebrity, someone with whom most Americans felt comfortable even if they did not always agree with his policies. He carried into the White House the same midwestern affability that had characterized his leadership of the Republicans in the House of Representatives, where he had served for twenty-five years. "For a nation that needed healing, and for an office that needed a calm and steady hand, Gerald Ford came along when we needed him most," President George W. Bush said in a statement broadcast on December 27, 2006. "During his time in office, the American people came to know President Ford as a man of complete integrity who led our country with common sense and kind instincts."

Ford's Early Years and Congressional Career

Ford's "common touch" may have stemmed from his middle-class, midwestern roots. Born in Omaha, Nebraska, on July 14, 1913, Ford was christened Leslie Lynch King Jr., after his father. When Ford was two years old, his mother divorced his father, moved to Grand Rapids, Michigan, and married paint salesman Gerald Rudolph Ford, who gave his adopted son his name. By all accounts, Ford was an outgoing and popular teenager who played center on the high school football team, was an Eagle Scout, worked at a local lunch counter, and had a modest academic record. Ford went to the University of Michigan, where he also played football; following graduation, he turned down an opportunity to play professionally for the Green Bay Packers and Chicago Bears in favor of attending Yale Law School. Back in Grand Rapids in 1941, he opened a law practice with his friend (and future White House legal counsel) Philip W. Buchen. A few months later, after Japan attacked Pearl Harbor and drew the United States into World War II, Ford joined the Navy as an ensign. He served until 1946, leaving with the rank of lieutenant commander.

Returning again to Grand Rapids, Ford resumed his law practice and began to explore a growing interest in Republican politics. His support for the Marshall Plan to rebuild postwar Europe brought him to the attention of Sen. Arthur Vandenberg, R-Mich., who persuaded the young war veteran to run in 1948 against Rep. Bartel J. Jonkman, a conservative Republican and isolationist who appeared to be well entrenched in his district. Despite the odds, Ford defeated Jonkman in the primary by nearly 10,000 votes and went on to win the general election with more than 60 percent of the vote—a margin he retained in each of his next twelve elections to the House of Representatives, thanks in large part to the personal attention he paid to his constituents.

In the House, Ford continued to support an internationalist approach to foreign policy and maintained a conservative Republican perspective on most domestic economic issues. His stance on civil rights issues was mixed: he voted for the Civil Rights Act of 1964 and the Voting Rights Act of 1965, but he opposed school busing to advance integration. During his House tenure, Ford never authored a major piece of legislation.

Ford was interested in party leadership and concentrated on climbing the ladder. From his position on the influential House Appropriations Committee, he gradually assumed a role helping shape Republican legislative strategy. In 1963, backed by other young moderates in the party, Ford successfully challenged the older, more conservative incumbent to become chairman of the House Republican Conference. Two years later, after the Republicans suffered massive losses at the hands of Lyndon Johnson and the Democrats, the younger Republicans completed the coup, replacing longtime leader Charles A. Halleck of Indiana with Ford.

As House minority leader, Ford ruled his caucus with gentle persuasion, forgoing as counterproductive harsh disciplinary tactics when a member strayed from the fold. "It's the damnedest thing," said Rep. Joe D. Waggonner Jr., an influential Democrat from Louisiana at the time. "Jerry just puts his arm around a colleague or looks him in the eye, says, 'I need your vote,' and gets it."

From Congress to the White House

Ford's greatest political ambition was to become Speaker of the House of Representatives, a dream that he gradually put behind him as it became apparent that Democrats were likely to maintain their strong majority in the House for many years to come. He told biographers that he had promised his wife in 1973 that he would run just once more, in 1974, in the small hope that Republicans could take control of the House, and then he would retire from politics in 1976.

That plan changed in October 1973, when Vice President Spiro T. Agnew pleaded no contest to a charge of federal income tax evasion and resigned his office. Nixon's first choice to replace Agnew under the new Twenty-fifth Amendment (ratified in 1967) was former Treasury secretary John B. Connally. Persuaded by several Republicans that Connally could not be confirmed, Nixon turned to Ford, who agreed to take the job. Ford later told the *New York Times* he thought, "Well, if I could be helpful, I knew I could be confirmed. And it was a nice way to end a career. I wasn't going to be Speaker." As anticipated, he was easily confirmed, by a vote of 92 to 3 in the Senate and 387 to 35 in the House, on December 6, 1973.

For the next eight months, Ford publicly defended Nixon against growing evidence that the president was personally involved in covering up the July 17, 1972, break-in, orchestrated by his reelection committee, at Democratic Party offices in the Watergate complex in Washington, D.C. Privately, however, Ford later said he began to have doubts about Nixon's innocence, especially after the president spurned his advice to demonstrate that innocence by turning over to Congress and the courts investigating the scandal the documents and White House tapes they were seeking. According to Ford, he received the first "concrete evidence" of Nixon's personal involvement in late July 1974 from White House chief of staff Gen. Alexander M. Haig Jr., who called to tell him that one of the tapes subpoenaed in the investigation clearly showed that the president's attempts to obstruct the investigation had begun just six days after the break-in. Forced by a Supreme Court decision to make the tapes public, Nixon announced on August 8, 1974, that he would resign the presidency the next day.

As Nixon left the White House on August 9, Ford was sworn in as the thirty-eighth president of the United States. "My fellow Americans, our long national nightmare is over," the new president reassured a country greatly relieved that it would be spared the virtually certain congressional impeachment and conviction of a sitting president for criminal conduct. "Our Constitution works; our great Republic is a government of laws and not of men. Here the people rule."

Ford's inauguration made him the first U.S. president elected neither to the presidency nor the vice presidency. Nonetheless, his openness and the openness with which he ran the White House made him highly popular with the public; his approval rating reached 71 percent soon after he took office. Public sentiment changed dramatically one month later, however, when he issued a full pardon to Nixon on September 8 for any crimes he "committed or may have committed" while in the White House. Ford explained that he believed the pardon was necessary to bring the Watergate era to a close. He felt that Nixon would eventually be pardoned and saw little reason to put the country though the pain of a public trial that would make the "healing process . . . much more difficult to achieve."

Ford had anticipated that the pardon would be unpopular, but he did not realize just how angry many Americans would be. He was chastised by editorial writers across the country, denounced by members of Congress of both parties, hurt by his friend and White House press secretary Jerald F. terHorst's resignation in protest, and accused by some of persuading Nixon to resign by promising a pardon. Ford responded to the last charge by taking an unusual step for any president: he testified before a House subcommittee. "There was no deal, period, under no circumstances," he declared. Although many politicians and others eventually came to view the pardon as the right thing done at the right time, its immediate effects were significant. Congress delayed confirmation of Ford's choice of former New York governor Nelson A. Rockefeller to become vice president, Ford's approval rating plunged, and Democrats achieved a landslide in the 1974 congressional elections. Many political observers also argued that it cost Ford his bid for election to the presidency in his own right in 1976—the only political election Ford ever lost.

A Lost War and a Difficult Economy

Even though he was president for only two and a half years, Ford dealt with several difficult issues. In the international arena, Ford's presidency witnessed the end of the Vietnam War. U.S. combat troops had been withdrawn by the time Ford entered the White House, but in 1975 he asked Congress for aid to help the South Vietnamese government resist an expected invasion from the North. Backed by public opinion, Congress refused, forcing Ford to stand by as South Vietnam fell to North Vietnam in April 1975, bringing to conclusion an undeclared war that had cost the United States more than 58,000 lives, billions of dollars, and social anguish that deeply affected a generation of young Americans. In

May 1975 Ford sent the U.S. Marines to free the crew of the *Mayaguez,* a U.S. merchant ship that Cambodia had seized. The crew was freed but forty-one of the rescuers were killed. Later that year, he signed the Helsinki human rights convention, a move that was sharply criticized by both his Republican opponent for the presidential nomination, Ronald Reagan, and his Democratic opponent in the general election, Jimmy Carter. In retrospect, several scholars have credited the accord with helping lay the groundwork for the protests in Eastern Europe and Russia that eventually led to the collapse of the Soviet Union.

On the domestic front, Ford faced a slew of economic problems, including chronic energy shortages that produced angry lines and skyrocketing costs at gas stations, double-digit inflation, and the highest unemployment levels since the 1930s. Ford initially attempted to fight inflation, which he called "public enemy No. 1," by vetoing some fifty spending bills, but the heavily Democratic Congress overturned many of the vetoes. He also announced a campaign to "Whip Inflation Now," or WIN, which won little traction either with the public or in Congress. By 1975 unemployment had become the more serious problem, and Ford compromised with Congress on a tax cut and spending plan designed to stimulate the sluggish economy. Although both inflation and the jobless rate remained high for the rest of his term, the economy did recover in late 1975 and 1976. Ford also narrowly escaped two assassination attempts in California in 1975.

As the 1976 election approached, Ford turned back a strong challenge from Reagan for the Republican presidential nomination. Polls showed Ford to be twenty-five points behind Carter at the start of the general election campaign, but he steadily closed that gap during the fall, until he misspoke during a debate with Carter, insisting that the Soviet Union did not dominate Eastern Europe. Ford's climb in the polls stopped, and he lost the election, taking 48 percent of the vote to Carter's 50 percent. Ford considered running for president again in 1980, but the party had already turned to Reagan. Reagan invited Ford to be his running mate, but the former president declined.

Ford spent his final years playing golf, building his presidential museum in Grand Rapids, and occasionally serving as an elder statesman. He was accompanied throughout his years in Washington and retirement by his wife, Betty, whose openness about her fight against breast cancer made her one of the most respected first ladies. She was also widely praised after acknowledging her dependency on drugs and alcohol (caused in part by pain from an inoperable pinched nerve), and she helped establish the Betty Ford Center for Drug and Alcohol Rehabilitation in Rancho Mirage, California.

It was at his home in Rancho Mirage that Ford died. His coffin was brought to Washington where it lay in state in the Capitol Rotunda from December 30, 2006, until January 2, 2007, when it was moved to Washington National Cathedral for a state funeral. Delivering one of several eulogies, Bush said that Ford's "time in office was brief, but history will long remember the courage and common sense that helped restore trust in the workings of our democracy." Ford was buried in a private ceremony in Grand Rapids on January 3. He was the longest-living president, followed by Ronald Reagan who died in 2004, also at age ninety-three.

Following is the text of the eulogy for former president Gerald R. Ford, delivered by President George W. Bush on January 2, 2007, at state funeral services at the National Cathedral in Washington, D.C.:

DOCUMENT

President Bush's Eulogy at the National Funeral Service for President Ford

January 2, 2007

Mrs. Ford; the Ford family; distinguished guests, including our Presidents and First Ladies; and our fellow citizens:

We are here today to say goodbye to a great man. Gerald Ford was born and reared in the American heartland. He belonged to a generation that measured men by their honesty and their courage. He grew to manhood under the roof of a loving mother and father. And when times were tough, he took part-time jobs to help them out. In President Ford, the world saw the best of America, and America found a man whose character and leadership would bring calm and healing to one of the most divisive moments in our Nation's history.

Long before he was known in Washington, Gerald Ford showed his character and his leadership. As a star football player for the University of Michigan, he came face to face with racial prejudice when Georgia Tech came to Ann Arbor for a football game. One of Michigan's best players was an African American student named Willis Ward. Georgia Tech said they would not take the field if a black man were allowed to play. Gerald Ford was furious at Georgia Tech for making the demand and for the University of Michigan for caving in. He agreed to play only after Willis Ward personally asked him to. The stand Gerald Ford took that day was never forgotten by his friend. And Gerald Ford never forgot that day either—and three decades later, he proudly supported the Civil Rights Act and the Voting Rights Act in the United States Congress.

Gerald Ford showed his character in the devotion to his family. On the day he became President, he told the Nation, "I am indebted to no man and only to one woman—to my dear wife." By then Betty Ford had a pretty good idea of what marriage to Gerald Ford involved. After all, their wedding had taken place less than 3 weeks before his first election to the United States Congress, and his idea of a honeymoon was driving to Ann Arbor with his bride so they could attend a brunch before the Michigan-Northwestern game the next day. [*Laughter*] And that was the beginning of a great marriage. The Fords would have four fine children. And Steve, Jack, Mike, and Susan know that, as proud as their dad was of being President, Gerald Ford was even prouder of the other titles he held: father and grandfather and great-grandfather.

Gerald Ford showed his character in the uniform of our country. When Pearl Harbor was attacked in December 1941, Gerald Ford was an attorney fresh out of Yale Law School, but when his Nation called, he did not hesitate. In early 1942 he volunteered for the Navy and, after receiving his commission, worked hard to get assigned to a ship headed into combat. Eventually his wish was granted, and Lieutenant Ford was assigned to the aircraft carrier USS *Monterey,* which saw action in some of the biggest battles of the Pacific.

Gerald Ford showed his character in public office. As a young Congressman, he earned a reputation for an ability to get along with others without compromising his principles. He was greatly admired by his colleagues, and they trusted him a lot. And so when President Nixon needed to replace a Vice President who had resigned in scandal, he naturally turned to a man whose name was a synonym for integrity: Gerald R. Ford. And

8 months later, when he was elevated to the Presidency, it was because America needed him, not because he needed the office.

President Ford assumed office at a terrible time in our Nation's history. At home, America was divided by political turmoil and wracked by inflation. In Southeast Asia, Saigon fell just 9 months into his Presidency. Amid all the turmoil, Gerald Ford was a rock of stability. And when he put his hand on his family Bible to take the Presidential oath of office, he brought grace to a moment of great doubt.

In a short time, the gentleman from Grand Rapids proved that behind the affability was firm resolve. When a U.S. ship called the *Mayaguez* was seized by Cambodia, President Ford made the tough decision to send in the Marines, and all the crew members were rescued. He was criticized for signing the Helsinki accords, yet history has shown that document helped bring down the Soviet Union, as courageous men and women behind the Iron Curtain used it to demand their God-given liberties. Twice assassins attempted to take the life of this good and decent man, yet he refused to curtail his public appearances. And when he thought that the Nation needed to put Watergate behind us, he made the tough and decent decision to pardon President Nixon, even though that decision probably cost him the Presidential election.

Gerald Ford assumed the Presidency when the Nation needed a leader of character and humility—and we found it in the man from Grand Rapids. President Ford's time in office was brief, but history will long remember the courage and common sense that helped restore trust in the workings of our democracy.

Laura and I had the honor of hosting the Ford family for Gerald Ford's 90th birthday. It's one of the highlights of our time in the White House. I will always cherish the memory of the last time I saw him this past year in California. He was still smiling, still counting himself lucky to have Betty at his side, and still displaying the optimism and generosity that made him one of America's most beloved leaders.

And so, on behalf of a grateful nation, we bid farewell to our 38th President. We thank the Almighty for Gerald Ford's life, and we ask for God's blessings on Gerald Ford and his family.

SOURCE: U.S. Executive Office of the President. "Eulogy at the National Funeral Service for Former President Gerald R. Ford." January 2, 2007. *Weekly Compilation of Presidential Documents* 43, no. 1 (January 8, 2007): 3–4. Washington, D.C.: National Archives and Records Administration. www.gpoaccess.gov/wcomp/v43no1.html (accessed September 22, 2007).

OTHER HISTORIC DOCUMENTS OF INTEREST

FROM PREVIOUS *HISTORIC DOCUMENTS*

- Reagan's death, *2004*, p. 320
- Ford's swearing-in as president, *1974*, p. 695
- Ford's pardon of Nixon, *1974*, p. 811

DOCUMENT IN CONTEXT

President Bush on the "Surge" of U.S. Troops to Iraq

JANUARY 10, 2007

Running against the political winds in the United States, President George W. Bush in January ordered a "surge" of additional U.S. troops to Iraq in an effort to curtail the sectarian violence threatening the U.S.-backed government in Baghdad. He then relied on his veto power and the Republican minority in the Senate to prevent Democrats from using their new majority in Congress to reverse course on Iraq policy. By year's end the stronger U.S. military presence had improved security in Iraq and created an opportunity for the government there to consolidate its political position. Whether Iraq's feuding politicians would be able to take advantage of that opportunity was an open question.

The president's decision to send more troops to Iraq was politically risky. Opinion polls show that the majority of Americans wanted to leave Iraq as soon as possible, and the president's party had lost control of Congress two months earlier in a wave of public discontent over the conduct of the war. Commentators across the political spectrum called Bush either foolish or courageous to take a step so contrary to the public mood. Regardless, by January 2007 it was clear that Bush's presidency would be judged in large part by the outcome of the war in Iraq; the success or failure of the surge would affect his legacy.

Steps to a Policy Shift

The U.S. intervention in Iraq appeared to reach a low point during 2006, when sectarian violence killed an estimated 34,000 Iraqis, sectarian militias ruled the streets in many areas, and the new government elected in December 2005 was so weak and divided that it could make few decisions more important than when to take a vacation. Many Americans believed the president's policies in Iraq had failed and that there was little chance of a turnaround. In December 2006 the bipartisan Iraq Study Group recommended several course corrections—without fundamental changes to Iraq policy—but Bush rejected the commission's modest proposals.

By late 2006 administration officials were suggesting that Bush had other plans. The president would add more troops in an attempt to bolster security, particularly in Baghdad, and to give the Iraqi government time to make politically difficult decisions. Secretary of Defense Robert M. Gates, who had recently replaced Donald H. Rumsfeld, the principal architect of the Iraq War plans, played a key role in the president's decision.

Replacing Rumsfeld was just the first step in Bush's overhaul of his Iraq team. The president on January 5 swept aside most of the people who had carried out his previous policy and appointed generals and diplomats who appeared to accept the need for change.

The key appointment was that of Lt. Gen. David H. Petraeus as the new overall U.S. military commander in Iraq. Petraeus, a widely respected commander who had served two previous tours in Iraq, succeeded Gen. George W. Casey Jr., who had commanded U.S. forces in Iraq since 2004 and previously had argued that he had enough troops. Bush also chose veteran diplomat Ryan C. Crocker as the new ambassador to Iraq. The former ambassador, Zalmay Khalilzad, was a native of Afghanistan and Sunni Muslim who had not been able to convince Iraq's Shiite leaders of his even-handed approach to the country's sectarian battles.

Taking the Case to the American People

Bush had addressed the American people to explain his reasoning and appeal for their support at numerous points since he decided in 2002 to invade Iraq. The challenge of winning that support had never been greater than on the evening of January 10, when the president stood at a lectern in the White House library and explained to the nation why he was sending more troops to Iraq.

In forthright language, Bush acknowledged that "the situation in Iraq is unacceptable to the American people, and it is unacceptable to me." The blame did not rest with U.S. troops, he added: "Where mistakes have been made, the responsibility rests with me."

The president said he was changing course in response to the rise of sectarian violence in Iraq during 2006. The principal focus in 2007, he said, would be to increase the level of security in the Baghdad region, which had experienced 80 percent of the sectarian violence. The Iraqi government planned to deploy eighteen army and police brigades in Baghdad, he said. To support that effort the United States would send more than 20,000 additional troops to Iraq—most of them to be based in Baghdad. "Our troops will have a well-defined mission: to help [Iraqi forces] protect the local population; and to help ensure that the Iraqi forces left behind are capable of providing the security that Baghdad needs." Although aides described the additional troops as a "surge" that would be "temporary," Bush did not indicate how long the extra troops would be stationed in Iraq.

Bush cited two reasons he believed this effort to secure Baghdad would be more successful than previous failed attempts. First, the additional troops would ensure that "we'll have the force levels we need to hold the areas that have been cleared" of insurgents. In the past, U.S. troops had cleared insurgents from neighborhoods, but they returned once the Americans moved to another area. Second, the Iraqi government had agreed to allow joint Iraq-U.S. forces to enter neighborhoods "that are home to those fueling the sectarian violence." In the past, he said, Iraqi political considerations had prevented U.S. troops from entering some neighborhoods.

The president said his decision offered an opportunity for the Iraqi government led by Prime Minister Nouri al-Maliki, who understood that "now is the time to act." Otherwise, Bush stated, Maliki's government would lose the support of both the Iraqi and the American populations.

Bush noted that Maliki's government had made other promises in conjunction with the U.S. surge: to "take responsibility for security" in all provinces by November, to pass legislation sharing oil revenue among all regions, to spend $10 billion of Iraqi funds on reconstruction and infrastructure projects, to hold provincial elections during 2007, and to revise the "de-Baathification" laws imposed by the United States in 2003–2004 to rid the government and security forces of top officials from Saddam Hussein's ruling Baath party. The de-Baathification laws had proved counterproductive because hundreds of thousands of government workers and soldiers had been forced out of work; some of them formed the core of the antigovernment insurgency.

Although Bush declared that his plan represented a "new strategy" for Iraq, the only new elements were the appointment of new commanders on the ground and the temporary expansion of U.S. forces. General Petraeus, the new commander, intended to implement a new counterinsurgency approach based on a revised U.S. Army doctrine developed under his supervision, but this was not an explicit component of the president's policy shift.

To create the surge, four combat brigades would be deployed to Iraq sooner than had been planned, one additional brigade would be sent to Iraq, and a brigade of the Minnesota National Guard would see return to the United States delayed by four months. The net result of the plan Bush announced on January 10 was that the U.S. presence in Iraq would rise by about 21,500 troops to 153,000, the highest level since more than 165,000 troops were present during parliamentary elections in December 2005. The total number of U.S. troops actually reached 167,000 by August, however, when a total of twenty combat brigades were in place. The first decrease in the U.S. presence occurred in early November, when one brigade of the First Cavalry Division returned to its home base in Texas. At that point, U.S. officials estimated that the number of troops would decline gradually to about 140,000 by mid-June 2008.

Democrats' Challenge Fails

Democrats in the Senate and House of Representatives were quick to denounce Bush's plan. Their joint statement, issued on January 10, argued that the surge "endangers our national security by placing additional burdens on our already overextended military, thereby making it even more difficult to respond to other crises." Some Republicans were less than enthusiastic about the idea of sending more troops to Iraq, but most managed to find good points in Bush's plan, such as its emphasis on additional steps required of the Iraqi government. Kansas Republican senator Sam Brownback, who was mulling a bid for the presidency in 2008 and was visiting Baghdad at the time of Bush's speech, was one of the most critical. "I do not believe that sending more troops to Iraq is the answer," he said in a statement. "Iraq requires a political rather than a military solution." Among the few members of Congress who enthusiastically supported Bush's plan was Arizona Republican senator John McCain, another presidential contender. McCain long had been critical of administration policy in Iraq, insisting that the United States had never deployed adequate forces to stabilize the country.

The Democrats were in a position to challenge Bush's policy more strongly than in the past, now that they held the leadership positions in both chambers of Congress. By no means did they control Congress, however, and they lacked the votes to override a presidential veto. Moreover, they were divided on tactics. Some wanted to use the power of the purse to try to impose limits on the president's policy, while others opposed any step that might be seen as denying funds for troops in the field. Ultimately, the Democrats, with support from a few Republicans, tried several different legislative tactics; they succeeded in forcing numerous debates on Iraq policy but failed to sway the course of policy to any serious extent. On the primary campaign trail, although all of the half-dozen Democratic presidential candidates advocated withdrawing U.S. troops sooner than the administration planned, none called for a complete withdrawal immediately or even by the end of 2007.

Over the course of the year, majorities in both chambers of Congress voted in favor of various legislative formulations opposing the president's Iraq policy. A small number of Republicans supported these proposals, but enough stood by the administration to prevent the measures from becoming law.

Only one proposal to change course reached the president's desk. Congress on May 1 sent the president a supplemental spending bill (HR 1591) that contained funding for the troop surge but required troop withdrawals to begin in six months, with the goal of completing the U.S. pullout from Iraq by March 2008. Democrats deliberately timed passage of the bill to mark the fourth anniversary of Bush's famous speech, aboard the aircraft carrier USS *Abraham Lincoln,* declaring the end of major combat in Iraq. Bush vetoed the bill immediately, and the House failed the next day to muster enough votes to override the veto. Congress then sent Bush a troop-funding measure without timelines for withdrawing from Iraq.

The high-water mark for House opposition to Bush's policy was reached earlier in the legislative session: on February 16 members voted 246 to 182 in favor of a nonbinding resolution stating disapproval of Bush's surge. Seventeen Republicans joined all but two Democrats in voting for the resolution. House sentiment was tested again on July 12 when Democrats mustered 223 votes in favor of a free-standing bill (HR 2956) establishing a time schedule for troop withdrawals; the bill stalled in the Senate.

In late summer, a majority of members in both chambers again voted in favor of a measure that would have changed Iraq policy. This time, rather than targeting spending, the bill required minimum rest periods for troops between deployments. Drafted by Democrats in reaction to the fifteen-month-long deployments that had become standard, the proposal was widely seen as a back-door attempt to force the drawdown of troop levels. The House adopted this language (HR 3159) in August, and fifty-six senators twice voted in favor of it, falling four votes short of the number needed to quash a Senate filibuster. The decisive Senate vote was taken on September 19; Democrats failed to pick up the votes of several Republicans who had been considered likely to support this approach.

Bush prevailed in all these legislative struggles even though several leading Republicans had joined the chorus of dissent. Prominent among the Republicans calling for a course shift were Sen. Richard G. Lugar of Indiana, the ranking Republican on the Foreign Relations Committee, and Sen. John Warner of Virginia, the ranking Republican on the Armed Services Committee. Lugar made his views known in an impassioned speech on the Senate floor on June 25. "In my judgment, the costs and risks of continuing down the current path outweigh the potential benefits that might be achieved," Lugar said. "Persisting indefinitely with the surge strategy will delay policy adjustments that have a better chance of protecting our vital interests over the long term." Warner, a former Navy secretary, made his strongest comments against the war on August 23 after meeting at the White House with several of the president's aides. Calling for some troops to be withdrawn by year's end, he said, "We simply cannot as a nation stand and continue to put troops at continuous risk of loss of life and limb without beginning to take some decisive action." Their statements aside, Lugar and Warner voted against the major Democratic-sponsored measures during 2007 that tried to force the president to shift course.

Over the summer Bush spoke before friendly audiences of military communities and veterans, giving a series of speeches linking U.S. presence in Iraq to the overall "war on terror" and warning of tragic results if Congress forced troop withdrawals. The president's August 22 speech to the annual convention of the Veterans of Foreign Wars in Kansas City raised some controversy. He warned that withdrawing from Iraq would have disastrous consequences, like the U.S. pullout from Vietnam thirty-five years earlier. "Then, as now, people argued that the real problem was America's presence, and that if we would just withdraw, the killing would end," he said. "The world would learn just how costly these misimpressions would be." Many Democrats and historians rejected Bush's analogy, saying the president had oversimplified complex situations in both Vietnam and Iraq.

Shifts in Public Opinion

Once it became clear in late 2003 that rapid success in Iraq was unlikely, public opinion about the Iraq War closely tracked developments on the ground in Iraq—particularly the rate at which U.S. service personnel were being killed. Except for a few periods of success, such as the three Iraqi elections in 2005, public opinion polls in the United States showed a downward trend in support for the war and for Bush's handling of it. Public dissatisfaction with the conduct of the war was the primary reason voters gave the Democrats control of Congress in the November 2006 elections.

Public opinion shifted noticeably during 2007; polls started with strongly negative reactions to the Iraq War, then gradually moved to a more supportive posture as the surge appeared to be working to improve security for civilians and military personnel in Iraq. Immediately after Bush's January 10 speech announcing the surge, a national survey by the Pew Research Center for the People and the Press found broad opposition to the plan: 61 percent of respondents opposed the surge and only 31 percent favored it. Opposition was strongest among Democrats (82 percent), and 62 percent of independents also opposed the surge. Only Republicans backed the plan, by a margin of 60 percent to 33 percent.

Additional surveys by the Pew Center and other polling firms found that opinion continued to be negative through the first half of the year, when there was a dramatic upsurge of violence in Iraq, including some of the highest monthly rates of U.S. combat deaths since the 2003 invasion. More than 100 U.S. soldiers and marines were killed in each month in April, May, and June, the first time three consecutive months had seen that many deaths. The violence tapered off by the summer, however, and many Americans started to believe that the surge was working. By November, a Pew Center survey found Americans evenly divided (at 48 percent on each side) as to whether the U.S. military effort was going "very well" or "fairly well" versus "not too well" or "not at all well." This was a sharp turnaround from earlier in the year. A slight majority of 54 percent, though, still said the United States should "bring the troops home" rather than keep them in Iraq. The survey also found that positive news in Iraq had not yet translated into good news for Bush: the president's approval rating slid from 33 percent in a February poll to 30 percent in November, partly because of concerns about the slowing economy.

Rising Costs of the War

One of the most complicated—and politically controversial—issues in 2007 about the Iraq War was its cost. Because the Bush administration often sought money for military operations in Iraq through "emergency" supplemental spending bills, rather than through the standard budgeting process, the total cost of the war for any given year generally did not appear in any single budget line item. Moreover, the administration combined spending for military operations in Afghanistan and Iraq, making it difficult to determine how much the United States spent on either war. By 2007 the various estimates of the cost of the wars had become part of the political debate, with opponents emphasizing the long-term costs, such as interest payments on money borrowed for funding, and supporters playing down such costs.

Perhaps the most definitive estimate of the costs to date was made on October 24 by the Congressional Budget Office (CBO), which examined the range of expenses incurred since 2001 for the "war on terror," which included the Iraq War. In testimony to the House Budget Committee, CBO director Peter Orszag stated that Congress had appropriated $602 billion for military operations in Afghanistan, Iraq, and other anti-terrorism efforts from September 2001 through fiscal year 2007, which ended September 30. Orzsag estimated that about $412 billion of that amount had been spent in Iraq (including $368 billion for military operations, $19 billion for aid to Iraqi security forces, and $25 billion for foreign aid and U.S. diplomacy). Interest payments on the money borrowed to fund the wars through fiscal 2007 would add another $415 billion through fiscal year 2017, Orzsag said, putting the total cost at just over $1 trillion.

At the Budget Committee's request, the CBO also prepared cost estimates for the next decade (through fiscal year 2017), using two scenarios. In one scenario, the U.S. troop level would be reduced to 30,000 by 2010, and the ten-year cost would be $570 billion. In the other, the troop level would be reduced to 75,000 by 2013, resulting in a ten-year cost of $1.055 trillion. The only definitive figure in either scenario was the administration's request of $196 billion for the first year, fiscal 2008.

Impact of Iraq on the U.S. Armed Forces

More evidence emerged in 2007 that the prolonged wars in Afghanistan and Iraq were putting enormous strains on the nation's military services, particularly the U.S. Army, the Marine Corps, and the National Guard and Reserve forces. Long tours of duty in the war zones and "stop-loss" orders that extended soldiers' enlistments past their stated expiration dates had angered many service personnel and their families and had made recruiting new service members more difficult.

The military's personnel problems became obvious on April 11, when the Pentagon announced that all active-duty army soldiers in or on their way to Iraq and Afghanistan would be required to serve fifteen-month tours of duty, rather than the standard twelve-month tours. Defense Secretary Gates called the order "a difficult and necessary interim step" and acknowledged that it asked "a lot" of service personnel and their families. The order did not apply to the Marine Corps (which had a standard seven-month tour of duty but shorter home leaves than did the Army), or the National Guard or Army Reserve.

Army officials reported in October that they had met their target of 80,000 new recruits during fiscal year 2007—but only by handing out ever-larger enlistment bonuses and reducing minimum requirements; in the latter case, the army enlisted recruits who would have been rejected previously because of past behavior problems or even criminal offenses, such as theft or drug abuse. One worrisome trend was that the army started its 2008 fiscal year on October 1 with a record-low number of recruits who had signed up for delayed entry into basic training. The army's goal was about 20,000 of these "delayed entry" recruits, but it started the year with only about 7,400.

Following is the text of a nationally televised speech by President George W. Bush on January 10, 2007, describing his order of some 20,000 additional U.S. troops to Iraq.

DOCUMENT

President Bush Explains the "Surge" of Additional U.S. Troops to Iraq

January 10, 2007

Good evening. Tonight in Iraq, the Armed Forces of the United States are engaged in a struggle that will determine the direction of the global war on terror and our safety here at home. The new strategy I outline tonight will change America's course in Iraq and help us succeed in the fight against terror.

When I addressed you just over a year ago, nearly 12 million Iraqis had cast their ballots for a unified and democratic nation. The elections of 2005 were a stunning achievement. We thought that these elections would bring the Iraqis together and that as we trained Iraqi security forces, we could accomplish our mission with fewer American troops.

But in 2006, the opposite happened. The violence in Iraq, particularly in Baghdad, overwhelmed the political gains the Iraqis had made. Al Qaida terrorists and Sunni insurgents recognized the mortal danger that Iraq's elections posed for their cause, and they responded with outrageous acts of murder aimed at innocent Iraqis. They blew up one of the holiest shrines in Shi'a Islam, the Golden Mosque of Samarra, in a calculated effort to provoke Iraq's Shi'a population to retaliate. Their strategy worked. Radical Shi'a elements, some supported by Iran, formed death squads. And the result was a vicious cycle of sectarian violence that continues today.

The situation in Iraq is unacceptable to the American people, and it is unacceptable to me. Our troops in Iraq have fought bravely. They have done everything we have asked them to do. Where mistakes have been made, the responsibility rests with me.

It is clear that we need to change our strategy in Iraq. So my national security team, military commanders, and diplomats conducted a comprehensive review. We consulted Members of Congress from both parties, our allies abroad, and distinguished outside experts. We benefited from the thoughtful recommendations of the Iraq Study Group, a bipartisan panel led by former Secretary of State James Baker and former Congressman Lee Hamilton. In our discussions, we all agreed that there is no magic formula for success in Iraq. And one message came through loud and clear: Failure in Iraq would be a disaster for the United States.

The consequences of failure are clear. Radical Islamic extremists would grow in strength and gain new recruits. They would be in a better position to topple moderate governments, create chaos in the region, and use oil revenues to fund their ambitions. Iran would be emboldened in its pursuit of nuclear weapons. Our enemies would have a safe haven from which to plan and launch attacks on the American people. On September the 11th, 2001, we saw what a refuge for extremists on the other side of the world could bring to the streets of our own cities. For the safety of our people, America must succeed in Iraq.

The most urgent priority for success in Iraq is security, especially in Baghdad. Eighty percent of Iraq's sectarian violence occurs within 30 miles of the capital. This violence is splitting Baghdad into sectarian enclaves and shaking the confidence of all Iraqis. Only Iraqis can end the sectarian violence and secure their people, and their Government has put forward an aggressive plan to do it.

Our past efforts to secure Baghdad failed for two principal reasons: There were not enough Iraqi and American troops to secure neighborhoods that had been cleared of terrorists and insurgents, and there were too many restrictions on the troops we did have. Our military commanders reviewed the new Iraqi plan to ensure that it addressed these mistakes. They report that it does. They also report that this plan can work.

Now let me explain the main elements of this effort. The Iraqi Government will appoint a military commander and two deputy commanders for their capital. The Iraqi Government will deploy Iraqi Army and National Police brigades across Baghdad's nine districts. When these forces are fully deployed, there will be 18 Iraqi Army and National Police brigades committed to this effort, along with local police. These Iraqi forces will operate from local police stations, conducting patrols and setting up checkpoints and going door to door to gain the trust of Baghdad residents.

This is a strong commitment, but for it to succeed, our commanders say the Iraqis will need our help. So America will change our strategy to help the Iraqis carry out their campaign to put down sectarian violence and bring security to the people of Baghdad. This will require increasing American force levels. So I've committed more than 20,000 additional American troops to Iraq. The vast majority of them—five brigades—will be deployed to Baghdad. These troops will work alongside Iraqi units and be embedded in their formations. Our troops will have a well-defined mission: to help Iraqis clear and secure neighborhoods; to help them protect the local population; and to help ensure that the Iraqi forces left behind are capable of providing the security that Baghdad needs.

Many listening tonight will ask why this effort will succeed when previous operations to secure Baghdad did not. Well, here are the differences. In earlier operations, Iraqi and American forces cleared many neighborhoods of terrorists and insurgents, but when our forces moved on to other targets, the killers returned. This time, we'll have the force levels we need to hold the areas that have been cleared. In earlier operations, political and sectarian interference prevented Iraqi and American forces from going into neighborhoods that are home to those fueling the sectarian violence. This time, Iraqi and American forces will have a green light to enter those neighborhoods, and Prime Minister Maliki has pledged that political or sectarian interference will not be tolerated.

I have made it clear to the Prime Minister and Iraq's other leaders that America's commitment is not open-ended. If the Iraqi Government does not follow through on its promises, it will lose the support of the American people and it will lose the support of the Iraqi people. Now is the time to act. The Prime Minister understands this. Here is what he told his people just last week: "The Baghdad security plan will not provide a safe haven for any outlaws, regardless of their sectarian or political affiliation."

This new strategy will not yield an immediate end to suicide bombings, assassinations, or IED [improvised explosive device] attacks. Our enemies in Iraq will make every effort to ensure that our television screens are filled with images of death and suffering. Yet over time, we can expect to see Iraqi troops chasing down murderers, fewer brazen acts of terror, and growing trust and cooperation from Baghdad's residents. When this happens, daily life will improve, Iraqis will gain confidence in their leaders, and the Government will have the breathing space it needs to make progress in other critical areas. Most of Iraq's Sunni and Shi'a want to live together in peace, and reducing the violence in Baghdad will help make reconciliation possible.

A successful strategy for Iraq goes beyond military operations. Ordinary Iraqi citizens must see that military operations are accompanied by visible improvements in their neighborhoods and communities. So America will hold the Iraqi Government to the benchmarks it has announced.

To establish its authority, the Iraqi Government plans to take responsibility for security in all of Iraq's Provinces by November. To give every Iraqi citizen a stake in the country's economy, Iraq will pass legislation to share oil revenues among all Iraqis. To show that it is committed to delivering a better life, the Iraqi Government will spend $10 billion of its own money on reconstruction and infrastructure projects that will create new jobs. To empower local leaders, Iraqis plan to hold provincial elections later this year. And to allow more Iraqis to reenter their nation's political life, the Government will reform de-Ba'athification laws and establish a fair process for considering amendments to Iraq's Constitution.

America will change our approach to help the Iraqi Government as it works to meet these benchmarks. In keeping with the recommendations of the Iraq Study Group, we will increase the embedding of American advisers in Iraqi Army units and partner a coalition brigade with every Iraqi Army division. We will help the Iraqis build a larger and better equipped army, and we will accelerate the training of Iraqi forces, which remains the essential U.S. security mission in Iraq. We will give our commanders and civilians greater flexibility to spend funds for economic assistance. We will double the number of Provincial Reconstruction Teams. These teams bring together military and civilian experts to help local Iraqi communities pursue reconciliation, strengthen the moderates, and speed the transition to Iraqi self-reliance. And Secretary Rice will soon appoint a reconstruction coordinator in Baghdad to ensure better results for economic assistance being spent in Iraq.

As we make these changes, we will continue to pursue Al Qaida and foreign fighters. Al Qaida is still active in Iraq. Its home base is Anbar Province. Al Qaida has helped make Anbar the most violent area of Iraq outside the capital. A captured Al Qaida document describes the terrorists' plan to infiltrate and seize control of the Province. This would bring Al Qaida closer to its goals of taking down Iraq's democracy, building a radical Islamic empire, and launching new attacks on the United States, at home and abroad.

Our military forces in Anbar are killing and capturing Al Qaida leaders, and they are protecting the local population. Recently, local tribal leaders have begun to show their willingness to take on Al Qaida. And as a result, our commanders believe we have an opportunity to deal a serious blow to the terrorists. So I have given orders to increase American forces in Anbar Province by 4,000 troops. These troops will work with Iraqi and tribal forces to keep up the pressure on the terrorists. America's men and women in uniform took away Al Qaida's safe haven in Afghanistan, and we will not allow them to reestablish it in Iraq.

Succeeding in Iraq also requires defending its territorial integrity and stabilizing the region in the face of extremist challenges. This begins with addressing Iran and Syria. These two regimes are allowing terrorists and insurgents to use their territory to move in and out of Iraq. Iran is providing material support for attacks on American troops. We will disrupt the attacks on our forces. We'll interrupt the flow of support from Iran and Syria, and we will seek out and destroy the networks providing advanced weaponry and training to our enemies in Iraq.

We're also taking other steps to bolster the security of Iraq and protect American interests in the Middle East. I recently ordered the deployment of an additional carrier strike group to the region. We will expand intelligence sharing and deploy Patriot air defense systems to reassure our friends and allies. We will work with the Governments of

Turkey and Iraq to help them resolve problems along their border. And we will work with others to prevent Iran from gaining nuclear weapons and dominating the region.

We will use America's full diplomatic resources to rally support for Iraq from nations throughout the Middle East. Countries like Saudi Arabia, Egypt, Jordan, and the Gulf States need to understand that an American defeat in Iraq would create a new sanctuary for extremists and a strategic threat to their survival. These nations have a stake in a successful Iraq that is at peace with its neighbors, and they must step up their support for Iraq's unity Government. We endorse the Iraqi Government's call to finalize an international compact that will bring new economic assistance in exchange for greater economic reform. And on Friday, Secretary Rice will leave for the region to build support for Iraq and continue the urgent diplomacy required to help bring peace to the Middle East.

The challenge playing out across the broader Middle East is more than a military conflict. It is the decisive ideological struggle of our time. On one side are those who believe in freedom and moderation; on the other side are extremists who kill the innocent and have declared their intention to destroy our way of life. In the long run, the most realistic way to protect the American people is to provide a hopeful alternative to the hateful ideology of the enemy by advancing liberty across a troubled region. It is in the interests of the United States to stand with the brave men and women who are risking their lives to claim their freedom and to help them as they work to raise up just and hopeful societies across the Middle East.

From Afghanistan to Lebanon to the Palestinian Territories, millions of ordinary people are sick of the violence and want a future of peace and opportunity for their children. And they are looking at Iraq. They want to know: Will America withdraw and yield the future of that country to the extremists, or will we stand with the Iraqis who have made the choice for freedom?

The changes I have outlined tonight are aimed at ensuring the survival of a young democracy that is fighting for its life in a part of the world of enormous importance to American security. Let me be clear: The terrorists and insurgents in Iraq are without conscience, and they will make the year ahead bloody and violent. Even if our new strategy works exactly as planned, deadly acts of violence will continue, and we must expect more Iraqi and American casualties. The question is whether our new strategy will bring us closer to success. I believe that it will.

Victory will not look like the ones our fathers and grandfathers achieved. There will be no surrender ceremony on the deck of a battleship. But victory in Iraq will bring something new in the Arab world—a functioning democracy that polices its territory, upholds the rule of law, respects fundamental human liberties, and answers to its people. A democratic Iraq will not be perfect, but it will be a country that fights terrorists instead of harboring them, and it will help bring a future of peace and security for our children and our grandchildren.

This new approach comes after consultations with Congress about the different courses we could take in Iraq. Many are concerned that the Iraqis are becoming too dependent on the United States, and therefore, our policy should focus on protecting Iraq's borders and hunting down Al Qaida. Their solution is to scale back America's efforts in Baghdad or announce the phased withdrawal of our combat forces. We carefully considered these proposals, and we concluded that to step back now would force a collapse of the Iraqi Government, tear the country apart, and result in mass killings on an unimaginable scale. Such a scenario would result in our troops being forced to stay in Iraq even longer and confront an enemy that is even more lethal. If we increase our support at this crucial moment and help the Iraqis break the current cycle of violence, we can hasten the day our troops begin coming home.

In the days ahead, my national security team will fully brief Congress on our new strategy. If Members have improvements that can be made, we will make them. If circumstances change, we will adjust. Honorable people have different views, and they will voice their criticisms. It is fair to hold our views up to scrutiny. And all involved have a responsibility to explain how the path they propose would be more likely to succeed.

Acting on the good advice of Senator Joe Lieberman and other key Members of Congress, we will form a new, bipartisan working group that will help us come together across party lines to win the war on terror. This group will meet regularly with me and my administration; it will help strengthen our relationship with Congress. We can begin by working together to increase the size of the active Army and Marine Corps, so that America has the Armed Forces we need for the 21st century. We also need to examine ways to mobilize talented American civilians to deploy overseas, where they can help build democratic institutions in communities and nations recovering from war and tyranny.

In these dangerous times, the United States is blessed to have extraordinary and selfless men and women willing to step forward and defend us. These young Americans understand that our cause in Iraq is noble and necessary and that the advance of freedom is the calling of our time. They serve far from their families, who make the quiet sacrifices of lonely holidays and empty chairs at the dinner table. They have watched their comrades give their lives to ensure our liberty. We mourn the loss of every fallen American, and we owe it to them to build a future worthy of their sacrifice.

Fellow citizens, the year ahead will demand more patience, sacrifice, and resolve. It can be tempting to think that America can put aside the burdens of freedom. Yet times of testing reveal the character of a nation, and throughout our history, Americans have always defied the pessimists and seen our faith in freedom redeemed. Now America is engaged in a new struggle that will set the course for a new century. We can and we will prevail.

We go forward with trust that the Author of Liberty will guide us through these trying hours. Thank you and good night.

SOURCE: U.S. Executive Office of the President. "Address to the Nation on the War on Terror in Iraq." January 10, 2007. *Weekly Compilation of Presidential Documents* 41, no. 2 (January 15, 2007): 19–23. Washington, D.C.: National Archives and Records Administration. www.gpoaccess.gov/wcomp/v43no2.html (accessed February 25, 2008).

OTHER HISTORIC DOCUMENTS OF INTEREST

FROM THIS VOLUME

- Intelligence estimate of Iran's nuclear threat, p. 112
- U.S. diplomacy with Iran concerning Iraq, p. 257
- CIA's "secret prisons" in Europe, p. 275
- Treatment of veterans at Walter Reed Army Hospital, p. 364

- National Intelligence Estimate on terrorism threats, p. 335
- Security situation in Iraq, p. 407
- Iraqi government, p. 477
- The situation in Afghanistan, p. 547

From previous *Historic Documents*

- UN secretary-general on situation in Iraq, *2006*, p. 702
- Iraq Study Group report on U.S. policy in Iraq, *2006*, p. 725

DOCUMENT IN CONTEXT

State of the Union Address and Democratic Response

JANUARY 23, 2007

When President George W. Bush stood before a joint session of Congress on January 23 to deliver his annual State of the Union address, he faced a Democratic majority in both chambers of Congress for the first time in his presidency. Although many legislators in the audience had campaigned on their opposition to the Iraq War and Americans had expressed their concern about the conduct of the war, including the president's recent decision to deploy more troops, the president confidently discussed the troop surge, arguing that it offered "the best chance for success." He asked Americans to give the new plan "a chance to work."

On the domestic front, perhaps reflecting Democratic control of Congress and his own lame-duck status, the president proposed only a modest package of economic, health, education, and energy initiatives. And, as he had in previous years, he called on Democrats and Republicans to work together. "Like many before us, we can work through our differences, and we can achieve big things for the American people," he said. "Our citizens don't much care which side of the aisle we sit on, as long as we're willing to cross that aisle when there is work to be done." Democrats were polite throughout the speech; one Republican senator commented it was the first time in his four years in the Senate that he "didn't see a hiss or a boo."

The joint session of Congress gathered to hear the State of the Union may have been the high-water mark for comity between the parties in 2007. Again and again, the slim Democratic majorities tried to push their ambitious legislative program through Congress, at times with the help of some powerful Republican allies. But time after time they were blocked by the determined Republican minorities who used procedural rules to defeat or force changes in Democratic measures and by a president who made effective use of vetoes and veto threats. Republicans explained that they were acting out of principle to block what they described as Democrats' liberal tax-and-spend policies and efforts to weaken national security in the face of terrorist threats.

The American public gave clear signs that it was tiring of the polarization and partisan politics in Washington, D.C. As the Iraq War continued and fears of economic recession mounted, approval ratings for both the president and Congress dropped to record lows in the late summer and autumn before recovering somewhat toward the end of the year. For example, an Associated Press-Ipsos poll released on October 4 found that just 31 percent of those surveyed approved of the job Bush was doing, while only 22 percent approved of Congress's performance. Only 26 percent said they thought the country was heading in the right direction.

The President's Domestic Agenda

Bush began his speech by congratulating the new Democratic Speaker of the House. "Tonight, I have the high privilege and distinct honor of my own, as the first president to begin the State of the Union message with these words: 'Madame Speaker,'" he said, then turned to shake the hand of Rep. Nancy Pelosi of California, the first woman to hold that post.

The president next offered a domestic agenda that was modest in comparison to past years, a nod perhaps to the realization that neither party would be likely to concede on major tenets of party principles going into a presidential election year. Having decided to address the state of the economy in a separate speech, Bush simply called for a balanced budget within five years and urged legislators to make permanent the income tax cuts enacted during his first term. He also encouraged members of Congress to "work together" to overhaul entitlement programs, such as Social Security, Medicare, and Medicaid, to make them financially sound, but he offered no new proposals for achieving that goal. Bush had suffered a major defeat in 2005 when he was forced to shelve his plan to partially privatize Social Security in face of opposition from members of both parties and the public.

The president's major new initiative was a plan to expand Americans' access to health insurance by offering a tax incentive. Under current law, those who obtain coverage through their employers do not pay taxes on employers' contributions to their health plans. People purchasing their own coverage must do so with after-tax income. Under the Bush plan, employer contributions would be considered taxable income. All taxpayers who purchase coverage, whether through their employers or on their own, would be eligible for tax deductions of $7,500 for individuals or $15,000 for families. Democrats gave the proposal short shrift, suggesting that while the plan might enable some uninsured families to buy health insurance, families whose policies cost more than $15,000 per year would see their taxes go up.

Bush also proposed working with states to shift Medicaid and Medicare funding from "safety net" hospitals, which treat large numbers of uninsured patients, to grants that would "give our nation's governors more money and more flexibility to get private health insurance to those most in need." That approach flew in the face of a Democratic initiative to expand federal health insurance for low-income children to cover some middle-income families. In what turned out to be one of the major legislative battles of the year, Democrats twice passed and Bush twice vetoed an expansion of the children's health program. They finally compromised on a short-term extension of the program.

Addressing energy issues, Bush proposed reducing gasoline consumption by 20 percent over ten years, largely by tweaking or expanding previous proposals. His plan would require that 35 billion gallons of renewable and alternative fuels be used annually by 2017, compared with the annual 7.5 billion gallons currently required by 2012. He also renewed his 2006 proposal to give the president the authority to overhaul fuel-efficiency standards for passenger cars. As in 2006, Democrats and some Republicans preferred to write specific fuel-efficiency standards into law. In December, after almost a year of negotiations, partisan bickering, and veto threats, Congress passed and Bush signed the first statutory increase in fuel-economy standards in thirty-two years. The measure (HR 6—PL 110-140) also required that at least 36 billion gallons of ethanol and other biofuels be incorporated into gasoline annually by 2022. Democrats had wanted a package of tax incentives to encourage development of alternative sources of energy and a renewable-electricity standard, but dropped it to win Senate passage and avoid a presidential veto.

On most other domestic issues, Bush's speech reiterated proposals from previous years. He called for a reauthorization of the landmark education law he signed in 2002, known as No Child Left Behind, but efforts foundered when lawmakers were unable to bridge concerns among members of both parties. Another top priority was immigration reform. Although House Republicans had killed an immigration reform bill in 2006, Bush asked again for Congress to approve an overhaul that would include a temporary guest-worker program and a pathway for some illegal immigrants to become citizens. "We need to uphold the great tradition of the melting pot that welcomes and assimilates new arrivals," he said. Despite the president's active support and months of bipartisan negotiations, a reform bill died in the Senate after leaders twice failed to get enough votes to cut off debate.

President Bush did not win much of what he wanted on domestic issues, but neither did the Democrats. Under Pelosi's leadership, Democrats in the House began the year by passing, with significant Republican support, the six measures they had promised voters during the midterm elections. Only three were signed into law: the first hike in the minimum wage in a decade, enactment of the September 11 commission's recommendations for protecting the country against future terrorist attacks, and lower interest rates on student loans.

The Iraq War

Bush saved discussion of his new Iraq strategy until the end of his speech. He acknowledged that the U.S. goal in the war had shifted from stopping a dictator from building weapons of mass destruction to preventing escalating sectarian violence from erupting into an all-out civil war that could spill over into neighboring countries and further destabilize the Middle East. "This is not the fight we entered in Iraq, but it is the fight we're in," Bush said, reiterating plans to send what turned out to be 35,000 additional troops to help quell the violence, particularly in Baghdad, and to root out insurgents in Anbar province. Only when Baghdad was secure, Bush explained, could the Iraqi government be held to its commitments to meet the United States' series of benchmarks. These included reconciliation of Iraq's Shiites, Sunnis, and Kurds; an equitable sharing of the country's oil revenues; and a rebuilding of the nation's infrastructure, which in many places was in worse shape than before Saddam Hussein was ousted in March and April 2003. "We went into this [war] largely united, in our assumption and our convictions," Bush told the lawmakers in both parties. "And whatever you voted for, you did not vote for failure. Our country is pursuing a new strategy in Iraq, and I ask you to give it a chance to work."

Giving the Democratic response to the address, newly elected senator Jim Webb of Virginia contended that the majority of Americans no longer supported "the way this war is being fought" and called for a new direction. "Not a precipitous withdrawal," he said, but "an immediate shift toward strong, regionally-based diplomacy, a policy that takes our soldiers off the streets of Iraq's cities, and a formula, that will, in short order, allow our combat forces to leave Iraq."

In the end, however, Democrats had no alternative but to give Bush's surge a chance. In April lawmakers cleared a supplemental war spending measure that would have required the administration to start withdrawing troops from Iraq in about six months, completing the process by March 2008 at the latest. Bush vetoed the bill on May 1, and the next day the House fell short of the two-thirds majority needed to override the veto. Democrats then had to decide whether to send him the same bill, risking another veto that they would not be able to reverse—along with charges that they were delaying supplies to troops at war. Democratic leaders backed down. Congress passed and the president signed a new version of the spending bill (PL 110-28) that did not contain a withdrawal timeline.

That standoff set the pattern for the rest of the year. Subsequent congressional efforts to legislate change in Iraq policy faltered repeatedly, usually in the Senate where the Democrats' slim majority was not large enough to overcome the sixty-vote hurdle to break a filibuster. The closest the Democrats came to changing war policy was in September, when an amendment by Senator Webb came within four votes of cutting off debate. Webb's amendment would have required minimum rest times for troops between deployments, effectively resulting in a drawdown of combat forces in Iraq. Webb lost some key support after U.S. generals said the amendment would tie their hands in Iraq and Afghanistan.

By all measures, the surge helped to reduce the violence in Iraq, at least temporarily. The number of civilian and military casualties had dropped markedly by the end of the year. American combat deaths numbered 899, compared with 821 in 2006. Most of the 2007 deaths occurred during the first part of the year, when the surge was just getting under way. The Bush administration also benefited when Sunni tribal chiefs in Anbar province turned against combatants affiliated with al Qaeda, a move that helped to reduce violent incidents in that region. There were few indications that the dysfunctional Iraqi government was meeting any of the benchmarks it had committed to, however, and some analysts warned that violence might start to erupt again once the temporary surge began to wind down in early 2008.

Following are the texts of the State of the Union address delivered by President George W. Bush to a joint session of Congress on January 23, 2007, and the Democratic response, delivered by Sen. Jim Webb of Virginia shortly afterward.

DOCUMENT

The State of the Union Address

January 23, 2007

Thank you very much. And tonight I have the high privilege and distinct honor of my own as the first President to begin the State of the Union message with these words: Madam Speaker.

In his day, the late Congressman Thomas D'Alesandro, Jr., from Baltimore, Maryland, saw Presidents Roosevelt and Truman at this rostrum. But nothing could compare with the sight of his only daughter, Nancy, presiding tonight as Speaker of the House of Representatives. Congratulations, Madam Speaker.

Two Members of the House and Senate are not with us tonight, and we pray for the recovery and speedy return of Senator Tim Johnson and Congressman Charlie Norwood.

Madam Speaker, Vice President Cheney, Members of Congress, distinguished guests, and fellow citizens: The rite of custom brings us together at a defining hour when decisions are hard and courage is needed. We enter the year 2007 with large endeavors underway and others that are ours to begin. In all of this, much is asked of us. We must have the will to face difficult challenges and determined enemies and the wisdom to face them together.

Some in this Chamber are new to the House and the Senate, and I congratulate the Democrat majority. Congress has changed, but not our responsibilities. Each of us is guided by our own convictions, and to these we must stay faithful. Yet we're all held to the same standards and called to serve the same good purposes: to extend this Nation's prosperity; to spend the people's money wisely; to solve problems, not leave them to future generations; to guard America against all evil; and to keep faith with those we have sent forth to defend us.

We're not the first to come here with a Government divided and uncertainty in the air. Like many before us, we can work through our differences, and we can achieve big things for the American people. Our citizens don't much care which side of the aisle we sit on, as long as we're willing to cross that aisle when there is work to be done. Our job is to make life better for our fellow Americans and to help them build a future of hope and opportunity, and this is the business before us tonight.

A future of hope and opportunity begins with a growing economy, and that is what we have. We're now in the 41st month of uninterrupted job growth, a recovery that has created 7.2 million new jobs so far. Unemployment is low; inflation is low; wages are rising. This economy is on the move, and our job is to keep it that way, not with more government but with more enterprise.

Next week, I'll deliver a full report on the state of our economy. Tonight I want to discuss three economic reforms that deserve to be priorities for this Congress.

First, we must balance the Federal budget. We can do so without raising taxes. What we need is spending discipline in Washington, DC. We set a goal of cutting the deficit in half by 2009 and met that goal 3 years ahead of schedule. Now let us take the next step. In the coming weeks, I will submit a budget that eliminates the Federal deficit within the next 5 years. I ask you to make the same commitment. Together we can restrain the spending appetite of the Federal Government, and we can balance the Federal budget.

Next, there is the matter of earmarks. These special interest items are often slipped into bills at the last hour—when not even C-SPAN is watching. [*Laughter*] In 2005 alone, the number of earmarks grew to over 13,000 and totaled nearly $18 billion. Even worse, over 90 percent of the earmarks never make it to the floor of the House and Senate. They are dropped into committee reports that are not even part of the bill that arrives on my desk. You didn't vote them into law; I didn't sign them into law; yet they're treated as if they have the force of law. The time has come to end this practice. So let us work together to reform the budget process, expose every earmark to the light of day and to a vote in Congress, and cut the number and cost of earmarks at least in half by the end of this session.

And finally, to keep this economy strong, we must take on the challenge of entitlements. Social Security and Medicare and Medicaid are commitments of conscience, and so it is our duty to keep them permanently sound. Yet we're failing in that duty. And this failure will one day leave our children with three bad options: huge tax increases; huge deficits; or huge and immediate cuts in benefits. Everyone in this Chamber knows this to be true, yet somehow we have not found it in ourselves to act. So let us work together and do it now. With enough good sense and good will, you and I can fix Medicare and Medicaid and save Social Security.

Spreading opportunity and hope in America also requires public schools that give children the knowledge and character they need in life. Five years ago, we rose above partisan differences to pass the No Child Left Behind Act, preserving local control, raising standards, and holding schools accountable for results. And because we acted, students are performing better in reading and math and minority students are closing the achievement gap.

Now the task is to build on the success without watering down standards, without taking control from local communities, and without backsliding and calling it reform. We can lift student achievement even higher by giving local leaders flexibility to turn around failing schools and by giving families with children stuck in failing schools the right to choose someplace better. We must increase funds for students who struggle and make sure these children get the special help they need. And we can make sure our children are prepared for the jobs of the future and our country is more competitive by strengthening math and science skills. The No Child Left Behind Act has worked for America's children, and I ask Congress to reauthorize this good law.

A future of hope and opportunity requires that all our citizens have affordable and available health care. When it comes to health care, Government has an obligation to care for the elderly, the disabled, and poor children, and we will meet those responsibilities. For all other Americans, private health insurance is the best way to meet their needs.

But many Americans cannot afford a health insurance policy, and so tonight I propose two new initiatives to help more Americans afford their own insurance. First, I propose a standard tax deduction for health insurance that will be like the standard tax deduction for dependents. Families with health insurance will pay no income or payroll tax—or payroll taxes on $15,000 of their income. Single Americans with health insurance will pay no income or payroll taxes on $7,500 of their income. With this reform, more than 100 million men, women, and children who are now covered by employer-provided insurance will benefit from lower tax bills. At the same time, this reform will level the playing field for those who do not get health insurance through their job. For Americans who now purchase health insurance on their own, this proposal would mean a substantial tax savings—$4,500 for a family of four making $60,000 a year. And for the millions of other Americans who have no health insurance at all, this deduction would help put a basic, private health insurance plan within their reach. Changing the Tax Code is a vital and necessary step to making health care affordable for more Americans.

My second proposal is to help the States that are coming up with innovative ways to cover the uninsured. States that make basic private health insurance available to all their citizens should receive Federal funds to help them provide this coverage to the poor and the sick. I have asked the Secretary of Health and Human Services to work with Congress to take existing Federal funds and use them to create Affordable Choices grants. These grants would give our Nation's Governors more money and more flexibility to get private health insurance to those most in need.

There are many other ways that Congress can help. We need to expand health savings accounts. We need to help small businesses through association health plans. We need to reduce costs and medical errors with better information technology. We will encourage price transparency. And to protect good doctors from junk lawsuits, we need to pass medical liability reform. In all we do, we must remember that the best health care decisions are not made by government and insurance companies but by patients and their doctors.

Extending hope and opportunity in our country requires an immigration system worthy of America, with laws that are fair and borders that are secure. When laws and borders are routinely violated, this harms the interests of our country. To secure our border, we're doubling the size of the Border Patrol and funding new infrastructure and technology.

Yet even with all these steps, we cannot fully secure the border unless we take pressure off the border, and that requires a temporary-worker program. We should establish a legal and orderly path for foreign workers to enter our country to work on a temporary basis. As a result, they won't have to try to sneak in, and that will leave border agents free to chase down drug smugglers and criminals and terrorists. We'll enforce our immigration laws at the worksite and give employers the tools to verify the legal status of their workers, so there's no excuse left for violating the law.

We need to uphold the great tradition of the melting pot that welcomes and assimilates new arrivals. We need to resolve the status of the illegal immigrants who are already in our country without animosity and without amnesty. Convictions run deep in this Capitol when it comes to immigration. Let us have a serious, civil, and conclusive debate, so that you can pass and I can sign comprehensive immigration reform into law.

Extending hope and opportunity depends on a stable supply of energy that keeps America's economy running and America's environment clean. For too long, our Nation has been dependent on foreign oil. And this dependence leaves us more vulnerable to hostile regimes and to terrorists who could cause huge disruptions of oil shipments and raise the price of oil and do great harm to our economy.

It's in our vital interest to diversify America's energy supply. The way forward is through technology. We must continue changing the way America generates electric power by even greater use of clean coal technology, solar and wind energy, and clean, safe nuclear power. We need to press on with battery research for plug-in and hybrid vehicles and expand the use of clean diesel vehicles and biodiesel fuel. We must continue investing in new methods of producing ethanol, using everything from wood chips to grasses to agricultural wastes.

We made a lot of progress, thanks to good policies here in Washington and the strong response of the market. And now even more dramatic advances are within reach. Tonight I ask Congress to join me in pursuing a great goal. Let us build on the work we've done and reduce gasoline usage in the United States by 20 percent in the next 10 years. When we do that, we will have cut our total imports by the equivalent of three-quarters of all the oil we now import from the Middle East.

To reach this goal, we must increase the supply of alternative fuels by setting a mandatory fuels standard to require 35 billion gallons of renewable and alternative fuels in 2017—and that is nearly five times the current target. At the same time, we need to reform and modernize fuel economy standards for cars the way we did for light trucks and conserve up to 8 1/2 billion more gallons of gasoline by 2017.

Achieving these ambitious goals will dramatically reduce our dependence on foreign oil, but it's not going to eliminate it. And so as we continue to diversify our fuel supply, we must step up domestic oil production in environmentally sensitive ways. And to further protect America against severe disruptions to our oil supply, I ask Congress to double the current capacity of the Strategic Petroleum Reserve.

America is on the verge of technological breakthroughs that will enable us to live our lives less dependent on oil. And these technologies will help us be better stewards of the environment, and they will help us to confront the serious challenge of global climate change.

A future of hope and opportunity requires a fair, impartial system of justice. The lives of our citizens across our Nation are affected by the outcome of cases pending in our Federal courts. We have a shared obligation to ensure that the Federal courts have enough judges to hear those cases and deliver timely rulings. As President, I have a duty to nominate qualified men and women to vacancies on the Federal bench, and the United States Senate has a duty as well, to give those nominees a fair hearing and a prompt up-or-down vote on the Senate floor.

For all of us in this room, there is no higher responsibility than to protect the people of this country from danger. Five years have come and gone since we saw the scenes and felt the sorrow that the terrorists can cause. We've had time to take stock of our situation. We've added many critical protections to guard the homeland. We know with certainty that the horrors of that September morning were just a glimpse of what the terrorists intend for us—unless we stop them.

With the distance of time, we find ourselves debating the causes of conflict and the course we have followed. Such debates are essential when a great democracy faces great questions. Yet one question has surely been settled: that to win the war on terror, we must take the fight to the enemy.

From the start, America and our allies have protected our people by staying on the offense. The enemy knows that the days of comfortable sanctuary, easy movement, steady financing, and free flowing communications are long over. For the terrorists, life since 9/11 has never been the same.

Our success in this war is often measured by the things that did not happen. We cannot know the full extent of the attacks that we and our allies have prevented, but here is some of what we do know. We stopped an Al Qaida plot to fly a hijacked airplane into the tallest building on the west coast. We broke up a Southeast Asian terror cell grooming operatives for attacks inside the United States. We uncovered an Al Qaida cell developing anthrax to be used in attacks against America. And just last August, British authorities uncovered a plot to blow up passenger planes bound for America over the Atlantic Ocean. For each life saved, we owe a debt of gratitude to the brave public servants who devote their lives to finding the terrorists and stopping them.

Every success against the terrorists is a reminder of the shoreless ambitions of this enemy. The evil that inspired and rejoiced in 9/11 is still at work in the world. And so long as that's the case, America is still a nation at war.

In the mind of the terrorists, this war began well before September the 11th and will not end until their radical vision is fulfilled. And these past 5 years have given us a much clearer view of the nature of this enemy. Al Qaida and its followers are Sunni extremists possessed by hatred and commanded by a harsh and narrow ideology. Take almost any principle of civilization, and their goal is the opposite. They preach with threats, instruct with bullets and bombs, and promise paradise for the murder of the innocent.

Our enemies are quite explicit about their intentions. They want to overthrow moderate governments and establish safe havens from which to plan and carry out new attacks on our country. By killing and terrorizing Americans, they want to force our country to retreat from the world and abandon the cause of liberty. They would then be free to impose their will and spread their totalitarian ideology. Listen to this warning from the late terrorist Zarqawi: "We will sacrifice our blood and bodies to put an end to your dreams, and what is coming is even worse." Usama bin Laden declared: "Death is better than living on this Earth with the unbelievers among us."

These men are not given to idle words, and they are just one camp in the Islamist radical movement. In recent times, it has also become clear that we face an escalating danger from Shi'a extremists who are just as hostile to America and are also determined to dominate the Middle East. Many are known to take direction from the regime in Iran, which is funding and arming terrorists like Hizballah—a group second only to Al Qaida in the American lives it has taken.

The Shi'a and Sunni extremists are different faces of the same totalitarian threat. Whatever slogans they chant when they slaughter the innocent, they have the same wicked purposes. They want to kill Americans, kill democracy in the Middle East, and gain the weapons to kill on an even more horrific scale.

In the sixth year since our Nation was attacked, I wish I could report to you that the dangers have ended. They have not. And so it remains the policy of this Government to use every lawful and proper tool of intelligence, diplomacy, law enforcement, and military action to do our duty, to find these enemies, and to protect the American people.

This war is more than a clash of arms; it is a decisive ideological struggle. And the security of our Nation is in the balance. To prevail, we must remove the conditions that inspire blind hatred and drove 19 men to get onto airplanes and to come and kill us. What every terrorist fears most is human freedom, societies where men and women make their own choices, answer to their own conscience, and live by their hopes instead of their resentments. Free people are not drawn to violent and malignant ideologies, and most will choose a better way when they're given a chance. So we advance our own security interests by helping moderates and reformers and brave voices for democracy. The great question of our day is whether America will help men and women in the Middle East to build free societies and share in the rights of all humanity. And I say, for the sake of our own security, we must.

In the last 2 years, we've seen the desire for liberty in the broader Middle East, and we have been sobered by the enemy's fierce reaction. In 2005, the world watched as the citizens of Lebanon raised the banner of the Cedar Revolution. They drove out the Syrian occupiers and chose new leaders in free elections. In 2005, the people of Afghanistan defied the terrorists and elected a democratic legislature. And in 2005, the Iraqi people held three national elections, choosing a transitional government, adopting the most progressive, democratic Constitution in the Arab world, and then electing a Government under that Constitution. Despite endless threats from the killers in their midst, nearly 12 million Iraqi citizens came out to vote in a show of hope and solidarity that we should never forget.

A thinking enemy watched all of these scenes, adjusted their tactics, and in 2006, they struck back. In Lebanon, assassins took the life of Pierre Gemayel, a prominent participant in the Cedar Revolution. Hizballah terrorists, with support from Syria and Iran, sowed conflict in the region and are seeking to undermine Lebanon's legitimately elected Government. In Afghanistan, Taliban and Al Qaida fighters tried to regain power by regrouping and engaging Afghan and NATO forces. In Iraq, Al Qaida and other Sunni extremists blew up one of the most sacred places in Shi'a Islam, the Golden Mosque of Samarra. This atrocity, directed at a Muslim house of prayer, was designed to provoke retaliation from Iraqi Shi'a, and it succeeded. Radical Shi'a elements, some of whom receive support from Iran, formed death squads. The result was a tragic escalation of sectarian rage and reprisal that continues to this day.

This is not the fight we entered in Iraq, but it is the fight we're in. Every one of us wishes this war were over and won. Yet it would not be like us to leave our promises unkept, our friends abandoned, and our own security at risk. Ladies and gentlemen, on this day, at this hour, it is still within our power to shape the outcome of this battle. Let us find our resolve and turn events toward victory.

We're carrying out a new strategy in Iraq, a plan that demands more from Iraq's elected Government and gives our forces in Iraq the reinforcements they need to complete their mission. Our goal is a democratic Iraq that upholds the rule of law, respects the rights of its people, provides them security, and is an ally in the war on terror.

In order to make progress toward this goal, the Iraqi Government must stop the sectarian violence in its capital. But the Iraqis are not yet ready to do this on their own. So we're deploying reinforcements of more than 20,000 additional soldiers and marines to Iraq. The vast majority will go to Baghdad, where they will help Iraqi forces to clear and secure neighborhoods and serve as advisers embedded in Iraqi Army units. With Iraqis in the lead, our forces will help secure the city by chasing down the terrorists, insurgents, and the roaming death squads. And in Anbar Province, where Al Qaida terrorists have gathered and local forces have begun showing a willingness to fight them, we're sending an additional 4,000 United States marines, with orders to find the terrorists and clear them out. We didn't drive Al Qaida out of their safe haven in Afghanistan only to let them set up a new safe haven in a free Iraq.

The people of Iraq want to live in peace, and now it's time for their Government to act. Iraq's leaders know that our commitment is not open-ended. They have promised to deploy more of their own troops to secure Baghdad, and they must do so. They pledged that they will confront violent radicals of any faction or political party, and they need to follow through and lift needless restrictions on Iraqi and coalition forces, so these troops can achieve their mission of bringing security to all of the people of Baghdad. Iraq's leaders have committed themselves to a series of benchmarks: to achieve reconciliation; to share oil revenues among all of Iraq's citizens; to put the wealth of Iraq into the rebuilding of Iraq; to allow more Iraqis to reenter their nation's civic life; to hold local elections; and to take responsibility for security in every Iraqi Province. But for all of this to happen, Baghdad must be secure, and our plan will help the Iraqi Government take back its capital and make good on its commitments.

My fellow citizens, our military commanders and I have carefully weighed the options. We discussed every possible approach. In the end, I chose this course of action because it provides the best chance for success. Many in this Chamber understand that America must not fail in Iraq, because you understand that the consequences of failure would be grievous and far-reaching.

If American forces step back before Baghdad is secure, the Iraqi Government would be overrun by extremists on all sides. We could expect an epic battle between Shi'a extremists backed by Iran and Sunni extremists aided by Al Qaida and supporters of the old regime. A contagion of violence could spill out across the country, and in time, the entire region could be drawn into the conflict.

For America, this is a nightmare scenario; for the enemy, this is the objective. Chaos is the greatest ally—their greatest ally in this struggle. And out of chaos in Iraq would emerge an emboldened enemy with new safe havens, new recruits, new resources, and an even greater determination to harm America. To allow this to happen would be to ignore the lessons of September the 11th and invite tragedy. Ladies and gentlemen, nothing is more important at this moment in our history than for America to succeed in the Middle East, to succeed in Iraq, and to spare the American people from this danger.

This is where matters stand tonight, in the here and now. I have spoken with many of you in person. I respect you and the arguments you've made. We went into this largely united, in our assumptions and in our convictions. And whatever you voted for, you did not vote for failure. Our country is pursuing a new strategy in Iraq, and I ask you to give it a chance to work, and I ask you to support our troops in the field and those on their way.

The war on terror we fight today is a generational struggle that will continue long after you and I have turned our duties over to others. And that's why it's important to work together so our Nation can see this great effort through. Both parties and both branches should work in close consultation. It's why I've proposed to establish a special advisory council on the war on terror, made up of leaders in Congress from both political parties. We will share ideas for how to position America to meet every challenge that confronts us. We'll show our enemies abroad that we are united in the goal of victory.

And one of the first steps we can take together is to add to the ranks of our military so that the American Armed Forces are ready for all the challenges ahead. Tonight I ask the Congress to authorize an increase in the size of our active Army and Marine Corps by 92,000 in the next 5 years. A second task we can take on together is to design and establish a volunteer civilian reserve corps. Such a corps would function much like our military reserve. It would ease the burden on the Armed Forces by allowing us to hire civilians with critical skills to serve on missions abroad when America needs them. It would give people across America who do not wear the uniform a chance to serve in the defining struggle of our time.

Americans can have confidence in the outcome of this struggle because we're not in this struggle alone. We have a diplomatic strategy that is rallying the world to join in the fight against extremism. In Iraq, multinational forces are operating under a mandate from the United Nations. We're working with Jordan and Saudi Arabia and Egypt and the Gulf States to increase support for Iraq's Government.

The United Nations has imposed sanctions on Iran and made it clear that the world will not allow the regime in Tehran to acquire nuclear weapons. With the other members of the Quartet—the U.N., the EU, and Russia—we're pursuing diplomacy to help bring peace to the Holy Land and pursuing the establishment of a democratic Palestinian state living side by side with Israel in peace and security. In Afghanistan, NATO has taken the lead in turning back the Taliban and Al Qaida offensive—the first time the Alliance has deployed forces outside the North Atlantic area. Together with our partners in China and Japan, Russia and South Korea, we're pursuing intensive diplomacy to achieve a Korean Peninsula free of nuclear weapons.

We will continue to speak out for the cause of freedom in places like Cuba, Belarus, and Burma, and continue to awaken the conscience of the world to save the people of Darfur.

American foreign policy is more than a matter of war and diplomacy. Our work in the world is also based on a timeless truth: To whom much is given, much is required. We hear the call to take on the challenges of hunger and poverty and disease, and that is precisely what America is doing. We must continue to fight HIV/AIDS, especially on the continent of Africa. Because you funded the Emergency Plan for AIDS Relief, the number of people receiving lifesaving drugs has grown from 50,000 to more than 800,000 in 3 short years. I ask you to continue funding our efforts to fight HIV/AIDS. And I ask you to provide $1.2 billion over 5 years so we can combat malaria in 15 African countries.

I ask that you fund the Millennium Challenge Account, so that American aid reaches the people who need it, in nations where democracy is on the rise and corruption is in retreat. And let us continue to support the expanded trade and debt relief that are the best hope for lifting lives and eliminating poverty.

When America serves others in this way, we show the strength and generosity of our country. These deeds reflect the character of our people. The greatest strength we have is the heroic kindness and courage and self-sacrifice of the American people. You see this spirit often if you know where to look, and tonight we need only look above to the gallery.

Dikembe Mutombo grew up in Africa amid great poverty and disease. He came to Georgetown University on a scholarship to study medicine, but Coach John Thompson took a look at Dikembe and had a different idea. [*Laughter*] Dikembe became a star in the NBA and a citizen of the United States, but he never forgot the land of his birth or the duty to share his blessings with others. He built a brand new hospital in his old hometown. A friend has said of this good-hearted man: "Mutombo believes that God has given him this opportunity to do great things." And we are proud to call this son of the Congo a citizen of the United States of America.

After her daughter was born, Julie Aigner-Clark searched for ways to share her love of music and art with her child. So she borrowed some equipment and began filming children's videos in her basement. The Baby Einstein Company was born, and in just 5 years, her business grew to more than $20 million in sales. In November 2001, Julie sold Baby Einstein to Walt Disney Company, and with her help, Baby Einstein has grown into a $200 million business. Julie represents the great enterprising spirit of America. And she is using her success to help others—producing child safety videos with John Walsh of the National Center for Missing and Exploited Children. Julie says of her new project: "I believe it is the most important thing I have ever done. I believe that children have the right to live in a world that is safe." And so tonight we are pleased to welcome this talented business entrepreneur and generous social entrepreneur, Julie Aigner-Clark.

Three weeks ago, Wesley Autrey was waiting at a Harlem subway station with his two little girls when he saw a man fall into the path of a train. With seconds to act, Wesley jumped onto the tracks, pulled the man into the space between the rails, and held him as the train passed right above their heads. He insists he's not a hero. He says: "We got guys and girls overseas dying for us to have our freedoms. We have got to show each other some love." There is something wonderful about a country that produces a brave and humble man like Wesley Autrey.

Tommy Rieman was a teenager pumping gas in Independence, Kentucky, when he enlisted in the United States Army. In December 2003, he was on a reconnaissance mission in Iraq when his team came under heavy enemy fire. From his Humvee, Sergeant Rieman returned fire. He used his body as a shield to protect his gunner. He was shot in the chest and arm and received shrapnel wounds to his legs, yet he refused medical attention and stayed in the fight. He helped to repel a second attack, firing grenades at the enemy's position. For his exceptional courage, Sergeant Rieman was awarded the Silver Star. And like so many other Americans who have volunteered to defend us, he has earned the respect and the gratitude of our entire country.

In such courage and compassion, ladies and gentlemen, we see the spirit and character of America. And these qualities are not in short supply. This is a decent and honorable country—and resilient too. We've been through a lot together. We've met challenges and faced dangers, and we know that more lie ahead. Yet we can go forward with confidence, because the State of our Union is strong, our cause in the world is right, and tonight that cause goes on. God bless.

See you next year. Thank you for your prayers.

SOURCE: U.S. Executive Office of the President. "Address Before a Joint Session of the Congress on the State of the Union." January 23, 2007. *Weekly Compilation of Presidential Documents* 43, no. 4 (January 29, 2007): 57–64. Washington, D.C.: National Archives and Records Administration. www.gpoaccess.gov/wcomp/v43no4.html (accessed September 23, 2007).

DOCUMENT

Democratic Response to the State of the Union Address by Sen. Jim Webb (Va.)

January 23, 2007

Good evening.

I'm Senator Jim Webb, from Virginia, where this year we will celebrate the 400th anniversary of the settlement of Jamestown—an event that marked the first step in the long journey that has made us the greatest and most prosperous nation on earth.

It would not be possible in this short amount of time to actually rebut the President's message, nor would it be useful. Let me simply say that we in the Democratic Party hope that this administration is serious about improving education and healthcare for all Americans, and addressing such domestic priorities as restoring the vitality of New Orleans.

Further, this is the seventh time the President has mentioned energy independence in his state of the union message, but for the first time this exchange is taking place in a Congress led by the Democratic Party. We are looking for affirmative solutions that will strengthen our nation by freeing us from our dependence on foreign oil, and spurring a wave of entrepreneurial growth in the form of alternate energy programs. We look forward to working with the President and his party to bring about these changes.

There are two areas where our respective parties have largely stood in contradiction, and I want to take a few minutes to address them tonight. The first relates to how we see the health of our economy—how we measure it, and how we ensure that its benefits are properly shared among all Americans. The second regards our foreign policy—how we might bring the war in Iraq to a proper conclusion that will also allow us to continue to fight the war against international terrorism, and to address other strategic concerns that our country faces around the world.

When one looks at the health of our economy, it's almost as if we are living in two different countries. Some say that things have never been better. The stock market is at an all-time high, and so are corporate profits. But these benefits are not being fairly shared. When I graduated from college, the average corporate CEO made 20 times what the average worker did; today, it's nearly 400 times. In other words, it takes the average worker more than a year to make the money that his or her boss makes in one day.

Wages and salaries for our workers are at all-time lows as a percentage of national wealth, even though the productivity of American workers is the highest in the world. Medical costs have skyrocketed. College tuition rates are off the charts. Our manufacturing base is being dismantled and sent overseas. Good American jobs are being sent along with them.

In short, the middle class of this country, our historic backbone and our best hope for a strong society in the future, is losing its place at the table. Our workers know this, through painful experience. Our white-collar professionals are beginning to understand it, as their jobs start disappearing also. And they expect, rightly, that in this age of globalization, their government has a duty to insist that their concerns be dealt with fairly in the international marketplace.

In the early days of our republic, President Andrew Jackson established an important principle of American-style democracy—that we should measure the health of our society not at its apex, but at its base. Not with the numbers that come out of Wall Street, but with the living conditions that exist on Main Street. We must recapture that spirit today.

And under the leadership of the new Democratic Congress, we are on our way to doing so. The House just passed a minimum wage increase, the first in ten years, and the Senate will soon follow. We've introduced a broad legislative package designed to regain the trust of the American people. We've established a tone of cooperation and consensus that extends beyond party lines. We're working to get the right things done, for the right people and for the right reasons.

With respect to foreign policy, this country has patiently endured a mismanaged war for nearly four years. Many, including myself, warned even before the war began that it was unnecessary, that it would take our energy and attention away from the larger war against terrorism, and that invading and occupying Iraq would leave us strategically vulnerable in the most violent and turbulent corner of the world.

I want to share with all of you a picture that I have carried with me for more than 50 years. This is my father, when he was a young Air Force captain, flying cargo planes during the Berlin Airlift. He sent us the picture from Germany, as we waited for him, back here at home. When I was a small boy, I used to take the picture to bed with me every night, because for more than three years my father was deployed, unable to live with us full-time, serving overseas or in bases where there was no family housing. I still keep it, to remind me of the sacrifices that my mother and others had to make, over and over again, as my father gladly served our country. I was proud to follow in his footsteps, serving as a Marine in Vietnam. My brother did as well, serving as a Marine helicopter pilot. My son has joined the tradition, now serving as an infantry Marine in Iraq.

Like so many other Americans, today and throughout our history, we serve and have served, not for political reasons, but because we love our country. On the political issues—those matters of war and peace, and in some cases of life and death—we trusted the judgment of our national leaders. We hoped that they would be right, that they would measure with accuracy the value of our lives against the enormity of the national interest that might call upon us to go into harm's way.

We owed them our loyalty, as Americans, and we gave it. But they owed us—sound judgment, clear thinking, concern for our welfare, a guarantee that the threat to our country was equal to the price we might be called upon to pay in defending it.

The President took us into this war recklessly. He disregarded warnings from the national security adviser during the first Gulf War, the chief of staff of the army, two former commanding generals of the Central Command, whose jurisdiction includes Iraq, the director of operations on the Joint Chiefs of Staff, and many, many others with great integrity and long experience in national security affairs. We are now, as a nation, held hostage to the predictable—and predicted—disarray that has followed.

The war's costs to our nation have been staggering.

Financially.

The damage to our reputation around the world.

The lost opportunities to defeat the forces of international terrorism.

And especially the precious blood of our citizens who have stepped forward to serve.

The majority of the nation no longer supports the way this war is being fought; nor does the majority of our military. We need a new direction. Not one step back from the war against international terrorism. Not a precipitous withdrawal that ignores the possibility of further chaos. But an immediate shift toward strong regionally-based diplomacy, a policy that takes our soldiers off the streets of Iraq's cities, and a formula that will in short order allow our combat forces to leave Iraq.

On both of these vital issues, our economy and our national security, it falls upon those of us in elected office to take action.

Regarding the economic imbalance in our country, I am reminded of the situation President Theodore Roosevelt faced in the early days of the 20th century. America was then, as now, drifting apart along class lines. The so-called robber barons were unapologetically raking in a huge percentage of the national wealth. The dispossessed workers at the bottom were threatening revolt.

Roosevelt spoke strongly against these divisions. He told his fellow Republicans that they must set themselves "as resolutely against improper corporate influence on the one hand as against demagogy and mob rule on the other." And he did something about it.

As I look at Iraq, I recall the words of former general and soon-to-be President Dwight Eisenhower during the dark days of the Korean War, which had fallen into a bloody stalemate. "When comes the end?" asked the General who had commanded our forces in Europe during World War Two. And as soon as he became President, he brought the Korean War to an end.

These Presidents took the right kind of action, for the benefit of the American people and for the health of our relations around the world. Tonight we are calling on this

President to take similar action, in both areas. If he does, we will join him. If he does not, we will be showing him the way.

Thank you for listening. And God bless America.

SOURCE: Democratic Party. "The Democratic Response to the State of the Union." January 23, 2007. www.democrats.org/a/2007/01/the_democratic_16.php (accessed January 24, 2007).

Other Historic Documents of Interest

From this volume

- Immigration reform, p. 158
- State of the U.S. economy, p. 449
- Vetoes of children's health care legislation, p. 585

From previous *Historic Documents*

- 2006 State of the Union and Democratic response, *2006,* p. 23
- President Bush on Social Security, *2005,* p. 111

February

DOCUMENT IN CONTEXT

Palestinian Peace Efforts and the Hamas Takeover of Gaza

FEBRUARY 8, JUNE 20, AND SEPTEMBER 19, 2007

The entire landscape of the long-running conflict between Israelis and Palestinians shifted dramatically in June 2007 with the violent takeover of the impoverished Gaza Strip by Hamas, the Islamic Resistance Movement. The victory by Hamas in the latest of numerous bouts of Palestinian infighting was a devastating blow to the secular Fatah faction, which had been dominant since the rise of Palestinian nationalism in the twentieth century. Through the Palestinian Authority government led by President Mahmoud Abbas, Fatah still ran those portions of the West Bank not controlled by Israel, but it faced increasing popular support for Hamas.

The takeover of Gaza by Hamas posed serious security challenges for Israel, which suddenly faced what its government called a "hostile territory" on its borders. Over the longer term, Hamas's control of Gaza also was certain to make much more difficult any negotiations toward a peace agreement to produce a legitimate Palestinian state.

The Hamas military victory, which came less than fifteen months after Hamas scored a landmark political victory in legislative elections, took place with incredible speed, and possibly without advance planning by Hamas leaders. In just one week, Hamas fighters routed their Fatah opponents in bloody fighting along the crowded streets and alleyways of the tiny territory bordered by Egypt, Israel, and the Mediterranean Sea. The fighting left 161 Gazans dead, including 41 civilians, and at least another 700 wounded, according to the Palestinian Center for Human Rights. This decisive round of intra-Palestinian violence was in addition to the deaths and destruction inflicted on Gaza by the ongoing conflict between the Palestinians and the Israeli government. By the end of 2007, more than 4,500 Palestinians had been killed since the onset of the second intifada (or "uprising") in September 2000. Israel counted some 1,100 dead during that same period.

Agreement at Mecca

The most recent starting point for the long series of events leading to the June 2007 takeover was the death of longtime Palestinian leader Yasir Arafat in November 2004. Mahmoud Abbas, a former aide to Arafat, assumed the leadership of Fatah, the faction led by Arafat, and won the election in January 2005 to head the Palestinian Authority government. The election of Abbas, considered a moderate, raised hopes in Israel, in Western capitals, and among many Palestinians for a government that could negotiate the long-sought peace agreement with Israel. These hopes were spurred later in 2005 by Israel's withdrawal of its settlements and military posts from the Gaza Strip, one of the territories it had seized from its Arab neighbors during the June 1967 War.

In January 2006, however, Hamas unexpectedly won a majority of seats in parliamentary elections, demonstrating that many Palestinians were fed up with the corruption and incompetence of the Fatah faction that Abbas represented. The election results meant that Palestinians had chosen a two-headed government, with Abbas of Fatah serving as president and Hamas leaders running the government ministries. Because Hamas refused to recognize Israel and had been responsible for numerous attacks against the Jewish state, the United States, most European countries, and Israel designated Hamas a terrorist organization and refused to have any direct contact with the Hamas-led government. The United States and the European Union cut off millions of dollars in aid to the Palestinians, and Israel halted the transfer of about $50 million in taxes it collected each month on behalf of the Palestinian Authority.

The two-headed government, run by factions long at odds with each other, proved untenable and led late in 2006 to an escalation of the long-running violence in Gaza between Fatah and Hamas loyalists. Abbas threatened in December 2006 to call new elections if the violence did not stop; this essentially was a negotiating tactic to win concessions from Hamas because neither side could be certain of the results of another round of voting.

The factional fighting between Fatah and Hamas continued into early 2007 despite several attempts at a cease-fire and an inconclusive meeting in late January between Abbas and Khaled Meshaal, the exiled political leader of Hamas living in Damascus. Saudi Arabia intervened in early February, summoning leaders from both Palestinian factions to the holy city of Mecca for face-to-face negotiations. These talks produced an agreement on February 8 calling for a unity government and an end to the factional conflict. The Mecca agreement did not address a central demand of Israel and Western governments: that Hamas recognize Israel as a state and declare an end to terrorism targeted against it.

The cease-fire between Fatah and Hamas was both fragile and temporary, one that papered over fundamental differences. The two sides needed five weeks to negotiate the precise terms and composition of a unity government, which was announced on March 14. Ismail Haniyeh, a relative moderate in the Hamas hierarchy, retained his post as prime minister, but most cabinet members in the previous Hamas government were replaced by members of Fatah and other political movements. Key among them was Salam Fayyad, a US.-educated economist and former official of the International Monetary Fund who regained the finance ministry portfolio he had held in an earlier Fatah-led government. Fayyad's role in the new government was important because he was one of the few Palestinian leaders implicitly trusted by the Western governments whose aid was necessary to keep the territories' economy functioning.

The new unity government adopted traditional, mainstream Palestinian positions on key issues, notably ending the Israeli occupation, and avoided the radical rhetoric favored by most Hamas politicians. Regarding Israel, the government's program called for "consolidating the calm and expanding it to become a comprehensive reciprocal truce"—a reference to the Israeli-Palestinian cease-fire dating from February 2005 that was in danger of collapsing. This last point was in keeping with recent Hamas policy; the group said it would honor a long-term, mutual truce (*hudna* in Arabic) but would neither recognize nor negotiate with Israel.

The unity agreement achieved two of its main goals, at least temporarily: it sustained the lull in the intra-Palestinian violence, and it offered a reason for the United

States and the European nations to loosen some of their restrictions on contact with and aid to the Palestinian government. Several Arab governments, which had been reluctant to subsidize the feuding Palestinians, also transferred millions of dollars to the government. Israel denounced the new government because it failed to recognize Israel's right to exist.

Within weeks, however, it became clear that the Mecca agreement and the formation of a unity government had not resolved the underlying conflicts between Fatah and Hamas, most importantly over control of the numerous Palestinian security services. Thousands of armed men were loyal to these various services; some were commanded by individual Fatah leaders (not necessarily by Abbas or his party), and others were run by Hamas, including the Executive Force and the Izzedine al-Qassam Brigades, the movement's underground armed wing that had launched suicide bombings, rockets, and other attacks against Israel.

One precipitating event for renewed conflict was Abbas's appointment on March 18 of Muhammad Dahlan as his national security adviser. A senior Fatah official based in Gaza, Dahlan was particularly disliked by Hamas officials because he used Palestinian security services aggressively to suppress Hamas in the mid-1990s. Hamas leaders also were angered by the news that security services loyal to Fatah would receive millions of dollars in U.S. aid that was intended to bolster Abbas.

An outbreak of fighting between Fatah and Hamas loyalists in mid-May left more than thirty Palestinians dead. In the midst of the intra-Palestinian fighting, Israel intervened with repeated air and artillery strikes against Hamas targets in Gaza. Israel said it was responding to an upsurge in rocket attacks launched from Gaza into Israeli territory, notably the border town of Sderot. The Israeli attacks also appeared aimed at weakening Hamas relative to Fatah.

The Palestinian factions on May 20 signed yet another agreement calling for a halt to the violence, but that accord lasted only about two weeks. Fighting resumed early in June, at first between Hamas members and Abbas's Presidential Guard, which some reports indicated had been reinforced in Gaza with U.S.-trained units. In a televised speech on June 6—marking the fortieth anniversary of the beginning of the June 1967 War during which Israel captured Gaza, the West Bank, East Jerusalem, and the Golan Heights—Abbas warned that a civil war was possible, caused by internal divisions that he said were as calamitous to the Palestinians as the forty-year Israeli occupation had been.

An Explosion of Factional Violence

More fighting among Palestinian factions broke out on June 7 and escalated rapidly while senior politicians bickered over control of the various security services. The precipitating events for a major explosion of violence occurred on June 10. That morning, Hamas gunmen threw a member of Abbas's Presidential Guard from the roof of a high-rise building, killing him. Angry Fatah supporters went on a rampage, attacking several Hamas members and killing a pro-Hamas imam at a Gaza mosque. Late in the day, Fatah gunmen threw a Hamas loyalist to his death from the twelfth floor of another building.

These assassinations sparked a wave of revenge killings through the crowded towns and refugee camps of the Gaza Strip, even inside hospitals. In several cases, senior Fatah or Hamas leaders were dragged from their homes or offices and executed in the streets, gangland style. Fatah and Hamas officials tried to halt the violence, saying they had negotiated another cease-fire, but the fighting continued. The home of Hamas leader Haniyeh and Abbas's Gaza City office were targeted by rocket attacks. Hamas appeared to gain the upper hand on June 12 as its fighters took control of several police stations and other facilities run by Fatah security services. By this point it appeared that many of Fatah's security personnel were sitting out the fight.

The tipping point occurred on June 13, when Hamas seized control of Fatah headquarters in Gaza City. UN aid agencies issued an appeal for calm and reported that at least fifty-nine people had been killed in more than a week of fighting, including two Palestinian employees of the UN refugee agency. The next day, June 14, Hamas took control of all remaining Fatah posts in Gaza, including the Preventive Security Services headquarters, the main police station, which fell after an intense battle with some 500 Fatah security agents defending it. The last bastion of Fatah authority to fall to Hamas was the seafront presidential compound, which was defended by members of the Presidential Guard. (Abbas was in the West Bank during the fighting and was never endangered by it, although he later accused Hamas of trying to kill him.) Fatah gunmen attempted a counterattack on June 16, storming the parliament and other government buildings, but they were quickly beaten back.

Dahlan—the man most wanted by Hamas—was nowhere in sight. He lived in Gaza but reportedly had gone to Egypt for knee surgery before the fighting broke out. Dahlan later surfaced in the West Bank, as did several other high-level Fatah leaders who had fled Gaza. After the Hamas takeover, approximately 3,000 other Gazans attempted to escape, most via the single border crossing into Egypt but several hundred via the main crossing into Israel. It was unclear how many actually made it.

Hamas officials proclaimed their victory the "second liberation" of Gaza; the first was Israel's withdrawal of its civilian settlements and military bases from the territory in 2005. The sudden Hamas triumph led many people to wonder if the violence had been part of a deliberate Hamas strategy to gain control of Gaza. Abbas and other Fatah leaders accused Hamas of deliberately staging a "coup," noting that Hamas fighters had been well armed and prepared. Some Hamas leaders insisted the opposite was true: Fatah had provoked the fighting with the intention of ousting Hamas from Gaza. In any event, many observers reported that the fighters on both sides took command of the situation while their political leaders stood by, unable to exercise much control.

In a letter to Arab leaders on June 23, Hamas officials said "we never wanted" a full takeover of Gaza. The group blamed the events on Israel and the United States, which were intent on "conducting a coup against the results of the legislative elections" that Hamas had won in 2006. Hamas also insisted on the "indivisibility of the nation," explaining that it did not want a political separation between Gaza and the West Bank.

One of the most curious aspects of the fighting was that neither side attempted to mobilize broad public support while the guns were blazing. During previous political crises, both Fatah and Hamas had staged public demonstrations with tens of thousands of people waving yellow (for Fatah) or green (for Hamas) banners. Few civilians ventured outside their homes during the hostilities of mid-June, however,

Two Palestinian States

Acting with uncharacteristic speed, Abbas on June 14 declared a state of emergency and dismissed the unity government that had been in office for just three months. He appointed Fayyad as prime minister of a new government.

In a somewhat conciliatory speech, dismissed prime minister Haniyeh said his side wanted reconciliation and "the road is open and wide to reformulating these relations on a firm nationalistic basis." Hamas's political leader, the Damascus-based Meshaal, acknowledged that Abbas remained the president of the Palestinian Authority. "There's no one who would question or doubt that, he is an elected president, and we will cooperate with him for the sake of national interest," he said.

The new cabinet headed by Fayyad took office on June 17, but its writ was limited to the West Bank. The new government gained international support quickly, both from Arab governments that disliked Hamas and from the United States and most European countries, many of which declared they would resume aid to the Palestinian Authority now that Hamas no longer was part of it.

In a speech on June 20 to the Central Committee of the Palestine Liberation Organization (PLO), which claimed to represent all Palestinians regardless of where they lived, Abbas vowed not to have any contact with Hamas: "We say, no dialogue with those murderers; I don't know in our history any force or group that kills its own people and violates their homes and properties and insults their national symbols like these groups did in Gaza during that black week." So far as was known, Abbas held to his pledge except for a meeting in early November with several West Bank members of Hamas who were considered more moderate than their colleagues in Gaza.

The net result of the turmoil in Gaza was that each of the Palestinian territories now had its own leadership: Hamas in Gaza and Abbas's Fatah-led government in the West Bank. Suddenly, the old discussion of a "two-state solution" to the Israeli-Palestinian conflict seemed outdated, and people began talking, some more seriously than others, about the prospect of a "three-state solution." with two of them being Palestinian.

Israeli Response

For Israel, the development of what many Israelis were calling "Hamastan" next door posed practical and security problems. In practical terms, Israel had to decide whether and how to fulfill its obligations as the main source of food, water, and other supplies for the Gaza Strip. Although it had withdrawn its settlements and bases, Israel maintained tight control over the border crossings through which Gaza received nearly all of its supplies from the outside world.

Even more worrisome from an Israeli viewpoint was the security situation. Members of Hamas, Islamic Jihad, and other Islamist organizations had been launching regular rocket attacks into Israeli territory. Few of the rockets caused casualties, but the regular barrages unnerved residents of border areas, particularly those in the town of Sderot. Israel feared that the Hamas takeover of Gaza would give militants free rein to launch even more attacks.

Israeli prime minister Ehud Olmert was visiting the United States in the days after the Hamas takeover in Gaza and had suggested that Israel do everything possible to bolster Abbas. After meeting with President George W. Bush at the White House on June 19, Olmert said Israel would make "every possible effort to cooperate" with Abbas. On June 24 Olmert's government agreed, in principle, to transfer to the new Palestinian government about $600 million in tax revenue that Israel had been withholding. More symbolic support was offered on June 25, when Olmert joined President Hosni Mubarak of Egypt and King Abdullah II of Jordan in meeting with Abbas at the Egyptian resort town Sharm el-Sheikh.

Israel took another concrete step on July 20 by releasing 255 Palestinian prisoners, most of whom were Fatah members. The release of the more than 10,000 Palestinians held in Israeli jails long had been a central Palestinian demand; the Israeli government periodically freed dozens or even hundreds of prisoners as a way of demonstrating goodwill toward Abbas. In addition, Abbas and Olmert worked out an arrangement under which Israel would grant amnesty to some 200 members of the Fatah-affiliated al-Aqsa Martyrs Brigade, a West Bank–centered armed group that Israel had accused of repeated terrorist attacks. The fighters turned in their weapons and agreed not to attack Israel in exchange for amnesty.

Israel toughened its line against Hamas on September 19, when the security cabinet (composed of Olmert and top security officials) formally declared that "Hamas is a terrorist organization that has taken control of the Gaza Strip and turned it into hostile territory." By making this declaration, Israel could tighten its restrictions on the movement of people and supplies into Gaza, including the delivery of electricity and fuel. Israel directly provided most of Gaza's electricity and was the sole source of fuel for the territory's electrical power plant. The first cutbacks in Gaza's fuel supplies came in late October.

Israeli officials said their intention in stepping up pressure on Gaza was to create a situation in which frustrated Palestinians would turn against Hamas. Representatives of Hamas said that the Israelis were misleading themselves because Palestinians would blame Israel, not Hamas, for their problems. UN officials and representatives of human rights organizations accused Israel of engaging in "collective punishment" of Palestinians, including those who did not support Hamas.

The year ended with no progress on one of Israel's highest priorities: the return of Gilad Shalit, a soldier who had been captured in Israel by Palestinians in June 2006 and presumably taken into Gaza. Israel had mounted a major military offensive against Gaza in the weeks after Shalit's capture, killing several hundred Palestinians but failing to secure his return. By 2007 it appeared that Hamas was holding him as a bargaining chip.

International Response

Many international leaders had been reluctant to accept the idea of a "unity" Palestinian government that included Hamas. Suddenly they were confronted with the diplomatic headache of dealing with two Palestinian entities, one controlled entirely by Hamas. Attempting to make the best of the situation, the United States and most of its European allies—along with Israel—decided to isolate Hamas while supporting Abbas as much as possible as a tactic to undermine popular support for Hamas.

In an early step, the United States on June 18 announced the release of millions of dollars in aid for the Palestinians that was frozen after Hamas won the January 2006 elections. "We are going to support President Abbas and what he wants to do," Secretary of State Condoleezza Rice said. European nations also announced the release to Abbas's government of aid money that had been held in limbo.

On June 19 the key Arab governments of Egypt, Jordan, and Saudi Arabia also signaled their support for Abbas, in Egypt's case by moving its embassy from Gaza to the West Bank town of Ramallah, where Abbas's government was headquartered. Even so, most Arab leaders were careful not to turn away entirely from Hamas, fearing a public backlash in their own countries if they were perceived as doing the bidding of Israel and the United States.

President Bush on July 16 offered additional support to Abbas by announcing that the United States would sponsor a meeting later in the year to revive peace negotiations between Israel and the Palestinians. Held in Annapolis, Maryland, on November 27, that meeting led to low-level sessions among Israeli and Palestinian diplomats in December, with the expectation of more diplomacy in 2008 and the goal of reaching an overall peace agreement. Abbas and Olmert also began a series of regular private meetings intended to demonstrate to Palestinians that Abbas had influence outside the West Bank. Yet another effort was made at a December 18 conference in Paris, where the United States, European nations, and several Arab countries pledged $7.4 billion in aid to the Palestinians over three years, nearly one-half of which was to be provided in 2008.

The Humanitarian Situation in Gaza

Before the serious fighting in Gaza in June, most of the 1.4 million Palestinians in the territory were already living in miserable conditions. In a report issued on April 25, the UN's Relief and Works Agency for Palestinian Refugees noted that 80 percent of Gazans lived on less than $1 a day and were heavily dependent on cash subsidies and food donations. The refugee agency and the UN's World Food Program were responsible for feeding and housing nearly 1.1 million of Gaza's residents. The UN blamed Israel's restrictions on border access to Gaza for much of the territory's troubles, a charge that Israel rejected, saying Palestinian terrorism and poor governance were the root causes.

Israel sealed its border crossings with Gaza after the Hamas takeover but did allow passage into Gaza of food and medical supplies. However, Israel refused to permit shipments of raw materials for the territory's many small factories that produced clothing, furniture, and other goods for the Israeli market. The result, according to the UN, was that more than 10,000 workers, constituting a large percentage of private-sector employees in Gaza, lost their jobs. Israel also blocked the shipment of cement and other construction material, fearing it would be used by Hamas to build tunnels underneath the border. As a consequence, the UN refugee agency had to halt major projects to upgrade schools and housing in Gaza.

The one positive outcome for Gazans, no matter what their political leanings, was that once the mid-June fighting was over, the streets were relatively safe for the first time in years. Hamas security officials imposed an unusual degree of order, in part by targeting criminal gangs that long had received protection from Fatah-led governments. The major exception to the relative security in Gaza came on November 12 when Hamas gunmen fired on a large pro-Fatah demonstration in Gaza City commemorating the third anniversary of Arafat's death. Six people were reported killed, and more than fifty wounded, by the shooting and armed clashes afterward. Hamas staged its own mass rally on December 16 to mark the twentieth anniversary of the movement's founding.

Meanwhile, on the West Bank

The West Bank had been the more prosperous of the two Palestinian territories, even before the fighting in Gaza. The release of Western aid in the second half of 2007 was intended to increase that disparity and by doing so improve Abbas's standing in the West Bank while undermining public support for Hamas in Gaza.

The West Bank remained relatively peaceful during the year, with only a few clashes between Palestinians and the Israeli army, which controlled the major cities. The major exception was in late July, when Hamas supporters staged demonstrations in Nablus and several other cities; these were suppressed by Abbas's government. Even so, there were signs that popular support for Hamas in the West Bank was growing, not diminishing.

The UN's humanitarian office caused a stir on August 31 when it published maps showing that 38 percent of all West Bank land was taken up with Israeli "infrastructure," including Jewish settlements, military bases, roads accessible only to Israelis, and military areas closed to Palestinians. The maps, and the accompanying analysis by the Office for the Coordination of Humanitarian Affairs, were the most detailed studies to date of how much of the West Bank was off-limits to Palestinians. Among other issues, the information suggested the difficulty Palestinians would face in building a territorially contiguous and viable independent state on the West Bank.

Following are the texts of three documents. First, "The Text of the Mecca Agreement," issued on February 8, 2007, calling for an end to Fatah and Hamas factional conflict and establishment of a unity government. Second, the unofficial translation of the "Speech of President Abbas before the PLO Central Council Meeting, Ramallah," issued on June 20, 2007. Third, "Security Cabinet declares Gaza hostile territory," issued on September 19, 2007, by Prime Minister Ehud Olmert and the Israel Ministry of Foreign Affairs, in which Hamas was described as a "terrorist organization."

DOCUMENT

Mecca Agreement between Hamas and Fatah on a Unity Palestinian Government

February 8, 2007

Based on the generous initiative announced by Saudi King Abdullah Ben Abdul Aziz and under the sponsorship of his majesty, Fatah and Hamas Movements held in the period February 6–8, 2007 in Holy Mecca the dialogues of Palestinian conciliation and agreement and these dialogues, thanks to God, ended with success and an agreement was reached on the following:

First: to stress on banning the shedding of the Palestinian blood and to take all measures and arrangements to prevent the shedding of the Palestinian blood and to stress on the importance of national unity as basis for national steadfastness and confronting the occupation and to achieve the legitimate national goals of the Palestinian people and adopt the language of dialogue as the sole basis for solving the political disagreements on the Palestinian arena.

Within this context, we offer gratitude to the brothers in Egypt and the Egyptian security delegation in Gaza who exerted tremendous efforts to calm the conditions in Gaza Strip in the past period.

Second: Final agreement to form a Palestinian national unity government according to a detailed agreement ratified by both sides and to start on an urgent basis to take the constitutional measures to form this government.

Third: to move ahead in measures to activate and reform the PLO [Palestine Liberation Organization] and accelerate the work of the preparatory committee based on the Cairo and Damascus Understandings [previous intra-Palestinian agreements].

It has been agreed also on detailed steps between both sides on this issue.

Fourth: to stress on the principle of political partnership on the basis of the effective laws in the PNA [Palestinian National Authority] and on the basis of political pluralism according to an agreement ratified between both parties.

We gladly announce this agreement to our Palestinian masses and to the Arab and Islamic nation and to all our friends in the world. We stress on our commitment to this agreement in text and spirit so that we can devote our time to achieve our national goals and get rid of the occupation and regain our rights and devote work to the main files, mainly Jerusalem, the refugees, the Aqsa Mosque, the prisoners and detainees and to confront the wall and settlements.

SOURCE: Jerusalem Media & Communication Centre. "The Text of the Mecca Agreement." February 8, 2007. www.jmcc.org/new/07/feb/meccaagree.htm (accessed February 26, 2008).

DOCUMENT

President Abbas's Speech to the PLO Central Council

June 20, 2007

My dear sisters and brothers,

We meet today in the context of the PLO Central Council in its capacity as the highest Palestinian authority in the absence of the Palestinian National Council. This meeting will have a historical role and importance as our national project is being subjected to a clear attack that aims to destroy it; this project which we laid down its pillars with blood and sacrifices for the last forty years.

It is the project of regaining the unity of the people and homeland through the return of the dispersed people to their homeland and the establishment of a Palestinian independent state with al-Quds al-Shareef as its capital and so that this state can have democratic pillars that allow our people to grow and develop and to keep up with the modern developments and progress in the world and to end ignorance and close-mindedness and to give the next generations a chance to live a free and dignified life. . . .

We are one homeland for one people. I tried using all methods, including the continuous dialogue before, during and after Mecca Agreement with patience to solve the internal problems and to prevent the dangers of the civil war and protect the Palestinian blood and institutions so that the democratic process can proceed and so that this homeland wont face another Nakba [the "catastrophe" of the creation of Israel in 1948] that can give the occupation another chance to continue with its schemes to prevent the establishment of an independent Palestinian state within the borders of 1967.

I always remained silent in front of the wounds they caused and I made concessions that many didn't approve and all of this just because I wanted to avoid reaching a dark future that can threaten our national soil and the future of our national project. We all remember how we worked since the national reconciliation document until we reached to Mecca Agreement in order to find common denominators that can open room for unity to prevail over disunity.

But the scheme of separating Gaza from the West Bank and the establishment of a semi-state consisting of one color controlled by one wing which is known for its fanaticism and extremism. This scheme was prepared for at the military and political levels through the establishment and expansion of an armed militia belonging to this wing alone which was planning to seize the authority in our beloved Gaza Strip. . . .

These groups attacked the headquarters of the national security forces, the intelligence apparatus, the preventive security apparatus and the presidential guard forces and atrocities were committed and these acts are strange to our heritage and traditions in terms of the killings and executions in the streets and throwing strugglers from rooftops. They also executed acts of looting against the security institutions and headquarters and against NGOs [nongovernmental organizations] and against Christian institutions and even churches were attacked; a church was looted and burnt in Gaza and this church is one of the oldest churches in Palestine and this church existed in Palestine even before our presence in Palestine; there were churches and Christians who lived there and they are our brothers, now we discover that they are our enemies and that they have to leave; this explained what happened in the churches and this is the least that we can say: this is disgraceful to the Palestinian people and it is disgraceful for those who committed these acts; it is shameful for those murderers who attacked homes of hundreds of cadres of the Authority and homes of innocent civilians and against the symbols of the national sovereignty, including the Presidential Headquarters and other posts. After they occupied and destroyed the looted the presidential headquarters, some thugs sat in the offices; they said it is possible to hand over al-Muntada to some people; but we say no dialogue with those murderers; I don't know in our history any force or group that kills its own people and violates their homes and properties and insults their national symbols like these groups did in Gaza during that black week.

Why did all this happen? And under what religion was it committed? God forbid to say that these acts were done in the name of religion; the religion is innocent of such acts and of such people because Islam calls for tolerance and compassion and calls for freedom; this is not Islam. This is strange and far away from Islam; Islam is innocent from such acts or cause or sect. This happened against the courageous and steadfast Gaza and it happened only to satisfy the sick dream of establishing the Emirate of darkness and ignorance and to control by force the lives and thoughts and future of 1.5 million members of our great people in Gaza Hashem.

Those who tried to simplify matters and picture this as a conflict between Fatah and Hamas are wrong. It is a conflict between the national project and the militias' project: between the one homeland project and the project of the Emirate or the semi-state; between a project that aims to impose its conditions by force and to establish their own closed system and a project that depended on democracy, dialogue and national participation as means to solve disputes; between those who resort to assassinations and killings and executions and planning conspiracies to achieve their factional purposes and those who resort to the rule of the law and who show concern and save the unity of the homeland and people. . . .

Today, I address our great people in Gaza and I tell them that the coup project won't last long and has no future and will become a painful part in the memory as soon as possible. But we will work with you our people in Gaza so that our people in Gaza can avoid the pain and suffering resulting from this treacherous coup. We will exert all efforts to prevent the economic, financial or humanitarian siege and we will continue to show our commitment towards our people who show concern for the unity of the people and homeland and who refuse to make any compromises on the unity. Don't feel sad; you will prevail eventually.

Within this context, I affirm our total rejection to any Israeli attempt that aims to exploit the acts committed by that group in Gaza and tighten the siege on our people in Gaza. This group that committed the coup placed their own factional interests as a priority at the expense of the interests of the people and homeland; they offered the best opportunity for all those who want to separate Gaza from the West Bank and they offered this chance for those who are working to punish an entire people because of the stupid and short-sighted acts of one group. . . .

We have reached a new juncture where the previous calls for dialogue became useless; this coup has to end in all its forms, including the dissolution of the Executive Force which is the tool that executed the coup and we announced in a presidential decree that this force is an outlaw. Hamas leadership must apologize to the Palestinian people and to the PLO for the bloody coup crime they committed. They have to hand over all the PA institutions and its centers and headquarters to the new legitimate government of one unified Palestine and to work according to the decisions of this government on the basis of the law with regards to the violations, crimes, executions, looting and other violations that are still happening until this day in Gaza Strip.

I want to affirm at the same time that we are moving in our path towards the future and this group will not obstruct our work and will not succeed in derailing us from working to accomplish the freedom to our people and to ensure the security and safety of our people and their children and to enable them to get rid of the chains of poverty, deprivation, economic deterioration and unemployment.

For this purpose and in order to prevent this group from achieving its goals in disuniting the homeland and in order to stop them from inciting part of the people against another part and to prevent them from planting the seeds of continuous and permanent civil war, we discharged the former government and we declared a state of emergency and we formed a Palestinian national government consisting of independent national competent people headed by my brother Sr. Salam Fayyad so that it can assume its responsibilities in implementing a national program with security of the citizens as its basis and to secure dignified living conditions to the entire people in all parts of the homeland in the West Bank, Gaza Strip and Jerusalem.

The formation of this government in this way came to affirm that we reject monopoly of the authority by one faction or group and to demonstrate that our Palestinian society which possesses potentials and qualified people is capable of producing a government that can protect its interests and that can end as soon as possible the crisis, collapse, poverty, famine and sabotage of the institutions. . . .

As long as we are in authority we are responsible for the entire people, including those who oppose us and those who support us, even those who fought against us and we are responsible even for those people to prevent them from their darkness and killings based on the law and Islamic Law.

Palestine will remain the homeland for all its people inside and abroad.

We will cling to this slogan: one homeland for one people.

Peace be upon you and thank you.

SOURCE: Jerusalem Media & Communication Centre. "Speech of President Abbas before the PLO Central Council Meeting, Ramallah." Unofficial translation. June 20, 2007. www.jmcc.org/documents/abbasspeechjun07.htm (accessed March 5, 2008).

DOCUMENT

Israeli Security Cabinet on Gaza as a "Hostile Territory"

September 19, 2007

Prime Minister Ehud Olmert today (Wednesday), 19 September 2007, convened the Security Cabinet, in continuation of its September 5 meeting, in order to discuss possibilities for taking action in light of both the continued Kassam rocket fire from the Gaza Strip at Sderot and other communities near the Gaza Strip and the continued terrorism in the Strip.

In today's unanimous decision, it was determined:

Hamas is a terrorist organization that has taken control of the Gaza Strip and turned it into hostile territory. This organization engages in hostile activity against the State of Israel and its citizens and bears responsibility for this activity.

In light of the foregoing, it has been decided to adopt the recommendations that have been presented by the security establishment, including the continuation of military and counter-terrorist operations against the terrorist organizations. Additional sanctions will be placed on the Hamas regime in order to restrict the passage of various goods to the Gaza Strip and reduce the supply of fuel and electricity. Restrictions will also be placed on the movement of people to and from the Gaza Strip. The sanctions will be enacted following a legal examination, while taking into account both the humanitarian aspects relevant to the Gaza Strip and the intention to avoid a humanitarian crisis.

SOURCE: Israel Ministry of Foreign Affairs. "Security Cabinet Declares Gaza Hostile Territory." September 19, 2007. www.mfa.gov.il/MFA/Terrorism+Obstacle+to+Peace/Hamas+war+against+Israel/Security+Cabinet+declares+Gaza+hostile+territory+19-Sep-2007.htm (accessed March 6, 2008).

Other Historic Documents of Interest

From this volume

- Winograd Report on Israel's conduct of the war in Lebanon, p. 200
- Statements at the Annapolis Middle East Peace Conference, p. 680

From previous *Historic Documents*

- Hamas victory in Palestinian elections, *2006*, p. 13
- Prime Minister Olmert on his new government, *2006*, p. 193
- UN Security Council resolution on the war between Israel and Hezbollah, *2006*, p. 454
- Abbas on his election as president of the Palestinian Authority, *2005*, p. 27
- Israeli withdrawal from Gaza, *2005*, p. 529

DOCUMENT IN CONTEXT

Federal Response to Imported Food Safety Concerns

FEBRUARY 8 AND DECEMBER 11, 2007

In 2007 high-profile cases of contaminated pet food, toothpaste, and seafood imported from China called attention to the lack of safety controls in that nation's vast food production system and to the potential dangers lurking in food and other imports from China and other countries that received little inspection or monitoring by the U.S. government. This put new pressure on the U.S. federal food safety system and ratcheted up warnings that the system was on the verge of breakdown.

Public outcry over the unsafe Chinese products—which also included counterfeit drugs and children's toys—led the Bush administration to announce new import safety and food protection programs in November. In December the administration signed an agreement with China designed to improve the quality and safety of certain foods and drugs intended for export.

Those actions did not, however, address increasing calls for reform from consumer organizations, legislators, and scientists. In February the Government Accountability Office (GAO) put "fragmented" federal oversight of food safety on its "high-risk" list of federal programs needing "urgent attention." In a report released on December 1, a scientific advisory board to the commissioner of the Food and Drug Administration (FDA) flatly stated that the agency—responsible for overseeing the safety of about 80 percent of the nation's food supply (as well as drugs, medical devices, and cosmetics)—did "not have the capacity to ensure the safety of food for the nation." The board placed much of the blame on inadequate funding—a theme that was echoed a few days later by a broad coalition of legislators, food industry representatives, and consumer groups, who called on President George W. Bush to double the FDA's funding over the next five years.

Threat from Imports

In two of the three incidents involving Chinese imports, the problems appeared to stem from companies that deliberately and knowingly used contaminants for material gain.

By far the largest scare involved pet food. On March 16 Menu Foods—an Ontario, Canada, firm that made pet food for brands such as Iams and Science Diet and for stores such as Wal-Mart, Safeway, and Kroger—recalled more than 60 million cans and pouches of wet foods after thousands of cats and dogs fell sick. Ultimately, at least six pet food manufacturers recalled more than 100 brands of wet and dry products involving about 1 percent of the U.S. pet food supply; a minimum of 300 pet deaths were linked to the contaminated food. Menu Foods told investigators it suspected the problem was a batch of wheat gluten, a protein added to animal feed, purchased from a new supplier in China. FDA scientists confirmed that the likely source of the contamination was melamine, an inexpensive chemical frequently used in fertilizer; the chemical and its compounds had been found in the Chinese wheat gluten. Because it mimics protein in chemical tests, melamine is sometimes added to animal feed to trick buyers into thinking the feed has a higher protein content.

Based on its findings, the FDA banned importation of wheat gluten from China and urged importers to screen nearly every other kind of food and feed additive coming into the United States from China, including corn gluten, soy protein, and rice protein concentrate. By that time several American companies had discovered that they had used melamine-contaminated additives in their animal feed, sparking fears that tainted pork, poultry, and fish had entered the human food supply. As melamine is not known to pose a threat to human health, the risk to humans of eating contaminated food was thought to be extremely low. Nonetheless, some hogs and poultry were quarantined until their tissue could be tested for melamine levels.

On June 1 the FDA reacted to a second threat, advising consumers against buying toothpaste labeled as made in China and issued an "import alert" to prevent toothpaste containing diethylene glycol (DEG) from entering the United States. Unscrupulous foreign manufacturers have sometimes substituted DEG, a chemical used in antifreeze, for glycerin, a more expensive chemical commonly used as a sweetener in drugs. More than 100 people in Panama had died in 2006 after taking cough syrup containing diethylene glycol from China that had been mislabeled as glycerin.

The third incident involved Chinese farm-raised seafood from producers using antibiotics and carcinogenic pesticides banned in the United States in their production facilities. China produced about 70 percent of the world's farm-raised seafood, often in overcrowded and polluted water; the antibiotics and other drugs helped keep the fish healthy, but residues of the drugs and pesticides remained in the fish. About 80 percent of all seafood consumed in the United States was imported, with China accounting for about one-fifth of the imports. In late June the FDA announced that it would detain imports of certain Chinese farm-raised seafood, including shrimp and catfish, until testing showed that the seafood was free of the contaminants.

These incidents highlighted the multitude of problems the Chinese faced in producing high-quality and safe food for export in 2007. Regulations governing food production were lax in China, and the government did not have the apparatus to enforce existing regulations. Foods, food ingredients, and chemicals were produced in hundreds of thousands of small farms and factories; the products were often mislabeled and sometimes counterfeit. It was nearly impossible to trace a contaminated product back to its point of origin. Although China strongly defended the quality and safety of its foods and drugs, it announced in June that it would revise its safety rules. In what was widely seen as a political move to show the world China was serious about cleaning up its food and drug industries, the government executed the former director of its top food and drug agency after he pleaded guilty to corruption and taking bribes. The Chinese government did not want to endanger its food export business, which had grown into a major industry. The value of its agriculture exports to the United States alone reached $2.26 billion in 2006, making it the third-largest exporter of food, by value, to the United States after Canada and Mexico.

On December 11 Health and Human Services Secretary Mike Leavitt and the head of China's General Administration of Quality Supervision, Inspection, and Quarantine (AQSIQ) signed a memorandum of agreement to ensure that China's food exports to the

United States would meet U.S. quality and safety standards. The agreement, which created a certification program for shipments to the United States, covered ten specific products, including preserved foods like canned mushrooms and olives, pet foods, some raw materials used in making manufactured foods, and farm-raised fish. The agreement also required certain exporters to register with Chinese officials and agree to annual inspections. China further agreed to establish a system for tracing a product from the source of production to its point of exportation. In addition, China would grant FDA inspectors greater access to food-growing and -processing sites, and the two countries would establish joint training programs. A similar agreement was signed to enhance the safety of Chinese drugs exported to the United States.

While China's food safety issues may have received the most media attention in the United States, it was not the only country that had problems with its food exports. According to the *New York Times*, FDA port inspectors stopped more food shipments from India and Mexico in 2006 than from China. Salmonella was the chief reason cited for blocking Indian products such as black pepper, coriander powder, and shrimp; while filth was reported as the main reason food from Mexico was stopped.

Shipments of food into the United States more than doubled from 2000 to 2007. The FDA, however, employed only enough inspectors to examine about 1 percent of the estimated 9.1 million food shipments, valued at nearly $80 billion, in 2007. Some monitoring was provided by large companies like Wal-Mart that required their foreign suppliers to certify that the food products were not contaminated, and by large food processors like Cargill that had set up quality assurance programs in their foreign-operated plants. Smaller companies and importers were less likely to have such requirements, and few of their goods were likely to be inspected before entering the United States. With few U.S. inspections and lax regulation in many exporting countries, it seemed inevitable that some contaminated food would find its way into the U.S. food chain.

Domestic Threats

In the United States, where regulation and inspection was much tighter, incidents of contamination still occurred. All domestic meat and poultry slaughterhouses and processing plants were monitored daily by inspectors for the U.S. Department of Agriculture (USDA), and beef processors spent about $350 million a year to prevent contamination. Still, there were twenty recalls of beef and beef products in 2007, the most since 2002; there were just five such recalls in 2005 and eight in 2006. In the largest recall in 2007—and the second largest in U.S. history—Topps Meat of New Jersey recalled 21.7 million pounds of frozen hamburger patties after forty people were sickened in eight states; the meat might have been contaminated with a deadly strain of bacteria known as E. coli O157:H7. E. coli bacteria are commonly found in human and animal digestive tracts, and most strains are harmless, but E. coli O157:H7 can cause illness and even death in humans. The Topps recall sent the company into bankruptcy and also closed down Rancher's Beef Ltd., of Alberta, Canada, which had supplied Topps with the tainted meat. In a separate incident, Cargill, Inc., of Minneapolis recalled nearly 2 million pounds of ground beef that was suspected of being contaminated with E. coli.

In another notable recall, ConAgra Foods, Inc., in February recalled all Peter Pan and Great Value peanut butter made at its plant in Sylvester, Georgia, due to possible salmonella contamination. A week later the Centers for Disease Control and Prevention confirmed the presence of salmonella in the tainted peanut butter, which sickened at least

425 people in forty-four states. The company eventually traced the contamination to a leaky roof and faulty sprinkler system. After correcting those problems, ConAgra resumed production at the Georgia plant in August 2007. In recent years, salmonella bacteria, typically found in soil and bird and animal feces, sickened approximately 40,000 people a year in the United States and killed about 600.

A System on the Verge of Breakdown

Many legislators, consumer and industry groups, and even some government officials argued that the myriad problems with domestic and imported food products indicated a much broader issue underlying the government's system for ensuring food safety. In testimony before a House Appropriations subcommittee on February 8, David M. Walker, the comptroller general of the United States, laid the blame squarely on the fragmentation of the food safety system that he said "resulted in inconsistent oversight, ineffective coordination, and inefficient use of resources."

In 2007 fifteen federal agencies had some oversight responsibility for the nation's food supply, but two of them were primary. USDA was responsible for the safety of meat, poultry, and processed egg products—about 20 percent of the nation's food. By law, the USDA's 7,600 inspectors were stationed in every slaughterhouse in the country and visually inspected every carcass as it was processed. Samples were tested in a laboratory. The USDA also limited imports of beef, chicken, lamb, and pork to about forty countries with similar food quality standards and that were certified by the USDA.

In contrast, the FDA had about half the funding and one-fifth the inspectors and lab technicians to monitor the safety of about 80 percent of the food supply. Consequently, it conducted inspections less frequently. Producers of foods susceptible to contamination, such as fresh fruits and vegetables, were supposed to be inspected once a year, although producers with good safety records might be inspected only once every two or three years. Other producers, whose risk of contamination was considered to be lower, were inspected on a rotating basis, depending on the agency's resources. Imports of food and food ingredients other than meat and poultry were not restricted. According to Consumers Union, the FDA's annual food safety budget was about $450 million, compared with USDA's $850 million. Some legislators had long called for merging all food safety programs under one agency's supervision, a proposal that has been rejected by both USDA and FDA.

The Bush administration on November 6 unveiled two plans designed to improve the food safety monitoring process. The Import Safety Action Plan would shift the focus from spot inspections of imports at the border to a "risk-based" strategy that would beef up the safety certification process for foreign manufacturers and exporters, especially those whose products posed the greatest risk of illness or injury to consumers. At the same time, the FDA announced a new "Food Protection Plan" that also focused on identifying potential hazards and preventing them from entering the food supply chain. The administration said it would make these changes without adding any new resources.

Many FDA critics, however, viewed the lack of new resources as the agency's underlying problem. Warning that "American lives were at risk," a subcommittee of the FDA Science Advisory Board said in a report released on December 1 that "the continued expansion of FDA responsibilities coupled with a dramatic decline in resources" represented a "dangerous trend" that the country could ill afford. Although Congress had enacted 125 statutes giving the FDA new or expanded responsibilities since 1988, the number of agency employees had remained about the same and its budget had lost ground to inflation. The FDA's total budget for fiscal 2007 was just over $1.7 billion. The Science and Technology Subcommittee warned in its report that the FDA was not able to recruit and retain leading scientists in key areas and could not to keep up with the scientific advances driving both the drug and food industries. The subcommittee quoted estimates indicating the agency needed at least an additional $390 million just to reform the food safety system.

A week later a coalition of legislators and food industry and consumer groups called on President Bush to commit to a significant increase in funding for FDA food safety programs. "For years Congress has pointed out that the FDA is understaffed and underfunded. Now the FDA's science advisors admit the agency is failing at its mission," Sen. Dick Durbin, D-Ill., said at a December 6 news conference. "I am calling on the administration to commit to doubling FDA funding over the next five years. We simply cannot leave American families vulnerable when it comes to food safety." Durbin was joined by fellow Democratic senator Edward M. Kennedy of Massachusetts and representatives from the Grocery Manufacturers Association, the Food Marketing Institute, the American Frozen Food Institute, the Coalition for a Stronger FDA and FDA Alliance, the Center for Science in the Public Interest, and the Consumer Federation of America.

Following are excerpts from two documents. First, testimony to House Appropriations Subcommittee on Agriculture, Rural Development, FDA, and Related Agencies on February 8, 2007, by David M. Walker, comptroller general of the United States. Walker designated the federal government's oversight of food safety as being at high risk of failure. Second, a press release from the U.S. Department of Health and Human Services, dated December 11, 2007, announcing a memorandum of agreement between the United States and China intended to improve the quality and safety of certain Chinese food exports to the United States.

DOCUMENT

Comptroller General Designates Food Safety System as a High-Risk Area

February 8, 2007

Madam Chairwoman and Members of the Subcommittee:

I am pleased to be here today to discuss the designation of federal oversight of food safety as a high-risk area in the January 2007 update to our High-Risk Series. Let me state at the outset that this nation enjoys a plentiful and varied food supply that is generally considered to be safe. However, each year, about 76 million people contract a foodborne illness in the United States; about 325,000 require hospitalization; and about 5,000 die, according to the Centers for Disease Control and Prevention. In addition, as we have repeatedly reported, our fragmented food safety system has resulted in inconsistent oversight, ineffective coordination, and inefficient use of resources. With 15 agencies collectively administering at least 30 laws related to food safety, the patchwork nature of the federal food safety oversight system calls into question whether the government can more efficiently and effectively protect our nation's food supply. As a result, we added the federal oversight of food safety to our list of programs needing urgent attention and transformation in order to ensure that our national government functions in the most economical, efficient, and effective manner possible.

Our high-risk status reports are provided at the start of each new Congress to help in setting congressional oversight agendas and to help in raising the priority and visibility of government programs needing transformation. These reports also help Congress and the executive branch carry out their responsibilities while improving the government's performance and enhancing its accountability for the benefit of the American people. . . .

Fragmented Federal Oversight of Food Safety Led to High-Risk Designation

For several years, we have reported on issues that suggest that food safety could be designated as a high-risk area because of the need to transform the federal oversight framework to reduce risks to public health as well as the economy. Specifically, the patchwork nature of the federal food oversight system calls into question whether the government can plan more strategically to inspect food production processes, identify and react more quickly to outbreaks of contaminated food, and focus on promoting the safety and the integrity of the nation's food supply. This challenge is even more urgent since the terrorist attacks of September 11, 2001, heightened awareness of agriculture's vulnerabilities to terrorism, such as the deliberate contamination of food or the introduction of disease to livestock, poultry, and crops.

An accidental or deliberate contamination of food or the introduction of disease to livestock, poultry, and crops could undermine consumer confidence in the government's ability to ensure the safety of the U.S. food supply and have severe economic consequences. Agriculture, as the largest industry and employer in the United States, generates more than $1 trillion in economic activity annually, or about 13 percent of the gross domestic product. The value of U.S. agricultural exports exceeded $68 billion in fiscal year 2006. An introduction of a highly infectious foreign animal disease, such as avian influenza or foot-and-mouth disease, would cause severe economic disruption, including substantial losses from halted exports. Similarly, food contamination, such as the recent *E. coli* outbreaks, can harm local economies. For example, industry representatives estimate losses from the recent California spinach *E. coli* outbreak to range from $37 million to $74 million.

While 15 agencies collectively administer at least 30 laws related to food safety, the two primary agencies are the U.S. Department of Agriculture (USDA), which is responsible for the safety of meat, poultry, and processed egg products, and the Food and Drug Administration (FDA), which is responsible for virtually all other foods. Among other agencies with responsibilities related to food safety, the National Marine Fisheries Service (NMFS) in the Department of Commerce conducts voluntary, fee-for-service inspections of seafood safety and quality; the Environmental Protection Agency (EPA) regulates the use of pesticides and maximum allowable residue levels on food commodities and animal feed; and the Department of Homeland Security (DHS) is responsible for coordinating agencies' food security activities.

The food safety system is further complicated by the subtle differences in food products that dictate which agency regulates a product as well as the frequency with which inspections occur. For example, how a packaged ham and cheese sandwich is regulated depends on how the sandwich is presented. USDA inspects manufacturers of packaged open-face meat or poultry sandwiches (e.g., those with one slice of bread), but FDA inspects manufacturers of packaged closed-face meat or poultry sandwiches (e.g., those with two slices of bread). Although there are no differences in the risks posed by these products, USDA inspects wholesale manufacturers of open-face sandwiches sold in interstate commerce daily, while FDA inspects manufacturers of closed-face sandwiches an average of once every 5 years.

This federal regulatory system for food safety, like many other federal programs and policies, evolved piecemeal, typically in response to particular health threats or economic crises. During the past 30 years, we have detailed problems with the current fragmented federal food safety system and reported that the system has caused inconsistent oversight, ineffective coordination, and inefficient use of resources. Our most recent work demonstrates that these challenges persist. Specifically:

- Existing statutes give agencies different regulatory and enforcement authorities. For example, food products under FDA's jurisdiction may be marketed without the agency's prior approval. On the other hand, food products under USDA's jurisdiction must generally be inspected and approved as meeting federal standards before being sold to the public. Under current law, thousands of USDA inspectors maintain continuous inspection at slaughter facilities and examine all slaughtered meat and poultry carcasses. They also visit each processing facility at least once during each operating day. For foods under FDA's jurisdiction, however, federal law does not mandate the frequency of inspections.

- Federal agencies are spending resources on overlapping food safety activities. USDA and FDA both inspect shipments of imported food at 18 U.S. ports of entry. However, these two agencies do not share inspection resources at these ports. For example, USDA officials told us that all USDA-import inspectors are assigned to, and located at, USDA-approved import inspection facilities and some of these facilities handle and store FDA-regulated products. USDA has no jurisdiction over these FDA-regulated products. Although USDA maintains a daily presence at these facilities, the FDA-regulated products may remain at the facilities for some time awaiting FDA inspection. In fiscal year 2003, USDA spent almost $16 million on imported food inspections, and FDA spent more than $115 million.

- Food recalls are voluntary, and federal agencies responsible for food safety have no authority to compel companies to carry out recalls—with the exception of FDA's authority to require a recall for infant formula. USDA and FDA provide guidance to companies for carrying out voluntary recalls. We reported that USDA and FDA can do a better job in carrying out their food recall programs so they can quickly remove potentially unsafe food from the marketplace. These agencies do not know how promptly and completely companies are carrying out recalls, do not promptly verify that recalls have reached all segments of the distribution chain, and use procedures that may not be effective to alert consumers to a recall.

- The terrorist attacks of September 11, 2001, have heightened concerns about agriculture's vulnerability to terrorism. The Homeland Security Act of 2002 assigned DHS the lead coordination responsibility for protecting the nation against terrorist attacks, including agroterrorism. Subsequent presidential directives further define agencies' specific roles in protecting agriculture and the food system against terrorist attacks. We reported that in carrying out these new responsibilities, agencies have taken steps to better manage the risks of agroterrorism, including developing national plans and adopting standard protocols. However, we also found several management problems that can reduce the effectiveness of the agencies' routine efforts to protect against agroterrorism. For example, there are weaknesses in the flow of critical information among key stakeholders and shortcomings in DHS's coordination of federal working groups and research efforts.

- More than 80 percent of the seafood that Americans consume is imported. We reported in 2001 that FDA's seafood inspection program did not sufficiently protect consumers. For example, FDA tested about 1 percent of imported seafood products. We subsequently found that FDA's program has improved: More foreign firms are inspected, and inspections show that more U.S. seafood importers are complying with its requirements. Given FDA officials' concerns about limited inspection resources, we also identified options, such as using personnel in the National Oceanic and Atmospheric Administration's (NOAA) Seafood Inspection Program to augment FDA's inspection capacity or state regulatory laboratories for analyzing imported seafood. FDA agreed with these options.
- In fiscal year 2003, four agencies—USDA, FDA, EPA, and NMFS—spent a total of $1.7 billion on food safety-related activities. USDA and FDA together were responsible for nearly 90 percent of federal expenditures for food safety. However, these expenditures were not based on the volume of foods regulated by the agencies or consumed by the public. The majority of federal expenditures for food safety inspection were directed toward USDA's programs for ensuring the safety of meat, poultry, and egg products; however, USDA is responsible for regulating only about 20 percent of the food supply. In contrast, FDA, which is responsible for regulating about 80 percent of the food supply, accounted for only about 24 percent of expenditures.

Federal Oversight of Food Safety Should Be Addressed as a 21st Century Challenge

We have cited the need to integrate the fragmented federal food safety system as a significant challenge for the 21st century, to be addressed in light of the nation's current deficit and growing structural fiscal imbalance. The traditional incremental approaches to budgeting will need to give way to more fundamental reexamination of the base of government. While prompted by fiscal necessity, such a reexamination can serve the vital function of updating programs to meet present and future challenges within current and expected resource levels. To help Congress review and reconsider the base of federal spending, we framed illustrative questions for decision makers to consider. While these questions can apply to other areas needing broad-based transformation, we specifically cited the myriad of food safety programs managed across several federal agencies. Among these questions are the following:

- How can agencies partner or integrate their activities in new ways, especially with each other, on crosscutting issues, share accountability for crosscutting outcomes, and evaluate their individual and organizational contributions to these outcomes?

- How can agencies more strategically manage their portfolio of tools and adopt more innovative methods to contribute to the achievement of national outcomes?

Integration can create synergy and economies of scale and can provide more focused and efficient efforts to protect the nation's food supply. Further, to respond to the nation's pressing fiscal challenges, agencies may have to explore new ways to achieve their missions. We have identified such opportunities. For example, as I already mentioned, USDA and FDA spend resources on overlapping food safety activities, and we have made recommendations designed to reduce this overlap. Similarly, regarding FDA's seafood inspection program, we have discussed options for FDA to use personnel at NOAA to augment FDA's inspection capacity.

Many of our recommendations to agencies to promote the safety and integrity of the nation's food supply have been acted upon. Nevertheless, as we discuss in the 2007 High-Risk Series, a fundamental reexamination of the federal food safety system is warranted. Such a reexamination would need to address criticisms that have been raised about USDA's dual mission as both a promoter of agricultural and food products and an overseer of their safety. Taken as a whole, our work indicates that Congress and the executive branch can and should create the environment needed to look across the activities of individual programs within specific agencies and toward the goals that the federal government is trying to achieve.

To that end, we have recommended, among other things, that Congress enact comprehensive, uniform, and risk-based food safety legislation and commission the National Academy of Sciences or a blue ribbon panel to conduct a detailed analysis of alternative organizational food safety structures. We also recommended that the executive branch reconvene the President's Council on Food Safety to facilitate interagency coordination on food safety regulation and programs. . . .

As I have discussed, GAO designated the federal oversight of food safety as a high-risk area that is in need of a broad-based transformation to achieve greater economy, efficiency, effectiveness, accountability, and sustainability. The high-risk designation raises the priority and visibility of this necessary transformation and thus can bring needed attention to address the weaknesses caused by a fragmented system. GAO stands ready to provide professional, objective, fact-based, and nonpartisan information and thereby assist Congress as it faces tough choices on how to fundamentally reexamine and transform the government. Lasting solutions to high-risk problems offer the potential to save billions of dollars, dramatically improve service to the American public, strengthen public confidence and trust in the performance and accountability of our national government, and ensure the ability of government to deliver on its promises. . . .

Source: U.S. Congress. Government Accountability Office. "Federal Oversight of Food Safety: High-Risk Designation Can Bring Needed Attention to Fragmented System." Testimony by Comptroller General David M. Walker before the House Subcommittee on Agriculture, Rural Development, FDA, and Related Agencies, Committee on Appropriations. GAO-07-449T. February 8, 2007. purl.access.gpo.gov/GPO/LPS78824 (accessed January 12, 2008).

DOCUMENT

U.S.-China Agreement on Food Imported from China

December 11, 2007

On December 11, 2007, the U.S. Department of Health and Human Services (HHS) and the General Administration of Quality Supervision, Inspection, and Quarantine (AQSIQ) of the People's Republic of China signed a Memorandum of Agreement (MOA) to enhance the safety of food and feed imported into the United States from China. HHS Secretary Mike Leavitt and the Honorable Li Changjiang, Minister of AQSIQ, signed the Agreement in Beijing in advance of the third session under the United States-China Strategic Economic Dialogue.

Specifically, the two countries are establishing a bilateral mechanism to provide greater information to ensure products imported into the United States from China meet standards for quality and safety. Implementation of the agreement will begin with a determined list of products, such as preserved foods (e.g. canned mushrooms, olives, various vegetables), pet food/pet treats of plant origin or animal origin, raw materials used in making manufactured foods (e.g. wheat gluten and rice protein used in canned and dry pet food for dogs and cats), and farm-raised fish (e.g. shrimp and catfish). The two sides can add additional products by mutual agreement.

Key Terms of Agreement

New Registration and Certification Requirements. To enhance the safety of products sold in the U.S., Chinese authorities will implement two programs, both subject to an audit by the Food and Drug Administration (FDA) within HHS. The first will require exporters to the United States to register with AQSIQ and agree to annual inspections to ensure their goods meet U.S. standards. AQSIQ will notify HHS/FDA of those that fail inspection, and why. HHS/FDA will maintain an online list of registered companies. AQSIQ will also notify HHS/FDA of all companies AQSIQ suspended or that have lost their registered status. To better contain and resolve safety problems, AQSIQ will implement a system to trace products from the source of production or manufacture to the point of exportation.

Second, new certification requirements will help ensure products exported from China to the United States meet our standards. Once AQSIQ's Inspection Bureau confirms a shipment meets HHS/FDA requirements, it will issue a certificate that carries a unique identifying number, which AQSIQ must also file with HHS/FDA. To avoid counterfeit certificates, technical experts from both countries will work together to implement a secure electronic system. AQSIQ will also develop a testing program that provides, as determined by HHS/FDA, a high level of statistical confidence in the quality of products exported to the United States.

HHS/FDA will explore mechanisms to notify AQSIQ when shipments of products exported to the United States are not certified, or come from a company not registered with AQSIQ.

Greater Information-Sharing. Each party commits to notify the other within 48 hours of the emergence of significant risks to public health related to product safety, recalls, and other situations. In the past, there was no system of notification. HHS/FDA can request a timely investigation regarding any covered product if there is reason to believe it could pose a health or safety risk.

Increased Access to Production Facilities. AQSIQ will assist and facilitate the inspection of manufacturing, cultivation, or processing sites in China by HHS/FDA. The two countries will develop joint training programs and activities, including in laboratory and risk-assessment methodologies and compliance and enforcement programs.

Implementation and Establishing Key Benchmarks. HHS/FDA and AQSIQ will create a Working Group to meet within 60 days to develop a plan that further details specific activities each will undertake to implement the agreement, and to establish performance measures to evaluate progress. To do so, HHS/FDA could rely on benchmarks such as the rate at which HHS/FDA refuses entry of covered products into the United States; the percentage of items exported to the United States that are uncertified or exported by companies not registered with Chinese authorities; and the volume, frequency and public-health significance of products recalled, including counterfeit goods, as compared to the previous year. The Working Group will meet annually. . . .

SOURCE: U.S. Department of Health and Human Services. "New Agreement Will Enhance the Safety of Food and Feed Imported from the People's Republic of China." Press release. December 11, 2007. www.hhs.gov/news/facts/foodfeed.html (accessed February 3, 2008).

OTHER HISTORIC DOCUMENTS OF INTEREST

FROM THIS VOLUME

- The safety of imported drugs, p. 635

FROM PREVIOUS *HISTORIC DOCUMENTS*

- FDA on foodborne illness and the safety of fresh produce, *2006,* p. 538
- GAO on food safety inspections, *2002,* p. 579

DOCUMENT IN CONTEXT

Russian President Putin on World Affairs and Russian Politics

FEBRUARY 10 AND OCTOBER 1, 2007

Russian president Vladimir Putin ended several years of speculation by announcing in December 2007 that he would relinquish the presidency when his second four-year term ended in 2008; under the constitution, he was ineligible to succeed himself. He would not, however, give up his political power. Putin anointed longtime aide Dmitri Medvedev as his candidate for the presidency in elections scheduled for March 2008. Medvedev immediately promised that, once elected, he would name Putin as his prime minister—a step almost certain to make Putin the first prime minister in modern Russian history to be the most important figure in the Kremlin.

The proposed job-swap capped a year in which Putin further consolidated his control over all levers of government by leading his political party to an overwhelming win in parliamentary elections. His victory marked the total defeat of the liberal opposition that had sought to steer Russia on a more democratic course. Putin also engaged in a year-long feud with the United States, largely over Washington's plan for a European component of a missile defense system, which he viewed as a potential threat to Russia.

The man who brought Putin to prominence, Boris Yeltsin, died on April 23 at the age of seventy-six. Yeltsin had been a Communist Party official in the Soviet Union but resigned from the leadership in 1987, accusing then–Soviet leader Mikhail Gorbachev of dragging his feet in making reforms. Four years later, after party hard-liners mounted a failed coup against Gorbachev, Yeltsin maneuvered to push Gorbachev out of power and the Soviet Union into the dustbin of history.

Yeltsin served as the first popularly elected president of Russia, which inherited most of the Soviet landmass, most of its nuclear weapons, and most of its problems. Russians remembered the Yeltsin years (1991–1999) for the chaos of the collapse of Communist Party authority, the transfer of government-owned industries (usually at fire-sale prices) to Yeltsin's cronies, and the resulting economic depression that ended when the government was forced in 1998 to suspend payment on foreign debts. Yeltsin's last major act was to pluck Putin from obscurity in 1999, naming him first as prime minister, then as his successor as president when Yeltsin stepped aside on the last day of that year. In the end, although Yeltsin created some of the necessary conditions for Russia's emergence as a modern democratic society, he was unable to abolish all of the country's historic tendencies toward authoritarianism.

Putin Takes on Washington

The year 2007 started with warning signs of escalating tension between Russia and the United States. Since the collapse of the Soviet Union, Moscow and Washington had been on remarkably good terms, largely because leaders in both countries saw more advantage in cooperating than competing. This was particularly true in the early relationship between Putin and U.S. president George W. Bush, whose terms coincided: Putin succeeded Yelstin in January 2000, one year before Bush took office, and he was scheduled to leave the presidency after the March 2008 elections, nearly one year before Bush would leave the White House.

After the two men first met on June 16, 2001, Bush, in a burst of enthusiasm, described their meeting: "I looked the man in the eye. I found him to be very straightforward and trustworthy. We had a very good dialogue. I was able to get a sense of his soul; a man deeply committed to his country and the best interests of his country." Critics then and later called the statement naïve. From all public appearances, and from comments by U.S. officials, the two men seemed to get along very well despite their differences on several issues, notably Bush's determination to build a missile defense system that the Russian military considered a threat.

The Bush-Putin relationship was reinforced after the September 11, 2001, attacks against the United States. Bush appealed to Putin for logistical and political support for the subsequent U.S. invasion of Afghanistan. Putin was happy to provide that backing because Russia itself had faced what he considered to be Islamist terrorism in the form of guerrillas fighting for independence in the Muslim-majority republic of Chechnya.

It was not until 2005, when Bush first made gentle comments about the shortcomings of Russian democracy, that relations between the two men became strained. At a meeting in Slovakia in February 2005, according to U.S. officials on the scene, Putin turned the tables and lectured Bush on the failures of American democracy, citing Bush's own rise to power as the result of the disputed elections in 2000. Vice President Dick Cheney ratcheted tensions up a notch in May 2006 with a speech explicitly criticizing Putin's crackdown on his domestic political opponents. Cheney's tough rhetoric raised hackles in Russia, even among some of Putin's critics, and appeared to signal a cooling of U.S.-Russian relations, if not the onset of the "second cold war" that one Russian newspaper predicted.

Despite the gradually escalating rhetoric, Putin's use of a high-profile appearance in Germany on February 10, 2007, to launch a bitter attack on nearly every aspect of U.S. foreign policy came as a surprise. Addressing an annual security conference in Munich—attended by German chancellor Angela Merkel, U.S. defense secretary Robert Gates, and other top officials on both sides of the Atlantic—Putin recited a long list of Russian grievances about the actions of Western nations, particularly the United States, and described a "unipolar world" in which Washington was seeking domination.

"Today we are witnessing an almost uncontained hyper use of force—military force—in international relations, force that is plunging the world into an abyss of permanent conflicts," he said. "One state and, of course, first and foremost the United States, has overstepped its national borders in every way. This is visible in the economic, political, cultural and educational policies it imposes on other nations." In response to what he called "the force's domination," several countries had sought weapons of mass destruction, and threats such as terrorism had taken on "a global character." As he had in the past, Putin also struck a resentful note about U.S. lectures on his steps to curb democracy. "Russia—

we—are constantly being taught about democracy. But for some reason those who teach us do not want to learn themselves," he said.

Putin's vehemence caught many of those attending the conference off-guard—even some European officials who shared a number of the Russian's concerns about the unilateral nature of U.S. foreign policy during the Bush administration. An immediate U.S. response came from Senator John McCain, R-Ariz., a regular attendee at the Munich conference, who suggested Putin was provoking "needless confrontation" because the world was "multipolar," not the U.S.-centric planet Putin had denounced.

Another U.S. response came the next day from recently appointed defense secretary Robert Gates, who was working hard to portray himself as the antithesis to his predecessor, Donald H. Rumsfeld, who was not beloved in Europe. Addressing the same forum as Putin had, Gates opened his remarks by reminding listeners that, as a former CIA official, he had been engaged in foreign affairs issues for many years, then commented, "Speaking of issues going back many years, as an old cold warrior, one of yesterday's speeches almost filled me with nostalgia for a less complex time. Almost," he emphasized. He added that "one cold war was quite enough." Gates went on to prod Putin, noting that although Russia was a "partner" with the West in many respects, "we wonder, too, about some Russian policies that seem to work against international stability, such as its arms transfers and its temptation to use energy resources for political coercion." Gates was referring to reports that Russia had provided weapons to Iran and that it had blocked natural gas supplies to Ukraine early in 2006 in an apparent attempt to destabilize the new pro-Western government in Kiev.

Frayed Russian-U.S. Relations

Less than a month after Putin's salvo in Munich, a panel of U.S. foreign policy experts, convened by the Council on Foreign Relations, issued on March 6 a report on U.S.-Russian relations concluding that the long-hoped for "partnership" between the two countries was "unfortunately not a realistic prospect" in the immediate future. Instead, the panel, chaired by former vice presidential candidates John Edwards (a Democrat) and Jack Kemp (a Republican), advised U.S. policymakers to strive "to make selective cooperation—and in some cases selective opposition—serve important international goals." On the same day the report was published, the *New York Times* quoted administration officials who acknowledged that Washington needed to be more receptive to Russian grievances and should emphasize the two countries' mutual interests rather than the matters on which they disagree.

For much of the year, the central issue of contention between Moscow and Washington was an old one: Bush's plan to create a missile defense system, part of it based in Eastern Europe and ostensibly aimed at blocking missiles from Iran. Russian defense officials long had criticized the proposal, saying its real purpose appeared to be protecting Europe from Russia rather than from Iran. Moscow's objections also reflected long-standing frustrations that former Soviet bloc countries, bordering Russia, had joined the U.S.-led North Atlantic Treaty Organization (NATO) alliance, posing what the Kremlin viewed as a direct threat to Russia.

Early in 2007 the United States moved closer to putting the missile defense system in place. In January the United States received permission from the Czech Republic to build a radar station there for the system. Washington also had asked Poland to allow use of its territory as a base for interceptor missiles to shoot down "enemy" missiles. The commander of Russian space forces, Lt. Gen. Vladimir Popovkin, said in January that the U.S. installations in the Czech Republic and Poland "are an obvious threat to us."

Making Russia's theoretical complaints real, Putin announced on April 26 that Russia would suspend its compliance with a 1990 treaty that established limits on the deployment of conventional forces in Europe. The treaty, which was strengthened in 1999, once had represented a high-water mark of collaboration between Moscow and the West but in recent years had become the subject of disputes. Most important, NATO nations refused to ratify the 1999 revision unless Russia withdrew its troops stationed in Georgia and Moldova. Putin's use of the treaty to protest the U.S. missile defense plan thus was a largely symbolic—but nevertheless politically potent—bit of diplomatic theater. He served formal notice to NATO of his treaty suspension in July and signed legislation to that effect in December.

More harsh words were exchanged in the weeks before the annual Group of Eight (G-8) summit, scheduled for early June in Germany. Diplomats from Washington and Moscow traded barbs, and the two presidents joined in with their own comments: Putin told reporters, in an interview published on June 4, that the United States had created a "horror" with its detentions of suspected terrorists; Bush responded the next day in a speech in Prague, asserting that Putin had "derailed" democratic reforms in post-Soviet Russia. The two leaders set aside the harsh rhetoric during the G-8 summit, and again in July when Putin visited the Bush family summer compound in Kennebunkport, Maine. Even so, it was clear that Putin continued to object to the U.S. plans for missile defense in Europe, and Bush was skeptical, at best, about Putin's counterproposal for a shared missile defense system based in Russia. By this time, however, Bush had dropped his public criticisms of Russian politics, reportedly having concluded that Putin was not listening.

The issue of Iran and nuclear weapons also came into play, even though it was not strictly related to missile defense or the other matters of dispute between Moscow and Washington. The United States needed continued Russian support for the United Nations Security Council to put pressure on Iran to provide honest answers to questions about its alleged ambitions for nuclear weapons. After years of hesitation, Russia in 2006 had supported the U.S. hard-line position, including sanctions, in response to Iranian foot-dragging. Continued Russian cooperation always was uncertain, though, largely because the Kremlin was reluctant to jeopardize large-scale commercial trade with Iran.

In October U.S. secretary of state Condoleezza Rice and defense secretary Gates traveled to Moscow to present a plan for a missile defense system planned and operated jointly by the United States, NATO, and Russia. The proposal failed to defuse Putin's criticisms of the planned U.S. missile defense system. At year's end, Moscow and Washington were still engaging in disputes over the missile defense system and suspicious of each other's intentions.

Consolidating Power in the Kremlin

Throughout his eight years as president, Putin strove both to reassert Russia's role on the world stage and to exert strong leadership at home. Many Russians appeared to believe the latter had been missing during the Yeltsin years, so they offered Putin broad popular support as he took steps to consolidate political power in his own hands.

Putin made his big move in late 2004 when he took control of two of the last remaining sources of independent authority in Russia: the parliament (called the State

Duma) and the eighty-nine provincial governors. He used one of the country's greatest tragedies in recent years as the pretext for asserting power. In September 2004 Chechen separatist guerrillas seized a school in the southern Russian city of Beslan, taking about 1,200 people hostage, most of them children. Nearly 400 hostages died when government troops stormed the school and killed the guerrillas. Although the rescue effort was bungled and was accompanied by many false reports from government officials, Putin was able to use the incident to justify a nationwide security crackdown and his plan to consolidate control over the parliament and the provincial governments.

Also in 2004, the parliament approved a proposal giving the president control over the committee that appointed judges to the federal courts. At various points during Putin's tenure, the Kremlin also secured effective control over all Russian television stations and over the oil and gas industry. The government also imposed tight restrictions on the operations of nongovernmental organizations (NGOs), particularly those that advocated on human rights and political issues. The restrictions led some international NGOs to withdraw from Russia; other groups found it difficult to publicize views of which the government disapproved.

Parliamentary Elections

Russia was due for parliamentary elections in 2007—the first to be conducted under the new system, installed after the Beslan tragedy in 2004, that posed new hurdles for opposition parties. Under the new system, voters would cast their ballots for a slate of candidates offered by the party of their choice, rather than for individual candidates. Further, to appear on the ballot, parties had to submit proof that they had at least 50,000 members, and to gain seats in parliament, a party would have to win at least 7 percent of the national vote. This system gave an overwhelming advantage to Putin's United Russia Party, which was the only party with national reach and the ability to get friendly news coverage from the state-controlled television stations. The government also used the 50,000-member requirement to disqualify several opposition parties, alleging that the lists they submitted contained too many invalid names. Another law, adopted in 2006 as an "anti-extremism" measure, barred candidates from criticizing their opponents or current officeholders. Officials used that law vigorously in prosecuting opposition groups, particularly political candidates who were critical of Putin.

In the weeks before the election, the Kremlin skirmished with one of the main pan-European bodies that monitored elections, particularly in former Soviet republics: the Organization for Security and Cooperation in Europe (OSCE), of which Russia was a member. Russia demanded restrictions on the number of observers for the elections, and on what they could do, arguing that its election system "is one of the most advanced in the world." When disagreements over these matters could not be resolved, Moscow simply refused to provide visas for the OSCE's fifty-seven observers, and the agency withdrew. Putin angrily denounced the OSCE action as a U.S.-sponsored attempt at "delegitimization" of the election.

In a televised speech three days before the voting, Putin appealed for support of his United Russia Party and attacked opposition parties, accusing them of drawing their principal support from foreign governments—a reference to "democracy" programs backed by the United States. The goal of those programs, he said, was to "change the course that the Russian people support and return to times of humiliation, dependence and dissolution" that Russia experienced during the 1990s.

No surprises emerged from the voting on December 2. United Russia won 64.3 percent of the vote nationwide, entitling it to 315 seats in the 450-seat lower house, the Duma. A party created by the Kremlin to foster the appearance of an opposition, called Just Russia, won another 7.8 percent of the vote, for 38 seats. Also clearing the 7 percent minimum threshold were the Communist Party, with 11.6 percent, and the ultranationalist Liberal Democratic Party, with 8.2 percent. The few remaining parties—composed of Western-oriented, pro-democracy liberals—were shut out of parliament for the first time since the fall of communism sixteen years earlier.

Most opposition leaders denounced the election as unfair, but Putin touted the results as a "sign of trust" in his government. The OSCE and the Council of Europe (a quasi-governmental human rights organization) issued a statement saying the election "failed to meet many OSCE and Council of Europe commitments and standards for democratic elections."

Medvedev's Rise to Prominence

Ever since former president Yeltsin made him prime minister in 1999, Putin had proven himself to be a master of bureaucratic intrigue. A former colonel in the KGB, he surrounded himself with numerous former KGB officers, creating a circle of insiders known as the *siloviki.* By early 2007 it was unclear whether he would keep his oft-stated promise not to change the constitution so he could run for another term as president. As a result, each of his personnel moves was scrutinized for possible clues as to who, if anyone, might succeed him. A cabinet shuffle in February drew particular interest: Putin on February 15 named his defense minister, Sergei B. Ivanov, as first deputy prime minister, the same post held by Medvedev, who had been an aide to Putin since the early 1990s. At least on paper, both Ivanov and Medvedev reported to prime minister Mikhail Y. Fradkov, but most Kremlin-watchers suggested that Putin likely would choose one of the two deputies as his favored presidential candidate if he did not run again himself. Ivanov seemed to have the inside track, according to some observers, because he was a member of the *siloviki,* and Medvedev was not.

Putin on April 26 made the clearest statement, to that date, indicating that he would, in fact, accept the two-term constitutional limit on his presidency. In his annual state of the nation speech, he said without elaboration, "The next state of the nation address will be given by another head of state."

Putin shuffled top government posts once again—in early September, three months before the parliamentary elections—ousting Fradkov as prime minister and replacing him with a little-known official, Viktor A. Zubkov, who, like Medvedev, had worked with Putin in the city government of St. Petersburg during the early 1990s. This move immediately generated new waves of speculation about Putin's intentions: Was he still planning to pick either Ivanov or Medvedev as his successor? Had he switched to Zubkov? Was he still deciding what to do? Putin fueled such speculation by remarking, on September 14, that "a minimum of five people" had a chance of being elected president. Putin noted he had not decided on his own future role, but added that he would "have an influence on events."

Putin cleared up some of the mystery on October 1, when he announced that he would head the list of parliamentary candidates put forward by United Russia. He also hinted that he would become prime minister once his term expired, saying that a "proposal [for him] to head the government is entirely realistic," but it was "too soon" for such talk. Some analysts suggested he might confine himself to a post as party chairman or that he might allow a successor to serve briefly as president, to satisfy the constitution, then take up the presidency again himself. Opinion polls showed that most Russians simply wanted to keep Putin in power; his approval ratings routinely hovered around 80 percent, bolstered in large part by perceptions that he was the strong leader that Russians historically desired.

All the questions were answered in a rush little more than a week after the parliamentary elections. On December 10 Putin announced that he would support Medvedev for the presidency: "I've been very close to him for more than seventeen years and fully and completely support this candidacy." Most Russian political analysts agreed that the announcement guaranteed Medvedev, who had no political base of his own other than Putin's backing, would win the election handily against a field of opposition candidates who could expect little coverage by the state-controlled broadcast media.

The next day, when Medvedev said he would ask Putin to be his prime minister "to ensure the continuity of the course of the past eight years," Putin made a show of considering the idea before announcing on December 18 that he would accept the offer. "If our people will trust Mr. Medvedev and elect him the president of the Russian Federation, I will be prepared to continue in our joint work, in this case in the position of premier of the government," he said.

Opposition politicians, and the country's remaining independent media commentators, expressed no surprise at the announcement. Switching from being president to being the prime minister while retaining power in his own hands would be "true to form" for Putin, according to a December 13 editorial in the English-language *Moscow Times.* "Despite the lip service Putin has paid to observing both the letter and spirit of the law, Putin has spent the last eight years managing democracy to bolster and defend the prerogatives of his own power."

Putin received a back-handed endorsement from the West on December 19, when *TIME* magazine named him as the "Person of the Year." The magazine's editor, Richard Stengel, said Putin was recognized for his "extraordinary feat of leadership in taking a country that was in chaos and bringing it stability."

Investigating the Politkovskaya and Litvinenko Deaths

Mysterious killings of opposition figures had long been part of the Russian political landscape, but the deaths of two people in 2006 continued to have ramifications in 2007. One was Anna Politkovskaya, the most prominent investigative journalist in Russia, who was shot to death in October 2006. The other was Alexander Litvinenko, a former KGB agent, who died in London in November 2006 of radiation poisoning. On his deathbed, Litvinenko accused Putin of ordering the poisoning, a charge the Kremlin hotly denied.

The Litvinenko case roiled relations between Britain and Russia during the first half of 2007 because Russian authorities refused to extradite the sole suspect: Andrei Lugovoi, another former KGB agent who had met with Litvinenko at a London bar just before Litvinenko fell ill. In the dispute over the case, Britain and Russia each expelled a handful of the other's diplomats. Lugovoi in December won a seat in parliament as a representative of the ultranationalist Liberal Democratic Party.

The killing of Politkovskaya, a critic of the harshness of Russia's war against separatists in Chechnya, generated a wide range of conspiracy theories, most of them featuring a role for the Russian secret services. Putin himself theorized about foreign involvement, while offering no specifics. The government's top prosecutor announced on August

27 that ten people had been arrested in connection with the case, but several of the suspects later were released. By mid-September the government had turned its attention to a former Chechen official as a prime suspect. The case remained unsolved at year's end.

Following are two documents. First, excerpts from a speech by Russian president Vladimir Putin, on February 10, 2007, to the Munich Conference on Security Policy, in Munich, Germany. Second, the text of Putin's concluding remarks to the annual congress of the United Russia Party on October 1, 2007.

DOCUMENT

Russian President Putin on World Affairs

February 10, 2007

. . . I am truly grateful to be invited to such a representative conference that has assembled politicians, military officials, entrepreneurs and experts from more than 40 nations.

This conference's structure allows me to avoid excessive politeness and the need to speak in roundabout, pleasant but empty diplomatic terms. This conference's format will allow me to say what I really think about international security problems. And if my comments seem unduly polemical, pointed or inexact to our colleagues, then I would ask you not to get angry with me. After all, this is only a conference. And I hope that after the first two or three minutes of my speech Mr Teltschik will not turn on the red light over there.

Therefore. It is well known that international security comprises much more than issues relating to military and political stability. It involves the stability of the global economy, overcoming poverty, economic security and developing a dialogue between civilisations.

This universal, indivisible character of security is expressed as the basic principle that "security for one is security for all". As Franklin D. Roosevelt said during the first few days that the Second World War was breaking out: "When peace has been broken anywhere, the peace of all countries everywhere is in danger."

These words remain topical today. Incidentally, the theme of our conference—global crises, global responsibility—exemplifies this.

Only two decades ago the world was ideologically and economically divided and it was the huge strategic potential of two superpowers that ensured global security.

This global stand-off pushed the sharpest economic and social problems to the margins of the international community's and the world's agenda. And, just like any war, the Cold War left us with live ammunition, figuratively speaking. I am referring to ideological stereotypes, double standards and other typical aspects of Cold War bloc thinking.

The unipolar world that had been proposed after the Cold War did not take place either.

The history of humanity certainly has gone through unipolar periods and seen aspirations to world supremacy. And what hasn't happened in world history?

However, what is a unipolar world? However one might embellish this term, at the end of the day it refers to one type of situation, namely one centre of authority, one centre of force, one centre of decision-making.

It is [a] world in which there is one master, one sovereign. And at the end of the day this is pernicious not only for all those within this system, but also for the sovereign itself because it destroys itself from within.

And this certainly has nothing in common with democracy. Because, as you know, democracy is the power of the majority in light of the interests and opinions of the minority.

Incidentally, Russia—we—are constantly being taught about democracy. But for some reason those who teach us do not want to learn themselves.

I consider that the unipolar model is not only unacceptable but also impossible in today's world. And this is not only because if there was individual leadership in today's—and precisely in today's—world, then the military, political and economic resources would not suffice. What is even more important is that the model itself is flawed because at its basis there is and can be no moral foundations for modern civilisation.

Along with this, what is happening in today's world—and we just started to discuss this—is a tentative to introduce precisely this concept into international affairs, the concept of a unipolar world.

And with which results?

Unilateral and frequently illegitimate actions have not resolved any problems. Moreover, they have caused new human tragedies and created new centres of tension. Judge for yourselves: wars as well as local and regional conflicts have not diminished. Mr [Horst] Teltschik [moderator of the Munich conference] mentioned this very gently. And no less people perish in these conflicts—even more are dying than before. Significantly more, significantly more!

Today we are witnessing an almost uncontained hyper use of force—military force—in international relations, force that is plunging the world into an abyss of permanent conflicts. As a result we do not have sufficient strength to find a comprehensive solution to any one of these conflicts. Finding a political settlement also becomes impossible.

We are seeing a greater and greater disdain for the basic principles of international law. And independent legal norms are, as a matter of fact, coming increasingly closer to one state's legal system. One state and, of course, first and foremost the United States, has overstepped its national borders in every way. This is visible in the economic, political, cultural and educational policies it imposes on other nations. Well, who likes this? Who is happy about this?

In international relations we increasingly see the desire to resolve a given question according to so-called issues of political expediency, based on the current political climate.

And of course this is extremely dangerous. It results in the fact that no one feels safe. I want to emphasise this—no one feels safe! Because no one can feel that international law is like a stone wall that will protect them. Of course such a policy stimulates an arms race.

The force's dominance inevitably encourages a number of countries to acquire weapons of mass destruction. Moreover, significantly new threats—though they were also well-known before—have appeared, and today threats such as terrorism have taken on a global character.

I am convinced that we have reached that decisive moment when we must seriously think about the architecture of global security.

And we must proceed by searching for a reasonable balance between the interests of all participants in the international dialogue. Especially since the international landscape is so varied and changes so quickly—changes in light of the dynamic development in a whole number of countries and regions.

Madam Federal Chancellor [German chancellor Angela Merkel] already mentioned this. The combined GDP [gross domestic product] measured in purchasing power parity of countries such as India and China is already greater than that of the United States. And a similar calculation with the GDP of the BRIC countries—Brazil, Russia, India and China—surpasses the cumulative GDP of the EU [European Union]. And according to experts this gap will only increase in the future.

There is no reason to doubt that the economic potential of the new centres of global economic growth will inevitably be converted into political influence and will strengthen multipolarity.

In connection with this the role of multilateral diplomacy is significantly increasing. The need for principles such as openness, transparency and predictability in politics is uncontested and the use of force should be a really exceptional measure, comparable to using the death penalty in the judicial systems of certain states.

However, today we are witnessing the opposite tendency, namely a situation in which countries that forbid the death penalty even for murderers and other, dangerous criminals are airily participating in military operations that are difficult to consider legitimate. And as a matter of fact, these conflicts are killing people—hundreds and thousands of civilians!

But at the same time the question arises of whether we should be indifferent and aloof to various internal conflicts inside countries, to authoritarian regimes, to tyrants, and to the proliferation of weapons of mass destruction? As a matter of fact, this was also at the centre of the question that our dear colleague Mr Lieberman [U.S. senator Joseph Lieberman] asked the Federal Chancellor. If I correctly understood your question (addressing Mr Lieberman), then of course it is a serious one! Can we be indifferent observers in view of what is happening? I will try to answer your question as well: of course not.

But do we have the means to counter these threats? Certainly we do. It is sufficient to look at recent history. Did not our country have a peaceful transition to democracy? Indeed, we witnessed a peaceful transformation of the Soviet regime—a peaceful transformation! And what a regime! With what a number of weapons, including nuclear weapons! Why should we start bombing and shooting now at every available opportunity? Is it the case when without the threat of mutual destruction we do not have enough political culture, respect for democratic values and for the law?

I am convinced that the only mechanism that can make decisions about using military force as a last resort is the Charter of the United Nations. And in connection with this, either I did not understand what our colleague, the Italian Defence Minister, just said or what he said was inexact. In any case, I understood that the use of force can only be legitimate when the decision is taken by NATO, the EU, or the UN. If he really does think so, then we have different points of view. Or I didn't hear correctly. The use of force can only be considered legitimate if the decision is sanctioned by the UN. And we do not need to substitute NATO or the EU for the UN. When the UN will truly unite the forces of the international community and can really react to events in various countries, when we will leave behind this disdain for international law, then the situation will be able to change. Otherwise the situation will simply result in a dead end, and the number of serious mistakes will be multiplied. Along with this, it is necessary to make sure that international law have a universal character both in the conception and application of its norms.

And one must not forget that democratic political actions necessarily go along with discussion and a laborious decision-making process.

Dear ladies and gentlemen!

The potential danger of the destabilisation of international relations is connected with obvious stagnation in the disarmament issue.

Russia supports the renewal of dialogue on this important question.

It is important to conserve the international legal framework relating to weapons destruction and therefore ensure continuity in the process of reducing nuclear weapons.

Together with the United States of America we agreed to reduce our nuclear strategic missile capabilities to up to 1700–2000 nuclear warheads by 31 December 2012. Russia intends to strictly fulfil the obligations it has taken on. We hope that our partners will also act in a transparent way and will refrain from laying aside a couple of hundred superfluous nuclear warheads for a rainy day. And if today the new American Defence Minister declares that the United States will not hide these superfluous weapons in warehouse or, as one might say, under a pillow or under the blanket, then I suggest that we all rise and greet this declaration standing. It would be a very important declaration.

Russia strictly adheres to and intends to further adhere to the Treaty on the Non-Proliferation of Nuclear Weapons as well as the multilateral supervision regime for missile technologies. The principles incorporated in these documents are universal ones.

In connection with this I would like to recall that in the 1980s the USSR and the United States signed an agreement on destroying a whole range of small- and medium-range missiles but these documents do not have a universal character.

Today many other countries have these missiles, including the Democratic People's Republic of Korea, the Republic of Korea, India, Iran, Pakistan and Israel. Many countries are working on these systems and plan to incorporate them as part of their weapons arsenals. And only the United States and Russia bear the responsibility to not create such weapons systems.

It is obvious that in these conditions we must think about ensuring our own security.

At the same time, it is impossible to sanction the appearance of new, destabilising high-tech weapons. Needless to say it refers to measures to prevent a new area of confrontation, especially in outer space. Star wars is no longer a fantasy—it is a reality. In the middle of the 1980s our American partners were already able to intercept their own satellite.

In Russia's opinion, the militarisation of outer space could have unpredictable consequences for the international community, and provoke nothing less than the beginning of a nuclear era. And we have come forward more than once with initiatives designed to prevent the use of weapons in outer space.

Today I would like to tell you that we have prepared a project for an agreement on the prevention of deploying weapons in outer space. And in the near future it will be sent to our partners as an official proposal. Let's work on this together.

Plans to expand certain elements of the anti-missile defence system to Europe cannot help but disturb us. Who needs the next step of what would be, in this case, an inevitable arms race? I deeply doubt that Europeans themselves do.

Missile weapons with a range of about five to eight thousand kilometres that really pose a threat to Europe do not exist in any of the so-called problem countries. And in the near future and prospects, this will not happen and is not even foreseeable. And any hypothetical launch of, for example, a North Korean rocket to American territory through western Europe obviously contradicts the laws of ballistics. As we say in Russia, it would be like using the right hand to reach the left ear.

And here in Germany I cannot help but mention the pitiable condition of the Treaty on Conventional Armed Forces in Europe.

The Adapted Treaty on Conventional Armed Forces in Europe was signed in 1999. It took into account a new geopolitical reality, namely the elimination of the Warsaw bloc. Seven years have passed and only four states have ratified this document, including the Russian Federation.

NATO countries openly declared that they will not ratify this treaty, including the provisions on flank restrictions [on deploying a certain number of armed forces in the flank zones], until Russia removed its military bases from Georgia and Moldova. Our army is leaving Georgia, even according to an accelerated schedule. We resolved the problems we had with our Georgian colleagues, as everybody knows. There are still 1,500 servicemen in Moldova that are carrying out peacekeeping operations and protecting warehouses with ammunition left over from Soviet times. We constantly discuss this issue with Mr [Javier] Solana [chief foreign policy representative of the EU] and he knows our position. We are ready to further work in this direction.

But what is happening at the same time? Simultaneously the so-called flexible frontline American bases with up to five thousand men in each. It turns out that NATO has put its frontline forces on our borders, and we continue to strictly fulfil the treaty obligations and do not react to these actions at all.

I think it is obvious that NATO expansion does not have any relation with the modernisation of the Alliance itself or with ensuring security in Europe. On the contrary, it represents a serious provocation that reduces the level of mutual trust. And we have the right to ask: against whom is this expansion intended? And what happened to the assurances our western partners made after the dissolution of the Warsaw Pact? Where are those declarations today? No one even remembers them. But I will allow myself to remind this audience what was said. I would like to quote the speech of NATO General Secretary Mr. [Manfred] Woerner in Brussels on 17 May 1990. He said at the time that: "the fact that we are ready not to place a NATO army outside of German territory gives the Soviet Union a firm security guarantee". Where are these guarantees?

The stones and concrete blocks of the Berlin Wall have long been distributed as souvenirs. But we should not forget that the fall of the Berlin Wall was possible thanks to a historic choice—one that was also made by our people, the people of Russia—a choice in favour of democracy, freedom, openness and a sincere partnership with all the members of the big European family.

And now they are trying to impose new dividing lines and walls on us—these walls may be virtual but they are nevertheless dividing, ones that cut through our continent. And is it possible that we will once again require many years and decades, as well as several generations of politicians, to dissemble and dismantle these new walls?

Dear ladies and gentlemen!

We are unequivocally in favour of strengthening the regime of non-proliferation. The present international legal principles allow us to develop technologies to manufacture nuclear fuel for peaceful purposes. And many countries with all good reasons want to create their own nuclear energy as a basis for their energy independence. But we also understand that these technologies can be quickly transformed into nuclear weapons.

This creates serious international tensions. The situation surrounding the Iranian nuclear programme acts as a clear example. And if the international community does not find a reasonable solution for resolving this conflict of interests, the world will continue to suffer similar, destabilising crises because there are more threshold countries than simply Iran. We both know this. We are going to constantly fight against the threat of the proliferation of weapons of mass destruction.

Last year Russia put forward the initiative to establish international centres for the enrichment of uranium. We are open to the possibility that such centres not only be created in Russia, but also in other countries where there is a legitimate basis for using civil nuclear energy. Countries that want to develop their nuclear energy could guarantee that they will receive fuel through direct participation in these centres. And the centres would, of course, operate under strict IAEA [International Atomic Energy Agency] supervision.

The latest initiatives put forward by American President George W. Bush are in conformity with the Russian proposals. I consider that Russia and the USA are objectively and equally interested in strengthening the regime of the non-proliferation of weapons of mass destruction and their deployment. It is precisely our countries, with leading nuclear and missile capabilities, that must act as leaders in developing new, stricter non-proliferation measures. Russia is ready for such work. We are engaged in consultations with our American friends.

In general, we should talk about establishing a whole system of political incentives and economic stimuli whereby it would not be in states' interests to establish their own capabilities in the nuclear fuel cycle but they would still have the opportunity to develop nuclear energy and strengthen their energy capabilities. . . .

In conclusion I would like to note the following. We very often—and personally, I very often—hear appeals by our partners, including our European partners, to the effect that Russia should play an increasingly active role in world affairs.

In connection with this I would allow myself to make one small remark. It is hardly necessary to incite us to do so. Russia is a country with a history that spans more than a thousand years and has practically always used the privilege to carry out an independent foreign policy.

We are not going to change this tradition today. At the same time, we are well aware of how the world has changed and we have a realistic sense of our own opportunities and potential. And of course we would like to interact with responsible and independent partners with whom we could work together in constructing a fair and democratic world order that would ensure security and prosperity not only for a select few, but for all. . . .

SOURCE: Russian Federation. President of Russia. "Speech and the Following Discussion at the Munich Conference on Security Policy." February 10, 2007. http://president.kremlin.ru/eng/speeches/2007/02/10/0138_type82912type82914type82917type84779_118123.shtml (accessed February 17, 2008).

DOCUMENT

Speech by Russian President Putin to the United Russia Party Congress

October 1, 2007

First of all, I would like to thank you for the analysis you made in your speeches and, of course, for your kind words and your serious employment proposals.

We all remember what state the country was in seven–eight years ago. A lot of people doubted that we would have the strength to combat terrorism and separatism and that we would even manage to hold the country together. We are still far from having solved all these problems. Russia still faces numerous threats and dangers, but we have succeeded in turning the situation around: constitutional order has been restored and our country's territorial integrity is once again intact and guaranteed. We will continue our uncompromising fight against terrorism until this scourge is totally eradicated.

It looked at one time as though we would never manage to free ourselves from powerful oligarchs [millionaire businessmen who, in the 1990s, controlled large corporations formerly owned by the government] whose influence was based on corruption, violence and information blackmail. We have accomplished a great deal in terms of cleaning out this illegitimate influence from the upper levels of state power, but, as has been said today, we still have much to do in fighting corruption. Our economy, and therefore our entire social sphere, was dependent on foreign debt and foreign borrowing, and this meant that we were not independent and enabled others to interfere in our domestic affairs. We turned a blind eye to this situation, sometimes even pretending to like it, and we were unable to effectively protect our own interests in international affairs.

We still face many problems in other areas too. As has already been said, there is a great divide between the incomes of rich and poor in our country. We need to work towards fairer provision for people in their old age and for those who are unable to work. We still have much work to do in the areas of health, education, science, strengthening our defence capability and rural development. But the situation in these areas is improving. Most important is not the positive trends that have emerged, but the fact that the policies we are pursuing ensure a high economic growth rate and growing investment and incomes, and this creates the conditions that enable us to be fully confident about our ability to solve all the problems we face today, above all the social problems, and to do so with respect for human rights and liberties and in a free and democratic system.

Now, regarding the ideas and proposals put forward today.

Thank you once again for your offers. Though I was one of the initiators of United Russia's founding, I, like the vast majority of people in our country, am not a member of any political party, and I would rather not change this status. I think it would be wrong to change the Constitution to suit one particular person, even if that person is someone I most certainly trust.

The proposal to head the Government is entirely realistic, but it is too soon to talk about this at the moment because at least two conditions would first need to be met. First, United Russia would have to win the State Duma [parliament] election on December 2, and second, our voters would have to elect a decent, effective and modern-thinking President with whom it would be possible to work together. What we need to talk about today is the fact that your party can and should be an instrument for guaranteeing social stability and ensuring that the next parliament and the state power system in general can function effectively. Your party should initiate development and provide support for the executive authorities in carrying out the plans that have been made. This is all something that can and should be the subject of discussion.

It is therefore with gratitude that I accept your proposal to head the United Russia list.

SOURCE: Russian Federation. President of Russia. "Concluding Remarks at the VIIIth United Russia Party Congress." October 1, 2007. http://president.kremlin.ru/eng/speeches/2007/10/01/2210_type82912type82913type84779_146510.shtml (accessed February 17, 2008).

OTHER HISTORIC DOCUMENTS OF INTEREST

FROM THIS VOLUME

- United Nations Security Council resolution on Iran, p. 112

FROM PREVIOUS *HISTORIC DOCUMENTS*

- Vice President Cheney and Russian foreign minister Lavrov on U.S.-Russian relations, *2006,* p. 202
- Russian president Putin on the state of the nation, *2005,* p. 299
- Russian president Putin on hostage-taking tragedy, *2004,* p. 564
- President Yeltsin on the Russian financial crisis, *1998,* p. 601

DOCUMENT IN CONTEXT

International Efforts to Denuclearize North Korea

FEBRUARY 13 AND OCTOBER 3, 2007

Significant progress was made during 2007 in the international campaign to persuade North Korea to give up its nuclear weapons and eliminate its capability to make new ones. Despite the progress, by the end of the year it was apparent that a great deal of hard bargaining would be required before North Korea could be considered a nuclear-free country. Bush administration officials said they hoped to come even closer to that goal in 2008.

In two landmark agreements developed during multilateral negotiations spearheaded by China, North Korea promised to disable a plant that had produced the fuel for a small nuclear weapon it had tested in 2006. North Korea also pledged to disclose every aspect of its nuclear weapons program. In return, the United States and other countries promised benefits to North Korea for its abandonment of nuclear weapons, including the provision of supplies of fuel oil and other aid, the start of negotiations toward normalization of relations with Japan and the United States, and the beginning of a process for the United States to drop some of its legal restrictions that effectively barred North Korea from access to the international financial system.

The reclusive North Korean government kept some of its promises but failed to meet a year-end deadline on others. Even so, negotiations continued, and it appeared that the two opposing sides had made fundamental decisions: the North Korean regime, to abandon its weapons programs in the hope of receiving benefits that would help it retain power; and the United States and its allies, to give the regime, notably leader Kim Jong Il, a bit of time and space to take the difficult steps necessary to fulfill its commitments. Whether or when North Korea would give up the weapons it had already built, along with the nuclear material it had used to make those weapons, remained unclear.

Background

The negotiations in 2007 represented a second attempt by the United States and other countries to persuade North Korea to back down from its determination to acquire nuclear weapons. The first attempt was made in the early 1990s, when North Korea built a nuclear plant near the city of Yongbyon (north of the capital, Pyongyang) that appeared to be intended to produce plutonium, a fuel for nuclear weapons. Under intense pressure from the United States, North Korea in 1994 signed an agreement (called the "Agreed Framework") that halted work at the plant in exchange for a promise by Japan, South Korea, and the United States to provide fuel for North Korea's electrical plants and to build two light-water nuclear reactors capable of producing electricity but not weapons-grade material.

In 2002 the Bush administration accused North Korea of violating that agreement by secretly operating a separate program to produce material for nuclear weapons via a different route: enriching uranium. To punish Pyongyang, the administration forced a halt to the construction of the two light-water reactors. North Korea reacted to the U.S. allegations by voiding the 1994 agreement and ousting inspectors from the International Atomic Energy Agency (IAEA), which had been monitoring that agreement. North Korea also began extracting plutonium from spent fuel rods at the Yongbyon reactor—a clear indication of its determination to produce weapons. U.S. officials later said they believed North Korea had extracted about 110 pounds of plutonium, enough to build six to eight small nuclear weapons.

China, North Korea's most important patron, intervened in the dispute in 2003 by sponsoring a series of multilateral negotiations in Beijing on the nuclear issue, known as the Six-Party Talks. The participants included China, Japan, North Korea, Russia, South Korea, and the United States.

These negotiations proceeded fitfully—and were suspended for nearly a year during 2004 and 2005—but produced an important accord. On September 19, 2005, North Korea joined its five negotiating partners in issuing the "Joint Statement," a pledge for the "verifiable denuclearization of the Korean Peninsula in a peaceful manner." North Korea pledged to abandon "all nuclear weapons and existing nuclear programs," while the United States promised not to station nuclear weapons on the peninsula and declared that it had "no intention" of attacking or invading North Korea.

Putting substance behind those promises involved dozens of negotiating sessions in various venues over the next seventeen months—talks that at several points appeared on the verge of collapse, as when on October 9, 2006, North Korea conducted its first-ever test of a nuclear weapon. That action was widely seen as a chest-beating exercise because the weapon was small and the test was only partly successful. Five days later, the United Nations Security Council issued a harsh rebuke and imposed several sanctions against North Korea. According to diplomats, the test proved to be a serious mistake for Pyongyang because it angered Chinese leaders, who had warned against such a test; once it occurred, they decided the time had come to pressure Kim Jong Il to show flexibility in the Six-Party Talks.

A separate, but clearly related, dispute between North Korea and the United States over a bank in the Chinese territory of Macao proved to be another stumbling block. U.S. officials had accused North Korea of using the bank (Banco Delta Asia) to launder money gained through narcotics trafficking and counterfeiting U.S. currency. The United States in October 2005 imposed sanctions against the bank, effectively freezing about $25 million in North Korean assets and blocking Pyongyang's access to the global financial system. This matter became an important element in subsequent negotiations: Washington used the bank freeze against Pyongyang, which in turn resisted making concessions on the nuclear issue until Washington would agree to unfreeze the money. This question reportedly was the main subject of a round of talks held in December 2006.

A Breakthrough Agreement

The first sign that real progress might be possible during 2007 came in January, when the lead U.S. and North Korean negotiators held bilateral talks in Berlin. U.S. officials, who had proposed the talks, apparently saw an opportunity to revive the negotiating process that had faltered since North Korea's nuclear test the previous October.

Washington was represented by Christopher R. Hill, assistant secretary of state for East Asian and Pacific Affairs; North Korea's negotiator was Kim Kye Gwan, a vice minister for foreign affairs. The two men reportedly reached a tentative understanding: North Korea would shut down its Yongbyon reactor and allow inspections by the IAEA in exchange for aid from the United States and its allies. Secretary of State Condoleezza Rice met with Hill while the Berlin negotiations were under way, approved the agreement, and then obtained President Bush's approval in Washington, according to subsequent news reports.

The year's first round of Six-Party Talks began in Beijing on February 8, and from the start it seemed that a major agreement would be possible. The diplomats worked from a draft agreement, prepared by China, calling for North Korea to shut down its Yongbyon reactor in exchange for incentives from the other parties. The main question was whether the North Korean side would insist on receiving the benefits before taking action—a position that U.S. officials called a nonstarter. Hill predicted "some rather hard bargaining" before an agreement could be reached.

The diplomats moved quickly, however, apparently because Hill and his counterpart from Pyongyang already had settled some of the basic issues. Early on February 13, the sixth day of talks, the negotiators reached an agreement based on the understanding discussed weeks earlier by Hill and Kim. China announced the agreement later in the day, after it was approved by each of the capitals represented at the talks.

Called "Initial Actions for the Implementation of the Joint Statement," the agreement laid out a series of specific steps toward the denuclearization goal in the September 2005 agreement. The document encompassed just the first steps necessary to guarantee that North Korea would stop producing plutonium at its Yongbyon plant.

Under the agreement, North Korea pledged to "shut down and seal" its facilities at Yongbyon within sixty days; these included an operating reactor and a facility to extract plutonium from fuel rods. The agreement had an explicit quid pro quo: In return for closing the Yongbyon facilities, North Korea would receive the equivalent of 50,000 tons of heavy fuel oil within sixty days. As it took other steps to disable the Yongbyon reactor and provide a complete list of its nuclear facilities, North Korea would receive additional assistance equal to 950,000 tons of heavy fuel oil.

In addition, the United States and Japan each agreed to begin direct talks with North Korea aimed at normalizing relations; neither country maintained diplomatic relations with Pyongyang at the time. In Japan's case, a serious obstacle to normal relations was Tokyo's demand that North Korea acknowledge the fate of Japanese citizens who had been abducted in recent decades, apparently as part of Pyongyang's effort to collect intelligence about Japan. Previously, North Korea had admitted abducting thirteen Japanese citizens from various locations; it released five of them in 2002 and said the others had died. The Japanese government contended that more than thirteen of its citizens had been abducted, rejected the claim that some of them had died, and insisted on a full accounting of all of those abducted.

Ultimately, the dispute between Japan and North Korea became linked implicitly to a separate issue between North Korea and the United States. Washington for years had listed North Korea as a state sponsor of terrorism because of its alleged sponsorship of several major terrorist attacks in the 1980s and its abductions of Japanese citizens. As part of the February 13 agreement, the United States promised to "begin the process" of removing North Korea from the terrorism list and to "advance the process" of eliminating restrictions, under the provisions of the World War I–era Trading with the Enemy Act, that barred U.S. financial transactions with North Korea. Later in the year, Japan pressed the United States to retain the terrorism designation until North Korea had met Tokyo's demands on the abducted citizens, and Washington complied, at least for the time being.

The February 13 agreement did not explicitly address the Bush administration's allegations about North Korean work to enrich uranium. It did, however, indicate that the six parties would discuss "all" of the North's nuclear programs, a term that U.S. officials argued included enrichment facilities even though North Korea denied having any. Hill praised the agreement as a "very solid step forward," and the White House issued a statement, in Bush's name, saying the negotiating process reflected "the common commitment of the participants to a Korean peninsula that is free of nuclear weapons."

The terms of the agreement angered hard-line neoconservatives in the United States—some of them current or former Bush administration officials. A number of conservatives long had been unhappy about the administration's willingness to negotiate with North Korea, and they were especially displeased by what they viewed as Hill's unnecessary flexibility in the talks. By far the most prominent and vocal critic was John Bolton, who had served earlier in the Bush administration as the State Department's top official on nonproliferation matters and then as ambassador to the United Nations until Senate Democrats blocked his confirmation in that position. Bolton said the agreement "sends exactly the wrong signal to would-be proliferators around the world: 'If we hold out long enough, wear down the State Department negotiators, eventually you get rewarded,' in this case with massive shipments of heavy fuel oil for only doing partially what needs to be done."

The *Washington Post* reported on February 15 that Elliott Abrams, another prominent conservative and a senior member of the National Security Council staff, fired off e-mails to other officials " expressing bewilderment" about the agreement. Abrams, the *Post* said, was particularly concerned about the United States' promise to begin action to remove North Korea from the list of terrorism-supporting nations; he argued that Pyongyang had done nothing to prove that it no longer supported terrorist groups.

Leading Democrats in Congress offered an entirely different criticism. If President Bush had been as willing to negotiate with North Korea when he came to office in 2001 as he was six years later, they said, he might have gotten the same agreement under better circumstances—before North Korea reprocessed plutonium from the Yongbyon reactor and tested a nuclear weapon.

More Diplomacy, then Real Action

Another round of talks opened in Beijing on March 19 but stalled three days later when North Korean representatives walked out, protesting the United States' failure—said to be for technical reasons—to lift the freeze on North Korean funds in the Macao bank. Apparently because of this dispute, North Korea failed to meet its April 14 deadline to shut down the Yongbyon reactor. The situation remained stuck for two more months: the talks remained on hold, North Korea kept the reactor open, and U.S. officials said they were resolving the technical difficulties. By this time, North Korean officials had laid out what they called an "action for action" strategy: Pyongyang would act on its commitments only when other parties acted on theirs.

In late June events occurred rapidly. Hill traveled to Pyongyang for the first time, North Korea reached agreement with the IAEA on methods to verify the shutdown of the Yongbyon reactor, and the United States released the North Korean money that had been frozen at the Macao bank. In July additional progress was made when South Korea sent the first shipment of 50,000 tons of fuel oil to North Korea, and the IAEA verified the shutdown of the working reactor and related facilities at Yongbyon between July 16 and July 18. Another round of negotiations, however, failed to make much headway, in part because of the continued dispute between Japan and North Korea over the abducted Japanese citizens.

More talks took place in August, including a bilateral session in Beijing between Hill and North Korean negotiator Kim and the first of a series of meetings among the six countries by "working groups" on technical issues. Early in September, though, an action in the Middle East momentarily raised questions about prospects for further progress in the talks. On September 6 Israel bombed a building in northeastern Syria. Although the Israeli government gave no justification, news organizations quoted Israeli and U.S. officials as saying that the building appeared to be a nuclear reactor under construction and that North Korean technicians were in the area at the time. This led to speculation that the talks could be endangered by new concerns about North Korean nuclear activities, but the negotiating process continued.

Another Major Agreement

The Six-Party Talks resumed in Beijing on September 27 and progressed rapidly, suggesting that all sides wanted to keep the process moving forward. On the second day of talks, President Bush approved the first shipment of U.S. fuel oil to North Korea under the February 13 agreement—an action intended to reassure Pyongyang that Washington would keep its part of the bargain. China said on September 30 that a tentative agreement had been reached, subject to approval by the six governments. Hill explained that the agreement laid out "an entire roadmap" for actions through the rest of the year. "We're into the nuts and bolts now of implementing denuclearization," he told reporters.

While the latest agreement was being reviewed in the six capitals, North Korean leader Kim Jong Il and outgoing South Korean president Roh Moo-hyun held three days of separate but related summit meetings in Pyongyang. These meetings, which started on October 2, appeared to produce little of substance but had great symbolic importance in that they restored the so-called sunshine process, intended to lead to eventual unification of the two Koreas, which had been on hold since 2000.

On October 3 China released the official text of the latest six-party agreement, formally titled the "Second-Phase Actions for the Implementation of the September 2005 Joint Statement." President Bush had approved the accord the day before, during a meeting at the White House with Hill, Vice President Dick Cheney, Secretary of State Rice, and other officials.

Key provisions of the new agreement required North Korea, by the end of 2007, to "disable" the shut down reactor and plutonium-reprocessing facilities at Yongbyon and to give the five other countries "a complete and correct declaration of all its nuclear programs." The United States was to provide the technicians and the funding for disabling the Yongbyon facilities. In return, Japan and the United States restated their commitments to improving relations with North Korea, including Washington's taking steps to remove its legal stranglehold on North Korean finances.

In a briefing for reporters, Hill noted that the language called for the Yongbyon facilities to be disabled rather than fully dismantled by year's end, as a practical matter to ensure that North Korea could no longer produce new plutonium for weapons. The goal was to put the facilities in such a state that "if the North Koreans ever wanted to change their mind, it would be quite difficult to restart the [weapons] program." Actual dismantling of the facilities—plus the question of getting North Korea to give up the nuclear weapons it had built—would be the subject of later negotiations. Hill noted that the end-of-year disclosure provision specifically applied to any work that North Korea had done to enrich uranium, even though Pyongyang continued to be vague about whether it ever had such a program. In another provision, North Korea promised not to transfer "nuclear materials, technology, or know-how" to other countries.

Although the agreement set a December 31 deadline for North Korean actions, it did not specify when the United States would lift its economic sanctions. Hill said the two countries had a "very clear understanding" of when this would happen but refused to give a precise date. He did acknowledge that removing North Korea from the list of terrorism-supporting countries was a matter of "political sensitivity," citing in particular Tokyo's requests that Washington not act until Pyongyang provided a complete account for the abducted citizens.

More Steps, More Delays

U.S. technicians began disabling the Yongbyon facilities on November 5. This momentous step reflected an extraordinary degree of cooperation between two countries that in recent years had been unrelentingly hostile toward each other. By early December, it became clear that the work would stretch into the early months of 2008 because of the labor required to remove some 8,000 spent fuel rods. In general, U.S. officials reported receiving good cooperation from their North Korean counterparts at the Yongbyon facility.

More problematic, however, was the parallel requirement that North Korea give a "complete and correct" accounting of all its nuclear facilities by year's end. Officials from Japan, South Korea, and the United States in November began expressing doubts that North Korea would meet the deadline, but no one was then prepared to claim that Pyongyang was backtracking on this commitment.

Hill returned to North Korea on December 3 for bilateral talks. It later emerged that he was carrying a letter from President Bush to Kim Jong Il urging him to fulfill his disclosure commitment. The full text of the letter was not made public, but the *New York Times* reported on December 6 that the president demanded precise information about how much plutonium and how many warheads the North Koreans had produced. He also stressed that the United States was serious about normalizing relations with North Korea once it had met all its obligations on the nuclear issue. Bush said on December 14: "I got his [Kim's] attention with a letter, and he can get my attention by fully disclosing his programs, including any plutonium he may have processed and converted. Whatever he's used it for, we just need to know." U.S. officials reported Kim had replied to Bush's letter with a statement that North Korea would keep its pledges so long as the United States met its obligations.

On December 19 South Korean voters elected a conservative candidate, Lee Myung-bak, as their new president, succeeding Roh at the end of his five-year term. This event had the potential to influence the nuclear issue, although it was unclear by year's end just what that influence would be. Lee, who was scheduled to take office in February 2008, had been harshly critical of North Korea during his campaign. In his first news conference after the election, he declared that "there will be a change from the past government's practice of avoiding criticism of North Korea and unilaterally flattering it."

As it became clear that North Korea would fail to meet the December 31 deadline, U.S. officials began saying that an accurate disclosure from Pyongyang was more important than the date it was given. Reuters on December 21 quoted one U.S. official as stating the requirement to disclose such sensitive information was causing understandable problems within the closed leadership of North Korea. "This is where you get into military secrets and, in a country that would keep a sweater size secret, you can imagine the difficulty in revealing military secrets," he said.

The December 31 deadline passed without the disclosure, and U.S. technicians continued to work on disabling the Yongbyon nuclear facilities. Hill and other U.S. officials said 2008 could be a decisive year, one that determined whether North Korea had, in fact, chosen to abandon its nuclear weapons—as well as its nuclear ambitions—in exchange for more normal diplomatic and economic relations with the rest of the world.

Following are the texts of two documents. First, "Initial Actions for the Implementation of the Joint Statement," issued on February 13, 2007, at the conclusion of a round of Six-Party Talks in Beijing, China, among representatives from China, Japan, North Korea, Russia, South Korea, and the United States. Second, "Second-Phase Actions for the Implementation of the September 2005 Joint Statement," issued by the Chinese foreign minister on October 3, 2007, following another round of Six-Party Talks.

DOCUMENT

Initial Implementation of the Joint Statement on Denuclearization

February 13, 2007

The Third Session of the Fifth Round of the Six-Party Talks was held in Beijing among the People's Republic of China [(PRC)], the Democratic People's Republic of Korea [(DPRK, North Korea)], Japan, the Republic of Korea [(ROK, South Korea)], the Russian Federation and the United States of America from 8 to 13 February 2007.

Mr. Wu Dawei, Vice Minister of Foreign Affairs of the PRC, Mr. Kim Gye Gwan, Vice Minister of Foreign Affairs of the DPRK; Mr. Kenichiro Sasae, Director-General for Asian and Oceanian Affairs, Ministry of Foreign Affairs of Japan; Mr. Chun Yung-woo, Special Representative for Korean Peninsula Peace and Security Affairs of the ROK Ministry of Foreign Affairs and Trade; Mr. Alexander Losyukov, Deputy Minister of Foreign Affairs of the Russian Federation; and Mr. Christopher Hill, Assistant Secretary for East Asian and Pacific Affairs of the Department of State of the United States attended the talks as heads of their respective delegations.

Vice Foreign Minister Wu Dawei chaired the talks.

I. The Parties held serious and productive discussions on the actions each party will take in the initial phase for the implementation of the Joint Statement of 19 September

2005. The Parties reaffirmed their common goal and will to achieve early denuclearization of the Korean Peninsula in a peaceful manner and reiterated that they would earnestly fulfill their commitments in the Joint Statement. The Parties agreed to take coordinated steps to implement the Joint Statement in a phased manner in line with the principle of "action for action".

II. The Parties agreed to take the following actions in parallel in the initial phase:

1. The DPRK will shut down and seal for the purpose of eventual abandonment the Yongbyon nuclear facility, including the reprocessing facility and invite back IAEA [International Atomic Energy Agency] personnel to conduct all necessary monitoring and verifications as agreed between IAEA and the DPRK.
2. The DPRK will discuss with other parties a list of all its nuclear programs as described in the Joint Statement, including plutonium extracted from used fuel rods, that would be abandoned pursuant to the Joint Statement.
3. The DPRK and the US will start bilateral talks aimed at resolving pending bilateral issues and moving toward full diplomatic relations. The US will begin the process of removing the designation of the DPRK as a state-sponsor of terrorism and advance the process of terminating the application of the Trading with the Enemy Act with respect to the DPRK.
4. The DPRK and Japan will start bilateral talks aimed at taking steps to normalize their relations in accordance with the Pyongyang Declaration, on the basis of the settlement of unfortunate past and the outstanding issues of concern.
5. Recalling Section 1 and 3 of the Joint Statement of 19 September 2005, the Parties agreed to cooperate in economic, energy and humanitarian assistance to the DPRK. In this regard, the Parties agreed to the provision of emergency energy assistance to the DPRK in the initial phase. The initial shipment of emergency energy assistance equivalent to 50,000 tons of heavy fuel oil (HFO) will commence within next 60 days.

The Parties agreed that the above-mentioned initial actions will be implemented within next 60 days and that they will take coordinated steps toward this goal.

III. The Parties agreed on the establishment of the following Working Groups (WG) in order to carry out the initial actions and for the purpose of full implementation of the Joint Statement:

1. Denuclearization of the Korean Peninsula
2. Normalization of DPRK-US relations
3. Normalization of DPRK-Japan relations
4. Economy and Energy Cooperation
5. Northeast Asia Peace and Security Mechanism

The WGs will discuss and formulate specific plans for the implementation of the Joint Statement in their respective areas. The WGs shall report to the Six-Party Heads of Delegation Meeting on the progress of their work. In principle, progress in one WG shall not affect progress in other WGs. Plans made by the five WGs will be implemented as a whole in a coordinated manner.

The Parties agreed that all WGs will meet within next 30 days.

IV. During the period of the Initial Actions phase and the next phase— which includes provision by the DPRK of a complete declaration of all nuclear programs and disablement of all existing nuclear facilities, including graphite-moderated reactors and reprocessing plant—economic, energy and humanitarian assistance up to the equivalent of 1 million tons of heavy fuel oil (HFO), including the initial shipment equivalent to 50,000 tons of HFO, will be provided to the DPRK.

The detailed modalities of the said assistance will be determined through consultations and appropriate assessments in the Working Group on Economic and Energy Cooperation.

V. Once the initial actions are implemented, the Six Parties will promptly hold a ministerial meeting to confirm implementation of the Joint Statement and explore ways and means for promoting security cooperation in Northeast Asia.

VI. The Parties reaffirmed that they will take positive steps to increase mutual trust, and will make joint efforts for lasting peace and stability in Northeast Asia. The directly related parties will negotiate a permanent peace regime on the Korean Peninsula at an appropriate separate forum.

VII. The Parties agreed to hold the Sixth Round of the Six-Party Talks on 19 March 2007 to hear reports of WGs and discuss on actions for the next phase.

SOURCE: U.S. Department of State. Under Secretary for Public Diplomacy and Public Affairs. Bureau of Public Affairs: Press Office. Office of the Spokesman. "North Korea—Denuclearization Action Plan. Initial Actions for the Implementation of the Joint Statement." Press release 2007/099. February 13, 2007. www.state.gov/r/pa/prs/ps/2007/february/80479.htm (accessed February 12, 2008).

DOCUMENT

Second-Phase Implementation of the September 2005 Joint Statement

October 3, 2007

The Second Session of the Sixth Round of the Six-Party Talks was held in Beijing among the People's Republic of China [(PRC)], the Democratic People's Republic of Korea [(DPRK, North Korea)], Japan, the Republic of Korea [(ROK, South Korea)], the Russian Federation and the United States of America from 27 to 30 September 2007.

Mr. Wu Dawei, Vice Minister of Foreign Affairs of the PRC, Mr. Kim Gye Gwan, Vice Minister of Foreign Affairs of the DPRK, Mr. Kenichiro Sasae, Director-General for Asian and Oceanian Affairs, Ministry of Foreign Affairs of Japan, Mr. Chun Yung-woo, Special Representative for Korean Peninsula Peace and Security Affairs of the ROK, Ministry of Foreign Affairs and Trade, Mr. Alexander Losyukov, Deputy Minister of Foreign Affairs of the Russian Federation, and Mr. Christopher Hill, Assistant Secretary for East Asian and Pacific Affairs of the Department of State of the United States, attended the talks as heads of their respective delegations.

Vice Foreign Minister Wu Dawei chaired the talks.

The Parties listened to and endorsed the reports of the five Working Groups, confirmed the implementation of the initial actions provided for in the February 13 agreement, agreed to push forward the Six-Party Talks process in accordance with the consensus reached at the meetings of the Working Groups and reached agreement on second-phase actions for the implementation of the Joint Statement of 19 September 2005, the goal of which is the verifiable denuclearization of the Korean Peninsula in a peaceful manner.

I. On Denuclearization of the Korean Peninsula

1. The DPRK agreed to disable all existing nuclear facilities subject to abandonment under the September 2005 Joint Statement and the February 13 agreement.

The disablement of the 5 megawatt Experimental Reactor at Yongbyon, the Reprocessing Plant (Radiochemical Laboratory) at Yongbyon and the Nuclear Fuel Rod Fabrication Facility at Yongbyon will be completed by 31 December 2007. Specific measures recommended by the expert group will be adopted by heads of delegation in line with the principles of being acceptable to all Parties, scientific, safe, verifiable, and consistent with international standards. At the request of the other Parties, the United States will lead disablement activities and provide the initial funding for those activities. As a first step, the US side will lead the expert group to the DPRK within the next two weeks to prepare for disablement.

2. The DPRK agreed to provide a complete and correct declaration of all its nuclear programs in accordance with the February 13 agreement by 31 December 2007.

3. The DPRK reaffirmed its commitment not to transfer nuclear materials, technology, or know-how.

II. On Normalization of Relations between Relevant Countries

1. The DPRK and the United States remain committed to improving their bilateral relations and moving towards a full diplomatic relationship. The two sides will increase bilateral exchanges and enhance mutual trust. Recalling the commitments to begin the process of removing the designation of the DPRK as a state sponsor of terrorism and advance the process of terminating the application of the Trading with the Enemy Act with respect to the DPRK, the United States will fulfill its commitments to the DPRK in parallel with the DPRK's actions based on consensus reached at the meetings of the Working Group on Normalization of DPRK-U.S. Relations.

2. The DPRK and Japan will make sincere efforts to normalize their relations expeditiously in accordance with the Pyongyang Declaration, on the basis of the settlement of the unfortunate past and the outstanding issues of concern. The DPRK and Japan committed themselves to taking specific actions toward this end through intensive consultations between them.

III. On Economic and Energy Assistance to the DPRK

In accordance with the February 13 agreement, economic, energy and humanitarian assistance up to the equivalent of one million tons of HFO (inclusive of the 100,000 tons of HFO already delivered) will be provided to the DPRK. Specific modalities will be finalized through discussion by the Working Group on Economy and Energy Cooperation.

IV. On the Six-Party Ministerial Meeting

The Parties reiterated that the Six-Party Ministerial Meeting will be held in Beijing at an appropriate time.

The Parties agreed to hold a heads of delegation meeting prior to the Ministerial Meeting to discuss the agenda for the Meeting.

SOURCE: U.S. Department of State. Under Secretary for Public Diplomacy and Public Affairs. Bureau of Public Affairs: Press Office. Office of the Spokesman. "Six-Party Talks—Second-Phase Actions for the Implementation of the September 2005 Joint Statement." Press release 2007/842. October 3, 2007. www.state.gov/r/pa/prs/ps/2007/oct/93217.htm (accessed February 12, 2008).

OTHER HISTORIC DOCUMENTS OF INTEREST

FROM THIS VOLUME

- Iranian nuclear weapons programs, p. 112

FROM PREVIOUS *HISTORIC DOCUMENTS*

- UN Security Council on nuclear weapons tests by North Korea, *2006,* p. 606
- Joint statement of Six-Party Talks on denuclearization, *2005,* p. 604
- North Korean nuclear agreement, *2004,* p. 602
- State Department official on the "dictatorship" in North Korea, *2003,* p. 592
- Korean unification talks, *2000,* p. 359

March

DOCUMENT IN CONTEXT

Acting Surgeon General Moritsugu on the Dangers of Underage Drinking

MARCH 6, 2007

The United States "can no longer ignore what alcohol is doing to our children," acting Surgeon General Kenneth P. Moritsugu said March 6, releasing a "call to action" for reducing underage drinking in the United States. Although fewer teenagers were smoking or using illegal drugs than in the past, alcohol use remained at consistently high levels, Moritsugu said. The 2005 National Survey on Drug Use and Health estimated there were 11 million underage drinkers (drinkers under age twenty-one) in the United States. About 7.2 million of those were considered binge drinkers, who consumed five drinks or more on a single occasion.

"Underage drinking is deeply embedded in the American culture, is often viewed as a rite of passage, is frequently facilitated by adults, and has proved stubbornly resistant to change," the "Surgeon General's Call to Action to Prevent and Reduce Underage Drinking" reported. The report outlined six goals aimed at reducing the incidence of underage drinking and spelled out dozens of strategies that parents, schools and colleges, communities, and others could take to discourage adolescents from drinking. The report also called for more research on adolescent alcohol use, including risk factors and ways to treat teenage alcohol use effectively.

Changing social attitudes about drinking was likely to be a struggle. According to a study released by the federal Substance Abuse and Mental Health Services Administration on February 28, 51.1 percent of Americans reported drinking alcohol in the previous month. The data were collected in interviews conducted with 136,100 people age twelve and older in 2004 and 2005. A separate study, published in the July 2 issue of *Archives of General Psychiatry*, reported that more than 30 percent of American adults had abused alcohol or suffered from alcoholism at some point in their lives.

Scope of the Problem

According to the surgeon general's report, which assembled data from dozens of studies, by age fifteen about half of all boys and girls surveyed had had at least one whole drink of alcohol; that percentage rose to about 90 percent for twenty-one-year-olds. The highest prevalence of alcohol dependence was among those ages eighteen through twenty. The drinking behaviors described by slightly more than 12 percent of this age group matched the definition of alcohol dependence adopted by the American Psychiatric Association. About 11 percent of those ages twenty-one through twenty-four were categorized as alcohol dependent, and the percentage declined markedly for older age groups.

More underage children used alcohol than smoked or used marijuana, the illegal drug most common among adolescents. More than 40 percent of twelfth graders reported having a drink in the previous month, compared with slightly more than 20 percent who said they had smoked and about 18 percent who said they had used marijuana. About one-third of tenth graders and more than 15 percent of eighth graders also reported having a drink in the previous month. Adolescents did not drink as often as adults, but when they did drink, they tended to be binge drinkers, consuming five or more drinks on a single occasion. For boys, the incidence of binge drinking steadily increased to nearly five days in any given month at age twenty; binge drinking was lower among girls, reaching a peak of just under three days a month by age eighteen before tapering down.

The surgeon general's report listed a number of adverse affects attributable to underage drinking. About 5,000 people under age twenty-one died every year from injuries involving underage drinking; of those deaths, about 1,900 were related to car crashes, while 1,600 involved homicides. Underage drinking also factored heavily in risky sexual behavior, which increased the risk for unplanned pregnancies and for contracting sexually transmitted diseases, including HIV, the virus that causes AIDS. Underage drinking was also associated with smoking, illegal drug use, and academic failure and had numerous physical consequences ranging from hangovers to death from alcohol poisoning. Recent research showed that underage drinking could adversely alter the structure and function of the brain, which continues to develop into a person's twenties.

Underage drinking was also reported to be a risk factor for heavy drinking later in life, which had its own range of social and health consequences, including increased risk of cancers and stroke, as well as pancreatitis and cirrhosis of the liver. Finally, underage drinking had what the report called "secondhand effects," such as unruly behavior, property destruction, violence, and death. About 45 percent of those killed in accidents involving a drinking underage driver were people other than the driver, according to the report.

Steps for Reducing Underage Drinking

Efforts to reduce underage drinking, according to the report, had to take into account three broad areas: the "dynamic developmental processes" of adolescence, the adolescent's environment, and the individual characteristics of each adolescent that affected the decision to drink. The report outlined several goals aimed at preventing or reducing underage drinking, including changing society's acceptance, norms, and expectations about adolescent drinking. The report also stressed the importance of engaging parents, schools, communities, and other social systems in a coordinated effort to prevent or delay adolescents from starting to drink. It then listed specific strategies that parents, schools, communities, and others could employ to help curb the incidence of underage drinking.

Even before the report was released, many schools, colleges, and other organizations were taking steps to combat adolescent drinking. Several colleges, for example, had begun to notify parents when their underage children were caught breaking campus policies regarding illegal drinking and drug use. The University of Georgia and the University of New Mexico were among those that alerted parents of their children's first offense. The University of Georgia also toughened its penalties, putting students on probation for a first offense and suspending them for a second offense committed while on probation. Some high schools suspended student athletes after photos showing them drinking at parties were posted on the Internet and e-mailed to school administrators. A growing number of high schools were using breathalyzers to test students at extracurricular activities such as dances and football games where underage drinking was considered a problem. Schools called parents or police if a student failed the test; in some cases, the student was suspended.

State legislatures were also stepping up their laws and regulations aimed at slowing underage drinking. Among other strategies, some states monitored online social networks and used hotlines to find out about and then stop underage drinking parties. Several states also tightened their policing of and penalties for adults who supplied alcohol to minors. At the urging of a group of teenagers, California officials voted to raise the tax on flavored malt beverages by nearly $2 per six-pack. "The ruling will send a signal to youth that these drinks are hard liquor because they have costs similar to hard liquor," said one of the officials.

In addition, state attorneys general challenged beer companies for targeting their advertising of flavored malt beverages to underage drinkers. On May 10 twenty-nine state attorneys general wrote to Anheuser-Busch, expressing concern about the marketing and packaging for its drinks Spykes, Tilt, and Bud Extra, which combined alcohol with caffeine. The state officials said the packaging was similar to that for non-alcoholic caffeinated "energy" drinks, such as Red Bull, that were hugely popular with teenagers. They also criticized Spykes because it was packaged in "tiny, attractive, brightly colored containers that can be easily concealed in a pocket or purse" and because the beverage was distributed in grocery and convenience stores "where it may be more readily seen and purchased by underage youth than it if were sold only in liquor stores." On May 18 the company announced that it was stopping sales of Spykes "due to limited volume potential and unfounded criticism."

In August a group of thirty attorneys general wrote to the federal Alcohol and Tobacco Tax and Trade Bureau asking it to determine whether the alcohol-caffeine drinks contained too much alcohol to be sold as malt beverages, which allowed them to be distributed, like beer, in more venues, including grocery stores and other places frequented by minors. The attorneys general also argued that the advertising for those alcoholic beverages was misleading and attractive to underage drinkers: "Alcoholic beverage manufacturers have taken advantage of the youth appeal (of non-alcoholic energy) drinks by engaging in aggressive marketing campaigns for pre-mixed alcoholic energy drinks."

Following are excerpts from the "Surgeon General's Call to Action to Prevent and Reduce Underage Drinking," released on March 6, 2007, by acting surgeon general Kenneth P. Moritsugu.

DOCUMENT

Surgeon General's Call to Action to Prevent and Reduce Underage Drinking

March 6, 2007

FOREWORD FROM THE ACTING SURGEON GENERAL U.S. DEPARTMENT OF HEALTH AND HUMAN SERVICES

Alcohol is the most widely used substance of abuse among America's youth. A higher percentage of young people between the ages of 12 and 20 use alcohol than use tobacco or illicit drugs. The physical consequences of underage alcohol use range from medical problems to death by alcohol poisoning, and alcohol plays a significant role in risky sexual behavior, physical and sexual assaults, various types of injuries, and suicide. Underage drinking also creates secondhand effects for others, drinkers and nondrinkers alike, including car crashes from drunk driving, that put every child at risk. Underage alcohol consumption is a major societal problem with enormous health and safety consequences and will demand the Nation's attention and committed efforts to solve.

For the most part, parents and other adults underestimate the number of adolescents who use alcohol. They underestimate how early drinking begins, the amount of alcohol adolescents consume, the many risks that alcohol consumption creates for adolescents, and the nature and extent of the consequences to both drinkers and nondrinkers. Too often, parents are inclined to believe, "Not my child." Yet, by age 15, approximately one-half of America's boys and girls have had a whole drink of alcohol, not just a few sips, and the highest prevalence of alcohol dependence in any age group is among people ages 18 to 20.

I have issued this *Surgeon General's Call to Action To Prevent and Reduce Underage Drinking* to focus national attention on this enduring problem and on new, disturbing research which indicates that the developing adolescent brain may be particularly susceptible to long-term negative consequences from alcohol use. Recent studies show that alcohol consumption has the potential to trigger long-term biological changes that may have detrimental effects on the developing adolescent brain, including neurocognitive impairment.

Fortunately, the latest research also offers hopeful new possibilities for prevention and intervention by furthering our understanding of underage alcohol use as a developmental phenomenon—as a behavior directly related to maturational processes in adolescence. New research explains why adolescents use alcohol differently from adults, why they react uniquely to it, and why alcohol can pose such a powerful attraction to adolescents, with unpredictable and potentially devastating outcomes.

Emerging research also makes it clear that an adolescent's decision to use alcohol is influenced by multiple factors. These factors include normal maturational changes that all adolescents experience; genetic, psychological, and social factors specific to each adolescent; and the various social and cultural environments that surround adolescents, including their families, schools, and communities. These factors—some of which protect adolescents from alcohol use and some of which put them at risk—change during the course of adolescence. Because environmental factors play such a significant role, responsibility for the prevention and reduction of underage drinking extends beyond the parents of adolescents, their schools, and communities. It is the collective responsibility of the Nation as a whole and of each of us individually.

The process of solving the public health problem of underage alcohol use begins with an examination of our own attitudes toward underage drinking—and our recognition of the seriousness of its consequences for adolescents, their families, and society as a whole. Adolescent alcohol use is not an acceptable rite of passage but a serious threat to adolescent development and health, as the statistics related to adolescent impairment, injury, and death attest.

A significant point of the *Call to Action* is this: Underage alcohol use is not inevitable, and schools, parents, and other adults are not powerless to stop it. The latest research demonstrates a compelling need to address alcohol use early, continuously, and in the context of human development using a systematic approach that spans childhood through adolescence into adulthood. Such an approach is described in this *Call to Action.* Such an approach can be effective when, as a Nation and individually, we commit ourselves to solving the problem of underage drinking in America. We owe nothing less to our children and our country.

Kenneth P. Moritsugu, M.D., M.P.H.

TAKING ACTION: A VISION FOR THE FUTURE

Underage alcohol use is a complex problem that has proved resistant to solution for decades. Established and emerging research, however, suggests a new evidence-based approach with considerable promise. It is that approach—and the possibilities it holds for the Nation's youth—that inspires the vision of *The Surgeon General's Call to Action To Prevent and Reduce Underage Drinking.*

Principles

The *Call to Action* is based on several overarching principles from which its goals and the means for achieving them were derived. These principles are:

1. *Underage alcohol use is a phenomenon that is directly related to human development.* Because of the nature of adolescence itself, alcohol poses a powerful attraction to adolescents, with unpredictable outcomes that can put any child at risk.
2. *Factors that protect adolescents from alcohol use as well as those that put them at risk change during the course of adolescence.* Internal characteristics, developmental issues, and shifting factors in the adolescent's environment all play a role.
3. *Protecting adolescents from alcohol use requires a comprehensive, developmentally based approach* that is initiated before puberty and continues throughout adolescence with support from families, schools, colleges, communities, the health care system, and government.
4. *The prevention and reduction of underage drinking is the collective responsibility of the Nation.* Scaffolding the Nation's youth is the responsibility of all people in all of the social systems in which adolescents operate: family, schools, communities, health care systems, religious institutions, criminal and juvenile justice systems,

all levels of government, and society as a whole. Each social system has a potential impact on the adolescent, and the active involvement of all systems is necessary to fully maximize existing resources to prevent underage drinking and its related problems. When all the social systems work together toward the common goal of preventing and reducing underage drinking, they create a powerful synergy that is critical to realize the vision.

5. *Underage alcohol use is not inevitable*, and parents and society are not helpless to prevent it.

Goals

The healthy development of America's youth is a national goal that is threatened by underage alcohol consumption and the adverse consequences it can bring. In sometimes subtle and sometimes dramatic ways, underage alcohol use can sidetrack the trajectory of a child's life—or end it. The freedom to fulfill one's potential and to develop without the impairment of alcohol's negative consequences is a significant part of the vision for the future described in this *Call to Action*. The fulfillment of that vision rests on the achievement of six goals that the Surgeon General has proposed for the Nation. . . .

Goal 1: Foster Changes in American Society That Facilitate Healthy Adolescent Development and That Help Prevent and Reduce Underage Drinking.

Rationale

Culture generally is thought to mean the set of attitudes, values, norms, customs, and beliefs that distinguishes one group of people from another group. Nations, communities, ethnic and religious groups, schools, and peer groups all have distinct cultures. The various cultures in which an adolescent lives have a significant influence on his or her decisions about alcohol use. The attitudes and values of an adolescent's community with regard to underage alcohol use and appropriate adult use form an important part of the social structure that protects youth from alcohol use—or puts them at risk—because they determine the extent to which the community itself and the adults within that community will encourage or discourage underage drinking.

Alcohol, in its many forms, is familiar to children and adolescents and often appears relatively benign, if not openly enticing. Yet it is not benign for underage drinkers. Reducing cultural forces that encourage or support underage alcohol consumption lessens both the attraction of alcohol and the likelihood that it will be consumed by youth.

A culture in which youth feel that underage drinking is accepted, acceptable, or even expected promotes underage drinking. Society as a whole needs to send the message that it strongly disapproves of underage alcohol use because of its potentially adverse consequences and that it will not condone or permit it. At the same time, it is necessary to work to increase those societal forces that facilitate and support an alcohol-free childhood and adolescence.

Challenges

Culture is complex, and changing it requires sustained efforts on the part of multiple segments of society. The culture around drinking in the United States is especially difficult to change because alcohol use is embedded in American society, is legal and acceptable for most adults, and is often regarded as a rite of passage for youth. Many young people believe that drinking is not only acceptable but expected of them and a way for them to feel more grown-up. Finally, alcohol holds a powerful attraction for adolescents because of the nature of adolescence itself.

Strategies

For parents and other caregivers: Parents have a responsibility to help shape the culture in which their adolescents are raised, particularly the culture of their schools and community. Parental strategies include the following:

- Partner with other parents in their child's network to ensure that parties and other social events do not allow underage alcohol consumption, much less facilitate its use or focus on it.
- Collaborate with other parents in coalitions designed to ensure that the culture in the schools and community support and reward an adolescent's decision not to drink.
- Serve as a positive role model for adolescents by not drinking excessively, by avoiding alcohol consumption in high-risk situations (e.g., when driving a motor vehicle, while boating, and while operating machinery), and by seeking professional help for alcohol-related problems.

For colleges and universities: Given the prevalence of underage drinking on college campuses, institutions of higher education should examine their policies and practices on alcohol use by their students and the extent to which they may directly or indirectly encourage, support, or facilitate underage alcohol use. Colleges and universities can change a campus culture that contributes to underage alcohol use. Some measures to consider are to:

- Establish, review, and enforce rules against underage alcohol use with consequences that are developmentally appropriate and sufficient to ensure compliance. This practice helps to confirm the seriousness with which the institution views underage alcohol use by its students.
- Eliminate alcohol sponsorship of athletic events and other campus social activities.
- Restrict the sale of alcoholic beverages on campus or at campus facilities, such as football stadiums and concert halls.
- Implement responsible beverage service policies at campus facilities, such as sports arenas, concert halls, and campus pubs.
- Hold all student groups on campus, including fraternities, sororities, athletics teams, and student clubs and organizations, strictly accountable for underage alcohol use at their facilities and during functions that they sponsor.
- Eliminate alcohol advertising in college publications.
- Educate parents, instructors, and administrators about the consequences of underage drinking on college campuses, including secondhand effects that range from interference with studying to being the victim of an alcohol-related assault

or date rape, and enlist their assistance in changing any culture that currently supports alcohol use by underage students.

- Partner with community stakeholders to address underage drinking as a community problem as well as a college problem and to forge collaborative efforts that can achieve a solution.
- Expand opportunities for students to make spontaneous social choices that do not include alcohol (e.g., by providing frequent alcohol-free late-night events, extending the hours of student centers and athletics facilities, and increasing public service opportunities).

For communities: Adolescents generally obtain alcohol from adults who sell it to them, purchase it on their behalf, or allow them to attend or give parties where it is served. Therefore, it is critical that adults refuse to provide alcohol to adolescents and that communities value, encourage, and reward an adolescent's commitment not to drink. A number of strategies can contribute to a culture that discourages adults from providing alcohol to minors and that supports an adolescent's decision not to drink. Communities can:

- Invest in alcohol-free youth friendly programs and environments.
- Widely publicize all policies and laws that prohibit underage alcohol use.
- Work with sponsors of community or ethnic holiday events to ensure that such events do not promote a culture in which underage drinking is acceptable.
- Urge the alcohol industry to voluntarily reduce outdoor alcohol advertising.
- Promote the idea that underage alcohol use is a local problem that local citizens can solve through concerted and dedicated action.
- Establish organizations and coalitions committed to establishing a local culture that disapproves of underage alcohol use, that works diligently to prevent and reduce it, and that is dedicated to informing the public about the extent and consequences of underage drinking.
- Work to ensure that members of the community are aware of the latest research on adolescent alcohol use and, in particular, the adverse consequences of alcohol use on underage drinkers and other members of the community who suffer from its secondhand effects. An informed public is an essential part of an overall plan to prevent and reduce underage drinking and to change the culture that supports it.
- Change community norms to decrease the acceptability of underage drinking, in part, through public awareness campaigns.
- Focus as much attention on underage drinking as on tobacco and illicit drugs, making it clear that underage alcohol use is a community problem. When the American people rejected the use of tobacco and illicit drugs as a culturally acceptable behavior, the use of those substances declined, and the culture of acceptance shifted to disapproval. The same change process is possible with underage drinking.

For the criminal and juvenile justice systems and law enforcement: The justice system and law enforcement can:

- Enforce uniformly and consistently all policies and laws against underage alcohol use and widely publicize these efforts.
- Gain public support for enforcing underage drinking laws by working with other stakeholders to ensure that the public understands that underage drinking affects both the public health and safety.
- Work with State, Tribal, and local coalitions to reduce underage drinking.

For the alcohol industry: The alcohol industry has a public responsibility relating to the marketing of its product, since its use is illegal for more than 80 million underage Americans. That responsibility can be fulfilled through products and advertising design and placement that meet these criteria:

- The message adolescents receive through the billions of dollars spent on industry advertising and responsibility campaigns does not portray alcohol as an appropriate rite of passage from childhood to adulthood or as an essential element in achieving popularity, social success, or a fulfilling life.
- The placement of alcohol advertising, promotions, and other means of marketing do not disproportionately expose youth to messages about alcohol.
- No alcohol product is designed or advertised to disproportionately appeal to youth or to influence youth by sending the message that its consumption is an appropriate way for minors to learn to drink or that any form of alcohol is acceptable for drinking by those under the age of 21.
- The content and design of industry Web sites and Internet alcohol advertising do not especially attract or appeal to adolescents or others under the legal drinking age.

For the entertainment and media industries: Because of their reach and potential impact, the entertainment and media industries have a responsibility to the public in the way they choose to depict alcohol use, especially by those under the age of 21, in motion pictures, television programming, music, and video games. That responsibility can be fulfilled by creating and distributing entertainment that:

- Does not glamorize underage alcohol use.
- Does not present any form of underage drinking in a favorable light, especially when entertainment products are targeted toward underage audiences or likely to be viewed or heard by them.
- Seeks to present a balanced portrayal of alcohol use, including its attendant risks.
- Avoids gratuitous portrayals of alcohol use in motion pictures and television shows that target children as a major audience. This is important because children's expectations toward alcohol and its use are, in part, based on what they see on the screen.

For governments and policymakers: Governments and policymakers can:

- Focus as much attention on underage drinking as on tobacco and illicit drugs, making it clear that underage alcohol use is an important public health problem.
- Ensure that all communications are clearly written and culturally sensitive.

Goal 2: Engage Parents and Other Caregivers, Schools, Communities, All Levels of Government, All Social Systems That Interface With Youth, and Youth Themselves in a Coordinated National Effort to Prevent and Reduce Underage Drinking and Its Consequences.

. . .

Goal 3: Promote an Understanding of Underage Alcohol Consumption in the Context of Human Development and Maturation That Takes Into Account Individual Adolescent Characteristics as Well as Ethnic, Cultural, and Gender Differences.

. . .

Goal 4: Conduct Additional Research on Adolescent Alcohol Use and Its Relationship to Development.

. . .

Goal 5: Work to Improve Public Health Surveillance on Underage Drinking and on Population-Based Risk Factors for This Behavior.

. . .

Goal 6: Work to Ensure That Policies at All Levels Are Consistent With the National Goal of Preventing and Reducing Underage Alcohol Consumption.

. . .

SOURCE: U.S. Department of Health and Human Services. Office of the Surgeon General. "The Surgeon General's Call to Action to Prevent and Reduce Underage Drinking." March 6, 2007. www.surgeongeneral.gov/topics/underagedrinking (accessed January 13, 2007).

OTHER HISTORIC DOCUMENTS OF INTEREST

FROM PREVIOUS *HISTORIC DOCUMENTS*

- National task force on college drinking, *2002*, p. 175

DOCUMENT IN CONTEXT

Federal Appeals Court on D.C. Handgun Ban

MARCH 9, 2007

A federal appeals court on March 9 struck down a District of Columbia ban on handguns, setting the stage for the Supreme Court to decide whether the Second Amendment to the Constitution confers the right to keep and bear arms on individuals as well as on organized militias like the National Guard. The Court, which was scheduled to hear arguments in the case in March 2008, had not examined the meaning of the Second Amendment for nearly seventy years. In a 1939 case, the Court strongly suggested, without explicitly stating, that the right was collective rather than individual. The Court's ruling, expected before the end of its session in June 2008, would likely be highly controversial and could potentially make gun control an important issue in the 2008 presidential campaign.

Meanwhile, in the wake of a mass shooting at Virginia Tech University in April 2007, Congress in December enacted the first major national gun control law since 1994. The measure, supported by the National Rifle Association (NRA), was designed to ensure that people with diagnosed mental health problems, such as the shooter at Virginia Tech, would be included in the national instant background check system and thereby prohibited from buying guns. But the NRA and its supporters blocked additional legislation that would have given law enforcement agencies greater access to information that would help them trace the movement of illegal weapons across state lines.

D.C. Handgun Ban Overturned

The Second Amendment states: "A well regulated Militia, being necessary to the security of a free State, the right of the people to keep and bear Arms, shall not be abridged." Through 2007, the Supreme Court had never ruled directly on whether the Second Amendment guarantees individuals the right to possess guns. The Court last addressed the issue in 1939 in *United States v. Miller,* and the majority opinion was widely seen as endorsing the view that the amendment applied only to a collective right, such as that of the National Guard, to "keep and bear arms."

Over the years, as the number of crimes committed with guns soared, federal, state, and local governments imposed restrictions and regulations on gun ownership, including outright bans in some jurisdictions. The District of Columbia's law, considered among the toughest in the nation, was adopted during a 1976 crime wave; it banned possession of virtually all handguns that were not registered before 1976. It also made it unlawful to carry an unlicensed weapon from one room to another and required that all registered guns be stored unloaded and disassembled or guarded by a trigger lock. While law enforcement officials considered such controls to be important tools to reduce violent crime, the NRA and other gun-rights advocates argued that such bans prevented only law-abiding citizens from obtaining guns and left them feeling unsafe and insecure in their own homes.

The D.C. handgun ban's success in controlling crime was a matter of some controversy by 2007. In recent years, more than 80 percent of the murders in the city had been committed with guns, and police seized more than 2,600 illegal weapons in 2006. Police indicated that most of the illegal weapons came into the city from surrounding jurisdictions with less stringent gun controls and that the problem would be worse without the ban.

The D.C. Second Amendment case began in May 2001, when Attorney General John Ashcroft, a long-time NRA member, told the politically powerful pro-gun lobby that he believed the Second Amendment protected the right of individuals to keep and bear arms. Later in the year, Ashcroft formalized that position as Justice Department policy, reversing the department's long-standing holding that the Second Amendment guarantees applied only to state and federal militias, not to individuals.

Soon after Ashcroft announced the new policy, Robert A. Levy, a wealthy libertarian lawyer and District native living in Florida, decided to challenge the D.C. handgun ban, seeking ultimately to get a definitive ruling on the Second Amendment from the Supreme Court. Levy, who was also funding the challenge, said he was not particularly concerned about owning a gun but was concerned about what he considered government infringement on personal liberties. He told the *Washington Post* in March that although he did not oppose "reasonable" gun controls, the District's "outright prohibition" on handguns "offends my constitutional sensibilities." Working with lawyers from the libertarian Cato Institute in Washington, Levy identified six District residents who were willing to join him in challenging the constitutionality of the ban; all six said they wanted to keep handguns in their homes for self-defense and argued that the D.C. law effectively denied them the use of a "functional" firearm for self-defense.

A federal judge dismissed the case in 2004, ruling that the Second Amendment did not confer an individual right but only a collective one. By a 2–1 vote, a three-judge panel for the District of Columbia Circuit Court of Appeals reversed the lower court on March 9, 2007, after first finding that only one of the six plaintiffs, Dick A. Heller, had legal standing to challenge the law. "We conclude that the Second Amendment protects an individual right to keep and bear arms," wrote Senior Judge Laurence H. Silberman, who was joined in his opinion by Judge Thomas B. Griffith. "That right existed prior to the formation of the new government under the Constitution and was premised on the private use of arms for activities such as hunting and self-defense." That individual right, he continued, "facilitated militia service by ensuring that citizens would not be barred from keeping the arms they would need when called forth for militia duty," but the individual right to keep and bear arms is "not limited to militia service." Judge Karen LeCraft Henderson dissented, arguing that the Second Amendment did not apply to the District because the capital was not a state. All three judges were appointed by Republican presidents.

D.C. mayor Adrian M. Fenty said he was "deeply disappointed and frankly outraged" by the decision, which "flies in the face of laws that helped decrease gun violence" in the city. In May the full appeals court refused Fenty's request to review the decision. Despite concerns among gun control proponents that a Supreme Court decision to uphold the appeals court ruling could jeopardize gun control laws throughout the country, Fenty announced on July 16 that the city would appeal to the high court: "We have made the determination that this law can and should be defended, and we are willing to take our case to the highest court in the land to protect the city's residents." In an unusual move, Heller's attorneys joined the city in asking the Court to hear the case.

On November 20 the Court announced that it would take the case and specifically address the question of whether the law violated "the Second Amendment rights of individuals who are not affiliated with any state-regulated militia, but who wish to keep handguns and other firearms for private use in their homes." The announcement was welcomed by many gun rights advocates, who judged that, given the Court's current conservative majority, the Court was more likely than it had been for decades to find an individual right to keep and bear arms in the Second Amendment. Arguments in the case were scheduled for March 2008 and a decision was expected by June, just five months before the 2008 presidential election.

Mental Health Checks for Gun Buyers

An unusual bipartisan congressional effort resulted in passage at the end of 2007 of the first significant federal gun control legislation in more than a decade. The measure, which President George W. Bush signed in January 2008, combined monetary incentives and penalties to encourage states to tighten their requirements to report people judged to be mentally ill for inclusion in the national database that gun sellers used to check the backgrounds of would-be gun purchasers. The 1968 Gun Control Act prohibited anyone found by a court to be "mentally defective" from purchasing a gun (it also barred gun purchases by felons, fugitives, drug addicts, and other law breakers), but most states failed to report many of their citizens who fell into this category.

Some legislators had been trying for years to plug this loophole, but their efforts were blocked by the NRA and other gun rights advocates. The matter came to a head in April, when news reports revealed that Cho Seung Hui, the student who shot and killed thirty-two people at Virginia Tech University in Blacksburg, Virginia, before committing suicide, had once been evaluated as being a danger to himself and had been ordered into treatment. Virginia law did not mesh with federal law, however, and Cho's name never made it into the national database. In the wake of the shootings, key gun rights and gun control advocates in Congress worked with the NRA to tighten the reporting loopholes. In exchange for its support, the NRA requested additional changes to the database to allow certain persons, including some veterans whose names had been put into the database for alleged mental health problems, to petition to have their names removed. "No law will prevent evildoers from doing evil acts, but this law will help ensure that those deemed dangerous by the courts will not be able to purchase a weapon" said Rep. John D. Dingell, D-Mich., in June before House passage. Dingell, chair of the House Energy and Commerce Committee and a former board member of the NRA, was a key negotiator on the measure.

Tracing Illegal Weapons

There was no agreement on another gun issue—making it easier for law enforcement officers to trace guns used in crimes. The federal Bureau of Alcohol, Tobacco, Firearms, and Explosives (ATF) operated a National Tracing Center, which collected information on gun sales around the country, including the make and serial number of each gun and its manufacturer, buyer, and seller. Since 2003 Congress had added language to the annual spending bill for the ATF that restricted the information on federal gun trace data that could be shared with law enforcement agencies, prevented its dissemination to city officials, and barred its use in civil lawsuits against firearms makers and sellers.

A coalition of 225 mayors, led by New York City mayor Michael R. Bloomberg, and some law enforcement groups unsuccessfully lobbied to repeal the language, arguing that it prevented police from taking illegal guns off the street. They cited statistics showing that about three-fifths of all illegal guns were sold by just 1 percent of gun dealers. But the amendment's sponsor, Rep. Todd Tiahrt, R-Kans., the NRA, and the Fraternal Order of Police argued that the language was necessary to protect the gun owners' privacy and that mayors wanted the gun trace information so they could sue out-of-state gun dealers. Gun rights advocates had sharply criticized Bloomberg for using private investigators to uncover crooked gun dealers in Virginia and other states. Although the repeal effort failed, the mayors and their supporters did succeed in blocking an amendment that would have made it a crime, punishable by prison time, for police officers to use federal gun trace data in ways other than in a specific criminal investigation.

Following are excerpts from the majority and minority opinions in the case of Parker v. District of Columbia, *in which a three-judge panel of the U.S. Court of Appeals for the District of Columbia Circuit ruled, 2-1, on March 9, 2007, that the D.C. handgun ban was an unconstitutional infringement on the Second Amendment.*

DOCUMENT

Shelley Parker v. District of Columbia

March 9, 2007

Before: HENDERSON and GRIFFITH, *Circuit Judges,* and SILBERMAN, *Senior Circuit Judge.*

Opinion for the Court filed by *Senior Circuit Judge* SILBERMAN.

Dissenting opinion filed by *Circuit Judge* HENDERSON.

SILBERMAN, *Senior Circuit Judge:* Appellants contest the district court's dismissal of their complaint alleging that the District of Columbia's gun control laws violate their Second Amendment rights. The court held that the Second Amendment ("A well regulated Militia, being necessary to the security of a free State, the right of the people to keep and bear Arms, shall not be infringed") does not bestow any rights on individuals except, perhaps, when an individual serves in an organized militia such as today's National Guard. We reverse.

I

Appellants, six residents of the District, challenge D.C. Code § 7-2502.02(a)(4), which generally bars the registration of handguns (with an exception for retired D.C. police officers); D.C. Code § 22-4504, which prohibits carrying a pistol without a license, insofar as that provision would prevent a registrant from moving a gun from one room to another within his or her home; and D.C. Code § 7-2507.02, requiring that all lawfully owned firearms be kept unloaded and disassembled or bound by a trigger lock or similar device. . . .

Essentially, the appellants claim a right to possess what they describe as "functional firearms," by which they mean ones that could be "readily accessible to be used effectively when necessary" for self-defense in the home. They are not asserting a right to carry such weapons outside their homes. Nor are they challenging the District's authority *per se* to require the registration of firearms.

Appellants sought declaratory and injunctive relief . . . , but the court below granted the District's motion to dismiss on the grounds that the Second Amendment, at most, protects an individual's right to "*bear* arms for service in the Militia.". . . And, by "Militia," the court concluded the Second Amendment referred to an organized military body—such as a National Guard unit.

II

[Omitted. In this section the court rules that only one of the plaintiffs, Dick A. Heller, has standing to challenge the D.C. law because he is the only one who applied for and was denied a permit to possess a gun in his home.]

III

As we noted, the Second Amendment provides:

> A well regulated Militia, being necessary to the security of a free State, the right of the people to keep and bear Arms shall not be infringed.

. . .

The provision's second comma divides the Amendment into two clauses; the first is prefatory, and the second operative. Appellants' argument is focused on their reading of the Second Amendment's operative clause. According to appellants, the Amendment's language flat out guarantees an individual right "to keep and bear Arms." Appellants concede that the prefatory clause expresses a civic purpose, but argue that this purpose, while it may inform the meaning of an ambiguous term like "Arms," does not qualify the right guaranteed by the operative portion of the Amendment.

The District of Columbia argues that the prefatory clause declares the Amendment's only purpose—to shield the state militias from federal encroachment—and that the operative clause, even when read in isolation, speaks solely to military affairs and guarantees a civic, rather than an individual, right. In other words, according to the District, the operative clause is not just limited by the prefatory clause, but instead both clauses share an explicitly civic character. The District claims that the Second Amendment "protects private possession of weapons only in connection with performance of civic duties as part of a well-regulated citizens militia organized for the security of a free state." Individuals may be able to enforce the Second Amendment right, but only if the law in question "will impair their participation in common defense and law enforcement when called to serve in the militia." But because the District reads "a well regulated Militia" to signify only the organized militias of the founding era—institutions that the District implicitly argues are no longer in existence today—invocation of the Second Amendment right is conditioned upon service in a defunct institution. Tellingly, we think, the District did not suggest what sort of law, if any, would violate the Second Amendment today—in fact, at oral argument, appellees' counsel asserted that it would be constitutional for the District to ban all firearms outright. In short, we take the District's position to be that the Second Amendment is a dead letter.

We are told by the District that the Second Amendment was written in response to fears that the new federal government would disarm the state militias by preventing men from bearing arms while in actual militia service, or by preventing them from keeping arms at home in preparation for such service. Thus the Amendment should be understood to check federal power to regulate firearms only when federal legislation was directed at the abolition of state militias, because the Amendment's *exclusive* concern was the preservation of those entities. At first blush, it seems passing strange that the able lawyers and statesmen in the First Congress (including James Madison) would have expressed a sole concern for state militias with the language of the Second Amendment. Surely there was a more direct locution, such as "Congress shall make no law disarming the state militias" or "States have a right to a well-regulated militia."

The District's argument—as strained as it seems to us—is hardly an isolated view. In the Second Amendment debate, there are two camps. On one side are the collective right theorists who argue that the Amendment protects only a right of the various state governments to preserve and arm their militias. So understood, the right amounts to an expression of militant federalism, prohibiting the federal government from denuding the states of their armed fighting forces. On the other side of the debate are those who argue that the Second Amendment protects a right of individuals to possess arms for private use. To these individual right theorists, the Amendment guarantees personal liberty analogous to the First Amendment's protection of free speech, or the Fourth Amendment's right to be free from unreasonable searches and seizures. However, some entrepreneurial scholars purport to occupy a middle ground between the individual and collective right models.

The most prominent in-between theory developed by academics has been named the "sophisticated collective right" model. The sophisticated collective right label describes several variations on the collective right theme. All versions of this model share two traits: They (1) acknowledge individuals could, theoretically, raise Second Amendment claims against the federal government, but (2) define the Second Amendment as a purely civic provision that offers no protection for the private use and ownership of arms.

The District advances this sort of theory and suggests that the ability of individuals to raise Second Amendment claims serves to distinguish it from the pure collective right model. But when seen in terms of its practical consequences, the fact that individuals have standing to invoke the Second Amendment is, in our view, a distinction without a difference. . . . Both the collective and sophisticated collective theories assert that the Second Amendment was written for the exclusive purpose of preserving state militias, and both theories deny that individuals *qua* individuals can avail themselves of the Second Amendment today. The latter point is true either because, as the District appears to argue, the "Militia" is no longer in existence, or, as others argue, because the militia's modern analogue, the National Guard, is fully equipped by the federal government, creating no need for individual ownership of firearms. It appears to us that for all its nuance, the sophisticated collective right model amounts to the old collective right theory giving a tip of the hat to the problematic (because ostensibly individual) text of the Second Amendment.

The lower courts are divided between these competing interpretations. Federal appellate courts have largely adopted the collective right model. Only the Fifth Circuit has interpreted the Second Amendment to protect an individual right. State appellate courts, whose interpretations of the U.S. Constitution are no less authoritative than those of our sister circuits, offer a more balanced picture. And the United States Department of Justice has recently adopted the individual right model. . . . The great legal treatises of the nineteenth century support the individual right interpretation, . . . as does Professor Laurence Tribe's leading treatise on constitutional law. Because we have no direct precedent—either in this court or the Supreme Court—that provides us with a square holding on the question, we turn first to the text of the Amendment.

A

We start by considering the competing claims about the meaning of the Second Amendment's operative clause: "the right of the people to keep and bear Arms shall not be infringed.". . .

In determining whether the Second Amendment's guarantee is an individual one, or some sort of collective right, the most important word is the one the drafters chose to describe the holders of the right—"the people." That term is found in the First, Second, Fourth, Ninth, and Tenth Amendments. It has never been doubted that these provisions were designed to protect the interests of *individuals* against government intrusion, interference, or usurpation. We also note that the Tenth Amendment—"The powers not delegated to the United States by the Constitution, nor prohibited by it to the states, are reserved to the states respectively, or to the people"—indicates that the authors of the Bill of Rights were perfectly capable of distinguishing between "the people," on the one hand, and "the states," on the other. The natural reading of "the right of the people" in the Second Amendment would accord with usage elsewhere in the Bill of Rights.

The District's argument, on the other hand, asks us to read "the people" to mean some subset of individuals such as "the organized militia" or "the people who are engaged in militia service," or perhaps not any individuals at all—e.g., "the states.". . . These strained interpretations of "the people" simply cannot be squared with the uniform construction of our other Bill of Rights provisions. . . .

In sum, the phrase "the right of the people" . . . leads us to conclude that the right in question is individual. . . .

The wording of the operative clause also indicates that the right to keep and bear arms was not created by government, but rather preserved by it. . . . Hence, the Amendment acknowledges "*the* right . . . to keep and bear Arms," a right that pre-existed the Constitution like "*the* freedom of speech." Because the right to arms existed prior to the formation of the new government, . . . the Second Amendment only guarantees that the right "shall not be infringed." . . .

To determine what interests this pre-existing right protected, we look to the lawful, private purposes for which people of the time owned and used arms. The correspondence and political dialogue of the founding era indicate that arms were kept for lawful use in self-defense and hunting. . . .

The pre-existing right to keep and bear arms was premised on the commonplace assumption that individuals would use them for these private purposes, in addition to whatever militia service they would be obligated to perform for the state. The premise that private arms would be used for self-defense accords with Blackstone's observation, which had influenced thinking in the American colonies, that the people's right to arms was auxiliary to the natural right of self-preservation. . . . The right of self-preservation, in turn, was understood as the right to defend oneself against attacks by lawless individuals, or, if absolutely necessary, to resist and throw off a tyrannical government. . . .

When we look at the Bill of Rights as a whole, the setting of the Second Amendment reinforces its individual nature. The Bill of Rights was almost entirely a declaration of individual rights, and the Second Amendment's inclusion therein strongly indicates that it, too, was intended to protect personal liberty. The collective right advocates ask us to imagine that the First Congress situated a *sui generis* states' right among a catalogue of cherished individual liberties without comment. We believe the canon of construction known as *noscitur a sociis* applies here. Just as we would read an ambiguous statutory term in light of its context, we should read any supposed ambiguities in the Second Amendment in light of *its* context. Every other provision of the Bill of Rights, excepting the Tenth, which speaks explicitly about the allocation of governmental power, protects rights enjoyed by citizens in their individual capacity. The Second Amendment would be an inexplicable aberration if it were not read to protect individual rights as well.

[The court discusses whether "the phrase 'keep and bear Arms' should be read as purely military language" and concludes that "it is not accurate to construe it exclusively so." The court then examines the meaning of the word *militia* at the time the Constitution was written.]

The crucial point is that the existence of the militia preceded its organization by Congress, and it preceded the implementation of Congress's organizing plan by the states. The District's definition of the militia is just too narrow. The militia was a large segment of the population—not quite synonymous with "the people," as appellants contend—but certainly not the organized "divisions, brigades, regiments, battalions, and companies" mentioned in the second Militia Act. . . .

* * *

As we observed, the District argues that *even if* one reads the operative clause in isolation, it supports the collective right interpretation of the Second Amendment. Alternatively, the District contends that the operative clause should not, in fact, be read in isolation, and that it is imbued with the civic character of the prefatory clause when the Amendment is read, correctly, as two interactive clauses. The District points to the singular nature of the Second Amendment's preamble as an indication that the operative clause must be restricted or conditioned in some way by the prefatory language. . . . However, the structure of the Second Amendment turns out to be not so unusual when we examine state constitutional provisions guaranteeing rights or restricting governmental power. It was quite common for prefatory language to state a principle of good government that was narrower than the operative language used to achieve it. . . .

We think the Second Amendment was similarly structured. The prefatory language announcing the desirability of a well-regulated militia—even bearing in mind the breadth of the concept of a militia—is narrower than the guarantee of an individual right to keep and bear arms. The Amendment does not protect "the right of militiamen to keep and bear arms," but rather "the right of the people." The operative clause, properly read, protects the ownership and use of weaponry beyond that needed to preserve the state militias. Again, we point out that if the competent drafters of the Second Amendment had meant the right to be limited to the protection of state militias, it is hard to imagine that they would have chosen the language they did. We therefore take it as an expression of the drafters' view that the people possessed a natural right to keep and bear arms, and that the preservation of the militia was the right's most salient political benefit—and thus the most appropriate to express in a political document. . . .

B

We have noted that there is no unequivocal precedent that dictates the outcome of this case. This Court has never decided whether the Second Amendment protects an individual or collective right to keep and bear arms. . . . The Supreme Court has not decided this issue either. . . . As we have said, the leading Second Amendment case in the Supreme Court is *United States v. Miller* [(1939)]. . . .

Few decisions of Second Amendment relevance arose in the early decades of the twentieth century. Then came *Miller,* the Supreme Court's most thorough analysis of the Second Amendment to date, and a decision that both sides of the current gun control debate have claimed as their own. . . . Although *Miller* did not explicitly accept the individual right position, the decision implicitly assumes that interpretation. . . .

On the question whether the Second Amendment protects an individual or collective right, the Court's opinion in *Miller* is most notable for what it omits. The government's first argument in its *Miller* brief was that "the right secured by [the Second Amendment] to the people to keep and bear arms is not one which may be utilized for private purposes but only one which exists where the arms are borne in the militia or some other military organization provided for by law and intended for the protection of the state.". . . This is a version of the collective right model. Like the Fifth Circuit, we think it is significant that the Court did not decide the case on this, the government's primary argument. . . . Rather, the Court followed the logic of the government's secondary position, which was that a short-barreled shotgun was not within the scope of the term "Arms" in the Second Amendment. . . .

. . . [T]he Court was focused only on what arms are protected by the Second Amendment . . . and not the collective or individual nature of the right. If the *Miller* Court intended to endorse the government's first argument, i.e., the collective right view, it would have undoubtedly pointed out that the two defendants were not affiliated with a state militia or other local military organization. . . .

. . . The term "Arms" was quite indefinite, but it would have been peculiar, to say the least, if it were designed to ensure that people had an individual right to keep weapons capable of mass destruction—e.g., cannons. Thus the *Miller* Court limited the term "Arms"—interpreting it in a manner consistent with the Amendment's underlying civic purpose. Only "Arms" whose "use or possession . . . has some reasonable relationship to the preservation or efficiency of a well regulated militia" . . . would qualify for protection.

Essential, then, to understanding what weapons qualify as Second Amendment "Arms" is an awareness of how the founding-era militia functioned. The Court explained its understanding of what the Framers had in mind when they spoke of the militia in terms we have discussed above. The members of the militia were to be "civilians primarily, soldiers on occasion.". . . When called up by either the state or the federal government, "these men were expected to appear *bearing arms supplied by themselves and of the kind in common use at the time*" (emphasis added). . . .

Miller's definition of the "Militia," then, offers further support for the individual right interpretation of the Second Amendment. Attempting to draw a line between the ownership and use of "Arms" for *private purposes* and the ownership and use of "Arms" for *militia* purposes would have been an extremely silly exercise on the part of the First Congress if indeed the very survival of the militia depended on men who would bring their commonplace, *private* arms with them to muster. A ban on the use and ownership of weapons for private purposes, if allowed, would undoubtedly have had a deleterious, if not catastrophic, effect on the readiness of the militia for action. We do not see how one could believe that the First Congress, when crafting the Second Amendment, would have engaged in drawing such a foolish and impractical distinction, and we think the *Miller* Court recognized as much.

* * *

To summarize, we conclude that the Second Amendment protects an individual right to keep and bear arms. That right existed prior to the formation of the new government under the Constitution and was premised on the private use of arms for activities such as hunting and self-defense, the latter being understood as resistance to either private lawlessness or the depredations of a tyrannical government (or a threat from abroad). In addition, the right to keep and bear arms had the important and salutary civic purpose of helping to preserve the citizen militia. The civic purpose was also a political expedient for the Federalists in the First Congress as it served, in part, to placate their Antifederalist opponents. The individual right facilitated militia service by ensuring that citizens would not be barred from keeping the arms they would need when called forth for militia duty. Despite the importance of the Second Amendment's civic purpose, however, the activities it protects are not limited to militia service, nor is an individual's enjoyment of the right contingent upon his or her continued or intermittent enrollment in the militia.

IV

[Omitted. This section dismisses the District's argument that the Second Amendment does not apply because the city is not a state.]

V

[Omitted. This section dismisses the District's argument "that modern handguns are not the sort of weapons covered by the Second Amendment."]

VI

For the foregoing reasons, the judgment of the district court is reversed and the case is remanded. Since there are no material questions of fact in dispute, the district court is ordered to grant summary judgment to Heller consistent with the prayer for relief contained in appellants' complaint.

KAREN LECRAFT HENDERSON, *Circuit Judge,* dissenting:

As has been noted by Fifth Circuit Judge Robert M. Parker in *United States v. Emerson* . . . (2001) . . . exhaustive opinions on the origin, purpose and scope of the Second

Amendment to the United States Constitution have proven to be irresistible to the federal judiciary. . . . The result has often been page after page of "dueling dicta"—each side of the debate offering law review articles and obscure historical texts to support an outcome it deems proper. Today the majority adds another fifty-plus pages to the pile. Its superfluity is even more pronounced, however, because the meaning of the Second Amendment in the District of Columbia (District) is purely academic. Why? As Judge Walton declared in *Seegars v. Ashcroft* . . . (2004), . . . "the District of Columbia is not a state within the meaning of the Second Amendment and therefore the Second Amendment's reach does not extend to it." For the following reasons, I respectfully dissent. . . .

SOURCE: United States Court of Appeals for the District of Columbia Circuit. *Shelly Parker v. District of Columbia.* No. 04-7041. March 9, 2007. http://pacer.cadc.uscourts.gov/docs/common/opinions/200703/04-7041a.pdf (accessed November 20, 2007).

OTHER HISTORIC DOCUMENTS OF INTEREST

FROM THIS VOLUME

- Virginia Tech shootings, p. 167

FROM PREVIOUS *HISTORIC DOCUMENTS*

- Federal appeals courts on the right to bear arms, *2002,* p. 959; *2001,* p. 722

DOCUMENT IN CONTEXT

UN Resolution and U.S. Intelligence Estimate on Iran's Nuclear Threat

MARCH 24 AND DECEMBER 3, 2007

Only rarely does the release of a government report, written in bureaucratic language, dramatically and immediately affect the course of world events. Such a thing happened in December, when the U.S. government released a new National Intelligence Estimate (NIE) declaring that Iran had halted its work to develop nuclear weapons in 2003. The report immediately undermined the increasingly intense efforts by the United States and some of its European allies to discourage Iran from its presumed intent to develop nuclear weapons. In March 2007 the United Nations Security Council had adopted its third resolution, since July 2006, imposing sanctions against the government of Iran, and Washington had been pressing for more punishment of Iran.

The intelligence report also ended, at least for the time being, suggestions by President George W. Bush's administration and senior Israeli officials of potential military action against Iran and its weapons program. Although the NIE indicated that Tehran probably was still interested in acquiring nuclear weapons, the document's unexpected assessment that Iran had stopped development work altered international dynamics.

Despite the high-stakes action and rhetoric on the nuclear issue, diplomats from the United States and Iran held their first direct, substantive talks in many years. During the three rounds of talks between May and August, centering on the situation in Iraq, Iran and the United States discussed their mutual interests in backing the weak Iraqi government, but Washington accused Tehran of fomenting violence by extremist militias in Iraq.

Unrelated to the weapons issue, Iran moved a step closer in 2007 to completion of a Russian-built and -supplied nuclear power plant. The long-delayed plant, in the southern port city of Bushehr, received its first shipment of enriched uranium from Russia on December 17. Russian officials estimated that the plant could begin producing electrical power by late 2008. U.S. officials continued to question why Iran, a major oil-producing country, required nuclear energy.

One More Security Council Resolution

International concerns about Iran's nuclear intentions emerged in 2002 with a revelation that the country was secretly building facilities that could be used to produce fissile material necessary for nuclear weapons. The most important of these facilities was a plant in Natanz, south of Teheran, to enrich uranium; that fuel could be used for power plants or, if enriched to a high enough level, for weapons. Iran maintained that its nuclear work was for civilian, not military, purposes.

The International Atomic Energy Agency (IAEA), the United Nations' nuclear watchdog, attempted to step up inspections of Iran's facilities. At the same time, diplomats from the European Union launched several rounds of negotiations aimed at getting Iran to drop its presumed plans to develop nuclear weapons. These inspections and diplomatic efforts appeared to have only limited success, and by 2006 Iran was clearly enriching uranium despite international pressure to halt the work. At the urging of the United States, the UN Security Council adopted two 2006 resolutions demanding that Iran halt its enrichment efforts and cooperate more fully with IAEA inspections. The second resolution, adopted in December, also imposed limited economic sanctions against the Iranian government, as well as companies, officials, and businessmen involved in the country's nuclear work.

On February 22, 2007, the IAEA reported that Iran was, or was on the verge of, operating about 1,000 centrifuges—fast-spinning machines used to create enriched uranium. Iran was therefore moving steadily toward its goal of having 3,000 centrifuges in operation—enough, under the right conditions, to produce sufficient fuel for one nuclear weapon within about one year.

Alarmed by the IAEA report, the UN Security Council on March 24 adopted its third resolution on Iran, number 1747, again demanding that Iran end its enrichment work within sixty days. If Iran did not comply, the nation would face new sanctions, including a ban on Iranian exports of weapons and a freeze on the assets of twenty-eight people and organizations linked to Iran's ballistic missile and nuclear programs. Unlike the two resolutions adopted in 2006, this one was remarkable because it had support from China and Russia, veto-bearing members of the Security Council that in previous years had been reluctant to punish Iran on the nuclear issue.

The ink was barely dry on the new resolution when, on April 9, Iranian president Mahmoud Ahmadinejad proudly announced that Iran was producing enriched uranium "at an industrial level," with 3,000 centrifuges in operation. As with many such statements by Iranian officials, this turned out to be an exaggeration. IAEA managing director Mohammed ElBaradei said three days later that Iran was running only "hundreds" of centrifuges, not 3,000. Later in April, IAEA officials determined that Iran had indeed begun enriching uranium, using some 1,300 centrifuges.

On May 12, Vice President Dick Cheney, the leading hawk in the Bush administration, ratcheted up U.S. rhetoric against Iran. Addressing sailors aboard the USS *John C. Stennis* aircraft carrier in the Persian Gulf just 150 miles off Iran's coast, Cheney issued a warning: the United States was ready to use naval power to prevent Iran from disrupting oil supplies or "gaining nuclear weapons and dominating this region." Cheney's tough line was reminiscent of his similar remarks in 2002 foreshadowing the U.S. invasion of Iraq in 2003.

International concern that a crisis might be in the offing arose after the IAEA reported on May 23 that Iran had not complied with the Security Council's sixty-day deadline for suspending its uranium enrichment, nor had it "agreed to any of the required transparency measures" to assure the world that it was not developing a weapons program. The agency confirmed that 1,312 centrifuges were in operation and being fed the uranium hexafluoride gas needed to make enriched uranium; another 328 had been built and tested, and 492 more were under construction. ElBaradei told the *New York Times* that Iran had solved key technical problems and "we believe they pretty much have the knowledge about how to enrich" uranium.

The United States, which customarily took the lead in imposing sanctions against Iran, began preparing to implement the restrictions the Security Council had authorized in March. Later in the year, however, U.S. officials acknowledged they had encountered difficulties in acting against some of the companies and individuals named by the Security Council because accurate information about them could not be found.

Agreement between Iran and the United Nations

Confrontation seemed less likely after August 21, when IAEA and Iranian officials agreed on a "work plan" under which Iran would answer, by December, the many questions the UN agency had been posing for four years about the nation's nuclear programs. The plan was harshly criticized by several Western governments and nonproliferation experts after it was made public on August 27. Critics argued it deprived the IAEA of needed access to weapons sites and documents and had prematurely ended discussion of some long-standing issues, notably Iran's efforts during the 1990s to produce plutonium, another fuel for nuclear weapons.

ElBaradei defended the work plan during an IAEA board meeting on September 10. "This is the first time that Iran has agreed on a plan to address all outstanding issues, with a defined timeline, and is therefore an important step in the right direction," he said.

More Tough Rhetoric

The work plan agreed to by the IAEA and Iran did not end discussion in some Western capitals, notably Paris and Washington, of taking additional steps to punish Iran. The new French government of President Nicholas Sarkozy took the lead, at first adopting the hawkish rhetoric of the White House. On September 16, French foreign minister Bernard Kouchner warned that if Iran were to develop a nuclear weapon, "we have to prepare for the worst, and the worst is war." This language alarmed other European leaders, who pressed Kouchner to tone down the rhetoric. He did so quickly, saying the French position was to "negotiate, negotiate, negotiate, and work with our European friends on credible sanctions."

Subsequent cross-Atlantic discussions centered on imposing new controls on Iran's access to international credit and technology. France, the United States, and Britain agreed on September 28 to consider additional sanctions if Iran had not made progress by November on the promises it had made to the IAEA in late August. In the meantime, President Bush again ratcheted up the rhetoric, saying at an October 17 news conference that Iran would be "a dangerous threat to world peace" if it built nuclear weapons. "So I told people that if you're interested in avoiding World War III, it seems like you ought to be interested in preventing them from having the knowledge necessary to make a nuclear weapon. I take the threat of Iran with a nuclear weapon very seriously," he said.

Such talk again heightened speculation that Bush was signaling a potential military strike against Iran. White House press secretary Dana Perino quickly sought to dampen the speculation, explaining that Bush "wasn't making any declarations" about war. Even so, Cheney on October 21 warned of "serious consequences" should Iran continue on its nuclear course. "We will not allow Iran to have a nuclear weapon," he said in a speech to a forum on Middle East policy.

The Bush administration took direct action against Iran on October 25, imposing unilateral U.S. sanctions against a significant component of Iran's military: the elite Quds Force of the Revolutionary Guards Corps. The administration labeled the division a "terrorist" organization, accusing it of providing weapons to Shiite militia groups in Iraq. The symbolic step that barred U.S. companies and citizens from doing business with the Quds Force marked the first time the United States had taken such action against the armed forces of a sovereign country. The administration's order also targeted four government-owned banks in Iran.

Iran seemed to take its own confrontational step in response, replacing its chief negotiator on the nuclear issue, Ali Larijani, a pragmatic diplomat generally considered a moderate within the context of Iran's fractured leadership circles. His successor, Saeed Jalili, appeared to be loyal to President Ahmadinejad and other hard-liners in the government.

On November 15 the IAEA issued a mixed report on Iran's cooperation, which it described as "selective and incomplete." Iran, the IAEA said, had failed to give the agency the promised detailed information it needed to evaluate Iran's nuclear programs "past and present." Because of Iran's continuing restrictions on IAEA's access to nuclear facilities, the agency reported its knowledge of the overall nuclear program was "diminishing."

The IAEA also reported that although Iran had 2,952 centrifuges in operation, the machines were not operating at full capacity and were enriching uranium only to a level of 4 percent, which was below the 4.8 percent figure Iran had claimed and well below the 90 percent level necessary for weapons-grade material. ElBaradei followed up with a verbal report to the IAEA board of governors on November 22, in which he explained he was "unable to provide credible assurance about the absence of the undeclared nuclear material and activities in Iran"—in other words, whether Iran was still hiding components of its nuclear program.

After this report, diplomats headed back to the negotiating table. On November 30, representatives from the European Union and Iran held unproductive talks in London that were widely considered to be a last-ditch effort to head off a new push for sanctions. Iran's new lead negotiator, Saeed Jalili, suggested that the negotiations were back to square one because Iran was no longer willing to consider the proposals discussed during the previous three years of talks with European diplomats. The next day, diplomats from the six countries most involved in discussions about Iran at the United Nations—Britain, China, France, Germany, Russia, and the United States—met in Paris and agreed to draft a new resolution to put more pressure on Tehran.

A Shocking Intelligence Estimate

On December 3, just three days after the stand-off in talks between Iran and the Europeans, intelligence agencies in Washington published an entirely unexpected assessment of Iran's nuclear program. In one stroke, the agencies undercut the most important of the Bush administration's arguments about Iran's intentions—even while suggesting that past pressure had played a role in Tehran's decision to suspend its drive for nuclear weapons.

In one of the most startling statements to come from any U.S. government agency in many years, the unclassified version of the National Intelligence Estimate (NIE) by the National Intelligence Council bluntly said: "We judge with high confidence that in the fall of 2003, Tehran halted its nuclear weapons program; we also assess with moderate-to-high confidence that Tehran at a minimum is keeping open the option to develop nuclear weapons." The assessment defined Iran's nuclear weapons program as work to design and create weapons, including secret work to enrich uranium to the level necessary for weapons. Enriching uranium to the level required for nonmilitary uses was defined as a "civilian" program. The full classified version of the report ran to 130 pages; its contents had not been reported publicly as of year's end.

Despite its bold statements, the NIE offered several caveats; the most important was that U.S. intelligence agencies did not have enough information to determine "whether Tehran is willing to maintain the halt of its nuclear weapons program indefinitely while it weighs its options, or whether it will or already has set specific deadlines or criteria that will prompt it to restart the program." The assessment also suggested that, under the right circumstances, Iran might be able to produce enough enriched uranium for weapons sometime between 2010 and 2015.

The bottom line, according to the NIE, was that Iran appeared to be "less determined to develop nuclear weapons than we have been judging since 2005," when the intelligence agencies had drafted their previous report on the matter. The intelligence community also changed its view of the effectiveness of international pressure during the two years between reports: in 2005 intelligence agencies suggested Iran would not give in to such pressure, while in 2007 they considered it likely that Iran suspended its nuclear weapons work "primarily in response to" international pressure. A key conclusion to be drawn from the 2007 report was that Iranian officials were more rational in their decision-making than many Western politicians had assumed. The report described Iran's decisions as "guided by a cost-benefit approach rather than a rush to a weapon irrespective of the political, economic and military costs."

The NIE mirrored, in some respects, the Russian position. On October 10, Russian president Vladimir Putin told French president Sarkozy, during the latter's first visit to Moscow, that Russia was reluctant to impose new sanctions against Iran because "we don't have information showing that Iran is striving to produce nuclear weapons. That's why we're proceeding on the basis that Iran does not have such plans."

U.S. officials emphasized that the report was a consensus assessment by the U.S. intelligence community—consisting of sixteen civilian and military agencies—and not by policymakers in the Bush administration. Even so, White House officials had agreed to the release of the NIE despite its likely impact on administration policy. Bush later said he had been told in August that the intelligence agencies were reevaluating their views on Iran's weapons, but he did not indicate whether he asked for more information at that time. The president received his first briefing on the new report on November 28.

Administration officials quickly sought to play down the impact of the NIE, saying it demonstrated the need for more, not less, pressure on Iran. To ensure that Iran did not develop weapons, national security adviser Stephen Hadley said, "the international community has to turn up the pressure on Iran with diplomatic isolation, United Nations sanctions, and with other financial pressure."

President Bush said the next day that the report offered a "warning signal" about Iran. "I have said Iran is dangerous," he stated, "and the NIE doesn't do anything to change my opinion about the danger Iran poses to the world. Quite the contrary." In a news conference, the president also rebuffed several opportunities to distance himself from the report and its findings, complimenting the intelligence agencies "for their good work."

Reaction to the Report

In the days before the NIE was made public, intelligence officials briefed key leaders of Congress about its findings, and so some members reacted to its publication promptly. Most Democrats appeared to be relieved that the report undercut what they had perceived as an attempt by some administration officials to build a justification for a military attack on Iran. Senate majority leader Harry Reid, D-Nev., said he hoped the administration "appropriately adjusts its rhetoric and policy," starting with a "diplomatic surge to effectively address the challenges posed by Iran"; this was a reference to Bush's military "surge" of 30,000 additional troops in Iraq. Some Republicans, however, strongly questioned the NIE findings and suggested that intelligence officials had not collected enough information to make informed judgments about Iran. A handful of Senate Republicans discussed creating a bipartisan commission to examine the issue, but they did not pursue the matter.

Iran's foreign minister, Manouchehr Mottaki, expressed his government's pleasure that "some of the same countries which had questions or ambiguities about our nuclear program are changing their views realistically." Ahmadinejad later declared the U.S. report to be "a victory for the Iranian nation in the nuclear issue, against all international powers."

Israel was also immediately affected by the report. Hawkish officials and politicians had been building a case for months for a military strike against Iran's weapons program. The U.S. report undermined prospects for such action, as noted the next day by commentators Shmuel Rosner and Aluf Benn in the liberal daily newspaper *Haaretz:* "However successful or flawed this report may be, there is a new, dramatic reality, in all aspects of the struggle against the Iranian bomb: The military option, American or Israeli, is off the table, indefinitely." Israeli defense minister Ehud Barak, however, said his government believed that even if Iran had stopped its weapons work in 2003, it had since resumed it.

Ironically, the IAEA—which the Bush administration long had accused of being too soft on Iran—was also skeptical of the NIE assessment. IAEA officials told reporters they were not yet willing to declare either that Iran had worked to build weapons up until 2003 or that it had stopped such work at that point. So much information about Iran's program remained hidden that definitive judgments were not yet possible, agency officials said.

In addition to stifling military action, the report threw into question the prospects for tougher sanctions against Iran. Several European officials said they were surprised by the timing of the report, which was made public just two days after diplomats had agreed to work on a fourth Security Council resolution on Iran. No such work had taken place as of year's end.

A near-universal reaction to the report was that it served as a reminder of recent failures by the U.S. intelligence community—notably its incorrect assessment in late 2002 that Iraq was working furiously to build weapons of mass destruction, including nuclear weapons. Now, intelligence officials were saying they had been wrong in 2005, when they insisted Iran was working to build nuclear weapons. Some conservatives in Washington cited this history to cast doubt on the new Iran assessment, which they said could be just as a wrong as the previous reports.

Following are two documents. First, the text of United Nations Security Council Resolution 1747, adopted on March 24, 2007, demanding Iranian compliance with UN requirements to provide information about its nuclear programs and imposing sanctions against Iran if it did not comply within sixty days. Second, excerpts from the unclassified version of "Iran: Nuclear Intentions and Capabilities," a National Intelligence Estimate drafted in November 2007 by the National Intelligence Council and made public on December 3, 2007.

DOCUMENT

UN Security Council Resolution 1747 Imposing Sanctions against Iran

March 24, 2007

The Security Council,

Recalling the Statement of its President, S/PRST/2006/15, of 29 March 2006, and its resolution 1696 (2006) of 31 July 2006, and its resolution 1737 (2006) of 23 December 2006, and *reaffirming* their provisions,

Reaffirming its commitment to the Treaty on the Non-Proliferation of Nuclear Weapons, the need for all States Party to that Treaty to comply fully with all their obligations, and recalling the right of States Party, in conformity with Articles I and II of that Treaty, to develop research, production and use of nuclear energy for peaceful purposes without discrimination,

Recalling its serious concern over the reports of the IAEA [International Atomic Energy Agency] Director General as set out in its resolutions 1696 (2006) and 1737 (2006),

Recalling the latest report by the IAEA Director General (GOV/2007/8) of 22 February 2007 and *deploring* that, as indicated therein, Iran has failed to comply with resolution 1696 (2006) and resolution 1737 (2006),

Emphasizing the importance of political and diplomatic efforts to find a negotiated solution guaranteeing that Iran's nuclear programme is exclusively for peaceful purposes, and *noting* that such a solution would benefit nuclear non-proliferation elsewhere, and *welcoming* the continuing commitment of China, France, Germany, the Russian Federation, the United Kingdom and the United States, with the support of the European Union's High Representative to seek a negotiated solution,

Recalling the resolution of the IAEA Board of Governors (GOV/2006/14), which states that a solution to the Iranian nuclear issue would contribute to global non-proliferation efforts and to realizing the objective of a Middle East free of weapons of mass destruction, including their means of delivery,

Determined to give effect to its decisions by adopting appropriate measures to persuade Iran to comply with resolution 1696 (2006) and resolution 1737 (2006) and with the requirements of the IAEA, and also to constrain Iran's development of sensitive technologies in support of its nuclear and missile programmes, until such time as the Security Council determines that the objectives of these resolutions have been met,

Recalling the requirement on States to join in affording mutual assistance in carrying out the measures decided upon by the Security Council,

Concerned by the proliferation risks presented by the Iranian nuclear programme and, in this context, by Iran's continuing failure to meet the requirements of the IAEA Board of Governors and to comply with the provisions of Security Council resolutions 1696 (2006) and 1737 (2006), *mindful* of its primary responsibility under the Charter of the United Nations for the maintenance of international peace and security,

Acting under Article 41 of Chapter VII of the Charter of the United Nations,

1. *Reaffirms* that Iran shall without further delay take the steps required by the IAEA Board of Governors in its resolution GOV/2006/14, which are essential to build confidence in the exclusively peaceful purpose of its nuclear programme and to resolve

outstanding questions, and, in this context, *affirms* its decision that Iran shall without further delay take the steps required in paragraph 2 of resolution 1737 (2006);

2. *Calls upon* all States also to exercise vigilance and restraint regarding the entry into or transit through their territories of individuals who are engaged in, directly associated with or providing support for Iran's proliferation sensitive nuclear activities or for the development of nuclear weapon delivery systems, and *decides* in this regard that all States shall notify the Committee established pursuant to paragraph 18 of resolution 1737 (2006) (herein "the Committee") of the entry into or transit through their territories of the persons designated in the Annex to resolution 1737 (2006) or Annex I to this resolution, as well as of additional persons designated by the Security Council or the Committee as being engaged in, directly associated with or providing support for Iran's proliferation of sensitive nuclear activities or for the development of nuclear weapon delivery systems, including through the involvement in procurement of the prohibited items, goods, equipment, materials and technology specified by and under the measures in paragraphs 3 and 4 of resolution 1737 (2006), except where such travel is for activities directly related to the items in subparagraphs 3 (b) (i) and (ii) of that resolution;

3. *Underlines* that nothing in the above paragraph requires a State to refuse its own nationals entry into its territory, and that all States shall, in the implementation of the above paragraph, take into account humanitarian considerations, including religious obligations, as well as the necessity to meet the objectives of this resolution and resolution 1737 (2006), including where Article XV of the IAEA Statute is engaged;

4. *Decides* that the measures specified in paragraphs 12, 13, 14 and 15 of resolution 1737 (2006) shall apply also to the persons and entities listed in Annex I to this resolution;

5. *Decides* that Iran shall not supply, sell or transfer directly or indirectly from its territory or by its nationals or using its flag vessels or aircraft any arms or related materiel, and that all States shall prohibit the procurement of such items from Iran by their nationals, or using their flag vessels or aircraft, and whether or not originating in the territory of Iran;

6. *Calls upon* all States to exercise vigilance and restraint in the supply, sale or transfer directly or indirectly from their territories or by their nationals or using their flag vessels or aircraft of any battle tanks, armoured combat vehicles, large calibre artillery systems, combat aircraft, attack helicopters, warships, missiles or missile systems as defined for the purpose of the United Nations Register on Conventional Arms to Iran, and in the provision to Iran of any technical assistance or training, financial assistance, investment, brokering or other services, and the transfer of financial resources or services, related to the supply, sale, transfer, manufacture or use of such items in order to prevent a destabilizing accumulation of arms;

7. *Calls upon* all States and international financial institutions not to enter into new commitments for grants, financial assistance, and concessional loans, to the Government of the Islamic Republic of Iran, except for humanitarian and developmental purposes;

8. *Calls upon* all States to report to the Committee within 60 days of the adoption of this resolution on the steps they have taken with a view to implementing effectively paragraphs 2, 4, 5, 6 and 7 above;

9. *Expresses* the conviction that the suspension set out in paragraph 2 of resolution 1737 (2006) as well as full, verified Iranian compliance with the requirements set out by the IAEA Board of Governors would contribute to a diplomatic, negotiated solution that guarantees Iran's nuclear programme is for exclusively peaceful purposes,

underlines the willingness of the international community to work positively for such a solution, *encourages* Iran, in conforming to the above provisions, to re-engage with the international community and with the IAEA, and *stresses* that such engagement will be beneficial to Iran;

10. *Welcomes* the continuous affirmation of the commitment of China, France, Germany, the Russian Federation, the United Kingdom and the United States, with the support of the European Union's High Representative, to a negotiated solution to this issue and *encourages* Iran to engage with their June 2006 proposals (S/2006/521), attached in Annex II to this resolution, which were endorsed by the Security Council in resolution 1696 (2006), and *acknowledges* with appreciation that this offer to Iran remains on the table, for a long-term comprehensive agreement which would allow for the development of relations and cooperation with Iran based on mutual respect and the establishment of international confidence in the exclusively peaceful nature of Iran's nuclear programme;

11. *Reiterates* its determination to reinforce the authority of the IAEA, strongly supports the role of the IAEA Board of Governors, *commends and encourages* the Director General of the IAEA and its secretariat for their ongoing professional and impartial efforts to resolve all outstanding issues in Iran within the framework of the IAEA, *underlines* the necessity of the IAEA, which is internationally recognized as having authority for verifying compliance with safeguards agreements, including the non-diversion of nuclear material for non-peaceful purposes, in accordance with its Statute, to continue its work to clarify all outstanding issues relating to Iran's nuclear programme;

12. *Requests* within 60 days a further report from the Director General of the IAEA on whether Iran has established full and sustained suspension of all activities mentioned in resolution 1737 (2006), as well as on the process of Iranian compliance with all the steps required by the IAEA Board and with the other provisions of resolution 1737 (2006) and of this resolution, to the IAEA Board of Governors and in parallel to the Security Council for its consideration;

13. *Affirms* that it shall review Iran's actions in light of the report referred to in paragraph 12 above, to be submitted within 60 days, and:

(a) that it shall suspend the implementation of measures if and for so long as Iran suspends all enrichment-related and reprocessing activities, including research and development, as verified by the IAEA, to allow for negotiations in good faith in order to reach an early and mutually acceptable outcome;

(b) that it shall terminate the measures specified in paragraphs 3, 4, 5, 6, 7 and 12 of resolution 1737 (2006) as well as in paragraphs 2, 4, 5, 6 and 7 above as soon as it determines, following receipt of the report referred to in paragraph 12 above, that Iran has fully complied with its obligations under the relevant resolutions of the Security Council and met the requirements of the IAEA Board of Governors, as confirmed by the IAEA Board;

(c) that it shall, in the event that the report in paragraph 12 above shows that Iran has not complied with resolution 1737 (2006) and this resolution, adopt further appropriate measures under Article 41 of Chapter VII of the Charter of the United Nations to persuade Iran to comply with these resolutions and the requirements of the IAEA, and underlines that further decisions will be required should such additional measures be necessary;

14. *Decides* to remain seized of the matter.

Annex I

Entities involved in nuclear or ballistic missile activities

1. Ammunition and Metallurgy Industries Group (AMIG) (aka Ammunition Industries Group) (AMIG controls 7th of Tir, which is designated under resolution 1737 (2006) for its role in Iran's centrifuge programme. AMIG is in turn owned and controlled by the Defence Industries Organisation (DIO), which is designated under resolution 1737 (2006))

2. Esfahan Nuclear Fuel Research and Production Centre (NFRPC) and Esfahan Nuclear Technology Centre (ENTC) (Parts of the Atomic Energy Organisation of Iran's (AEOI) Nuclear Fuel Production and Procurement Company, which is involved in enrichment-related activities. AEOI is designated under resolution 1737 (2006))

3. Kavoshyar Company (Subsidiary company of AEOI, which has sought glass fibres, vacuum chamber furnaces and laboratory equipment for Iran's nuclear programme)

4. Parchin Chemical Industries (Branch of DIO, which produces ammunition, explosives, as well as solid propellants for rockets and missiles)

5. Karaj Nuclear Research Centre (Part of AEOI's research division)

6. Novin Energy Company (aka Pars Novin) (Operates within AEOI and has transferred funds on behalf of AEOI to entities associated with Iran's nuclear programme)

7. Cruise Missile Industry Group (aka Naval Defence Missile Industry Group) (Production and development of cruise missiles. Responsible for naval missiles including cruise missiles)

8. Bank Sepah and Bank Sepah International (Bank Sepah provides support for the Aerospace Industries Organisation (AIO) and subordinates, including Shahid Hemmat Industrial Group (SHIG) and Shahid Bagheri Industrial Group (SBIG), both of which were designated under resolution 1737 (2006))

9. Sanam Industrial Group (subordinate to AIO, which has purchased equipment on AIO's behalf for the missile programme)

10. Ya Mahdi Industries Group (subordinate to AIO, which is involved in international purchases of missile equipment)

Iranian Revolutionary Guard Corps entities

1. Qods Aeronautics Industries (Produces unmanned aerial vehicles (UAVs), parachutes, para-gliders, para-motors, etc. Iranian Revolutionary Guard Corps (IRGC) has boasted of using these products as part of its asymmetric warfare doctrine)

2. Pars Aviation Services Company (Maintains various aircraft including MI-171, used by IRGC Air Force)

3. Sho'a' Aviation (Produces micro-lights which IRGC has claimed it is using as part of its asymmetric warfare doctrine)

Persons involved in nuclear or ballistic missile activities

1. Fereidoun Abbasi-Davani (Senior Ministry of Defence and Armed Forces Logistics (MODAFL) scientist with links to the Institute of Applied Physics, working closely with Mohsen Fakhrizadeh-Mahabadi, designated below)

2. Mohsen Fakhrizadeh-Mahabadi (Senior MODAFL scientist and former head of the Physics Research Centre (PHRC). The IAEA have asked to interview him about the activities of the PHRC over the period he was head but Iran has refused)

3. Seyed Jaber Safdari (Manager of the Natanz Enrichment Facilities)

4. Amir Rahimi (Head of Esfahan Nuclear Fuel Research and Production Center, which is part of the AEOI's Nuclear Fuel Production and Procurement Company, which is involved in enrichment-related activities)

5. Mohsen Hojati (Head of Fajr Industrial Group, which is designated under resolution 1737 (2006) for its role in the ballistic missile programme)

6. Mehrdada Akhlaghi Ketabachi (Head of SBIG, which is designated under resolution 1737 (2006) for its role in the ballistic missile programme)

7. Naser Maleki (Head of SHIG, which is designated under resolution 1737 (2006) for its role in Iran's ballistic missile programme. Naser Maleki is also a MODAFL official overseeing work on the Shahab-3 ballistic missile programme. The Shahab-3 is Iran's long range ballistic missile currently in service)

8. Ahmad Derakhshandeh (Chairman and Managing Director of Bank Sepah, which provides support for the AIO and subordinates, including SHIG and SBIG, both of which were designated under resolution 1737 (2006))

Iranian Revolutionary Guard Corps key persons

1. Brigadier General Morteza Rezaie (Deputy Commander of IRGC)
2. Vice Admiral Ali Akbar Ahmadian (Chief of IRGC Joint Staff)
3. Brigadier General Mohammad Reza Zahedi (Commander of IRGC Ground Forces)
4. Rear Admiral Morteza Safari (Commander of IRGC Navy)
5. Brigadier General Mohammad Hejazi (Commander of Bassij resistance force)
6. Brigadier General Qasem Soleimani (Commander of Qods force)
7. General Zolqadr (IRGC officer, Deputy Interior Minister for Security Affairs)

Annex II

Elements of a long-term agreement

Our goal is to develop relations and cooperation with Iran, based on mutual respect and the establishment of international confidence in the exclusively peaceful nature of the nuclear programme of the Islamic Republic of Iran. We propose a fresh start in the negotiation of a comprehensive agreement with Iran. Such an agreement would be deposited with the International Atomic Energy Agency (IAEA) and endorsed in a Security Council resolution.

To create the right conditions for negotiations,

We will:

- Reaffirm Iran's right to develop nuclear energy for peaceful purposes in conformity with its obligations under the Treaty on the Non-Proliferation of

Nuclear Weapons (hereinafter, NPT), and in this context reaffirm our support for the development by Iran of a civil nuclear energy programme.

- Commit to support actively the building of new light water reactors in Iran through international joint projects, in accordance with the IAEA statute and NPT.
- Agree to suspend discussion of Iran's nuclear programme in the Security Council upon the resumption of negotiations.

Iran will:

- Commit to addressing all of the outstanding concerns of IAEA through full cooperation with IAEA.
- Suspend all enrichment-related and reprocessing activities to be verified by IAEA, as requested by the IAEA Board of Governors and the Security Council, and commit to continue this during these negotiations.
- Resume the implementation of the Additional Protocol.

Areas of future cooperation to be covered in negotiations on a long-term agreement

1. Nuclear

We will take the following steps:

Iran's rights to nuclear energy

- Reaffirm Iran's inalienable right to nuclear energy for peaceful purposes without discrimination and in conformity with articles I and II of NPT, and cooperate with Iran in the development by Iran of a civil nuclear power programme.
- Negotiate and implement a Euratom/Iran nuclear cooperation agreement.

Light water reactors

- Actively support the building of new light water power reactors in Iran through international joint projects, in accordance with the IAEA statute and NPT, using state-of-the-art technology, including by authorizing the transfer of necessary goods and the provision of advanced technology to make its power reactors safe against earthquakes.
- Provide cooperation with the management of spent nuclear fuel and radioactive waste through appropriate arrangements.

Research and development in nuclear energy

- Provide a substantive package of research and development cooperation, including possible provision of light water research reactors, notably in the fields of radioisotope production, basic research and nuclear applications in medicine and agriculture.

Fuel guarantees

- Give legally binding, multilayered fuel assurances to Iran, based on:
    - Participation as a partner in an international facility in Russia to provide enrichment services for a reliable supply of fuel to Iran's nuclear reactors. Subject to negotiations, such a facility could enrich all uranium hexaflouride (UF_6) produced in Iran.
    - Establishment on commercial terms of a buffer stock to hold a reserve of up to five years' supply of nuclear fuel dedicated to Iran, with the participation and under supervision of IAEA.
    - Development with IAEA of a standing multilateral mechanism for reliable access to nuclear fuel, based on ideas to be considered at the next meeting of the Board of Governors.

Review of moratorium

The long-term agreement would, with regard to common efforts to build international confidence, contain a clause for review of the agreement in all its aspects, to follow:

- Confirmation by IAEA that all outstanding issues and concerns reported by it, including those activities which could have a military nuclear dimension, have been resolved;
- Confirmation that there are no undeclared nuclear activities or materials in Iran and that international confidence in the exclusively peaceful nature of Iran's civil nuclear programme has been restored.

2. Political and economic

Regional security cooperation

Support for a new conference to promote dialogue and cooperation on regional security issues.

International trade and investment

Improving Iran's access to the international economy, markets and capital, through practical support for full integration into international structures, including the World Trade Organization and to create the framework for increased direct investment in Iran and trade with Iran (including a trade and economic cooperation agreement with the European Union). Steps would be taken to improve access to key goods and technology.

Civil aviation

Civil aviation cooperation, including the possible removal of restrictions on United States and European manufacturers in regard to the export of civil aircraft to Iran, thereby widening the prospect of Iran renewing its fleet of civil airliners.

Energy partnership

Establishment of a long-term energy partnership between Iran and the European Union and other willing partners, with concrete and practical applications.

Telecommunications infrastructure

Support for the modernization of Iran's telecommunication infrastructure and advanced Internet provision, including by possible removal of relevant United States and other export restrictions.

High technology cooperation

Cooperation in fields of high technology and other areas to be agreed upon.

Agriculture

Support for agricultural development in Iran, including possible access to United States and European agricultural products, technology and farm equipment.

SOURCE: United Nations. Security Council. "Resolution 1747 (2007)." S/RES/1747 (2007). March 24, 2007. http://daccess-ods.un.org/TMP/9899216.html (accessed January 23, 2008).

DOCUMENT

U.S. National Intelligence Estimate on Iran's Nuclear Intentions and Capabilities

December 3, 2007

SCOPE NOTE

This National Intelligence Estimate (NIE) assesses the status of Iran's nuclear program, and the program's outlook over the next 10 years. This time frame is more appropriate for estimating capabilities than intentions and foreign reactions, which are more difficult to estimate over a decade. In presenting the Intelligence Community's assessment of Iranian nuclear intentions and capabilities, the NIE thoroughly reviews all available information on these questions, examines the range of reasonable scenarios consistent with this information, and describes the key factors we judge would drive or impede nuclear progress in Iran. This NIE is an extensive reexamination of the issues in the May 2005 assessment.

This Estimate focuses on the following key questions:

- What are Iran's intentions toward developing nuclear weapons?
- What domestic factors affect Iran's decisionmaking on whether to develop nuclear weapons?
- What external factors affect Iran's decisionmaking on whether to develop nuclear weapons?
- What is the range of potential Iranian actions concerning the development of nuclear weapons, and the decisive factors that would lead Iran to choose one course of action over another?
- What is Iran's current and projected capability to develop nuclear weapons? What are our key assumptions, and Iran's key chokepoints/vulnerabilities?

This NIE does *not* assume that Iran intends to acquire nuclear weapons. Rather, it examines the intelligence to assess Iran's capability and intent (or lack thereof) to acquire nuclear weapons, taking full account of Iran's dual-use uranium fuel cycle and those nuclear activities that are at least partly civil in nature.

This Estimate does assume that the strategic goals and basic structure of Iran's senior leadership and government will remain similar to those that have endured since the death of Ayatollah Khomeini in 1989. We acknowledge the potential for these to change during the time frame of the Estimate, but are unable to confidently predict such changes or their implications. This Estimate does not assess how Iran may conduct future negotiations with the West on the nuclear issue.

This Estimate incorporates intelligence reporting available as of 31 October 2007. . . .

Key Judgments

A. We judge with high confidence that in fall 2003, Tehran halted its nuclear weapons program (For the purposes of this Estimate, by "nuclear weapons program" we mean Iran's nuclear weapon design and weaponization work and covert uranium conversion-related and uranium enrichment-related work; we do not mean Iran's declared civil work related to uranium conversion and enrichment.); we also assess with moderate-to-high confidence that Tehran at a minimum is keeping open the option to develop nuclear weapons. We judge with high confidence that the halt, and Tehran's announcement of its decision to suspend its declared uranium enrichment program and sign an Additional Protocol to its Nuclear Non-Proliferation Treaty Safeguards Agreement, was directed primarily in response to increasing international scrutiny and pressure resulting from exposure of Iran's previously undeclared nuclear work.

- We assess with high confidence that until fall 2003, Iranian military entities were working under government direction to develop nuclear weapons.
- We judge with high confidence that the halt lasted at least several years. (Because of intelligence gaps discussed elsewhere in this Estimate, however, DOE [the Department of Energy] and the NIC [National Intelligence Council] assess with only moderate confidence that the halt to those activities represents a halt to Iran's entire nuclear weapons program.)
- We assess with moderate confidence Tehran had not restarted its nuclear weapons program as of mid-2007, but we do not know whether it currently intends to develop nuclear weapons.
- We continue to assess with moderate-to-high confidence that Iran does not currently have a nuclear weapon.
- Tehran's decision to halt its nuclear weapons program suggests it is less determined to develop nuclear weapons than we have been judging since 2005. Our assessment that the program probably was halted primarily in response to international pressure suggests Iran may be more vulnerable to influence on the issue than we judged previously.

B. We continue to assess with low confidence that Iran probably has imported at least some weapons-usable fissile material, but still judge with moderate-to-high confidence it has not obtained enough for a nuclear weapon. We cannot rule out that Iran has acquired from abroad—or will acquire in the future—a nuclear weapon or enough fissile material for a weapon. Barring such acquisitions, if Iran wants to have nuclear weapons it would need to produce sufficient amounts of fissile material indigenously—which we judge with high confidence it has not yet done.

C. We assess centrifuge enrichment is how Iran probably could first produce enough fissile material for a weapon, if it decides to do so. Iran resumed its declared centrifuge enrichment activities in January 2006, despite the continued halt in the nuclear weapons program. Iran made significant progress in 2007 installing centrifuges at Natanz [the location of a nuclear facility], but we judge with moderate confidence it still faces significant technical problems operating them.

- We judge with moderate confidence that the earliest possible date Iran would be technically capable of producing enough HEU [highly enriched uranium, a fuel for nuclear weapons] for a weapon is late 2009, but that this is very unlikely.
- We judge with moderate confidence Iran probably would be technically capable of producing enough HEU for a weapon sometime during the 2010–2015 time frame. (INR [the State Department's Bureau of Intelligence and Research] judges Iran is unlikely to achieve this capability before 2013 because of foreseeable technical and programmatic problems.) All agencies recognize the possibility that this capability may not be attained until *after* 2015.

D. Iranian entities are continuing to develop a range of technical capabilities that could be applied to producing nuclear weapons, if a decision is made to do so. For example, Iran's civilian uranium enrichment program is continuing. We also assess with high confidence that since fall 2003, Iran has been conducting research and development projects with commercial and conventional military applications—some of which would also be of limited use for nuclear weapons.

E. We do not have sufficient intelligence to judge confidently whether Tehran is willing to maintain the halt of its nuclear weapons program indefinitely while it weighs its options, or whether it will or already has set specific deadlines or criteria that will prompt it to restart the program.

- Our assessment that Iran halted the program in 2003 primarily in response to international pressure indicates Tehran's decisions are guided by a cost-benefit approach rather than a rush to a weapon irrespective of the political, economic, and military costs. This, in turn, suggests that some combination of threats of intensified international scrutiny and pressures, along with opportunities for Iran to achieve its security, prestige, and goals for regional influence in other ways, might—if perceived by Iran's leaders as credible—prompt Tehran to extend the current halt to its nuclear weapons program. It is difficult to specify what such a combination might be.
- We assess with moderate confidence that convincing the Iranian leadership to forgo the eventual development of nuclear weapons will be difficult given the linkage many within the leadership probably see between nuclear weapons development and Iran's key national security and foreign policy objectives, and given Iran's considerable effort from at least the late 1980s to 2003 to develop such weapons. In our judgment, only an Iranian political decision to abandon

a nuclear weapons objective would plausibly keep Iran from eventually producing nuclear weapons—and such a decision is inherently reversible.

F. We assess with moderate confidence that Iran probably would use covert facilities—rather than its declared nuclear sites—for the production of highly enriched uranium for a weapon. A growing amount of intelligence indicates Iran was engaged in covert uranium conversion and uranium enrichment activity, but we judge that these efforts probably were halted in response to the fall 2003 halt, and that these efforts probably had not been restarted through at least mid-2007.

G. We judge with high confidence that Iran will not be technically capable of producing and reprocessing enough plutonium for a weapon before about 2015.

H. We assess with high confidence that Iran has the scientific, technical and industrial capacity eventually to produce nuclear weapons if it decides to do so.

Key Differences Between the Key Judgments of This Estimate on Iran's Nuclear Program and the May 2005 Assessment

2005 IC Estimate	2007 National Intelligence Estimate
Assess with high confidence that Iran currently is determined to develop nuclear weapons despite its international obligations and international pressure, but we do not assess that Iran is immovable.	Judge with high confidence that in fall 2003, Tehran halted its nuclear weapons program. Judge with high confidence that the halt lasted at least several years. (DOE and the NIC have moderate confidence that the halt to those activities represents a halt to Iran's entire nuclear weapons program.) Assess with moderate confidence Tehran had not restarted its nuclear weapons program as of mid-2007, but we do not know whether it currently intends to develop nuclear weapons. Judge with high confidence that the halt was directed primarily in response to increasing international scrutiny and pressure resulting from exposure of Iran's previously undeclared nuclear work. Assess with moderate-to-high confidence that Tehran at a minimum is keeping open the option to develop nuclear weapons.
We have moderate confidence in projecting when Iran is likely to make a nuclear weapon; we assess that it is unlikely before early-to-mid next decade.	We judge with moderate confidence that the earliest possible date Iran would be technically capable of producing enough highly enriched uranium (HEU) for a weapon is late 2009, but that this is very unlikely. We judge with moderate confidence Iran probably would be technically capable of producing enough HEU for a weapon sometime during the 2010–2015

2005 IC Estimate	2007 National Intelligence Estimate
	time frame. (INR judges that Iran is unlikely to achieve this capability before 2013 because of foreseeable technical and programmatic problems.)
Iran could produce enough fissile material for a weapon by the end of this decade if it were to make more rapid and successful progress than we have seen to date.	We judge with moderate confidence that the earliest possible date Iran would be technically capable of producing enough highly enriched uranium (HEU) for a weapon is late 2009, but that this is very unlikely.

SOURCE: U.S. Director of National Intelligence. National Intelligence Council. "Iran: Nuclear Intentions and Capabilities." National Intelligence Estimate. December 3, 2007. www.dni.gov/press_releases/20071203_releases.pdf (accessed January 23, 2008).

OTHER HISTORIC DOCUMENTS OF INTEREST

FROM THIS VOLUME

- President Bush on the "surge" in Iraq, p. 9
- North Korea's pledge to dismantle its nuclear weapons, p. 77
- Nicolas Sarkozy takes office as president of France, p. 240

FROM PREVIOUS *HISTORIC DOCUMENTS*

- Iranian president's comments to the United States, *2006*, p. 212
- United Nations Security Council on sanctions against Iran, *2006*, p. 781
- U.S. intelligence findings on Iraq's weapons of mass destruction, *2004*, p. 711; *2003*, p. 874

DOCUMENT IN CONTEXT

World Health Organization on Male Circumcision and HIV

MARCH 28, 2007

There was good news and bad news in 2007 in the worldwide effort to reverse the spread of AIDS (acquired immunodeficiency syndrome) and HIV (human immunodeficiency virus), the virus that causes the incurable disease. The Joint United Nations Programme on HIV/AIDS (UNAIDS), the agency responsible for coordinating the international response to the pandemic, substantially lowered its estimates of the number of people living with and dying from HIV/AIDS and said the epidemic had shown signs of slowing for nearly a decade. Largely due to of better data, particularly from India, the estimate of the number of people worldwide living with HIV in 2007 was 33.2 million, down from the 39.5 million estimated in 2006 before the new data were available.

The new data also led to lower estimates for new HIV infections—about 2.5 million in 2007, nearly 40 percent fewer than estimated in 2006. UN officials said the number of new infections probably peaked in the late 1990s. For the second year in a row, the number of deaths attributed to the disease declined; for 2007 the estimate was 2.1 million people worldwide. UNAIDS officials attributed much of the decline in mortality to the greater availability in the developing world of antiretroviral drugs, which were keeping AIDS sufferers alive longer.

The lower estimates helped offset some of the disappointment resulting from halted clinical trials of what had been thought to be a promising vaccine against HIV. Not all new prevention efforts resulted in failure, however. In March the World Health Organization (WHO) officially recommended male circumcision as a way to prevent heterosexual transmission of HIV after studies found that the procedure could reduce the risk of infection in men by about 60 percent. WHO's recommendation meant that international donors would be likely to advocate for and fund the procedure as part of their AIDS prevention programs.

Lowering the Estimates

UNAIDS announced its new estimates on November 20. The biggest downward revision was for India, where the number of people living with HIV (the prevalence rate) was lowered from 5.7 million to 2.5 million. The revision resulted largely from better surveillance methods, including a nationwide household survey and greater use of HIV testing. The new data showed that the epidemic was not as pervasive in India as UNAIDS had previously warned and that it was mainly confined to high-risk populations, such as injecting-drug users and sex workers. Other major downward revisions resulting from better data occurred in Angola, Kenya, Mozambique, Nigeria, and Zimbabwe. Together these six countries accounted for 70 percent of the reduction in estimated HIV prevalence from 2006 and 2007.

UNAIDS had long been criticized by some experts who argued that the agency exaggerated the scope of the epidemic to ensure continuing political and financial support for combating the disease. "There was a tendency toward alarmism, and that fit perhaps a certain fundraising agenda," Helen Epstein, one of those critics and the author of a book on the fight against AIDS in Africa, told the *Washington Post* in November. "I hope these new numbers will help refocus the response in a more pragmatic way." James Chin, a former WHO expert on the disease who had long argued that the worldwide HIV prevalence rate was around 25 million people, welcomed the new estimate of 33 million. "It's a little high, but it's not outrageous anymore," he said.

At a November 20 news conference on the revised estimates, Dr. Paul De Lay, director of monitoring and policy at UNAIDS, said any notion that the agency had deliberately been inflating the numbers was "absurd." Dr. Kevin M. De Cock, director of WHO's HIV/AIDS department, agreed. He said that it was not clear before 2003 that the estimates were probably too high and attributed the large decline in prevalence primarily to the new data from India, which had only become available in 2006.

The new estimates of HIV prevalence substantially changed the picture of the global spread of AIDS. China and India no longer appeared to be on the brink of tipping into an HIV epidemic that would affect the general population. In those countries, and elsewhere outside of Africa, the disease was primarily concentrated in high-risk groups such as injecting-drug users, sex workers, and men who had sex with other men. Africa remained the epicenter of the HIV/AIDS epidemic, with 22.5 million people infected and nearly 1.7 million new cases in 2007. The epidemic was largely concentrated in eight southern countries, where more than 15 percent of the adult population was infected. In both East and West Africa, the prevalence rates were lower—considerably lower in some countries. A key factor in the spread of the disease in some parts of Africa was the number of sexual partners; in some African countries, as many as 40 percent of adults had multiple partners at the same general time.

The revised picture of the AIDS crisis led some experts to suggest that it was time to adjust the distribution of AIDS funds; many of these experts had been advocating a similar readjustment for years to little avail. For example, they said, aid programs directed at the broad population might not be particularly useful in countries where the primary problem remained among high-risk populations. "The take-home message" from the new estimates, Daniel Halperin, an AIDS specialist at the Harvard School of Public Health, told the *Chicago Tribune* in November, "is that in countries like India, and in virtually all of Asia, [controlling AIDS] is really about doing effective HIV prevention with high-risk groups—condom promotion with sex workers, needle interventions with drug users."

Halperin and others said that the large amounts of money devoted to combating HIV/AIDS may have skewed some countries' budgets and health delivery systems toward fighting one disease while resources were diverted from other common diseases and health problems, such as malaria, tuberculosis, diarrhea, and maternal mortality, which also have devastating effects. "It would be really a mistake to say we don't need to fund AIDS anymore, that it's not a problem," Halperin said. "But in countries that have a low prevalence and no strong indication the problem is getting worse, spending shouldn't be disproportionate for one problem."

A New Intervention: Circumcision

The WHO and UNAIDS officially enlisted a new intervention in the fight against HIV/AIDS, strongly urging countries with high rates of heterosexual infection to offer male circumcision either free of charge or at subsidized rates. Some AIDS researchers had long noted that countries, particularly in Africa, with high HIV rates tended to have relatively low circumcision rates. Those arguments were not taken seriously until late in 2006, when randomized controlled tests in Kenya, South Africa, and Uganda showed that circumcision could reduce the risk of heterosexually acquired HIV infection in males by about 60 percent. "The evidence is really now quite conclusive that male circumcision is effective at preventing HIV among men," De Cock said March 28 as the WHO issued its recommendation.

De Cock and other health officials emphasized that circumcision alone would not provide complete protection; it should be done as part of comprehensive HIV prevention package including counseling, HIV testing, and the promotion of safer sex practices. To avoid infection, circumcised men would have to continue to use other forms of prevention such as condoms, abstinence, and a reduced number of sexual partners. WHO officials also warned that it would take some years to realize the benefits of circumcision in reducing the incidence of HIV. Nonetheless, modeling studies showed that male circumcision in sub-Saharan Africa could prevent 5.7 million new cases of HIV infection and 3 million deaths over twenty years.

The WHO and UNAIDS recommended that countries with high rates of generalized heterosexual HIV "urgently" scale up their ability to provide clean, safe circumcisions. They indicated, however, that there would be "limited public health impact" from promoting circumcision in the general population in countries where HIV was concentrated among sex workers or injecting-drug users. The head of President George W. Bush's anti-AIDS program, the President's Emergency Plan for AIDS Relief (PEPFAR), issued a statement on March 28 saying the president's program would "support safe male circumcision services when host country governments include this new HIV prevention recommendation as part of an expanded approach to reduce HIV infections." The Global Fund to Fight AIDS, Tuberculosis and Malaria and other large donors also said they were willing to underwrite the cost of the procedure.

In 2007 international donors had yet to sign onto another, more controversial intervention that researchers said held great promise for reducing the incidence of HIV in newborns—promoting contraceptive measures such as birth control pills, intrauterine devices (IUDs), or sterilization. According to UNAIDS estimates, about 90 percent of the 2.5 million children in the world with HIV lived in sub-Saharan Africa. The vast majority of children acquired the infection from their mothers. Despite sustained efforts to ensure that every pregnant woman with HIV received the antiretroviral drugs that could prevent transmission of the infection to her newborn, only about one in every ten infected mothers actually received them. That failure led some activists to argue that birth control measures would be cheaper, easier, and more effective.

Failed Vaccine Tests

Overall, the number of HIV sufferers gaining access to antiretroviral drugs increased significantly in 2007. According to the UN, more than 2 million people living in low- and middle-income countries received the drugs, which can ease the symptoms of AIDS while prolonging life and improving its quality. In sub-Saharan Africa, 28 percent of those in need of the treatments were receiving them in 2007, compared with just 2 percent of that population three years earlier. In addition, several new antiretroviral medications were approved for use in 2007 that showed great promise in treating strains of HIV that had developed tolerance to the older drugs.

But efforts to develop a vaccine to prevent infection came to a disappointing—and disturbing—end in September, when the pharmaceutical company Merck announced that it was stopping a large clinical trial of a promising vaccine because the vaccine had failed to work. The vaccine, which was funded in part by the National Institutes of Health (NIH), was one of several in development that represented a new strategy in fighting the virus; scientists had held high hopes for it. The study's participants were 3,000 uninfected but high-risk volunteers, primarily living in the United States and Latin America, and final results had been expected no earlier than the end of 2008. When the board monitoring the trial looked at an interim analysis of 1,500 volunteers, however, it saw that of the 741 people receiving the vaccine, 24 had been infected, compared with 21 infections in the 762 people taking the placebo. As the vaccine was obviously not effective, Merck halted the trial, as well as another one in South Africa and two smaller ones elsewhere.

About six weeks later researchers announced that the vaccine may have actually increased the risks of infection for some of the volunteers in the study. At the end of the year, researchers from Merck, the NIH, and other laboratories were still trying to determine why the vaccinated test group had a higher rate of infection than the control group.

Following is the text of a press release issued by the World Health Organization on March 28, 2007, officially recommending the use of male circumcision in some situations to reduce the risk of HIV infection.

DOCUMENT

WHO and UNAIDS Recommend Male Circumcision for Prevention of HIV

March 28, 2007

In response to the urgent need to reduce the number of new HIV infections globally, WHO and the UNAIDS Secretariat convened an international expert consultation to determine whether male circumcision should be recommended for the prevention of HIV infection.

Based on the evidence presented, which was considered to be compelling, experts attending the consultation recommended that male circumcision now be recognized as an additional important intervention to reduce the risk of heterosexually acquired HIV infection in men. The international consultation, which was held 6–8 March 2007 in Montreux, Switzerland, was attended by participants representing a wide range of stakeholders, including governments, civil society, researchers, human rights and women's health advocates, young people, funding agencies and implementing partners.

"The recommendations represent a significant step forward in HIV prevention," said Dr Kevin De Cock, Director, HIV/AIDS Department in WHO. "Countries with high rates of heterosexual HIV infection and low rates of male circumcision now have an additional intervention which can reduce the risk of HIV infection in heterosexual men. Scaling up male circumcision in such countries will result in immediate benefit to individuals. However, it will be a number of years before we can expect to see an impact on the epidemic from such investment."

There is now strong evidence from three randomized controlled trials undertaken in Kisumu, Kenya; Rakai District, Uganda (funded by the US National Institutes of Health); and Orange Farm, South Africa (funded by the French National Agency for Research on AIDS) that male circumcision reduces the risk of heterosexually acquired HIV infection in men by approximately 60%. This evidence supports the findings of numerous observational studies that have also suggested that the geographical correlation long described between lower HIV prevalence and high rates of male circumcision in some countries in Africa, and more recently elsewhere, is, at least in part, a causal association. Currently, 665 million men, or 30% of men worldwide, are estimated to be circumcised.

Male circumcision should be part of a comprehensive HIV prevention package

Male circumcision should always be considered as part of a comprehensive HIV prevention package, which includes

- the provision of HIV testing and counselling services;
- treatment for sexually transmitted infections;
- the promotion of safer sex practices; and
- the provision of male and female condoms and promotion of their correct and consistent use.

Counselling of men and their sexual partners is necessary to prevent them from developing a false sense of security and engaging in high-risk behaviours that could undermine the partial protection provided by male circumcision. Furthermore, male circumcision service provision was seen as a major opportunity to address the frequently neglected sexual health needs of men.

"Being able to recommend an additional HIV prevention method is a significant step towards getting ahead of this epidemic," said Catherine Hankins, Associate Director, Department of Policy, Evidence and Partnerships at UNAIDS. "However, we must be clear: Male circumcision does not provide complete protection against HIV. Men and women who consider male circumcision as an HIV preventive method must continue to use other forms of protection such as male and female condoms, delaying sexual debut and reducing the number of sexual partners."

Need for quality and safe services

Health services in many developing countries are weak and there is a shortage of skilled health professionals. There is a need, therefore, to ensure that male circumcision services for HIV prevention do not unduly disrupt other health care programmes, including other HIV/AIDS interventions. In order to both maximize the opportunity afforded by male circumcision and ensure longer-term sustainability of services, male circumcision should, wherever possible, be integrated with other services.

The risks involved in male circumcision are generally low, but can be serious if circumcision is undertaken in unhygienic settings by poorly trained providers or with inadequate instruments. Wherever male circumcision services are offered, therefore, training and certification of providers, as well as careful monitoring and evaluation of programmes, will be necessary to ensure that these meet their objectives and that quality services are provided safely in sanitary settings, with adequate equipment and with appropriate counselling and other services.

Male circumcision has strong cultural connotations implying the need also to deliver services in a manner that is culturally sensitive and that minimizes any stigma that might be associated with circumcision status. Countries should ensure that male circumcision is undertaken with full adherence to medical ethics and human rights principles, including informed consent, confidentiality and absence of coercion.

Maximizing public health benefit

A significant public health impact is likely to occur most rapidly if male circumcision services are first provided where the incidence of heterosexually acquired HIV infection is high. It was therefore recommended that countries with high prevalence, generalized heterosexual HIV epidemics that currently have low rates of male circumcision consider urgently scaling up access to male circumcision services. A more rapid public health benefit will be achieved if age groups at highest risk of acquiring HIV are prioritized, although providing male circumcision services to younger age groups will also have public health impact over the longer term. Modelling studies suggest that male circumcision in sub-Saharan Africa could prevent 5.7 million new cases of HIV infection and 3 million deaths over 20 years.

Experts at the meeting agreed that the cost-effectiveness of male circumcision is acceptable for an HIV prevention measure and that, in view of the large potential public health benefit of expanding male circumcision services, countries should also consider providing the services free of charge or at the lowest possible cost to the client, as for other essential services.

In countries where the HIV epidemic is concentrated in specific population groups such as sex workers, injecting drug users or men who have sex with men, there would be limited public health impact from promoting male circumcision in the general population. However, there may be an individual benefit for men at high risk of heterosexually acquired HIV infection. . . .

SOURCE: United Nations. World Health Organization. "WHO and UNAIDS Announce Recommendations from Expert Consultation on Male Circumcision for HIV Prevention." Press release. March 28, 2007. www.who.int/mediacentre/news/releases/2007/pr10/en/index.html (accessed March 25, 2008).

OTHER HISTORIC DOCUMENTS OF INTEREST

FROM PREVIOUS *HISTORIC DOCUMENTS*

- Federal guidelines on drugs to prevent HIV infection, *2005,* p. 50

April

DOCUMENT IN CONTEXT

Supreme Court on EPA Regulation of Greenhouse Gases

APRIL 2, 2007

Tensions over federal policy on climate change continued to spark heated debate in 2007, as all three branches of the federal government, the states, and numerous advocacy groups weighed in on the issue. Most importantly, on April 2 a divided Supreme Court ruled that the Environmental Protection Agency (EPA) has authority under the Clean Air Act to regulate carbon dioxide and other heat-trapping gases in vehicle tailpipe emissions—the greenhouse gases that many scientists say contribute to global warming. The EPA had declined to regulate greenhouse gases, arguing that they are not air pollutants within the meaning of the Clean Air law.

The Court's decision allowed California, joined by several other states, to press the EPA to rule on its request for permission to set state tailpipe emissions standards that would effectively require greater gasoline fuel economy than that mandated by federal standards. At the same time, Congress was considering legislation to raise federal fuel economy standards for the first time since 1985. For years, the powerful auto industry and its influential supporters in Congress had been able to block any increases in those standards, but in the face of growing public pressure to do something about global warming and with gasoline prices rising to around $3.00 a gallon, the political climate was changing. On December 19 President George W. Bush signed into law a measure raising federal fuel economy standards to an average of 35 miles per gallon by 2020, compared with the current average of about 25 miles per gallon. That same day, the EPA rejected California's request, an announcement that critics immediately denounced as a political decision designed to help the auto industry. As the year ended, Democrats in Congress were investigating allegations that the EPA administrator had ignored the advice of his staff and that the White House had unduly interfered in the administrator's decision, while California was preparing to challenge to the EPA decision in the courts.

Supreme Court on Greenhouse Gases

In the face of the federal government's reluctance to regulate greenhouse gas emissions, several environmental groups sought to force the issue, petitioning the EPA to use its authority under the Clean Air Act to regulate the gases. The agency denied the petition, saying it had no such authority and, even if it did, it would not take action because the link between greenhouse gases and climate change had not been firmly established. The EPA also responded that such regulation would interfere with the president's policy of encouraging private businesses to reduce carbon dioxide emissions voluntarily and with his ability to persuade key developing countries such as China to reduce their greenhouse gas emissions. Joined by Massachusetts and eleven other states, as well as American Samoa, the District of Columbia, the city of Baltimore, and New York City, the environmentalists appealed the EPA denial to the courts.

The Supreme Court's April 2 decision in *Massachusetts v. Environmental Protection Agency* represented the first time it had ruled on an issue related to global warming, and the 5–4 decision came down on the side of environmental advocates. Writing for the majority, Justice John Paul Stevens said that the EPA had provided "no reasoned explanation for its refusal to decide whether greenhouse gases cause or contribute to climate change." By offering "nothing more than a laundry list of reasons not to regulate," the agency had defied the Clean Air Act's "clear statutory command" to regulate air pollutants, he said. The EPA could only "avoid taking further action" now, Stevens wrote, by "determin[ing] that greenhouse gases do not contribute to climate change" or by providing a "reasoned judgment for declining to form a scientific judgment." Justices Anthony M. Kennedy, David H. Souter, Ruth Bader Ginsburg, and Stephen G. Breyer joined the majority opinion.

In a dissent joined by Justices Antonin Scalia, Clarence Thomas, and Samuel A. Alito Jr., Chief Justice John G. Roberts argued that Massachusetts and the other plaintiffs should not have been granted legal standing to sue the EPA in the first place. "Global warming might be 'the most pressing environmental problem of our time,' " Roberts wrote, but it is an issue that should be resolved by Congress and the chief executive, not the federal courts. Joined by Roberts, Thomas, and Alito, Scalia wrote a separate dissent, arguing that the majority had substituted "its own desired outcome for the reasoned judgment of the responsible agency."

Many of the reactions to the ruling seemed to acknowledge that the political climate had shifted toward those who wanted the government to take meaningful steps to address climate change. Rep. John D. Dingell, D-Mich., the chairman of the House Energy and Commerce Committee and a longtime backer of the auto industry, issued a statement: "While I still believe Congress did not intend for the Clean Air Act to regulate greenhouse gases, the Supreme Court has made clear its decision and the matter is now settled. Today's ruling provides another compelling reason why Congress must enact, and the President must sign, comprehensive climate change legislation." Dave McCurdy, the president of the Alliance of Automobile Manufacturers, which had backed the EPA decision not to regulate, said that his organization "believes that there needs to be a national, federal, economy-wide approach to addressing greenhouse gases."

California Waiver Denied

California governor Arnold Schwarzenegger said that he was "very encouraged" by the decision and that he expected the EPA "to move quickly now" to grant California its requested waiver to allow it to impose tough new fuel economy standards. Under the Clean Air Act, California was given the authority to adopt air pollution standards more stringent than those of the federal government provided it obtained approval, or a waiver, from the EPA. Other states could choose to follow California's regulations or the federal standards. Since the late 1960s, California had been granted more than fifty waivers, under which it developed several innovations to reduce air pollution, including the catalytic converter to reduce tailpipe emissions, tighter fuel caps to reduce gasoline evaporation, and computerized detectors to warn when a car's smog controls were not working.

The waiver currently at issue was for a plan adopted by the California Air Resources Board in 2004 to reduce tailpipe emissions of greenhouse gases by 30 percent between 2009 and 2016. By the end of 2007, another sixteen states, accounting for about 45 percent of the U.S. auto market, had either adopted the California standards or pledged to do so. Because the primary way of cutting greenhouse gas emissions from gasoline-powered vehicles was to cut fuel use, the California standard was effectively a fuel economy standard. Estimates varied about the fuel efficiency necessary to meet the standard. The EPA said the standard would require an average of 33.9 miles per gallon (mpg) by 2016, while California officials believed the average would be at least 36 mpg. Automakers, who opposed the waiver, worried it could be as high as 43 mpg.

Although California applied for the waiver in 2005, the EPA put the decision on hold to await the Supreme Court's ruling on the agency's authority to regulate greenhouse gases. Also pending was a suit brought in federal court in 2004 by a group of auto dealers in California's San Joaquin Valley, backed by the Alliance of Auto Manufacturers, which challenged the state's legal authority to regulate greenhouse gas emissions. On December 13 Judge Anthony W. Ishii ruled against the automakers, saying that both California and the EPA had authority to regulate tailpipe emissions.

That ruling was a major defeat for automakers, who were also embroiled in a tense lobbying battle in Washington over raising federal fuel-economy standards. The federal corporate average fuel economy (CAFE) standards were first introduced in 1975, in the midst of an oil shortage. Under existing rules, each manufacturer's fleet of passenger cars had to average at least 27.5 miles per gallon, a standard that had not changed since 1985. The standard for light-truck fleets—minivans, pickups, and sports utility vehicles (SUVs)—was 21.3 mpg in 2007 and was set to increase to 22.5 mpg in 2008.

Domestic automakers had long—and successfully—opposed any increase in the CAFE standards, arguing that raising them would up production costs and force them to lay workers off. They also argued that the standards failed to take into account the safety advantages of SUVs and light trucks. But with the Democratic takeover of Congress in January, widespread concern over climate change, and the rising price of gasoline, the political calculus changed. In June the automakers abruptly switched their strategy, from complete opposition to support of a mild increase in the mileage standards in an effort to ward off more stringent federal regulations, and possibly the California regulations as well.

The legislation that Congress cleared on December 18 and that the president signed into law (HR 6—PL 110-140) on December 19 was tougher than the automakers had wanted but not as tough as some Democrats and environmentalists would have liked. The new law set a single standard for cars and light trucks and required that a manufacturer's entire fleet average 35 mpg in the 2020 model year.

Hours after President Bush signed the bill, EPA administrator Stephen L. Johnson announced that he was denying California's request for a waiver for its tailpipe emissions regulation. "The Bush administration is moving forward with a clear national solution, not a confusing patchwork of state rules, to reduce America's climate footprint from vehicles," Johnson said. "President Bush and Congress have set the bar high, and when fully implemented, our federal fuel-economy standard will achieve significant benefits by applying to all 50 states." Johnson noted that earlier waivers granted to California covered "pollutants that predominantly impacted local and regional air quality." But "climate change," he continued, "affects everyone regardless of where greenhouse gases occur, so California is not exclusive" and therefore does not meet the "compelling and extraordinary circumstances" for obtaining a waiver.

Johnson's decision and his explanation were immediately decried by California state officials and environmentalists. "It is disappointing that the federal government is standing in our way and ignoring the will of tens of millions of people across the nation," Schwarzenegger said. "We will continue to fight this battle." One environmental activist pointed out that the EPA had never before distinguished between local, national, and international air pollution. The head of the California Air Resources Board, which had written the state standards, said that the federal fuel economy standards were "not the same thing" as the state's "comprehensive program for reducing greenhouse gases."

Many of the critics, and much of the editorial comment, said the decision was dictated by politics. A December 21 *New York Times* editorial called Johnson's decision "an indefensible act of executive arrogance that can only be explained as the product of ideological blindness and as a political payoff to the automobile industry."

The *Los Angeles Times* reported on December 21 that EPA staff had urged Johnson to grant the waiver. "California met every criteria . . . on the merits. The same criteria we have used for the last 40 years on all the other waivers," one EPA staffer told the paper. "We told him [Johnson] that. All the briefings we have given him laid out the facts." Technical and legal staff at the EPA also reportedly advised Johnson that he would likely lose a legal challenge. Several news reports indicated that several EPA staff members believed Johnson made his decision after Vice President Dick Cheney met with auto executives in November to hear their arguments against allowing either the EPA or California to regulate greenhouse gases. The *St. Petersburg Times* reported that during the late summer and fall, Transportation Secretary Mary Peters had quietly urged members of Congress and some governors to let the EPA know they opposed the waiver request.

The House Oversight and Government Reform Committee and the Senate Environment and Public Works Committee, led by California Democrats Henry A. Waxman and Barbara Boxer, announced on December 20 that they would investigate Johnson's decision and requested all related internal EPA documents, including any communications with the White House. The EPA indicated that it would comply with the request. Meanwhile, Schwarzenegger and several other governors declared their intent to challenge the decision in the U.S. Court of Appeals for the District of Columbia Circuit.

Other EPA Actions during 2007

Although the EPA had not responded to the Supreme Court decision on greenhouse gases by the end of the year, the agency took other actions, some as a result of court orders. Following are a few of the more notable actions.

Smog standards

On June 21 EPA administrator Johnson announced that the agency was proposing new rules to reduce current smog standards, although not by as much as the EPA's science advisers and staff had recommended. The standards were last changed in 1997, and the EPA was under court order to review them. Smog, or ground-level ozone, is produced when nitrogen oxides and other gases found in car exhaust, industrial emissions, and gasoline vapors are aggravated by heat and sunlight. Smog has been shown to cause several respiratory diseases, including asthma, and to lead to premature deaths. Smog is measured by calculating the concentration of ozone molecules in the atmosphere over an eight-hour period; the standard in 2007 was 84 parts per billion. The EPA proposed decreasing that to between 70 and 75 parts per billion but took public comment on reducing it to 65 parts per billion or keeping the current standard. On August 2 the EPA released documents showing that at the behest of the Office of Management and Budget (OMB), it would also consider a less stringent standard of 79 parts per billion. The final rule was scheduled to be published in March 2008.

Many environmental advocacy groups, as well as the EPA's Clean Air Scientific Advisory Committee, had urged that the new standard be no higher than 70 parts per billion. At 65 parts per billion, an EPA analysis showed that the costs of controlling smog could be as high as $46 billion in 2020, with a savings of between 530 and 2,400 premature deaths each year. Reducing the current standard to 79 parts per billion would cost up to $3.3 billion by 2020 and prevent 19 to 85 premature deaths every year. Business and industry groups had lobbied for no change in the standard, citing the costs involved, what they considered to be the uncertainty of the health benefits to be gained, and the fact that about eighty counties around the country had not yet been able to meet the existing standard.

Benzene

In a related development, the EPA issued a final rule on February 9 that by 2030 would reduce the amount of benzene in gasoline substantially below 1990 levels. Benzene, a carcinogen, is the second most toxic air pollutant after diesel fuel. Found naturally in crude oil, it is increased during refining to raise gasoline's octane rating. Under the new rule, scheduled to take effect in 2011, the amount of benzene in gasoline would drop from an average of about 1 percent to 0.62 percent. Oil refineries could trade pollution credits among themselves provided the entire industry stayed below the limit, but the EPA put a cap of 1.3 percent benzene on any given product. The agency also tightened rules on benzene emissions from cars starting in cold temperatures and from fuel containers such as gasoline cans. The ruling was published under a court-ordered deadline issued after environmental advocacy organizations successfully sued the EPA for failing to promulgate a benzene rule by 2004. The new rule was tighter than the one first proposed by the EPA and was generally applauded by environmentalists, although some expressed skepticism about the trading scheme.

Toxic chemicals reporting

Twelve states filed suit against the EPA in federal district court in New York City, challenging the agency's decision to roll back the amount of data that companies were required to report on toxic chemicals that they used, stored, and released into the environment. The regulation was adopted in 1986, two years after the release of deadly chemical gas at a Union Carbide plant in Bhopal, India, killed thousands of people. The Toxics Release Inventory program (TRI) required companies to provide a detailed report whenever they stored or emitted 500 pounds of specific toxic chemicals. The reports were frequently used by communities, watchdog groups, and government agencies to monitor public exposure to potentially dangerous chemical releases.

In December 2006, however, the EPA relaxed the rule to allow companies to file an abbreviated form if they used less than 5,000 pounds of toxic chemicals or released less than 2,000 pounds annually. Companies using the most dangerous chemicals would still have to use the long reporting form unless they certified that they were not releasing the chemicals into the environment. Announcing the lawsuit on November 28, 2007, New

York attorney general Andrew Cuomo said the new regulations "rob New Yorkers—and people across the country—of their right to know about toxic dangers in their own backyard. Along with eleven other states, we will restore the public's right to information about chemical hazards, despite the Bush administration's best attempts to hide it."

Two weeks later, on December 12, the Government Accountability Office (GAO) issued a report entitled "Toxic Chemical Releases," saying that the new rules would "significantly reduce" information available to the public on toxic chemicals used in their neighborhoods. The GAO estimated that the new rule would permit about 3,500 facilities to file abbreviated reports and that the number of detailed TRI reports would be cut by about 22,000, a decline of 25 percent. GAO also reported that the EPA administrator had "expedited" the rule-making process under pressure from OMB to reduce the paperwork burden on industry and, as a result, had not conducted a required cost-benefit analysis or solicited input from the agency's own offices that relied on the TRI data for decision making.

Following are excerpts from the majority opinion, written by Justice John Paul Stevens, in the case of Massachusetts v. Environmental Protection Agency, *in which the Supreme Court ruled 5–4 on April 2, 2007, that the Environmental Protection Agency has the authority under the Clean Air Act of 1970 to regulate greenhouse gases.*

DOCUMENT

Massachusetts v. Environmental Protection Agency

No. 05–1120

Massachusetts, et al., Petitioners *v.* Environmental Protection Agency et al.	On writ of certiorari to the United States Court of Appeals for the District of Columbia Circuit

[April 2, 2007]

JUSTICE STEVENS delivered the opinion of the Court.

A well-documented rise in global temperatures has coincided with a significant increase in the concentration of carbon dioxide in the atmosphere. Respected scientists believe the two trends are related. For when carbon dioxide is released into the atmosphere, it acts like the ceiling of a greenhouse, trapping solar energy and retarding the escape of reflected heat. It is therefore a species—the most important species—of a "greenhouse gas."

Calling global warming "the most pressing environmental challenge of our time," a group of States, local governments, and private organizations, alleged in a petition for certiorari that the Environmental Protection Agency (EPA) has abdicated its responsibility under the Clean Air Act to regulate the emissions of four greenhouse gases, including carbon dioxide. Specifically, petitioners asked us to answer two questions . . . : whether EPA has the statutory authority to regulate greenhouse gas emissions from new motor vehicles; and if so, whether its stated reasons for refusing to do so are consistent with the statute. . . .

[Sections I through V, giving the legal background of the case and ruling that Massachusetts did have legal standing to bring the suit, are omitted.]

VI

On the merits, the first question is whether §202(a)(1) of the Clean Air Act authorizes EPA to regulate greenhouse gas emissions from new motor vehicles in the event that it forms a "judgment" that such emissions contribute to climate change. We have little trouble concluding that it does. In relevant part, §202(a)(1) provides that EPA "shall by regulation prescribe . . . standards applicable to the emission of any air pollutant from any class or classes of new motor vehicles or new motor vehicle engines, which in [the Administrator's] judgment cause, or contribute to, air pollution which may reasonably be anticipated to endanger public health or welfare." . . Because EPA believes that Congress did not intend it to regulate substances that contribute to climate change, the agency maintains that carbon dioxide is not an "air pollutant" within the meaning of the provision.

The statutory text forecloses EPA's reading. The Clean Air Act's sweeping definition of "air pollutant" includes "*any* air pollution agent or combination of such agents, including *any* physical, chemical . . . substance or matter which is emitted into or otherwise enters the ambient air. . . ." (emphasis added). On its face, the definition embraces all airborne compounds of whatever stripe, and underscores that intent through the repeated use of the word "any." Carbon dioxide, methane, nitrous oxide, and hydrofluorocarbons are without a doubt "physical [and] chemical . . . substance[s] which [are] emitted into . . . the ambient air." The statute is unambiguous.

Rather than relying on statutory text, EPA invokes postenactment congressional actions and deliberations it views as tantamount to a congressional command to refrain from regulating greenhouse gas emissions. Even if such postenactment legislative history could shed light on the meaning of an otherwise-unambiguous statute, EPA never identifies any action remotely suggesting that Congress meant to curtail its power to treat greenhouse gases as air pollutants. That subsequent Congresses have eschewed enacting binding emissions limitations to combat global warming tells us nothing about what Congress meant when it amended §202(a)(1) in 1970 and 1977. And unlike EPA, we have no difficulty reconciling Congress' various efforts to promote interagency collaboration and research to better understand climate change with the agency's pre-existing mandate to regulate "any air pollutant" that may endanger the public welfare. . . . Collaboration and research do not conflict with any thoughtful regulatory effort; they complement it.

EPA's reliance on *Brown & Williamson Tobacco Corp.* [2000] . . . is similarly misplaced. In holding that tobacco products are not "drugs" or "devices" subject to Food and Drug Administration (FDA) regulation pursuant to the Food, Drug and Cosmetic Act (FDCA), . . . we found critical at least two considerations that have no counterpart in this case.

First, we thought it unlikely that Congress meant to ban tobacco products, which the FDCA would have required had such products been classified as "drugs" or "devices." . . . Here, in contrast, EPA jurisdiction would lead to no such extreme measures. EPA would only *regulate* emissions, and even then, it would have to delay any action "to permit the development and application of the requisite technology, giving appropriate consideration to the cost of compliance." . . . However much a ban on tobacco products clashed with the "common sense" intuition that Congress never meant to remove those products from circulation, . . . there is nothing counterintuitive to the notion that EPA can curtail the emission of substances that are putting the global climate out of kilter.

Second, in *Brown & Williamson* we pointed to an unbroken series of congressional enactments that made sense only if adopted "against the backdrop of the FDA's consistent and repeated statements that it lacked authority under the FDCA to regulate tobacco." . . . We can point to no such enactments here: EPA has not identified any congressional action that conflicts in any way with the regulation of greenhouse gases from new motor vehicles. Even if it had, Congress could not have acted against a regulatory "backdrop" of disclaimers of regulatory authority. Prior to the order that provoked this litigation, EPA had never disavowed the authority to regulate greenhouse gases, and in 1998 it in fact affirmed that it *had* such authority. . . . There is no reason, much less a compelling reason, to accept EPA's invitation to read ambiguity into a clear statute.

EPA finally argues that it cannot regulate carbon dioxide emissions from motor vehicles because doing so would require it to tighten mileage standards, a job (according to EPA) that Congress has assigned to DOT. . . . But that DOT sets mileage standards in no way licenses EPA to shirk its environmental responsibilities. EPA has been charged with protecting the public's "health" and "welfare," . . . a statutory obligation wholly independent of DOT's mandate to promote energy efficiency. . . . The two obligations may overlap, but there is no reason to think the two agencies cannot both administer their obligations and yet avoid inconsistency.

While the Congresses that drafted §202(a)(1) might not have appreciated the possibility that burning fossil fuels could lead to global warming, they did understand that without regulatory flexibility, changing circumstances and scientific developments would soon render the Clean Air Act obsolete. The broad language of §202(a)(1) reflects an intentional effort to confer the flexibility necessary to forestall such obsolescence. . . . Because greenhouse gases fit well within the Clean Air Act's capacious definition of "air pollutant," we hold that EPA has the statutory authority to regulate the emission of such gases from new motor vehicles.

VII

The alternative basis for EPA's decision—that even if it does have statutory authority to regulate greenhouse gases, it would be unwise to do so at this time—rests on reasoning divorced from the statutory text. While the statute does condition the exercise of EPA's authority on its formation of a "judgment," . . . that judgment must relate to whether an air pollutant "cause[s], or contribute[s] to, air pollution which may reasonably be anticipated to endanger public health or welfare." . . . Put another way, the use of the word "judgment" is not a roving license to ignore the statutory text. It is but a direction to exercise discretion within defined statutory limits.

If EPA makes a finding of endangerment, the Clean Air Act requires the agency to regulate emissions of the deleterious pollutant from new motor vehicles. . . . EPA no doubt has significant latitude as to the manner, timing, content, and coordination of its regulations with those of other agencies. But once EPA has responded to a petition for rulemaking, its reasons for action or inaction must conform to the authorizing statute. Under the clear terms of the Clean Air Act, EPA can avoid taking further action only if it determines that greenhouse gases do not contribute to climate change or if it provides some reasonable explanation as to why it cannot or will not exercise its discretion to determine whether they do. . . . To the extent that this constrains agency discretion to pursue other priorities of the Administrator or the President, this is the congressional design.

EPA has refused to comply with this clear statutory command. Instead, it has offered a laundry list of reasons not to regulate. For example, EPA said that a number of voluntary executive branch programs already provide an effective response to the threat of global warming, . . . that regulating greenhouse gases might impair the President's ability to negotiate with "key developing nations" to reduce emissions, . . . and that curtailing motor-vehicle emissions would reflect "an inefficient, piecemeal approach to address the climate change issue." . . .

Although we have neither the expertise nor the authority to evaluate these policy judgments, it is evident they have nothing to do with whether greenhouse gas emissions contribute to climate change. Still less do they amount to a reasoned justification for declining to form a scientific judgment. In particular, while the President has broad authority in foreign affairs, that authority does not extend to the refusal to execute domestic laws. In the Global Climate Protection Act of 1987, Congress authorized the State Department—not EPA—to formulate United States foreign policy with reference to environmental matters relating to climate. . . . EPA has made no showing that it issued the ruling in question here after consultation with the State Department. Congress did direct EPA to consult with other agencies in the formulation of its policies and rules, but the State Department is absent from that list. . . .

Nor can EPA avoid its statutory obligation by noting the uncertainty surrounding various features of climate change and concluding that it would therefore be better not to regulate at this time. . . . If the scientific uncertainty is so profound that it precludes EPA from making a reasoned judgment as to whether greenhouse gases contribute to global warming, EPA must say so. That EPA would prefer not to regulate greenhouse gases because of some residual uncertainty . . . is irrelevant. The statutory question is whether sufficient information exists to make an endangerment finding.

In short, EPA has offered no reasoned explanation for its refusal to decide whether greenhouse gases cause or contribute to climate change. Its action was therefore "arbitrary, capricious, . . . or otherwise not in accordance with law." . . . We need not and do not reach the question whether on remand EPA must make an endangerment finding, or whether policy concerns can inform EPA's actions in the event that it makes such a finding. . . . We hold only that EPA must ground its reasons for action or inaction in the statute.

VIII

The judgment of the Court of Appeals is reversed, and the case is remanded for further proceedings consistent with this opinion.

It is so ordered.

Source: Supreme Court of the United States. *Massachusetts v. EPA*. 549 U.S. ___ (2007). Docket 05–1120. April 2, 2007. www.supremecourtus.gov/opinions/06pdf/05-1120.pdf (accessed March 6, 2008).

Other Historic Documents of Interest

From this volume

- Scientists on climate change, p. 657
- Gore speech accepting the Nobel Prize, p. 691

From previous *Historic Documents*

- British Treasury report on the economics of climate change, *2006,* p. 615
- EPA inspector general on mercury pollution regulations, *2005,* p. 95
- Government panel on science and technology appointments, *2004,* p. 841

DOCUMENT IN CONTEXT

Catholic Bishops on Human Rights Abuses in Zimbabwe

APRIL 5, 2007

Zimbabwe slipped deeper into crisis in 2007, just ahead of elections scheduled for 2008 that longtime president Robert Mugabe appeared determined to win at any cost. The country's economy, formerly one of the most productive in Africa, had all but collapsed, with inflation running at an annual rate of upwards of 60,000 percent and with an estimated three-quarters of the working population unemployed. In March the government launched a fierce crackdown on the opposition; its acts included the beating of a leading politician, who was jailed and emerged bloodied but defiant. A mediation effort by Mugabe's colleagues in southern Africa floundered, in part because African leaders were reluctant to intervene in each other's affairs.

During the 1970s, Mugabe led the main black guerrilla movement that fought white minority rule in the country then called Rhodesia. He came to power in 1980, renamed the country Zimbabwe, suppressed the opposition, and ruled with an increasingly authoritarian hand over the next two decades. Zimbabwe's economy began a downhill slide in the early 1990s that worsened after 2000, when the government instituted a "land reform" program that seized millions of acres of the most productive farmland from white farmers and handed most of it to government supporters, including former members of Mugabe's rebel group.

Mugabe faced his first serious political challenge in 2000, when voters rejected constitutional changes intended to keep him in power indefinitely. Two years later an opposition candidate, Morgan Tsvangirai, insisted he had beaten Mugabe in the presidential election, but the government ruled for Mugabe and later charged Tsvangirai with treason, claiming he had conspired to kill Mugabe. Tsvangirai was acquitted, but the government pressed other charges against him, still pending in 2007, of attempting to overthrow the president.

A Downward Economic Spiral

It was difficult to see how the country's economy could get any worse during 2007, but it did so as the year and political crisis wore on. Any recitation of economic statistics had to reach for superlatives, all of the wrong kind. According to the International Monetary Fund, Zimbabwe had the worst economic record of any country in the world; its gross domestic product had fallen by about half since 2000.

Agricultural production, once the underpinning of the economy, had come nearly to a halt in some regions because of a prolonged drought and inefficiencies the government had introduced when it confiscated nearly all of the choicest land from minority white farmers. The United Nations (UN) Food and Agriculture Organization estimated, in a June 2007 report, that black farmers who had settled on the former white estates were using less than half the land because they lacked the necessary equipment, seeds, fertilizers, and other resources. The same report estimated that at least 2 million, and potentially as many as 4 million, Zimbabweans would depend totally on donated food during the year.

Faced with high inflation and food shortages, the government responded on June 25 by imposing price controls on food, fuel, and other essential commodities, then began arresting businesspeople who charged prices higher than the state-mandated levels. One predictable result was a sudden shortage of many basic commodities because businesses stopped producing and consumers hoarded supplies—when they could get them. In a pessimistic September 25 report describing "food security" issues in Zimbabwe, the UN World Food Programme noted that shortages "have been most profound in urban markets, where sporadic deliveries of basic goods are met with long lines, and not everyone makes it into stores before stocks run out."

The price controls were eased slightly after two months, but not before they had the perverse impact of boosting inflation, which had soared annually by several hundred percent in early 2007 and reached 14,000 percent by October, according to official estimates. Unofficial estimates by Western economists put the inflation rate much higher—close to 60,000 percent at year's end, making Zimbabwe's dollar all but worthless. A local correspondent for the BBC reported in July that the price for one banana in local currency had reached fifteen times what she had paid for a four-bedroom house seven years earlier. In late July the government issued its highest denomination bill, worth 200,000 Zimbabwe dollars—about enough to buy a quart of gasoline. On December 20 the government withdrew that bill and replaced it with new ones with face values of up to 750,000 dollars.

Estimates of unemployment generally fell in the range of 70 percent to 80 percent. The government repeatedly mandated pay raises for those who still worked, but the increases lagged well behind inflation, meaning that millions of people could no longer afford food, housing, and other essentials of daily life. The government's Central Statistics Office said in May that the basic "family food basket" for a family of six cost the equivalent of 100 U.S. dollars per month—at least five times the average monthly salary for a Zimbabwean teacher or nurse, according to the UN Children's Fund (UNICEF). Many people survived by bartering goods and services or by relying on remittances, money sent home by family members who had fled to South Africa or other countries. According to UN figures, at least one-quarter of Zimbabwe's 12.5 million people had left the country in recent years. Refugees streamed into neighboring countries at the rate of about 3,000 per day, most of them only to be forced back into Zimbabwe.

Mugabe and his government officials denied that their policies had anything to do with Zimbabwe's economic collapse; indeed, they denied that the country was even in serious economic difficulty. To the extent that Zimbabwe did have troubles, they said, the blame lay with Western countries, Britain in particular, that had imposed economic sanctions against Zimbabwe. "Our nation faces continued socioeconomic challenges from the illegal sanctions [imposed] on us by our detractors as punishment for repossessing our land," Mugabe said in a February 24 speech marking his eighty-third birthday. This analysis ignored the fact that all of the Western sanctions specifically targeted Mugabe and his fellow leaders—for example, by banning their travel to Europe or the United States and by freezing their international bank accounts—and not the Zimbabwean economy as a whole.

Carrying its land reform program one step further, the government in 2007 adopted legislation to "indigenize" foreign-owned businesses in the country. The stated goal of the plan was to require that every company operating in Zimbabwe was majority-owned by black Zimbabweans.

Crackdown on the Opposition

Political events in Zimbabwe in 2007 centered around numerous public demonstrations by human rights groups, teachers, students, and opposition political parties—most of which were suppressed by the Mugabe government, often with violence. The first major event of the year along this line came on February 17–18, when police disrupted rallies in Harare, the capital, by two competing factions of the opposition party, the Movement for Democratic Change (MDC). The police acted against the rallies even though the nation's high court had ruled that they should be allowed. More than 125 opposition supporters were arrested. The police later banned all public demonstrations, at first just in parts of Harare, then nationwide.

Four weeks later, on March 11, a coalition of civil society and political organizations, organized as the Save Zimbabwe Campaign, sponsored what they called a prayer meeting in the Highfield suburb of Harare. Participants arrived to find the rally site surrounded by security forces, who, according to most witnesses, launched an unprovoked attack on the protesters. In response to police beatings, some of the protesters threw stones, and the police reacted by firing tear gas and live ammunition, killing MDC member Gift Tandare. Police arrested several dozen people, including the leaders of both MDC factions: Tsvangirai, a former union official, and Arthur Mutambara, a research scientist who had challenged Tsvangirai for the party leadership.

Some of the arrested protesters later said they were beaten at police stations, in some cases for several hours, but were not taken for medical treatment until the next day. Among them was Tsvangirai, who appeared in court on March 13 with a bloody gash on the side of his head and a swollen eye. On the same day, police attacked crowds attending the funeral for Tandare.

In subsequent days, police arrested several hundred people, many of them leaders and members of opposition political parties and other civil society organizations. Human rights organizations also documented numerous cases in which police attacked people who had no connection with opposition groups but happened to live in neighborhoods where the opposition was strong.

Responding to international outrage over the attack on Tsvangirai, Mugabe said Western critics "can go hang." Some of Mugabe's fellow African leaders broke their long silence on Zimbabwe's troubles. Among them was Levy Mwanawasa, the president of neighboring Zambia, who called Zimbabwe a "sinking Titanic" and expressed concern about the exodus of Zimbabweans to neighboring countries, including his.

One of the year's most important political attacks against Mugabe came during Easter week, when the Zimbabwe Catholic Bishops' Conference issued a pastoral letter saying that the country was in a crisis of "extreme danger and difficulty." The letter, which was given to parishioners during the first week of April, described in detail the economic and political agonies of the country, which the bishops attributed to a "crisis of leadership." In particular, the bishops traced the nation's current travails to the years after independence in 1980 when, they said, the "power and wealth of the tiny white Rhodesian elite was appropriated by an equally exclusive black elite"—a reference to Mugabe, the leaders of his party, and a small number of black businessmen who supported Mugabe and, in turn, benefited from government policies.

The bishops' letter was an extraordinary event in Zimbabwe. Previously, only a few church leaders had spoken out individually, and several had been supporters of Mugabe's until recently; further, they had never before taken such a strong political position as a group. About one-quarter of Zimbabweans adhered to Catholic or other Christian faiths, while the majority adopted a mixture of Christian and traditional African religious beliefs.

Mugabe, a Catholic, denounced the letter in an interview with the pro-government *Herald* newspaper published on April 5: "Once [the bishops] turn political, we regard them as no longer spiritual and our relations with them would be conducted as if we are dealing with political entities, and this is quite a dangerous path they have chosen for themselves." Mugabe got some revenge in September when Archbishop Pius Ncube, his harshest critic among the bishops, was forced by the Vatican to step down after Zimbabwe government-run news media published photographs that appeared to show him having sex with a married woman.

Enforcing the government ban on public demonstrations, police repeatedly broke up protests throughout the year, in most cases arresting the protest leaders and in some cases beating demonstrators on the spot. Amnesty International, Human Rights Watch, local human rights groups in Zimbabwe, and other advocacy groups reported that they had counted dozens of cases in which police tortured political detainees. The government denied such charges.

There were two main questions about Zimbabwean politics during the second half of the year: Would the opposition party be able to unite behind a single presidential candidate? Would opposition to Mugabe emerge from within his party? At year's end, the answers appeared to be maybe, and maybe. The MDC remained split between the factions led by Tsvangirai and Mutambara. Late in December, however, party leaders said the factions probably would unite behind one candidate, likely Tsvangirai.

Mugabe's ruling party, the Zimbabwe African National Union-Patriotic Front (ZANU-PF), faced its own internal dissension, generated largely by the president's refusal to give way to a younger generation. One possible candidate from within the party was a previous Mugabe insider, former finance minister Simba Makoni. Mugabe received backing from his party in March and again in December; on the latter occasion, he said he expected a "thunderous" victory in the elections scheduled for March 2008.

International Diplomacy

By 2007 it was clear that Western diplomatic pressure on Mugabe over the years not only had failed but had, in fact, given him a convenient foil to blame for Zimbabwe's troubles. The imposition of limited sanctions against Mugabe and other top leaders, and the suspension of Zimbabwe's membership in the British Commonwealth starting in 2002, had no apparent impact on Mugabe except to toughen his resolve.

Zimbabwe's neighbors—under the auspices of a regional economic authority called the Southern African Development Council (SADC)—in 2003 launched what they called "quiet diplomacy" to persuade Mugabe to reconcile with his domestic opponents. This approach appeared to be no more productive than the heated rhetoric of Western leaders, however. Mugabe's colleagues were reluctant to pressure him publicly for two reasons: his legacy as a liberation hero during the struggle against white minority rule in Zimbabwe (and his support in the 1980s for South African blacks against their white rulers), and African leaders' general resistance to outside "interference" of any kind in their internal affairs.

The man at the center of international attention on the Zimbabwe question was the region's most important leader, South African president Thabo Mbeki, who had been elected democratically in 1999 and pledged to step down when his term ended in 2009. Mbeki appeared deeply reluctant to challenge Mugabe openly about his authoritarian practices and plans to retain power indefinitely.

SADC leaders held several rounds of talks with Mugabe and his agents during the year, starting with an "emergency" session on March 28, following the crackdown on the opposition rallies. Expectations that this session might lead to pressure on Mugabe went unfulfilled, however. The leaders issued a statement expressing "solidarity" with Mugabe and calling on Western countries to end sanctions against him and his fellow government officials. Mugabe boasted later that he had received "full backing" from his fellow presidents. Even so, the SADC leaders authorized Mbeki to meet with the various political factions in Zimbabwe, their first formal recognition that something might be amiss in that country.

Another summit in Lusaka, Zambia, on August 18 produced a call for the people of Zimbabwe to be allowed to elect their leaders "in an atmosphere of peace and tranquility" but offered no specific steps toward that goal. Subsequent mediation efforts by Mbeki produced rumors of a deal that would open Zimbabwe's political process. However, the only practical result appeared to be an agreement by the government in mid-December to modify several laws that had enabled the police to ban public protests and crack down on the few remaining independent news organizations. Although opposition leaders welcomed the government's amendment of the laws on December 18, they considered the step insufficient preparation for fair elections in March 2008.

The slow pace and apparent ineffectiveness of Mbeki's diplomacy frustrated Western leaders, who had hoped that an African-centered approach would have more influence on Mugabe than their own diplomacy had. Even so, leaders in London, Washington, and elsewhere in the West had few alternatives by 2007. In a December 3 speech in Washington, Jendayi Frazer, the U.S. assistant secretary of state for African affairs, acknowledged the shortcomings of SADC diplomacy, but said "we support without reservation the SADC initiative to bring together the ruling and opposition parties for talks on resolving the country's political and economic crisis. Specifically, we commend President Mbeki for his leadership and public commitment to deliver free and fair elections in Zimbabwe."

Following are excerpts from "God Hears the Cry of the Oppressed," the pastoral letter by the Zimbabwe Catholic Bishops' Conference concerning "the crisis of our country." The letter was posted on the Internet on March 30, 2007, distributed in Zimbabwe's Catholic churches during the week of Easter, and officially dated April 5, 2007. The letter was signed by Robert C. Ndlovu, archbishop of Harare; Pius Alec M. Ncube, archbishop of Bulawayo; Alexio Churu Muchabaiwa, bishop of Mutare; Michael D. Bhasera, bishop of Masvingo; Angel Floro, bishop of Gokwe; Martin Munyanyi, bishop of Gweru; Dieter B. Scholz SJ, bishop of Chinhoyi; Albert Serrano, bishop of Hwange; and Patrick M. Mutume, auxiliary bishop of Mutare.

DOCUMENT

Pastoral Letter by Catholic Bishops of Zimbabwe

April 5, 2007

As your Shepherds we have reflected on our national situation and, in the light of the Word of God and Christian Social Teaching, have discerned what we now share with you, in the hope of offering guidance, light and hope in these difficult times.

The Crisis

The people of Zimbabwe are suffering. More and more people are getting angry, even from among those who had seemed to be doing reasonably well under the circumstances. The reasons for the anger are many, among them, bad governance and corruption. A tiny minority of the people have become very rich overnight, while the majority are languishing in poverty, creating a huge gap between the rich and the poor. Our Country is in deep crisis. A crisis is an unstable situation of extreme danger and difficulty. Yet, it can also be turned into a moment of grace and of a new beginning, if those responsible for causing the crisis repent, heed the cry of the people and foster a change of heart and mind especially during the imminent Easter Season, so our Nation can rise to new life with the Risen Lord.

In Zimbabwe today, there are Christians on all sides of the conflict; and there are many Christians sitting on the fence. Active members of our Parish and Pastoral Councils are prominent officials at all levels of the ruling party. Equally distinguished and committed office-bearers of the opposition parties actively support church activities in every parish and diocese. They all profess their loyalty to the same Church. They are all baptised, sit and pray and sing together in the same church, take part in the same celebration of the Eucharist and partake of the same Body and Blood of Christ. While the next day, outside the church, a few steps away, Christian State Agents, policemen and soldiers assault and beat peaceful, unarmed demonstrators and torture detainees. This is the unacceptable reality on the ground, which shows much disrespect for human life and falls far below the dignity of both the perpetrator and the victim.

In our prayer and reflection during this Lent, we have tried to understand the reasons why this is so. We have concluded that the crisis of our Country is, in essence, a crisis of governance and a crisis of leadership apart from being a spiritual and moral crisis.

A Crisis of Governance

The national health system has all but disintegrated as a result of prolonged industrial action by medical professionals, lack of drugs, essential equipment in disrepair and several other factors.

In the educational sector, high tuition fees and levies, the lack of teaching and learning resources, and the absence of teachers have brought activities in many public schools and institutions of higher education to a standstill. The number of students forced to terminate their education is increasing every month. At the same time, Government interference with the provision of education by private schools has created unnecessary tension and conflict.

Public services in Zimbabwe's towns and cities have crumbled. Roads, street lighting, water and sewer reticulation are in a state of severe disrepair to the point of constituting an acute threat to public health and safety, while the collection of garbage has come to a complete standstill in many places. Unabated political interference with the work of democratically elected Councils is one of the chief causes of this breakdown.

The erosion of the public transport system has negatively affected every aspect of our Country's economy and social life. Horrific accidents claim the lives of dozens of citizens each month.

Almost two years after the Operation Murambatsvina, thousands of victims are still without a home. That inexcusable injustice has not been forgotten.

Following a radical land reform programme seven years ago, many people are today going to bed hungry and wake up to a day without work. Hundreds of companies were forced to close. Over 80 per cent of the people of Zimbabwe are without employment. Scores risk their lives week after week in search of work in neighbouring countries.

Inflation has soared to over 1,600 per cent, and continues to rise, daily. It is the highest in the world and has made the life of ordinary Zimbabweans unbearable, regardless of their political preferences. We are all concerned for the turnaround of our economy but this will remain a dream unless corruption is dealt with severely irrespective of a person's political or social status or connections.

The list of justified grievances is long and could go on for many pages.

The suffering people of Zimbabwe are groaning in agony: "*Watchman, how much longer the night*"? *(Is 21:11)*

A Crisis of Moral Leadership

The crisis of our Country is, secondly, a crisis of leadership. The burden of that crisis is borne by all Zimbabweans, but especially the young who grow up in search of role models. The youth are influenced and formed as much by what they see their elders doing as by what they hear and learn at school or from their peers.

If our young people see their leaders habitually engaging in acts and words which are hateful, disrespectful, racist, corrupt, lawless, unjust, greedy, dishonest and violent in order to cling to the privileges of power and wealth, it is highly likely that many of them will behave in exactly the same manner. The consequences of such overtly corrupt leadership as we are witnessing in Zimbabwe today will be with us for many years, perhaps decades, to come. Evil habits and attitudes take much longer to rehabilitate than to acquire. Being elected to a position of leadership should not be misconstrued as a licence to do as one pleases at the expense of the will and trust of the electorate.

A Spiritual and Moral Crisis

Our crisis is not only political and economic but first and foremost a spiritual and moral crisis. As the young independent nation struggles to find its common national spirit, the people of Zimbabwe are reacting against the "structures of sin" in our society. Pope John Paul II says that the "structures of sin" are "rooted in personal sin, and thus always linked to the concrete acts of individuals who introduce these structures, consolidate them and make them difficult to remove. And thus they grow stronger, spread, and become the source of other sins, and so influence people's behaviour." The Holy Father stresses that in order to understand the reality that confronts us, we must "give a name to the root of the evils which afflict us." That is what we have done in this Pastoral Letter.

The Roots of the Crisis

The present crisis in our Country has its roots deep in colonial society. Despite the rhetoric of a glorious socialist revolution brought about by the armed struggle, the colonial structures and institutions of pre-independent Zimbabwe continue to persist in our society. None of the unjust and oppressive security laws of the Rhodesian State have been repealed; in fact, they have been reinforced by even more repressive legislation, the Public Order and Security Act and the Access to Information and Protection of Privacy Act, in particular. It almost appears as though someone sat down with the Declaration of Human Rights and deliberately scrubbed out each in turn.

Why was this done? Because soon after Independence, the power and wealth of the tiny white Rhodesian elite was appropriated by an equally exclusive black elite, some of whom have governed the country for the past 27 years through political patronage. Black Zimbabweans today fight for the same basic rights they fought for during the liberation struggle. It is the same conflict between those who possess power and wealth in abundance, and those who do not; between those who are determined to maintain their privileges of power and wealth at any cost, even at the cost of bloodshed, and those who demand their democratic rights and a share in the fruits of independence; between those who continue to benefit from the present system of inequality and injustice, because it favours them and enables them to maintain an exceptionally high standard of living, and those who go to bed hungry at night and wake up in the morning to another day without work and without income; between those who only know the language of violence and intimidation, and those who feel they have nothing more to lose because their Constitutional rights have been abrogated and their votes rigged. Many people in Zimbabwe are angry, and their anger is now erupting into open revolt in one township after another.

The confrontation in our Country has now reached a flashpoint. As the suffering population becomes more insistent, generating more and more pressure through boycotts, strikes, demonstrations and uprisings, the State responds with ever harsher oppression through arrests, detentions, banning orders, beatings and torture. In our judgement, the situation is extremely volatile. In order to avoid further bloodshed and avert a mass uprising the nation needs a new people-driven Constitution that will guide a democratic leadership chosen in free and fair elections that will offer a chance for economic recovery under genuinely new policies.

Our Message of Hope: God Is Always on the Side of the Oppressed

The Bible has much to say about situations of confrontation. The conflict between the oppressor and the oppressed is a central theme throughout the Old and New Testaments. Biblical scholars have discovered that there are no less than twenty different root words in Hebrew to describe oppression. . . .

Generations of Zimbabweans, too, throughout their own long history of oppression and their struggle for liberation, have remembered, prayed and sung these texts from the Old and New Testaments and found strength, courage and perseverance in their faith that

Jesus is on their side. That is the message of hope we want to convey in this Pastoral Letter: God is on your side. He always hears the cry of the poor and oppressed and saves them.

Conclusion

We conclude our Pastoral Letter by affirming with a clear and unambiguous Yes our support of morally legitimate political authority. At the same time we say an equally clear and unambiguous No to power through violence, oppression and intimidation. We call on those who are responsible for the current crisis in our Country to repent and listen to the cry of their citizens. To the people of Zimbabwe we appeal for peace and restraint when expressing their justified grievances and demonstrating for their human rights.

Words call for concrete action, for symbols and gestures which keep our hope alive. We therefore invite all the faithful to a Day of Prayer and Fasting for Zimbabwe, on Saturday, 14 April 2007. This will be followed by a Prayer Service for Zimbabwe, on Friday, every week, in all parishes of our Country. As for the details, each Diocese will make known its own arrangements.

May the Peace and Hope of the Risen Lord be with you always. Happy Easter.

SOURCE: Zimbabwe Catholic Bishops' Conference. "God Hears the Cry of the Oppressed." Harare, Zimbabwe. April 5, 2007. Released March 30, 2007. www.usccb.org/sdwp/international/Zmbabwebishopstatement.pdf (accessed March 13, 2008).

Other Historic Documents of Interest

From this volume

- World Bank report on Africa aid programs, p. 592

From previous *Historic Documents*

- British Commonwealth leaders on Zimbabwe, *2003,* p. 1110
- Commonwealth Commission on elections in Zimbabwe, *2002,* p. 133

DOCUMENT IN CONTEXT

President Bush on Border Security and Immigration Reform

APRIL 9, 2007

For the second year in a row, Congress failed to enact comprehensive immigration reform, ensuring that the topic would be a major factor in the 2008 presidential campaign. The failure was a blow to President George W. Bush, who actively worked for passage of a bipartisan compromise measure that called for beefed-up border security, a temporary guest-worker program, and a path toward citizenship for many of the estimated 12 million illegal immigrants already in the country.

As in 2006, the debate pitted those who thought immigrants who held jobs and paid taxes had earned a chance to apply for citizenship even if they had come into the country illegally against those who were concerned about border security, the strains and costs immigrants placed on social services, and rewarding illegal actions. The discourse in 2007, however, was much harsher, at times even ugly, in tone. Partly fueled by conservative talk show hosts who railed against granting "amnesty" to the millions of illegal immigrants who were "invading" America across its border with Mexico, opponents of the measure deluged their legislators with petitions, phone calls, and e-mails.

In contrast, immigrant communities and their supporters were more subdued than they had been in 2006, when Hispanic immigrants had rallied in many cities across the country in support of that year's legislation. In 2007 rallies were smaller and fewer, and immigrants, both legal and illegal, told reporters and pollsters they felt intimidated not only by the government's stepped-up efforts to stop illegal immigrants from entering into the country and to deport those living here illegally, but also by the sometimes vitriolic rhetoric of those leading the opposition to the legislation. A survey released on December 13 by the Pew Hispanic Center found that about half the Hispanics polled said their lives were made more difficult by the political fight over immigration reform. About 12 percent said they had experienced difficulty finding or keeping a job, 15 percent said they had a harder time finding housing, and around 20 percent said they had been asked more frequently in the past year to show proof of their immigration status. Nonetheless, 71 percent described their overall lives as good or excellent. Pollsters did not ask Hispanics about their immigration status.

Republican political operatives feared that the heated rhetoric against immigration reform emanating from conservative circles would have negative repercussions for the party's efforts to draw more Hispanic voters. President Bush won about 40 percent of the Hispanic vote in 2004, but in the 2006 midterm elections, GOP candidates won only about 25 percent. Many observers predicted the percentage would drop even lower in 2008. "The tone of the debate, and the way it was framed in sort of an 'us against them' way, has done great harm in wooing Hispanics to the party," Linda Chavez, chairwoman of the Center for Equal Opportunity, a conservative public policy organization, told the *New York Times.*

Some voters in 2008 were likely to be newly naturalized citizens. In the first six months of 2007 the number of legal immigrants, largely Hispanics, who applied for citizenship rose sharply, in large part because of an impending fee increase; the cost of applying for citizenship was scheduled to jump from $400 to $675 on July 30. Many immigrants also told reporters they were motivated to become citizens so they could vote. "I realized that I want to be able to vote and speak up for my people because they are not getting enough support," one Hispanic applicant for citizenship told the *Times.*

Stepping Up Enforcement

Bush had championed immigration reform as a presidential candidate in 2000 and made it one of his administration's top priorities until the attacks of September 11, 2001, turned his attention to Afghanistan and then Iraq. Returning to the issue in 2006, Bush had hoped to bolster his domestic legacy with passage of a comprehensive reform measure. He actively worked to win passage of a compromise measure in the Senate that won broad support from Democrats and moderate Republicans. The measure died in the House, however, after Republicans refused to accept what many of them described as an amnesty bill. (House Republicans had initially won approval of a measure that would have made anyone in the United States without a valid visa a felon.)

To show Americans the administration was serious about controlling the borders— and to help win support for immigration reform— Bush took steps to increase the number of border patrol agents. Toward the end of the year, immigration authorities began to conduct raids of immigrant communities and workplaces looking for illegal aliens, who were then deported. Congress also sought to increase border security. After the reform measure died, Congress passed legislation calling for the construction of 700 miles of fencing along the border, as well as a "virtual fence" of cameras, sensors, unmanned aerial vehicles, and other surveillance technology to detect and turn back illegal immigrants trying to sneak into the country.

As a result of the increased enforcement efforts, the border patrol in 2006 caught 1.1 million illegal immigrants, mostly Mexicans who were promptly sent back to Mexico. Non-Mexicans were supposed to be detained until their status was determined and then deported if they were found to be in the country illegally. Because of a shortage of detention facilities, for most of the year these detainees were almost immediately released, and few of them ever returned for their status hearings. Starting in August 2006, however, the government deported almost all non-Mexicans who were caught; the only people allowed to stay were those seeking political asylum. Citing a drop in the number of non-Mexicans detained, the administration claimed the captures and quick deportations had discouraged many would-be illegal immigrants: after quadrupling over the four previous years, the number of non-Mexicans caught fell 35 percent, to 108,026 in 2006. The number of Mexicans captured and sent home had also dropped, in some cases, such as around Yuma, Arizona, by nearly two-thirds in just over four months. Whether the slowdown in attempted border crossings was permanent remained unknown.

Bush touted these achievements in his remarks at the Border Patrol's offices in Yuma on April 9, 2007, and again called for comprehensive immigration reform similar to the legislation that collapsed in 2006. "It's important that we get a bill done," Bush said. "We deserve a system that secures our borders and honors our proud history as a nation of immigrants."

A Second Attempt at Federal Reform

In Washington, Homeland Security Secretary Michael Chertoff and Commerce Secretary Carlos Gutierrez convened closed-door meetings with a group of senators who called themselves the "grand bargainers." Led by Massachusetts Democrat Edward M. Kennedy and Arizona Republican John Kyl, formerly a vocal critic of the president's immigration reform proposals, the group produced a bipartisan bill that was made public on May 17. The legislation combined border security and enforcement measures with a temporary worker program to allow illegal immigrants to remain and work in the United States. Under the proposal, illegal aliens would be able to get on a path to citizenship—which they could receive after eight years and if they paid fines, passed an English and civics exam, made a return trip to their native country, and remained employed and in good legal standing. The measure also sought to change the future distribution of green cards by moving away from the current system, which rewarded family ties, toward a merit point system based on U.S. economic needs.

The good will that existed behind the scenes never caught on beyond the conference room. From the moment the nearly 800-page bill was made public, its contents inflamed grassroots conservatives and divided liberals. Republicans continued to argue that earned citizenship was equivalent to amnesty for those who had broken the law. Democrats, backed by labor groups and civil rights organizations, rejected the green card point system and criticized the guest worker program for not offering its own path to citizenship. After several weeks of procedural maneuvering by both sides, supporters of the measure on June 7 lost three votes to cut off debate, and the legislation appeared to be all but dead.

Not willing to let go of what was likely his last opportunity as president to get a comprehensive reform bill passed, Bush attended the Senate Republican policy lunch on June 12 and urged GOP senators to give the bill a second chance. In response to their suggestions, Bush and the bill's backers agreed to introduce new overhaul legislation that included $4.4 billion in mandatory spending for additional border security and enforcement. Supporters, however, were unable to overcome partisan bickering over the procedure for voting on amendments—or the thousands of phone calls and e-mails urging senators to vote against the legislation. The June 28 vote to cut off debate on the new overhaul legislation failed even to win a majority in the Senate.

Aftermath

Congress's failure to pass comprehensive federal legislation meant that states and localities would likely continue to enact a myriad of conflicting laws, regulations, and ordinances regarding employment, housing, health, education, and other services for the illegal immigrant population. On July 3, just days after the federal legislation died, Arizona governor Janet Napolitano signed a bill setting harsh sanctions for employers who knowingly hired illegal immigrants. Napolitano said the measure had problems that the state legislature should fix but that she signed it "because Congress has failed miserably" to reform federal immigration laws. Under the new Arizona law, considered one of the toughest in the nation, employers had to verify the legal status of their employees; failure to do so could result in a suspension of their state business license, whereas a second offense could result in permanent revocation of the license and shut down the business. The law was scheduled to take effect on January 1, 2008.

The Arizona law was just one of the 244 laws related to immigration that were enacted by forty-six state legislatures in 2007, up from 84 in 2006, according to a report by the National Conference of State Legislators issued in late November. Most of the laws dealt with whether illegal immigrants could obtain driver's and other licenses or, like the Arizona law, set penalties for employers who hired illegal immigrants. At least eleven states passed laws prohibiting illegal immigrants from receiving state assistance. In the last two years several towns had also passed ordinances fining employers and landlords who hired or rented to illegal aliens; some were later repealed after federal and state courts upheld challenges to them.

The federal government also sought to crack down on illegal immigrants who used fraudulent documents to obtain jobs and on employers who knowingly hired them. On August 10 Secretaries Chertoff and Gutierrez announced a new rule that required employers to fire workers within ninety days if the federal government could not verify their Social Security numbers, as illegal immigrants often used phony Social Security numbers. An unusual coalition that included the AFL-CIO and the Chamber of Commerce of the United States challenged the rule, fearing it would lead employers in low-wage industries to lay off any Hispanic employee, legal or illegal, and lead to massive work disruptions.

On October 10 a federal district judge in San Francisco, Charles R. Breyer, issued an order delaying the new rule indefinitely. Breyer, the brother of Supreme Court justice Stephen G. Breyer, said that the government had failed to follow the proper procedures for issuing a new rule, including offering a legal explanation for its action or conducting a required survey of the costs to and impact on small businesses. Breyer also said that the Social Security database was so riddled with errors that legal immigrants might be told their Social Security numbers could not be verified, leading their employers to fire them rather than risk being penalized for hiring illegal workers. The rule could cause "irreparable harm to innocent workers and employers," Breyer concluded. In late November the administration said it would revise the rule to try to meet Breyer's concerns. The new rule was expected by the end of March 2008.

Late in the year, the *Washington Post* reported data from the Department of Homeland Security showing that, despite promises to penalize employers who knowingly hired illegal workers, the government had made criminal arrests of only ninety-two employers in fiscal 2007; only seventeen companies faced criminal fines or other penalties. Yet there were approximately 7.5 million illegal immigrants working in the United States. Some Democrats were quick to suggest the low number of arrests might reflect the Bush administration's reluctance to target corporate business and farm interests. "Why is it that hundreds of bar owners can be sanctioned in Missouri every year for letting somebody with a fake ID have a beer, but we can't manage to sanction hundreds of employers for letting people use fake identities to obtain a job?" asked Sen. Claire McCaskill of Missouri, a former prosecutor.

Following are excerpts from the text of remarks on border security and immigration reform made by President George W. Bush in Yuma, Arizona, on April 9, 2007.

DOCUMENT

President Bush's Call for Comprehensive Immigration Reform

April 9, 2007

Thank you all. Thank you all very much. Please be seated. . . .

. . . I'm glad to be back here in Yuma [Arizona]. Thank you so very much for your hospitality. Thanks for your service to the country. I appreciate so very much the work you're doing day and night to protect these borders. And the American people owe you a great debt of gratitude. . . .

Last May, I visited this section of the border, and it was then that I talked about the need for our Government to give you the manpower and resources you need to do your job. We were understaffed here. We weren't using enough technology to enable those who work here to be able to do the job the American people expect. I returned to check on the progress, to make sure that the check wasn't in the mail—it, in fact, had been delivered.

I went to a neighborhood that abuts up against the border when I was here in May. It's the place where a lot of people came charging across. One or two agents would be trying to do their job and stopping a flood of folks charging into Arizona, and they couldn't do the job—just physically impossible. Back at this site, there's now infrastructure; there's fencing. And the amount of people trying to cross the border at that spot is down significantly.

. . . The efforts are working. This border is more secure, and America is safer as a result.

Securing the border is a critical part of a strategy for comprehensive immigration reform. It is an important part of a reform that is necessary so that the Border Patrol agents down here can do their job more effectively. Congress is going to take up the legislation on immigration. It is a matter of national interest, and it's a matter of deep conviction for me. I've been working to bring Republicans and Democrats together to resolve outstanding issues so that Congress can pass a comprehensive bill and I can sign it into law this year. . . .

I hope by now the American people understand the need for comprehensive immigration reform is a clear need. Illegal immigration is a serious problem—you know it better than anybody. It puts pressure on the public schools and the hospitals, not only here in our border States but States around the country. It drains the State and local budgets. I was talking to the Governor [Janet Napolitano of Arizona] about how it strained the budgets. Incarceration of criminals who are here illegally strains the Arizona budget. But there's a lot of other ways it strains the local and State budgets. It brings crime to our communities.

It's a problem, and we need to address it aggressively. This problem has been growing for decades, and past efforts to address it have failed. These failures helped create a perception that America was not serious about enforcing our immigration laws and that they could be broken without consequence. Past efforts at reform did not do enough to secure our Nation's borders. As a result, many people have been able to sneak into this country.

If you don't man your borders and don't protect your borders, people are going to sneak in, and that's what's been happening for a long time. Past efforts at reform failed to address the underlying economic reasons behind illegal immigration. People will make great sacrifices to get into this country to find jobs and provide for their families.

When I was the Governor of Texas, I used to say, family values did not stop at the Rio Grande River. People are coming here to put food on the table, and they're doing jobs Americans are not doing. And the farmers in this part of the world understand exactly what I'm saying. But so do a lot of other folks around the country. People are coming to work, and many of them have no lawful way to come to America, and so they're sneaking in.

Past efforts at reform also failed to provide sensible ways for employers to verify the legal status of the workers they hire. It's against the law to knowingly hire an illegal alien. And as a result, because they couldn't verify the legal status, it was difficult for employers to comply. It was difficult for the government to enforce the laws at the worksite, and yet it is a necessary part of a comprehensive plan. You see, the lessons of all these experiences—the lesson of these experiences is clear: All elements of the issue must be addressed together. You can't address just one aspect and not be able to say to the American people that we're securing our borders.

We need a comprehensive bill, and that's what I'm working with Members of Congress on, a comprehensive immigration bill. And now is the year to get it done. The first element, of course, is to secure this border. That's what I'm down here for, to remind the American people that we're spending their taxpayer—their money, taxpayers' money, on securing the border. And we're making progress. This border should be open to trade and lawful immigration and shut down to criminals and drug dealers and terrorists and *coyotes* and smugglers, people who prey on innocent life.

We more than doubled the funding for border security since I've been the President. In other words, it's one thing to hear people come down here and talk; it's another thing for people to come down and do what they say they're going to do. And I want to thank Congress for working on this issue. The funding is increasing manpower. The additional funding is increasing infrastructure, and it's increasing technology.

When I landed here at the airport, the first thing I saw was an unmanned aerial vehicle. It's a sophisticated piece of equipment. You can fly it from inside a truck, and you can look at people moving at night. It's the most sophisticated technology we have, and it's down here on the border to help the Border Patrol agents do their job. We've expanded the number of Border Patrol agents from about 9,000 to 13,000, and by the end of 2008, we're going to have a total of more than 18,000 agents. . . .

This new technology is really important to, basically, leverage the manpower. Whether it be the technology of surveillance and communication—we're going to make sure the agents have got what is necessary to be able to establish a common picture and get information out to the field as quickly as possible so that those 18,000 agents, when they're finally on station, can do the job the American people expect.

But manpower can't do it alone. In other words, there has to be some infrastructure along the border to be able to let these agents do their job. And so I appreciate the fact that we've got double fencing, all-weather roads, new lighting, mobile cameras. The American people have no earthly idea what's going on down here. One of the reasons I've come is to let you know—let the taxpayers know, the good folks down here are making progress.

We've worked with our Nation's Governors to deploy 6,000 National Guard members to provide the Border Patrol with immediate reinforcements. In other words, it takes time to train the Border Patrol, and until they're fully trained, we've asked the Guard to come down. It's called Operation Jump Start, and the Guard down here is serving nobly.

. . . More than 600 members of the Guard are serving here in the Yuma Sector. And I thank the Guard, and equally importantly, I thank their families for standing by the men and women who wear the uniform during this particular mission. And you e-mail them back home and tell them how much I appreciate the fact they're standing by you.

I appreciate very much the fact that illegal border crossings in this area are down. In the months before Operation Jump Start, an average of more than 400 people a day were apprehended trying to cross here. The number has dropped to fewer than 140 a day. In other words, one way that the Border Patrol can tell whether or not we're making progress is the number of apprehensions. When you're apprehending fewer people, it means fewer are trying to come across. And fewer are trying to come across because we're deterring people from attempting illegal border crossings in the first place.

. . . And that's part of the effort we're doing. We're saying, we're going to make it harder for you, so don't try in the first place.

We're seeing similar results all across the southern border. The number of people apprehended for illegally crossing our southern border is down by nearly 30 percent this year. We're making progress. And thanks for your hard work. It's hard work but necessary work.

Another important deterrent to illegal immigration is to end what was called catch-and-release. I know how this discouraged some of our Border Patrol agents; I talked to them personally. They worked hard to find somebody sneaking in the country; they apprehended them; the next thing they know, they're back in society on our side of the border.

There's nothing more discouraging than have somebody risk their life or work hard and have the fruits of their labor undermined. And that's what was happening with catch-and-release. In other words, we'd catch people, and we'd say, "Show up for your court date," and they wouldn't show up for their court date. That shouldn't surprise anybody, but that's what was happening. And the reason why that was happening is because we didn't have enough beds to detain people.

Now, most of the people we apprehend down here are from Mexico. About 85 percent of the illegal immigrants caught crossing into—crossing this border are Mexicans—crossing the southern border are Mexicans. And they're sent home within 24 hours. It's the illegal immigrants from other countries that are not that easy to send home.

For many years, the government didn't have enough [detention] space, and so [Homeland Security Secretary] Michael [Chertoff] and I worked with Congress to increase the number of beds available. So that excuse was eliminated. The practice has been effectively ended. Catch-and-release for every non-Mexican has been effectively ended. And I want to thank the Border Patrol and the leaders of the Border Patrol for allowing me to stand up and say that's the case.

And the reason why is, not only do we have beds; we've expedited the legal process to cut the average deportation time. Now, these are non-Mexican, illegal aliens that we've caught trying to sneak into our country. We're making it clear to foreign governments that they must accept back their citizens who violate our immigration laws. I said we're going to effectively end catch-and-release, and we have. And I appreciate your hard work in doing that.

The second element of a comprehensive immigration reform is a temporary-worker program. You cannot fully secure the border until we take pressure off the border. And that requires a temporary-worker program. It seems to make sense to me that if you've got people coming here to do jobs Americans aren't doing, we need to figure out a way that they can do so in a legal basis for a temporary period of time. And that way our Border Patrol can chase the criminals and the drug runners, potential terrorists, and not have to try to chase people who are coming here to do work Americans are not doing.

If you want to take the pressure off your border, have a temporary-worker program. It will help not only reduce the number of people coming across the border, but it will do something about the inhumane treatment that these people are subjected to. There's a whole smuggling operation—you know this better than I do. There's a bunch of smugglers that use the individual as a piece of—as a commodity. And they make money off these poor people, and they stuff them in the back of 18-wheelers, and they find hovels for them to hide in. And there's a whole industry that has sprung up. And it seems like to me that since this country respects human rights and the human condition, that it would be a great contribution to eliminate this thuggery, to free these people from this kind of extortion that they go through. And one way to do so is to say, "You can come and work in our country for jobs Americans aren't doing, for a temporary period of time."

The third element of a comprehensive reform is to hold employers accountable for the workers they hire. In other words, if you want to make sure that we've got a system in which people are not violating the law, then you've got to make sure we hold people to account, like employers. Enforcing immigration is a vital part of any successful reform. And so Chertoff and his department are cracking down on employers who knowingly violate the law.

But not only are there *coyotes* smuggling people in, there are document forgers that are making a living off these people. So, in other words, people may want to comply with the law, but it's very difficult at times to verify the legal status of their employees. And so to make the worksite enforcement practical on a larger scale, we have got to issue a tamper-proof identification card for illegal—for legal foreign workers. And we must create a better system for employers to verify the legality of the workers. In other words, we got work to do. And part of a comprehensive bill is to make sure worksite enforcement is effective.

Fourth, we've got to resolve the status of millions of illegal immigrants already here in the country. People who entered our country illegally should not be given amnesty. Amnesty is the forgiveness of an offense without penalty. I oppose amnesty, and I think most people in the United States Congress oppose amnesty. People say, "Why not have amnesty?" Well, the reason why is because you—10 years from now, you don't want to have a President having to address the next 11 million people who might be here illegally. That's why you don't want amnesty. And secondly, we're a nation of law, and we expect people to uphold the law.

And so we're working closely with Republicans and Democrats to find a practical answer that lies between granting automatic citizenship to every illegal immigrant and deporting every illegal immigrant. It is impractical to take the position that, oh, we'll just find the 11 million or 12 million people and send them home. That's just an impractical position; it's not going to work. It may sound good. It may make nice sound-bite news. It won't happen.

And therefore, we need to work together to come up with a practical solution to this problem, and I know people in Congress are working hard on this issue. Illegal immigrants who have roots in our country and want to stay should have to pay a meaningful penalty for breaking the law and pay their taxes and learn the English language and show work—show that they've worked in a job for a number of years. People who meet a reasonable number of conditions and pay a penalty of time and money should be able to apply for citizenship. But approval would not be automatic, and they would have to wait in line behind those who played by the rules and followed the law. What I've described is a way for those who've broken the law to pay their debt to society and demonstrate the character that makes a good citizen.

Finally, we have got to honor the tradition of the melting pot and help people assimilate into our society by learning our history, our values, and our language. Last June, I created a new task force to look for ways to help newcomers assimilate and succeed in our country. Many organizations, from churches to businesses to civic associations, are working to answer this call, and I'm grateful for their service. . . .

It's important that we address this issue in good faith. And it's important for people to listen to everybody's positions. And it's important for people not to give up, no matter how hard it looks from a legislative perspective. It's important that we get a bill done. We deserve a system that secures our borders and honors our proud history as a nation of immigrants. . . .

SOURCE: U.S. Executive Office of the President. "Remarks on Border Security and Immigration Reform in Yuma." April 9, 2007. *Weekly Compilation of Presidential Documents* 43, no. 15 (April 16, 2007): 431–435. Washington, D.C.: National Archives and Records Administration. www.gpoaccess.gov/wcomp/v43no15.html (accessed January 8, 2008).

OTHER HISTORIC DOCUMENTS OF INTEREST

FROM PREVIOUS *HISTORIC DOCUMENTS*

- Overhaul of U.S. immigration law, *2006,* p. 230
- Census Bureau on U. S. population growth, *2006,* p. 601

DOCUMENT IN CONTEXT

Remarks and Findings on the Shootings at Virginia Tech

APRIL 17 AND AUGUST 30, 2007

Just after 7:00 a.m. on April 16, a deeply disturbed and angry student shot and killed two other students in a dormitory on the Blacksburg campus of Virginia Polytechnic Institute and State University, known informally as Virginia Tech. He then drove to the local post office, where he mailed a package of videos and writings ranting against his "oppressors." Returning to campus, he entered a classroom building, chained the doors shut, and moved from classroom to classroom, methodically shooting as many people as he could. In little more than ten minutes, Cho Seung Hui killed another thirty students and teachers and wounded seventeen before killing himself, apparently when he heard police enter the building.

It was the deadliest school shooting in the nation's history; the death toll surpassed that of the Columbine High School shooting in Littleton, Colorado, which had occurred almost exactly eight years earlier, on April 20, 1999. Although Cho's motivations were unclear, the Columbine shooting, in which two high school students killed thirteen and wounded twenty-three before killing themselves, apparently had a significant effect on Cho, who was in eighth grade at the time. The deadliest previous college campus shooting occurred on August 1, 1966, when Charles Whitman opened fire from the observation deck of a clock tower at the University of Texas in Austin, killing sixteen and wounding thirty-one before being shot and killed by police.

The horrific killings plunged the Virginia Tech campus and the country into grief and mourning. President George W. Bush visited the campus the day after the shootings to offer his and the nation's condolences to the "Hokie" community of Virginia Tech. "It's impossible to make sense of such violence and suffering," the president said at a memorial convocation in Cassell Coliseum on April 17. "Those whose lives were taken did nothing to deserve their fates. They were simply in the wrong place at the wrong time."

The slayings also raised difficult questions: Why did the campus police not alert students to the first murders sooner? How did Cho's mental instability go unrecognized even though he gave clear signs that he was a deeply troubled young man? How was Cho able to buy two handguns when a record of his mental illness should have placed him on the federal background check list that would have prohibited his buying the guns? In the aftermath of the shootings, the university, Virginia governor Tim Kaine, and the federal government set up separate review panels to investigate what went wrong and what could have been done better during and immediately after the emergency. All the panels had issued their reports by the end of August.

By the end of the year, both Virginia and Congress had taken steps to tighten requirements for placing people with certain mental health problems on the Federal Bureau of Investigation's background check list for gun purchases. But unlike the response after the Columbine shootings, there were few public demands for, and virtually no talk on Capitol Hill of, tightening federal gun control laws. Congress may have been reluctant to take action in light of the political influence of the pro-gun National Rifle Association and in the wake of a recent federal appeals court ruling that called the constitutionality of gun control laws into question. In that ruling, a federal court held for the first time that the Second Amendment right to "keep and bear arms" pertains to individuals, not just to organized militias such as the National Guard. The Supreme Court agreed in September to review the case; arguments were scheduled for March 2008.

A Student Runs Amok

The Virginia Tech slayings began shortly after 7:00 a.m., when Cho shot and killed a female student in her dorm room along with a resident adviser, who was apparently investigating the noise. Cho, who left the dormitory unseen, had no known connection to the female student or any obvious motive for attacking her. Police arrived on the scene quickly and immediately began to try to track down the girl's boyfriend, who lived off campus and was said to be an avid gun user. Meanwhile, at 9:01 a.m., Cho mailed a package of videos and a manifesto he had made in the preceding weeks to NBC News in New York. In the videos, an armed Cho delivered a rambling, incoherent message raging against his "oppressors" and alluded to the massacre he was planning. Cho then returned to campus, where witnesses later said they saw him first outside and then inside Norris Hall, an engineering building. At 9:26 a.m., two hours after the first shootings, university administrators sent e-mails to staff, faculty, and students, notifying them of the two murders at the dormitory. At about the same time, Cho was chaining shut the three main entrances to Norris Hall.

At approximately 9:40 a.m., Cho entered a classroom on the second floor and shot the professor and several students. For the next eleven minutes, Cho went from classroom to classroom, shooting as many people as he could. In some cases, he shot through doors that students and teachers were barricading. He then went back to classrooms he had already been in, shooting—sometimes multiple times—anyone who looked like he or she was still alive. Alerted by a cell phone call from students in one of the classrooms, police arrived at Norris Hall within the first five minutes of the shootings but took another five minutes to enter the building because of the chained entrances. As police reached the second floor, Cho shot himself in the head. Investigators later determined that he had fired off 174 rounds and had more than 200 rounds remaining when he died.

Because of Cho's wounds, it took some time to identify the senior English major, who had emigrated from South Korea in 1992 at the age of eight with his parents and older sister. Once his identity was determined, Cho's history of mental instability began to emerge. He had been diagnosed with "selective mutism" in 1997. In 1999, shortly after the Columbine shootings, Cho's eighth grade teachers reported suicidal and homicidal thoughts in his writings. Cho received therapy, which continued into his high school years, and presented no more behavioral problems.

Cho seemed to do well during his first two years at Virginia Tech, but in his junior year, some of his English teachers became so troubled by the violent cast of his creative writings and by his occasionally explosive temper that they warned university officials about him. He had also come to the attention of campus police on two occasions when female students accused him of harassing them. After the second encounter, in 2005, Cho sent his suitemate an instant message, saying "I might as well kill myself now." The suitemate notified the campus police, who ordered Cho to a psychiatric hospital, where he was evaluated and ordered to get follow-up treatment as an outpatient. Cho did not seek the treatment, and the Cook Counseling Center on the Virginia Tech campus never followed up with him. In 2006 another concerned teacher asked a dean about Cho, but the dean found no mention of mental health problems or police reports in Cho's records.

According to investigators' reconstruction of events leading to the shootings, Cho bought the two guns and the multiple rounds of ammunition between February 2 and April 14, 2007. On April 15 he made his usual Sunday evening call to his family in Fairfax County, Virginia; they later said Cho said nothing to cause them concern. Hours later, Cho had killed thirty-two people and himself, leaving behind great grief and anguish but very few answers about his motivations.

Follow-up Investigations

Along with the shock and grief came questions. Might at least some of the deaths have been prevented if students and faculty had been alerted sooner about the first two shootings? Why did relevant school officials seem unaware of Cho's troubled mental health history? On April 19 Governor Kaine appointed an eight-member independent review panel, led by a former state superintendent of police and including former homeland security secretary Tom Ridge, as well as experts on health, security, and education. On April 20 President Bush asked cabinet officials to meet with educators, mental health specialists, and government officials across the country to determine how the federal government could help avert such tragedies in the future. And in early May, Virginia Tech president Charles W. Steger set up three internal review panels to look at campus security, communications, and procedures for dealing with high-risk students like Cho.

The first to conclude its review was the federal panel, which issued its formal report on June 13. The panel found widespread "confusion and differing interpretations" about state and federal privacy laws that often prevented the sort of information sharing about a student's mental health condition that might make it easier to recognize when a student was in trouble. On October 30 the Department of Education released three brochures giving guidelines to school and college educators and parents on the types of information about a student that could and could not be shared among government agencies and parents. The panel also urged state and local governments to review and clarify their own privacy laws. Other recommendations involved ensuring people received necessary mental health services—an effort that was likely to require a considerable expenditure of money as well as additional personnel given the need. For example, a 2006 survey conducted by the American College Health Association found that 8.5 percent of college students had seriously considered suicide and 15 percent had been diagnosed with depression. A survey conducted in 2007 by the Anxiety Disorders Association of America found that 13 percent of students at major universities, and nearly double that percentage at liberal arts colleges, were using campus mental health services.

The three internal committees set up by Steger reported back on August 22. They called for modifications of physical infrastructure, including locks on classroom doors, to provide greater security; a fully integrated digital campus telecommunications system; and a series of steps to expand the campus health center's capacity to identify and treat students experiencing mental health problems. In October the university began to test an expanded alert system that incorporated cell phone text messages, e-mail, online instant messaging, and Internet announcements in addition to traditional warning sirens.

The Virginia Tech Review Panel was the last to weigh in. Governor Kaine released the major conclusions at a news conference on August 30. While the panel credited campus and Blacksburg police and emergency medical personnel for responding quickly to the shootings at both the dormitory and Norris Hall, it also faulted the police and university administrators for not issuing a campus-wide notification of the first two killings sooner. "It might be argued that the total toll would have been less if the university had canceled classes and announced it was closed for business immediately after the first shooting; or if the earlier alert message had been stronger and clearer," the panel's report stated. But the panel also determined that even a lockdown of the campus after the first shootings was unlikely to have stopped Cho's second rampage. "There does not seem to be a plausible scenario of university response to the double homicide that could have prevented a tragedy of considerable magnitude on April 16," the panel concluded. "Cho had started on a mission of fulfilling a fantasy of revenge."

The panel also faulted the campus and state mental health care system for failing to recognize Cho's need for help, despite "clear warnings." Although "various individuals and departments within the university" knew about separate incidents when Cho showed signs of distress, "no one knew all the information and no one connected all the dots." Misunderstandings about federal privacy prohibitions on sharing information prevented connection of the dots, while lack of resources, concerns about privacy, and "passivity" contributed to the failure of the university health team and counseling center to provide Cho with needed services. Beyond the campus, the report said, Virginia's mental health laws were "flawed" and mental health services "inadequate," resulting in "gaps in the mental health system including short term crisis stabilization and comprehensive outpatient services."

Although Cho was ineligible to purchase guns because he had been judged a danger to himself and ordered into treatment, that information never found its way into the federal database used to conduct background checks on would-be gun purchasers. Kaine issued an executive order on April 30 requiring that Virginia report all people involuntarily committed for outpatient mental health treatment to the federal database; the state legislature codified that requirement later in the year.

Kaine and members of the panel appeared to be most struck by the failures identified in the mental health system. "I think this is a huge issue," Kaine told the *Washington Post* on August 31. "I know it's not just a problem in Virginia. It's a problem elsewhere.... We need to fix this issue of follow-up [when someone has come in for or been ordered into treatment]. We need accountability." Kaine expected to make recommendations for improvements to both state and federal officials after reviewing the panel's report in more detail.

The report did not satisfy some parents and others, who angrily called for university president Steger's resignation. "It's hard to believe anybody could read that report and not find that people should be held accountable for the many, many mistakes that were identified," one family member said. But at the August 30 news conference, Kaine said he was satisfied with the report, calling it "comprehensive and thorough, objective and in many instances hard-hitting." He saw no point in firing anyone, suggesting that Steger and others had already suffered enough. "This is not something where the university officials, faculty, administrators have just been very blithe. There has been deep grieving about this, and it's torn the campus up," Kaine said, adding that he wanted to focus his energies on fixing the problems the report identified to "reduce the chance of anything like this every happening again."

Following are two documents. First, the text of President George W. Bush's remarks on April 17, 2007, at the Virginia Tech Memorial Convocation for the victims of a mass shooting on the campus the day before; Bush delivered his remarks at Cassell Coliseum at the Virginia Tech campus in Blacksburg. Second, the "Summary of Key Findings" issued on August 30, 2007, by the Virginia Tech Review Panel, an independent commission appointed by Virginia governor Tim Kaine to determine what went right and what went wrong in connection with the shootings.

DOCUMENT

President Bush's Remarks at Virginia Tech Memorial Convocation

April 17, 2007

Governor [Tim Kaine], thank you. [Virginia Tech] President [Charles] Steger, thank you very much. Students, and faculty, and staff, and grieving family members, and members of this really extraordinary place.

Laura and I have come to Blacksburg today with hearts full of sorrow. This is a day of mourning for the Virginia Tech community—and it is a day of sadness for our entire nation. We've come to express our sympathy. In this time of anguish, I hope you know that people all over this country are thinking about you, and asking God to provide comfort for all who have been affected.

Yesterday began like any other day. Students woke up, and they grabbed their backpacks and they headed for class. And soon the day took a dark turn, with students and faculty barricading themselves in classrooms and dormitories—confused, terrified, and deeply worried. By the end of the morning, it was the worst day of violence on a college campus in American history—and for many of you here today, it was the worst day of your lives.

It's impossible to make sense of such violence and suffering. Those whose lives were taken did nothing to deserve their fate. They were simply in the wrong place at the wrong time. Now they're gone—and they leave behind grieving families, and grieving classmates, and a grieving nation.

In such times as this, we look for sources of strength to sustain us. And in this moment of loss, you're finding these sources everywhere around you. These sources of strength are in this community, this college community. You have a compassionate and resilient community here at Virginia Tech. Even as yesterday's events were still unfolding, members of this community found each other; you came together in dorm rooms and dining halls and on blogs. One recent graduate wrote this: "I don't know most of you guys, but we're all Hokies, which means we're family. To all of you who are okay, I'm happy for that. For those of you who are in pain or have lost someone close to you, I'm sure you can call on anyone of us and have help any time you need it."

These sources of strength are with your loved ones. For many of you, your first instinct was to call home and let your moms and dads know that you were okay. Others took on the terrible duty of calling the relatives of a classmate or a colleague who had been wounded or lost. I know many of you feel awfully far away from people you lean on and people you count on during difficult times. But as a dad, I can assure you, a parent's love is never far from their child's heart. And as you draw closer to your own families in the coming days, I ask you to reach out to those who ache for sons and daughters who will never come home.

These sources of strength are also in the faith that sustains so many of us. Across the town of Blacksburg and in towns all across America, houses of worship from every faith have opened their doors and have lifted you up in prayer. People who have never met you are praying for you; they're praying for your friends who have fallen and who are injured. There's a power in these prayers, real power. In times like this, we can find comfort in the grace and guidance of a loving God. As the Scriptures tell us, "Don't be overcome by evil, but overcome evil with good."

And on this terrible day of mourning, it's hard to imagine that a time will come when life at Virginia Tech will return to normal. But such a day will come. And when it does, you will always remember the friends and teachers who were lost yesterday, and the time you shared with them, and the lives they hoped to lead. May God bless you. May God bless and keep the souls of the lost. And may His love touch all those who suffer and grieve.

Source: U.S. Executive Office of the President. Office of the Press Secretary. "President Bush Offers Condolences at Virginia Tech Memorial Convocation." Blacksburg, Virginia. April 17, 2007. www.white-house.gov/news/releases/2007/04/20070417-1.html (accessed November 20, 2007).

DOCUMENT

Summary of Key Findings of the Virginia Tech Review Panel

August 30, 2007

Summary of Key Findings

On April 16, 2007, Seung Hui Cho, an angry and disturbed student, shot to death 32 students and faculty of Virginia Tech, wounded 17 more, and then killed himself.

The incident horrified not only Virginians, but people across the United States and throughout the world.

Tim Kaine, Governor of the Commonwealth of Virginia, immediately appointed a panel to review the events leading up to this tragedy; the handling of the incidents by public safety officials, emergency services providers, and the university; and the services subsequently provided to families, survivors, care-givers, and the community.

The Virginia Tech Review Panel reviewed several separate but related issues in assessing events leading to the mass shootings and their aftermath:

- The life and mental health history of Seung Hui Cho, from early childhood until the weeks before April 16.
- Federal and state laws concerning the privacy of health and education records.
- Cho's purchase of guns and related gun control issues.
- The double homicide at West Ambler Johnston (WAJ) residence hall and the mass shootings at Norris Hall, including the responses of Virginia Tech leadership and the actions of law enforcement officers and emergency responders.
- Emergency medical care immediately following the shootings, both onsite at Virginia Tech and in cooperating hospitals.
- The work of the Office of the Chief Medical Examiner of Virginia.
- The services provided for surviving victims of the shootings and others injured, the families and loved ones of those killed and injured, members of the university community, and caregivers.

The panel conducted over 200 interviews and reviewed thousands of pages of records, and reports the following major findings:

1. Cho exhibited signs of mental health problems during his childhood. His middle and high schools responded well to these signs and, with his parents' involvement, provided services to address his issues. He also received private psychiatric treatment and counseling for selective mutism and depression.

   In 1999, after the Columbine shootings, Cho's middle school teachers observed suicidal and homicidal ideations in his writings and recommended psychiatric counseling, which he received. It was at this point that he received medication for a short time. Although Cho's parents were aware that he was troubled at this time, they state they did not specifically know that he thought about homicide shortly after the 1999 Columbine school shootings.
2. During Cho's junior year at Virginia Tech, numerous incidents occurred that were clear warnings of mental instability. Although various individuals and departments within the university knew about each of these incidents, the university did not intervene effectively. No one knew all the information and no one connected all the dots.
3. University officials in the office of Judicial Affairs, Cook Counseling Center, campus police, the Dean of Students, and others explained their failures to communicate with one another or with Cho's parents by noting their belief that such communications are prohibited by the federal laws governing the privacy of health and education records. In reality, federal laws and their state counterparts afford ample leeway to share information in potentially dangerous situations.
4. The Cook Counseling Center and the university's Care Team failed to provide needed support and services to Cho during a period in late 2005 and early 2006. The system failed for lack of resources, incorrect interpretation of privacy laws, and passivity. Records of Cho's minimal treatment at Virginia Tech's Cook Counseling Center are missing.
5. Virginia's mental health laws are flawed and services for mental health users are inadequate. Lack of sufficient resources results in gaps in the mental health system

including short term crisis stabilization and comprehensive outpatient services. The involuntary commitment process is challenged by unrealistic time constraints, lack of critical psychiatric data and collateral information, and barriers (perceived or real) to open communications among key professionals.

6. There is widespread confusion about what federal and state privacy laws allow. Also, the federal laws governing records of health care provided in educational settings are not entirely compatible with those governing other health records.
7. Cho purchased two guns in violation of federal law. The fact that in 2005 Cho had been judged to be a danger to himself and ordered to outpatient treatment made him ineligible to purchase a gun under federal law.
8. Virginia is one of only 22 states that report any information about mental health to a federal database used to conduct background checks on would-be gun purchasers. But Virginia law did not clearly require that persons such as Cho—who had been ordered into out-patient treatment but not committed to an institution—be reported to the database. Governor Kaine's executive order to report all persons involuntarily committed for outpatient treatment has temporarily addressed this ambiguity in state law. But a change is needed in the Code of Virginia as well.
9. Some Virginia colleges and universities are uncertain about what they are permitted to do regarding the possession of firearms on campus.
10. On April 16, 2007, the Virginia Tech and Blacksburg police departments responded quickly to the report of shootings at West Ambler Johnston residence hall, as did the Virginia Tech and Blacksburg rescue squads. Their responses were well coordinated.
11. The Virginia Tech police may have erred in prematurely concluding that their initial lead in the double homicide was a good one, or at least in conveying that impression to university officials while continuing their investigation. They did not take sufficient action to deal with what might happen if the initial lead proved erroneous. The police reported to the university emergency Policy Group that the "person of interest" probably was no longer on campus.
12. The VTPD erred in not requesting that the Policy Group issue a campus-wide notification that two persons had been killed and that all students and staff should be cautious and alert.
13. Senior university administrators, acting as the emergency Policy Group, failed to issue an all-campus notification about the WAJ killings until almost 2 hours had elapsed. University practice may have conflicted with written policies.
14. The presence of large numbers of police at WAJ led to a rapid response to the first 9-1-1 call that shooting had begun at Norris Hall.
15. Cho's motives for the WAJ or Norris Hall shootings are unknown to the police or the panel. Cho's writings and videotaped pronouncements do not explain why he struck when and where he did.
16. The police response at Norris Hall was prompt and effective, as was triage and evacuation of the wounded. Evacuation of others in the building could have been implemented with more care.
17. Emergency medical care immediately following the shootings was provided very effectively and timely both onsite and at the hospitals, although providers from different agencies had some difficulty communicating with one another. Communication of accurate information to hospitals standing by to receive the

wounded and injured was somewhat deficient early on. An emergency operations center at Virginia Tech could have improved communications.

18. The Office of the Chief Medical Examiner properly discharged the technical aspects of its responsibility (primarily autopsies and identification of the deceased). Communication with families was poorly handled.
19. State systems for rapidly deploying trained professional staff to help families get information, crisis intervention, and referrals to a wide range of resources did not work.
20. The university established a family assistance center at The Inn at Virginia Tech, but it fell short in helping families and others for two reasons: lack of leadership and lack of coordination among service providers. University volunteers stepped in but were not trained or able to answer any questions and guide families to the resources they needed.
21. In order to advance public safety and meet public needs, Virginia's colleges and universities need to work together as a coordinated system of state-supported institutions.

As reflected in the body of the report, the panel has made more than 70 recommendations directed to colleges, universities, mental health providers, law enforcement officials, emergency service providers, law makers, and other public officials in Virginia and elsewhere.

SOURCE: Commonwealth of Virginia. Official Site of the Governor of Virginia. "Report of the Virginia Tech Review Panel." August 30, 2007. www.governor.virginia.gov/TempContent/techPanelReport.cfm (accessed November 20, 2007).

OTHER HISTORIC DOCUMENTS OF INTEREST

FROM THIS VOLUME

- Federal appeals court on Washington, D.C., handgun ban, p. 101

FROM PREVIOUS *HISTORIC DOCUMENTS*

- Columbine shootings, *2001*, p. 347; *1999*, p. 179

DOCUMENT IN CONTEXT

Supreme Court on the "Partial-Birth" Abortion Ban

APRIL 18, 2007

On a day long awaited by abortion opponents, the Supreme Court on April 18 upheld a federal ban on so-called partial-birth abortion. For the first time since its 1973 decision in *Roe v. Wade,* recognizing a woman's constitutional right to obtain an abortion, the Court upheld a ban on a specific abortion procedure. The 5–4 decision was also the first in which the Court approved a restriction on abortion that did not include an exception to protect the woman's health (although the legislation did include an exception to protect the life of the mother). The ruling was thus widely seen as shifting the weight of the abortion debate toward opponents of the practice and as clearing the way for states and Congress to place more restrictions on abortion.

The pivotal vote was cast by the newest justice, Samuel A. Alito Jr., who voted with the majority to uphold the controversial ban. In 2000 his predecessor, Justice Sandra Day O'Connor, had voted to strike down a similar state law, in part because it did not contain an exception to protect the health of the woman. Alito, nominated to the Court by President George W. Bush, had refused to state his position on abortion during his confirmation hearing in January 2006—as is typical for nominees—but his earlier writings and decisions as a federal appellate court judge strongly indicated his opposition to the practice. With the possibility that the next nominee to the Court could represent the deciding vote to uphold or overturn *Roe v. Wade,* activists on both sides of the issue were carefully following both parties' 2008 presidential contenders.

Background: A State Ban Ruled Unconstitutional

The term *partial-birth abortion* refers to a procedure generally described in the medical terminology as intact dilation and extraction (intact D & E). In this procedure, used in the second trimester of pregnancy, the woman's cervix is dilated and the body of the fetus brought intact by instruments from the uterus into the vagina. Because the head is too large to pass through the cervix, the doctor pierces the fetal skull, suctions out the contents, and crushes or collapses the skull to complete removal of the fetus. The procedure is a variant of a more common second-trimester procedure known as dilation and evacuation, in which the fetus is dismembered as it is brought through the cervical opening.

No reliable records have been kept on the number of such abortions performed each year, but by most estimates the number is fewer than 5,000. Between 85 and 90 percent of the 1.3 million annual abortions are performed in the first trimester, when the fetus is typically suctioned from the uterus. Abortions are rare in the third trimester, when a fetus is generally capable of living on its own.

Anti-abortion activists had been seeking a ban on intact D&E since the mid-1990s, when it became apparent that the Court as then constituted was unlikely to overturn *Roe v. Wade* outright. Using graphic pictures of nearly intact fetuses with their brains spilling out of collapsed skulls, anti-abortion organizations called the procedure "partial-birth" abortion, likened it to infanticide, and mounted lobbying campaigns asking state legislatures and the U.S. Congress to ban the operation. Congress twice passed such bans, in 1996 and 1997, but failed both times to override President Bill Clinton's veto. Thirty states also passed laws banning this type of abortion; a challenge to the constitutionality of Nebraska's law came before the Supreme Court in 2000.

In that case, *Stenberg v. Carhart,* the Court struck down the Nebraska law by a 5–4 vote. The majority held that the law was unconstitutional both because it was too broad, in that it did not distinguish clearly between the intact D & E procedure and the more common dilation and evacuation operation, and because it did not contain an exception to protect the woman's health.

Upholding a Federal Ban

In 2003 Congress again passed a measure banning intact D & E, and this time President George W. Bush signed it into law. Like the Nebraska law, the legislation banned the procedure in all instances except to protect the life of the woman, and it did not contain an exception to protect the woman's health. Instead it relied on congressional findings showing that the procedure was never medically necessary. Under the law, doctors who performed the banned procedure would be subject to fines and up to two years in prison.

Less than an hour after Bush signed the measure on November 5, 2003, a federal judge in Nebraska issued a temporary restraining order prohibiting its enforcement. Federal judges in New York and California issued similar orders the next day, effectively blocking the ban from taking effect pending the outcome of lawsuits challenging its constitutionality. In all three states, the district judges ruled that the federal ban was unconstitutional, and those rulings were later affirmed by appeals courts.

In February 2006 the Supreme Court agreed to hear the Nebraska case, *Gonzales v. Carhart.* The case came from the Eight Circuit Court of Appeals, which ruled that the failure to include a health exception for the woman invalidated the entire law. In June the Court agreed to expand its review, adding a second case, *Gonzales v. Planned Parenthood.* In this case the Ninth Circuit Court of Appeals ruled that the ban violated due process because it was so vague that it could have the effect of criminalizing the dilation and evacuation procedure as well as the intact D & E procedure.

Writing for the majority, Justice Anthony M. Kennedy rejected arguments that the federal law is unconstitutionally vague, saying that unlike the Nebraska statute, the federal ban not only describes the precise procedure it outlaws but applies only to those doctors who deliberately and intentionally perform the procedure. On the failure of the federal ban to create a health exception, the majority accepted Congress's finding that there is "never" a medical need for the controversial procedure and noted the availability of alternative, safe procedures. "The medical uncertainty over whether the Act's prohibition creates significant health risks provides a sufficient basis to conclude . . . that the Act does not impose an undue burden," Kennedy said, adding that "the government may use its voice and regulatory authority to show its profound respect for the life within the woman." Moreover, Kennedy said, a woman or her doctor may challenge the health exception on an individual basis.

Perhaps more notable than the ruling itself was the tone of Kennedy's opinion. Kennedy clearly showed his abhorrence for late-term abortion by describing both the banned method and the permitted dilation and evacuation procedure in minute and graphic detail, as well as by referring to fetuses as unborn children and babies and to the doctors who perform abortions as "abortion doctors" rather than by their medical specialties. Speaking through Kennedy, the majority also argued that the federal ban protects the moral and emotional health of women, a controversial stance seen as patronizing by many women and abortion rights supporters. "Respect for human life finds an ultimate expression in the bond of love the mother has for her child," Kennedy wrote. He acknowledged the lack of conclusive data showing that abortion in general or the banned method in particular harms women's emotional and mental health. Nonetheless, he wrote, "it seems unexceptional to conclude some women come to regret their choice to abort the infant life they once created and sustained."

Joining Kennedy and Alito in the majority were Chief Justice John G. Roberts Jr. and Justices Antonin Scalia and Clarence Thomas. Thomas and Scalia also filed a separate opinion stating their long-expressed belief that Roe "has no basis in the Constitution."

Writing for the four dissenters, Justice Ruth Bader Ginsburg called the majority opinion "alarming," its "hostility" to the right guaranteed by Roe unconcealed, and its reasoning "bewildering," "flimsy," and "transparent." Ginsburg rejected the majority's claim that the federal ban would advance a legitimate and substantial government interest in protecting fetal life, arguing that the ban targets only a method for performing abortion. "In short," she wrote, "the Court upholds a law that while doing nothing to 'preserve fetal life,' bars a woman from choosing" the medical procedure that offers the best protection of her health. Leaving little doubt that she thought the ruling resulted solely from a change in the Court's composition, Ginsburg said the decision "cannot be understood as anything other than an effort to chip away at a right declared again and again by this court—and with increasing comprehension of its centrality to women's lives." Joining Ginsburg in dissent were Justices John Paul Stevens, David H. Souter, and Stephen G. Breyer

Reaction

In a statement issued later that day, President Bush said the ruling "affirms that the Constitution does not stand in the way of the people's representatives enacting laws reflecting the compassion and humanity of America." He added that the decision also attested to the "progress we have made over the past six years in protecting human dignity and upholding the sanctity of life."

Anti-abortion activists were overjoyed. "This is the most monumental win on the abortion issue that we have ever had," said Jay Sekulow, chief counsel for the American Center for Law and Justice. Roberta Combs, president of the Christian Coalition of America, predicted that "it is just a matter of time before the infamous *Roe v. Wade* . . . will also be struck down by the court." Pro-choice advocates were concerned that Combs might be right. "The impact of Sandra Day O'Connor's retirement is painfully clear," said Nancy Northrup, president of the Center for Reproductive Rights. "It took just a year for this new Court to overturn three decades of established constitutional law." Cecile Richards, president of the Planned Parenthood Federation of America, concurred: "Until this decision, I think a lot of people were skeptical about whether *Roe* could be overturned. But there clearly is no longer a presumption that women's health will be protected by the courts."

Some doctors and medical organizations expressed dismay that the Court and Congress would tell them how to practice medicine. The American College of Obstetricians and Gynecologists called the decision "shameful and incomprehensible" and said it denied the use of the "safest" procedure for women who suffer from certain conditions that make the more common dilation and evacuation procedure "especially dangerous." Some physicians interviewed by news media said the decision left abortion providers in legal jeopardy, but others indicated most legal difficulties could be avoided by injecting the fetus with a drug that stops its heart before performing the abortion. The ruling "will not stop any abortions from taking place," Dr. Nancy Stanwood, an obstetrics professor at the University of Rochester who delivered babies and performed abortions, told the *Los Angeles Times*. "We physicians will make some slight changes in our practice. An injection for the fetus adds another risk to woman's health [if the injection misses its target], and it means added time and money," she said. "But if that's what's necessary, that's what we'll do."

Virtually all sides in the debate agreed that the decision would prompt anti-abortion activists to lobby state legislatures and Congress for a variety of additional restrictions, including a ban on abortions of viable fetuses unless the woman's life is endangered and a ban on mid- and late-term abortions for fetal abnormalities such as Down's syndrome or brain malformations. If adopted, such restrictions would be almost certain to be challenged. Measures requiring doctors to explain abortion procedures in graphic detail and to extend waiting periods were also expected to be introduced in those states that did not already have such laws. In addition, some states were already moving to impose tighter regulations on clinics providing abortions, with the clear intention of making it too expensive and burdensome for them to remain in business.

Following are excerpts from the majority and minority opinions in the cases of Gonzales v. Carhart *and* Gonzales v. Planned Parenthood Federation of America, *in which the Supreme Court upheld, on a 5–4 vote announced April 18, 2007, the constitutionality of a federal ban barring a medical procedure often called "partial-birth" abortion.*

DOCUMENT

Gonzales v. Carhart

Nos. 05–380 and 05–1382

Alberto R. Gonzales, Attorney General, Petitioner *v.* Leroy Carhart et al.	On writ of certiorari to the United States Court of Appeals for the Eighth Circuit
Alberto R. Gonzales, Attorney General, Petitioner *v.* Planned Parenthood Federation of America, Inc., et al.	On writ of certiorari to the United States Court of Appeals for the Ninth Circuit

[April 18, 2007]

JUSTICE KENNEDY delivered the opinion of the Court.

These cases require us to consider the validity of the Partial-Birth Abortion Ban Act of 2003 (Act), 18 U. S. C. §1531 (2000 ed., Supp. IV), a federal statute regulating abortion procedures. In recitations preceding its operative provisions the Act refers to the Court's opinion in *Stenberg v. Carhart* . . . (2000), which also addressed the subject of abortion procedures used in the later stages of pregnancy. Compared to the state statute at issue in *Stenberg,* the Act is more specific concerning the instances to which it applies and in this respect more precise in its coverage. We conclude the Act should be sustained against the objections lodged by the broad, facial attack brought against it. . . .

I

[Omitted. This section describes the abortion procedures prohibited by the federal legislation and the findings of various appeals courts on the constitutionality of the law.]

II

The principles set forth in the joint opinion in *Planned Parenthood of Southeastern Pa. v. Casey* . . . (1992) did not find support from all those who join the instant opinion. . . . Whatever one's views concerning the *Casey* joint opinion, it is evident a premise central to its conclusion—that the government has a legitimate and substantial interest in preserving and promoting fetal life—would be repudiated were the Court now to affirm the judgments of the Courts of Appeals.

Casey involved a challenge to *Roe v. Wade* . . . (1973). The opinion contains this summary:

> "It must be stated at the outset and with clarity that *Roe*'s essential holding, the holding we reaffirm, has three parts. First is a recognition of the right of the woman to choose to have an abortion before viability and to obtain it without undue interference from the State. Before viability, the State's interests are not strong enough to support a prohibition of abortion or the imposition of a substantial obstacle to the woman's effective right to elect the procedure. Second is a confirmation of the State's power to restrict abortions after fetal viability, if the law contains exceptions for pregnancies which endanger the woman's life or health. And third is the principle that the State has legitimate interests from the outset of the pregnancy in protecting the health of the woman and the life of the fetus that may become a child. These principles do not contradict one another; and we adhere to each." . . .

Though all three holdings are implicated in the instant cases, it is the third that requires the most extended discussion; for we must determine whether the Act furthers the legitimate interest of the Government in protecting the life of the fetus that may become a child.

To implement its holding, *Casey* rejected both *Roe*'s rigid trimester framework and the interpretation of *Roe* that considered all previability regulations of abortion unwarranted. . . . On this point *Casey* overruled the holdings in two cases because they undervalued the State's interest in potential life. . . .

We assume the following principles for the purposes of this opinion. Before viability, a State "may not prohibit any woman from making the ultimate decision to terminate her pregnancy." . . . It also may not impose upon this right an undue burden, which exists if a regulation's "purpose or effect is to place a substantial obstacle in the path of a woman seeking an abortion before the fetus attains viability." . . . On the other hand, "[r]egulations which do no more than create a structural mechanism by which the State, or the parent or guardian of a minor, may express profound respect for the life of the unborn are permitted, if they are not a substantial obstacle to the woman's exercise of the right to choose." . . .

Casey, in short, struck a balance. The balance was central to its holding. We now apply its standard to the cases at bar.

III

We begin with a determination of the Act's operation and effect. A straightforward reading of the Act's text demonstrates its purpose and the scope of its provisions: It regulates and proscribes, with exceptions or qualifications to be discussed, performing the intact D&E procedure.

Respondents agree the Act encompasses intact D&E, but they contend its additional reach is both unclear and excessive. Respondents assert that, at the least, the Act is void for vagueness because its scope is indefinite. In the alternative, respondents argue the Act's text proscribes all D&Es. Because D&E is the most common second-trimester abortion method, respondents suggest the Act imposes an undue burden. In this litigation the Attorney General does not dispute that the Act would impose an undue burden if it covered standard D&E.

We conclude that the Act is not void for vagueness, does not impose an undue burden from any overbreadth, and is not invalid on its face.

A

The Act punishes "knowingly perform[ing]" a "partial birth abortion." . . . It defines the unlawful abortion in explicit terms. . . .

First, the person performing the abortion must "vaginally delive[r] a living fetus." . . . The Act does not restrict an abortion procedure involving the delivery of an expired fetus. The Act, furthermore, is inapplicable to abortions that do not involve vaginal delivery (for instance, hysterotomy or hysterectomy). The Act does apply both previability and postviability because, by common understanding and scientific terminology, a fetus is a living organism while within the womb, whether or not it is viable outside the womb. . . . We do not understand this point to be contested by the parties.

Second, the Act's definition of partial-birth abortion requires the fetus to be delivered "until, in the case of a head-first presentation, the entire fetal head is outside the body of the mother, or, in the case of breech presentation, any part of the fetal trunk past the navel is outside the body of the mother." . . . The Attorney General concedes, and we agree, that if an abortion procedure does not involve the delivery of a living fetus to one of these "anatomical 'landmarks' " . . . —the prohibitions of the Act do not apply. . . .

Third, to fall within the Act, a doctor must perform an "overt act, other than completion of delivery, that kills the partially delivered living fetus." . . . For purposes of criminal liability, the overt act causing the fetus' death must be separate from delivery. And the overt act must occur after the delivery to an anatomical landmark. This is because the Act proscribes killing "the partially delivered" fetus, which, when read in context, refers to a fetus that has been delivered to an anatomical landmark. . . .

Fourth, the Act contains scienter requirements concerning all the actions involved in the prohibited abortion. To begin with, the physician must have "deliberately and intentionally" delivered the fetus to one of the Act's anatomical landmarks. . . . If a living fetus is delivered past the critical point by accident or inadvertence, the Act is inapplicable. In addition, the fetus must have been delivered "for the purpose of performing an overt act that the [doctor] knows will kill [it]." . . . If either intent is absent, no crime has occurred. This follows from the general principle that where scienter is required no crime is committed absent the requisite state of mind. . . .

B

Respondents contend the language described above is indeterminate, and they thus argue the Act is unconstitutionally vague on its face. "As generally stated, the void-for-vagueness doctrine requires that a penal statute define the criminal offense with sufficient definiteness that ordinary people can understand what conduct is prohibited and in a manner that does not encourage arbitrary and discriminatory enforcement." . . . The Act satisfies both requirements. . . .

C

We next determine whether the Act imposes an undue burden, as a facial matter, because its restrictions on second-trimester abortions are too broad. A review of the statutory text discloses the limits of its reach. The Act prohibits intact D&E; and, notwithstanding respondents' arguments, it does not prohibit the D&E procedure in which the fetus is removed in parts.

1

The Act prohibits a doctor from intentionally performing an intact D&E. The dual prohibitions of the Act, both of which are necessary for criminal liability, correspond with the steps generally undertaken during this type of procedure. First, a doctor delivers the fetus until its head lodges in the cervix, which is usually past the anatomical landmark for a breech presentation. . . . Second, the doctor proceeds to pierce the fetal skull with scissors or crush it with forceps. This step satisfies the overt-act requirement because it kills the fetus and is distinct from delivery. . . . The Act's intent requirements, however, limit its reach to those physicians who carry out the intact D&E after intending to undertake both steps at the outset.

The Act excludes most D&Es in which the fetus is removed in pieces, not intact. If the doctor intends to remove the fetus in parts from the outset, the doctor will not have the requisite intent to incur criminal liability. A doctor performing a standard D&E procedure can often "tak[e] about 10–15 'passes' through the uterus to remove the entire fetus." . . . Removing the fetus in this manner does not violate the Act because the doctor will not have delivered the living fetus to one of the anatomical landmarks or committed an additional overt act that kills the fetus after partial delivery. . . .

[Justice Kennedy then set out several ways in which he said the federal law differed from the Nebraska statute.]

The canon of constitutional avoidance, finally, extinguishes any lingering doubt as to whether the Act covers the prototypical D&E procedure. " '[T]he elementary rule is that every reasonable construction must be resorted to, in order to save a statute from unconstitutionality.' " . . . It is true this longstanding maxim of statutory interpretation has, in the past, fallen by the wayside when the Court confronted a statute regulating abortion. . . . *Casey* put this novel statutory approach to rest. . . . *Stenberg* need not be interpreted to have revived it. We read that decision instead to stand for the uncontroversial proposition that the canon of constitutional avoidance does not apply if a statute is not "genuinely susceptible to two constructions." . . . In *Stenberg* the Court found the statute covered D&E. . . . Here, by contrast, interpreting the Act so that it does not prohibit standard D&E is the most reasonable reading and understanding of its terms.

2

Contrary arguments by the respondents are unavailing. Respondents look to situations that might arise during D&E, situations not examined in *Stenberg*. They contend—relying on the testimony of numerous abortion doctors—that D&E may result in the delivery of a living fetus beyond the Act's anatomical landmarks in a significant fraction of cases. This is so, respondents say, because doctors cannot predict the amount the cervix will dilate before the abortion procedure. It might dilate to a degree that the fetus will be removed largely intact. To complete the abortion, doctors will commit an overt act that kills the partially delivered fetus. Respondents thus posit that any D&E has the potential to violate the Act, and that a physician will not know beforehand whether the abortion will proceed in a prohibited manner. . . .

This reasoning, however, does not take account of the Act's intent requirements, which preclude liability from attaching to an accidental intact D&E. If a doctor's intent at the outset is to perform a D&E in which the fetus would not be delivered to either of the Act's anatomical landmarks, but the fetus nonetheless is delivered past one of those points, the requisite and prohibited scienter is not present. . . . When a doctor in that situation completes an abortion by performing an intact D&E, the doctor does not violate the Act. It is true that intent to cause a result may sometimes be inferred if a person "knows that that result is practically certain to follow from his conduct." . . . Yet abortion doctors intending at the outset to perform a standard D&E procedure will not know that a prohibited abortion "is practically certain to follow from" their conduct. . . . A fetus is only delivered largely intact in a small fraction of the overall number of D&E abortions. . . .

The evidence also supports a legislative determination that an intact delivery is almost always a conscious choice rather than a happenstance. . . .

Many doctors who testified on behalf of respondents, and who objected to the Act, do not perform an intact D&E by accident. On the contrary, they begin every D&E abortion with the objective of removing the fetus as intact as possible. . . . This does not prove, as respondents suggest, that every D&E might violate the Act and that the Act therefore imposes an undue burden. It demonstrates only that those doctors who intend to perform a D&E that would involve delivery of a living fetus to one of the Act's anatomical landmarks must adjust their conduct to the law by not attempting to deliver the fetus to either of those points. Respondents have not shown that requiring doctors to intend dismemberment before delivery to an anatomical landmark will prohibit the vast majority of D&E abortions. The Act, then, cannot be held invalid on its face on these grounds.

IV

Under the principles accepted as controlling here, the Act, as we have interpreted it, would be unconstitutional "if its purpose or effect is to place a substantial obstacle in the path of a woman seeking an abortion before the fetus attains viability." . . . The abortions affected by the Act's regulations take place both previability and postviability; so the quoted language and the undue burden analysis it relies upon are applicable. The question is whether the Act, measured by its text in this facial attack, imposes a substantial obstacle to late-term, but previability, abortions. The Act does not on its face impose a substantial obstacle, and we reject this further facial challenge to its validity.

A

The Act's purposes are set forth in recitals preceding its operative provisions. A description of the prohibited abortion procedure demonstrates the rationale for the congressional enactment. The Act proscribes a method of abortion in which a fetus is killed just inches before completion of the birth process. Congress stated as follows: "Implicitly approving such a brutal and inhumane procedure by choosing not to prohibit it will further coarsen society to the humanity of not only newborns, but all vulnerable and innocent human life, making it increasingly difficult to protect such life." . . . The Act expresses respect for the dignity of human life.

Congress was concerned, furthermore, with the effects on the medical community and on its reputation caused by the practice of partial-birth abortion. The findings in the Act explain:

> "Partial-birth abortion . . . confuses the medical, legal, and ethical duties of physicians to preserve and promote life, as the physician acts directly against the physical life of a child, whom he or she had just delivered, all but the head, out of the womb, in order to end that life." . . .

There can be no doubt the government "has an interest in protecting the integrity and ethics of the medical profession." . . . Under our precedents it is clear the State has a significant role to play in regulating the medical profession.

Casey reaffirmed these governmental objectives. The government may use its voice and its regulatory authority to show its profound respect for the life within the woman. A central premise of the opinion was that the Court's precedents after *Roe* had "undervalue[d] the State's interest in potential life." . . . The plurality opinion indicated "[t]he fact that a law which serves a valid purpose, one not designed to strike at the right itself, has the incidental effect of making it more difficult or more expensive to procure an abortion cannot be enough to invalidate it." . . . This was not an idle assertion. The three premises of *Casey* must coexist. . . . The third premise, that the State, from the inception of the pregnancy, maintains its own regulatory interest in protecting the life of the fetus that may become a child, cannot be set at naught by interpreting *Casey*'s requirement of a health exception so it becomes tantamount to allowing a doctor to choose the abortion method he or she might prefer. Where it has a rational basis to act, and it does not impose an undue burden, the State may use its regulatory power to bar certain procedures and substitute others, all in furtherance of its legitimate interests in regulating the medical profession in order to promote respect for life, including life of the unborn.

The Act's ban on abortions that involve partial delivery of a living fetus furthers the Government's objectives. No one would dispute that, for many, D&E is a procedure itself laden with the power to devalue human life. Congress could nonetheless conclude that the type of abortion proscribed by the Act requires specific regulation because it implicates additional ethical and moral concerns that justify a special prohibition. Congress determined that the abortion methods it proscribed had a "disturbing similarity to the killing of a newborn infant," . . . and thus it was concerned with "draw[ing] a bright line that clearly distinguishes abortion and infanticide." . . . The Court has in the past confirmed the validity of drawing boundaries to prevent certain practices that extinguish life and are close to actions that are condemned. . . .

Respect for human life finds an ultimate expression in the bond of love the mother has for her child. The Act recognizes this reality as well. Whether to have an abortion requires a difficult and painful moral decision. . . . While we find no reliable data to measure the phenomenon, it seems unexceptionable to conclude some women come to regret their choice to abort the infant life they once created and sustained. . . . Severe depression and loss of esteem can follow. . . .

In a decision so fraught with emotional consequence some doctors may prefer not to disclose precise details of the means that will be used, confining themselves to the required statement of risks the procedure entails. From one standpoint this ought not to be surprising. Any number of patients facing imminent surgical procedures would prefer not to hear all details, lest the usual anxiety preceding invasive medical procedures become the more intense. This is likely the case with the abortion procedures here in issue. . . .

It is, however, precisely this lack of information concerning the way in which the fetus will be killed that is of legitimate concern to the State. . . . The State has an interest in ensuring so grave a choice is well informed. It is self-evident that a mother who comes to regret her choice to abort must struggle with grief more anguished and sorrow more profound when she learns, only after the event, what she once did not know: that she allowed a doctor to pierce the skull and vacuum the fast-developing brain of her unborn child, a child assuming the human form.

It is a reasonable inference that a necessary effect of the regulation and the knowledge it conveys will be to encourage some women to carry the infant to full term, thus reducing the absolute number of late-term abortions. The medical profession, furthermore, may find different and less shocking methods to abort the fetus in the second trimester, thereby accommodating legislative demand. The State's interest in respect for life is advanced by the dialogue that better informs the political and legal systems, the medical profession, expectant mothers, and society as a whole of the consequences that follow from a decision to elect a late-term abortion.

It is objected that the standard D&E is in some respects as brutal, if not more, than the intact D&E, so that the legislation accomplishes little. What we have already said, however, shows ample justification for the regulation. Partial-birth abortion, as defined by the Act, differs from a standard D&E because the former occurs when the fetus is partially outside the mother to the point of one of the Act's anatomical landmarks. It was reasonable for Congress to think that partial-birth abortion, more than standard D&E, "undermines the public's perception of the appropriate role of a physician during the delivery process, and perverts a process during which life is brought into the world." . . . There would be a flaw in this Court's logic, and an irony in its jurisprudence, were we first to conclude a ban on both D&E and intact D&E was overbroad and then to say it is irrational to ban only intact D&E because that does not proscribe both procedures. In sum, we reject the contention that the congressional purpose of the Act was "to place a substantial obstacle in the path of a woman seeking an abortion." . . .

B

The Act's furtherance of legitimate government interests bears upon, but does not resolve, the next question: whether the Act has the effect of imposing an unconstitutional burden on the abortion right because it does not allow use of the barred procedure where " 'necessary, in appropriate medical judgment, for [the] preservation of the . . . health of the mother.' " . . . The prohibition in the Act would be unconstitutional, under precedents we here assume to be controlling, if it "subject[ed] [women] to significant health risks." . . . [W]hether the Act creates significant health risks for women has been a contested factual question. The evidence presented in the trial courts and before Congress demonstrates both sides have medical support for their position.

Respondents presented evidence that intact D&E may be the safest method of abortion, for reasons similar to those adduced in *Stenberg*. . . . Abortion doctors testified, for example, that intact D&E decreases the risk of cervical laceration or uterine perforation because it requires fewer passes into the uterus with surgical instruments and does not require the removal of bony fragments of the dismembered fetus, fragments that may be sharp. Respondents also presented evidence that intact D&E was safer both because it reduces the risks that fetal parts will remain in the uterus and because it takes less time to complete. Respondents, in addition, proffered evidence that intact D&E was safer for women with certain medical conditions or women with fetuses that had certain anomalies. . . .

These contentions were contradicted by other doctors who testified in the District Courts and before Congress. They concluded that the alleged health advantages were based on speculation without scientific studies to support them. They considered D&E always to be a safe alternative. . . .

There is documented medical disagreement whether the Act's prohibition would ever impose significant health risks on women. . . . The three District Courts that considered the Act's constitutionality appeared to be in some disagreement on this central factual question. . . .

The question becomes whether the Act can stand when this medical uncertainty persists. The Court's precedents instruct that the Act can survive this facial attack. The Court has given state and federal legislatures wide discretion to pass legislation in areas where there is medical and scientific uncertainty. . . .

Medical uncertainty does not foreclose the exercise of legislative power in the abortion context any more than it does in other contexts. . . . The medical uncertainty over whether the Act's prohibition creates significant health risks provides a sufficient basis to conclude in this facial attack that the Act does not impose an undue burden.

The conclusion that the Act does not impose an undue burden is supported by other considerations. Alternatives are available to the prohibited procedure. As we have noted, the Act does not proscribe D&E. . . . If the intact D&E procedure is truly necessary in some circumstances, it appears likely an injection that kills the fetus is an alternative under the Act that allows the doctor to perform the procedure. . . .

. . . When standard medical options are available, mere convenience does not suffice to displace them; and if some procedures have different risks than others, it does not follow that the State is altogether barred from imposing reasonable regulations. The Act is not invalid on its face where there is uncertainty over whether the barred procedure is ever necessary to preserve a woman's health, given the availability of other abortion procedures that are considered to be safe alternatives.

V

. . . As the previous sections of this opinion explain, respondents have not demonstrated that the Act would be unconstitutional in a large fraction of relevant cases. . . . We note that the statute here applies to all instances in which the doctor proposes to use the prohibited procedure, not merely those in which the woman suffers from medical complications. It is neither our obligation nor within our traditional institutional role to resolve questions of constitutionality with respect to each potential situation that might develop. . . .

The Act is open to a proper as-applied challenge in a discrete case. . . . No as-applied challenge need be brought if the prohibition in the Act threatens a woman's life because the Act already contains a life exception. . . .

* * *

Respondents have not demonstrated that the Act, as a facial matter, is void for vagueness, or that it imposes an undue burden on a woman's right to abortion based on its overbreadth or lack of a health exception. For these reasons the judgments of the Courts of Appeals for the Eighth and Ninth Circuits are reversed.

It is so ordered.

JUSTICE THOMAS, with whom JUSTICE SCALIA joins, concurring.

I join the Court's opinion because it accurately applies current jurisprudence, including *Planned Parenthood of Southeastern Pa. v. Casey* . . . (1992). I write separately to reiterate my view that the Court's abortion jurisprudence, including *Casey* and *Roe v. Wade* . . . (1973), has no basis in the Constitution. . . . I also note that whether the Act constitutes a permissible exercise of Congress' power under the Commerce Clause is not before the Court. The parties did not raise or brief that issue; it is outside the question presented; and the lower courts did not address it. . . .

JUSTICE GINSBURG, with whom JUSTICE STEVENS, JUSTICE SOUTER, and JUSTICE BREYER join, dissenting.

In *Planned Parenthood of Southeastern Pa. v. Casey,* . . . (1992), the Court declared that "[l]iberty finds no refuge in a jurisprudence of doubt." There was, the Court said, an "imperative" need to dispel doubt as to "the meaning and reach" of the Court's 7-to-2 judgment, rendered nearly two decades earlier in *Roe v. Wade* . . . (1973). Responsive to that need, the Court endeavored to provide secure guidance to "[s]tate and federal courts as well as legislatures throughout the Union," by defining "the rights of the woman and the legitimate authority of the State respecting the termination of pregnancies by abortion procedures." . . .

Taking care to speak plainly, the *Casey* Court restated and reaffirmed *Roe*'s essential holding. . . . First, the Court addressed the type of abortion regulation permissible prior to fetal viability. It recognized "the right of the woman to choose to have an abortion before viability and to obtain it without undue interference from the State." . . . Second, the Court acknowledged "the State's power to restrict abortions *after fetal viability,* if the law contains exceptions for pregnancies which endanger the woman's life *or health.*" . . . (emphasis added). Third, the Court confirmed that "the State has legitimate interests from the outset of the pregnancy in protecting *the health of the woman* and the life of the fetus that may become a child." . . . (emphasis added).

In reaffirming *Roe,* the *Casey* Court described the centrality of "the decision whether to bear . . . a child," . . . to a woman's "dignity and autonomy," her "personhood" and "destiny," her "conception of . . . her place in society." . . . Of signal importance here, the *Casey* Court stated with unmistakable clarity that state regulation of access to abortion procedures, even after viability, must protect "the health of the woman." . . .

Seven years ago, in *Stenberg v. Carhart* . . . , the Court invalidated a Nebraska statute criminalizing the performance of a medical procedure that, in the political arena, has been dubbed "partial-birth abortion." With fidelity to the *Roe-Casey* line of precedent, the Court held the Nebraska statute unconstitutional in part because it lacked the requisite protection for the preservation of a woman's health. . . .

Today's decision is alarming. It refuses to take *Casey* and *Stenberg* seriously. It tolerates, indeed applauds, federal intervention to ban nationwide a procedure found necessary and proper in certain cases by the American College of Obstetricians and Gynecologists (ACOG). It blurs the line, firmly drawn in *Casey,* between previability and postviability abortions. And, for the first time since *Roe,* the Court blesses a prohibition with no exception safeguarding a woman's health.

I dissent from the Court's disposition. Retreating from prior rulings that abortion restrictions cannot be imposed absent an exception safeguarding a woman's health, the

Court upholds an Act that surely would not survive under the close scrutiny that previously attended state-decreed limitations on a woman's reproductive choices.

[The remainder of Justice Ginsburg's dissent is omitted.]

SOURCE: Supreme Court of the United States. *Gonzales v. Carhart.* 550 U.S. ___ (2007). Docket 05-380. April 18, 2007. www.supremecourtus.gov/opinions/06pdf/05-380.pdf (accessed September 23, 2007).

OTHER HISTORIC DOCUMENTS OF INTEREST

FROM PREVIOUS *HISTORIC DOCUMENTS*

- Alito nomination and confirmation, *2006,* p. 41; *2005,* p. 563
- Federal ban on partial-birth abortion, *2003,* p. 995
- *Stenberg v. Carhart* decision, *2000,* p. 429

DOCUMENT IN CONTEXT

Turkey's General Staff and the European Union on Turkish Politics

APRIL 27 AND NOVEMBER 6, 2007

Turkey's military, long the country's self-proclaimed guardian of secularism, attempted in 2007 to prevent a leader of the ruling Islamic-oriented party from being elected president. The military's intervention backfired, leading instead to new parliamentary elections in July that increased the political muscle of the ruling party, which then elected its candidate, Foreign Minister Abdullah Gül, as president in August. Gül's election to the office in August was a particularly bitter pill for the military because the presidency had long been held by the nation's founder, former general Mustafa Kemal ("Atatürk"), whose legacy the generals insisted they were attempting to preserve.

From the perspective of the generals and the Turks who agreed with them, the year's events represented a step back from some of the secularist, Western-oriented policies, known as Kemalism, that Atatürk had imposed during the 1920s and 1930s. From a different perspective, the end result of Turkey's political crisis in 2007 was the strengthening of democracy—an attribute of governing that Atatürk had embraced in theory but not in practice. The ruling Justice and Development Party (AKP, in its Turkish acronym) demonstrated at the ballot box that it had broad public support, even among many secular Turks. Under the leadership of Prime Minister Recep Tayyip Erdoğan, the AKP was pushing to modernize many aspects of Turkish society and government in the hope of winning entry into the European Union (EU), a goal that had eluded Turkey's secular leaders in previous decades.

At year's end Turkey was engaging in military action involving neighboring Iraq. A militant group, the Kurdish Workers Party, had been launching attacks into Turkey from mountainous Kurdish regions of Iraq. Turkey retaliated with bombings along the border but also threatened to invade Iraq in pursuit of the guerrillas. Such an invasion, however, would risk a confrontation with Turkey's ally, the United States, which was backing the fragile Iraqi government with some 150,000 troops.

Turkey and the United States had avoided another potential confrontation in October when the U.S. House of Representatives stepped back from action on a resolution condemning the killings of more than 1 million ethnic Armenians in Turkey during World War I. Turkish governments for decades had vehemently denied responsibility for the deaths of the Armenians and denounced attempts by other countries to label the event as a genocide.

The Military Intervenes

Turkey essentially was a one-party state (the party being Atatürk's Republican People's Party) until 1950, when an opposition party swept into power. The military intervened directly in politics four times during the next five decades to force civilian leaders from office; in each case the generals said they were acting to ensure stability and returned the government to civilian control after a period of time. The last of these coups occurred in 1997, when the generals pressured the government—Turkey's first to be led by an Islamic party, the Welfare Party—to resign. This was called a "soft coup" because it took place over a period of months.

Several leaders of the Welfare Party later formed the new Justice and Development Party (AKP). Its leader, Erdoğan, was a former mayor of Istanbul and a controversial figure because he had been convicted of "religious incitement" for publicly reading a poem the authorities considered to be overtly Islamic. Erdoğan was one of Turkey's most charismatic politicians in decades, however, and he led his party to victory in 2002 parliamentary elections. The AKP won only about one-third of the popular vote, but because of the electoral system it gained just shy of a two-thirds majority in parliament. Turkey's mainstream secular parties were effectively shut out of power.

Defying widespread expectations, Erdoğan's government did not attempt to roll back Atatürk's secular policies that had imposed a strict separation between the mosque and the government. Instead, Erdoğan emphasized free market economic policies and pushed for Turkey's admission into the EU, winning the long-sought invitation from EU leaders in 2004. The prospect of Turkish membership remained controversial in Europe and even in Turkey, however, and the earliest Turkey could adopt the necessary economic and legal reforms was expected to be around 2015.

The main event on the political calendar for 2007 was the election of a new president to replace Ahmet Necdet Sezer, whose term was supposed to expire in May. Early in the year, speculation centered around whether Erdoğan himself would seek the presidency, which was a largely ceremonial post but did have great prestige, stemming from the Atatürk era, as well as the power to veto legislation. The prospect of Erdoğan as president energized both his supporters and critics; the latter mounted protests, starting with a demonstration of some 300,000 in Ankara, the capital, on April 14. "I would not like to have an Islamic regime in Turkey," one female demonstrator told the BBC.

After suggesting for several weeks that he might be a candidate, Erdoğan stepped aside in favor of Gül, his closest political ally. A former economics professor, Gül had helped promote many of the AKP's economic reforms and had become a familiar figure on the world stage as an articulate advocate for Turkey. He was not as controversial among Turkey's secularists as was Erdoğan, but he did come from the same Islamic background, and his wife, like Erdoğan's, wore a headscarf—a symbol of piety that was immensely divisive in Turkey. The AKP nominated Gül as its presidential candidate on April 24, and he declared that he would be "loyal to secular principles."

Parliament took its first vote on a new president on April 27, but Gül fell 10 votes short of the two-thirds majority (367 votes) needed to win. This vote was marred by a controversy between the ruling AKP and the main opposition party, the Republican People's Party (CHP), over what constituted a quorum. The CHP boycotted the vote and then appealed to Turkey's top court, the Constitutional Court, to require a two-thirds quorum in parliament for the presidential vote; such a requirement would allow the minority party, by boycotting a vote, effectively to veto a decision.

Later on April 27, the general staff of the Turkish military posted on its Web site an explicitly political message warning that "certain circles are waging a relentless struggle to erode the founding principles of the Turkish Republic starting with secularism." Citing the presidential election, the generals warned that they "will make their position and stance perfectly clear if needs be. Let nobody have any doubt about this." Although the generals, led by Chief of Staff General Yaşar Büyükanit, did not specifically name Gül as the target of their concern, there was no need for them to do so. Commentators across the political spectrum said it was unclear whether the generals intended to mount a coup to force Erdoğan's government from office; such a prospect was seen as dangerous because of the broad popular support the government enjoyed. Despite the uncertainty, many Turks referred to this event as the country's first "Internet coup."

The controversy then moved into the streets and into the court room. First, on April 29, a crowd of several hundred thousand people gathered in central Istanbul to rally on behalf of secularism; many in the crowd chanted slogans calling on the government to resign. This demonstration led to numerous protests, reflecting both sides of the issue, in Istanbul, Ankara, and other cities over the next several days.

On May 1 the Constitutional Court sided with the opposition party, ruling that parliament did not have a quorum when it took its first vote on Gül. This decision was significant because it effectively gave the minority a veto power. Erdoğan denounced the ruling as "firing a bullet at democracy" and said he would seek to move up parliamentary elections, which had been scheduled for November 4, to June or July. At this delicate moment, the EU and the United States got involved, issuing separate statements on May 2, calling on the Turkish military to stay out of politics. In Turkey these interventions were widely seen as important because of the country's pending application to join the EU and its status as ally of the United States in the North Atlantic Treaty Organization (NATO).

Parliament voted on the presidency again on May 6. Opposition parties boycotted the proceedings, and Gül once more fell short of a two-thirds majority. Gül promptly withdrew his candidacy, saying the votes "have depreciated politicians in the eyes of the people."

The ruling party did win an important vote on May 10, when parliament approved a constitutional amendment providing for the president to be chosen in a direct popular election; it would take effect with the 2014 election. President Sezer, who was continuing to serve because his successor had not yet been chosen, vetoed this proposal on May 25. Parliament approved it again, and Sezer appealed to the Constitutional Court, which narrowly upheld the legislation on July 5. The measure was submitted to a public referendum on October 21 and won overwhelming approval.

July Elections

The early parliamentary elections called by Erdoğan were scheduled for July 22. In the weeks beforehand, opinion polls and other evidence suggested that a significant shift was taking place in Turkish public opinion: Many secular Turks, particularly young people, liberal intellectuals, and businesspeople hoping to increase the country's ties with Europe, were siding with Erdoğan's AKP. These new supporters of the Islamic party had differing reasons for their changes of heart, but in general they expressed anger at the military for its latest political intervention and at the old secular parties that were widely viewed as having failed Turkey for decades. "This election is a power struggle between those who want change and those who don't," Zafer Üskül, a constitutional lawyer who had jumped to the AKP and was running for parliament under its banner, told the *New York Times* several days before the election. "Religion is just an excuse" to oppose change, he added.

The advance signs of this political shift proved accurate, indeed. On July 22 the AKP won 46.7 percent of the national vote, an improvement of 13 percentage points over its showing five years earlier. This was an overwhelming vote of support in Turkish politics, given that ruling parties had not won outright majorities since the advent of multiparty democracy in 1950. The strong vote also demonstrated that the AKP was able to draw support from a diverse constituency beyond its original base of the rural poor. Even so, the AKP ended up with thirteen fewer seats than in the previous parliament because more minority parties won seats in this election than had been the case five years earlier.

The CHP, the successor to Atatürk's old party, won 20.8 percent of the vote, about 2 points better than its previous showing. One potentially significant change from prior elections was that the Nationalist Action Party, an extreme right-wing party that played to Turkish nationalist fears about the ethnic Kurdish minority, won 14.3 percent of the vote, giving it representation in parliament for the first time.

A triumphant Erdoğan vowed to "press ahead with reforms and the economic development that we have been following so far." He also pledged that the question of selecting a new president would be settled "without causing tensions."

Gül Elected President

The new parliament took office on August 4; one of its first orders of business was the election of a president. On August 13 Erdoğan once again nominated Gül. Two days later, he described Gül as "the ideal candidate" and warned the military not to interfere with the election. "If there is a backsliding in democracy, it would have serious effects on the economy," he said. "The will of the people that was reflected in the ballot boxes must be respected." General Büyükanit declined at this point to discuss the military's position on the matter except to say that the new president should adhere "in earnest and not just in words to the ideal of a secular state."

Gül received slightly less than the two-thirds majority needed for election in the first two rounds of voting, held on August 20 and August 24, but under the rules he was assured victory in the third round, when he needed only a simple majority to win. On the night of August 27, just before the third round was scheduled to take place, the military issued another statement, warning against "furtive plans which aim to undo the modern advances and ruin the Turkish republic's secular and democratic structure." Unlike the statement in April, this one contained no veiled threats of a coup.

Parliament acted on August 28, giving Gül a victory with 339 of 550 votes in parliament. He was immediately sworn into office, but none of the generals attended the proceedings, a step interpreted as a snub of their new commander in chief. In his first speech to parliament as president, Gül pledged to uphold all of Turkey's traditions and to be a nonpartisan leader. "Secularism is one of the basic principles of the republic," he said of the main topic on the minds of many people.

The peaceful conclusion of Turkey's long-running political crisis brought praise from an important outside source: the EU. In a November 6 "progress report" on Turkey's membership application, EU staff praised the outcome as a victory for democracy. "Overall, through free and fair parliamentary elections, Turkey resolved the political and constitutional crisis which followed the April presidential elections," the report said. "The elections were fully in line with the rule of law and international democratic standards." The report indicated that Turkey still fell short of EU membership by many other measures, but it made clear that lack of democracy no longer was among these failings.

Following are two documents. First, the text of a statement posted on April 27, 2007, on the Web site of the Turkish general staff, discussing the country's political situation. Second, excerpts from the "Turkey 2007 Progress Report," adopted by the European Commission on November 6, 2007.

DOCUMENT

Turkey's General Staff on Political Developments

April 27, 2007

It is being observed that certain circles that are waging a relentless struggle to erode the founding principles of the Turkish Republic starting with secularism have recently increased their efforts. These activities, which are constantly being brought to the attention of the pertinent authorities in an appropriate manner, encompass a broad spectrum of activities ranging from their wish to question and redefine the founding principles to the creation of alternative celebrations to our national holidays, which are the symbol of our state's independence and the unity and integrity of our nation.

The people engaged in these activities do not hesitate to exploit our people's sacred religious sentiments, and they work to conceal their true aims by dressing up these efforts, which have become an open challenge to the state, in the apparel of religion. The way they use women and children in particular in the front line of their activities carries striking similarities to the destructive and separatist activities being conducted against our country's unity and integrity.

In connection with this:

A Koran recital competition had been organized in Ankara for the same day as the National Sovereignty and Children's Holiday on 23 April, but was cancelled thanks to pressure brought to bear by the sensitive media and public.

In Sanliurfa on 22 April 2007 a chorus made up of young girls dressed in outmoded clothing that was inappropriate to their ages and at an hour at which they should have long been in their beds was made to recite Islamic hymns with the participation of certain groups from the provinces of Mardin, Gaziantep and Diyarbakir. The true motives and intentions of the people organizing the evening became apparent when they attempted to take down pictures of Ataturk and the Turkish flag.

Furthermore, there has been disturbing news to the effect that all the school principals in Ankara's Altindag district were ordered to attend a "Sacred Birth Celebration," that Islamic hymns were sung by primary school girls wearing Islamic headscarves during an event arranged jointly by the Denizli Provincial Muftu's Office and a political party, and that despite there being four mosques in the borough of Nikfer in Denizli's Tavas district a sermon and religious talk directed at women were held in the Ataturk Primary School.

The activities to be celebrated in schools are specified in the pertinent Ministry of Education directives. However, despite the fact that these kinds of events were arranged according to non-directive instructions and the fact that the General Staff notified the pertinent authorities it has been observed that no preventative measures have been taken.

The fact that the above mentioned activities took place with the permission and knowledge of the authorities whose duty it is to intercede in and prevent them makes the issue all the more serious. It is possible to list more examples.

This reactionary mindset, which is opposed to our Republic and has no other aim than to undermine the founding principles of our state, has been encouraged by certain developments and rhetoric in recent days and is broadening the scope of its activities.

Developments in our region are replete with examples that should be heeded of the disasters that can be caused by playing with religion and exploiting faith for political rhetoric and ends. It can be seen in both our country and abroad that when a sacred faith is used to try and carry political rhetoric or ideology it changes into something by taking faith out of the picture. The incident in Malatya can be said to be a striking example of this. It goes without saying that the only condition under which the state of the Turkish Republic may live in peace and stability as a modern democracy is to stand up for the founding qualities of the state as specified in the Constitution.

It is a clear fact that this behavior and these actions contradict entirely the principle of "being loyal to the Republic regime in spirit and not in word and of acting in such a way as to show this" as stated by the Chief of Staff in a news conference on 12 April 2007, and that they violate the founding qualities and provisions of the Constitution.

The question that has come to the fore in the recent run up to the presidential elections is focused on the secularism debate. This situation is being watched in trepidation by the Turkish Armed Forces. It must not be forgotten that the Turkish Armed Forces do take sides in this debate and are the sure and certain defenders of secularism. Moreover, the Turkish Armed Forces are definitely on the receiving end of the debates being argued and the negative commentary, and they will make their position and stance perfectly clear if needs be. Let nobody have any doubt about this.

In short, anybody who opposes the idea as stated by the founder of the Republic the Great Leader Ataturk of "Happy is the man who says I am a Turk!" is an enemy of the Turkish Republic and will stay that way.

The Turkish Armed Forces remain steadfast in their unwavering commitment to carry out in full the duties given to them by law to protect these qualities. Its allegiance to and faith in this commitment is certain.

The public has been respectfully informed.

SOURCE: Republic of Turkey. Office of the Chief of the General Staff. "On Reactionary Activities, Army's Duty." April 27, 2007. Statement No. BA-08/07. Obtained from the United States Embassy, Ankara, Turkey. Not available online.

DOCUMENT

European Union Report on Turkey's Progress toward Membership

November 6, 2007

. . .

2. Political Criteria and Enhanced Political Dialogue

This section examines progress made by Turkey towards meeting the Copenhagen political criteria which require stability of institutions guaranteeing democracy, the rule of law, human rights and respect for and protection of minorities. It also monitors the respect for international obligations, regional cooperation, and good neighbourly relations with enlargement countries and Member States.

2.1. Democracy and the rule of law

Constitution

On 10 May 2007, the Turkish Grand National Assembly adopted a package of constitutional reforms proposed by the majority Justice and Development Party (AKP). The package introduces the election of the President by popular vote for a renewable term of five years, the shortening of the government's term of office from five to four years and the establishment of a quorum of one third for all sessions and decisions of parliament. A referendum held on 21 October endorsed these reforms.

In a separate constitutional amendment, the minimum age for a person to be elected to parliament was lowered from 30 to 25 years. The new rules will not be applicable until the next parliamentary elections.

Parliament

Parliamentary elections were held on 22 July 2007. Voter turnout was over 83%. Following an invitation from the Turkish authorities, the OSCE [Organization for Security and Co-operation in Europe] Office for Democratic Institutions and Human Rights, OSCE/ODIHR, carried out an election assessment mission. In a press statement, the OSCE/ODIHR stressed that the electoral process was characterized by pluralism and a high level of public confidence underscored by the transparent, professional and efficient performance of the election administration. A delegation from the Parliamentary Assembly of the Council of Europe (PACE) also observed the elections and came to similar conclusions.

Three parties crossed the 10% threshold of the national vote required to be represented in Parliament. These were the Justice and Development Party (AKP) with 46.6%, resulting in 341 seats, the Republican People's Party (CHP) with 20.9% (99 seats) and the Nationalist Movement Party (MHP) which obtained 14.3% (70 seats). 26 independent candidates were also elected. 20 of these, from the Democratic Society Party (DTP), formed their own political group. This brought the number of political groups to four. Additional parties represented in parliament are the Democratic Left Party (DSP), with 13 Members of Parliament, the Grand Unity Party (BBP) and the Freedom and Democracy Party (ÖDP) with one seat each.

The newly-elected parliament is now more representative of the country's political diversity. Nevertheless, the debate continued on reducing the 10% threshold, which is the highest among European parliamentary systems. This issue was also brought to the European Court of Human Rights (ECtHR), which ruled in January 2007 that the threshold does not violate the right to free elections. However, it also noted that it would be desirable for the threshold to be lowered in order to ensure optimal representation, while preserving the objective of achieving stable parliamentary majorities. The issue was referred to the Grand Chamber.

President

In view of the expiration of the presidential term of President Sezer in May, in April the Parliament convened to elect a new President of the Republic. The first round of voting, held in the Parliament on 27 April 2007, was boycotted by opposition parties, and the sole candidate, Foreign Minister and Deputy Prime Minister Abdullah Gül, failed to obtain the required two-thirds majority. On the same day the General Staff interfered with the presidential elections by issuing a memorandum.

Following an application of the main opposition party CHP the Constitutional Court ruled on 1 May that the vote be invalidated because a two-thirds quorum of participants was lacking. A new vote was held, but the two-thirds quorum was not achieved. Mr. Gül then withdrew his candidacy and the entire procedure was cancelled. This triggered early elections, as provided for in the Turkish Constitution.

In August the newly-elected Parliament elected Minister Gül President at the third round with 339 votes.

President Sezer exercised his right of veto on several laws related to political reforms, notably the law on the Ombudsman, the law on Foundations, and the law on private education institutions. The President also appealed to the Constitutional Court against the law on the Ombudsman. Strained relations between the President and the government contributed to slowing work on political reforms.

Government

Following the general election a single-party AKP government was formed by Prime Minister Erdoğan and endorsed by Parliament on 5 September. The government programme includes a commitment to continue reforms. The government plans to carry out extensive constitutional reforms aimed in particular at fully aligning Turkey to international standards in the area of fundamental rights. The government reiterated its intention to push forward the implementation of the Turkish road map for EU accession presented in April 2007. The Road Map provides internal guidance to line Ministries on alignment with the *acquis* and covers alignment of primary and secondary legislation to be adopted and implemented between 2007 and 2013. The inter-ministerial Reform Monitoring Group met in September.

In the new government the Foreign Ministry will continue to be in charge of accession negotiations with the EU. The Secretariat General for EU Affairs (EUSG) was placed under the Ministry of Foreign Affairs and will continue to play a coordinating role, in particular on the political criteria, financial cooperation and negotiations on individual chapters. In September 2007 it was announced that the EUSG and the State Planning Organisation will exercise a quarterly progress review of the road map implementation.

However, given its significant role there is a need to strengthen the EUSG staff and resources. Only limited action was taken in this respect.

Overall, through free and fair parliamentary elections Turkey resolved the political and constitutional crisis which followed the April presidential elections. The elections were fully in line with the rule of law and international democratic standards. Participation was high and the new Parliament is highly representative of Turkish political diversity. Elections of the President in August took place smoothly and in accordance with the Constitution. A new government was formed and presented an EU-oriented reform agenda. . . .

Civilian oversight of the security forces

Despite public comments from the army and attempts to interfere in the political process, the outcome of the spring 2007 constitutional crisis reaffirmed the primacy of the democratic process.

The National Security Council (NSC) continued to meet in line with its revised role. Ambassador Burcuoğlu was appointed as new Secretary-General in September. The total staff of the NSC decreased from 408 to 224, and the number of military personnel from 26 to 12.

However, the armed forces continued to exercise significant political influence. Senior members of the armed forces have stepped up their public comments on domestic and foreign policy questions including Cyprus, secularism and Kurdish issues. On a number of occasions, the General Staff reacted publicly to government statements or decisions. The General Staff directly interfered with the April 2007 presidential election by publishing a memorandum on its website expressing concern at the alleged weakening of secularism in the country.

There were several attempts from senior members of the armed forces to restrict academic research and public debate in Turkey, in particular on security and minority rights issues. Furthermore, the military targeted the press on various occasions.

The 1997 *EMASYA* secret protocol on Security, Public Order and Assistance Units remains in force. The protocol, signed by the General Staff and the Ministry of Interior, allows for military operations to be carried out for internal security matters under certain conditions without a request from the civilian authorities.

No change has been made to the Turkish Armed Forces Internal Service Law and the law on the National Security Council. These laws define the role and duties of the Turkish military and grant the military a wide margin of manoeuvre by providing a broad definition of national security. No progress has been made in enhancing civilian control over the Gendarmerie when engaged in civilian activities.

No progress has been made in terms of strengthening parliamentary oversight of the military budget and expenditure. The Parliamentary Planning and Budget Committee reviews the military budget only in a general manner. It does not examine programmes and projects. Furthermore, extra-budgetary funds are excluded from parliamentary scrutiny.

As regards auditing, according to the Constitution the Court of Auditors can carry out external ex-post audit of military expenditures and properties. However, the Court remains unable to audit military properties, pending the adoption of the Law on the Court of Auditors. Furthermore, the 2003 Law on Public Financial Management and Control providing for the internal audit of security institutions has yet to be properly implemented.

Overall, no progress has been made in ensuring full civilian supervisory functions over the military and parliamentary oversight of defence expenditure. On the contrary, the tendency for the military to make public comments on issues going beyond its remit, including on the reform agenda, has increased.

Judicial system

Some progress has been made in terms of the efficiency of the judiciary, including through amendments to the Turkish Criminal Code (CC) and the Criminal Procedure Code (CPC) adopted in December 2006. These amendments extend the discretion of the prosecutor as regards decisions not to prosecute, while the provisions regarding mediation are simplified. Judicial supervision—introduced in the CPC as an alternative to arrest for offences requiring imprisonment of three years or less—has started functioning satisfactorily. Probation is an area where progress has been achieved in implementation: 133 probation centres employing 1,298 staff have become fully operational since November 2006.

Efforts to modernise the judiciary through the use of information technology continued. Judges have reported positive results as regards the National Judicial Network Project (UYAP) on court proceedings while the lawyers' portal was integrated into this network in March. 864 judges and 476 prosecutors were appointed during the reporting period. The funds for the judiciary have increased from 409 million in 2005 to 482 million in 2006 and are planned to reach 865 million by the end of 2007. In May 2007, nine locations for regional courts of appeal were identified and their geographical areas of jurisdiction defined in line with legal requirements.

However, concerns remain as regards the independence and the impartiality of the judiciary. In the context of the election of the new president in April, the Constitutional Court ruled by a majority of seven to four that a quorum of two thirds (367 deputies) is necessary for the first and second rounds of presidential elections in Parliament, and annulled the first round of voting. This decision led to strong political reactions and allegations that the Constitutional Court had not been impartial when reaching this decision. In the event, and as regards the election of the President of the Republic by Parliament, the Court introduced a one-third blocking minority. . . .

SOURCE: European Union. Commission of the European Communities. "Turkey 2007 Progress Report." SEC(2007) 1436. Commission Staff Working Document. November 6, 2007. http://ec.europa.eu/enlargement/pdf/key_documents/2007/nov/turkey_progress_reports_en.pdf (accessed March 25, 2008).

OTHER HISTORIC DOCUMENTS OF INTEREST

FROM THIS VOLUME

- U.S. reports on the security situation in Iraq, p. 407
- European Union treaty, p. 721

FROM PREVIOUS *HISTORIC DOCUMENTS*

- European Union president on a membership invitation to Turkey, *2004,* p. 973
- Parliamentary program of the Turkish Justice and Development Party, *2002,* p. 906

DOCUMENT IN CONTEXT

Report on Israel's Handling of Its 2006 War with Hezbollah

APRIL 30, 2007

Israeli prime minister Ehud Olmert survived investigations that condemned nearly every aspect of his government's handling of the 2006 war with the Hezbollah militia in Lebanon, but he was severely damaged politically. He managed to remain in office through 2007 largely because key elements of Israel's political establishment disliked a probable alternative: new elections that would bring to power former prime minister Binyamin Netanyahu, one of the country's most divisive figures. Two other senior figures were forced from office, however: Defense Minister Amir Peretz and Lt. Gen. Dan Halutz, the chief of staff of the army.

Israel fought a month-long war with Hezbollah in July and August 2006 to secure release of two soldiers whom Hezbollah had captured during a cross-border raid. The war quickly escalated into a much broader conflict in which Olmert and other Israeli officials pledged to "destroy" Hezbollah, which had become the most potent military force in Lebanon. In a month of bombing and a shorter ground invasion, Israel caused serious damage to much of southern Lebanon but failed to win return of its soldiers or to destroy Hezbollah, which fired some 4,000 rockets into northern Israel in retaliation. The war killed about 1,100 Lebanese (civilians and Hezbollah fighters), 120 Israeli soldiers, and 39 Israeli civilians.

For Israel, the war ended at best in a draw. However, many Israelis grudgingly acknowledged that their army—the most powerful and sophisticated in the Middle East—had suffered a stinging strategic defeat. Hezbollah's leader, Hassan Nasrallah, claimed to have won a "divine victory," an assessment that was widely shared in the Middle East, even among moderate Arabs repelled by Nasrallah's Islamist rhetoric.

In addition to the controversies over the war, Israel was bombarded with political scandals during 2007. Olmert survived several investigations into his personal conduct, but Israeli president Moshe Katsav did not. Katsav was forced to resign his largely ceremonial position in late June after an investigation into charges of sexual abuse of his staff. The Knesset (parliament) elected as his successor Shimon Peres, a senior political figure who had held just about every other post in the Israeli government.

Winograd Report

Responding to public anger over the outcome of the war, Olmert in September 2006 appointed an independent commission to examine his government's actions. The panel was headed by Eliyahu Winograd, a retired judge; its other members were two retired generals, a professor of law, and a professor of political science. Olmert gave closed-door testimony to the panel on February 1, 2007; according to news reports, he defended his actions and insisted that Israel had achieved important objectives in the war.

As the Winograd panel was finishing its work, retired general Dan Shomron was conducting a separate official investigation that examined specific details of Israeli military operations during the war. Shomron issued part of his report in December 2006 and the balance in January 2007. His report sharply criticized the military's organization of the war and raised serious questions about the judgment of senior commanders, in particular General Halutz. In response to the Shomron report, and in anticipation of similar conclusions by the Winograd panel, Halutz submitted his resignation on January 16. An air force officer, Halutz had been widely criticized for relying too heavily on bombs and rockets to destroy Hezbollah positions, many of which were located in deep underground bunkers.

Winograd on April 30 made public an unclassified summary of his panel's findings covering the first phase of the war: from July 12, 2006, when Hezbollah conducted its raid into Israel, until July 17, when Olmert appeared before the Knesset and outlined his objectives, which included the destruction of Hezbollah. Winograd said the panel would continue its investigation into the rest of the war and issue a final report by year's end. That report was delayed until early in 2008.

Although the April 30 report covered just six days of the war, it offered an unvarnished critique of nearly every important action and decision by Israel's military and political leaders. The report focused primarily on Olmert, Peretz, and Halutz, but other leaders did not escape censure; the cabinet, for example, was faulted for failing to ask necessary questions about the decision to go to war.

Winograd's panel charged Olmert with "a serious failure in exercising judgment, responsibility, and prudence." The prime minister, who had no military command experience, made the decision to go to war "hastily" after failing to consult with others who did have military experience and disregarding the "political and professional reservations" that some officials did express to him, the report said. Olmert also failed to match his goals in the war with the military means he was employing; as a result, he was partly responsible for the fact "that the declared goals were over-ambitious and not feasible."

The panel had similar criticisms of Defense Minister Peretz, who also lacked military experience despite his position. According to the report, Peretz also failed to consult with others, ask appropriate questions, and carefully assess the goals and the means of achieving them. In a single damning sentence, the panel concluded that Peretz, in his service as defense minister during the war, "impaired Israel's ability to respond well to its challenges."

The Winograd panel depicted Halutz as the "dominant" figure in decision making during the early phase of the war because of the inexperience of Olmert and Peretz. Halutz acted "impulsively" and failed to alert political leaders to the military's shortcomings or its internal debates about the war. Halutz, the panel said, "failed in his duties" as commander and "exhibited flaws in professionalism, responsibility, and judgment."

Beyond these specific criticisms of the main leaders, the panel offered a pointed critique of Israel's military preparedness and strategic planning, particularly concerning threats from Lebanon. This critique was directed at the government's lack of intelligence information about the extent and capability of Hezbollah's arsenal of rockets. Before the war, the Israeli government reportedly believed that Hezbollah had only short-range rockets capable of reaching just a few miles into Israeli territory. Hezbollah's launching of some 4,000 rockets—many of them deep into northern Israel—demonstrated flaws in Israel's intelligence agencies, widely assumed to be among the best in the world. These flaws predated Olmert and the current administration, the panel said. Even so, the panel did not specifically name Olmert's predecessor: former prime minister Ariel Sharon, who was a former general and one of Israel's most important military heroes.

Another tough report about the government's handling of the war was released on December 31 by the Knesset's Foreign Affairs Committee. That panel focused on military operations and cited many of the same failures noted in the Winograd report. The bottom line, the committee said, was that "Israel did not succeed in defeating the enemy."

Political Fallout

The Winograd panel's harsh findings prompted calls for Olmert and Peretz to resign. Such calls came not only from predictable quarters, such as opposition politicians, but even from some of Olmert's allies; among the latter was Foreign Minister Tzipi Livni (who also served as a deputy prime minister). Livni said on May 2 that she had told Olmert that "resignation would be the right thing for him to do," and offered herself as a replacement. Olmert rebuffed such calls, saying immediately after release of the Winograd report: "It would not be correct to resign, and I have no intention of resigning."

Israel's political situation often was complex and divisive, and this certainly was the case in the period during which the Winograd report was published. Olmert sat atop a coalition government that, in theory, was solid, representing 77 of the 120 seats in the Knesset. Behind this facade of a strong majority, however, were numerous fractures. Olmert's Kadima Party was a middle-of-the-road, breakaway faction of the right-wing Likud Party, which had governed Israel for much of the previous three decades. Sharon founded Kadima at the end of 2005 but suffered an incapacitating stroke in January 2006, leaving his deputy, Olmert, in charge. Although Olmert was a lifelong public servant and had been mayor of Jerusalem, he was an unpopular figure, in part because of numerous allegations of improper business dealings and in part because many Israelis considered him opportunistic.

The other major party in the coalition was the center-left Labor Party, which had not won a national election since 1999 and had undergone numerous leadership changes, the most recent of which put Peretz in charge. The Kadima-Labor coalition also depended on the support of smaller parties that did not share the priorities of either Kadima or Labor but did receive political benefits, including cabinet posts and public spending for their own priorities, in return for their votes. The fragility of Olmert's coalition had been demonstrated, even before the Winograd report was published, in polls showing Kadima and Labor both lagging behind Likud and even a new ultra-right party, Yisrael Beiteinu ("Israel Is Our Home"), dominated by immigrants from the former Soviet Union.

Olmert survived major tests in early May. Late on May 2, fellow legislators from the Kadima Party offered him their support after what reportedly was a fractious closed-door meeting. At this meeting, Livni and several colleagues called on Olmert to resign. When he refused, other members rallied around him. On the evening of May 3, more than 100,000 protesters gathered in Tel Aviv to demand Olmert's resignation. By this point, Olmert's most dangerous political rival, Likud leader Netanyahu, had broken his silence after the Winograd report and called on the government to "return to the people and let them speak their minds." Likud lacked the votes in the Knesset to force Olmert from office, however. The Olmert government beat back three no-confidence motions on May 7.

Defense Minister Peretz was in an even more untenable situation than Olmert because he had led the Labor Party to a second-place finish in the most recent elections, in March 2006, and had no prospect—after Winograd's description of his performance—of doing any better in the future. He announced on May 5 that he would leave the defense ministry and take over the finance ministry if his party reelected him as leader in a primary scheduled for May 29. Peretz came in third in that vote, however, and he stepped aside.

The Labor leadership contest was won on June 13 by former prime minister Ehud Barak, a retired general and Israeli military hero. Barak had won elections in 1999 but lost to Sharon in February 2001 after the failure of U.S.-sponsored peace negotiations with the Palestinians. Although Barak had called on Olmert to resign following the Winograd report, he joined the cabinet later in June as defense minister.

Olmert's coalition stayed in office through the rest of the year despite defections by minority parties. Israeli analysts said most political figures had nothing to gain by upsetting the fragile coalition and bringing on new elections. The one exception was Netanyahu, whose Likud Party led the polls all through the latter part of the year. Netanyahu had few allies outside his party who wanted to see him in office again, however.

Olmert enjoyed a bump in popularity after September 6, when he ordered Israeli jets to bomb a target in Syria. Israeli officials refused to comment publicly on the raid, but subsequent news reports said the target had been a nuclear installation under construction, possibly with help from North Korea. The validity of these reports was not confirmed by year's end.

During much of his career Olmert had been dogged by charges that he had used his official positions to benefit himself and political associates. At least four separate investigations into such allegations were under way in 2007, none of which resulted in criminal charges being filed against him. Judicial authorities on November 29 closed an investigation into one of the most controversial of the allegations: that Olmert, as finance minister in 2005, had improperly tried to steer the sale of a state-owned bank to a friend.

Peres as President

Israel did not get a new prime minister in 2007, but it did get a new president as a result of a personal scandal involving Katsav, who had held the post since 2000. Attorney General Menachem Mazuz on January 24 announced his intention to indict Katsav on several charges, including rape, sexual harassment, and abuse of power. These charges stemmed from complaints by several women who had worked for Katsav over the years that he had pressured them to engage in sex. Insisting on his innocence and refusing calls for his resignation, Katsav asked the Knesset for a leave of absence, which was refused. Katsav submitted his resignation on June 29 as part of an agreement with prosecutors under which he pleaded guilty to one count of indecent assault.

With Katsav's term due to expire anyway in July, the Knesset on June 13 took up the matter of choosing his successor. Shimon Peres, the leading candidate, fell just three votes short of the required majority on the first round and then won an overwhelming vote when two competing candidates withdrew. This election broke a long string of political defeats for eighty-three-year-old Peres, who had served in parliament since 1959. One of the most embarrassing of those defeats had come when he sought the presidency in 2000, only to lose to the lesser-known Katsav. A lifetime member of the Labor Party, Peres in late 2005 had jumped to the new Kadima Party to join Sharon, who had been both a close friend and longtime political rival.

Following is the summary of the Interim Report of the Winograd Commission, which examined Israel's handling of its July–August 2006 war with the Hezbollah militia in Lebanon. This unclassified summary of the report was made public on April 30, 2007.

DOCUMENT

Winograd Commission Report on the 2006 Israel-Hezbollah War

April 30, 2007

1. On September 17th 2006, the Government of Israel decided, under section 8A of Basic Law: The Government 2001, to appoint a governmental commission of examination "To look into the preparation and conduct of the political and the security levels concerning all the dimensions of the Northern Campaign which started on July 12th 2006". Today we have submitted to the Prime Minister and the Minister of Defense the classified interim report, and we are now presenting the unclassified report to the public.

2. The Commission was appointed due to a strong sense of a crisis and deep disappointment with the consequences of the campaign and the way it was conducted. We regarded accepted [sic] this difficult task both as a duty and a privilege. It is our belief that the larger the event and the deeper the feeling of crisis—the greater the opportunity to change and improve matters which are essential for the security and the flourishing of state and society in Israel. We believe Israeli society has great strength and resilience, with a robust sense of the justice of its being and of its achievements. These, too, were expressed during the war in Lebanon and after it. At the same time, we must not underrate deep failures among us.

3. This conception of our role affected the way we operated. No one underestimates the need to study what happened in the past, including the imposition of personal responsibility. The past is the key for learning lessons for the future. Nonetheless, learning these lessons and actually implementing them are the most [important] of the conclusions of the Commission.

4. This emphasis on learning lessons does not only follow from our conception of the role of a public Commission. It also follows from our belief that one [of] Israeli society's greatest sources of strength is its being free, open and creative. Together with great achievements, the challenges facing it are existential. To cope with them, Israel must be a learning society—a society which examines its achievements and, in particular, its failures, in order to improve its ability to face the future.

5. Initially we hoped that the appointment of the Commission will serve as an incentive to accelerate learning processes in the relevant systems, while we are working, so that we could devote our time to study all of the materials in depth, and present the public with a comprehensive picture. However, learning processes have been limited. In some ways an opposite, and worrying, process emerged—a process of 'waiting' for the Commission's Report before energetic and determined action is taken to redress failures which have been revealed.

6. Therefore we decided to publish initially an Interim Report, focusing on the decisions related to starting the war. We do this in the hope that the relevant bodies will act urgently to change and correct all that it implies. We would like to reiterate and emphasize that we hope that this Partial Report, which concentrates on the functioning of the highest political and military echelons in their decision to move into the war will not divert attention from the overall troubling complete picture revealed by the war as a whole.

7. The interim report includes a number of chapters dealing with the following subjects:

a. The Commission's conception of its role, and its attitude to recommendations in general and to recommendations dealing with specific persons in particular. (chapter 2): We see as the main task of a public commission of inquiry (or investigation) to determine findings and conclusions, and present them—with its recommendations—before the public and decision makers so that they can take action. A public commission should not—in most cases—replace the usual political decision-making processes and determine who should serve as a minister or senior military commander. Accordingly, we include personal conclusions in the interim report, without personal recommendations. However, we will reconsider this matter towards our Final Report in view of the depiction of the war as a whole.

b. The way we balanced our desire to engage in a speedy and efficient investigation with the rights of those who may be negatively affected to 'natural justice' (chapter 3): The special stipulations of the Commissions of Inquiry Act in this regard do not apply to a governmental commission of Examination, but we regard ourselves, naturally, as working under the general principles of natural justice. The commission notified those who may be affected by its investigation, in detailed letters of invitation, of the ways in which they may be negatively affected, and enabled them to respond to allegations against them, without sending "notices of warning" and holding a quasi-judicial hearing before reaching our conclusions. We believe that in this way we provided all who may be negatively affected by our report with a full opportunity to answer all allegations against them.

c. The processes and developments in the period between the withdrawal of the IDF [Israeli Defense Forces] from Lebanon until July 11, 2006 which contributed to the background of the Lebanon War (Chapter 4): These processes created much of the factual background against which the decision-makers had to operate on July 12th, and they are thus essential to both the understanding and the evaluation of the events of the war. Understanding them is also essential for drawing lessons from the events, whose significance is often broader than that of the war itself.

8. The core of the interim report is a detailed examination of the decisions of senior political and military decision-makers concerning the decision to go to war at the wake of the abduction of the two soldiers on the morning of July 12th. We start with the decision of the government on the fateful evening of the 12th to authorize a sharp military response, and end with the speech of the Prime Minister in the Knesset on July 17th, when he officially presented the campaign and its goals. These decisions were critical and constitutive, and therefore deserve separate investigation. We should note that these decisions enjoyed broad support within the government, the Knesset and the public throughout this period.

9. Despite this broad support, we determine that there are very serious failings in these decisions and the way they were made. We impose the primary responsibility for these

failures on the Prime Minister, the minister of defense and the (outgoing) Chief of Staff. All three made a decisive personal contribution to these decisions and the way in which they were made. However, there are many others who share responsibility for the mistakes we found in these decisions and for their background conditions.

10. The main failures in the decisions made and the decision-making processes can be summed up as follows:

a. The decision to respond with an immediate, intensive military strike was not based on a detailed, comprehensive and authorized military plan, based on careful study of the complex characteristics of the Lebanon arena. A meticulous examination of these characteristics would have revealed the following: the ability to achieve military gains having significant political-international weight was limited; an Israeli military strike would inevitably lead to missiles fired at the Israeli civilian north; there was not [an]other effective military response to such missile attacks than an extensive and prolonged ground operation to capture the areas from which the missiles were fired—which would have a high "cost" and which did not enjoy broad support. These difficulties were not explicitly raised with the political leaders before the decision to strike was taken.

b. Consequently, in making the decision to go to war, the government did not consider the whole range of options, including that of continuing the policy of 'containment', or combining political and diplomatic moves with military strikes below the 'escalation level', or military preparations without immediate military action—so as to maintain for Israel the full range of responses to the abduction. This failure reflects weakness in strategic thinking, which derives the response to the event from a more comprehensive and encompassing picture.

c. The support in the cabinet for this move was gained in part through ambiguity in the presentation of goals and modes of operation, so that ministers with different or even contradictory attitudes could support it. The ministers voted for a vague decision, without understanding and knowing its nature and implications. They authorized to commence a military campaign without considering how to exit it.

d. Some of the declared goals of the war were not clear and could not be achieved, and in part were not achievable by the authorized modes of military action.

e. The IDF did not exhibit creativity in proposing alternative action possibilities, did not alert the political decision-makers to the discrepancy between its own scenarios and the authorized modes of action, and did not demand—as was necessary under its own plans—early mobilization of the reserves so they could be equipped and trained in case a ground operation would be required.

f. Even after these facts became known to the political leaders, they failed to adapt the military way of operation and its goals to the reality on the ground. On the contrary, declared goals were too ambitious, and it was publicly stated that fighting will continue till they are achieved. But the authorized military operations did not enable their achievement.

11. The primary responsibility for these serious failings rests with the Prime Minister, the minister of defense and the (outgoing) Chief of Staff. We single out these three because it is likely that had any of them acted better—the decisions in the relevant period and the ways they were made, as well as the outcome of the war, would have been significantly better.

12. Let us start with the Prime Minister.

a. The Prime Minister bears supreme and comprehensive responsibility for the decisions of 'his' government and the operations of the army. His responsibility for the failures in the initial decisions concerning the war stem from both his position and from his behavior, as he initiated and led the decisions which were taken.

b. The Prime Minister made up his mind hastily, despite the fact that no detailed military plan was submitted to him and without asking for one. Also, his decision was made without close study of the complex features of the Lebanon front and of the military, political and diplomatic options available to Israel. He made his decision without systematic consultation with others, especially outside the IDF, despite not having experience in external-political and military affairs. In addition, he did not adequately consider political and professional reservations presented to him before the fateful decisions of July 12th.

c. The Prime Minister is responsible for the fact that the goals of the campaign were not set out clearly and carefully, and that there was no serious discussion of the relationships between these goals and the authorized modes of military action. He made a personal contribution to the fact that the declared goals were over-ambitious and not feasible.

d. The Prime Minister did not adapt his plans once it became clear that the assumptions and expectations of Israel's actions were not realistic and were not materializing.

e. All of these add up to a serious failure in exercising judgment, responsibility and prudence.

13. The Minister of Defense is the minister responsible for overseeing the IDF, and he is a senior member in the group of leaders in charge of political-military affairs.

a. The Minister of Defense did not have knowledge or experience in military, political or governmental matters. He also did not have good knowledge of the basic principles of using military force to achieve political goals.

b. Despite these serious gaps, he made his decisions during this period without systemic consultations with experienced political and professional experts, including outside the security establishment. In addition, he did not give adequate weight to reservations expressed in the meetings he attended.

c. The Minister of Defense did not act within a strategic conception of the systems he oversaw. He did not ask for the IDF's operational plans and did not examine them; he did not check the preparedness and fitness of IDF; and did not examine the fit between the goals set and the modes of action presented and authorized for achieving them. His influence on the decisions made was mainly pointillist and operational. He did not put on the table—and did not demand presentation—of serious strategic options for discussion with the Prime Minister and the IDF.

d. The Minister of Defense did not develop an independent assessment of the implications of the complexity of the front for Israel's proper response, the goals of the campaign, and the relations between military and diplomatic moves within it. His lack of experience and knowledge prevented him from challenging in a competent way both the IDF, over which he was in charge, and the Prime Minister.

e. In all these ways, the Minister of Defense failed in fulfilling his functions. Therefore, his serving as Minister of Defense during the war impaired Israel's ability to respond well to its challenges.

14. The Chief of Staff (COS) is the supreme commander of the IDF, and the main source of information concerning the army, its plans, abilities and recommendations presented to the political echelon. Furthermore, the COS's personal involvement with decision making within the army and in coordination with the political echelon were dominant.

a. The army and the COS were not prepared for the event of the abduction despite recurring alerts. When the abduction happened, he responded impulsively. He did not alert the political leaders to the complexity of the situation, and did not present information, assessments and plans that were available in the IDF at various levels of planning and approval and which would have enabled a better response to the challenges.

b. Among other things, the COS did not alert the political echelon to the serious shortcomings in the preparedness and the fitness of the armed forces for an extensive ground operation, if that became necessary. In addition, he did not clarify that the military assessments and analyses of the arena were that a military strike against Hezbollah will with a high probability make such a move necessary.

c. The COS's responsibility is aggravated by the fact that he knew well that both the Prime Minister and the Minister of Defense lacked adequate knowledge and experience in these matters, and by the fact that he had led them to believe that the IDF was ready and prepared and had operational plans fitting the situation.

d. The COS did not provide adequate responses to serious reservation about his recommendations raised by ministers and others during the first days of the campaign, and he did not present to the political leaders the internal debates within the IDF concerning the fit between the stated goals and the authorized modes of actions.

e. In all these the Chief of Staff failed in his duties as commander in chief of the army and as a critical part of the political-military leadership, and exhibited flaws in professionalism, responsibility and judgment.

15. Concomitantly we determine that the failures listed here, and in the outcomes of the war, had many other partners.

a. The complexity of the Lebanon scene is basically outside Israel's control.

b. The ability of Hezbollah to sit 'on the border', its ability to dictate the moment of escalation, and the growth of its military abilities and missile arsenal increased significantly as a result of Israel's unilateral withdrawal in May 2000 (which was not followed, as had been hoped, by The Lebanese Army deploying on the border with Israel.)

c. The shortcomings in the preparedness and the training of the army, its operational doctrine, and various flaws in its organizational culture and structure, were all the responsibility of the military commanders and political leaders in charge years before the present Prime Minister, Minister of Defense and Chief of Staff took office.

d. On the political-security strategic level, the lack of preparedness was also caused by the failure to update and fully articulate Israel's security strategy doctrine, in the fullest sense of that term, so that it could not serve as a basis for coping comprehensively [with] all the challenges facing Israel. Responsibility for this lack of an updated national security strategy lies with Israel's governments over the years. This omission made it difficult to devise an immediate proper response to the abduction, because it led to stressing an immediate and sharp military strike. If the response had been derived from a more comprehensive security strategy, it would have been easier to take into account Israel's overall balance of strengths and vulnerabilities, including the preparedness of the civil population.

e. Another factor which largely contributed to the failures is the weakness of the high staff work available to the political leadership. This weakness existed under all previous Prime Ministers and this continuing failure is the responsibility of these PMs and their cabinets. The current political leadership did not act in a way that could compensate for this lack, and did not rely sufficiently on other bodies within and outside the security system that could have helped it.

f. Israel's government in its plenum failed in its political function of taking full responsibility for its decisions. It did not explore and seek adequate response for various reservations that were raised, and authorized an immediate military strike that was not thought-through and suffered from over-reliance on the judgment of the primary decision-makers.

g. Members of the IDF's general staff who were familiar with the assessments and intelligence concerning the Lebanon front, and the serious deficiencies in preparedness and training, did not insist that these should be considered within the army, and did not alert the political leaders concerning the flaws in the decisions and the way they were made.

16. As a result of our investigation, we make a number of structural and institutional recommendations, which require urgent attention:

a. The improvement of the quality of discussions and decision making within the government through strengthening and deepening staff work; strict enforcement of the prohibition of leaks; improving the knowledge base of all members of the government on core issues of Israel's challenges, and orderly procedures for presentation of issues for discussion and resolution.

b. Full incorporation of the Ministry of Foreign Affairs in security decisions with political and diplomatic aspects.

c. Substantial improvement in the functioning of the National Security Council, the establishment of a national assessment team, and creating a center for crises management in the Prime Minister's Office.

17. We regard it is of great importance to make findings, reach conclusions and present recommendations on the other critical issues which emerged in this war. We will cover them in the final report, which we strive to conclude soon. These subjects include, among others, the direction of the war was led [sic] and its management by the political echelon; the conduct of the military campaign by the army; the civil-military relationship in the war; taking care of Israel's civilian population under missile attack; the diplomatic negotiations by the Prime Minister's office and the Ministry of Foreign Affairs; censorship, the media and secrecy; the effectiveness of Israel's media campaign; and the discussion of various social and political processes which are essential for a comprehensive analysis of the events of the war and their significance.

18. Let us add a few final comments: It took the government till March 2007 to name the events of the summer of 2006 'The Second Lebanon War'. After 25 years without a war, Israel experienced a war of a different kind. The war thus brought back to center stage some critical questions that parts of Israeli society preferred to avoid.

19. The IDF was not ready for this war. Among the many reasons for this we can mention a few: Some of the political and military elites in Israel have reached the conclusion that Israel is beyond the era of wars. It had enough military might and superiority to deter others from declaring war against her; these would also be sufficient to send a painful reminder to anyone who seemed to be undeterred; since Israel did not intend to initiate a war, the conclusion was that the main challenge facing the land forces would be low intensity asymmetrical conflicts.

20. Given these assumptions, the IDF did not need to be prepared for 'real' war. There was also no urgent need to update in a systematic and sophisticated way Israel's overall security strategy and to consider how to mobilize and combine all its resources and sources of strength—political, economic, social, military, spiritual, cultural and scientific—to address the totality of the challenges it faces.

21. We believe that—beyond the important need to examine the failures of conducting the war and the preparation for it, beyond the need to identify the weaknesses (and strengths) in the decisions made in the war—these are the main questions raised by the Second Lebanon war. These are questions that go far beyond the mandate of this or that commission of inquiry; they are the questions that stand at the center of our existence here as a Jewish and democratic state. It would be a grave mistake to concentrate only on the flaws revealed in the war and not to address these basic issues.

We hope that our findings and conclusions in the interim report and in the final report will not only impel taking care of the serious governmental flaws and failures we examine and expose, but will also lead towards a renewed process in which Israeli society, and its political and spiritual leaders will take up and explore Israel's long-term aspirations and the ways to advance them.

SOURCE: State of Israel. Ministry of Foreign Affairs. "Winograd Commission Submits Interim Report." April 30, 2007. www.mfa.gov.il/MFA/Government/Communiques/2007/Winograd+Inquiry+Commission+submits+Interim+Report+30-Apr-2007.htm (accessed March 20, 2008).

OTHER HISTORIC DOCUMENTS OF INTEREST

FROM THIS VOLUME

- Upheavals in the Palestinian territories, p. 39
- United Nations reports on Lebanon, p. 624
- Middle East peace conference, p. 680

FROM PREVIOUS *HISTORIC DOCUMENTS*

- Prime Minister Olmert on the new Israeli government, *2006*, p. 193
- UN Security Council resolution on war between Israel and Hezbollah, *2006*, p. 454

May

DOCUMENT IN CONTEXT

Leaders of Northern Ireland on Their New Government

MAY 8, 2007

It was a sight that few people in Northern Ireland had believed they would ever see: key leaders of the opposing hard-line factions appearing together in the Great Hall of the parliament building in Belfast as the top officials in a new government for the long-troubled British province. On May 8 Rev. Ian Paisley, leader of the Democratic Unionist Party, took his post as the "first minister" (equivalent to prime minister) of a newly restored regional government for Northern Ireland; Martin McGuinness, the number-two leader of Sinn Féin, the political wing of the Irish Republican Army (IRA), assumed the role of Paisley's chief deputy.

Each man made clear in his inaugural speech that he had not abandoned his long-term goals. Paisley insisted that Northern Ireland should remain within the United Kingdom (the "unionist" or "loyalist" position of most Protestants, who were the majority of the population), while McGuinness held on to the opposing goal of joining Northern Ireland with the independent Republic of Ireland to the south (the "republican" or "nationalist" position of most Catholics). But the act of these two formerly bitter rivals joining together in government suggested, more than any other step in recent years, that peace finally was a reality, not just a dream, for Northern Ireland.

The creation of a coalition government in Northern Ireland was a final achievement for British prime minister Tony Blair, who announced his long-expected resignation two days later. Blair had devoted much of his decade as prime minister to bringing peace to the province.

Background

The installation of a new provincial government headed by Paisley and McGuinness was the culmination of long-running, always-fragile efforts to end sectarian violence—known locally as "the Troubles"—in Northern Ireland. The violence broke out in 1969 when the IRA launched attacks against the British government. By 2007 more than 3,600 people had died as the result of bombings, shootings, and other forms of violence on both sides. Most of the victims were civilians, but 763 were British soldiers.

A landmark peace accord in 1998, known as the Good Friday Agreement, sought to end the violence through a series of compromises that kept Northern Ireland within the United Kingdom but with greater political rights and economic benefits for the Catholic minority. Hopes for peace suffered setbacks in 2002, when the provincial assembly stopped functioning because of partisan disagreements, and again in 2003, when hard-liners, notably Paisley, dominated parliamentary elections. Important progress toward peace was again made in 2005, when the IRA announced that it had destroyed ("decommissioned") all or nearly all of its weapons. Paisley and other militant Protestants voiced skepticism about the IRA's claim, but the move was verified by an independent committee, thus putting enormous pressure on Protestant militias to follow suit and on Paisley and other Protestant politicians to negotiate with the IRA's political leaders.

Another significant step forward was taken in October 2006, when Blair and Irish prime minister Bertie Ahern summoned Northern Ireland's key leaders to a conference in St. Andrews, Scotland. That conference produced a power-sharing agreement under which Paisley would head a new Northern Ireland government.

The St. Andrews Agreement also dealt with what long had been a central impediment to long-term peace: allegations by Catholics that the provincial police force, the Royal Ulster Constabulary, routinely targeted Catholics and had been infiltrated by members of Protestant militias. The British government had mandated reforms in the police force, including the addition of Catholic officers; at St. Andrews, Sinn Féin leaders agreed to support the new force, renamed the Police Service of Northern Ireland. Sinn Féin party members endorsed that position at a Dublin meeting on January 28, 2007, thus removing the last major barrier to reconciliation between Catholic and Protestant political leaders. "Today, you have created the potential to change the political landscape of this island forever," Sinn Féin leader Gerry Adams told about 2,000 of his fellow party members after they voted.

Putting the St. Andrews Agreement in Place

Throughout the past four decades, the prospect of peace in Northern Ireland arose repeatedly, only to be crushed by the refusal of one or both sides to compromise. By 2007 all the lost opportunities, particularly the 1998 Good Friday Agreement, appeared to have been necessary stepping stones that enabled the leaders of both factions to advance toward each another.

For Paisley, time was running out to achieve his long-term ambition to head the government of Northern Ireland. At age eighty-one, Paisley had set aside some of his earlier radical positions but still, according to his associates, retained his desire to be the dominant politician in Northern Ireland.

On the opposing side, Adams and other Sinn Féin leaders had conceded years earlier that violence had brought them no closer to their goal of uniting Northern Ireland with the Republic of Ireland. For them, long-term peace offered instead the hope that Catholics would share more fully in the recent economic prosperity seen in the Republic of Ireland and by most Protestants in the north. Northern Ireland was segregated, with nearly three-quarters of its 1.6 million residents living in communities known either as Catholic or as Protestant, and Catholics remained an underclass.

In this context, the St. Andrews Agreement, followed by Sinn Féin's endorsement of police reforms, set the stage for a rush of political change in Northern Ireland during the first five months of 2007. The next major step was the holding of elections for the 108 seats in the provincial assembly. The voting, on March 7–8, produced strong showings for the two leading hard-line parties: Paisley's Democratic Unionist Party won 30 percent of the total, and Sinn Féin won 26 percent. The balance was spread among smaller parties, including the more moderate parties whose leaders had been responsible for the 1998 Good Friday Agreement.

In theory, the strong showings by the two hard-line parties could have been a setback for the peace process. In practice, though, the voting gave leaders of both parties the political leverage they needed to take the final steps toward compromise. Political commentators across the spectrum said this was particularly true in Paisley's case because he now had full control of his party and could shunt aside those who were even more opposed to compromise than he had been.

Following the election, the leaders were working under a deadline: the British government had said it would abandon plans for a power-sharing deal in Northern Ireland unless the two sides agreed by March 26 to form a government. Although accompanied by the usual threats and ultimatums, negotiations to meet that deadline represented progress because they brought together bitter foes who long had refused to deal with each other.

Ultimately, the negotiations produced a historic agreement on the day of the March 26 deadline. Meeting in Belfast at Stormont, the provincial parliament building, Paisley and Adams announced that their parties would form a joint government for Northern Ireland on May 8. It was the first time the two men had met. Even the location of the meeting was important; for decades Stormont was perhaps the main symbol of Protestant domination in Northern Ireland, a place where Protestant factions held their rallies and where Catholics felt unwelcome.

Paisley, who for years had seemed to relish the nickname "Dr. No" for his heated denunciations of Catholicism and his vehement opposition to any compromise, suddenly was the voice of conciliation. "We must not allow our justified loathing of the horrors and tragedies of the past to become a barrier to creating a better and more stable future," he said while seated across a table from Adams, whom he had previously called a "terrorist." "In looking to the future we must never forget those who have suffered during the dark period from which we are, please God, emerging."

Adams, a former fighter for the IRA, expressed similar sentiments. "We are very conscious of the many people who have suffered," he said. "We owe it to them to build the best possible future. It is a time for generosity, a time to be mindful of the common good and the future of all our people."

Despite the conciliatory rhetoric, Paisley did not seem able to bring himself to shake hands with Adams, who by offering his hand appeared more willing to establish personal contact across the sectarian divide. At the conclusion of their historic meeting, Paisley gathered up his papers and left the room, avoiding even eye contact with his former enemy. Britain's chief representative in Northern Ireland, Peter Hain, called the meeting between Adams and Paisley "a graphic manifestation of the power of politics over bigotry and conflict, bitterness and horror."

A New Government Takes Office

In the weeks after the March 26 announcement, officials of the parties headed by Adams and Paisley negotiated the details of running a government together. While they did so, another significant event occurred. On May 3 the Ulster Volunteer Force, the oldest and most violent of the Protestant militias, renounced violence and said it had disbanded as a military force. The British government had blamed this group and its associated commando groups for the murders of more than 500 people, including the first murder in June 1967 that sparked the four decades of the Troubles. In some ways, the disbanding of the Ulster Volunteer Force matched earlier steps by the IRA, which had declared its own cease-fire and laid down its weapons. The two groups had something else in common:

over the decades of violence, the people of Northern Ireland had come to consider both more as criminal gangs than as legitimate warriors for political causes.

The ultimate step toward reconciliation was taken on May 8, as the newly elected parliament met at Stormont and formally installed Paisley as first minister and McGuinness as his deputy. McGuinness, a former IRA commander who once was banned from Britain as a "terrorist," took the leadership post rather than Sinn Féin chairman Adams, who was concentrating his efforts on the party's campaign for parliamentary elections in the Republic of Ireland.

Paisley and McGuinness had met for the first time only a few weeks earlier, but it appeared that they had already established the beginnings of a working relationship. In back-to-back speeches, both said the time for peace had come, and each pledged to do his part to ensure that peace stayed alive in Northern Ireland. "We are making this declaration, we are all aiming to build a Northern Ireland in which all can live together in peace, being equal under the law and equally subject to the law," Paisley said. For his part, McGuinness noted that the peace process had been encouraged by "friends" of Ireland worldwide: "We will continue to rely on that support as we strive toward a society moving from division and disharmony to one which celebrates our diversity and is determined to provide a better future for our people."

Despite these words, each man also insisted that he had not given up his ultimate goal: Paisley that Northern Ireland would remain within the United Kingdom, and McGuinness that the province one day would join a "united Ireland."

Among those in attendance at the ceremony were prime ministers Blair and Ahern, men who had staked much of their own political careers on bringing a final settlement to the troubles of Northern Ireland. In their speeches, both emphasized the historic nature of the occasion. Blair considered the past and future: "Look back, and we see centuries pockmarked by conflict, hardship, even hatred among the people of these islands. Look forward, and we see the chance to shake off these heavy chains of history." Ahern recalled the thousands who had been killed or wounded: "We cannot undo our sad and turbulent past and none of us can forget the many victims of the Troubles, but we can, and are, shaping our future in a new and better way. And in doing so we can put the divisions of the past behind us forever. Northern Ireland is now a place of peace and promise."

One of the new government's first, and symbolically most important, steps was to nominate members for a new Policing Board to oversee the province's revamped police force. For the first time, four representatives of Sinn Féin were to serve on the nineteen-member board, a move intended to assure Catholics that the police were no longer aligned against them.

Another significant moment came on August 1, when the British government officially drew down its military contingent in Northern Ireland to just 5,000 troops—less than one-fifth of the troop level at the height of the Troubles in the early 1970s. Nominally, British troops had been assigned to the province to "support" the police, but in reality the military found itself playing an active role in combating violence, at first by defending Catholics against Protestant militias and later by targeting IRA bombers. In recent years the British government had closed barracks and outposts throughout the province, notably posts in several Catholic-dominated areas that often had been flashpoints for violence.

In perhaps one of the greatest ironies of the history of Northern Ireland's Troubles, key leaders found themselves advising politicians from strife-torn Iraq on resolving sectarian differences. At a conference in Finland early in September, McGuiness and senior

Unionist Party official Jeffrey Donaldson helped Shiite and Sunni Muslim leaders from Iraq develop a framework for political harmony. "That we had come not as academics with theories but from the university of hard-life knocks and real-life conflicts, I think that made a powerful statement to them," Donaldson later told the *USA Today* newspaper.

Following are speeches delivered on May 8, 2007, by Rev. Ian Paisley and Martin McGuinness, following their installation as first minister and deputy first minister, respectively, of the provincial government of Northern Ireland.

DOCUMENT

Paisley's Address to the Northern Ireland Assembly

May 8, 2007

How true are the words of Holy Scripture, "We know not what a day may bring forth."

If anyone had told me that I would be standing here today to take this office, I would have been totally unbelieving. I am here by the vote of the majority of the electorate of our beloved Province. During the past few days I have listened to many very well placed people from outside Northern Ireland seeking to emphasise the contribution they claim to have made in bringing it about. However, the real truth of the matter is rather different.

If those same people had only allowed the Ulster people to settle the matter without their interference and insistence upon their way and their way alone, we would all have come to this day a lot earlier.

I remember well the night the Belfast Agreement was signed, I was wrongfully arrested and locked up on the orders of the then Secretary of State for Northern Ireland. It was only after the Assistant Chief of Police intervened that I was released. On my release I was kicked and cursed by certain loyalists who supported the Belfast Agreement. But that was yesterday, this is today, and tomorrow is tomorrow.

Today at long last we are starting upon the road—I emphasise starting—which I believe will take us to lasting peace in our Province. I have not changed my unionism, the union of Northern Ireland within the United Kingdom, which I believe is today stronger than ever.

We are making this declaration, we are all aiming to build a Northern Ireland in which all can live together in peace, being equal under the law and equally subject to the law. I welcome the pledge we have all taken to that effect today. That is the rock foundation upon which we must build.

Today we salute Ulster's honoured and unaging dead—the innocent victims, that gallant band, members of both religions, Protestant and Roman Catholic, strong in their allegiance to their differing political beliefs, Unionist and Nationalist, male and female, children and adults, all innocent victims of the terrible conflict. In the shadows of the evenings and in the sunrise of the mornings we hail their gallantry and heroism. It cannot and will not be erased from our memories.

Nor can we forget those who continue to bear the scars of suffering and whose bodies have been robbed of sight, robbed of hearing, robbed of limbs. Yes, and we must all shed the silent and bitter tear for those whose loved ones' bodies have not yet been returned.

Let me read to you the words of Deirdre Speer who lost her Police officer father in the struggle:

Remember me! Remember me!
My sculptured glens where crystal rivers run,
My purple mountains, misty in the sun,
My coastlines, little changed since time begun,
I gave you birth.
Remember me! Remember me!
Though battle-scarred and weary I abide.
When you speak of history say my name with pride.
I am Ulster.

In politics, as in life, it is a truism that no-one can ever have one hundred percent of what they desire. They must make a verdict when they believe they have achieved enough to move things forward. Unlike at any other time I believe we are now able to make progress.

Winning support for all the institutions of policing has been a critical test that today has been met in pledged word and deed. Recognising the significance of that change from a community that for decades demonstrated hostility for policing, has been critical in Ulster turning the corner.

I have sensed a great sigh of relief amongst all our people who want the hostility to be replaced with neighbourliness. The great king Solomon said, "To everything there is a season, and a time to every purpose under heaven.

A time to be born and a time to die.
A time to plant and a time to pluck up that which is planted.
A time to kill and a time to heal.
A time to break down and a time to build up.
A time to get and a time to lose.
A time to keep and a time to cast away.
A time to love and a time to hate.
A time of war and a time of peace."

I believe that Northern Ireland has come to a time of peace, a time when hate will no longer rule. How good it will be to be part of a wonderful healing in our Province. Today we have begun to plant and we await the harvest.

Source: United Kingdom. Northern Ireland Executive. Office of the First Minister and Deputy First Minister. "Speeches Delivered by First Minister and Deputy First Minister Today in Parliament Buildings, Belfast." May 8, 2007. www.northernireland.gov.uk/news/news-ofmdfm/news-ofmdfm-may-2007/news-ofmdfm-080507-speeches-delivered-by.htm (accessed October 3, 2007).

DOCUMENT

McGuinness's Address to the Northern Ireland Assembly

May 8, 2007

I am proud to stand here today as an Irish Republican who believes absolutely in a United Ireland. I too wish to welcome the Prime Minister, Tony Blair, and the Taoiseach [prime minister of the Republic of Ireland], Bertie Ahern, and all our friends from around the world whose encouragement and support helped us reach this day.

Many people in this hall today played an important part in our peace process. Many others could not be with us today. I want to send our warmest thanks to them. We will continue to rely on that support as we strive towards a society moving from division and disharmony to one which celebrates our diversity and is determined to provide a better future for all our people.

One which cherishes the elderly, the vulnerable, the young and all of our children equally. Which welcomes warmly those from other lands and cultures who wish to join us and forge a future together.

A society which remembers those who have lost their lives. Last Saturday I spent time with families in County Tyrone who had lost loved ones. They and many others throughout our community have suffered and continue to suffer as a result of our difficult and painful past. So we must look to the future to find the means to help them heal.

We must also focus on the practical. To build we need the tools and as I have said we look to our friends on these islands and beyond to provide the practical support we need.

As joint heads of the Executive the First Minister and I pledge to do all in our power to ensure it makes a real difference to the lives of all our people by harnessing their skills through a first rate education system, caring for our sick in the best health service we can provide and building our economy through encouraging investment and improving our infrastructure.

We know that this will not be easy and the road we are embarking on will have many twists and turns. It is however a road which we have chosen and which is supported by the vast majority of our people. In the recent elections they voted for a new political era based on peace and reconciliation.

On the evening of the Assembly election results I received a phone call from a 100 years young woman, Molly Gallagher, in County Donegal. She told me she was very happy with the election results and that she was looking forward to seeing Ian Paisley and myself together. I'm sure she is watching us today. Hello Molly!

As for Ian Paisley, I want to wish you all the best as we step forward towards the greatest yet most exciting challenge of our lives.

Ireland's greatest living poet, a fellow Derry man, Seamus Heaney, once told a gathering that I attended at Magee University that for too long and too often we speak of the others or the other side and that what we need to do is to get to a place of through otherness. The Office of the First and Deputy First Ministers is a good place to start. This will only work if we collectively accept the wisdom and importance of Seamus Heaney's words.

Since March 26 much work has been done which has confounded critics and astounded the sceptics.

Like these talented people from Sky's the Limit, who entertained us so wonderfully today, we must overcome the difficulties which we face in order to achieve our goals and seize the opportunities that exist. This, and future generations expect and deserve no less from us.

SOURCE: United Kingdom. Northern Ireland Executive. Office of the First Minister and Deputy First Minister. "Speeches Delivered by First Minister and Deputy First Minister Today in Parliament Buildings, Belfast." May 8, 2007. www.northernireland.gov.uk/news/news-ofmdfm/news-ofmdfm-may-2007/news-ofmdfm-080507-speeches-delivered-by.htm (accessed October 3, 2007).

OTHER HISTORIC DOCUMENTS OF INTEREST

FROM THIS VOLUME

- Tony Blair steps down as British prime minister, p. 221
- Politics in Iraq, p. 477

FROM PREVIOUS *HISTORIC DOCUMENTS*

- Irish Republican Army on an end to violence, *2005,* p. 507

DOCUMENT IN CONTEXT

Outgoing and Incoming Prime Ministers on British Politics

MAY 10 AND JUNE 27, 2007

Tony Blair, who revived Britain's sickly Labour Party in the mid-1990s, then won acclaim and suffered derision as prime minister, stepped down from power in June after serving more than ten years in office. Blair was Britain's second-longest-serving prime minister in modern times—just behind the also controversial Margaret Thatcher, who governed the country with an iron hand during the 1980s. Blair was succeeded by Chancellor of the Exchequer Gordon Brown, who had waited, with diminishing patience in recent years, for the turn in office that Blair had promised him before they came to power together in 1997.

The contrast between the glib Blair and the dour Brown could not have been greater. When Blair entered office, critics called him "Tony Blur" for his tendency to make flashy speeches that seemed carefully tailored, by polling and the use of focus groups, to enhance his popularity. During his years in power, however, Blair cast himself as a leader unafraid to confront grand questions of domestic and foreign policy. It was his stance on one such issue—his unwavering support for the U.S. invasion and subsequent occupation of Iraq—that undermined his popularity and was likely to affect his legacy.

Brown, on the other hand, seemed comfortable with his public image as a workman-like, sometimes brooding politician, from solid Scottish stock, who would focus on the unglamorous aspects of governing that Blair had been accused of neglecting. For a while, Brown's publicists portrayed him as "Not Flash, Just Gordon," an ironic comparison with the 1930s comic book and movie hero.

Although Blair left the prime minister's official London residence at Number 10 Downing Street in June, he did not disappear from the public eye. He quietly campaigned for, and won, an appointment as the chief envoy for the Quartet, the diplomatic assemblage of the European Union (EU), Russia, the United Nations (UN), and the United States, that was promoting peace between the Israelis and the Palestinians. Blair appeared to have higher ambitions, however. At year's end he was mounting another, thinly disguised campaign, this time to win appointment as the first EU president under a new charter adopted by EU leaders in December.

Blair's Legacy

The Labour Party had been out of power for eighteen years when a brash forty-three-year-old politician named Tony Blair seized control of it in 1994 and came up with the marketing slogan of "New Labour." The word *new* was intended to distinguish Blair's revamped party from the tired organization, long dominated by labor unions, that had won elections only occasionally and could never stay in office for long. Blair also portrayed himself, and his party, as belonging to a "third way" of progressive politics in Western democracies—selecting policies from both conservatives and liberals— championed by then president Bill Clinton in the United States.

Describing the Labour Party as new also put it in contrast with the Conservative Party, which, under Thatcher and then John Major, had been in office since 1979; by the early 1990s, that party had run out of political petrol and was afflicted by scandals. Blair won an unexpectedly large landslide victory on May 1, 1997, making him prime minister the next day with a huge parliamentary majority of 179 seats. Blair and Labour won another landslide victory in June 2001, then held onto office with a sharply diminished majority after the next elections in May 2005.

Blair had many achievements as prime minister, both in domestic affairs and in foreign policy, which increasingly became his focus of attention. He had been in office just four months when voters in Scotland and Wales supported "devolution" plans, which Blair had advocated, that transferred some powers from the central government in London to local officials—a move that eased nationalist pressures that had been building because of Scottish and Welsh resentment of English authority. Over the years, Blair's government also poured money into Britain's schools and into the National Health Service—once the crown jewel of the country's social service system but by the 1990s known better for forcing patients to wait weeks or even months to receive indifferent medical care. Britain's economy also grew at a steady pace during the Blair years, with low inflation and unemployment, in marked contrast to the sluggish economies of France, Germany, and Italy, the other major European powers.

Blair even managed to shake up one of the oldest of all British institutions, the House of Lords. In October 1999 he negotiated a deal with other parties that essentially ended the right of some 600 hereditary peers to sit in that chamber. Subsequent attempts to set a new formula for selecting members of the upper house fell short, however, and the matter was still pending when Blair stepped down.

What was perhaps Blair's greatest achievement involved another matter with a long history: the violent conflict between Catholics and Protestants in Northern Ireland. With the help of important partners—including former president Clinton, former U.S. senator George Mitchell, and Irish prime minister Bertie Ahern—Blair pushed and pushed until the central figures in the conflict gradually caught up with the public's desire for peace. It was no coincidence that Blair announced his resignation just two days after his peacemaking culminated, on May 8, in the installation of a unity government for Northern Ireland, something that had been unimaginable for many years.

Despite his successes, Blair's final years in office were beset by two challenges that tarnished his image and affected his popularity. The more consequential challenge was Blair's embrace of Bush's decision to invade Iraq in 2003. Perhaps less important in historic terms, but no less damaging politically, was a scandal over the Labour Party's alleged awarding of peerages and other honors in exchange for campaign contributions.

Blair was an early convert, in 2002, to the argument by "neoconservatives" in the United States that Iraqi leader Saddam Hussein had to be deposed. The Bush administration linked Hussein, without offering evidence, to the al Qaeda network responsible for the September 11, 2001, attacks against the United States. Bush also repeatedly insisted, with support from Blair and several other European leaders, that the Iraqi regime possessed stores of biological and chemical weapons and was working to build nuclear weapons.

During the months before the March 2003 U.S.-led invasion of Iraq, Blair made the case for the invasion with just as much fervor as Bush. More importantly, Blair contributed 40,000 British service personnel to participate in the invasion, making Britain the cornerstone of what U.S. officials called a "coalition of the willing." After Saddam was ousted, Britain also played a crucial role in the occupation of Iraq by securing Basra, the country's second largest city.

Unfortunately for Blair, his fellow Britons gave him only lukewarm support for the Iraq War at the beginning; as the occupation dragged on and violence increased in Iraq, opposition mounted steadily. Blair managed to win reelection to a third term in May 2005 despite growing anti-war sentiment, but he owed that victory as much to weak leadership of the opposition Conservative Party as to public support for his policies.

Two months after that election, Britain faced one of its most serious terrorist incidents of modern times. On July 7, 2005, four suicide bombers attacked the London transport system, killing fifty-two people, plus themselves. Britons were particularly shocked by the attack because three of the bombers were British-born men whose families had emigrated from Pakistan—they had lived all their lives in Britain, attending local schools and supposedly absorbing British culture. Blair's government, which had already adopted stringent antiterrorism policies, tightened controls even further, resulting in complaints that traditional British civil liberties were being undermined.

The scandal over the alleged sale of peerages cut just as deeply into Blair's popularity as had his support for the Iraq War, in part because it reinforced public perceptions that he was a politician willing to do or say anything to win an election. The scandal erupted in 2006 with reports that, before the 2005 elections, the Labour Party had accepted large loans from wealthy supporters in exchange for promises that the donors would be given honors, such as knighthoods or "life peerages" (the nonhereditary title of Lord or Lady for the person's lifetime). It long had been assumed that prime ministers awarded such peerages to political supporters, but these allegations involved specific instances of quid pro quo arrangements. Several senior Labour Party officials and top aides to Blair were investigated during 2006 and early 2007, and police even questioned Blair about the matter twice, making him the first sitting prime minister to be questioned as a witness in a criminal inquiry. Ultimately, the investigation led nowhere; British prosecutors announced on July 20 that no charges would be brought in the case because of "insufficient evidence."

Blair Steps Down

Months before standing for reelection for a third time, in May 2005, Blair said he would not seek an unprecedented fourth term as prime minister. This stance immediately raised questions about when he would step down in favor of Brown, his presumed successor. Blair parried such questions until September 7, 2006, two days after fifteen Labour Party parliamentarians sent him a letter calling for an "urgent" change of party leadership. Blair finally said he intended to step down within one year. He again offered no firm date, but it was clear that he wanted to serve at least until early May 2007, when he would reach his tenth anniversary in office.

Almost daily reports about the cash-for-honors scandal added to the drumbeat of questions about Blair's departure date. The announcement was finally made on May 10, two days after Blair presided over the ceremony in Belfast that marked the triumphal conclusion to his dogged efforts to bring peace to Northern Ireland. For the bittersweet moment, Blair returned to his constituency in Sedgefield, in northeastern England, where he addressed more than two hundred supporters at a crowded Labour Party clubhouse.

In an emotional, reflective speech announcing that he would resign on June 27, Blair noted that he had just reached his tenth anniversary: "In this job, in the world today, that is long enough, for me, but more especially for the country." Blair admitted that he had not fulfilled all of the "great expectations" created by his landslide electoral wins, but he insisted his government had solid accomplishments nonetheless. "There is only one government since 1945 that can say all of the following: 'More jobs, fewer unemployed, better health and education results, lower crime and economic growth in every quarter,'— this one," he said. Blair also offered this simple response to his many critics: "Accept one thing, hand on heart, I did what I thought was right."

Recalling that his decision to support the invasion of Iraq had not met with universal approval in Britain, and that there had been "blowback" in the form of increased terrorism, the prime minister insisted that he had made the right choice. "For me, I think we must see it through," he said. "They, the terrorists, who threaten us here and round the world, will never give up if we give up."

As could be expected, Blair's counterpart in the Conservative Party, David Cameron, offered an unsympathetic assessment. "I think a lot of people will look back on the last ten years of dashed hopes and big disappointments, of so much promised, so little delivered," he said.

Before he stepped down, Blair made one last official trip to Washington, D.C., for a visit at the White House with Bush on May 17. Despite their many differences on matters other than the Iraq War, the two men had formed a close bond. Bush clearly relished the support he received from Blair, one of the few world leaders to embrace his Iraq mission wholeheartedly.

Brown Takes Over

Although Blair and Brown were former office mates and political allies from their early days in Parliament in the 1980s and had weathered many storms together, in recent years they had grown apart. Less than a year after Blair became prime minister, the two men had their first major public feud, and a source close to Blair was quoted as saying Brown was "psychologically flawed." Their relationship reportedly soured from that point on.

Over the years, several Labour Party members put themselves forward as alternatives to Brown as Blair's eventual successor. One by one they dropped away until Brown stood out as the man who, finally, would get the post he believed Blair had promised him years earlier. Unfortunately for Brown, by the time his chance came, the public's restiveness with Tony Blair had rubbed off on the Labour Party, which now faced stiff competition from a rejuvenated Conservative Party led by Cameron, who had Blair's gift for gab. "His dread must be that he will finally get to sup from the chalice only to find that it is poisoned," commentator Andrew Rawnsley wrote in *The Observer* magazine about Brown's ascendancy.

Brown announced his candidacy for the Labour Party leadership on May 11, having secured the expected formal endorsement from Blair. Attempting to demonstrate that he was his own man, Brown suggested subtle changes, such as calling for government to be "more open" and promising Parliament a greater voice in decisions to go to war, and declared that politics was not "about celebrity" but about "what you believe in and what you want to achieve."

The Labour Party formally elected Brown as its leader, and thus as the next prime minister, on June 24. The transfer of power took place on June 27, after Blair engaged in one last "question time" in the House of Commons, the weekly exercise during which prime ministers and their top cabinet members are subjected to often-brutal questions and comments from fellow legislators. He received a rare standing ovation, even from Conservatives across the aisle. Later in the day, diplomats meeting at the UN officially named Blair as special envoy for the Quartet's negotiations in the Middle East peace process.

After a ceremonial meeting with Queen Elizabeth II at Buckingham Palace, Brown made his first statement at Downing Street as prime minister and emphasized that changes were on the way; in a brief speech of less than ten minutes, he used the word "change" eight times. "This will be a new government with new priorities," he said. Brown introduced his "new" cabinet the following day. Most of its members were holdovers from Blair's last cabinet, although all but one of them were in new jobs.

The new team faced its first challenges almost immediately. On June 30 police discovered two bomb-laden cars in downtown London, and on July 1 two men attempted to crash a sports utility vehicle into Glasgow airport, setting the vehicle ablaze. These events raised fears of another major terrorist attack, just two years after the bombings in the London transportation system. Brown succeeded in calming the jittery nation and won high marks from a usually combative press corps, with a somber televised speech and a subsequent interview on the BBC.

Brown won more kudos in mid-July when he responded with concern and sympathy to widespread flooding in rural areas, the worst to hit Britain in more than 150 years. Polls showed Britons warming to their new leader and, correspondingly, to the Labour Party, which surged ahead of the Conservatives for the first time in months.

There was much speculation, on both sides of the Atlantic, about how well Brown would get along with Bush. Although Brown had loyally supported Blair's Iraq policy, he had appeared to do so with diminishing enthusiasm, leading some observers to expect a frosty meeting between the new prime minister and the president. The two men appeared to get along well during a meeting at the Camp David presidential retreat on July 30, although Brown was not as effusive in his praise of Bush as Blair had been. Brown signaled no dramatic changes in British policy toward Iraq, a stance he maintained at a London news conference on September 4. Brown visited Iraq the following month, however, and subsequently announced that the British force level there would be cut in half, to about 2,500 troops, by the spring of 2008. In November Britain carried out a previously scheduled handover to the Iraqis of responsibility for security in the Basra area.

Just as Brown was basking in the warm glow of a political honeymoon, he did something that raised doubts about his decisiveness. After a successful Labour Party conference in late September, Brown's aides suggested he would call a "snap" election to capitalize on the party's new lead in the polls. Brown allowed expectations for such an election to grow but on October 7 dashed them, saying it was more important that his government "got on with the job" than it was to hold early elections. One apparent explanation for the turnabout was a sudden rise in the polls for the Conservatives, who at their annual meeting pledged to abolish the unpopular inheritance tax on estates worth under one million pounds (about $2 million). Cameron, the Conservative leader, was quick to jump on the offensive: addressing Brown during the October 10 House of Commons question time, he asked, "Do you realize what a phony you now look?"

As often happens when a political leader shows weakness, a series of problems cascaded down on Brown. Among the most serious were a run on a major bank (Northern

Rock) as the result of a global liquidity crisis that originated in the United States, the revelation that the government had lost computer disks containing detailed personal information on some 25 million Britons, and reports that the Labour Party had improperly accepted about $1 million from a real estate developer. The fundraising scandal was especially damaging to Brown because it echoed the political scandals of Blair's last years in office. In this case, Brown's seriousness may have been a disadvantage. Unlike Blair, he seemed unable to brush off criticism with a loquacious and disarming comeback. At year's end, Britons once again saw Gordon Brown scowling, with his brow furrowed.

Following are two documents. First, the text of a speech by British prime minister Tony Blair, on May 10, 2007, announcing his intention to resign his office. Second, a statement by new prime minister Gordon Brown at 10 Downing Street on June 27, 2007.

DOCUMENT

Prime Minister Blair on His Resignation

May 10, 2007

I have come back here, to Sedgefield, to my constituency. Where my political journey began and where it is fitting it should end.

Today I announce my decision to stand down from the leadership of the Labour Party. The Party will now select a new Leader. On 27 June I will tender my resignation from the office of Prime Minister to The Queen.

I have been Prime Minister of this country for just over 10 years. In this job, in the world today, that is long enough, for me but more especially for the country. Some times the only way you conquer the pull of power is to set it down.

It is difficult to know how to make this speech today. There is a judgment to be made on my premiership. And in the end that is, for you, the people to make.

I can only describe what I think has been done over these last 10 years and perhaps more important why.

I have never quite put it like this before.

I was born almost a decade after the Second World War. I was a young man in the social revolution of the 60s and 70s. I reached political maturity as the Cold War was ending, and the world was going through a political, economic and technological revolution.

I looked at my own country.

A great country.

Wonderful history.

Magnificent traditions.

Proud of its past.

But strangely uncertain of its future. Uncertain about the future. Almost old-fashioned.

All of that was curiously symbolized in its politics.

You stood for individual aspiration and getting on in life or social compassion and helping others.

You were liberal in your values or conservative.

You believed in the power of the State or the efforts of the individual. Spending more money on the public realm was the answer or it was the problem.

None of it made sense to me. It was 20th century ideology in a world approaching a new millennium. Of course people want the best for themselves and their families but in an age where human capital is a nation's greatest asset, they also know it is just and sensible to extend opportunities, to develop the potential to succeed, for all not an elite at the top.

People are today open-minded about race and sexuality, averse to prejudice and yet deeply and rightly conservative with a small 'c' when it comes to good manners, respect for others, treating people courteously.

They acknowledge the need for the state and the responsibility of the individual.

They know spending money on our public services matters and that it is not enough. How they are run and organized matters too.

So 1997 was a moment for a new beginning; for sweeping away all the detritus of the past.

Expectations were so high. Too high. Too high in a way for either of us.

Now in 2007, you can easily point to the challenges, the things that are wrong, the grievances that fester.

But go back to 1997. Think back. No, really, think back. Think about your own living standards then in May 1997 and now.

Visit your local school, any of them round here, or anywhere in modern Britain.

Ask when you last had to wait a year or more on a hospital waiting list, or heard of pensioners freezing to death in the winter unable to heat their homes.

There is only one Government since 1945 that can say all of the following:

More jobs

Fewer unemployed

Better health and education results

Lower crime;

And economic growth in every quarter.

This one.

But I don't need a statistic. There is something bigger than what can be measured in waiting lists or GCSE results [General Certificate of Secondary Education, national education tests] or the latest crime or jobs figures.

Look at our economy. At ease with globalization. London the world's financial centre. Visit our great cities and compare them with 10 years ago.

No country attracts overseas investment like we do.

Think about the culture of Britain in 2007. I don't just mean our arts that are thriving. I mean our values. The minimum wage. Paid holidays as a right. Amongst the best maternity pay and leave in Europe. Equality for gay people.

Or look at the debates that reverberate round the world today. The global movement to support Africa in its struggle against poverty. Climate change. The fight against terrorism. Britain is not a follower. It is a leader. It gets the essential characteristic of today's world: its interdependence.

This is a country today that for all its faults, for all the myriad of unresolved problems and fresh challenges, is comfortable in the 21st Century.

At home in its own skin, able not just to be proud of its past but confident of its future.

I don't think Northern Ireland would have been changed unless Britain had changed. Or the Olympics won if we were still the Britain of 1997.

As for my own leadership, throughout these 10 years, where the predictable has competed with the utterly unpredicted, right at the outset one thing was clear to me.

Without the Labour Party allowing me to lead it, nothing could ever have been done. But I knew my duty was to put the country first. That much was obvious to me when just under 13 years ago I became Labour's Leader.

What I had to learn, however, as Prime Minister was what putting the country first really meant.

Decision-making is hard. Every one always says: listen to the people. The trouble is they don't always agree.

When you are in Opposition, you meet this group and they say why can't you do this? And you say: it's really a good question. Thank you. And they go away and say: its great, he really listened.

You meet that other group and they say: why can't you do that? And you say: it's a really good question. Thank you. And they go away happy you listened.

In Government you have to give the answer, not an answer, the answer.

And, in time, you realise putting the country first doesn't mean doing the right thing according to conventional wisdom or the prevailing consensus or the latest snapshot of opinion.

It means doing what you genuinely believe to be right.

Your duty is to act according to your conviction.

All of that can get contorted so that people think you act according to some messianic zeal.

Doubt, hesitation, reflection, consideration and re-consideration these are all the good companions of proper decision-making.

But the ultimate obligation is to decide.

Sometimes the decisions are accepted quite quickly. Bank of England independence was one, which gave us our economic stability.

Sometimes like tuition fees or trying to break up old monolithic public services, they are deeply controversial, hellish hard to do, but you can see you are moving with the grain of change round the world.

Sometimes like with Europe, where I believe Britain should keep its position strong, you know you are fighting opinion but you are content with doing so.

Sometimes as with the completely unexpected, you are alone with your own instinct.

In Sierra Leone and to stop ethnic cleansing in Kosovo, I took the decision to make our country one that intervened, that did not pass by, or keep out of the thick of it.

Then came the utterly unanticipated and dramatic. September 11th 2001 and the death of 3,000 or more on the streets of New York.

I decided we should stand shoulder to shoulder with our oldest ally.

I did so out of belief.

So Afghanistan and then Iraq.

The latter, bitterly controversial.

Removing Saddam and his sons from power, as with removing the Taliban, was over with relative ease.

But the blowback since, from global terrorism and those elements that support it, has been fierce and unrelenting and costly. For many, it simply isn't and can't be worth it.

For me, I think we must see it through. They, the terrorists, who threaten us here and round the world, will never give up if we give up.

It is a test of will and of belief. And we can't fail it.

So: some things I knew I would be dealing with.

Some I thought I might be.

Some never occurred to me on that morning of 2 May 1997 when I came into Downing Street for the first time.

Great expectations not fulfilled in every part, for sure.

Occasionally people say, as I said earlier, they were too high, you should have lowered them.

But, to be frank, I would not have wanted it any other way. I was, and remain, as a person and as a Prime Minister an optimist. Politics may be the art of the possible; but at least in life, give the impossible a go.

So of course the vision is painted in the colours of the rainbow; and the reality is sketched in the duller tones of black, white and grey.

But I ask you to accept one thing. Hand on heart, I did what I thought was right.

I may have been wrong. That's your call. But believe one thing if nothing else. I did what I thought was right for our country.

I came into office with high hopes for Britain's future. I leave it with even higher hopes for Britain's future.

This is a country that can, today, be excited by the opportunities not constantly fretful of the dangers.

People often say to me: it's a tough job.

Not really.

A tough life is the life the young severely disabled children have and their parents, who visited me in Parliament the other week.

Tough is the life my Dad had, his whole career cut short at the age of 40 by a stroke.

I have been very lucky and very blessed.

This country is a blessed nation.

The British are special.

The world knows it.

In our innermost thoughts, we know it.

This is the greatest nation on earth.

It has been an honour to serve it. I give my thanks to you, the British people, for the times I have succeeded, and my apologies to you for the times I have fallen short.

Good Luck.

SOURCE: United Kingdom. Labour Party. "Tony Blair's Statement on 10th May." May 10, 2007. www.labour.org.uk/leadership/tony_blair_resigns (accessed January 29, 2008).

DOCUMENT

Prime Minister Brown on the New British Government

June 27, 2007

I have just accepted the invitation of Her Majesty The Queen to form a Government. This will be a new Government with new priorities and I have been privileged to have been granted the great opportunity to serve my country and at all times I will be strong in purpose, steadfast in will, resolute in action in the service of what matters to the British people, meeting the concerns and aspirations of our whole country.

I grew up in the town that I now represent in Parliament. I went to the local school. I wouldn't be standing here without the opportunities that I received there and I want the best of chances for everyone. That is my mission, that if we can fulfil the potential and realise the talents of all our people then I am absolutely sure that Britain can be the great global success story of this century.

As I have travelled round the country, and as I have listened I have learnt from the British people—and as Prime Minister I will continue to listen and learn from the British people—I have heard the need for change, change in our NHS [National Health Service], change in our schools, change with affordable housing, change to build trust in Government, change to protect and extend the British way of life. And this need for change cannot be met by the old politics. So I will reach out beyond narrow Party interests, I will build a government that uses all the talents, I will invite men and women of goodwill to contribute their energies in a new spirit of public service to make our nation what it can be. And I am convinced that there is no weakness in Britain today that cannot be overcome by the strengths of the British people.

On this day I remember words that have stayed with me since my childhood and which matter a great deal to me today, my school motto: "I will try my utmost". This is my promise to all of the people of Britain and now let the work of change begin.

Thank you.

Source: United Kingdom. 10 Downing Street. "Statement at Downing Street." June 27, 2007. www.pm.gov.uk/output/Page12155.asp (accessed January 12, 2008).

Other Historic Documents of Interest

From this volume

- U.S. policy in Iraq, p. 9
- Mortgage and housing crisis, p. 514
- Middle East peace negotiations, p. 680
- New European Union charter, p. 721

FROM PREVIOUS *HISTORIC DOCUMENTS*

- Queen Elizabeth's speech to Parliament, *1997,* p. 277
- European leaders on the prospect of a war in Iraq, *2003,* p. 40
- Bombings in London, *2005,* p. 393

DOCUMENT IN CONTEXT

Chrysler Executives on Takeover, Export Deal, and Job Reductions

MAY 14, JULY 3, AND NOVEMBER 1, 2007

Two significant developments—successful labor negotiations and the takeover of Chrysler by a private equity firm—breathed new life into the "Big Three" American automakers in 2007. Whether the developments were enough to stave off further financial problems remained to be seen. Continuing high gasoline prices, coupled with the crisis in the housing market and the increasingly gloomy outlook for the economy, slowed car sales toward the end of the year. Automakers predicted that sales were likely to worsen in 2008.

Meanwhile, General Motors (GM) barely held onto its position as the world's number one automaker. Throughout 2007 GM vied with Toyota to be first in global sales; when the final count was in, GM had sold 9,369,524 vehicles, about 3,000 more than Toyota had. It was GM's second-best sales year on record, behind 1978, and was based on strong sales overseas. Sales in the domestic market fell 6 percent, however. Even Toyota had a weak year in U.S. sales, gaining only 3 percent over 2006, after posting three straight years of double-digit annual sales growth.

Long-standing Troubles in Detroit

Many of the problems facing U.S. car manufacturers had been in the making for decades. Critics had long complained that the American companies were hurting themselves by making the cars they wanted to make, rather than the cars consumers wanted to buy. During the oil crisis of the 1970s and the economic slowdown that followed, American buyers turned to the more fuel-efficient, reliable Japanese cars that were just beginning to be imported into the United States. The Big Three found themselves in financial trouble as they scrambled to deal with new fuel economy, clean air, and safety regulations while also retooling their factories to follow the "lean production" patterns of their foreign rivals.

With the general pickup in the economy in the mid-1990s, domestic automakers increasingly left the markets for small and luxury cars to their foreign competitors and concentrated on building and selling popular and highly profitable pickup trucks and sports utility vehicles. When spikes in world oil prices pushed gasoline prices above $2.00 per gallon in 2004 and then to near or above $3.00 per gallon in parts of 2005 and much of 2006, the U.S. car manufacturers had few smaller, more fuel-efficient cars to offer buyers.

Another problem for the American automakers was their wage and benefit structure. Union wages at the domestic car companies' factories were roughly twice as high as those at nonunion U.S. factories run by foreign makers. Moreover, the companies' generous pension and health benefit costs had mounted as their workforce aged and retired. By the mid-2000s, GM, Ford, and Chrysler had run up enormous "legacy" costs, which they said added about $1,800 to the cost of every car. Much of this obligation was unfunded, leading to a lowering of their credit ratings, which in turn made it more expensive for them to borrow money to invest in plants and new development.

Between 2005 and early 2007, each of the Big Three had announced restructuring plans; together, they would cut about 70,000 jobs and close dozens of factories in the United States and Canada. In 2005 GM and Ford won special concessions on health care for retirees, and in 2006 the two companies negotiated buyouts that led to early retirements for tens of thousands of their workers. But with the three companies losing a combined total of $16 billion in 2006, negotiations scheduled to start in July with the United Auto Workers (UAW) for new four-year contracts were expected to take on added urgency.

Labor Negotiations

Despite some tough talk going into the negotiations, UAW leaders and many union workers seemed resigned to granting some significant concessions in an effort to keep their jobs. The focal point of the talks, which formally began on July 20, was a proposal by the automakers that would lift the burden of their legacy costs. Under the plan, each automaker would contribute a lump-sum payment to a special trust fund, called a voluntary employee benefit association (VEBA). The contribution would effectively wipe out the car makers' health care liabilities for retired workers. That responsibility would be assumed by the UAW, which would invest the assets and pay out health care and other benefits to retired workers.

The UAW typically bargained with one automaker at a time, with the expectation that the deal reached with the first company would set the pattern for the remaining negotiations. In 2007 the union chose to negotiate first with General Motors. The two parties reached a landmark agreement on September 26—but not before union workers struck GM plants nationwide for two days. As anticipated, the union agreed to establish a VEBA, and GM agreed to make a payment in cash, stock, and other assets, worth about 70 percent of its $55 billion liability for health care benefits for current and future retirees. The initial investment was set at $24.1 billion, with an additional contribution of $5.8 billion to follow. GM also pledged to add a maximum of $1.6 billion if the trust did not earn as much on its investments as GM anticipated. As the VEBA required court and regulatory approval, which was likely to take two years, it was scheduled to take effect on January 1, 2010. Until then, GM would maintain its current retiree health care plan. Establishment of the VEBA was expected to save GM as much as $3 billion per year.

In an important concession to union workers, GM agreed to maintain its current union workforce at 73,000, providing them with about $13,000 in bonuses and cost of living increases over the life of the contract. In a major concession to GM, the union agreed to lower wages for new hires in what were called "noncore" positions, such as subassembly and materials handling. These workers would be paid $14.00–14.61 an hour, roughly half as much as the workers who currently held the jobs. Estimates showed that after an anticipated round of worker buyouts, about one-third of GM's union workforce might be new hires paid at the new, lower rate. Union workers ratified the new contract on October 10, with about two-thirds of them approving it.

On October 10 the UAW announced it had reached a similar agreement with Chrysler negotiators—six hours after the union began a national strike at the automaker's plants across the country. That impasse was reportedly over job guarantees—the company's new owner, Cerberus Capital Management, was said to be reluctant to commit to a specific number of new products or jobs while it was in the middle of a restructuring. In the end, Chrysler officials eased back on some of their plans for plant closings.

Like GM, Chrysler shed a hefty proportion of its legacy costs, agreeing to contribute $8.8 billion to fund a VEBA. The UAW and Chrysler also identified some 11,000 of the company's 45,000 union jobs as noncore, for which new hires would be paid about half as much as workers currently holding those jobs. That provision, along with a lack of job guarantees at some plants, jeopardized ratification of the deal. Workers at several Chrysler assembly plants voted against the contract, but in the end a narrow majority voted for it, and the new contract was ratified on October 28.

Less than a week later, Ford and the UAW announced on November 3 that they had reached agreement on a new four-year contract that closely followed the pattern of the earlier two. Ratified by nearly four-fifths of Ford's 54,000 union workers on November 14, the contract called for a VEBA that would allow Ford to shed about $23 billion in future liabilities and set up a two-tier wage scale for noncore jobs. Ford also agreed to keep open six plants that it had planned to close.

Sale of the Chrysler Group

The second major development in the U.S. auto industry in 2007 was the sale of the Chrysler Group to Cerberus Capital Management, a private equity firm. In 1998 Chrysler had merged with the German company Daimler Benz to become DaimlerChrysler, in a move to expand the markets of both companies by exploiting each other's strengths. Largely because of differences in management and marketing philosophies, the merged entity never ran very smoothly, and Chrysler's sagging sales in its North American market threatened the rest of the company. Chrysler restructured at the start of the twenty-first century and was the only one of the Big Three to begin 2006 with an operating profit. But the company ended that year with a $1.5 billion loss, a large excess inventory, and the possibility of yet another restructuring.

On February 14, 2007, Chrysler announced that it was undertaking another restructuring that would eliminate 13,000 jobs in the United States and Canada and close all or parts of four plants. On the same day, DaimlerChrysler said all options were open regarding the Chrysler Group's future. Three months later, on May 14, DaimlerChrysler announced it was selling an 80.1 percent share of the Chrysler Group to Cerberus. Under the deal, Cerberus agreed to make a capital contribution of $7.4 billion to the new Chrysler holding company and to pay Daimler $1.35 billion. Cerberus also agreed to take over all of Chrysler's health care and pension liabilities, valued at $18 billion. Daimler reported that it expected to pay out $1.6 billion before the deal closed to cover Chrysler's negative cash flow, and the German company also agreed to pay closing costs. In sum, Daimler effectively paid $650 million to sell a company it had bought for $36 billion just nine years earlier.

Cerberus, which was chaired by former Treasury secretary John W. Snow, was one of the largest private equity firms in the United States, with holdings in companies in a variety of markets, such as pharmaceuticals and paper products, retail and grocery stores, and financial services. The purchase of Chrysler was its most ambitious undertaking to date, and many observers questioned whether Cerberus would have the time or the patience to see through a turnaround of the troubled carmaker. UAW president Ron

Gettelfinger had initially opposed the sale, believing Cerberus would "strip and flip" Chrysler to earn a quick profit. He reportedly changed his mind when Chrysler officials informed him that the company had no other options to stay afloat. He commented that he thought the transaction was "in the best interest" of everyone involved, including union members.

Before the sale was complete, Chrysler signed a deal on July 4 with Chery Automobile, China's largest automaker, to produce small cars for export under the Dodge brand—first to Latin America and Asia and then, probably around 2010, to the United States. It was the first such export agreement between U.S. and Chinese auto companies; major automakers already had factories in China, but production was aimed at the booming Chinese market. The deal gave Chery an opportunity to reach new markets and gave Chrysler a source of small cars it could sell under its brand names.

Once Cerberus took charge of Chrysler on August 3, it moved quickly to assert its control. Within a month, Chrysler had hired Robert L. Nardelli as the new chairman and chief executive officer of Chrysler and brought on James Press as co-president. Nardelli, a former General Electric (GE) executive, had been forced out as chief executive of Home Depot in January by shareholders disgruntled over his very generous pay package. Press, the former head of Toyota in North America, was well respected in the auto industry for his sales and marketing acumen. On September 8 Nardelli announced that Chrysler's new owner was concerned that the upheaval in the housing market could hurt car and truck sales and that the company was "going to look hard at the next three years financially" and possibly revise the restructuring plan the previous owners had put in place in February.

On November 1, just four days after Chrysler's UAW workers narrowly approved a new four-year contract, Nardelli announced that Chrysler would cut as many as 10,000 additional hourly jobs by the end of 2008, close down some shifts at five assembly plants, and replace four slow-selling models with four new ones, including two hybrids. Nardelli stated that the cuts were a necessary response to dramatic changes in the market situation. Not only was Chrysler facing lower sales, but its owner, Cerberus, was also facing unanticipated losses in another recently acquired company—GMAC Financial Services, once General Motors' lending arm. GMAC was reportedly suffering serious losses in connection with the collapse of the subprime loan market.

After the *Wall Street Journal* quoted Nardelli as saying that Chrysler was "operationally bankrupt," Nardelli issued a statement on December 24, explaining that his comments were meant only to "convey a sense of urgency" to employees about the car company's situation. In the statement, Nardelli said Chrysler was "fully funded with working capital to meet our present and future needs and objectives. We have a solid strategic direction to return the company to long-term profitability. We are on target and have the unwavering support of Cerberus, as well as our other key partner, Daimler, AG."

Following are the texts of three press releases issued by Chrysler announcing the purchase of Chrysler by Cerberus Capital Management (May 14, 2007), the cooperative agreement between Chrysler and China's Chery Automobile Co. (July 3, 2007), and the new round of job reductions and shift eliminations (November 1, 2007).

DOCUMENT

Announcement of Purchase of Chrysler by Cerberus Capital Management

May 14, 2007

We are confident that this transaction will create a standalone Chrysler that is financially stronger, with a winning combination of people, industry know-how, operational expertise and spirit of innovation that will accelerate the company's recovery, and help us regain our position as a competitive industry leader.

Cerberus is the right strategic buyer for Chrysler, with a long-term commitment to Chrysler's growth and success. They are committed to working constructively with both union leadership and Chrysler's management team to help Chrysler realize its full potential. There are no new job cuts planned in connection with this transaction announced today.

As a private company, Chrysler will be better positioned to focus on its long-term plan for recovery, rather than just short-term results. It will allow Chrysler to renew its focus on what has always made us special—our passion, creativity and commitment to delivering exciting Chrysler, Jeep and Dodge vehicles and quality Mopar parts to our customers, along with unparalleled customer service.

With strong backing from Cerberus and a continued relationship with Daimler, Chrysler must demonstrate once and for all that we can win in this global marketplace. It is ours to win. And Chrysler has it in its DNA to do just that.

SOURCE: Chrysler LLC. "DaimlerChrysler Definitive Agreement to Sell Chrysler Group (including Chrysler Financial Corporation) to Cerberus Capital Management, L.P." Press release. Statement attributed to Tom LaSorda, President and CEO of Chrysler Group. May 14, 2007. http://cgcomm.daimler-chrysler.com/documents.do?method=display&docType=pressrelease&docId=6891 (accessed February 1, 2008).

DOCUMENT

Chrysler Announces Deal with Chinese Automaker

July 3, 2007

Chrysler Group President and CEO Tom LaSorda joined Chery Automobile Co. Chairman and President Yin Tongyue today to finalize the highly anticipated cooperative agreement between the two automakers.

Chinese governmental authorities from the State Development and Reform Commission officially approved the agreement and marked the occasion by hosting a first-of-its-kind signing event. The ceremony was held at Beijing's Diaoyutai State Guesthouse.

Under the agreement, Chery, based in Wuhu, Anhui Province, China and Chrysler, based in Auburn Hills, Michigan, USA, will work together to develop, manufacture and distribute Chery-made small and sub-compact cars in North America, Europe and other major automotive markets under the Chrysler Group brands.

"This is a win-win for both of our companies, and I am confident this will be a successful relationship," said Yin. "Chrysler brands are very well known in the U.S. and Europe. We're prepared to work with Chrysler Group to expand their small-vehicle lineup with competitive products and accelerate both our companies' international competitiveness."

Chrysler will identify several small-car models now being developed by Chery in China and work collaboratively to make any necessary branding and regulatory modifications prior to their entry into other markets. Both companies also will jointly develop new globally competitive products based on future Chery small-car platforms.

Strategic growth in international markets—while defending market share in North America—is an important part of Chrysler Group's Recovery and Transformation Plan.

"This is the start of a very long relationship between Chrysler and Chery," said LaSorda. "Chery's participation in this agreement and their focus on small and sub-compact cars will have a nearly immediate effect on Chrysler Group's offerings in the small-vehicle segments. This strategic partnership is part of a new business model that is allowing us to introduce all-new products more quickly, with less capital spending."

Today's announcement is not the first milestone for Chrysler in China. Chrysler's relationship with China began 25 years ago when it formed Beijing Jeep® Corp., the first international automotive joint venture in the country.

The DaimlerChrysler Supervisory Board approved the framework for the Agreement earlier this year.

Profile of Chery Automobile Co., Ltd.

Since the foundation of Chery in 1997, Chery has always been pursuing independent innovation, after 10 years of development, Chery now is endowed with systematic new products research and development, manufacturing and sales. Its products have been sold to more than 50 countries and regions, and it has become the pacesetter of Chinese independent auto brands with its average annual growing rate at 169 percent.

Profile of Chrysler Group

The Chrysler Group is the Auburn Hills, Michigan-based unit of DaimlerChrysler AG. Employing more than 84,000 people worldwide, Chrysler Group designs, manufactures, markets, distributes and—through its Chrysler, Jeep and Dodge dealers—sells cars, trucks, minivans and sport-utility vehicles to customers in more than 125 countries worldwide. Its brands feature some of the world's most recognizable vehicles, including the Dodge Viper, Jeep Wrangler, Chrysler PT Cruiser and Chrysler 300C. In 2006, Chrysler Group sold 2.8 million vehicles globally.

SOURCE: Chrysler LLC. "Chery and Chrysler Group Finalize Cooperative Agreement." Press release. July 3, 2007. http://cgcomm.daimlerchrysler.com/documents.do?method=display&docType=pressrelease&docId=7002 (accessed February 1, 2008).

DOCUMENT

Chrysler Announces Job, Product, and Shift Eliminations

November 1, 2007

Chrysler LLC today announced that it would make volume-related reductions at several of its North American assembly and powertrain plants, and eliminate four products from its line-up.

Shifts will be eliminated at five North American assembly plants which, combined with other volume-related manufacturing actions, will lead to a reduction of 8,500–10,000 additional hourly jobs through 2008.

Additional actions include reductions of salaried employment by 1,000 and supplemental (contract) employment by 37 percent. The Company also plans to eliminate hourly and salaried overtime and reduce purchased services due to reduction in volume.

The volume-related actions are in addition to 13,000 jobs eliminated by the three-year Recovery and Transformation Plan (RTP) announced in February. The objectives of the RTP remain the same.

"The market situation has changed dramatically in the eight months since Chrysler established the Recovery and Transformation Plan as its blueprint," said Bob Nardelli, Chairman and Chief Executive Officer. "Annual industry volume (U.S. market) then was running at a 17.2 million clip. Now, we expect a seasonally adjusted annual volume for 2007 to be significantly lower and carry over into 2008."

"We have to move now to adjust the way our company looks and acts to reflect a smaller market," added Tom LaSorda, Vice Chairman and President. "That means a cost base that is right-sized and an appropriate level of plant utilization."

LaSorda added that third-shift operations at assembly plants usually reflect a high demand after a product is launched. Three of the five plants affected by this action are the result of elimination of third shifts—in Belvidere, Illinois; Toledo, Ohio, and Brampton, Ontario.

In contract negotiations just concluded with the United Auto Workers, Chrysler committed to spending more than $15 billion on products, plants and engineering during the life of the contract through 2011.

The company announced that it will eliminate four models through 2008, including Dodge Magnum, the convertible version (only) of Chrysler PT Cruiser, Chrysler Pacifica and Chrysler Crossfire. In the same time frame, Chrysler will add two all-new products to its portfolio: the Dodge Journey and Dodge Challenger, along with two new hybrid models, the Chrysler Aspen and Dodge Durango.

"These actions reflect our new customer-driven philosophy and allow us to focus our resources on new, more profitable and appealing products," added Jim Press, Vice Chairman and President. "Further, these product actions are all in response to dealer requests."

Manufacturing Actions

Chrysler will eliminate shifts at five assembly plants, and take further volume-related actions at several other facilities. It will:

Drop third-shift operations at Belvidere (Ill.) Assembly Plant in the first quarter 2008. Belvidere builds the Dodge Caliber, Jeep Patriot and Jeep Compass.

Drop second-shift operations at its Jefferson North (Detroit, Mich.) Assembly Plant in the first quarter 2008. It's expected that the plant will return to two shifts in first quarter 2010 with the introduction of the next generation of sport-utility vehicles. The addition of a third shift will remain an option, depending on market demand. Jefferson North builds the Jeep Grand Cherokee and Jeep Commander.

Drop third-shift operations at the Toledo (Ohio) North Assembly Plant in the first quarter 2008. Toledo North builds the Jeep Liberty and Dodge Nitro.

Drop third-shift operations at Brampton (Ontario) Assembly Plant in first quarter 2008. Brampton will build the Chrysler 300, Dodge Charger and Dodge Challenger. The Dodge Magnum will be discontinued.

Drop second shift operations at Sterling Heights (Mich.) Assembly Plant in first quarter 2008. Sterling Heights builds the Dodge Avenger and Chrysler Sebring sedans and Chrysler Sebring Convertible.

In addition, Mack Avenue (Detroit) Engine Plant II will return to a traditional two-shift/two-crew operation in the first quarter 2008 after operating on a three-crew, two-shift, 120-hour-per-week (3/2/120) schedule. Mack II builds the 3.7-liter V-6 engine.

"I'm confident that we have the right team in place and a business plan that doesn't need to be re-written," concluded Nardelli. "Like all good plans, the RTP has built-in flexibility that allows us to stay one step ahead of market change. And that is the way to long-term sustained profitability."

Source: Chrysler LLC. "Chrysler Announces Product and Plant Changes." Press release. November 1, 2007. http://cgcomm.daimlerchrysler.com/documents.do?method=display&docType=pressrelease&docId=7414 (accessed February 1, 2008).

Other Historic Documents of Interest

From this volume

- Slump in housing market, p. 514

From previous *Historic Documents*

- Problems facing American automakers, *2006*, p. 293.

DOCUMENT IN CONTEXT

Sarkozy on His Inauguration as President of France

MAY 16, 2007

France in 2007 elected its first new president in twelve years—a politician named Nicolas Sarkozy who promised a break with the past and set out to keep his word, not the least by improving relations with U.S. president George W. Bush. Sarkozy, who was often described as "hyperactive" and "hard-edged," had little trouble defeating Ségolène Royal, the first woman to make a serious run for the presidency in France.

Upon his inauguration on May 16, Sarkozy assumed leadership of a country in the economic doldrums, with low growth, high unemployment, and a sharply diminished sense of its greatness. Ever since the devastation of World War II, French leaders, notably Charles de Gaulle, had played up France's historic grandeur and glory—even as the country sank into the second tier of nations by all measures except food, wine, and tourism. By 2007 many commentators were speaking of a nostalgic "malaise" in France, one that had no obvious cure and had stubbornly resisted the rhetorical flourishes of politicians. Sarkozy's platform for dealing with France's ills emphasized hard work, particularly for the country's heavily unionized laborers, who enjoyed 35-hour work weeks and generous public services. Some of his first proposals sparked one of the most extensive transit strikes in recent French history.

Chirac's Legacy

The twelve years that Sarkozy's predecessor, Jacques Chirac, spent at the Élysée Palace (the president's official residence and office) were marked by nearly as many failures as successes—and Chirac's failures were extraordinary. One of the first was his decision to call early parliamentary elections in 1997, apparently on the assumption that his center-right supporters would win. They lost to a resurgent Socialist Party, however, forcing Chirac to share power with Socialist prime minister Lionel Jospin, an uncomfortable arrangement the French called "cohabitation."

In 2003 Chirac voiced the sentiments of most of his constituents by opposing, in harsh terms, President Bush's decision to invade Iraq. However, he angered fellow leaders in eastern European nations, many of whom supported Bush's decision, by telling them they were supposed to "keep quiet" in the face of his leadership.

Chirac lost another high-risk gamble in 2005 when he submitted a new European Union (EU) constitution to a national referendum, a step he was not required to take but one that he seemed to believe would produce a ringing endorsement of his leadership. In the campaign leading up to the vote, Chirac often appeared incapable of understanding public concerns about the complex constitution, and his public speeches undermined, rather than reinforced, support for the measure. The defeat of the referendum, coupled with a similar outcome weeks earlier in the Netherlands, deeply embarrassed Chirac and forced European leaders to put the constitution on ice. The EU crafted a less ambitious replacement in 2007.

Also in 2005, France was rocked by three weeks of rioting in Paris suburbs, mostly by immigrant Muslim youth protesting the lack of job opportunities and public services. The government was forced to declare a state of emergency to quell the disturbances, which reminded the people of France that the revolutionary slogan of "liberty, equality, fraternity" had not become a reality even after two centuries. Serving as Chirac's interior minister, Sarkozy heightened tensions during the rioting by referring to those responsible as "scum."

Throughout his presidency, Chirac was dogged by numerous allegations of corruption, many of them stemming from his service as mayor of Paris (1977–1995). While in office, he was protected from prosecution by the immunity accorded the president, but once he stepped down, eager investigating judges reopened old inquiries. One such judge on November 22, 2007, placed Chirac under formal investigation in connection with allegations that, as mayor, Chirac had improperly given jobs at the Paris city hall to political sympathizers. Chirac insisted that the allegations were false and pledged to fight for "truth and honor."

The 2007 Election Campaign

The son of a Hungarian immigrant, Sarkozy had been an ambitious junior official in the government until Chirac named him to head the interior ministry—considered the most important domestic post in the cabinet—in 2002. Two years later, Chirac also put Sarkozy in charge of the ruling party, the Union for a Popular Movement (UMP), which Chirac had created from an assortment of center-right and Gaullist parties as the base for his 2002 reelection campaign. Sarkozy's personal ambition marked him as a man destined for greater things, but it also made him many enemies, notably his chief rival in Chirac's government, Dominique de Villepin, who served as foreign minister and interior minister before taking over as prime minister in 2006. But by the time 2007 rolled around, Sarkozy had already brushed aside de Villepin for the UMP's nomination. De Villepin endorsed Sarkozy for the presidency on March 12, one day after Chirac formally announced that he would not seek a third term.

Ségolène Royal, the Socialist candidate, had a more bruising time winning her party's endorsement. She was one of a half-dozen serious candidates for the nomination; one of her principal rivals also happened to be her partner at home, party chief François Hollande, with whom she had four children. Royal shot up in the polls during the second half of 2006, buoyed in large part by her charisma and a widespread desire for fresh leadership unencumbered by the stagnant politics of the François Mitterrand and Chirac eras. She committed a number of verbal gaffes—both in campaigning around France and during foreign trips intended to burnish her foreign policy credentials—that raised serious questions about her knowledge of public affairs and her judgment. By early 2007 Royal had fallen behind Sarkozy in the polls and her frantic efforts to catch up merely increased doubts about her capabilities.

By March an estimated one-third to one-half of French voters were looking for an alternative to Sarkozy and Royal, and many found it in the person of François Bayrou.

A longtime politician from the south of France, Bayrou headed a centrist party, the Union for French Democracy, which was founded by former president Valéry Giscard d'Estaing. Bayrou presented himself as a compromise candidate capable of bridging the old left-right divide in French politics. As the campaign progressed, that stance won him increasing attention and support—so much so that Royal's advisers began to worry that Bayrou could edge her out in the first round of voting.

The major contrast with the previous French election, in 2002, was that far-right candidate Jean-Marie Le Pen was not considered to be a major factor. Le Pen had pulled off an astounding upset in 2002, coming in ahead of Socialist party candidate Lionel Jospin in the first round of balloting before being overwhelmed by Chirac in the second round. Five years later, the anti-immigration sentiment that fueled Le Pen's appeal remained strong in France, but after four previous runs at the presidency, the aging nationalist leader was no longer able to portray himself as a fresh voice of dissent. Also, Sarkozy, with his criticisms of immigrants who refused to learn French or adopt French lifestyles, had captured the support of many voters concerned about the impact of Muslim immigrants from North Africa.

The first round of voting on April 22 produced results that had been widely expected: Sarkozy finished first with 31.2 percent, Royal was second with 25.9 percent, and Bayrou came in third with 18.3 percent. Le Pen captured 11.3 percent, and other candidates shared the rest of the vote. Because no candidate achieved a majority, as had been expected with a crowded field, a second round of voting was required. Although Sarkozy clearly was out in front, the results did give Royal reason to hope that she could win the second round if she captured the bulk of the votes that had gone to Bayrou. As a result, the two weeks between the first and second rounds were dominated by both Sarkozy and Royal appealing to Bayrou's supporters. Bayrou refused to endorse either of the final two candidates, although he did make clear his intense dislike for Sarkozy's style of governing and his pro-business economic policies.

On April 28, midway between the first and second rounds, Royal took the unusual step of engaging in a 100-minute, televised debate with Bayrou. This marked the first time in France that a candidate defeated in the first round played such a role in the second round. Sarkozy and Royal held their one direct debate of the campaign on May 2; it was a heated affair in which a suddenly combative Royal tried repeatedly to put the usually aggressive Sarkozy on the defensive. Surprisingly, Sarkozy came across as the calmer candidate, thus easing fears among some voters about his temper and heightening concerns about Royal's leadership skills.

In the end, a majority of French voters chose to go with Sarkozy despite any concerns about his personality. In the second-round voting on May 6, Sarkozy defeated Royal by a margin of 53.1 percent to 46.9 percent, with turnout estimated at 85 percent of the nation's 44.5 million registered voters. "The French people have chosen change," he said on election night. "I will implement that change. Because that is the mandate I received and because France needs change." Royal's defeat was devastating for the Socialists, who had lost three straight presidential elections and the previous two parliamentary elections.

Strengthening a Hold on Power

Sarkozy took office on May 16, accepting power from Chirac in an elaborate ceremony at the Élysée Palace. In a relatively brief inaugural address, Sarkozy offered no new specific proposals and did not lay out a legislative agenda. Instead, he repeated the principle themes of his campaign, including increasing national unity and restoring hard work and individual merit—rather than entitlements—as virtues in French society. Perhaps the central focus of his speech was an emphasis on the "imperative of change because never has inertia been so dangerous for France as in this world in flux where everyone strives to change faster than the others, where any delay can be fatal and quickly becomes irretrievable."

The new president quickly made it clear that some elements of his administration would be unconventional, certainly by French standards. To begin with, he reached out to the opposition, asking for direct support from the Socialists and appointing a senior leader of the defeated party as his new foreign minister: Bernard Kouchner, who had gained worldwide fame as a founder of the humanitarian group Médicins sans Frontières (Doctors Without Borders). For prime minister, Sarkozy named his chief campaign aide and former cabinet minister, François Fillon. His choice for the important post of finance minister also was widely seen as heralding change: Christine Lagarde, who had served as Chirac's trade minister, was also widely known as the first female head of a major American law firm, Baker & McKenzie. Often called "l'Americaine" because of her years spent in the United States, Lagarde was a staunch advocate of the free market economic policies Sarkozy had pledged to implement in France.

Sarkozy cemented his control over government—and, not incidentally, the center-right political party that Chirac founded and he now led—in two rounds of parliamentary elections on June 10 and 17. Although his UMP actually lost 45 seats from the previous parliament, it still ended up with a strong majority of 314 of the 577 seats. A major setback for Sarkozy was the defeat of Alain Juppe, a former prime minister whom Sarkozy had named to head a newly combined ministry of the environment, transportation, and energy. Juppe's loss in his constituency forced him to step down from the cabinet.

Despite the importance of the elections, much of the country's attention was quickly diverted to a personal matter. Shortly after the polls closed, aides to Royal revealed that she had left her long-time partner, François Holland. This de facto divorce raised new questions about who would lead the Socialist Party. Later in the year, Sarkozy and his glamorous wife, Cécilia, divorced; she was a former model who had made clear her discomfort with life as the wife of a politician.

Revamping the Economy

In his election campaign, Sarkozy never directly accused his compatriots of laziness, but he did have a clear message: French workers had been spoiled by a thirty-five-hour work week and needed to work harder so the national economy would grow. Adopting the first of his economic policies, the parliament in July abolished taxes on overtime pay—taxes that Sarkozy said discouraged workers from putting in extra effort—and cut the maximum income tax rate to 50 percent.

The government followed these steps with new proposals in September and October, including more tax cuts, incentives for employees to work more than 35 hours, and some cutbacks in benefits for government workers. Among these proposed cutbacks was the scrapping of a pension plan that gave particularly generous terms to about 1 million employees in physically demanding jobs, such as miners and train drivers. The country's powerful transport unions called strikes to protest this proposal—a brief one on October 18 and a more extensive one that started on November 13 and lasted for nine days until the government and unions agreed to negotiations. Sarkozy vowed not to back down, and polls showed that the public continued to support his economic reform plans.

But French unions had succeeded in blocking similar benefit cutbacks in the past, and at year's end it remained unclear whether the new president would have more success than had his predecessors.

Improved Relations with Washington

Of all Sarkozy's unconventional moves during his first months in office, perhaps none was more unusual for contemporary France than his warm embrace of the United States and its president, who was deeply unpopular in France. In sharp contrast to Chirac and most other French politicians, Sarkozy made clear in his election campaign that he admired the United States, wanted France to be more like the United States, and considered President Bush to be an outstanding leader. In electing Sarkozy, French voters apparently ignored these positions or placed greater importance on other matters.

Sarkozy wasted no time following through on his admiration of all things American. In early August, he and his family headed to the United States for a vacation at a lakeside resort in New Hampshire—an almost unheard-of destination for a senior French politician. While in New England, Sarkozy on August 11 had lunch with the president on the Maine coast at the Bush family compound owned by Bush's parents, former president George H. W. and Barbara Bush.

Sarkozy returned to the United States in November for a visit that included a formal dinner at the White House, a speech to a joint meeting of Congress, and a symbolic trip with Bush to Mount Vernon, the home of George Washington. At the last locale, the two leaders basked in Revolutionary War symbolism of Franco-American friendship.

Following is the text of a speech delivered by Nicolas Sarkozy in Paris on May 16, 2007, following his inauguration as president of France.

DOCUMENT

Inaugural Address of Nicolas Sarkozy

May 16, 2007

Ladies and gentlemen,

On this day when I officially take up my duties as President of the French Republic, I'm thinking of France, this venerable country which has gone through so many ordeals and always picked itself up, which has always spoken for all mankind and which I now have the weighty task of representing in the eyes of the world.

I'm thinking of all the presidents of the Fifth Republic who have preceded me. I'm thinking of General [Charles] de Gaulle who twice saved the Republic, who gave France back her sovereignty and the State its dignity and authority. I'm thinking of Georges Pompidou and Valéry Giscard d'Estaing who, each in his own way, did so much to take France into the modern era.

I'm thinking of François Mitterrand, who found the way to safeguard the institutions and embody the changeover of political power at a time when it was becoming necessary for the Republic to belong to all the French.

I'm thinking of Jacques Chirac who, for 12 years, worked for peace and projected France's universal values throughout the world. I'm thinking of his role in making mankind aware of the imminence of the ecological disaster and of everyone's responsibility to the coming generations.

But at such a solemn moment, my thoughts go first to the French people, a great people with a great history and who stood up and declared their faith in democracy, said they no longer wanted to have no say. I'm thinking of the French people who have always been able to overcome ordeals courageously and find in themselves the strength to transform the world.

I'm thinking, with emotion, about this expectation, this hope, this need to believe in a better future which were voiced so strongly during the campaign which has just ended. I'm thinking solemnly about the mandate the French people have entrusted to me and the extremely high imperative expressed by them—I have no right to disappoint them.

Imperative of bringing the French together because France is strong only when she is united, and today she needs to be strong to take up the challenges confronting her.

Imperative of keeping promises and honouring commitments because trust has never been as shaken, as fragile. A moral imperative because never has the crisis of values been as deep, because never has the need for people to regain their bearings been as strong.

Imperative of restoring the value of work, of effort, of merit, of respect, because these values underpin human dignity and requirement for social progress.

Imperative of tolerance and opening-up because never have intolerance and sectarianism been so destructive, because never has it been so necessary for all women and all men of goodwill to pool their talents, their intellectual skill, their ideas for conceiving the future.

Imperative of change because never has inertia been so dangerous for France as in this world in flux where everyone strives to change faster than the others, where any delay can be fatal and quickly becomes irretrievable.

Imperative of security and protection because it has never been so necessary to fight the fear of the future and feeling of vulnerability which discourage initiative and risk-taking.

Imperative for order and authority because we have too often given in to disorder and violence from which those who suffer the greatest are the most vulnerable and humble.

Imperative to deliver results because the French have had enough of nothing in their daily lives ever improving, because the French have had enough of their lives becoming ever tougher, ever harder, because the French have had enough of sacrifices being imposed on them with no result.

Imperative of justice because for a very long time so many French have not felt such a strong sense of injustice, and had the feeling that the sacrifices weren't fairly shared, that everyone did not have equal rights.

Imperative of breaking with past behaviours, ways of thinking and intellectual conformism because never have the problems to be resolved been so completely new.

The people have entrusted me with a mandate. I shall fulfil it. I shall fulfil it scrupulously, with the determination to be worthy of the trust the French have placed in me.

I shall defend France's independence and identity.

I shall ensure respect for the State's authority and its impartiality.

I shall strive to build a Republic founded on genuine rights and an irreproachable democracy.

I shall fight for a Europe which protects, for the unity of the Mediterranean and for the development of Africa.

I shall make the defence of human rights and battle against climate warming the priorities of France's diplomatic action in the world.

The task will be difficult and will have to be long-term.

Every one of you in your official position in the State and all citizens in their positions in society are destined to contribute to it.

I want to express my conviction that in the service of France there are no sides. There is only the goodwill of those who love their country. There are only the skills, ideas and convictions of those fired by their passion for serving the general interest.

To all those who want to serve their country, I say that I am ready to work with you and that I shall not ask them to renounce their beliefs, betray their friendships or forget their history. It is for them to decide, in all conscience as free men and women, how they want to serve France.

On 6 May there was only one victory, that of the France who doesn't want to die, who wants order but also movement, who wants progress, but wants fraternity, who wants efficiency, but wants justice, who wants identity, but wants an opening-up.

On 6 May there was only one victor, the French people who don't want to give up, who don't want to be confined to inertia and conservatism, who no longer want others to decide for them, think for them.

Well, to this France who wants to go on living, to this people who don't want to give up, who deserve our love and our respect, I want to express my determination not to disappoint them.

Long live the Republic!

Long live France!

Source: French Republic. Embassy of France in the United States. "Speech by Nicolas Sarkozy, President of the Republic, at the Investiture Ceremony." Paris, France. May 16, 2007. www.ambafrance-us.org/news/statmnts/2007/sarkozy_inaugural_speech051607.asp (accessed October 3, 2007).

Other Historic Documents of Interest

From previous *Historic Documents*

- French vote on the EU constitution, *2005,* p. 339
- European leaders on a potential war in Iraq, *2003,* p. 40

DOCUMENT IN CONTEXT

World Bank Statements on the Resignation of President Wolfowitz

MAY 17, 2007

In a year of exceptional economic and financial uncertainty, two of the world's most important financial institutions underwent leadership changes. Paul Wolfowitz, a controversial figure who helped plan the U.S. invasion of Iraq, was forced in May to step down after serving two years as the president of the World Bank; his resignation was the result of a scandal over his arranging a high-paying State Department job for his girlfriend. He was succeeded by Robert B. Zoellick, who also had held top posts in the current Bush administration but was considered less controversial than Wolfowitz.

Over at the bank's sister institution, the International Monetary Fund (IMF), Managing Director Rodrigo de Rato stepped down in October for personal reasons after serving just three years. He was succeeded by French economist and politician Dominique Strauss-Kahn.

The World Bank and the IMF were created by the United States and its allies at the end of World War II as international institutions, aligned with the United Nations (UN), to help rebuild damaged economies after the war and stabilize the global financial situation. Both institutions succeeded at some of their tasks but faced a range of criticism: they were viewed as too cumbersome and bureaucratic (particularly in the bank's case) and too wedded to Western-style free market economic theories that did not always meet the immediate needs of impoverished developing countries (mainly in the fund's case). Some critics also had suggested since the mid-1990s that both institutions were no longer as relevant as they once had been. The IMF itself was searching for a new role, and one of Strauss-Kahn's major tasks would be to devise and implement major reforms.

The coincidental turnover of leadership at both institutions in 2007 also raised questions about the sixty-three-year-old tradition for choosing their leaders: the United States chooses the president of the bank, while European nations select the managing director of the fund. There were widespread calls, especially from developing countries, for new procedures to select the leaders, but the practice was unchanged for the selections in 2007.

An Embattled Wolfowitz Resigns

As deputy defense secretary during President George W. Bush's first term, Wolfowitz was one of the most prominent of a group of neoconservative intellectuals in the administration who had built the case for ousting Iraqi leader Saddam Hussein. By 2005 it was clear that Wolfowitz and his colleagues had not been as careful in developing a plan for what the

United States would do in Iraq once Hussein had been removed. Wolfowitz and his boss, Defense Secretary Donald H. Rumsfeld, became the two administration officials, aside from Bush himself, most identified with the U.S. enterprise in Iraq, particularly its failures.

In March 2005, early in Bush's second term, the president nominated Wolfowitz for the presidency of the World Bank. The nomination was controversial for two reasons. First, Wolfowitz was a divisive figure because of his reputation as a primary exponent of the U.S. invasion and occupation of Iraq, which remained deeply unpopular in most of the world. Second, there was growing resistance to the tradition of an American serving as president of the World Bank. Critics argued that this tradition was outdated, not least because the world economy had become more globalized. Moreover, many international leaders—including some U.S. allies in Europe—were still smarting from what they considered the Bush administration's unilateral approach to foreign policy, an approach that Wolfowitz had strongly advocated while at the Pentagon.

Despite these concerns, and after extensive lobbying by Bush administration officials, Wolfowitz won approval from the bank's board. He quickly set about the task of demonstrating to developing countries that he had a deep personal interest in economic development and poverty reduction. He won over many doubters in developing countries—particularly in Africa—with speeches emphasizing the need for rich countries, such as the United States, to be more generous to their poorer brethren. Wolfowitz encountered widespread resistance, however, when he made fighting corruption in developing countries one of the bank's chief priorities. He insisted that deeply rooted corruption, notably in autocratic societies, was a major barrier to economic development. Some of the resistance to this campaign came from within the bank, where many longtime staff members resented what they viewed as heavy-handed behavior by aides Wolfowitz had brought with him from the Bush administration.

It was in the context of unease within the bank that news reports in late March revealed that Wolfowitz had arranged a job at the State Department for a woman he described as his "domestic partner." Shaha Ali Riza was a bank employee when Wolfowitz became president. Because the bank's anti-nepotism regulations barred employees from supervising anyone with whom they had a personal relationship, the issue of Wolfowitz's relationship with Riza arose during negotiations over the terms of his contract. The eventual solution was to transfer Riza to a job at the State Department, while the World Bank continued to pay her salary and other benefits—which, according to news reports, by early 2007 reached $193,500, slightly more than the salary of Secretary of State Condoleezza Rice. In addition, Riza was guaranteed a senior-level position at the bank when Wolfowitz stepped down.

Between late March and late May, reporting by news organizations and an internal investigation by the bank revealed the extent of Wolfowitz's actions during 2005 to find employment for Riza. At first, Wolfowitz insisted that he had recused himself from the matter, but it quickly became clear that he had been deeply involved; in fact, he had negotiated the terms under which Riza obtained her job at the State Department. Hoping to stem a tide of news stories, Wolfowitz on April 12 issued a statement apologizing for his actions. "I made a mistake, for which I am sorry," he said. At this point, the bank's board was investigating the matter, but the bank's staff association said Wolfowitz should resign because his actions had "compromised the integrity and effectiveness" of the institution.

Wolfowitz dug in and refused to resign, even as reports emerged almost daily about his efforts to find a job for Riza and his strong-arm approach to dealing with subordinates at the bank. The Bush administration stood by Wolfowitz during the controversy, but other key players—especially the bank's other major donors in Europe and Japan—were notable by their public silence, their wait-and-see attitude, or their behind-the-scenes appeals to the Bush administration to drop its backing for him. On April 30 Wolfowitz said he had been the victim of a "smear campaign," and he insisted he would not resign "in the face of a plainly bogus charge of conflict of interest."

The slippery slope got much steeper for Wolfowitz on May 7, when a committee of bank directors informed him that he had been guilty of a conflict of interest. Presented to the full board and released publicly one week later, the directors' report said Wolfowitz should have withdrawn from all decision making about Riza and argued that the controversy had jeopardized the bank's own "reputation and credibility." Although the committee did not say directly that Wolfowitz should resign, the report left little doubt that panel members had reached that conclusion. Wolfowitz sent a statement to the board, calling the report "unbalanced and flawed." The Bush administration, which had continued to support Wolfowitz, began shifting its position on May 15, suggesting that he could be allowed to resign voluntarily.

Wolfowitz's rocky tenure at the bank ended on May 17, when he submitted his resignation, effective on June 30, without acknowledging any wrongdoing. The bank board formally accepted his assurances that he had "acted ethically and in good faith." This settlement, reportedly negotiated during heated board meetings in previous days, gave Wolfowitz the right to claim that he left on his own terms. For the bank, the settlement also lanced a controversy that had threatened its international support, in particular a forthcoming round of requests to big donors for money to support the bank's International Development Association (IDA), which makes concessional loans to the world's poorest countries. At the White House, President Bush expressed regret at the outcome and added, "I believe all parties in this matter have acted in good faith."

Choosing a Successor

Wolfowitz's resignation prompted much speculation about a possible fight over his successor if President Bush insisted on retaining the U.S. privilege of naming the World Bank's president. A group of more than 200 experts on economic development, including the heads of several major international aid organizations, sent the boards of the World Bank and the IMF a letter calling for the abandonment of the tradition of the Americans and Europeans naming the heads, respectively, of the two institutions. Instead, the letter said, the institutions' leaders should be chosen by a procedure emphasizing "transparency of process and competence of prospective leadership without regard to national origin."

Bush cut short the discussion of selection procedures and qualifications, however, by nominating Robert B. Zoellick, a man who had broad experience in international diplomacy and finance and, more importantly, who was viewed as a consensus builder. A long-time U.S. government official, Zoellick had served as U.S. trade representative during Bush's first term, then as deputy secretary of state. Since mid-2006, he had been a senior executive at the global investment firm Goldman Sachs. During his government-service days, Zoellick won international respect for his diplomatic skills, both as a trade negotiator and as a top diplomat. One of his main responsibilities at the State Department was to mediate difficult negotiations between the government of Sudan and rebel leaders in the troubled Darfur region. His tenacity had been a key factor leading to a peace agreement in May 2006—an agreement that had little impact in ending violence in Darfur but did demonstrate that a diplomatic solution might be possible.

European officials quickly made clear that they would accept Bush's nomination of Zoellick, thus leaving intact—at least for the immediate future—the arrangement for choosing the heads of the World Bank and the IMF. The leading critics of Zoellick's nomination were antiglobalization activists who charged that during his tenure as the U.S. trade negotiator, he had struck deals favoring multinational corporations at the expense of developing countries and low-income workers in the United States.

The World Bank board formally approved Zoellick's appointment on June 25 and issued a statement saying that he brought "strong leadership and managerial qualities as well as a proven track record in international affairs and the drive required to enhance the credibility and effectiveness of the bank." Zoellick started work on July 1, one day after Wolfowitz officially left the bank. Wolfowitz immediately assumed a position as a visiting scholar at the American Enterprise Institute, a Washington think tank that had become a center of neoconservative advocacy.

On September 12, Zoellick accepted, and said he would implement, the recommendations of a panel of experts, headed by former Federal Reserve Board chairman Paul A. Volker, who had investigated charges that the bank had overlooked fraudulent spending of the money it lent to developing countries. The panel reported that losses from bribes, bid-rigging, shoddy services, and other problems were "substantial" but could not be put into specific dollar amounts.

Zoellick spent much of his early tenure at the bank healing wounds left raw by the battles over Wolfowitz. He acknowledged this just before the annual meeting of world finance ministers in mid-October, when he told reporters, "I believe we've been able to calm some of the waters while starting to navigate a course ahead." A significant part of that course was the bank's request for money for the latest replenishment for IDA. Zoellick announced on December 15 that $41.6 billion had been raised for IDA lending to the poorest countries—a record amount that included $25.1 billion in contributions from major donor countries (Britain, the United States, Japan, and Germany, in descending order) and $16.5 billion from the bank's own funds. For the first time, the United States was not the largest single donor. Even some critics of the World Bank said Zoellick's success in raising the money showed that he had quickly ended the turmoil of the brief Wolfowitz era.

A New IMF Chief

Just days before Zoellick succeeded Wolfowitz at the World Bank, IMF chief Rodrigo de Rato announced on June 28 that he would step down in October. In his surprise announcement, he cited "personal reasons," including the education of his children. A former Spanish finance minister, Rato had served just three years of his five-year term.

Rato's announcement, like Wolfowitz's at the World Bank, prompted numerous calls, particularly from developing countries, for abandoning the tradition for choosing the fund's leader. Remarkably, some European officials seemed to agree: Britain's new chancellor of the exchequer, Alistair Downing, suggested broad consultation among IMF-member countries, and Italian finance minister Tommaso Padoa-Schioppa was quoted as saying "the passport shouldn't be the determining thing" in selecting a head for the fund.

France's newly elected president, Nicolas Sarkozy, did not want to abandon the old tradition. He apparently saw the appointment as an opportunity to reassert French importance in world affairs and to bolster his domestic political standing in one stroke.

The conservative Sarkozy had astonished friends and foes, once he took office in May, by appointing leaders of the opposition Socialist party to several high-level posts, including foreign minister. Sarkozy announced in July that he was supporting Dominique Strauss-Kahn for the IMF post—a startling turn of events because the latter in 2006 had unsuccessfully sought the Socialist Party nomination to oppose Sarkozy and had once called him a "danger" to democracy. An academic economist, Strauss-Kahn had served as a cabinet minister in past Socialist governments but was forced to step down as finance minister in 1999 when he was caught up in a broader corruption scandal. A court later cleared him of any wrongdoing.

Most other European countries quickly endorsed Sarkozy's nomination of Strauss-Kahn, as did the United States. Russia sought to intervene by nominating Josef Tošovský, a former governor of the central bank in the Czech Republic, but Strauss-Kahn won broad acceptance by touring developing countries to make the case that he was the better choice. Part of Strauss-Kahn's appeal was his socialist background, which helped him to convince leaders of poor countries that he was not a creature of multinational corporations. The IMF board approved his appointment on September 29, and he took office on November 1.

In several interviews with news organizations, as well as in his formal interview with the IMF board, Strauss-Kahn said he sympathized with developing countries' view that the fund did not meet their needs and, in some cases, provided counterproductive advice. For example, the IMF had been criticized harshly after it demanded economic reforms in East Asian countries during a massive financial crisis in 1998; the fund forced some countries to make cuts in social programs that deepened poverty and contributed to social unrest. In another case four years later, Argentina cancelled payments on its foreign debts and rejected IMF advice for getting out of its crisis; the country subsequently experienced strong economic growth. In more recent years, the IMF struggled to assert a role for itself in the face of dramatic changes in global financial systems, notably the accumulation of enormous foreign exchange reserves by China and oil-producing countries that far outweighed the fund's own resources. "There is a risk that more and more countries say that we are not interested in the Fund, that our voice is not listened to," Strauss-Kahn told the Bloomberg news service on August 31. "Emerging countries are legitimate in asking for more voice."

Following is the text of two statements released by the World Bank on May 17, 2007, regarding the resignation of Paul Wolfowitz as president of the bank. First, a statement by the bank's board of executive directors. Second, a statement by Wolfowitz.

Statements of Executive Directors and President Wolfowitz

DOCUMENT

May 17, 2007

Statement of Executive Directors

Over the last three days we have considered carefully the report of the ad hoc group, the associated documents, and the submissions and presentations of Mr. Wolfowitz. Our deliberations were greatly assisted by our discussion with Mr. Wolfowitz. He assured us that he acted ethically and in good faith in what he believed were the best interests of the institution, and we accept that. We also accept that others involved acted ethically and in good faith. At the same time, it is clear from this material that a number of mistakes were made by a number of individuals in handling the matter under consideration, and that the Bank's systems did not prove robust to the strain under which they were placed. One conclusion we draw from this is the need to review the governance framework of the World Bank Group, including the role as well as procedural and other aspects of the Ethics Committee. The Executive Directors accept Mr. Wolfowitz's decision to resign as President of the World Bank Group, effective end of the fiscal year (June 30, 2007). The Board will start the nomination process for a new President immediately.

We are grateful to Mr. Wolfowitz for his service at the Bank. Much has been achieved in the last two years, including the Multilateral Debt Relief Initiative, the Clean Energy Investment Framework, the Africa Action Plan, and the Avian Flu Initiative. 2006 was a record year for IDA [International Development Association] lending, especially in Africa. The Bank has launched emergency action programmes in Liberia, the Democratic Republic of the Congo and the Central African Republic, and played a key role in the Lebanon and Afghanistan donors conference. In March, after an unprecedented global consultation process, we adopted a new strategy for the Bank's work on Governance and Anti-Corruption. And we have new strategies for Rapid Response in Fragile States, for the Health Sector and for the Financial Sector. We thank Mr. Wolfowitz for his leadership and for championing the Bank's work across so many areas.

It is regrettable that these achievements have been overshadowed by recent events. Mr. Wolfowitz has stressed his deep support for and attachment to the World Bank and his responsibility, as its President, to act at all stages in the best interests of the institution. This sense of duty and responsibility has led him to his announcement today. We thank him for this and underscore our appreciation for his commitment to development and his continuing support for the World Bank and its mission.

Statement of Paul Wolfowitz

I am pleased that after reviewing all the evidence the Executive Directors of the World Bank Group have accepted my assurance that I acted ethically and in good faith in what I believed were the best interests of the institution, including protecting the rights of a valued staff member.

The poorest people of the world, especially in Sub-Saharan Africa deserve the very best that we can deliver. Now it is necessary to find a way to move forward.

To do that, I have concluded that it is in the best interests of those whom this institution serves for that mission to be carried forward under new leadership. Therefore, I am announcing today that I will resign as President of the World Bank Group effective at the end of the fiscal year (June 30, 2007).

The World Bank Group is a critical institution with a noble mission, that of enabling the world's poor—and particularly the more than a billion men, women and children who struggle to survive on less than a dollar a day—to escape the shackles of poverty. I have had the privilege of visiting World Bank Group staff and programs in some 25 developing countries in the last two years. I've had a chance to see with my own eyes and hear with my own ears how eager people are to work hard if they have a chance for a good job, how excited children are to have a chance for the first time to go to school, and how willing parents are to sacrifice so that their children can have a better future.

It has been truly inspirational to be able to help them achieve their goals and it is a privilege for all of us in the World Bank Group to have a chance, every day that we come to work, to make a difference in the lives of those who are less fortunate. I am grateful to have enjoyed that privilege for nearly two years and I am proud of what we have accomplished together as a team.

We provided record levels of support last year to the poorest countries of the world, $9.5 billion, through the International Development Agency [sic] (IDA) and we are headed to a new record this year. Half of that support is going to Sub-Saharan African countries, also setting new records;

- We are further increasing support to the poorest countries through the Multilateral Debt Relief Initiative completed last year which canceled $38 billion of debt owed by the HIPC countries [highly-indebted poor countries] to IDA, along with specific commitments by the IDA donors to provide additional support to make up for the lost reflows to IDA on a dollar-for-dollar basis;
- And last year we transferred a record amount of Bank Group income, $950 million, to IDA, including the first-ever transfer from the IFC [International Finance Corp.] to IDA;

We have not only increased the quantity of resources available to the poorest countries through IDA, we are also making those resources more effective, and we are providing greater assurance to donors that they are being used properly:

- By helping developing countries strengthen systems of governance and supporting their efforts to fight corruption and to recover stolen assets;
- By placing greater emphasis on measuring the results our support is producing, although much more work needs to be done in this area; and
- By strengthening cooperation among donors, and particularly among the Multilateral Development Banks in such areas as fighting corruption and averting unsustainable debt burdens;

We have also strengthened our work significantly in a number of important specific sectors, particularly:

- Infrastructure—which was a major concern of the Finance Ministers from Africa when I first met with them two years ago;
- Combating malaria, a preventable disease that is killing 3,000 people a day, most of them children and most of them in Sub-Saharan Africa. In the last 18 months we have approved over $360 million in assistance for anti-malaria programs compared to $50 million in the first five years of this decade.
- Here, too, we are emphasizing quality as well as quantity, pressing the development of a "malaria scorecard" to track results and effectively coordinate the work of the many donors so that gaps can be identified and filled.

Some of the work which has been most inspiring to me has been the Bank Group's response to countries emerging from conflict, countries with new leadership which urgently need assistance to consolidate peace and jumpstart recovery:

- We have responded with unprecedented speed to help fragile states with new leadership, such as Liberia, the Central African Republic and the Democratic Republic of the Congo;
- We have adopted a new Rapid Response and Fragile States policy to enable us to move faster in situations with new opportunity and to encourage more of our staff to work in fragile states.
- We have helped lead successful donor conferences for many post-conflict countries, including Afghanistan, Lebanon and Liberia.

Our work is important, however, to more than just the poorest countries. Indeed, the majority of the world's poor live in the more successful developing countries, our partners in middle income countries, which borrow from the IBRD [International Bank for Reconstruction and Development, the main arm of the World Bank].

- These countries still seek help to deal with their large challenges to fight poverty and preserve their environment, but the World Bank Group needs to be increasingly innovative and flexible if we are to be useful to these countries which are already highly sophisticated and have access to many other sources of funds. To do that we developed a new "Middle Income Country Strategy" last year and we are working hard on implementing it.

Some of our most important work has been strengthening the development of the private sector, which is the most important source of the growth and jobs that people need to escape poverty:

- The International Finance Corporation, which works with the private sector, has been setting impressive records, including $8 billion in new commitments this year.
- What should inspire us even more than the numbers is the greatly increased emphasis the IFC is placing on the development impact of their work and on expansion into "frontier markets." Indeed, Africa is the fastest growing region for IFC work—a five-fold increase in five years—and the IFC has greatly expanded its field staff in Africa.

- Perhaps most of all, I am proud of the innovative work the IFC is doing, through the "Doing Business" report, to help developing countries identify the obstacles to private sector growth and I have been delighted at how eager many governments have been to remove those obstacles once we help identify them.

This is not an exhaustive description of the work of the World Bank Group—or even just the part that I have been involved in—but I need to mention one more thing: the importance of the World Bank partnership with the developed countries to promote sustainable global development:

- The Bank helps rich countries carry out their obligation and their interest to help the world's poor.
- We support the interest of the developed countries to mobilize global resources for common purposes, such as containing the spread of Avian flu—where the Bank has played a leadership role—or to preserving the planet's environmental heritage, as we are doing in Brazil and the DRC [Democratic Republic of Congo], by supporting Amazon Basin and Congo River Basin initiatives.
- Most important of all has been the Bank's development of the Clean Energy Investment Framework which we were first asked to do by the Gleneagles Summit of the G-8 in July 2005. As the world mobilizes resources to diversify energy sources, reduce carbon emissions, avoid deforestation and help countries deal with the effects of climate change, most of those resources have to come from the developed countries. The most productive place to invest them will often be in developing countries. The World Bank Group has been and continues to be in a unique position to facilitate those investment flows and the Global Environment Facility and the Clean Energy Investment Framework form the foundation on which the Bank Group can build.

All of that work—and much more—is only possible because of the dedicated efforts of very hard-working staff. I am particularly impressed by our staff in country offices, including remarkable local staff members, many of whom face daily risks to their health and security in order to help the poor whom we strive to serve. They too have been treated unfairly by much of the press coverage of the past weeks and they deserve better. I hope that can happen now.

I have made many strong appointments both from inside and outside the Bank of which I am personally proud. My Senior Management Team is an exceptional group of talented managers and devoted international public servants who it has been an honor to have as friends and colleagues.

But, I am particularly proud to have appointed two African women as Vice-Presidents in key positions, each of them a former cabinet minister with real-world experience in solving problems in democratically elected Sub-Saharan governments. Only when African voices with African experiences are fully empowered at the Bank, will the Bank be seen as a center for solutions in that part of the world. We need senior leaders who have real-world experience in tackling the toughest challenges in the poorest countries.

I am also grateful for the dedicated professionalism of the many staff throughout the World Bank Group who have stayed focused on their work during the recent controversy. In the month of April alone, they delivered nearly $1 billion of support for Africa, an innovative new strategy for Bank work in the health sector, and a strategy for Bank Group support for financial sector work in developing countries, and much more. I am particularly grateful to the entire staff of the President's office who have given me such strong professional support throughout the last two years and particularly during the last month.

It is inspiring to work with people like those and I will miss them.

Finally, I want to say a special word of thanks to the many people inside and outside the Bank who have publicly or privately expressed their support for me and asked me to stay. One of the most moving was a phone call I received from the democratically elected President of a Sub-Saharan African country. It was a private call so I will not quote him by name. But he thanked me for doing so much, in his words, to make the World Bank an institution "that listens, that cares, that understands and that takes action." If that is true, and if I have "touched the hearts of Africans," as he told me, then the last two years have been worth it.

I hope I can continue working with him and with the many other Africans, official and non-official, who have been such an inspiration to me—although I will have to find other ways to do so. They are the ones who have convinced me that Africa has a real chance to turn a corner and join the progress that we have seen in many other parts of the developing world in recent decades. It is those Africans who are stepping up—often at great personal sacrifice and even risk, to bring peace, good governance and sound policies to their countries that are the reason for hope. They deserve all the support that the World Bank Group can give them and I hope they get it.

The next President will have my full support. Hopefully the difficulties of the last few weeks can actually strengthen the Bank by identifying some of the areas of governance and human resource management where reform is needed.

Change should not be feared, it is something to welcome. It is the key to keeping this important institution relevant and effective in the future and meeting the needs of the world's poor, and of humanity as a whole.

SOURCE: World Bank. "Statements of Executive Directors and President Wolfowitz." Press release. May 17, 2007. go.worldbank.org/NDB91EQRJ0 (accessed May 17, 2007).

OTHER HISTORIC DOCUMENTS OF INTEREST

FROM THIS VOLUME

- Sarkozy on his inauguration as president of France, p. 240
- World Bank agriculture programs in Africa, p. 592
- International Monetary Fund's world economic outlook, p. 603

FROM PREVIOUS *HISTORIC DOCUMENTS*

- UN Security Council on Darfur crisis in Sudan, *2006,* p. 497
- Wolfowitz approved as World Bank president, *2005,* p. 414
- Argentine financial crisis, *2002,* p. 80
- Asian financial crisis, *1998,* p. 722

DOCUMENT IN CONTEXT

Ambassador Crocker on U.S.-Iranian Diplomacy

MAY 28, 2007

The United States and Iran during 2007 held their first direct and detailed diplomatic meetings in more than two decades. The talks, which centered on concerns about security and stability in Iraq, broke what had been a diplomatic taboo in both Tehran and Washington ever since the end of the Iran hostage crisis in 1981. By year's end it was unclear whether the diplomacy had affected the situation in Iraq, but it did create the possibility that the two outside countries with the most influence there might eventually find some way to focus on shared interests in a peaceful and stable Iraq.

Diplomacy between the United States and Iran broke down in November 1979, when Iranian students seized more than sixty diplomats and others at the U.S. embassy in Tehran and held them hostage for 444 days. Although Washington severed relations with Iran, over the course of 1980 and into January 1981, U.S. and Iranian diplomats negotiated—with Algerian mediation—the terms under which the hostages would be released. After the hostages were freed on January 20, 1981, all direct contact between the two countries was halted, with one significant exception. Between late 1985 and early 1986, the administration of President Ronald Reagan negotiated secretly with Iran in an effort to win the freedom of several American citizens then being held hostage in Lebanon. This failed diplomacy became public in late 1986 and sparked numerous investigations into what became known as the Iran-contra affair.

In subsequent years, it became politically impossible for successive governments in both countries to resume any kind of formal diplomacy. Numerous attempts by one side or the other to reach out failed, largely because the timing was not right. This was particularly true between 1998 and 2000, when U.S. president Bill Clinton and relatively moderate Iranian president Mohammad Khatami each initiated tentative moves toward a rapprochement only to find that the other side was not yet ready.

Relations between the United States and Iran worsened again early in President George W. Bush's first term, notably after Bush in January 2002 described Iran (along with Iraq and North Korea) as part of an "axis of evil." This diplomatic standoff continued after the United States and its European allies in 2004 began putting pressure on Iran to halt its presumed work to build nuclear weapons. European diplomats met repeatedly with their Iranian counterparts on the nuclear issue, while the United States remained on the sidelines and engaged in harsh rhetoric. The closest the two countries came to diplomatic contact was at a regional meeting in Egypt in 2004, when Secretary of State Colin Powell sat next to his Iranian counterpart; the two men reportedly exchanged only pleasantries.

Powell's successor, Condoleezza Rice, in May 2006 announced that the United States would be willing to participate in the European talks with Iran—but only after Iran suspended the nuclear work that was at the heart of those negotiations. Iranian president Mahmoud Ahmadinejad appeared to try to reach past the diplomatic impasse in 2006 with two public letters, one addressed directly to Bush and the other to the American people. The Bush administration dismissed both letters as propaganda. Ahmadinejad visited the United States in September 2007, for the annual opening sessions of the United Nations (UN) General Assembly. He made two speeches to American audiences but apparently had no formal contact with the U.S. government.

Hardline Statements, then an Opening

Two imperatives, both involving Iraq, seemed to be at work in the Bush administration's decision to reach out to Iran in 2007. First, the administration wanted to show the nation that Bush's decision to send a "surge" of about 30,000 additional troops to Iraq was part of a broader effort to stem sectarian violence there. Among the nonmilitary components of that effort was securing cooperation from Iraq's immediate neighbors, particularly Iran and Syria. A second and related goal was to convince Iran to stop supplying weapons, including roadside bombs, to Shiite militias and other insurgent groups in Iraq that were destabilizing the Iraqi government and attacking U.S. troops.

Getting regional support for stability in Iraq had been the chief recommendation of the December 2006 report by the Iraq Study Group, a commission of respected public policy figures mandated by Congress to develop a bipartisan policy for Iraq. Noting that Arab insurgents from elsewhere in the Middle East were entering Iraq through Syria and that key leaders of the Baghdad government had ties with Iran, the study group recommended that Bush drop his reluctance to deal with Damascus and Tehran, preferably in the setting of a regional conference. Bush and his aides at first rejected this advice and instead focused on military escalation as the primary solution to the imperiled security of Iraq.

The administration's position appeared to harden in late January and early February 2007, when U.S. troops arrested alleged Iranian intelligence operatives in Iraq and the Pentagon made repeated claims that Iran was backing antigovernment militias in Iraq. Bush said on January 26 that he had authorized U.S. troops in Iraq to capture or kill any Iranians in Iraq who were involved in "killing our soldiers or hurting Iraqi people."

In early February Pentagon officials said they had proof that Iran had given Shiite militias in Iraq the sophisticated munitions used to make particularly deadly versions of the roadside bombs known as improvised explosive devices (IEDs). Bush stepped up his rhetoric against Iran on February 14, saying he was certain that the IED munitions had been supplied by the Quds (Jerusalem) Force of the Iranian Revolutionary Guard. This sudden flurry of allegations against Iran led some observers to believe the administration was laying the political groundwork for an eventual war against Iran, while others suggested the administration simply was putting verbal pressure on Tehran in the absence of direct diplomacy.

Suddenly, on February 27, the Bush administration announced that the United States would participate in a "Neighbors Conference" in Baghdad, to which Iran and Syria also had been invited. The announcement provoked commentary about whether Bush had reversed his position against negotiating with Iran—something the White House heatedly denied. "You guys are getting it wrong," White House press secretary Tony Snow told reporters who questioned whether Bush had changed course.

David M. Satterfield, the State Department's top adviser on Iraq, and Zalmay Khalilzad, the U.S. ambassador to Iraq, represented the United States at the conference, which was held in the fortified Green Zone in central Baghdad on March 10. The Iranian delegation was led by Deputy Foreign Minister Abbas Araghchi. In addition, representatives from Iraq and several other Middle Eastern and European countries attended. Khalilzad reported later that he and his fellow U.S. diplomats shook hands with their counterparts from Iran but had not engaged in any direct substantive discussions with them. The principal outcome of the meeting reportedly was the creation of several "working groups" to discuss specific issues, such as border security and the plight of Iraqi refugees in neighboring countries.

A more important follow-up meeting took place on May 3–4 in the Egyptian resort city of Sharm el-Sheikh at a regional conference chaired by Iraqi prime minister Nouri al-Maliki and UN secretary-general Ban Ki-moon; more than thirty foreign ministers from Iraq's neighbors and other key countries participated, including Secretary of State Rice and Iranian foreign minister Manuchehr Mottaki. On the first day, the conference officially launched a UN- and World Bank–sponsored reconstruction plan, the International Compact with Iraq, in which donor nations pledged increased aid to Iraq in return for pledges by the Baghdad government to carry out economic and political reforms.

The second day focused on security issues in Iraq, with Maliki demanding that "neighboring countries stop the infiltration of terrorist groups inside Iraq, and prevent them from getting any funds and political and media support." As at previous meetings, the U.S. and Iranian representatives reportedly exchanged pleasantries but did not engage in substantive discussions. Rice brushed aside questions about why she did not meet directly with her Iranian counterpart, saying the regional conference "is not about the United States and Iran. It's about Iraq and I think we can find different ways to deal with and reinforce the message" that Iran should stop meddling in Iraq.

U.S.-Iranian Diplomacy

The security situation in Iraq continued to deteriorate in the days after the high-level conference in Egypt, perhaps increasing the incentives for both the United States and Iran to discuss their mutual concerns about Iraq directly. The diplomatic breakthrough took place at Iraqi prime minister Maliki's office on May 28, when, for the first time in a generation, diplomats from Iran and the United States attended a publicly announced meeting and discussed serious matters face to face. The U.S. delegation was led by Ambassador Ryan C. Crocker, who had arrived in Baghdad the previous month to replace Ambassador Khalilzad. Ambassador Hassan Kazemi Qumi headed the Iranian delegation. Maliki opened the session with a short statement, then left the room; his national security adviser, Mowaffak al-Rubaie, remained behind as a moderator to create the appearance of a three-sided discussion. The U.S. and Iranian delegations sat at a long polished conference table, facing each other.

The history-making talks lasted for about four hours; afterward, Crocker told reporters the atmosphere was "business-like." Crocker said each side described its overall policies regarding Iraq, for which "there was pretty good congruence right down the line—support for a secure, stable, democratic, federal Iraq, in control of its own security, at peace with its neighbors." At a separate news conference on the same day, Crocker said he found it "somewhat encouraging" that the Iranians had started their presentation with a "positive policy declaration" on the issues of Iraqi security.

Despite the agreement on generalities, each side used the meeting to air its grievances against the other. Crocker said he laid out "a number of our direct specific concerns about their behavior in Iraq, their support for militias that are fighting both the Iraqi security forces and coalition forces." At his own news conference after the talks, Iranian ambassador Qumi said he told the Americans that Iraq's problems stemmed from the U.S. "occupation" of the country: "We made clear the role of occupiers in Iraq and reminded them [of] their mistakes in running the affairs in Iraq." Qumi also reported that his side gave the U.S. diplomats "appropriate responses" to the allegations about Iranian military support for insurgents in Iraq. The Iranians also proposed what Qumi called a "trilateral security committee" to coordinate security issues in Iraq.

The two sides met again on July 24, this time for about seven hours at Maliki's official residence. This longer meeting produced an agreement on the earlier Iranian proposal to create a working group, or what Crocker called a "security subcommittee" of experts, to discuss such matters as the Iraqi militias and insurgent groups.

Both sides said the July meeting brought out the underlying tensions between the United States and Iran and featured what Crocker described as "some heated moments" when he made allegations about Iranian behavior and the Iranians responded in kind. "We've got a lot of problems with the Iranians and, you know, face to face we're not going to pull any punches. They've got problems of their own, so I don't read anything too negative into that," he said.

Crocker added that his complaints centered on continued Iranian aid to the insurgent groups and militias in Iraq—aid that he said had increased, rather than decreased, since the May 28 meeting. The most significant U.S. allegation was that Iranian security forces were working with the Lebanese militia group Hezbollah to train Shiite militia fighters in Iraq.

The security working group held its first session on August 6 but did not announce any definitive results. Another meeting was scheduled for mid-December but postponed until early 2008, reportedly because of disputes over the agenda and an Iranian request for continued meetings at the ambassadorial, rather than working group, level.

Despite these procedural disagreements, U.S. officials in late December said they had seen evidence that Iran had decided to curtail its support for militias in Iraq. In an interview with the *Washington Post* published on December 23, State Department adviser Satterfield said the Tehran government had decided "at the most senior levels" to cut back munitions and other supplies for the Shiite militias targeting U.S. troops in Iraq. He said the decision seemed to coincide with the August 29 cease-fire announcement by the "Mahdi army" militia loyal to Shiite cleric Moqtada al-Sadr, who had been living in Iran. Sadr made his announcement one day after his militia and a rival Shiite militia clashed in the holy city of Karbala, south of Baghdad, causing widespread outrage among Iraq's Shiite majority community.

Ahmadinejad in New York

The U.S.-Iranian diplomacy over Iraq came in the midst of continuing verbal warfare between the Bush administration and Iranian president Ahmadinejad. In 2006 Ahmadinejad had made several efforts to reach out to the United States, including writing a letter to President Bush, offering to debate Bush, and penning an open letter to the American people. The Bush administration dismissed these advances as posturing intended to ease international pressure against Iran's weapons program.

Ahmadinejad followed up in September 2007 with an extended visit to New York City during the annual opening session of the United Nations General Assembly. On September 24 he participated in two programs that enabled Americans to ask him questions, providing rare opportunities for interactions between an Iranian leader and Americans. The first was a noontime speech, by video link, to the National Press Club in Washington; the second was an appearance at a World Leaders Forum at Columbia University. Nearly as remarkable as Ahmadinejad's presence on the campus was the lengthy and hostile introductory speech by Columbia president Lee C. Bollinger, who defended the school's controversial decision to invite the Iranian leader but also denounced him for behaving like "a petty and cruel dictator." Ahmadinejad bristled at the "insult" and adopted a confrontational tone in response to several harsh questions. He refused to give direct answers to several questions concerning previous statements in which he appeared to call for the destruction of Israel and cast doubt on whether millions of European Jews really died at the hands of Nazi Germany in the 1930s and 1940s. Rather than discuss those matters, he focused on the Palestinians, whose fate, he said, was an "old wound." Ahmadinejad also received a scornful reaction from the audience when he responded to a question about Iran's treatment of women and homosexuals by saying, "In Iran, we don't have homosexuals like in your country."

Ahmadinejad adopted a combative tone in his speech to the UN General Assembly on September 25, accusing "arrogant powers"—by which he meant the United States and European countries—of attempting to deny Iran its right to the peaceful use of nuclear power. Citing Iran's work plan agreement with the IAEA one month earlier, he said, "I officially announce that in our opinion the nuclear issue of Iran is now closed and has turned into an ordinary agency [IAEA] matter."

Americans Detained in Iran

Also in 2007, the Iranian government detained four Iranian Americans for varying periods, charging at least three of them with endangering Iranian national security and espionage. Most international attention centered on the detention of Haleh Esfandiari, director of the Middle East Program at the Woodrow Wilson Center for Scholars. She was prevented from leaving Iran in December 2006 after visiting her elderly mother and was jailed in early May. After high-profile appeals on her behalf by U.S. officials and private individuals, Esfandiari was allowed to leave the country on September 3, but only after her mother's apartment in Iran was posted as bail.

Other Iranian Americans who were temporarily jailed or prevented from leaving Iran during the year were Kian Tajbakhsh, an urban planning consultant with the Open Society Institute founded by financier George Soros; journalist Parnaz Azima of the U.S.-financed Radio Farda; and Ali Shakri of the Center for Citizen Peacebuilding at the University of California at Irvine. Iranian officials suggested that all four of these Iranian Americans were part of a U.S. conspiracy to overthrow the Tehran government; Iran based this complaint on the Bush administration's $75 million annual program, funded through the State Department, that sought to promote democracy in Iran.

Following are excerpts from a statement and news conference by Ryan C. Crocker, U.S. ambassador to Iraq, on May 28, 2007, following negotiations among U.S., Iranian, and Iraqi diplomats in Baghdad.

DOCUMENT

U.S. Ambassador to Iraq on His Meeting with Iranian Officials

May 28, 2007

Ambassador Crocker: Good Afternoon. I apologize for being a little late. The talks ran a little longer than expected. I'll just make a few comments to characterize them and then I'll be happy to take your questions.

As you know, U.S. and Iranian delegations met this morning at the [Iraqi] Prime Minister's office—hosted by the Prime Minister—for talks that ran in total for about four hours. I would characterize the atmosphere of the talks as business-like.

The Iranians as well as ourselves laid out the principles that guide our respective policies toward Iraq. There was pretty good congruence right down the line—support for a secure, stable, democratic, federal Iraq, in control of its own security, at peace with its neighbors.

We both laid out our support for the governor of—[Iraqi] Prime Minister [Nouri al-]Maliki—as he undertakes a number of very difficult challenges and we all agreed—Iraqis, Americans, and Iranians, that the focus of our discussions was on Iraq and Iraq only and how we might support as effectively as possible Iraq, its people and its government in restoring security and stability to the country and furthering a political reconciliation process.

We also made it clear from the American point of view that this is about actions not just principles and I laid out before the Iranians a number of our direct specific concerns about their behavior in Iraq, their support for militias that are fighting both the Iraqi security forces and coalition forces.

The fact that a lot of the explosives and ammunition that are used by these groups are coming in from Iran, that such activities led by the IRGC Quds Force [Iranian Revolutionary Guards Corps Quds, or Jerusalem Force] needed to cease, and that we would be looking for results. The Iranians did not respond directly to that. They did again emphasize that their policy is support of the government and the Government of Iraq itself thanked both delegations for the statements of support that were laid out today and we'll see what happens next. . . .

Question: What were some of the concerns the Iranians had and what your reaction was to those?

Ambassador Crocker: The Iranians did not go into any great detail. They made the assertion that the Coalition [the U.S.-led multinational force in Iraq] presence was an occupation and that the effort to train and equip the Iraqi security forces had been inadequate to the challenges faced. We of course responded on both points, making clear that coalition forces are here at the Iraqi government's invitation and under security council authorities and that we have put literally billions of dollars into training and equipping an increasingly capable set of Iraqi security forces.

They—the Iranians—did propose a trilateral mechanism to coordinate on security matters. That of course would be a decision for Washington. The point that I made in the meeting is that the purpose of the meeting was not to discuss further meetings. It was to lay out concrete concerns as we did and our expectation that action would be taken on them.

Question: [In Arabic]. Your Excellency. Did you find the Iranian side in their negotiation today, positive and will open the road for other negotiations that could be useful to the Americans?

Ambassador Crocker: The Iranians, again, laid out their policy toward Iraq—their aims and goals in terms very similar to our own policy and very similar to what the Iraqi government has set as its own set of guiding principles. From that point of view I would say that the talks proceeded positively.

What we underscored to the Iranians, though, is that beyond principle there is practice and what we need to see is Iranian actions on the ground come into harmony with their stated principles because right now the action that I described to them and that I just described to you are running at cross purposes to their own policy.

So, in terms of what happens next I think we're going to want to wait and see not what is said next, but what happens next on the ground—whether we start to see indications of a change in Iranian behavior.

Question: [In Arabic]. Did you agree with the Iranians on other meetings or will the meetings end today? We heard from some sources that the American requests to Iran—Iran responded by giving American incentives. Is this true? What are these American incentives that will be provided to Iran? (Inaudible).

Ambassador Crocker: The Iraqi government said it would extend an invitation in the period ahead for another meeting. We'll obviously consider that invitation when we receive it. With respect to, again, the substance of the exchange—the Iranian side did not respond in detail to the points I laid out nor did they have specific issues to put on the table themselves beyond those that I mentioned and which we dealt with in the discussion concerning the status of Coalition Forces and our efforts at training and equipping Iraqi security forces. Their main focus was on mechanisms—if you will—and principles rather than the detailed security substance that we need to see improvement in the future for the sake of Iraq, its government and its people. . . .

Question: Mr. Ambassador, thank you. Were there any concrete agreements that were reached that could lower or improve the security in the country in the short term—anything at all. And what was the mechanism—the security trilateral mechanism they talked about, if you could expand on that.

Ambassador Crocker: Again, at the level of principle and policy there was broad agreement—Iraqis, Iranian and Americans. In terms of security specifics we laid out a number of them. The Iranians did not offer any detailed response. They did say they rejected such allegations but, again, there was no detailed exchange. The mechanism they propose for trilateral security cooperation did not go beyond pretty much that simple characterization.

My observation on it was, again, that the purpose of a meeting should not be simply to arrange other meetings, but also that it seemed to me that what we were doing today and the structure we had to do it—Iraqis, Americans, Iranians—that effectively was a security committee. Because again, on the level of policy, there isn't a great deal to argue about. We all are pretty much in the same place in terms of declaratory policy. The problem lies, in our view, with the Iranians not bringing their behavior on the ground into line with their own policy. . . .

Question: [In Arabic]. (Inaudible). What was the position of the Iraqi side? Did they intervene or propose anything in the meeting?

Ambassador Crocker: As you know, Prime Minister Maliki himself hosted these talks. And I think you all saw his televised statement at the opening session. The Iraqis, led by the Prime Minister, were not only present but very active really in overseeing the discussion. You could say, if you will, that this was an Iraqi-led process. Iraqis were in the chair for this. Both my Iranian colleague and I deferred to the Iraqi chair either the Prime Minister, or in the follow on sessions, the national security advisor, Dr. [Mowaffak al-] Rubaie. So, I think it's fair to say that the Iraqi government was not only present and was not only a participant to a very important degree, it was Iraq that led the discussions.

Question: [In Arabic]. Arabiya Channel. Your Excellency. Iran perhaps is looking for an assuring message for nonintervention by America in Iran. During this meeting today with the Iranians, did you mention that you are willing to provide such a comforting message?

Ambassador Crocker: Again, one of the points of agreement among all the three of us—Iraq, Iran and the United States—was that the subject of discussion would be the situation in Iraq and how the United States and Iran can work to improve conditions here. We both respected that so there was no discussion of issues outside of that framework including the one you mentioned.

Question: Steven Farrell, *The London Times.* Did you, at any stage, present evidence—pictorial, data or otherwise—about the claims you've been making about Iranian supply of weapons attacking Coalition forces? And it sounds like, from what you're saying, if you were speaking in detail and they were in generalities—did most of the talking come from your side as it were?

Ambassador Crocker: Again, the purpose of our effort in this meeting was not to build a legal case, presumably the Iranians know what they're doing. Our point was simply to say we know as well this is dangerous for Iraq, it contravenes Iran's own stated policy, and it is dangerous for the region because it can produce widespread instability. So, we were not there for the purpose of trying to lay out a judicial case on this. Simply to say, look, this is what's happening, this is what needs to stop. I really can't give you a word count except to say that, as you surely know among diplomats, you don't need a lot of substance to take up a lot of time. . . .

Source: U.S. Department of State. Embassy of the United States, Baghdad, Iraq. "Remarks by Ambassador Ryan C. Crocker at the Press Availability after Meeting with Iranian Officials." Press release. May 28, 2007. http://iraq.usembassy.gov/iraq/20070528_remarks_by_ambassador_ryan_c._crocker_at_the_press_availability_after_meeting_with_iranian_officials.html (accessed March 3, 2008).

Other Historic Documents of Interest

From this volume

- President Bush on the "surge" of U.S. troops to Iraq, p. 9
- UN Security Council resolution and U.S. National Intelligence Estimate on Iran's nuclear programs, p. 112

From previous *Historic Documents*

- Letter from Iranian president Ahmadinejad to President Bush, *2006,* p. 212
- Iraq Study Group report on U.S. policy in Iraq, *2006,* p. 725
- UN Security Council on sanctions against Iran, *2006,* p. 781

DOCUMENT IN CONTEXT

Yar'Adua on His Inauguration as President of Nigeria

MAY 29, 2007

Deeply flawed elections in April resulted in Nigeria's first-ever transfer of power from one elected civilian government to another. But disputes related to the election, combined with the country's ingrained corruption and widespread violence, called the new government's capabilities into question. Although Nigeria had the potential to be one of its continent's richest nations because of its oil resources—it was already the most populous, with 140 million people— it was as poor as most other African countries. Despite economic growth fueled by high oil prices, Nigeria's cities and towns lacked reliable electricity and other public services. Good-paying jobs were scarce, and many people considered themselves worse off under democracy than they had been under the previous decades of military dictatorship.

As expected, the state and national elections in April were won by the ruling People's Democratic Party, including its candidate for president, Umaru Yar'Adua. Yar'Adua, a little-known politician who had been the governor of the northern state of Katsina, was selected by outgoing president Olusegun Obasanjo, a military ruler in the 1970s who had returned to public life as a civilian and won questionable elections in 1999 and 2003. The 2007 elections that put Yar'Adua in office were even more heavily criticized domestically and internationally than the previous ones, suggesting that Nigeria had not yet developed the institutional capacity of a genuine democracy.

Before being elected president, Yar'Adua was one of the few sitting governors who had not been indicted on corruption charges—a status that enabled him to campaign as "Mr. Integrity." At year's end, however, it was unclear whether, as president, Yar'Adua had the political clout to reform the corrupt, ineffective national government he inherited.

Background

Nigeria achieved independence from Great Britain in 1960 and promised its citizens democratic rule. Like many other African countries, however, Nigeria quickly fell into the hands of generals or entrenched politicians who used power for their own ends. In its first four decades as an independent state, Nigeria was ruled by the military for nearly thirty years.

A new chance for democracy came in 1999 when Obasanjo campaigned and won elections on his promises to curb corruption, keep the military out of government, and bridge the country's many sectarian divides; Nigeria had some 250 ethnic and tribal groups and was roughly split between the mostly Muslim north and the mostly Christian south. When he came to office, Obasanjo was widely respected both in Nigeria and internationally because he was Nigeria's only military ruler who had voluntarily stepped down, in 1979, in favor of an elected government.

Obasanjo had some successes as president. Largely because world oil prices started to rise in 2005, Nigeria's overall economy boomed during his second term, enabling him to pay off all of the $32 billion in national debt. Obasanjo established Nigeria as a leading political force in Africa, in part by contributing thousands of troops to peacekeeping forces that helped end conflicts elsewhere on the continent. In addition, he guaranteed that elections—however flawed—would be the method for choosing Nigeria's leaders.

In other respects, conditions in Nigeria had not improved—and perhaps had even deteriorated—after Obasanjo's eight years in office. Despite the president's anticorruption drive, government officials at all levels continued to use their positions to line their pockets at the expense of ordinary citizens, who lived on an average income of $2 per day. Lack of investment left the country's oil industry outdated and inefficient; without adequate refining capacity, Nigeria had to import all of its gasoline, which was sold to the public at prices heavily subsidized by the government. One of the government's most striking failures was the lack of reliable electricity. According to most estimates, Nigeria's aging power industry generated only about 10 percent of the output necessary for twenty-four-hour service, even in metropolitan areas.

Obasanjo's years in office also saw the growth of a rebellion in the oil-producing Niger Delta region, where government opponents demanded a greater share of the oil wealth. Sporadic fighting in that region, and frequent attacks on the oil industry, killed thousands of people—some estimates put the figure at more than 10,000 since 1999—and curtailed oil production by an estimated 25 percent. By 2007 oil exports were down to about 2.1 million barrels per day, which was still enough to make Nigeria the eighth-largest producer in the world and the fifth-most-important source of imported oil for the United States.

Although Obasanjo brought an end to military rule, he did not succeed in cementing into place the procedures for valid elections. The elections that put him, and his People's Democratic Party, into office in 1999 and 2003 were widely seen as flawed; international observers refused to certify the validity of either poll. By the end of his eight years in power, many Nigerians complained that democracy had brought no improvements to their daily lives. "It doesn't matter who runs this country," Lagos shopkeeper Joyce Kadiri told the *Guardian*, a British newspaper, in May. "They are the elites. They get rich, we suffer."

April Elections

The 2007 election campaign started with disputes in February, when the Economic and Financial Crimes Commission, the country's anticorruption agency, ruled that 135 candidates for national and state posts were unfit for office. The most prominent of these was national vice president Atiku Abubakar, who had broken with Obasanjo in 2006 over the president's desire to amend the constitution to allow him a third four-year term in office. The parliament ultimately voted against amending the constitution on Obasanjo's behalf—an action widely seen as ratifying constitutional rule—but the rift between the two men never healed. Obasanjo's party chose Yar'Adua as its candidate, and an angry Abubakar sought the presidency under the banner of an opposition coalition, the Action Congress.

Abubakar had been accused of funneling millions of dollars from government agencies into his own business interests, a charge he denied while asserting that the anticorruption commission was a tool used by Obasanjo against his enemies. The commission's ruling was advisory and did not have the force of law, but the National Electoral Commission ruled on March 14 that Abubakar could not run for the presidency because of the corruption allegations.

While Abubakar was appealing that ruling, elections for state governors and legislatures were held on April 14. The voting was marred in many areas by logistical problems, violence that killed at least fifty people, and ballot stuffing. The ruling party won twenty-seven of the thirty-six governorships, prompting opposition parties to call unsuccessfully for annulment of the results.

Just two days after that election, and five days before the presidential election, the Supreme Court of Nigeria on April 16 ruled that Abubakar should appear on the presidential ballot. Some 61 million ballots had already been printed without Abubakar's name and sent to polling places. New ballots were printed at the last minute, but they showed only the names of parties, not candidates, and many polling places reportedly never received them. The other main opposition presidential candidate was former general Muhammadu Buhari, one of the country's military rulers in the 1980s.

As voters were preparing to go to the polls for the national election, Obasanjo acknowledged flaws in election procedures but insisted democracy was the key to Nigeria's future. "Do we have an alternative to democracy, no matter what difficulties we are encountering?" he asked on April 20. "The answer surely is no. Then let us continue to improve on the structure of the house rather than pull it down because it is leaking in part."

On election day, April 21, journalists and election monitors encountered numerous cases of vote rigging, most of them on behalf of the ruling party. In one example, a *Washington Post* reporter found young men filling out stacks of ballots at a polling station near Port Harcourt, the center of the oil industry. Election monitors reported other cases in which poll workers efficiently filled out all the ballots before any voters arrived, or failed to open the polling stations at the appointed hour, or did not have anywhere near the necessary number of ballots. At dozens of polling stations, men with bags of money were seen handing out cash to voters.

The Transition Monitoring Group, a coalition of Nigerian civil society groups that sent thousands of monitors to observe polling places, demanded the next day that the election be annulled because of widespread fraud and irregularities. "We cannot allow these sham elections to continue," Innocent Chukwuma, the group's chairman, said. International election monitoring groups also issued reports indicating that the voting failed to meet basic standards. Martin van den Berg, chief of the European Union's observer mission, said the election process "cannot be considered to have been credible." A later report by the U.S.-based advocacy group Human Rights Watch charged that, with "brazenly rigged polls, government officials have denied millions of Nigerians any real voice in selecting their political leaders."

The government's election commission rejected these criticisms and announced on April 23 that Yar'Adua had won the election with 24.6 million votes; Buhari finished second with 6.6 million votes, and Abubakar was third with 2.6 million. With 72 percent of the total vote, Yar'Adua easily met constitutional standards under which a presidential candidate could be declared the winner only by receiving a plurality of all votes cast nationwide and at least 25 percent of the votes in two-thirds of the thirty-six states. These requirements were intended to ensure that Nigeria's president drew broad support nationwide, not just from voters of the candidate's own region and ethnic background.

In a televised address, Yar'Adua said he had won "because my party is strong and we enjoy the goodwill of the Nigerian people." Obasanjo also went on television and acknowledged irregularities in the election, which he attributed to parties and candidates "who have employed thugs and violent means to secure what they consider electoral victory." Obasanjo urged those who were unhappy with the results to pursue their complaints through the courts—a step Abubakar had already said he was planning to take.

The violence that accompanied both the state and national elections—killing as many as 300 people, according to some estimates—led many observers to fear an even greater outbreak of violence after the final results were announced. That did not happen, however. Nigeria's customary violence—particularly in the Niger Delta—continued unabated, but there was no nationwide upsurge that could be attributed to widespread anger over the presidential election.

Yar'Adua in Office

Yar'Adua took office on May 29, amid a two-day strike called by several of the country's main unions to protest the flawed election. In an inaugural address that was much more conciliatory than Obasanjo's recent rhetoric, the new president reached out to political opponents, including the guerrillas in the Niger Delta region. Yar'Adua pledged to give "urgent attention" to that impoverished area and called on "all aggrieved communities, groups, and individuals to immediately suspend all violent activities and respect the law."

For the first time, Yar'Adua acknowledged problems in the electoral process that brought him to power, saying the elections "were not perfect and had lapses and shortcomings." Echoing Obasanjo's statement right after the voting, Yar'Adua urged "anyone aggrieved" by the voting to pursue complaints in the "well-established legal avenues of redress" and pledged to establish a panel to examine the electoral process and make recommendations. He later appointed an electoral reform panel and asked to receive its recommendations within one year.

The new president moved cautiously, taking nearly three months to assemble his thirty-two-member cabinet. In a nod to the opposition, he awarded cabinet posts to figures from other parties. Addressing his new cabinet members on July 26, Yar'Adua bluntly warned them against continuing Nigeria's customary corruption: "As an administration, we have absolute zero tolerance for corruption in all its ramifications. We must never abuse public trust either through misappropriation, misapplication, or outright stealing of public funds. Anyone who does so will have the full weight of the law to reckon with."

An action by the government late in the year raised questions about how effective Yar'Adua's anticorruption campaign could be. Officials announced on December 29 that Nuhu Ribadu, a former police officer and the zealous chairman of the anticorruption commission, had been relieved so he could spend a year studying. The government portrayed this as a routine move, but reform advocates noted that the announcement came shortly after Ribadu had arrested one of the president's key allies, and they suggested that Ribadu was being stopped before he reached the core of corruption in the ruling party.

Following is the text of the address by Umaru Yar'Adua upon his inauguration as president of Nigeria on May 29, 2007.

DOCUMENT

Inaugural Address of Umaru Yar'Adua

May 29, 2007

H.E. Vice President Goodluck Jonathan, President of the Senate, the Speaker House of Representatives, my Lord Chief Justice of Nigeria, [outgoing] President [Olusegun] Obasanjo, distinguished Presidents and Heads of governments who have graciously honored us with their presence today, leaders of our nation, guests from far and near, fellow citizens.

This is a historic day for our nation, for it marks an important milestone in our march towards a maturing democracy.

For the first time since we cast off the shackles of colonialism almost a half-century ago, we have at last managed an orderly transition from one elected government to another.

We acknowledge that our elections had some shortcomings. Thankfully, we have well-established legal avenues of redress, and I urge anyone aggrieved to pursue them.

I also believe that our experiences represent an opportunity to learn from our mistakes. Accordingly, I will set up a panel to examine the entire electoral process with a view to ensuring that we raise the quality and standard of our general elections, and thereby deepen our democracy.

This occasion is historic also because it marks another kind of transitional generational shift when the children of independence assume the adult responsibility of running the country at the heart of Africa.

My fellow citizens, I am humbled and honored that you have elected me and Vice President Jonathan to represent that generation in the task of building a just and humane nation, where its people have a fair chance to attain their fullest potential.

Luckily we are not starting from scratch. We are fortunate to have been led the past eight years by one of our nation's greatest patriots, President Obasanjo. On behalf of all our people, I salute you, Mr. President, for your vision, your courage and your boundless energy in creating the roadmap toward that united and economically thriving Nigeria that we seek.

Many of us may find it hard to believe now, but before you assumed the presidency eight years ago, the national conversation was about whether Nigeria deserved to remain one country at all.

Today we are talking about Nigeria's potential, to become one of the 20 largest economies in the world by the year 2020. That is a measure of how far we have come. And we thank you.

The administration of President Obasanjo has laid the foundation upon which we can build our future prosperity.

Over the past eight years Nigerians have reached a national consensus in at least four areas: to deepen democracy and the rule of law; build an economy driven primarily by the private sector, not government; display zero tolerance for corruption in all its forms, and, finally, restructure and staff our government to ensure efficiency and good governance.

I commit myself to these tasks.

Our goal now is to build on the greatest accomplishments of the past few years. Relying on the 7-point agenda that formed the basis of our compact with voters during the recent campaigns, we will concentrate on rebuilding our physical infrastructure and human capital in order to take our country forward.

We will focus on accelerating economic and other reforms in a way that makes a concrete and visible difference to ordinary people.

Our economy already has been set on the path of growth. Now we must continue to do the necessary work to create more jobs, lower interest rates, reduce inflation, and maintain a stable exchange rate. All this will increase our chances for rapid growth and development.

Central to this is rebuilding our basic infrastructure. We already have comprehensive plans for mass transportation, especially railroad development. We will make these plans a reality.

Equally important, we must devote our best efforts to overcoming the energy challenge. Over the next four years we will see dramatic improvements in power generation, transmission and distribution.

These plans will mean little if we do not respect the rule of law. Our government is determined to strengthen the capacity of law enforcement agencies, especially the police. The state must fulfill its constitutional responsibility of protecting life and property.

The crisis in the Niger Delta commands our urgent attention. Ending it is a matter of strategic importance to our country. I will use every resource available to me, with your help, to address this crisis in a spirit of fairness, justice, and cooperation.

We have a good starting point because our predecessor already launched a master plan that can serve as a basis for a comprehensive examination of all the issues. We will involve all stakeholders in working out a solution.

As part of this effort, we will move quickly to ensure security of life and property, and to make investments safe.

In the meantime, I appeal to all aggrieved communities, groups and individuals to immediately suspend all violent activities, and respect the law. Let us allow the impending dialogue to take place in a conducive atmosphere. We are all in this together, and we will find a way to achieve peace and justice.

As we work to resolve the challenges of the Niger Delta, so must we also tackle poverty throughout the country.

By fighting poverty, we fight disease. We will make advances in public health, to control the scourge of HIV/AIDS, malaria, and other diseases that hold back our population and limit our progress.

We are determined to intensify the war against corruption, more so because corruption is itself central to the spread of poverty. Its corrosive effect is all too visible in all aspects of our national life. This is an area where we have made significant progress in recent years, and we will maintain the momentum.

We also are committed to rebuilding our human capital, if we are to support a modern economy. We must revive education in order to create more equality, and citizens who can function more productively in today's world.

To our larger African family, you have our commitment to the goal of African integration. We will continue to collaborate with fellow African states to reduce conflict and free our people from the leg chains of poverty.

To all our friends in the international community, we pledge our continuing fidelity to the goals of progress in Africa and peace in the world.

Fellow citizens, I ask you all to march with me into the age of restoration. Let us work together to restore our time-honored values of honesty, decency, generosity, modesty, selflessness, transparency and accountability. These fundamental values determine societies that succeed or fail. We must choose to succeed.

I will set a worthy personal example as your president.

No matter what obstacles confront us, I have confidence and faith in our ability to overcome them. After all, we are Nigerians! We are a resourceful and enterprising people, and we have it within us to make our country a better place.

To that end I offer myself as a servant-leader. I will be a listener and doer, and serve with humility.

To fulfill our ambitions, all our leaders at all levels whether a local government councilor or state governor, senator or cabinet minister must change our style and our attitude. We must act at all times with humility, courage, and forthrightness.

I ask you, fellow citizens, to join me in rebuilding our Nigerian family, one that defines the success of one by the happiness of many.

I ask you to set aside negative attitudes, and concentrate all our energies on getting to our common destination.

All hands must be on deck.

Let us join together to ease the pains of today while working for the gains of tomorrow.

Let us set aside cynicism, and strive for the good society that we know is within our reach.

Let us discard the habit of low expectations of ourselves as well as of our leaders.

Let us stop justifying every shortcoming with that unacceptable phrase, "the Nigerian factor", as if to be a Nigerian is to settle for less.

Let us recapture the mood of optimism that defined us at the dawn of independence, that legendary can-do spirit that marked our Nigerianness.

Let us join together, now, to build a society worthy of our children.

We have the talent. We have the intelligence. We have the ability.

The challenge is great. The goal is clear. The time is now.

I thank you, and God bless you.

SOURCE: Federal Republic of Nigeria. Federal Ministry of Information and Communications. "Inaugural Address of Umaru Musa Yar'Adua, President of the Federal Republic of Nigeria and Commander-in-Chief of the Armed Forces." Abuja, Nigeria. May 29, 2007. www.nigeria.gov.ng/NR/exeres/B30C7654-36D9-4C57-837C-1EAB4A103091.htm?id=992#NewsSection (accessed October 21, 2007).

OTHER HISTORIC DOCUMENTS OF INTEREST

FROM PREVIOUS *HISTORIC DOCUMENTS*

- Obasanjo elected president of Nigeria, *1999*, p. 428

June

DOCUMENT IN CONTEXT

Council of Europe Investigation into the CIA's "Secret Prisons" in Europe

JUNE 7, 2007

Two years of controversy over the Central Intelligence Agency's use of "secret prisons" in Europe and elsewhere came to a close in 2007, only to be supplanted by revived debate over the agency's techniques of interrogating alleged terrorists. Both controversies resulted in numerous legal cases, most of which were quashed on procedural grounds by courts in the United States and Europe.

The secret prisons controversy jumped into the headlines in December 2005 when the *Washington Post* reported that the CIA had detained dozens of terrorism suspects at secret facilities around the world, including in Europe. The *Post* article gave substance to numerous previous reports about the CIA's handling of alleged terrorists—and its mention of facilities in Europe generated a firestorm of criticism on the continent, where public backing of Washington's foreign policies had sagged since the U.S.-led invasion of Iraq in 2003. Two European bodies launched formal investigations of CIA activities: the European Parliament established a special committee and the Council of Europe (a quasi-governmental human rights agency, separate from the European Union) appointed a special investigator. Both bodies issued preliminary reports in 2006 and final reports in 2007.

The administration of President George W. Bush had brushed aside stories about the CIA prisons until September 6, 2006, when Bush acknowledged that the CIA had held a "small number" of terrorism suspects at locations "outside the United States." Gen. Michael Hayden, the CIA director, said in November 2007 that the agency had held "fewer than 100 people" in its secret detention program, which he said started in 2002.

Controversy about the CIA's role in the Bush administration's "war on terror" gained new traction in December 2007 when newspapers reported that the agency had destroyed videotapes that documented its interrogation in 2002 of two alleged al Qaeda operatives. The destruction of the tapes raised questions of whether the CIA had tried to cover up abusive treatment of the detainees. Two congressional committees, the Justice Department, and the CIA itself launched their own investigations.

Reports on CIA Prisons in Europe

The first of the year's final reports about CIA activities in Europe came from a special committee established by the European Parliament. The panel published its report on January 30, then submitted it to the full parliament, which approved it, with changes, on February 14 by a vote of 382–256, with 76 abstentions. In general, members of leftist and center-left parties voted for the document, while members of rightist and center-right parties voted against it.

The report documented more than 1,245 CIA-operated flights over European air space from 2001 to 2005, but noted that "not all those flights" involved the transport of terrorism suspects to and from CIA bases. The report also took an ambiguous position on human rights groups' contention that the CIA had operated one or more secret prisons in Poland, concluding that "it is not possible to acknowledge or deny that secret detention centers were based in Poland."

The report took a somewhat stronger position on allegations that the CIA had operated a secret prison in Romania, saying that "no definitive evidence has been provided to contradict any of the allegations concerning the running of a secret detention facility on Romanian soil." The Romanian government had denied these allegations, and the Romanian senate established a special commission to investigate the matter. That group reported on February 21 that its yearlong investigation had "found no solid argument to make us believe that the CIA was running illegal transports of prisoners" in the country.

A more detailed and heavily documented report on the CIA prisons was issued on June 8 by a Council of Europe investigative panel, led by Swiss senator and lawyer Dick Marty. This panel specifically identified Poland and Romania as being among the countries where the CIA housed and interrogated terrorism suspects between 2003 and 2005; it based its claim on the flight records of CIA-operated aircraft and information from current and former intelligence officials.

In the case of Poland, the report noted numerous clandestine flights by CIA aircraft to Szymany, in northeastern Poland, near a former Soviet-era military base at the village of Stare Kiejkuty. Between eight and twelve high-level terrorism suspects were held at that facility, several of whom had been transferred directly from Kabul, Afghanistan, on flights that were deliberately disguised to hide CIA involvement, according to the report. Among them, the report said, were Khalid Sheikh Mohammed, the alleged mastermind of the September 11, 2001, attacks, and Abu Zubaydah, alleged to be a top al Qaeda official.

Marty's report provided fewer details about the CIA activities in Romania but did suggest that detainees were flown to the Mihail Kogalniceanu Airfield, near the Black Sea port city of Constanta. Numerous news reports since 2005 had cited that airport, which had both civilian and military uses, as the likely location of CIA operations in Romania.

CIA spokesman Paul Gimigliano issued a carefully worded response to Marty's report, calling it biased and distorted but not directly challenging key findings. "The CIA's counter-terrorism operations have been lawful, effective, closely reviewed, and of benefit to many people, including Europeans, by disrupting plots and saving lives," he said.

Officials in Poland and Romania flatly denied claims that the CIA had housed terrorism suspects in their countries. Alexander Kwasniewski, Poland's president from 1995 to 2005, insisted that "there were no secret prisons in Poland." A Romanian senator who investigated the matter, Norica Nicolai, called Marty's allegations "totally groundless."

A separate investigation by a British parliamentary committee also raised concerns about U.S. handling of terrorism suspects. The report by Parliament's Intelligence and Security Committee, published on July 27, focused on the relationship between British and U.S. intelligence services and suggested that British officials were not forceful enough in expressing their concerns about the actions of their American counterparts. Much of the report dealt with the case of two British residents who had been identified by British intelligence in Gambia in 2002 as possible terrorism suspects. The CIA arrested the men, even though British authorities said they should not be arrested. "The case shows a lack of concern, on the part of the U.S., for UK concerns," the report said.

CIA director Hayden made the most extensive U.S. response to these reports in a September 7 speech to the Council on Foreign Relations. He said that the agency had held fewer than 100 people in overseas facilities since the capture of Zubaydah in 2002, and fewer than that number had been transferred to or from foreign governments. He stated that the renditions had been conducted "lawfully, responsibly, and with a clear and simple purpose: to get terrorists off the streets and gain intelligence on those still at large."

Cases of Alleged Abuse

Legal action continued during 2007 on several cases involving alleged abuse by U.S. officials of terrorism suspects at the CIA's secret prisons and other venues. Among these cases were the following:

- Italian authorities launched proceedings in 2007 against Italian and U.S. officials allegedly involved in the case of Abu Omar (also known as Osama Moustafa Hassan Nasr), a Muslim cleric who claimed that the CIA abducted him in Milan, Italy, in February 2003 and flew him to Germany, then to Cairo, where he said he was tortured. An Italian judge on February 16 ordered a trial of twenty-six Americans (one lieutenant colonel in the air force and twenty-five CIA operatives, including Robert Seldon Lady, the former CIA station chief in Milan) and several Italian officials, including Nicolo Pollari, the former chief of the Italian military intelligence service. The case was suspended in June after the Italian government charged that the Milan prosecutor, Armando Spataro, violated state secrecy laws during his investigation.

- German prosecutors sought during 2007 to take action against thirteen people (some of them said to be CIA operatives) in connection with the mistaken kidnapping and jailing of Khaled el-Masri, a German national of Lebanese descent. Masri was seized in Macedonia in December 2003, flown by the United States to Afghanistan, where he was imprisoned for five months and, he said, beaten, then released without charges. A German court blocked an attempt to extradite the CIA operatives, reportedly on the grounds of maintaining good relations with the U.S. government, which had said it would refuse to cooperate in the matter.

  Masri also failed in his own effort to gain legal satisfaction in the United States. A U.S. federal court in May 2005 dismissed a suit Masri had filed against the CIA, and the Fourth Circuit Court of Appeals affirmed that decision on March 2, 2007, citing the need to protect U.S. "state secrets." The Supreme Court on October 9 refused to hear his appeal, thus ending his legal recourse in the United States. The Court's action brought a tart editorial denunciation by the *New York Times*, which said the Court effectively "has left an innocent person without any remedy for his wrongful imprisonment and torture."

- Abd al-Rahim al-Nashiri, a Saudi Arabian national of Yemeni descent, said in a hearing at Guantanamo Bay that he had been held in CIA secret prisons, and tortured, for five years before being transferred to Guantanamo Bay in September 2006. (The U.S. government transferred him and thirteen other suspects to

Guantanamo in conjunction with Bush's statement acknowledging the CIA program.) A heavily censored transcript of his hearing was released on March 30. In it, he claimed that, to satisfy his interrogators, he had falsely confessed to involvement in several terrorist acts, including the 2000 bombing of the U.S.S *Cole* in Yemen.

- Abd al-Hadi al-Iraqi, an Iraqi Kurd and alleged close advisor to al Qaeda leader Osama bin Laden, reportedly became the first terrorism suspect known to be placed in a CIA jail following Bush's September 2006 acknowledgment of the secret prisons. News reports said the Iraqi was captured by U.S. authorities in 2006, taken into CIA custody and questioned for several months, then transferred to Guantanamo Bay in April 2007. The *New York Times* quoted officials as saying he had provided valuable information when subjected to standard, nonabusive interrogation methods approved by the Pentagon.
- Secretary of State Condoleezza Rice on October 24 became the first U.S. official to acknowledge that the government mishandled the case of Maher Arar, a Canadian man who was detained by U.S. authorities in New York in 2002, then shipped to Syria, where, he said, he was tortured. The incident had strained relations between the Canadian government and the United States, which had refused to accept any responsibility for Arar's treatment. Canada apologized to Arar in 2005 and paid him about $10 million in compensation.

  At a congressional hearing, Rice said: "We do not think that this case was handled as it should have been. We do absolutely not wish to transfer anyone to any place in which they might be tortured." She did not, however, explicitly apologize to Arar or acknowledge that he had been tortured. A week earlier, on October 18, Arar had testified before two House of Representative panels via televised link because he was still banned from entering the United States. Several members of the House of Representatives did apologize to Arar for the ordeal he had suffered.

  Arar filed a civil rights lawsuit against the United States, but it was dismissed by a federal court in 2006 on national security grounds. His appeal of that dismissal was pending at the end of 2007. The 2007 movie *Rendition* was based loosely on the Arar case.

Following is the section entitled "Introductory Remarks—An Overview" from the report "Secret Detentions and Illegal Transfers of Detainees Involving Council of Europe Member States: Second Report" submitted on June 7, 2007, to the Committee on Legal Affairs and Human Rights of the Parliamentary Assembly of the Council of Europe, by Dick Marty, the committee's special rapporteur.

DOCUMENT

Council of Europe Report on CIA "Secret Prisons" in Europe

June 7, 2007

Introductory Remarks—An Overview

1. What was previously just a set of allegations is now proven: large numbers of people have been abducted from various locations across the world and transferred to countries where they have been persecuted and where it is known that torture is common practice. Others have been held in arbitrary detention, without any precise charges levelled against them and without any judicial oversight—denied the possibility of defending themselves. Still others have simply disappeared for indefinite periods and have been held in secret prisons, including in member states of the Council of Europe, the existence and operations of which have been concealed ever since.

2. Some individuals were kept in secret detention centres for periods of several years, where they were subjected to degrading treatment and so-called "enhanced interrogation techniques" (essentially a euphemism for a kind of torture), in the name of gathering information, however unsound, which the United States claims has protected our common security. Elsewhere, others have been transferred thousands of miles into prisons whose locations they may never know, interrogated ceaselessly, physically and psychologically abused, before being released because they were plainly not the people being sought. After the suffering they went through, they were released without a word of apology or any compensation—with one remarkable exception owing to the ethical and responsible approach of the Canadian authorities—and also have to put up with the opprobrium of doubts surrounding their innocence and, right here in Europe, racist harassment fuelled by certain media outlets. These are the terrible consequences of what in some quarters is called the "war on terror."

3. While the strategy in question was devised and put in place by the current United States administration to deal with the threat of global terrorism, it has only been made possible by the collaboration at various institutional levels of America's many partner countries. As was already shown in my report of 12 June 2006 (PACE Doc 10957), these partners have included several Council of Europe member states. Only exceptionally have any of them acknowledged their responsibility—as in the case of Bosnia and Herzegovina, for instance—while the majority have done nothing to seek out the truth. Indeed many governments have done everything to disguise the true nature and extent of their activities and are persistent in their unco-operative attitude. Moreover, only very few countries have responded favourably to the proposals made by the Secretary General of the Council of Europe at the end of the procedure initiated under Article 52 of the European Convention of Human Rights ("ECHR") (see document SG(2006)01).

4. The rendition, abduction and detention of terrorist suspects have always taken place outside the territory of the United States, where such actions would no doubt have been ruled unlawful and unconstitutional. Obviously, these actions are also unacceptable under the laws of European countries, who nonetheless tolerated them or colluded actively in carrying them out. This export of illegal activities overseas is all the more shocking in that it shows fundamental contempt for the countries on whose territories it was decided to

commit the relevant acts. The fact that the measures only apply to non-American citizens is just as disturbing: it reflects a kind of "legal apartheid" and an exaggerated sense of superiority. Once again, the blame does not lie solely with the Americans but also, above all, with European political leaders who have knowingly acquiesced in this state of affairs.

5. Some European governments have obstructed the search for the truth and are continuing to do so by invoking the concept of "state secrets". Secrecy is invoked so as not to provide explanations to parliamentary bodies or to prevent judicial authorities from establishing the facts and prosecuting those guilty of offences. This criticism applies to Germany and Italy, in particular. It is striking to note that state secrets are invoked on grounds almost identical to those advanced by the authorities in the Russian Federation in its crackdown on scientists, journalists and lawyers, many of whom have been prosecuted and sentenced for alleged acts of espionage. The same approach led the authorities of "the former Yugoslav Republic of Macedonia" to hide the truth and give an obviously false account of the actions of its own national agencies and the CIA in carrying out the secret detention and rendition of Khaled El-Masri [a German national seized by the CIA in Macedonia in 2003 then imprisoned for five months in Afghanistan, where he said he was beaten and tortured before being released].

6. Invoking state secrets in such a way that they apply even years after the event is unacceptable in a democratic state based on the rule of law. It is frankly all the more shocking when the very body invoking such secrets attempts to define their concept and scope, as a means of shirking responsibility. The invocation of state secrets should not be permitted when it is used to conceal human rights violations and it should, in any case, be subject to rigorous oversight. Here again, Canada seems to demonstrate the right approach, as will be seen later in this report.

7. There is now enough evidence to state that secret detention facilities run by the CIA did exist in Europe from 2003 to 2005, in particular in Poland and Romania. These two countries were already named in connection with secret detentions by Human Rights Watch in November 2005. At the explicit request of the American government, the Washington Post simply referred generically to "eastern European democracies", although it was aware of the countries actually concerned. It should be noted that ABC did also name Poland and Romania in an item on its website, but their names were removed very quickly in circumstances which were explained in our previous report. We have also had clear and detailed confirmation from our own sources, in both the American intelligence services and the countries concerned, that the two countries did host secret detention centres under a special CIA programme established by the American administration in the aftermath of 11 September 2001 to "kill, capture and detain" terrorist suspects deemed to be of "high value". Our findings are further corroborated by flight data of which Poland, in particular, claims to be unaware and which we have been able to verify using various other documentary sources.

8. The secret detention facilities in Europe were run directly and exclusively by the CIA. To our knowledge, the local staff had no meaningful contact with the prisoners and performed purely logistical duties such as securing the outer perimeter. The local authorities were not supposed to be aware of the exact number or the identities of the prisoners who passed through the facilities—this was information they did not "need to know." While it is likely that very few people in the countries concerned, including in the governments themselves, knew of the existence of the centres, we have sufficient grounds to declare that the highest state authorities were aware of the CIA's illegal activities on their territories.

9. We are not an investigating authority: we have neither the powers nor the resources. It is not therefore our aim to pass judgments, still less to hand down sentences. However, our task is clear: to assess, as far as possible, allegations of serious violations of human rights committed on the territory of Council of Europe member states, which therefore involve violations of the European Convention on Human Rights. We believe we have shown that the CIA committed a whole series of illegal acts in Europe by abducting individuals, detaining them in secret locations and subjecting them to interrogation techniques tantamount to torture.

10. In most cases, the acts took place with the requisite permissions, protections or active assistance of government agencies. We believe that the framework for such assistance was developed around NATO authorisations agreed on 4 October 2001, some of which are public and some of which remain secret. According to several concurring sources, these authorisations served as a platform for bilateral agreements, which—of course—also remain secret.

11. In our view, the countries implicated in these programmes have failed in their duty to establish the truth: the evidence of the existence of violations of fundamental human rights is concrete, reliable and corroborative. At the very least, it is such as to require the authorities concerned at last to order proper independent and thorough inquiries and stop obstructing the efforts under way in judicial and parliamentary bodies to establish the truth. International organisations, in particular the Council of Europe, the European Union and NATO, must give serious consideration to ways of avoiding similar abuses in future and ensuring compliance with the formal and binding commitments which states have entered into in terms of the protection of human rights and human dignity.

12. Without investigative powers or the necessary resources, our investigations were based solely on astute use of existing materials—for instance, the analysis of thousands of international flight records—and a network of sources established in numerous countries. With very modest means, we had to do real "intelligence" work. We were able to establish contacts with people who had worked or still worked for the relevant authorities, in particular intelligence agencies. We have never based our conclusions on single statements and we have only used information that is confirmed by other, totally independent sources. Where possible we have cross-checked our information both in the European countries concerned and on the other side of the Atlantic or through objective documents or data. Clearly, our individual sources were only willing to talk to us on the condition of absolute anonymity. At the start of our investigations, the Committee on Legal Affairs and Human Rights authorised us to guarantee our contacts strict confidentiality where necessary. This willingness to grant confidentiality to potential "whistleblowers" was also communicated to Mr Franco Frattini, Vice-President of the European Commission with responsibility for the area of freedom, security and justice, so that he could also notify the relevant ministers in EU countries. Guarantees of confidentiality undoubtedly contributed to a climate of trust and made it possible for many sources to agree to talk to us. The individuals concerned are not prepared at present to testify in public, but some of them may be in the future if the circumstances were to change.

13. The Polish authorities recently criticised us for not travelling to their country to visit the facility suspected of having housed a detention centre. However, we see no point in visiting the site: we are not forensic science experts and we have no doubts about the capability of those who would have removed any traces of the prisoners' presence. Moreover, a meeting at the site would only have been worthwhile if the Polish authorities

had first replied to the questions we put to them on numerous occasions and to which we are still awaiting replies.

14. We are fully aware of the seriousness of the terrorist threat and the danger it poses to our societies. However, we believe that the end does not justify the means in this area either. The fight against terrorism must not serve as an excuse for systematic recourse to illegal acts, massive violation of fundamental human rights and contempt for the rule of law. I hold this view not only because methods of this nature conflict with the constitutional order of all civilised countries and are ethically unacceptable, but also because they are not effective from the perspective of a genuine long-term response to terrorism.

15. We have said it before and others have said it much more forcefully, but we must repeat it here: having recourse to abuse and illegal acts actually amounts to a resounding failure of our system and plays right into the hands of the criminals who seek to destroy our societies through terror. Moreover, in the process, we give these criminals a degree of legitimacy—that of fighting an unfair system—and also generate sympathy for their cause, which cannot but serve as an encouragement to them and their supporters.

16. The fact is that there is no real international strategy against terrorism, and Europe seems to have been tragically passive in this regard. The refusal to establish and recognise a functioning international judicial and prosecution system is also a major weakness in our efforts to combat international terrorism. We also agree with the view expressed by Amnesty International in its recent annual report: governments are taking advantage of the fear generated by the terrorist threat to impose arbitrary restrictions on fundamental freedoms. At the same time, they are paying no attention to developments in other areas that claim many more lives, or they display a disconcerting degree of passivity. We need only cast our minds to human trafficking or the arms trade: how is it possible, for example, that aeroplanes full of weapons continue to land regularly in Darfur, where a human tragedy with tens of thousands of victims is unfolding?

17. In our view, it is also necessary to draw attention to an aspect we believe to be very dangerous: the legitimate fight against terrorism must not serve as a pretext for provoking racist and Islamophobic reactions among the public. The Council of Europe has rightly recognised the fundamental importance of intercultural and interfaith dialogue. The member states and observers really should carry these efforts forward and maintain the utmost of vigilance on the issue. Any excesses in this respect could have disastrous consequences in terms of an expanded future terrorist threat.

18. In the course of our investigations and through various specific circumstances, we have become aware of certain special mechanisms, many of them covert, employed by intelligence services in their counter-terrorist activities. It is not for us to judge these methods, although in this area, too, great liberties appear to be taken with lawfulness. Many of these methods give rise to chain reactions of blackmail and lies between different agencies and institutions in individual states, as well as between states. Therein may lie at least a partial explanation for certain governments' fierce opposition to revealing the truth. We cannot go into the details of this phenomenon without putting human lives at risk. Let me reiterate that we are fully convinced of the strategic importance of the work of intelligence services in combating terrorism. However, we believe equally strongly that the relevant agencies need to be subject to codes of conduct, accompanied by robust and thorough supervision.

19. With the mandate assigned to us, we believe that the Assembly has reached the limits of its possibilities. The resources at our disposal to address the issues presented to us are totally inadequate for the task. The Council of Europe should give serious consid-

eration to equipping itself with more effective and more binding instruments for dealing with such grave instances of massive and systematic violations of human rights. This is more necessary now than ever before, since it is clear that we are facing a worrying process of the erosion of fundamental freedoms and rights.

20. We must condemn the attitude of the many countries that did not deem it necessary to reply to the questionnaire we sent them through their national delegations. Similarly, NATO has never replied to our correspondence. . . .

SOURCE: Council of Europe. Parliamentary Assembly. Committee on Legal Affairs and Human Rights. "Secret Detentions and Illegal Transfers of Detainees Involving Council of Europe Member States: Second Report." June 7, 2007. http://assembly.coe.int/CommitteeDocs/2007/Emarty_20070608_NoEmbargo.pdf (accessed January 7, 2008).

OTHER HISTORIC DOCUMENTS OF INTEREST

FROM THIS VOLUME

- National Intelligence Estimate on terrorism, p. 335
- CIA interrogation techniques, p. 354

FROM PREVIOUS *HISTORIC DOCUMENTS*

- Sentencing of Moussaoui for his role in the September 11 attacks, *2006*, p. 182
- President Bush on the secret detention of terrorism suspects, *2006*, p. 511
- Secretary of state on U.S. policy concerning torture of detainees, *2005*, p. 905
- Justice Department opinions on the use of torture in terrorism interrogations, *2004*, p. 336
- U.S. Army report on abuses of Iraqi prisoners, *2004*, p. 207

DOCUMENT IN CONTEXT

Chinese President, Rights Activists on Preparation for the Olympic Games

JUNE 25 AND AUGUST 8, 2007

For China, 2007 was a year of anticipation and preparation for what many people considered its coming-out party as the next superpower. The 2008 Summer Olympics to be held in Beijing offered China an extraordinary opportunity to showcase the transformation of some of its society from backwardness and poverty to the modernity of glass-and-steel urban office towers and the glittering nightlife of Shanghai. The focus of world attention on China as it prepared for the Olympics exposed just how incomplete the transformation still was. China remained mostly rural and poor and under the grip of a single political party that, by 2007, was known as much for its inefficiency and corruption as for its ruthlessness in previous decades. China also was experiencing the negative consequences that accompanied its forced march to industrial development, notably extreme environmental devastation and heightened economic inequality.

On the economic front, China in 2007 had yet another year of blazing economic growth, reporting a near-record expansion. The global slowdown toward the end of the year suggested China's growth would slow modestly in 2008, however. By contrast, the glacial pace of political change also continued. The Communist Party held its seventeenth congress in October and signaled that President Hu Jintao and Prime Minister Wen Jiabao would remain in office for another five years. The party also elevated two new officials to its top ranks, sparking speculation that they were the top candidates to succeed Hu if he were to retire, as expected, in 2013.

Getting Ready for the Olympics

Hosting the international Olympics is a major endeavor for any country because of the attention, prestige, and business that accompanies the arrival of the world's best athletes, along with thousands of fans and journalists. For China, the 2008 Summer Olympics appeared to hold special significance as an opportunity for its government to demonstrate the enormous changes taking place there.

By 2007 the Chinese government had rebuilt large sections of its capital, clearing out thousands of acres of ancient neighborhoods known as the *hutongs,* as well as hundreds of squatter settlements in and around the city. According to Chinese dissidents, this urban renewal project dislocated more than 1 million people. The hardest hit were hundreds of thousands of migrants from rural areas who had moved to Beijing in search of work; few of them, if any, received compensation for their losses.

In place of these old neighborhoods and shantytowns, the government built hundreds of high-rise apartment and office complexes similar to those found in urban areas worldwide. A new opera house (intended to be the world's largest) and a modernistic stadium resembling a giant bird's nest were among the architectural gems being built in time for the Olympics.

Domestic human rights activists in China attempted to use the Olympics to draw international attention to their concerns. On August 8, exactly one year before the start of the Olympics, thirty-seven of China's best-known writers, intellectuals, and dissidents posted on the Internet a highly unusual open letter to Chinese and world leaders, calling on the government to live up to its promises to respect human rights. Rather than feeling pride about the advent of the Olympics, they said, "we feel disappointment and doubt as we witness the continuing systematic denial of the human rights of our fellow citizens even while—and sometimes because—Olympic preparations are moving forward."

Two of the signers of that letter, human rights activist Hu Jia and lawyer Teng Biao, published another letter on September 10. In "The Real Olympics and China," they told prospective visitors that they would see a shiny new Beijing, but "you may not know that the flowers, smiles, harmony and prosperity are built on a base of grievances, tears, imprisonment, torture and blood." In December Hu Jia gave testimony via the Internet to the European Parliament describing government oppression of dissent. He was arrested on December 27 and charged, early in 2008, with attempting to "subvert state power."

The Olympics and Dirty Air

Another matter that drew increasing attention because of the approaching Olympics was environmental degradation in China. Three decades of rapid industrialization had severely damaged nearly every aspect of China's environment, leaving hundreds of millions of Chinese dependent on toxics-laden air, land, and water. Of China's many environmental woes, the most obvious to any casual observer was the generally terrible air quality in Beijing and other urban areas. According to the United Nations, sixteen of the twenty cities with the world's worst air pollution were in China—among them Beijing. The capital's average concentration of fine-particle air pollution, the type of pollution most damaging to human respiratory systems, was seven times the level recommended by the U.S. Environmental Protection Agency. A World Bank study published in 2007 cited estimates that air pollution caused more than 400,000 premature deaths in China each year.

Although China had set up special pig and vegetable farms to provide toxin-free food for the international athletes, there was little the country could do by August 2008 to clean the air in Beijing. The government set up monitoring stations to report on the number of "blue sky" days, but the blue sky standard fell well below those used by the United States and Europe. In a visit to China in August, International Olympic Committee president Jacques Rogge suggested that some events at the Beijing Olympics, particularly those involving outdoor sports, might have to be postponed to avoid endangering the health of athletes. Sports medicine authorities and some athletes expressed similar concerns, thus drawing more international attention to a problem that the capital's approximately 12 million residents experienced regularly.

Economic Boom, but Clouds on Horizon

China's economy continued to grow at a rapid-fire pace, exceeding all records for other high-growth countries in modern history, notably Japan after World War II. According to the International Monetary Fund, the Chinese economy grew by 11.4 percent in 2007, its highest rate in more than a decade. This also was the fifth consecutive year of double-digit growth and part of three decades of sustained growth.

China achieved an important landmark in 2007: for the first time in modern history, it contributed more than any other country to world economic growth, exceeding the United States, whose economy was three times as big. China accomplished this feat through its high rate of growth, compared with the relative weakness of the U.S. economy, and the fact that by 2007 it had reached number three on the global economic totem pole, below the United States and Japan. Depending on what economic measures were used, China was expected to overtake the United States as the world's largest economy some time after 2020.

Economists and other experts, in China and other countries, pointed to several near- and long-term problems that, they said, could take some of the gloss off of China's golden economy. One short-term issue was the threat of inflation; China's leaders had worried for several years that their overheated economy would generate inflation, but the annual inflation rate had stayed low at about 3 percent. Most estimates for 2008, however, suggested China's inflation would reach or even exceed 5 percent, possibly signaling a problem. Also of some concern in late 2007 was the prospect of a recession in the United States, which was China's largest single export market (just ahead of Japan). If Americans bought fewer Chinese goods, Chinese producers would have to find alternative markets at a time of global economic uncertainty. These global economic worries led most economists to reduce their estimates of Chinese economic growth to under 10 percent for 2008 and down to about 9 percent for 2009.

Another challenge, with potential long-term negative consequences, that emerged in 2007 involved revelations about the toxic contamination of many Chinese export products, including seafood, toothpaste, pet food, and toys. Surveys showed that numerous recalls during the year had undermined public confidence, in the United States and other countries, about the safety of Chinese products. However, the extent to which this would translate into actual resistance to Chinese imports was still unclear at year's end. Although the Chinese government accused the foreign news media of sensationalizing the matter, authorities appeared to take some aspects of it seriously. Among the steps Beijing took in direct response to the revelations were announcing the closing of 180 food plants in June, stepping up inspections of export products, naming a high-level committee to develop improved standards for product safety, and—most dramatically—executing the former head of the national food and drug administration on July 10 after he pleaded guilty to bribery and corruption charges.

Public "Participation," but Not Democracy

China's current leaders—the fourth generation of leaders since the 1949 Communist revolution—had come to power in 2002 and 2003 with the retirement of former president Jiang Zemin and his colleagues. Chinese politics ran on a five-year schedule that approximated the timing for meetings of the National Congress of the Communist Party of China, the ultimate source of political power. The seventeenth such congress was set for

October 2007. Its main business was the selection of the senior committees, from which the ranks of top government officials would be selected (as least in theory) the following year by the parliament, the National People's Congress.

Behind the scenes, President Hu and other party leaders reportedly spent much of 2007 preparing the groundwork for the party congress, most importantly settling on candidates for the top leadership positions. Few hints of what transpired in these meetings emerged publicly, but one of the most important events appeared to be a high-profile speech given by Hu on June 25 to the party's leadership school, known as the Central Party School, in Beijing. In his speech, made public the next day, Hu seemed to steer a middle-of-the-road course between party elders who wanted to go back to the old, absolutely authoritarian, ways of doing things and others who advocated a somewhat quicker pace of political change to allow greater public participation.

Hu made clear that changes would continue, giving repeated references to the mantra "reform and opening up," advocated by former leader Deng Xiaoping, that had led to China's embrace of capitalism. But in one sentence, Hu appeared to summarize a philosophy of keeping any political changes within the strict bounds of Communist Party rule: "Our country's political structural reform must adhere to the correct political orientation, it must continuously move forward along with economic and social development, and it must be commensurate with the continuous rise of our people's enthusiasm for political participation." The key phrases in this sentence, according to close observers of Chinese politics, were "correct political orientation" and "political participation." In Chinese Communist dogma, the only "correct political orientation" is one in which the party is the ultimate authority; this phrase banished any thoughts that Hu was advocating the emergence of a multiparty democracy. The reference to popular "political participation" intrigued analysts, however, because it suggested Hu was looking for ways of satisfying the growing public desire for more influence on decision making in politics and government. Modest experiments in democracy already were under way in some local areas, but increasing public participation at higher levels was unlikely to occur soon. The party distributed Hu's speech to many of its 70 million members, and in the following weeks major Chinese newspapers ran lengthy articles analyzing it.

The seventeenth Communist Party congress began on October 15; most attention focused on the question of who was up, and who was down, in the leadership. Like its predecessors, this congress was shrouded in secrecy (its opening date was not even announced until mid-September) and generated no startling news. Hu opened the congress with a lengthy oration expanding on the themes of his June 25 speech. On the question of political reform, he told the 2,200 delegates that "citizens' participation in political affairs will expand in an orderly way"—in other words, any changes would be gradual and would not threaten the party's hold on power.

The most often-asked question in advance of the congress was whether party leaders would use it to signal a possible successor to Hu, who was expected to step down in 2013 at the end of his second five-year term as president. The few reports that emerged from high-level deliberations suggested former leader Jiang still had some influence and had prevented Hu from unilaterally choosing his successor. One sign of this lingering power struggle, according to experts, was the elevation of two new members to the Politburo Standing Committee, the nine-member panel that runs the party, and thus the country, on a daily basis. The more senior of these two new members was Xi Jingping, a protégé of Jiang's who had been named party chief in Shanghai in 2006 after the dramatic ousting of his predecessor, Chen Liangyu, on corruption charges. The other was Li

Keqiang, the head of Liaoning province, in the northeast, who had worked with Hu in the Communist Youth organization. Many observers agreed that one of these two men, both in their early fifties, might be well-positioned to emerge as Hu's successor in the "fifth generation" of China's leaders at the next Congress in 2012.

Communist Party–watchers also looked for the adoption of slogans as a sign of successful leadership. At the previous congress, in 2002, Jiang had inserted his favorite slogan, the "Three Represents," into the party constitution. That slogan meant that the party represented the workers, society at large, and "advanced forces" (a vague formulation that appeared to include the capitalists who were building the new economy).

By that standard, Hu achieved a success at the 2007 congress with the incorporation of his slogan of a "scientific outlook on development" into the party constitution. While vague to Western eyes, the slogan, along with a companion slogan of "harmonious society," seemed to be intended by Hu as advocating a holistic approach to China's economic development—one that took into account such things as the impact on the environment and the need to deal with economic inequalities that had worsened with the race toward national economic growth. Hu and Prime Minister Wen had both expressed concern about the widening gap between the rapidly developing eastern provinces and the interior rural areas where more than half of China's people still lived. Some observers said Hu and Wen simply were attempting to return to the egalitarian roots of the Communist Party. Others suggested the two leaders recognized the inherent dangers of a deeply polarized society and were struggling to improve the lives of the impoverished majority.

Following are excerpts from two documents. First, a speech on June 25, 2007, by Chinese president Hu Jintao to leaders of the Communist Party assembled at the Central Party School in Beijing. Second, an open letter, headed "One World, One Dream and Universal Human Rights" and addressed to "Chinese and World Leaders," posted on the Internet on August 8, 2007, and signed by thirty-seven Chinese human rights activists, intellectuals, lawyers, and others.

DOCUMENT

Address to the Chinese Party School by President Hu Jintao

June 25, 2007

Hu Jintao stressed: Socialism with Chinese characteristics is the banner of development and progress in contemporary China and the banner of struggle in unity for the whole party and for the people of all ethnic groups in China. We must unswervingly uphold Deng Xiaoping Theory and the important thinking of the "Three representations" as guide, deepen implementation of the scientific development concept, and unwaveringly uphold and develop socialism with Chinese characteristics.

Hu Jintao pointed out: Emancipating the mind, an essential requirement of the party's ideological line and a magic weapon of ours in dealing with all kind of new situations and problems lying on the road ahead of us and in our continuous efforts to create a new phase in our cause, must be upheld firmly. Reform and opening up, which are essential to the institutional mechanism for liberating and developing the productive forces and for invigorating itself through continuous innovation and which constitute a powerful motive force for developing socialism with Chinese characteristics, must be firmly pushed forward. Scientific development and social harmony, which are basic requirements for developing socialism with Chinese characteristics and intrinsic requirements for achieving sound and fast economic and social development, must be firmly put in place. Building a well-off society in an all-round way, which is the goal our party and country will work to attain by 2020 and where the fundamental interest of the people of all ethnic groups across the country lies, must be firmly strived for. Firmly accomplishing these four things is vitally important to the overall interest of maintaining smooth development of the cause of the party and country.

Hu Jintao pointed out: Under the present international and domestic circumstances, our country faces unprecedented development opportunities and equally unprecedented challenges. We have many favourable conditions as well as unfavourable factors. The important thing is how we did our work. We must be soberly aware of the major trend of development in today's world and in contemporary China, comprehensively grasp the new requirements of our country's development and the new expectations of the masses of the people, conscientiously sum up our party's practice and experience in administering the country, scientifically formulate action programmes and major policies and guidelines that accord with the requirements of the times and the people's wishes, proceed from the new historical starting point, and lead the people to continue building a well-off society and accelerating socialist modernization in an all-round way to accomplish the lofty mission bestowed on us by the times.

Hu Jintao stressed: Reform and opening up are a new great revolution waged by the people under our party's leadership against the backdrop of the new times. Over the last 29 years in the new period, our country has scored achievements in reform, opening up, and socialist modernization that have captured the attention of the whole world.

Facts have eloquently proved that reform and opening up are the road we must follow to develop socialism with Chinese characteristics and accomplish the Chinese nation's great rejuvenation. Since the beginning of reform and opening up, our party has led the people in blazing a socialist road with Chinese characteristics. This road is correct and is able to lead China to development and progress primarily because we have upheld the basic tenet of scientific socialism and give it distinct Chinese characteristics in light of our country's reality. We must continue deepening exploration, research, and experiment with socialism with Chinese characteristics and strive to broaden the socialist road with Chinese characteristics as we march forward.

Hu Jintao pointed out: Comrades throughout the party and especially the party's senior cadres must closely bear in mind our basic national conditions of being in the initial stage of socialism and clearly recognize the importance and the protracted and arduous nature of our endeavour to build a well-off society in an all-round way, realize our country's basic modernization, and consolidate and develop socialist system. We must enhance our resolve to focus intensely and whole-heartedly on construction and development, raise our awareness that we must not deviate from reality when pondering issues and handling things, and soberly and tenaciously work to accomplish the party's historical mission in a down-to-earth manner.

Hu Jintao stressed: Our country's development is at a new historical starting point in the new century and new period. We must scientifically analyse the new opportunities and challenges in our country's all-round participation in economic globalization;

profoundly grasp the new tasks and contradictions facing our country as it pushes forward development in various areas against the backdrop of deep[en]ing industrialization, urbanization, marketization, and internationalization; deepen implementation of the scientific development concept; more conscientiously promote scientific development; and do all we can to broaden the bright prospects of developing socialism with Chinese characteristics.

Hu Jintao stressed: Over a long period of time, the party's three-generation central leading collectives with Comrade Mao Zedong, Comrade Deng Xiaoping, and Comrade Jiang Zemin, respectively, as the core, led our party to continuously explore, experiment with, and study the major issue of building socialism and scored important achievements. Since the party's 16th national congress, the party Central Committee has inherited and developed the three-generation central leading collectives' important thinking on development and put forward the scientific development concept. Development is the scientific development concept's foremost important thing, putting people first is its central theme, all-round coordination for sustainable development is its basic requirement, and unified planning with due consideration for all concerned is its basic approach. Development is of decisive significance to building a well-off society in an all-round way and to accelerating socialist modernization while liberating and developing the productive forces has always been a fundamental task of socialism. We must firmly seize economic construction, which is the central task, and lay a solid material foundation for developing socialism with Chinese characteristics. Our party's fundamental goal is to serve the people wholeheartedly. All our party's endeavour and work are to benefit the people. We must always fully realize, safeguard, and develop the fundamental interest of the broad masses of the people, regarding it the starting point and the purpose of all the work of the party and country. We must be able to develop for the people, rely on the people to attain development, and let the people share the fruits of development.

Hu Jintao pointed out: Implementing the scientific development concept requires us to always uphold the basic line of "one central task and two basic points" [economic development and adherence to the four cardinal principles and implementation of reform and the open policy], actively build a harmonious socialist society, firmly push forward reform and opening up, and conscientiously strengthen and improve party building. Unswervingly upholding the party's basic line is the most reliable safeguard ensuring that our cause is capable of standing the test and potential dangers and that we can smoothly arrive at our destination. We must uphold the "one central task and two basic points" and integrate it into the great practice of developing socialism with Chinese characteristics. We must not waver in this at any time.

Speaking of promoting sound and fast development of the national economy, Hu Jintao said: To achieve sound and fast development of the national economy, the important thing is to achieve significant and new progress in changing the mode of economic development and perfecting the structure of our socialist market economy. Changing the mode of economic development is an important principle based on our experiment with and our grasp of the laws of our country's economic development as well as a major strategy put forward in light of the reality of our country's economic development. We must more profoundly and consciously grasp the economic development laws and adopt more effective measures with greater resolve to raise the quality and benefits of economic development. We must deepen implementation of the strategy of rejuvenating the country through science and education, accelerate the strategic adjustment of economic structure, conscientiously strengthen agriculture as the foundation of the national economy, earnestly build a resource conserving and environment friendly society, and continue implementing the overall strategy of regional development. We must uphold and perfect the fundamental economic system whereby public ownership remains predominant and different forms of ownership develop side by side; unwaveringly consolidate and develop public sector of the economy; unwaveringly encourage, support, and guide the development of non-public sector of the economy; and create a new setup in which different forms of ownership compete on an equal footing and promote each other. We must deepen reform of the finance, tax, banking, and planning structure to create a macroeconomic regulatory and control system conducive to scientific development. We must raise the standards of our open economy and develop new competitive edges for our participation in international economic cooperation and competition under the conditions of economic globalization.

Speaking of developing socialist democracy [minzhu zhengzhi], Hu Jintao said: Developing socialist democracy has always been a consistent goal of our party. Since the beginning of reform and opening up, we have consistently pressed ahead with political structural reform [zhengzhi tizhi gaige] in an active but prudent way in pace with the entire process of reform and development, and we have scored major success in building socialist democracy. Our country's political structural reform must adhere to the correct political orientation, it must continuously move forward along with economic and social development, and it must be commensurate with the continuous rise of our people's enthusiasm for political participation. We must uphold the party's leadership, make the people the masters of the country, rule the country by law, and bring the three into organic harmony; and we must continue to push forward the self-improvement and self-development of our socialist political system. We must continue to expand orderly political participation of our citizens, perfect the democratic system, enrich the form of democracy, and broaden the democratic channel; we must press on with our efforts to make decisionmaking more scientific and democratic and perfect decisionmaking information and intellectual support systems; we must develop grass-roots democracy and ensure that the people can exercise their democratic rights directly and in accordance with the law; we must comprehensively implement the basic plan for running the country according to law, foster the spirit of rule of law, and safeguard social equity and justice; and we must accelerate reform of the administration management structure and strengthen government's social management and public service functions. . . .

Speaking of accelerating society building, Hu Jintao said: Society building, which is closely related to the personal interests of the broad masses of the people, must be given more attention. To strengthen society building, we must focus on issues people care about most and issues that concern their most immediate and most realistic interest problems. We must apply the fruits of economic development more to improving the people's livelihood. We must pay particular attention to developing education on a priority basis, implement a development strategy that expands employment, deepen reform of the income distribution system, establish a basic social security system covering urban and rural residents, set up a basic medicare and public health system, improve the health of all the people, perfect social management, and safeguard social stability and unity.

Hu Jintao stressed: For our party to lead the people in winning new victories in building a well-off society in an all-round way and creating a new situation in building socialism with Chinese characteristics, the important thing is to carry out the party's self-construction. We must uphold the principle that the party must exercise control of its members and enforce strict discipline, and we must continue pushing forward the new great endeavour of party building. We must promote in-depth study of Marxism-Leninism-Mao Zedong Thought and Deng Xiaoping Theory as well as the important thinking of the "Three representations," and we must study the scientific development concept in a deep-going way. We must strengthen the party's organizational building and develop better-quality leading bodies, cadres' ranks, and the ranks of party members. We must continue to push forward the building of inner-party democracy actively but prudently, uphold democratic centralism, uphold the status of party members as the principal component of the party, and perfect the inner-party democratic system in order to widely enhance the sense of inner-party democracy, perfect the inner-party democratic system, and bring into full play the party's creative vigour. We must strengthen building the party' s ideological style, study style, work style, leadership style, and cadre's lifestyle; vigorously improve the study style and the style of writing; oppose formalism, bureaucratism, fraud practice, and extravagant and wasteful practice; and enable comrades of the whole party and especially leading cadres at all levels to more consciously uphold the spirit of seeking truth and being pragmatic, more conscientiously uphold the purpose of serving the people wholeheartedly, and more conscientiously uphold the party's mass line. We must insist on exercising power for the people, feeling as the people feel, and working for their interests, and we must always listen carefully to the masses' voice, truthfully reflect their wishes, genuinely concern ourself with their sufferings, and perform more good, concrete deeds for them.

Hu Jintao stressed: Party committees at all levels must fully realize the protracted, complicated, and arduous nature of the struggle against corruption; place combating corruption and promoting clean government in an even more prominent place; uphold the principle of addressing both the immediate and underlying causes of problems, comprehensively tackling the problem, using both punishment and preventive measures, and emphasizing prevention; and establish a sound system for cracking down on and preventing corruption, a system that puts equal emphasis on education, rules and regulations, and supervision. While taking resolute steps to crack down on corruption, we must pay more attention to seeking a permanent cure of the problem, more attention to prevention, and more attention to institutional building. We must strengthen the work of making leading cadres honest and self-disciplined, and we must resolutely investigate and deal with violations of law and discipline. . . .

SOURCE: China Elections and Governance. "Hu Jintao's Speech at Party School, Full Text." June 25, 2007. www.chinaelections.net/newsinfo.asp?newsid=5842 (accessed March 21, 2008).

DOCUMENT

Open Letter by Chinese Human Rights Activists and Dissidents on Olympics

August 8, 2007

Respected Leaders and Fellow World Citizens:

Upholding the fundamental principles of the Olympic spirit, including "respect for universal fundamental ethical principles" and "the harmonious development of man, with a view to promoting a peaceful society concerned with the preservation of human dignity" (Olympic Charter, Preamble);

Taking note of the Chinese government's official 2008 Olympic theme "One World, One Dream" and the Beijing Olympic Committee's stated objectives of hosting an "Open, Green, and Humane Olympics"; and

Mindful of the growing number of questions and criticisms in our own society and from around the world about the violations of the human rights of Chinese citizens in the name of the Beijing Olympics;

We, the undersigned citizens of the People's Republic of China, here voice our concerns and to propose changes in the ways in which our government is handling its preparations for the Olympics.

Today, August 8, 2007, marks the start of the one-year count-down to the 2008 Summer Olympics, a mega-event for China and the world. We, as citizens of the People's Republic of China, ought to be feeling pride in our country's glory in hosting the Games, whose purposes include the symbolization of peace, friendship, and fairness in the world community. We also ought to feel uplifted by the watchword chosen by the Beijing Olympic Committee: "One World, One Dream." Instead we feel disappointment and doubt as we witness the continuing systematic denial of the human rights of our fellow citizens even while—and sometimes because—Olympic preparations are moving forward. We hear "One World" and wonder: What kind of world will this be? "One Dream"? Whose dream is it that is coming true? We are gravely concerned about the question of whether authorities in our country can successfully host the Olympic Games in an authentic Olympic spirit so that the 2008 Beijing games can become an event of which China and the world community can be proud.

As the one world that we share "globalizes," lives and dreams are becoming increasingly intertwined. One person's "world dream," especially if it is implemented with unchecked power, and with endorsement from the world community, can turn into misery and nightmare for others. "One world" can still be a world where people suffer discrimination, political and religious persecution, and deprivation of liberty, as well as poverty, genocide, and war. Millions of people who survived such miseries and disasters in the 20th century have come to appreciate, and to pursue, human rights. Universal human rights have become the bedrock concept in pursuing lasting peace, sustainable development, and justice.

If "one dream" is truly to belong to all cultures and communities, it must involve protection of basic rights and liberties for all. Even the powerful, the rich and privileged might be punished unjustly tomorrow if fundamental rights are not assured today.

The government that rules our country has pledged to the Chinese people and to the world to protect human rights. It has acceded to obligations under numerous international human rights conventions and treaties, including the Universal Declaration of Human Rights, and it has amended the Chinese Constitution to include guarantees of human rights.

In order to avoid misunderstanding, and in order to alert the international community to un-Olympic conduct that tarnishes the true spirit of the Games, we, the undersigned citizens of the People's Republic of China, endorse the government's Olympic slogan with the following vital addition:

"One World, One Dream, and Universal Human Rights."

Without promoting human rights, which are the fundamental principle of universal ethics in China and elsewhere, it is gratuitous to promote "One World." Without the protection of the human rights of all Chinese citizens equally—i.e., without abolition of the rural-urban residential control system, without an end to discrimination against women and sexual, ethnic, and faith minorities, and without ending the suppression of political dissent—it is senseless to talk about "One Dream" for all of China.

China's government has promised the International Olympic Committee to "promote human rights" and has pledged to the United Nations Human Rights Council to "uphold the highest standard of human rights." On paper it has taken certain steps toward improving human rights—in 2003, for example, abolishing the arbitrary detention system known as "Custody and Repatriation" and in 2004 adding "human rights protection" as an amendment to China's Constitution. We believe that the government should be able to do much more.

Little has been done, in practice, to carry out the promises that have been made on paper. On the contrary we have experienced and witnessed violations of human rights many times—in press censorship and control of the Internet, in the persecution of human rights defenders and of people who expose environmental or public health disasters, in the exploitation of poor or disadvantaged social groups and in retaliation against them when they protest, and even in abuses by corrupt officials who are involved in the construction of Olympic facilities and city beautification projects that are aimed to prepare for the Olympics. All of these actions violate not only international standards but provisions of the Chinese constitution as well.

We find no consolation or comfort in the rise of grandiose sports facilities, or a temporarily beautified Beijing city, or the prospect of Chinese athletes winning medals. We know too well how these glories are built on the ruins of the lives of ordinary people, on the forced removal of urban migrants, and on the sufferings of victims of brutal land grabbing, forced eviction, exploitation of labor, and arbitrary detention.

Out of deep affection for our motherland and our sense of duty as citizens of the world, we will do our best, and urge leaders in China and in the world community to join hands with us, to make the Beijing Olympics a turning point in China's rise to greatness. China has the opportunity to use the Games to build true harmony on the basis of respect for human dignity and freedom and to become a respectable member of the community of civilized nations—not by loud rhetoric or brute force, but by taking actions to promote human rights at home and in the world.

In the "one world" in which we live, the dreams that are coming true in China today will significantly shape everyone's future. Therefore, in order to promote a successful Olympics consistent with human rights, we propose the following measures:

1. **Declare amnesty for all prisoners of conscience** so that they can enjoy the Olympic games in freedom.
2. **Open China's borders** to all Chinese citizens who have been forced into exile for their beliefs, expression, or faith, so that they can re-unite with their loved ones and celebrate the glory of the Olympics in their motherland;
3. **Implement the government ordinance to allow foreign journalists** to conduct interviews and reporting without pre-approval by authorities before October 17, 2008, granting **Chinese journalists** the same access and independence.
4. **Provide fair compensation to the victims of forced evictions and land appropriations** that have been done in order to construct Olympic facilities, and release

people who have been detained or imprisoned (often violently) for protesting or resisting such actions.

5. **Protect the rights of workers** on all Olympic construction sites, including their right to organize independent labor unions; end discrimination against rural migrant laborers and give them fair compensation.
6. **End police operations** intended to intercept, detain, or send home petitioners who try to travel to Beijing to complain about local officials' misconduct; abolish illegal facilities used for incarcerating, interrogating, and terrorizing petitioners; end the "clean up" operations aimed at migrants that demolish their temporary housing and close down schools for their children.
7. **Establish a system of citizen oversight over Olympics spending** and provide public accounting and independent auditing of Olympics-related expenditures; make the process of awarding contracts to businesses transparent, and hold legally accountable any official who embezzles or wastes public funds.

We further suggest setting up an independent Beijing Olympics Watch Committee, composed of independent experts and representatives of non-governmental organizations and affected communities such as migrant laborers and people who have been forcibly relocated. This Committee would oversee the implementation of the above proposals. It should be allowed to operate independently, to examine plans, to interview freely, and to release its findings to the public. Citizen participation is key to a successful Olympics.

If proposals even as straightforward as the foregoing cannot be adopted, we feel certain that the Beijing Olympics will not go down in history as the glorious events that everyone wishes them to be. We do not want to "politicize" the Olympic movement. However, pushing the Games through in ways that violate human rights and that hurt people who are forced into silence, all in the name of a "dream" that belongs only to "some" people, not our whole world, will only plant seeds of resentment that will exacerbate the crises in China and affect the future of the world.

SOURCE: Human Rights Watch. "Letter by Liu Xiaobo, Ding Zilin, Bao Tong and Other Activists One Year Before the Beijing Games." August 8, 2007. http://china.hrw.org/press/other_news/open_letter_by_42_chinese_signatories_on_the_need_for_human_rights_reform (accessed March 21, 2008).

OTHER HISTORIC DOCUMENTS OF INTEREST

FROM THIS VOLUME

- Imported food safety, p. 51
- North Korea denuclearization agreements, p. 77
- United Nations Security Council on peacekeeping in Darfur, p. 385

FROM PREVIOUS *HISTORIC DOCUMENTS*

- Chinese president Hu on relations between China and Africa, *2006,* p. 629
- State Department on U.S.-China relations, *2005,* p. 612
- Jiang Zemin on Chinese leadership transition, *2002,* p. 850

DOCUMENT IN CONTEXT

Supreme Court on Regulation of Issue Ads

JUNE 25, 2007

By the end of 2007, the 2008 presidential election campaign was shaping up to be one of superlatives. More candidates than in recent memory were offering themselves as contenders. Together they were breaking records for the most money raised by presidential campaigns. With a Supreme Court decision in June easing restrictions on election advertising, it appeared likely that special interest groups would raise and spend more money than ever before trying to influence the outcome. The Democratic race featured the first woman and the first African American man with a serious chance of winning the party's nomination and perhaps the general election. Primary elections were scheduled for earlier dates than ever before, and the campaign was certain to be the longest on record. Although American voters complained about the likely length of the campaign, more of them appeared to be paying closer attention than usual to the early candidate speeches, debates, and issue positions.

The Candidates

For the first time since 1928, neither a sitting president nor vice president was running for office, which left the field wide open to contenders from both parties. On the Republican side, the early front-runners in the contest to succeed President George W. Bush were Rudolph Giuliani, the former mayor of New York City who was widely acclaimed for his handling of the chaos that followed the attacks of September 11, 2001; Mitt Romney, the former governor of Massachusetts; and Arizona senator John McCain, known for his independent streak. None of the three candidates was particularly popular among the party's religious conservative base—Giuliani and McCain because they were not as supportive of key social issues as religious right leaders preferred, and Romney because of his changing position on those issues as well as his Mormon faith. Two candidates with more appeal to the Republican base were former Tennessee senator and actor Fred Thompson, who entered the race comparatively late in the season and engaged in relatively little campaigning, and former Arkansas governor Mike Huckabee. By year's end, Huckabee was mounting a surprisingly strong challenge against Romney in the upcoming Iowa caucuses. McCain was focusing his attention on New Hampshire, while Giuliani was making only a token effort in the early primary states and instead concentrating on more populous states such as New York and Florida.

Several Republicans entered the race in 2007 only to drop out later in the year as they found it difficult to raise enough money to fund their campaigns. In this group were

Tommy Thompson, the secretary of health and human services during Bush's first term; Kansas senator Sam Brownback; former Virginia governor Jim Gilmore; and Colorado representative Tom Tancredo, who ran on an anti-immigration platform. A few other candidates remained in the race at the end of the year: Alan Keyes, a former ambassador from Maryland and a perennial presidential candidate, as well as Rep. Duncan Hunter of California and Rep. Ron Paul of Texas. None of them was given any chance of winning the nomination, although Paul had surprised many observers by raising nearly $30 million for his campaign, most of it from donors contributing less than $200 each.

On the Democratic side, the early frontrunner was generally considered to be New York senator Hillary Rodham Clinton, followed closely by Illinois senator Barack Obama and former North Carolina senator John Edwards, who had been the party's vice presidential nominee in 2004. Running behind them, both in the polls and in fundraising, were Delaware senator Joseph R. Biden Jr., Connecticut senator Chris Dodd, New Mexico governor Bill Richardson, Ohio representative Dennis J. Kucinich, and former Alaska senator Mike Gravel.

Two notable Democrats who did not enter the race were Massachusetts senator John Kerry, the party's losing standard-bearer in 2004, and former vice president Al Gore, who won the popular vote against Bush in 2000 only to lose the Electoral College vote after the Supreme Court ended a dispute over a recount of Florida ballots. Another often-mentioned potential candidate for the presidency was New York City mayor Michael Bloomberg. Bloomberg's announcement in June that he was leaving the Republican Party fueled speculation that he might head an independent third-party ticket, but that had not occurred by the end of the year. Consumer advocate Ralph Nader had once again declared himself a candidate for the presidency. Many Democrats still blamed Nader's third-party candidacy in 2000 for siphoning off enough votes in Florida to cost Gore the election.

Early Primaries, Early Fundraising

One factor in the early start to the presidential campaign season was the number of states that were moving their primary elections to the front of the election-year calendar. For decades, Iowa's party caucuses had been the first test of the presidential field, followed closely by New Hampshire's primary. Iowa's event was typically held in the middle of January, while New Hampshire's was held toward the end of that month. For years, however, candidates in both parties had complained that these small, relatively rural, and predominantly white states carried too much influence in determining the parties' nominees. Finally responding to these complaints, the Democratic National Committee agreed to a new schedule for 2008 that sought to add geographic and ethnic diversity by allowing South Carolina and Nevada to hold their nominating events after Iowa's and New Hampshire's contests but before February 5, which was the earliest date that that party was allowing any other state to hold its primary or caucus. The Republican National Committee barred any state from electing convention delegates before February 5, although straw polls such as the one held in Iowa were permissible.

The primary calendar remained in flux for much of the year as state legislatures debated their options. Several of the bigger states, including California and New York, eventually moved up their events to February 5. Florida and Michigan, however, decided to defy national party rules. In late August, after Florida set its event for January 29, the Democratic National Committee (DNC) voted to strip the state of all its delegates to the national nominating convention. Despite that move and a pledge signed by all the major

Democratic candidates not to campaign in the state, Michigan set its primary for January 15. The DNC voted in December to strip Michigan of its delegates. Loss of those delegates could conceivably make a difference in a close contest between two Democratic candidates. Republicans penalized five states—New Hampshire, Florida, South Carolina, Michigan, and Wyoming—for scheduling nominating events before February 5; each state lost half its delegates.

Meanwhile, the traditional first events were pushed even closer to the beginning of the year. Iowa's Democratic caucus was scheduled for January 3, 2008; the New Hampshire primary was set for January 8. More than twenty states' primary or caucus events were scheduled for February 5, a full month earlier and about seven more states than had participated in Super Tuesday in 2004.

The earlier primary schedule, and the crowding of half the states' contests into the first five weeks of the year, meant that to be viable candidates, presidential hopefuls had to start earlier than ever before to organize their campaigns, map out their strategies, and, perhaps most important, line up campaign contributions. By the end of 2007, according to the nonpartisan Campaign Finance Institute, the presidential contenders had raised a combined total of nearly $552 million for the 2008 primaries, more than double the record amount raised in 2003 for the 2004 primaries. The top Democratic fundraisers were Obama, who raised $99.6 million, nearly one-third of which came from individuals donating less than $200 apiece; and Clinton, who raised $98.7 million. Edwards was a distant third, with $41.3 million. By far the top fundraiser among Republicans was Romney, who brought in $90.1 million for the primaries, $35.4 million of which was money he contributed or lent to his campaign. Giuliani raised $56.1 million, and McCain came in at $39.9 million.

Eased Restrictions on Political Advertising

Campaign finance experts predicted that spending on behalf of presidential and congressional candidates would come close to $5 billion in 2008. Much of that spending was expected to come from corporations, unions, and other special interests that would mount get-out-the-vote campaigns and run television and print ads supporting or opposing candidates. On June 25, the Supreme Court made it a little easier for some organizations to fund such ads when it effectively overturned key provisions of the Bipartisan Campaign Reform Act of 2002. Those provisions were aimed at curbing controversial "issue ads"—ads that often praise or scorn a candidate's position on a specific issue without directly instructing the viewer to vote for or against the candidate. Supporters of such ads said they were meant to be informative and educational; critics said the ads too often purveyed erroneous or misleading information and gave well-heeled special interests too large a role in campaigns. The 2002 act—better known as McCain-Feingold for its chief Senate sponsors, John McCain and Russ Feingold, D-Wis.—placed some restrictions on issue ads by barring corporations, unions, and corporate-funded nonprofits from directly paying for such ads to run in the sixty days before a general election and the thirty days before a primary. In 2003 the Supreme Court ruled 5–4 that the McCain-Feingold act was constitutional on its face. Of the restrictions on issue ads, the majority wrote that they were "a regulation of, not a ban on, expression."

That decision left open the question of whether the issue ad restriction might be found unconstitutional as applied to a specific advertisement. Wisconsin Right to Life raised that question, arguing that it should have been allowed to run advertisements in

2004 urging the state's two senators to oppose Senate filibusters of President Bush's judicial nominees. One of the senators was Feingold, who was running for reelection and had been targeted for defeat by the anti-abortion group in some of its other campaign advertising. Wisconsin Right to Life argued that its ads were legitimate issue ads protected by the First Amendment. Its position was supported by a diverse group of organizations on both the left and right, ranging from the American Civil Liberties Union and the AFL-CIO to the United States Chamber of Commerce, the National Rifle Association, and Focus on the Family.

The group's argument persuaded a majority of the Court, which ruled 5–4 in the case of *Federal Election Commission v. Wisconsin Right to Life, Inc.* that the McCain-Feingold restriction on issue advertising was unconstitutional as applied to the organization's ads. "Discussion of issues cannot be suppressed simply because the issues may also be pertinent in an election," Chief Justice John G. Roberts wrote. "Where the First Amendment is implicated, the tie goes to the speaker, not the censor." In a portion of his opinion joined only by Justice Samuel A. Alito Jr., Roberts set out a test for courts to use in determining whether an issue ad was permissible. "A court should find that an ad is the functional equivalent of express advocacy [for or against a particular candidate] only if the ad is susceptible of no reasonable interpretation other than as an appeal to vote for or against a specific candidate," he wrote. The three other justices making up the majority, Antonin Scalia, Clarence Thomas, and Anthony M. Kennedy, said Roberts's test was too vague; they would have struck down the McCain-Feingold provision altogether.

In dissent, Justice David H. Souter warned that "after today the ban on contributions by corporations and unions and the limitation on their corrosive spending when they enter the political arena are open to easy circumvention, and the possibilities for regulating corporate and union campaign money are unclear." He was joined by Justices John Paul Stevens, Stephen G. Breyer, and Ruth Bader Ginsburg, who together with then justice Sandra Day O'Connor had formed the majority upholding the constitutionality of the provision in 2003.

The Federal Election Commission (FEC), which oversees federal campaign financing, issued new advertising rules on November 20 accommodating the Supreme Court decision. Any organization supporting a federal candidate for office could use corporate or union money to fund independent ads in the weeks before an election provided the ads focused on an issue of public policy and did not mention an election, political party, or opposing candidate. Any advertising paid for by unions or corporations had to be publicly disclosed. Only candidates' campaign committees and political action committees would be allowed to air ads expressly advocating the election or defeat of a particular candidate. Contributions to such committees were limited. Richard L. Hasen, a campaign finance expert at Loyola Law School in Los Angles posted on his Web site examples of the kinds of ads that would be permissible under the new FEC regulation. One urged: "Call Mitt Romney and tell him more of our soldiers shouldn't die in an unnecessary war in Iraq." (Romney supported the war effort.)

A week before the Court's decision, John McCain on November 13 asked his supporters and donors to "cease and desist immediately" from funding advertising efforts that supported him but that were not connected to his presidential campaign. He referred specifically to the Foundation for a Secure and Prosperous America, a nonprofit set up by a media strategist and former aide to McCain, which was touting McCain's position on national security and government spending in ads running in South Carolina, one of the early primary states. The nonprofit was set up under the tax laws as a 501(c)4 organization, which meant that it could raise funds in unlimited amounts from donors who did not have to publicly disclose their identities. "Anyone who believes they could assist my campaign by exploiting a loophole in campaign finance laws is doing me and our country a disservice," McCain said. "If you respect me or my principles, I urge you to refrain from using my name and image in any ads or other activities."

Following are excerpts from the opinion written by Chief Justice John G. Roberts in the case of Federal Election Commission v. Wisconsin Right to Life Inc., *in which the Court ruled 5–4 on June 25, 2007, that provisions of the Bipartisan Campaign Reform Act of 2002 restricting issue ads were unconstitutional as applied to ads produced by the Wisconsin organization.*

DOCUMENT

Federal Election Commission v. Wisconsin Right to Life, Inc.

Nos. 06–969 and 06–970

Federal Election Commission, Appellant 06–969 *v.* Wisconsin Right to Life Inc.	On appeals from the United States District Court for the District of Columbia
Senator John McCain, et al., Appellants 06–970 *v.* Wisconsin Right to Life, Inc.	

[June 25, 2007]

CHIEF JUSTICE ROBERTS announced the judgment of the Court and delivered the opinion of the Court with respect to Parts I and II, and an opinion with respect to Parts III and IV, in which JUSTICE ALITO joins.

Section 203 of the Bipartisan Campaign Reform Act of 2002 (BCRA) . . . makes it a federal crime for any corporation to broadcast, shortly before an election, any communication that names a federal candidate for elected office and is targeted to the electorate. In *McConnell v. Federal Election Comm'n* . . . (2003), this Court considered whether §203 was facially overbroad under the First Amendment because it captured within its reach not only campaign speech, or "express advocacy," but also speech about public issues more generally, or "issue advocacy," that mentions a candidate for federal office. The Court concluded that there was no overbreadth concern to the extent the speech in question was the "functional equivalent" of express campaign speech. . . . On the other hand, the Court "assume[d]" that the interests it had found to "justify the regulation of campaign speech might not apply to the regulation of genuine issue ads." . . . The Court nonetheless determined that §203 was not facially overbroad. Even assuming §203 "inhibit[ed] some constitutionally protected corporate and union speech," the Court concluded that those challenging the law on its face had failed to carry their "heavy burden" of establishing that *all* enforcement of the law should therefore be prohibited. . . .

Last Term, we reversed a lower court ruling, arising in the same litigation before us now, that our decision in *McConnell* left "no room" for as-applied challenges to §203. . . . We held on the contrary that "[i]n upholding §203 against a facial challenge, we did not purport to resolve future as-applied challenges." *Wisconsin Right to Life, Inc. v. Federal Election Comm'n* . . . (2006). . . .

We now confront such an as-applied challenge. Resolving it requires us first to determine whether the speech at issue is the "functional equivalent" of speech expressly advocating the election or defeat of a candidate for federal office, or instead a "genuine issue a[d]." . . We have long recognized that the distinction between campaign advocacy and issue advocacy "may often dissolve in practical application. Candidates, especially incumbents, are intimately tied to public issues involving legislative proposals and governmental actions." . . Our development of the law in this area requires us, however, to draw such a line, because we have recognized that the interests held to justify the regulation of campaign speech and its "functional equivalent" "might not apply" to the regulation of issue advocacy. . . .

In drawing that line, the First Amendment requires us to err on the side of protecting political speech rather than suppressing it. We conclude that the speech at issue in this as-applied challenge is not the "functional equivalent" of express campaign speech. We further conclude that the interests held to justify restricting corporate campaign speech or its functional equivalent do not justify restricting issue advocacy, and accordingly we hold that BCRA §203 is unconstitutional as applied to the advertisements at issue in these cases.

I

. . . Appellee Wisconsin Right to Life, Inc. (WRTL), is a nonprofit, nonstock, ideological advocacy corporation recognized by the Internal Revenue Service as tax exempt. . . . On July 26, 2004, as part of what it calls a "grassroots lobbying campaign," . . . WRTL began broadcasting a radio advertisement entitled "Wedding." The transcript of "Wedding" reads as follows:

> " 'PASTOR: And who gives this woman to be married to this man?
>
> " 'BRIDE'S FATHER: Well, as father of the bride, I certainly could. But instead, I'd like to share a few tips on how to properly install drywall. Now you put the drywall up . . .
>
> " 'VOICE-OVER: Sometimes it's just not fair to delay an important decision.
>
> " 'But in Washington it's happening. A group of Senators is using the filibuster delay tactic to block federal judicial nominees from a simple "yes" or "no" vote. So qualified candidates don't get a chance to serve.
>
> " 'It's politics at work, causing gridlock and backing up some of our courts to a state of emergency.
>
> " 'Contact Senators Feingold and Kohl and tell them to oppose the filibuster.
>
> " 'Visit: BeFair.org

" 'Paid for by Wisconsin Right to Life (befair.org), which is responsible for the content of this advertising and not authorized by any candidate or candidate's committee.' " . . .

On the same day, WRTL aired a similar radio ad entitled "Loan." It had also invested treasury funds in producing a television ad entitled "Waiting," which is similar in substance and format to "Wedding" and "Loan." . . .

[Section II omitted. In this section the majority concluded that it had jurisdiction to hear the case.]

III

WRTL rightly concedes that its ads are prohibited by BCRA §203. Each ad clearly identifies Senator [Russ] Feingold, who was running (unopposed) in the Wisconsin Democratic primary on September 14, 2004, and each ad would have been "targeted to the relevant electorate" . . . during the BCRA blackout period. WRTL further concedes that its ads do not fit under any of BCRA's exceptions to the term "electioneering communication." . . . The only question, then, is whether it is consistent with the First Amendment for BCRA §203 to prohibit WRTL from running these three ads.

[Sections A and B omitted.]

C

"The freedom of speech . . . guaranteed by the Constitution embraces at the least the liberty to discuss publicly and truthfully all matters of public concern without previous restraint or fear of subsequent punishment." . . . To safeguard this liberty, the proper standard for an as-applied challenge to BCRA §203 must be objective, focusing on the substance of the communication rather than amorphous considerations of intent and effect. . . . It must entail minimal if any discovery, to allow parties to resolve disputes quickly without chilling speech through the threat of burdensome litigation. . . . And it must eschew "the open-ended rough-and-tumble of factors," which "invit[es] complex argument in a trial court and a virtually inevitable appeal." . . . In short, it must give the benefit of any doubt to protecting rather than stifling speech. . . .

In light of these considerations, a court should find that an ad is the functional equivalent of express advocacy only if the ad is susceptible of no reasonable interpretation other than as an appeal to vote for or against a specific candidate. Under this test, WRTL's three ads are plainly not the functional equivalent of express advocacy. First, their content is consistent with that of a genuine issue ad: The ads focus on a legislative issue, take a position on the issue, exhort the public to adopt that position, and urge the public to contact public officials with respect to the matter. Second, their content lacks indicia of express advocacy: The ads do not mention an election, candidacy, political party, or challenger; and they do not take a position on a candidate's character, qualifications, or fitness for office.

Despite these characteristics, appellants assert that the content of WRTL's ads alone betrays their electioneering nature. Indeed, the FEC suggests that *any* ad covered by §203 that includes "an appeal to citizens to contact their elected representative" is the "functional equivalent" of an ad saying defeat or elect that candidate. . . . We do not agree. To take just one example, during a blackout period the House considered the proposed Universal National Service Act. . . . There would be no reason to regard an ad supporting or opposing that Act, and urging citizens to contact their Representative about it, as the equivalent of an ad saying vote for or against the Representative. Issue advocacy conveys information and educates. An issue ad's impact on an election, if it exists at all, will come only after the voters hear the information and choose—uninvited by the ad—to factor it into their voting decisions.

The FEC and intervenors try to turn this difference to their advantage, citing *McConnell*'s statements "that the most effective campaign ads, like the most effective commercials for products . . . avoid the . . . magic words [expressly advocating the election or defeat of a candidate]," . . . and that advertisers "would seldom choose to use such words even if permitted." . . An expert for the FEC in these cases relied on those observations to argue that WRTL's ads are especially effective electioneering ads because they are "subtl[e]," focusing on issues rather than simply exhorting the electorate to vote against Senator Feingold. . . . Rephrased a bit, the argument perversely maintains that the *less* an issue ad resembles express advocacy, the more likely it is to be the functional equivalent of express advocacy. This "heads I win, tails you lose" approach cannot be correct. It would effectively eliminate First Amendment protection for genuine issue ads, contrary to our conclusion in *WRTL I* that as-applied challenges to §203 are available, and our assumption in *McConnell* that "the interests that justify the regulation of campaign speech might not apply to the regulation of genuine issue ads." . . Under appellants' view, there can be no such thing as a genuine issue ad during the blackout period—it is simply a very effective electioneering ad.

Looking beyond the content of WRTL's ads, the FEC and intervenors argue that several "contextual" factors prove that the ads are the equivalent of express advocacy. First, appellants cite evidence that during the same election cycle, WRTL and its Political Action Committee (PAC) actively opposed Senator Feingold's reelection and identified filibusters as a campaign issue. This evidence goes to WRTL's subjective intent in running the ads, and we have already explained that WRTL's intent is irrelevant in an as-applied challenge. Evidence of this sort is therefore beside the point, as it should be—WRTL does not forfeit its right to speak on issues simply because in other aspects of its work it also opposes candidates who are involved with those issues.

Next, the FEC and intervenors seize on the timing of WRTL's ads. They observe that the ads were to be aired near elections but not near actual Senate votes on judicial nominees, and that WRTL did not run the ads after the elections. To the extent this evidence goes to WRTL's subjective intent, it is again irrelevant. To the extent it nonetheless suggests that the ads should be interpreted as express advocacy, it falls short. That the ads were run close to an election is unremarkable in a challenge like this. *Every* ad covered by BCRA §203 will by definition air just before a primary or general election. If this were enough to prove that an ad is the functional equivalent of express advocacy, then BCRA would be constitutional in all of its applications. This Court unanimously rejected this contention in *WRTL I*.

That the ads were run shortly after the Senate had recessed is likewise unpersuasive. Members of Congress often return to their districts during recess, precisely to determine the views of their constituents; an ad run at that time may succeed in getting more constituents to contact the Representative while he or she is back home. In any event, a group can certainly choose to run an issue ad to coincide with public interest rather than a floor vote. Finally, WRTL did not resume running its ads after the BCRA blackout period because, as it explains, the debate had changed. . . . The focus of the Senate was on whether a majority would vote to change the Senate rules to eliminate the filibuster—not whether individual Senators would continue filibustering. Given this change, WRTL's decision not to continue running its ads after the blackout period does not support an inference that the ads were the functional equivalent of electioneering.

The last piece of contextual evidence the FEC and intervenors highlight is the ads' "specific and repeated cross-reference" to a website. . . . In the middle of the website's homepage, in large type, were the addresses, phone numbers, fax numbers, and email addresses of Senators Feingold and Kohl. Wisconsinites who viewed "Wedding," "Loan," or "Waiting" and wished to contact their Senators—as the ads requested—would be able to obtain the pertinent contact information immediately upon visiting the website. This is fully consistent with viewing WRTL's ads as genuine issue ads. The website also stated both Wisconsin Senators' positions on judicial filibusters, and allowed visitors to sign up for "e-alerts," some of which contained exhortations to vote against Senator Feingold. These details lend the electioneering interpretation of the ads more credence, but again, WRTL's participation in express advocacy in other aspects of its work is not a justification for censoring its issue-related speech. Any express advocacy on the website, already one step removed from the text of the ads themselves, certainly does not render an interpretation of the ads as genuine issue ads unreasonable.

Given the standard we have adopted for determining whether an ad is the "functional equivalent" of express advocacy, contextual factors of the sort invoked by appellants should seldom play a significant role in the inquiry. Courts need not ignore basic background information that may be necessary to put an ad in context—such as whether an ad "describes a legislative issue that is either currently the subject of legislative scrutiny or likely to be the subject of such scrutiny in the near future," . . . but the need to consider such background should not become an excuse for discovery or a broader inquiry of the sort we have just noted raises First Amendment concerns.

At best, appellants have shown what we have acknowledged . . . : that "the distinction between discussion of issues and candidates and advocacy of election or defeat of candidates may often dissolve in practical application.". . . Under the test set forth above, that is not enough to establish that the ads can only reasonably be viewed as advocating or opposing a candidate in a federal election. "Freedom of discussion, if it would fulfill its historic function in this nation, must embrace all issues about which information is needed or appropriate to enable the members of society to cope with the exigencies of their period.". . . Discussion of issues cannot be suppressed simply because the issues may also be pertinent in an election. Where the First Amendment is implicated, the tie goes to the speaker, not the censor. . . .

Because WRTL's ads may reasonably be interpreted as something other than as an appeal to vote for or against a specific candidate, we hold they are not the functional equivalent of express advocacy, and therefore fall outside the scope of *McConnell's* holding.

[Section IV omitted. In this section Justice Roberts found no compelling government interest in regulating ads that are "neither express advocacy nor its functional equivalent."]

* * *

These cases are about political speech. The importance of the cases to speech and debate on public policy issues is reflected in the number of diverse organizations that have joined in supporting WRTL before this Court: the American Civil Liberties Union, the National Rifle Association, the American Federation of Labor and Congress of Industrial Organizations, the Chamber of Commerce of the United States of America, Focus on the Family, the Coalition of Public Charities, the Cato Institute, and many others.

Yet, as is often the case in this Court's First Amendment opinions, we have gotten this far in the analysis without quoting the Amendment itself: "Congress shall make no law . . . abridging the freedom of speech." The Framers' actual words put these cases in proper perspective. Our jurisprudence over the past 216 years has rejected an absolutist interpretation of those words, but when it comes to drawing difficult lines in the area of pure political speech—between what is protected and what the Government may ban—it is worth recalling the language we are applying. *McConnell* held that express advocacy of a candidate or his opponent by a corporation shortly before an election may be prohibited, along with the functional equivalent of such express advocacy. We have no occasion to revisit that determination today. But when it comes to defining what speech qualifies as the functional equivalent of express advocacy subject to such a ban—the issue we *do* have to decide—we give the benefit of the doubt to speech, not censorship. The First Amendment's command that "Congress shall make no law . . . abridging the freedom of speech" demands at least that.

The judgment of the United States District Court for the District of Columbia is affirmed.

It is so ordered.

SOURCE: Supreme Court of the United States. *Federal Election Commission v. Wisconsin Right to Life, Inc.* Slip opinion. Docket 06-969. June 25, 2007. www.supremecourtus.gov/opinions/06pdf/06-969.pdf (accessed March 26, 2008).

OTHER HISTORIC DOCUMENTS OF INTEREST

FROM PREVIOUS *HISTORIC DOCUMENTS*

- Governor Vilsack and Senator Bayh on presidential campaign decisions, *2006*, p. 672
- Supreme Court on federal campaign finance law, *2003*, p. 1155

DOCUMENT IN CONTEXT

Supreme Court on Voluntary School Desegregation

JUNE 28, 2007

In a controversial ruling in which each side called on the historic school desegregation decision in *Brown v. Board of Education* (1954) to justify its position, a narrowly divided Supreme Court on June 28 held that race cannot be the determining factor in assigning students to schools, even if the intention is to achieve racial diversity. Led by Chief Justice John G. Roberts Jr., four of the justices would have gone further to state that schools could never consider race to achieve diversity. The majority's position was tempered somewhat by Justice Anthony M. Kennedy, who agreed that the two plans being challenged—one from Seattle, Washington, and the other from Louisville, Kentucky—violated the Fourteenth Amendment's equal protection clause. But Kennedy also said race can be taken into account in some circumstances. In a lengthy and angry dissent, the remaining four justices called the ruling a "cruel distortion" of the meaning of the *Brown* decision, in which a unanimous Court called for an end to segregated schools.

The decision had potentially widespread ramifications for schools across the country. By some counts, about 1,000 of the nations' 15,000 school districts had plans in place that used race in some way to try to ensure racial diversity and to prevent the school segregation or resegregation that occurred in areas where housing patterns were largely segregated. How many of these plans might be challenged in court or voluntarily changed or abandoned as a result of the ruling was unknown.

The deep divisions on the Court, together with the strongly worded opinions filed by five of the nine justices, reflected continuing differences in American society about issues of racial diversity, equality, and fairness. Several racially charged incidents in 2007 also reflected those differences. Two prominent figures, a radio shock jock and a Nobel Prize–winning biologist, were forced out of their jobs after making public racial slurs. And the largest civil rights demonstration in recent years gathered in Jena, Louisiana, to protest what many saw as racially biased treatment of six black students accused of beating up a white student after white students hung nooses—long a symbol of racial hatred—from a schoolyard tree.

At the same time, there was evidence of changing racial attitudes. A *Washington Post*–ABC News poll conducted immediately after the Supreme Court ruling was issued found that 56 percent of those surveyed disapproved of the decision, while only 40 percent supported it. Perhaps the most visible sign of changing attitudes, however, was the 2008 presidential election campaign, in which Barack Obama, an African American senator from Illinois, was mounting a strong challenge to Sen. Hillary Clinton of New York for the Democratic nomination.

Different Meanings Assigned to Same Words

The decision in *Parents Involved in Community Schools v. Seattle School District No. 1* consolidated challenges to two separate racial diversity plans. The Seattle school district had voluntarily adopted a plan that classified children as white or nonwhite and then used the classification as a "tiebreaker" to allocate slots in particular high schools. The Jefferson County, Kentucky, school district, which incorporates Louisville, had been under a court-ordered desegregation plan until 2000, when a court found that all practical vestiges of past segregation had been removed. To prevent the resegregation that was likely to recur from housing patterns in the county, the Kentucky school district in 2001 adopted a plan that classified children as black or "other" and then used those classifications to make certain assignments or transfers to elementary schools. In both cases, parents of white children challenged the plans, complaining that by making school assignments based solely on race, the plans violated the equal protection clause of the Fourteenth Amendment. In both cases, the federal district courts upheld the racial diversity plans, finding them narrowly tailored to meet a compelling government interest, and federal appellate courts agreed.

The Supreme Court overruled the lower courts by a 5–4 vote. Writing the court's opinion, Chief Justice John G. Roberts Jr. said that under its precedents, the Court recognized only two interests compelling enough to justify the use of racial classifications. One compelling interest was to remedy the effects of past segregation imposed by law. But, Roberts noted, Seattle had never been segregated, and a court in Kentucky had found that Jefferson County had already remedied its past segregation. The other compelling interest, Roberts said, was diversity, where racial classifications were "part of a broader assessment of diversity, and not simply an effort to achieve racial balance." In the two school district plans in this case, however, race "is decisive by itself. It is not simply one factor weighed with others in reaching a decision [about admission to a particular school], it is *the* factor," Roberts said. Justices Antonin Scalia, Clarence Thomas, Samuel A. Alito Jr., and Kennedy endorsed this portion of Roberts's opinion.

Roberts, joined by Scalia, Thomas, and Alito, went on to argue that the two school plans were aimed not at racial diversity but at racial balancing, "an objective this Court has repeatedly condemned as illegitimate." Roberts then chastised the dissenting justices for misinterpreting the meaning of *Brown v. Board of Education*. In that case, Roberts wrote, "we held that segregation deprived black children of equal educational opportunities regardless of whether school facilities and other tangible factors were equal, because government classification and separation on grounds of race themselves denoted inferiority. It was not the inequality of the facilities but the fact of legally separating children on the basis of race on which the Court relied" in 1954. "Before Brown," Roberts continued, "schoolchildren were told where they could and could not go to school based on the color of their skin. The school districts in these cases have not carried the heavy burden of demonstrating that we should allow this once again—even for very different reasons. For schools that never segregated on the basis of race . . . or that have removed the vestiges of past discrimination . . . the way to stop discrimination on the basis of race is to stop discriminating on the basis of race."

Kennedy agreed with the ruling in the case but said that parts of Roberts's opinion implied "an all-too-unyielding insistence that race cannot be a factor in instances when, in my view, it may be taken into account. The plurality opinion is too dismissive of the legitimate interest government has in ensuring all people have equal opportunity regardless of their race." Noting specifically the problem of de facto school segregation resulting from housing patterns, Kennedy suggested that school districts would not violate the equal protection clause if they were to adopt "race-conscious measures to address the problem in a general way and without treating each student in different fashion solely on the basis of a systematic, individual typing by race." Examples of permissible race-conscious measures, Kennedy said, would include strategic site selection of new schools, attendance zones drawn with general recognition of neighborhood demographics, allocating resources for magnet schools, and tracking enrollments, performance, and other statistics by race.

In a lengthy and passionate dissent, Justice Stephen G. Breyer, joined by Justices John Paul Stevens, David Souter, and Ruth Bader Ginsburg, argued that the four justices in the plurality misinterpreted the spirit of the rulings in *Brown* and subsequent decisions. "The plurality pays inadequate attention to this law, to past opinions' rationales, their language, and the contexts in which they arise," Breyer wrote. "As a result it reverses course and reaches the wrong conclusion. In doing so, it distorts precedent, it misapplies the relevant constitutional principles, it announces legal rules that will obstruct efforts by state and local governments to deal effectively with the growing resegregation of public schools, it threatens to substitute for present calm a disruptive round of race-related litigation, and it undermines Brown's promise of integrated primary and secondary education that local communities have sought to make a reality. This cannot be justified in the name of the Equal Protection Clause."

Alternatives to Race-Based Plans

It was unclear how schools with diversity plans similar to those struck down by the Court would react to the decision. Sharon L. Browne, an attorney with the Pacific Legal Foundation, a conservative group that supported the parents who sued the Seattle and Louisville school districts, issued a statement declaring that "with these decisions, an estimated 1,000 school districts around the country that are sending the wrong message about race to kids will have to stop." Her organization had already filed suits in Los Angeles and Berkeley, California, challenging those school districts' use of race to make student assignments, and Browne indicated that the foundation might also bring similar challenges elsewhere, including in Lynn, Massachusetts.

Browne, like many other attorneys and educators, looked to plans using income levels, which often correlate with race, as a race-neutral way to achieve racial diversity. Such plans would not only avoid the legal difficulties of race-based plans, but some educators believed they would also promote achievement by putting low-income children in schools where teachers were more experienced, children had high expectations for their own achievement, and more parents were actively involved in their children's education. "There is a large body of evidence going back several years that probably the most important thing you can do to raise the achievement of low-income students is to provide them with middle-class schools," said Richard D. Kahlenberg, a senior fellow at the Century Foundation, a nonpartisan public policy research group, and author of an analysis of diversity plans based on socioeconomic status.

About forty school districts already were using income levels to make school assignments in 2007. The plans varied considerably in their scope, and by most accounts the most successful one was in Wake County, North Carolina, which includes the city of Raleigh. For several years the district had sought to cap the proportion of low-income students in each of its 143 schools at 40 percent. Students were encouraged to attend, and sometimes assigned to, schools outside their neighborhoods, and magnet schools were used to attract suburban students to city schools. Early results showing improved achievement were encouraging—the proportion of African American children in grades three through eight who could read at grade level rose from 40 percent in 1995, before the plan was put in place, to 82 percent in 2006.

Others, especially many African Americans, argued that the real issue in schooling was not racial diversity but lack of adequate funding for schools in poor neighborhoods. "A lot of black folks say, 'give us the resources, give us the money, we're tired of chasing white folks, and we don't need integrated schools to have good education,' " Ted Shaw, president of the NAACP Legal Defense and Educational Fund Inc., told the *New York Times* after the Court handed down its decision. "It's hard to tell how much of that is weariness and cynicism with respect to efforts to get racially integrated schools, and I can understand that."

Racial Slurs

In 2007 two prominent public figures were forced to apologize and lost their jobs after making public comments that were viewed by many as denigrating African Americans. Two of the most prominent were radio shock jock Don Imus and noted biologist James D. Watson; the latter shared the 1962 Nobel Prize for describing the double-helix structure of DNA. In April both CBS radio and cable news channel MSNBC suspended and then canceled the show *Imus in the Morning* after host Imus called the players on Rutgers University women's basketball team "nappy-headed hos." Eight of the ten members of the team, which had just lost to Tennessee in the NCAA championship title game, were black. Imus met privately with the women on April 13 to apologize for what their coach, C. Vivian Stringer, referred to as "deplorable, despicable, and unconscionable remarks." He returned to the airwaves on December 8 for Citadel Radio.

Watson was forced to resign his position as chancellor of the Cold Spring Harbor Laboratory in New York after suggesting that blacks were not as intelligent as whites. In an interview published October 14 in the *Times* (London), Watson was quoted as saying he was "inherently gloomy about the prospect of Africa" because "all our social policies are based on the fact that their intelligence is the same as ours—whereas all the testing says not really." In a statement released October 19, Watson said he could not understand "how I could have said what I am quoted as having said," but he apologized "unreservedly," adding that "there is no scientific basis for such a belief."

Righting Past Wrongs

A former member of the Ku Klux Klan was tried and convicted for his role in the murders of two young black men in Mississippi in May 1964. On June 14, 2007, James Ford Seale was convicted in federal court of conspiring with other Klansmen to kidnap and murder Hezekiah Dee and Charlie Eddie Moore, who Seale and his cohorts had mistakenly believed were plotting an armed uprising. The Klansmen picked the two nineteen-year-olds up as they were hitchhiking and drove them into a national forest, where they interrogated them at gunpoint and beat them. The two men were then thrown into the Mississippi River and drowned. Seale, who was seventy-two in 2007, was sentenced to three life terms in prison.

Seale was the latest in a string of former Klansmen to be tried for murders committed during the Civil Rights era. In June 2005 Edgar Ray Killen was convicted for his role in the brutal murders of three civil rights workers, also in 1964. The deaths of the three—Michael Schwerner, James Chaney, and Andrew Goodman—galvanized the drive that resulted in passage of the Voting Rights Act of 1965. Additional trials could be forthcoming. The Justice Department in February announced that it was reexamining nearly 100 cases of what were thought to be racially inspired murders from as far back as the 1940s.

A recent case highlighted one of the many racial disparities between blacks and whites—black youths were more likely that their white counterparts to be arrested and tried as adults. The incident in rural Jena, Louisiana, began in 2006 when white high school students hung nooses from a tree in the school yard after a black student sat under it; the tree was a usual gathering place for white students. Offended by the blatant racist threat, black parents asked that the white students be expelled, but the students were instead suspended for three days. Weeks later, in December, six black youths beat up a white student, knocking him unconscious but doing no permanent damage.

The local district attorney first charged the six black teenagers with attempted second-degree murder but later reduced the charges for five of the boys. The sixth, Mychal Bell, was tried as an adult, even though he was sixteen at the time of the assault, and was convicted by an all-white jury of aggravated second-degree battery. An appeals court threw out that verdict in September and ordered that Bell be tried as a juvenile. In December Bell agreed to plead guilty to a juvenile charge of second-degree battery and accepted a sentence of eighteen months, ten of which he had already served. The "Jena Six" incident led in September to one of the largest civil rights demonstrations in recent history, as civil rights leaders, college campus activists, and others descended on Jena to protest what they saw as racial bias and inequality in the justice system.

Following are excerpts from opinions filed by Chief Justice John G. Roberts Jr. and Justice Anthony M. Kennedy in the case of Parents Involved in Community Schools v. Seattle School District No. 1, *in which the Supreme Court on June 28, 2007, ruled 5–4 to restrict the use of race-based plans for assigning students to elementary and secondary schools to achieve racial diversity.*

DOCUMENT

Parents Involved in Community Schools v. Seattle School District No. 1

Nos. 05–908 and 05–915

Parents Involved in Community Schools, Petitioner 05–908
v.
Seattle School District No. 1 et al.

On writ of certiorari to the United States Court of Appeals for the Ninth Circuit

Crystal D. Meredith, Custodial Parent and Next Friend of Joshua Ryan McDonald, Petitioner 05–915
v.
Jefferson County Board of Education et al.

On writ of certiorari to the United States Court of Appeals or the Sixth Circuit

[June 28, 2007]

CHIEF JUSTICE ROBERTS announced the judgment of the Court, and delivered the opinion of the Court with respect to Parts I, II, III–A, and III–C, and an opinion with respect to Parts III–B and IV, in which JUSTICES SCALIA, THOMAS, and ALITO join.

The school districts in these cases voluntarily adopted student assignment plans that rely upon race to determine which public schools certain children may attend. The Seattle school district classifies children as white or nonwhite; the Jefferson County school district as black or "other." In Seattle, this racial classification is used to allocate slots in oversubscribed high schools. In Jefferson County, it is used to make certain elementary school assignments and to rule on transfer requests. In each case, the school district relies upon an individual student's race in assigning that student to a particular school, so that the racial balance at the school falls within a predetermined range based on the racial composition of the school district as a whole. Parents of students denied assignment to particular schools under these plans solely because of their race brought suit, contending that allocating children to different public schools on the basis of race violated the Fourteenth Amendment guarantee of equal protection. The Courts of Appeals below upheld the plans. We granted certiorari, and now reverse.

I

Both cases present the same underlying legal question—whether a public school that had not operated legally segregated schools or has been found to be unitary may choose to classify students by race and rely upon that classification in making school assignments. Although we examine the plans under the same legal framework, the specifics of the two plans, and the circumstances surrounding their adoption, are in some respects quite different.

A

Seattle School District No. 1 operates 10 regular public high schools. In 1998, it adopted the plan at issue in this case for assigning students to these schools. . . . The plan allows incoming ninth graders to choose from among any of the district's high schools, ranking however many schools they wish in order of preference.

Some schools are more popular than others. If too many students list the same school as their first choice, the district employs a series of "tiebreakers" to determine who will fill the open slots at the oversubscribed school. The first tiebreaker selects for admission students who have a sibling currently enrolled in the chosen school. The next tiebreaker depends upon the racial composition of the particular school and the race of the individual student. In the district's public schools approximately 41 percent of enrolled students are white; the remaining 59 percent, comprising all other racial groups, are classified by Seattle for assignment purposes as nonwhite. . . . If an oversubscribed school is not within 10 percentage points of the district's overall white/nonwhite racial balance, it is what the district calls "integration positive," and the district employs a tiebreaker that selects for assignment students whose race "will serve to bring the school into balance.". . . If it is still necessary to select students for the school after using the racial tiebreaker, the next tiebreaker is the geographic proximity of the school to the student's residence. . . .

Seattle has never operated segregated schools—legally separate schools for students of different races—nor has it ever been subject to court-ordered desegregation. It nonetheless employs the racial tiebreaker in an attempt to address the effects of racially identifiable housing patterns on school assignments. . . .

For the 2000–2001 school year, five of these schools were oversubscribed—Ballard, Nathan Hale, Roosevelt, Garfield, and Franklin—so much so that 82 percent of incoming ninth graders ranked one of these schools as their first choice. . . . Three of the oversubscribed schools were "integration positive" because the school's white enrollment the previous school year was greater than 51 percent—Ballard, Nathan Hale, and Roosevelt. Thus, more nonwhite students (107, 27, and 82, respectively) who selected one of these three schools as a top choice received placement at the school than would have been the case had race not been considered, and proximity been the next tiebreaker. . . .

Petitioner Parents Involved in Community Schools (Parents Involved) is a nonprofit corporation comprising the parents of children who have been or may be denied assignment to their chosen high school in the district because of their race. . . . Parents Involved commenced this suit in the Western District of Washington, alleging that Seattle's use of race in assignments violated the Equal Protection Clause of the Fourteenth Amendment, Title VI of the Civil Rights Act of 1964, and the Washington Civil Rights Act. . . .

The District Court granted summary judgment to the school district, finding that state law did not bar the district's use of the racial tiebreaker and that the plan survived strict scrutiny on the federal constitutional claim because it was narrowly tailored to serve a compelling government interest. . . .

A panel of the Ninth Circuit . . . reversed the District Court. . . . The panel determined that while achieving racial diversity and avoiding racial isolation are compelling government interests, . . . Seattle's use of the racial tiebreaker was not narrowly tailored to achieve these interests. . . . The Ninth Circuit granted rehearing en banc, . . . and overruled the panel decision, affirming the District Court's determination that Seattle's plan was narrowly tailored to serve a compelling government interest. . . . We granted certiorari. . . .

B

Jefferson County Public Schools operates the public school system in metropolitan Louisville, Kentucky. In 1973 a federal court found that Jefferson County had maintained a segregated school system. . . . Jefferson County operated under this decree until 2000, when the District Court dissolved the decree after finding that the district had achieved unitary status by eliminating "[t]o the greatest extent practicable" the vestiges of its prior policy of segregation. . . .

In 2001, after the decree had been dissolved, Jefferson County adopted the voluntary student assignment plan at issue in this case. . . . Approximately 34 percent of the district's 97,000 students are black; most of the remaining 66 percent are white. . . . The plan requires all nonmagnet schools to maintain a minimum black enrollment of 15 percent, and a maximum black enrollment of 50 percent. . . .

At the elementary school level, based on his or her address, each student is designated a "resides" school to which students within a specific geographic area are assigned; elementary resides schools are "grouped into clusters in order to facilitate integration." . . The district assigns students to nonmagnet schools in one of two ways: Parents of kindergartners, first-graders, and students new to the district may submit an application indicating a first and second choice among the schools within their cluster; students who do not submit such an application are assigned within the cluster by the district. "Decisions to assign students to schools within each cluster are based on available space within the schools and the racial guidelines in the District's current student assignment plan." . . If a school has reached the "extremes of the racial guidelines," a student whose race would contribute to the school's racial imbalance will not be assigned there. . . . After assignment, students at all grade levels are permitted to apply to transfer between nonmagnet schools in the district. Transfers may be requested for any number of reasons, and may be denied because of lack of available space or on the basis of the racial guidelines. . . .

When petitioner Crystal Meredith moved into the school district in August 2002, she sought to enroll her son, Joshua McDonald, in kindergarten for the 2002–2003 school year. His resides school was only a mile from his new home, but it had no available space—assignments had been made in May, and the class was full. Jefferson County assigned Joshua to another elementary school in his cluster, Young Elementary. This school was 10 miles from home, and Meredith sought to transfer Joshua to a school in a different cluster, Bloom Elementary, which—like his resides school—was only a mile from home. . . . Space was available at Bloom, and inter cluster transfers are allowed, but Joshua's transfer was nonetheless denied because, in the words of Jefferson County, "[t]he transfer would have an adverse effect on desegregation compliance" of Young. . . .

Meredith brought suit in the Western District of Kentucky, alleging violations of the Equal Protection Clause of the Fourteenth Amendment. The District Court found that Jefferson County had asserted a compelling interest in maintaining racially diverse schools, and that the assignment plan was (in all relevant respects) narrowly tailored to serve that compelling interest. . . . The Sixth Circuit affirmed in a *per curiam* opinion relying upon the reasoning of the District Court, concluding that a written opinion "would serve no useful purpose." . . We granted certiorari. . . .

[Section II omitted. In this section, the majority held that the parents in both school districts had standing to sue.]

III

A

It is well established that when the government distributes burdens or benefits on the basis of individual racial classifications, that action is reviewed under strict scrutiny.... In order to satisfy this searching standard of review, the school districts must demonstrate that the use of individual racial classifications in the assignment plans hereunder review is "narrowly tailored" to achieve a "compelling" government interest....

Without attempting in these cases to set forth all the interests a school district might assert, it suffices to note that our prior cases, in evaluating the use of racial classifications in the school context, have recognized two interests that qualify as compelling. The first is the compelling interest of remedying the effects of past intentional discrimination.... Yet the Seattle public schools have not shown that they were ever segregated by law, and were not subject to court-ordered desegregation decrees. The Jefferson County public schools were previously segregated by law and were subject to a desegregation decree entered in 1975. In 2000, the District Court that entered that decree dissolved it, finding that Jefferson County had "eliminated the vestiges associated with the former policy of segregation and its pernicious effects," and thus had achieved "unitary" status.... Jefferson County accordingly does not rely upon an interest in remedying the effects of past intentional discrimination in defending its present use of race in assigning students....

Nor could it.... Once Jefferson County achieved unitary status, it had remedied the constitutional wrong that allowed race-based assignments. Any continued use of race must be justified on some other basis.

The second government interest we have recognized as compelling for purposes of strict scrutiny is the interest in diversity in higher education upheld in *Grutter [v. Bollinger (2003)]*. The specific interest found compelling in *Grutter* was student body diversity "in the context of higher education."... The diversity interest was not focused on race alone but encompassed "all factors that may contribute to student body diversity."...

The entire gist of the analysis in *Grutter* was that the admissions program at issue there focused on each applicant as an individual, and not simply as a member of a particular racial group.... The point of the narrow tailoring analysis in which the *Grutter* Court engaged was to ensure that the use of racial classifications was indeed part of a broader assessment of diversity, and not simply an effort to achieve racial balance, which the Court explained would be "patently unconstitutional."...

In the present cases, by contrast, race is not considered as part of a broader effort to achieve "exposure to widely diverse people, cultures, ideas, and viewpoints"...; race, for some students, is determinative standing alone. The districts argue that other factors, such as student preferences, affect assignment decisions under their plans, but under each plan when race comes into play, it is decisive by itself. It is not simply one factor weighed with others in reaching a decision...; it is *the* factor....

Prior to *Grutter,* the courts of appeals rejected as unconstitutional attempts to implement race-based assignment plans—such as the plans at issue here—in primary and secondary schools.... After *Grutter,* however, the two Courts of Appeals in these cases, and one other, found that race-based assignments were permissible at the elementary and secondary level, largely in reliance on that case....

In upholding the admissions plan in *Grutter,* though, this Court relied upon considerations unique to institutions of higher education.... The Court in *Grutter* expressly articulated key limitations on its holding—defining a specific type of broad-based diversity and noting the unique context of higher education—but these limitations were largely disregarded by the lower courts in extending *Grutter* to uphold race-based assignments in elementary and secondary schools. The present cases are not governed by *Grutter.*

B

Perhaps recognizing that reliance on *Grutter* cannot sustain their plans, both school districts assert additional interests, distinct from the interest upheld in *Grutter,* to justify their race-based assignments. In briefing and argument before this Court, Seattle contends that its use of race helps to reduce racial concentration in schools and to ensure that racially concentrated housing patterns do not prevent nonwhite students from having access to the most desirable schools. . . . Jefferson County has articulated a similar goal, phrasing its interest in terms of educating its students "in a racially integrated environment." . . Each school district argues that educational and broader socialization benefits flow from a racially diverse learning environment, and each contends that because the diversity they seek is racial diversity—not the broader diversity at issue in *Grutter*—it makes sense to promote that interest directly by relying on race alone.

The parties and their *amici* dispute whether racial diversity in schools in fact has a marked impact on test scores and other objective yardsticks or achieves intangible socialization benefits. The debate is not one we need to resolve, however, because it is clear that the racial classifications employed by the districts are not narrowly tailored to the goal of achieving the educational and social benefits asserted to flow from racial diversity. In design and operation, the plans are directed only to racial balance, pure and simple, an objective this Court has repeatedly condemned as illegitimate. . . .

Accepting racial balancing as a compelling state interest would justify the imposition of racial proportionality throughout American society, contrary to our repeated recognition that "[a]t the heart of the Constitution's guarantee of equal protection lies the simple command that the Government must treat citizens as individuals, not as simply components of a racial, religious, sexual or national class." . . Allowing racial balancing as a compelling end in itself would "effectively assur[e] that race will always be relevant in American life, and that the 'ultimate goal' of 'eliminating entirely from governmental decision making such irrelevant factors as a human being's race' will never be achieved." . .

The validity of our concern that racial balancing has "no logical stopping point" . . . is demonstrated here by the degree to which the districts tie their racial guidelines to their demographics. As the districts' demographics shift, so too will their definition of racial diversity. . . .

The principle that racial balancing is not permitted is one of substance, not semantics. Racial balancing is not transformed from "patently unconstitutional" to a compelling state interest simply by relabeling it "racial diversity." While the school districts use various verbal formulations to describe the interest they seek to promote—racial diversity, avoidance of racial isolation, racial integration—they offer no definition of the interest that suggests it differs from racial balance. . . .

. . . However closely related race-based assignments may be to achieving racial balance, that itself cannot be the goal, whether labeled "racial diversity" or anything else. To the extent the objective is sufficient diversity so that students see fellow students as individuals rather than solely as members of a racial group, using means that treat students solely as members of a racial group is fundamentally at cross-purposes with that end.

C

The districts assert, as they must, that the way in which they have employed individual racial classifications is necessary to achieve their stated ends. The minimal effect these classifications have on student assignments, however, suggests that other means would be effective. Seattle's racial tiebreaker results, in the end, only in shifting a small number of students between schools. . . . [T]he district could identify only 52 students who were ultimately affected adversely by the racial tiebreaker in that it resulted in assignment to a school they had not listed as a preference and to which they would not otherwise have been assigned. . . .

Similarly, Jefferson County's use of racial classifications has only a minimal effect on the assignment of students. . . .

While we do not suggest that *greater* use of race would be preferable, the minimal impact of the districts' racial classifications on school enrollment casts doubt on the necessity of using racial classifications. . . .

The districts have also failed to show that they considered methods other than explicit racial classifications to achieve their stated goals. Narrow tailoring requires "serious, good faith consideration of workable race-neutral alternatives" . . . , and yet in Seattle several alternative assignment plans—many of which would not have used express racial classifications—were rejected with little or no consideration. . . . Jefferson County has failed to present any evidence that it considered alternatives, even though the district already claims that its goals are achieved primarily through means other than the racial classifications. . . .

IV

JUSTICE BREYER's dissent takes a different approach to these cases, one that fails to ground the result it would reach in law. Instead, it selectively relies on inapplicable precedent and even dicta while dismissing contrary holdings, alters and misapplies our well-established legal framework for assessing equal protection challenges to express racial classifications, and greatly exaggerates the consequences of today's decision.

To begin with, JUSTICE BREYER seeks to justify the plans at issue under our precedents recognizing the compelling interest in remedying past intentional discrimination. . . . Not even the school districts go this far, and for good reason. The distinction between segregation by state action and racial imbalance caused by other factors has been central to our jurisprudence in this area for generations. . . . The dissent elides this distinction between *de jure* and *de facto* segregation, casually intimates that Seattle's school attendance patterns reflect illegal segregation, . . . and fails to credit the judicial determination—under the most rigorous standard—that Jefferson County had eliminated the vestiges of prior segregation. The dissent thus alters in fundamental ways not only the facts presented here but the established law.

[Text omitted in which Chief Justice Roberts explains where, in his view, the dissenting justices have misinterpreted facts and established law.]

At the same time it relies on inapplicable desegregation cases, misstatements of admitted dicta, and other non controlling pronouncements, JUSTICE BREYER's dissent candidly dismisses the significance of this Court's repeated *holdings* that all racial classifications must be reviewed under strict scrutiny. . . , arguing that a different standard of review should be applied because the districts use race for beneficent rather than malicious purposes. . . .

This Court has recently reiterated, however, that " '*all* racial classifications [imposed by government] . . . must be analyzed by a reviewing court under strict scrutiny.' ". . . JUSTICE BREYER nonetheless relies on the good intentions and motives of the school districts, stating that he has found "no case that . . . repudiated this constitutional asymmetry between that which seeks to *exclude* and that which seeks to *include* members of minority races.". . . We have found many. Our cases clearly reject the argument that motives affect the strict scrutiny analysis. . . .

This argument that different rules should govern racial classifications designed to include rather than exclude is not new; it has been repeatedly pressed in the past . . . and has been repeatedly rejected. . . .

JUSTICE BREYER's position comes down to a familiar claim: The end justifies the means. He admits that "there is a cost in applying 'a state-mandated racial label,' " . . . but he is confident that the cost is worth paying. Our established strict scrutiny test for racial classifications, however, insists on "detailed examination, both as to ends *and* as to means.". . . Simply because the school districts may seek a worthy goal does not mean they are free to discriminate on the basis of race to achieve it, or that their racial classifications should be subject to less exacting scrutiny.

Despite his argument that these cases should be evaluated under a "standard of review that is not 'strict' in the traditional sense of that word,". . . JUSTICE BREYER still purports to apply strict scrutiny to these cases. . . . It is evident, however, that JUSTICE BREYER's brand of narrow tailoring is quite unlike anything found in our precedents. Without any detailed discussion of the operation of the plans, the students who are affected, or the districts' failure to consider race-neutral alternatives, the dissent concludes that the districts have shown that these racial classifications are necessary to achieve the districts' stated goals. This conclusion is divorced from any evaluation of the actual impact of the plans at issue in these cases—other than to note that the plans "often have no effect.". . . Instead, the dissent suggests that some combination of the development of these plans over time, the difficulty of the endeavor, and the good faith of the districts suffices to demonstrate that these stark and controlling racial classifications are constitutional. The Constitution and our precedents require more. . . .

* * *

If the need for the racial classifications embraced by the school districts is unclear, even on the districts' own terms, the costs are undeniable. . . . Government action dividing us by race is inherently suspect because such classifications promote "notions of racial inferiority and lead to a politics of racial hostility," . . . "reinforce the belief, held by too many for too much of our history, that individuals should be judged by the color of their skin," . . . and "endorse race-based reasoning and the conception of a Nation divided into racial blocs, thus contributing to an escalation of racial hostility and conflict.". . . As the Court explained in *Rice v. Cayetano* . . . (2000), "[o]ne of the principal reasons race is treated as a forbidden classification is that it demeans the dignity and worth of a person to be judged by ancestry instead of by his or her own merit and essential qualities."

All this is true enough in the contexts in which these statements were made—government contracting, voting districts, allocation of broadcast licenses, and electing state officers—but when it comes to using race to assign children to schools, history will be heard. In *Brown v. Board of Education* . . . (1954) (*Brown I*), we held that segregation deprived black children of equal educational opportunities regardless of whether school facilities and other tangible factors were equal, because government classification and separation on grounds of race themselves denoted inferiority. . . . It was not the inequality of the facilities but the fact of legally separating children on the basis of race on which the Court relied to find a constitutional violation in 1954. . . . The next Term, we accordingly stated that "full compliance" with *Brown I* required school districts "to achieve a system of determining admission to the public schools *on a nonracial basis.*" . . .

The parties and their *amici* debate which side is more faithful to the heritage of *Brown,* but the position of the plaintiffs in *Brown* was spelled out in their brief and could not have been clearer: "[T]he Fourteenth Amendment prevents states from according differential treatment to American children on the basis of their color or race." . . . What do the racial classifications at issue here do, if not accord differential treatment on the basis of race? As counsel who appeared before this Court for the plaintiffs in *Brown* put it: "We have one fundamental contention which we will seek to develop in the course of this argument, and that contention is that no State has any authority under the equal-protection clause of the Fourteenth Amendment to use race as a factor in affording educational opportunities among its citizens." . . . There is no ambiguity in that statement. And it was that position that prevailed in this Court, which emphasized in its remedial opinion that what was "[a]t stake is the personal interest of the plaintiffs in admission to public schools as soon as practicable *on a nondiscriminatory basis,*" and what was required was "determining admission to the public schools *on a nonracial basis.*" . . . What do the racial classifications do in these cases, if not determine admission to a public school on a racial basis?

Before *Brown,* schoolchildren were told where they could and could not go to school based on the color of their skin. The school districts in these cases have not carried the heavy burden of demonstrating that we should allow this once again—even for very different reasons. For schools that never segregated on the basis of race, such as Seattle, or that have removed the vestiges of past segregation, such as Jefferson County, the way "to achieve a system of determining admission to the public schools on a nonracial basis, . . . is to stop assigning students on a racial basis. The way to stop discrimination on the basis of race is to stop discriminating on the basis of race.

The judgments of the Courts of Appeals for the Sixth and Ninth Circuits are reversed, and the cases are remanded for further proceedings.

It is so ordered.

[Justice Kennedy wrote a concurring opinion in which he concurred in part and concurred in the judgment of the majority.]

JUSTICE BREYER, with whom JUSTICE STEVENS, JUSTICE SOUTER, and JUSTICE GINSBURG join, dissenting.

These cases consider the longstanding efforts of two local school boards to integrate their public schools. The school board plans before us resemble many others adopted in the last 50 years by primary and secondary schools throughout the Nation. All of those plans represent local efforts to bring about the kind of racially integrated education that

Brown v. Board of Education . . . (1954), long ago promised—efforts that this Court has repeatedly required, permitted, and encouraged local authorities to undertake. This Court has recognized that the public interests at stake in such cases are "compelling." We have approved of "narrowly tailored" plans that are no less race-conscious than the plans before us. And we have understood that the Constitution *permits* local communities to adopt desegregation plans even where it does not *require* them to do so.

The plurality pays inadequate attention to this law, to past opinions' rationales, their language, and the contexts in which they arise. As a result, it reverses course and reaches the wrong conclusion. In doing so, it distorts precedent, it misapplies the relevant constitutional principles, it announces legal rules that will obstruct efforts by state and local governments to deal effectively with the growing resegregation of public schools, it threatens to substitute for present calm a disruptive round of race-related litigation, and it undermines *Brown*'s promise of integrated primary and secondary education that local communities have sought to make a reality. This cannot be justified in the name of the Equal Protection Clause.

I

Facts

The historical and factual context in which these cases arise is critical. In *Brown,* this Court held that the government's segregation of schoolchildren by race violates the Constitution's promise of equal protection. The Court emphasized that "education is perhaps the most important function of state and local governments.". . . And it thereby set the Nation on a path toward public school integration.

In dozens of subsequent cases, this Court told school districts previously segregated by law what they must do at a minimum to comply with *Brown*'s constitutional holding. The measures required by those cases often included race-conscious practices, such as mandatory busing and race-based restrictions on voluntary transfers. . . .

Beyond those minimum requirements, the Court left much of the determination of how to achieve integration to the judgment of local communities. Thus, in respect to race-conscious desegregation measures that the Constitution *permitted,* but did not *require* (measures similar to those at issue here), this Court unanimously stated:

> "School authorities are traditionally charged with broad power to formulate and implement educational policy and might well conclude, for example, that in order to prepare students to live in a pluralistic society each school should have a prescribed ratio of Negro to white students reflecting the proportion for the district as a whole. *To do this as an educational policy is within the broad discretionary powers of school authorities.*" *Swann v. Charlotte-Mecklenburg Bd. of Ed.* . . . (1971) (emphasis added).

As a result, different districts—some acting under court decree, some acting in order to avoid threatened lawsuits, some seeking to comply with federal administrative orders, some acting purely voluntarily, some acting after federal courts had dissolved earlier orders—adopted, modified, and experimented with hosts of different kinds of plans, including race-conscious plans, all with a similar objective: greater racial integration of public schools. . . . The techniques that different districts have employed range "from voluntary transfer programs to mandatory reassignment.". . . And the design of particular plans has been "dictated by both the law and the specific needs of the district." . . .

Overall these efforts brought about considerable racial integration. More recently, however, progress has stalled. Between 1968 and 1980, the number of black children attending a school where minority children constituted more than half of the school fell from 77% to 63% in the Nation (from 81% to 57% in the South) but then reversed direction by the year 2000, rising from 63% to 72% in the Nation (from 57% to 69% in the South). Similarly, between 1968 and 1980, the number of black children attending schools that were more than 90% minority fell from 64% to 33% in the Nation (from 78% to 23% in the South), but that too reversed direction, rising by the year 2000 from 33% to 37% in the Nation (from 23% to 31% in the South). As of 2002, almost 2.4 million students, or over 5% of all public school enrollment, attended schools with a white population of less than 1%. Of these, 2.3 million were black and Latino students, and only 72,000 were white. Today, more than one in six black children attend a school that is 99–100% minority. . . . In light of the evident risk of a return to school systems that are in fact (though not in law) resegregated, many school districts have felt a need to maintain or to extend their integration efforts.

The upshot is that myriad school districts operating in myriad circumstances have devised myriad plans, often with race-conscious elements, all for the sake of eradicating earlier school segregation, bringing about integration, or preventing retrogression. Seattle and Louisville are two such districts, and the histories of their present plans set forth typical school integration stories.

I describe those histories at length in order to highlight three important features of these cases. First, the school districts' plans serve "compelling interests" and are "narrowly tailored" on any reasonable definition of those terms. Second, the distinction between *de jure* segregation (caused by school systems) and *de facto* segregation (caused, *e.g.*, by housing patterns or generalized societal discrimination) is meaningless in the present context, thereby dooming the plurality's endeavor to find support for its views in that distinction. Third, real-world efforts to substitute racially diverse for racially segregated schools (however caused) are complex, to the point where the Constitution cannot plausibly be interpreted to rule out categorically all local efforts to use means that are "conscious" of the race of individuals.

In both Seattle and Louisville, the local school districts began with schools that were highly segregated in fact. In both cities plaintiffs filed lawsuits claiming unconstitutional segregation. In Louisville, a federal district court found that school segregation reflected pre-*Brown* state laws separating the races. In Seattle, the plaintiffs alleged that school segregation unconstitutionally reflected not only generalized societal discrimination and residential housing patterns, but also *school board policies and actions* that had helped to create, maintain, and aggravate racial segregation. In Louisville, a federal court entered a remedial decree. In Seattle, the parties settled after the school district pledged to undertake a desegregation plan. In both cities, the school boards adopted plans designed to achieve integration by bringing about more racially diverse schools. In each city the school board modified its plan several times in light of, for example, hostility to busing, the threat of resegregation, and the desirability of introducing greater student choice. And in each city, the school boards' plans have evolved over time in ways that progressively *diminish* the plans' use of explicit race-conscious criteria. . . .

[Breyer's histories of the Seattle and Louisville plans omitted.]

C

The histories I have set forth describe the extensive and ongoing efforts of two school districts to bring about greater racial integration of their public schools. In both cases the efforts were in part remedial. Louisville began its integration efforts in earnest when a federal court in 1975 entered a school desegregation order. Seattle undertook its integration efforts in response to the filing of a federal lawsuit and as a result of its settlement of a segregation complaint filed with the federal OCR [Office of Civil Rights].

The plans in both Louisville and Seattle grow out of these earlier remedial efforts. Both districts faced problems that reflected initial periods of severe racial segregation, followed by such remedial efforts as busing, followed by evidence of resegregation, followed by a need to end busing and encourage the return of, *e.g.*, suburban students through increased student choice. When formulating the plans under review, both districts drew upon their considerable experience with earlier plans, having revised their policies periodically in light of that experience. Both districts rethought their methods over time and explored a wide range of other means, including non-race-conscious policies. Both districts also considered elaborate studies and consulted widely within their communities.

Both districts sought greater racial integration for educational and democratic, as well as for remedial, reasons. Both sought to achieve these objectives while preserving their commitment to other educational goals, *e.g.*, district wide commitment to high quality public schools, increased pupil assignment to neighborhood schools, diminished use of busing, greater student choice, reduced risk of white flight, and so forth. Consequently, the present plans expand student choice; they limit the burdens (including busing) that earlier plans had imposed upon students and their families; and they use race-conscious criteria in limited and gradually diminishing ways. In particular, they use race-conscious criteria only to mark the outer bounds of broad population-related ranges.

The histories also make clear the futility of looking simply to whether earlier school segregation was *de jure* or *de facto* in order to draw firm lines separating the constitutionally permissible from the constitutionally forbidden use of "race-conscious" criteria. JUSTICE THOMAS suggests that it will be easy to identify *de jure* segregation because "[i]n most cases, there either will or will not have been a state constitutional amendment, state statute, local ordinance, or local administrative policy explicitly requiring separation of the races.". . . But our precedent has recognized that *de jure* discrimination can be present even in the absence of racially explicit laws. . . .

No one here disputes that Louisville's segregation was *de jure*. But what about Seattle's? Was it *de facto*? *De jure*? A mixture? Opinions differed. Or is it that a prior federal court had not adjudicated the matter? Does that make a difference? Is Seattle free on remand to say that its schools were *de jure* segregated, just as in 1956 a memo for the School Board admitted? The plurality does not seem confident as to the answer. . . .

A court finding of *de jure* segregation cannot be the crucial variable. After all, a number of school districts in the South that the Government or private plaintiffs challenged as segregated *by law* voluntarily desegregated their schools *without a court order*—just as Seattle did. . . .

Moreover, Louisville's history makes clear that a community under a court order to desegregate might submit a race-conscious remedial plan *before* the court dissolved the order, but with every intention of following that plan even *after* dissolution. How could such a plan be lawful the day before dissolution but then become unlawful the very next day? On what legal ground can the majority rest its contrary view? . . .

Are courts really to treat as merely *de facto* segregated those school districts that avoided a federal order by voluntarily complying with *Brown*'s requirements? . . . This Court has previously done just the opposite, permitting a race-conscious remedy without any kind of court decree. . . . Because the Constitution emphatically does not forbid the use of race-conscious measures by districts in the South that voluntarily desegregated their schools, on what basis does the plurality claim that the law forbids Seattle to do the same? . . .

The histories also indicate the complexity of the tasks and the practical difficulties that local school boards face when they seek to achieve greater racial integration. The boards work in communities where demographic patterns change, where they must meet traditional learning goals, where they must attract and retain effective teachers, where they should (and will) take account of parents' views and maintain *their* commitment to public school education, where they must adapt to court intervention, where they must encourage voluntary student and parent action—where they will find that their own good faith, their knowledge, and their understanding of local circumstances are always necessary but often insufficient to solve the problems at hand.

These facts and circumstances help explain why in this context, as to means, the law often leaves legislatures, city councils, school boards, and voters with a broad range of choice, thereby giving "different communities" the opportunity to "try different solutions to common problems and gravitate toward those that prove most successful or seem to them best to suit their individual needs.". . .

With this factual background in mind, I turn to the legal question: Does the United States Constitution prohibit these school boards from using race-conscious criteria in the limited ways at issue here? . . .

[Section II, interpreting the legal standard for judging these cases, Section III, on applying the legal standard, and Section IV, on direct precedents, omitted.]

V

Consequences

The Founders meant the Constitution as a practical document that would transmit its basic values to future generations through principles that remained workable over time. Hence it is important to consider the potential consequences of the plurality's approach, as measured against the Constitution's objectives. To do so provides further reason to believe that the plurality's approach is legally unsound.

For one thing, consider the effect of the plurality's views on the parties before us and on similar school districts throughout the Nation. . . .

The districts' past and current plans are not unique. They resemble other plans, promulgated by hundreds of local school boards, which have attempted a variety of desegregation methods that have evolved over time in light of experience. . . .

A majority of these desegregation techniques explicitly considered a student's race. . . . Transfer plans, for example, allowed students to shift from a school in which they were in the racial majority to a school in which they would be in a racial minority. Some districts, such as Richmond, California, and Buffalo, New York, permitted only "one-way" transfers, in which only black students attending predominantly black schools were permitted to transfer to designated receiver schools. . . . Fifty-three of the 125 studied districts used transfers as a component of their plans. . . .

At the state level, 46 States and Puerto Rico have adopted policies that encourage or require local school districts to enact interdistrict or intradistrict open choice plans. Eight of those States condition approval of transfers to another school or district on whether the transfer will produce increased racial integration. Eleven other States require local boards to deny transfers that are not in compliance with the local school board's desegregation plans. . . .

At a minimum, the plurality's views would threaten a surge of race-based litigation. Hundreds of state and federal statutes and regulations use racial classifications for educational or other purposes. . . . In many such instances, the contentious force of legal challenges to these classifications, meritorious or not, would displace earlier calm.

The wide variety of different integration plans that school districts use throughout the Nation suggests that the problem of racial segregation in schools, including *de facto* segregation, is difficult to solve. The fact that many such plans have used explicitly racial criteria suggests that such criteria have an important, sometimes necessary, role to play. The fact that the controlling opinion would make a school district's use of such criteria often unlawful (and the plurality's "colorblind" view would make such use always unlawful) suggests that today's opinion will require setting aside the laws of several States and many local communities.

As I have pointed out, . . . *de facto* resegregation is on the rise. . . . It is reasonable to conclude that such resegregation can create serious educational, social, and civic problems. . . . Given the conditions in which school boards work to set policy, . . . they may need all of the means presently at their disposal to combat those problems. Yet the plurality would deprive them of at least one tool that some districts now consider vital—the limited use of broad race-conscious student population ranges.

I use the words "may need" here deliberately. The plurality, or at least those who follow JUSTICE THOMAS' " 'color-blind' " approach, . . . may feel confident that, to end invidious discrimination, one must end *all* governmental use of race-conscious criteria including those with inclusive objectives. . . . By way of contrast, I do not claim to know how best to stop harmful discrimination; how best to create a society that includes all Americans; how best to overcome our serious problems of increasing *de facto* segregation, troubled inner city schooling, and poverty correlated with race. But, as a judge, I do know that the Constitution does not authorize judges to dictate solutions to these problems. Rather, the Constitution creates a democratic political system through which the people themselves must together find answers. And it is for them to debate how best to educate the Nation's children and how best to administer America's schools to achieve that aim. The Court should leave them to their work. And it is for them to decide, to quote the plurality's slogan, whether the best "way to stop discrimination on the basis of race is to stop discriminating on the basis of race.". . . That is why the Equal Protection Clause outlaws invidious discrimination, but does not similarly forbid all use of race-conscious criteria.

Until today, this Court understood the Constitution as affording the people, acting through their elected representatives, freedom to select the use of "race-conscious" criteria from among their available options. . . . Today, however, the Court restricts (and some Members would eliminate) that leeway. I fear the consequences of doing so for the law, for the schools, for the democratic process, and for America's efforts to create, out of its diversity, one Nation.

VI

Conclusions

To show that the school assignment plans here meet the requirements of the Constitution, I have written at exceptional length. . . .

. . . But [my] conclusion is short: The plans before us satisfy the requirements of the Equal Protection Clause. And it is the plurality's opinion, not this dissent that "fails to ground the result it would reach in law." . . .

Four basic considerations have led me to this view. *First,* the histories of Louisville and Seattle reveal complex circumstances and a long tradition of conscientious efforts by local school boards to resist racial segregation in public schools. Segregation at the time of *Brown* gave way to expansive remedies that included busing, which in turn gave rise to fears of white flight and resegregation. For decades now, these school boards have considered and adopted and revised assignment plans that sought to rely less upon race, to emphasize greater student choice, and to improve the conditions of all schools for all students, no matter the color of their skin, no matter where they happen to reside. The plans under review—which are less burdensome, more egalitarian, and more effective than prior plans—continue in that tradition. And their history reveals school district goals whose remedial, educational, and democratic elements are inextricably intertwined each with the others. . . .

Second, since this Court's decision in *Brown,* the law has consistently and unequivocally approved of both voluntary and compulsory race-conscious measures to combat segregated schools. The Equal Protection Clause, ratified following the Civil War, has always distinguished in practice between state action that excludes and thereby subordinates racial minorities and state action that seeks to bring together people of all races. From *Swann* to *Grutter,* this Court's decisions have emphasized this distinction, recognizing that the fate of race relations in this country depends upon unity among our children, "for unless our children begin to learn together, there is little hope that our people will ever learn to live together." . . .

Third, the plans before us, subjected to rigorous judicial review, are supported by compelling state interests and are narrowly tailored to accomplish those goals. Just as diversity in higher education was deemed compelling in *Grutter,* diversity in public primary and secondary schools—where there is even more to gain—must be, *a fortiori,* a compelling state interest. Even apart from *Grutter,* five Members of this Court agree that "avoiding racial isolation" and "achiev[ing] a diverse student population" remain today compelling interests. . . .

Fourth, the plurality's approach risks serious harm to the law and for the Nation. Its view of the law rests either upon a denial of the distinction between exclusionary and inclusive use of race-conscious criteria in the context of the Equal Protection Clause, or upon such a rigid application of its "test" that the distinction loses practical significance. Consequently, the Court's decision today slows down and sets back the work of local school boards to bring about racially diverse schools. . . .

Indeed, the consequences of the approach the Court takes today are serious. Yesterday, the plans under review were lawful. Today, they are not. Yesterday, the citizens of this Nation could look for guidance to this Court's unanimous pronouncements concerning desegregation. Today, they cannot. Yesterday, school boards had available to them a full range of means to combat segregated schools. Today, they do not. . . .

* * *

Finally, what of the hope and promise of *Brown*? For much of this Nation's history, the races remained divided. It was not long ago that people of different races drank from separate fountains, rode on separate buses, and studied in separate schools. In this Court's finest hour, *Brown v. Board of Education* challenged this history and helped to change it. For *Brown* held out a promise. It was a promise embodied in three Amendments designed to make citizens of slaves. It was the promise of true racial equality—not as a matter of fine words on paper, but as a matter of everyday life in the Nation's cities and schools. It was about the nature of a democracy that must work for all Americans. It sought one law, one Nation, one people, not simply as a matter of legal principle but in terms of how we actually live.

Not everyone welcomed this Court's decision in *Brown*. Three years after that decision was handed down, the Governor of Arkansas ordered state militia to block the doors of a white schoolhouse so that black children could not enter. The President of the United States dispatched the 101st Airborne Division to Little Rock, Arkansas, and federal troops were needed to enforce a desegregation decree. . . . Today, almost 50 years later, attitudes toward race in this Nation have changed dramatically. Many parents, white and black alike, want their children to attend schools with children of different races. Indeed, the very school districts that once spurned integration now strive for it. The long history of their efforts reveals the complexities and difficulties they have faced. And in light of those challenges, they have asked us not to take from their hands the instruments they have used to rid their schools of racial segregation, instruments that they believe are needed to overcome the problems of cities divided by race and poverty. The plurality would decline their modest request.

The plurality is wrong to do so. The last half-century has witnessed great strides toward racial equality, but we have not yet realized the promise of *Brown*. To invalidate the plans under review is to threaten the promise of *Brown*. The plurality's position, I fear, would break that promise. This is a decision that the Court and the Nation will come to regret.

I must dissent.

SOURCE: Supreme Court of the United States. *Parents Involved in Community Schools v. Seattle School District No. 1*. 551 U.S. __ (2007). Docket 05-908. June 28, 2007. www.supremecourtus.gov/opinions/06pdf/05-908.pdf (accessed September 30, 2007).

OTHER HISTORIC DOCUMENTS OF INTEREST

FROM THIS VOLUME

- The 2008 presidential race, p. 296

FROM PREVIOUS *HISTORIC DOCUMENTS*

- Trial for 1964 slayings of three civil rights workers, *2005*, p. 353
- Trial in 1963 Birmingham church bombing, *2002*, 239

July

DOCUMENT IN CONTEXT

President Bush on Clemency for Former Aide in the CIA Leak Case

JULY 2, 2007

A federal jury on March 6, 2007, convicted Vice President Dick Cheney's former chief of staff, I. Lewis "Scooter" Libby, on four counts of perjury and obstruction of justice in a case related to the leaking of the name of a covert CIA official. Libby, who did not take the stand in his own defense, was sentenced to two and a half years in prison and a $250,000 fine. On July 2, after another court ruled that Libby would have to remain in prison while appealing his conviction, President George W. Bush granted Libby executive clemency, freeing him of the obligation to serve any prison time. The clemency order did not extend to the fine or affect the conviction itself; the former aide would remain on probation for two years.

The trial confirmed that the original sources for the published leak were former deputy secretary of state Richard Armitage and Karl Rove, Bush's chief political strategist. Neither man was indicted for his role in the complicated affair. Many of Libby's supporters argued that it was unfair to penalize him when he had not been the original source of the leak. In his clemency statement, Bush agreed that the prison penalty was "excessive" for a "first-time offender with years of exceptional public service." However, he added, a jury weighed the evidence and found Libby guilty of perjury. "[O]ur entire system of justice relies on people telling the truth," Bush said. "And if a person does not tell the truth, particularly if he serves in government and holds the public trust, he must be held accountable."

Background

The events leading to the case began in February 2002, when the CIA sent Joseph C. Wilson IV to Niger to determine the truth of information the agency had received indicating that Iraq had tried to buy raw uranium from the African country in the late 1990s to use in its alleged nuclear weapons development program. Wilson was a former U.S. diplomat who had served in Niger and still had high-level connections there; he was also married to Valerie Plame, an undercover official in the CIA's bureau that analyzed attempts by foreign governments to build or acquire so-called weapons of mass destruction. Wilson reported back to the CIA in March 2002 that documents linking Iraq to Niger were "bogus." The CIA sent a report of its findings to the White House, but even years later it remained unclear who there had read it. None of these events were made public at the time.

Nearly a year after the CIA filed its report, Bush used his 2003 State of the Union address to make his case for going to war against Iraq. A key rationale was to stop Iraq from trying to build weapons of mass destruction, including nuclear weapons. The key piece of evidence Bush offered to show that Iraq had such a program was a British intelligence report about the purported Iraq-Niger uranium connection. "The British government has learned that Saddam Hussein recently sought significant quantities of uranium from Africa," Bush said in his speech on January 28. He did not mention that the British government had made no effort to verify the information or that Wilson had determined the information to be false. The Bush administration also brushed aside a March 7, 2003, statement by Mohammed ElBaradei, the head of the International Atomic Energy Agency, saying that the documents showing the Iraq-Niger connection had been forged. Two weeks later, on March 20, the United States led the invasion of Iraq that toppled Saddam Hussein's government in three weeks and then maintained a large military presence there that at the end of 2007 showed no signs of ceasing.

The Iraq-Niger link did not resurface until some weeks after the invasion, when it was becoming increasingly apparent that Hussein had not been engaged in building nuclear weapons. In a column that appeared in the July 6 *New York Times,* Wilson denounced the administration's "twisted" use of intelligence on Iraq's alleged weapons to justify the invasion. Two days later, White House spokesman Ari Fleischer acknowledged that the Iraq-Niger claim should not have been included in the State of the Union address.

These events set the stage for the CIA leak case. On July 14, 2003, newspapers around the country carried a column by syndicated writer Robert Novak, a conservative ally of the administration, who revealed that two "senior administration officials" had told him that Wilson's mission to Niger had been recommended by his wife, Valerie Plame, a covert CIA operative. The Novak column created a furor in Washington, in part because it was widely seen as part of an administration campaign to discredit Wilson, who had emerged as a key critic of Bush's justification for going to war in Iraq. But the column also raised speculation over which officials had disclosed Plame's identity as a covert intelligence official. Such disclosures were a felony offense under a law adopted by Congress two decades earlier. (The law applied only to those who disclosed the information and not to the journalists who published the disclosures.)

At the request of the CIA, the Justice Department on September 30, 2003, announced that it was conducting an investigation into the leaking of Plame's name. On December 30, 2003, the Justice Department appointed Patrick J. Fitzgerald, a U.S. district attorney in Chicago, as a special counsel to investigate the possible commission of a crime in the leak.

After nearly two years of investigation, a federal grand jury in Washington indicted Libby on October 28, 2005, on one count of obstruction of justice, two counts of perjury, and two counts of making false statements to investigators. The indictment, which was accompanied by transcripts of Libby's testimony, accused him of lying to FBI agents during interviews on October 14 and November 26, 2004, about his statements to journalists about Plame; of committing perjury in his sworn testimony to the grand jury on March 5 and March 24, 2004, again in connection with his statements about what he did or did not say to reporters about Plame; and of obstructing justice by his statements and actions that impeded the grand jury's investigation into the case. Libby resigned just before the indictment was made public; he said through his attorney that he expected to be "completely and totally exonerated." He pleaded not guilty to all five counts at a November 3, 2005, arraignment.

Much of the news reporting in the wake of the indictment centered on the question of why Libby, in his statements to the FBI and the grand jury, had exposed himself by making allegedly misleading and erroneous statements about who said what to whom about Plame. In a report on November 13, the *Washington Post* quoted unidentified "critics" as saying that the timing of Libby's statements "suggests an attempt to obscure Cheney's role, and possibly his legal culpability." In particular, the *Post* noted, Libby's interviews with the FBI agents in October 2004 and testimony to the grand jury in March 20005 took place as it was becoming clear that reporters would likely be forced to testify—and would in some cases "provide starkly different and damning accounts to the prosecutor" of what White House officials had said.

There was no evidence in the public record that Cheney directly told reporters about Plame's identity. However, the indictment contained several references to Cheney's knowledge of Plame's role at the CIA and his possible role in developing the White House media strategy to rebut Wilson's statements about the Iraq-Niger connection. In particular, the indictment said Libby had learned from Cheney on June 12, 2003, that Wilson's wife worked at the CIA and that Cheney had "learned this information from the CIA." According to the indictment, Libby testified that he had forgotten that Cheney had told him about Plame and thought that he had learned of her identity from reporters nearly a month later.

Only two things of note happened in 2006 in connection with the case. In June Fitzgerald told Rove, according to Rove's attorney, that he did not expect to charge Rove with any crime in connection with the case. In September former deputy secretary of state Richard Armitage admitted that he had revealed Plame's identity to both Novak and *Washington Post* reporter Bob Woodward. Armitage said he did not realize the Plame's job was undercover.

The Trial

Libby's trial began on January 16, 2007, in Washington before U.S. district judge Reggie Walton. In methodical fashion, Fitzgerald laid out his case against Libby point by point, conversation by conversation. He began by displaying charts to the jury showing that Libby first learned of Plame's identity from federal administration officials, including Cheney, and had conversations with several reporters about her from early June through early July. He then played a tape of Libby's grand jury testimony in which Libby said he believed he first learned about Plame in a conversation with Tim Russert of NBC on July 10 and that he was surprised by the information. "You can't be startled about something on Thursday that you told other people about on Monday and Tuesday," Fitzgerald said.

Over the next few days, Fitzgerald called to the witness stand administration officials and reporters who testified that they had learned of Plame's identity directly from Libby days before he said he had learned it from Russert. Former White House press secretary Ari Fleischer, who, like Armitage, had been granted immunity in the case, testified that he had learned about Plame from Libby on July 7 and in turn discussed her with several reporters. Judith Miller, a former *New York Times* reporter, also testified that Libby had discussed Plame with her on June 23, 2003, and again on July 8, 2003. Miller, who never wrote a story about Plame, had spent eighty-five days in jail in 2005 for refusing to tell the grand jury who had leaked the information to her. She agreed to testify after Libby released her from her pledge of confidentiality.

The most damaging testimony, however, probably came from Russert, who said that Libby had called him to complain about a news story and that Plame never came up during the conversation. He told the jury that he first learned about Plame when he read Novak's July 14 column, four days after his conversation with Libby.

For its part, Libby's defense team portrayed the vice presidential aide as a busy man dealing with important matters who could not be expected to remember every conversation he had. The attorneys also sought to undermine the credibility of the key prosecution witnesses. After broadly hinting that they would put Libby, and possibly Cheney, on the witness stand, Libby's lawyers chose not to do so, leaving unrebutted many of the questions raised by the prosecution.

Verdict, Sentencing, and Clemency

After ten days of deliberations, the jury on March 6 returned guilty verdicts on four of the five counts against Libby. He was found not guilty on one count of lying to an FBI agent. On June 5 Judge Walton sentenced Libby to thirty months in prison and a $250,000 fine. "Evidence in this case overwhelmingly indicated Mr. Libby's culpability," Walton said. Later in the day Cheney released a statement praising Libby for his long public service and expressing hope that his conviction would be overturned.

A week later, on June 14, Walton denied a request from Libby to delay the start of his prison term until his appeal had been heard. Walton said he thought there was very little chance that Libby's conviction would be overturned, "and I just think blue-collar criminals are entitled to the same kind of justice as white-collar criminals." Libby appealed Walton's decision to the U.S. Circuit Court of Appeals for the District of Columbia Circuit, which announced on July 2 that it was denying the request. Five hours later, Bush commuted Libby's prison sentence while keeping the fine intact. "I respect the jury's verdict," Bush said in a statement. "But I have concluded that the prison term given to Mr. Libby is excessive."

On December 10, Libby announced through his attorneys that he was dropping his appeal. The statement said the attorneys believed in Libby's innocence but that the "burden on Mr. Libby and his young family of continuing to pursue his complete vindication are too great to ask them to bear."

Following is the text of a statement issued by President George W. Bush on July 2, 2007, commuting the prison sentence given to I. Lewis "Scooter" Libby for perjury and obstruction of justice in connection with a CIA leak case.

DOCUMENT

Bush Statement on Granting Executive Clemency to I. Lewis Libby

July 2, 2007

The United States Court of Appeals for the DC Circuit today rejected Lewis Libby's request to remain free on bail while pursuing his appeals for the serious convictions of perjury and obstruction of justice. As a result, Mr. Libby will be required to turn himself over to the Bureau of Prisons to begin serving his prison sentence.

I have said throughout this process that it would not be appropriate to comment or intervene in this case until Mr. Libby's appeals have been exhausted. But with the denial of bail being upheld and incarceration imminent, I believe it is now important to react to that decision.

From the very beginning of the investigation into the leaking of Valerie Plame's name, I made it clear to the White House staff and anyone serving in my administration that I expected full cooperation with the Justice Department. Dozens of White House staff and administration officials dutifully cooperated.

After the investigation was underway, the Justice Department appointed United States Attorney for the Northern District of Illinois Patrick Fitzgerald as a special counsel in charge of the case. Mr. Fitzgerald is a highly qualified, professional prosecutor who carried out his responsibilities as charged.

This case has generated significant commentary and debate. Critics of the investigation have argued that a special counsel should not have been appointed, nor should the investigation have been pursued after the Justice Department learned who leaked Ms. Plame's name to columnist Robert Novak. Furthermore, the critics point out that neither Mr. Libby nor anyone else has been charged with violating the Intelligence Identities Protection Act or the Espionage Act, which were the original subjects of the investigation. Finally, critics say the punishment does not fit the crime: Mr. Libby was a first-time offender with years of exceptional public service and was handed a harsh sentence based in part on allegations never presented to the jury.

Others point out that a jury of citizens weighed all the evidence and listened to all the testimony and found Mr. Libby guilty of perjury and obstructing justice. They argue, correctly, that our entire system of justice relies on people telling the truth. And if a person does not tell the truth, particularly if he serves in government and holds the public trust, he must be held accountable. They say that had Mr. Libby only told the truth, he would have never been indicted in the first place.

Both critics and defenders of this investigation have made important points. I have made my own evaluation. In preparing for the decision I am announcing today, I have carefully weighed these arguments and the circumstances surrounding this case.

Mr. Libby was sentenced to 30 months of prison, 2 years of probation, and a $250,000 fine. In making the sentencing decision, the district court rejected the advice of the probation office, which recommended a lesser sentence, and the consideration of factors that could have led to a sentence of home confinement or probation.

I respect the jury's verdict. But I have concluded that the prison sentence given to Mr. Libby is excessive. Therefore, I am commuting the portion of Mr. Libby's sentence that required him to spend 30 months in prison.

My decision to commute his prison sentence leaves in place a harsh punishment for Mr. Libby. The reputation he gained through his years of public service and professional work in the legal community is forever damaged. His wife and young children have also suffered immensely. He will remain on probation. The significant fines imposed by the judge will remain in effect. The consequences of his felony conviction on his former life as a lawyer, public servant, and private citizen will be long-lasting.

The Constitution gives the President the power of clemency to be used when he deems it to be warranted. It is my judgment that a commutation of the prison term in Mr. Libby's case is an appropriate exercise of this power.

SOURCE: U.S. Executive Office of the President. "Statement on Granting Executive Clemency to I. Lewis Libby." July 2, 2007. *Weekly Compilation of Presidential Documents* 43, no. 27 (July 9, 2007): 901-902. Washington, D.C.: National Archives and Records Administration. www.gpoaccess.gov/wcomp/v43no27.html (accessed March 20, 2008).

OTHER HISTORIC DOCUMENTS OF INTEREST

FROM THIS VOLUME

- The death of Gerald R. Ford, p. 3

FROM PREVIOUS *HISTORIC DOCUMENTS*

- Libby indictment, *2005*, p. 699

DOCUMENT IN CONTEXT

U.S. National Intelligence Estimate on Terrorist Threats

JULY 17, 2007

Six years after the September 11, 2001, attacks against the United States, the organization that sponsored those attacks—al Qaeda—remained a potent threat and was rebuilding its capabilities to strike the U.S. homeland again. This was the root assessment of several statements and reports issued by the U.S. government in mid-2007, including the unclassified summary of a new National Intelligence Estimate released by the Bush administration on July 17. President George W. Bush and his allies said the latest information proved the importance of continuing the fight against insurgents in Iraq, including militants associated with a branch of al Qaeda. The president's Democratic foes insisted the intelligence estimate, plus other evidence, demonstrated that the United States was losing ground against terrorism in large part because of its own actions, particularly the invasion and subsequent occupation of Iraq.

Experts on terrorism, including U.S. and foreign intelligence officials, said al Qaeda had survived and been able to rebuild despite six years of extensive action against it under the "global war on terrorism" that President Bush proclaimed after the September 11 attacks. By most estimates, al Qaeda remained a loose confederation of extremist groups around the world that were united by radical interpretations of Islam, inspired but probably not directed by al Qaeda leader Osama bin Laden, and energized by the continued U.S. occupation of Iraq.

Bin Laden appeared in a video broadcast on September 7, his first in nearly three years, reading a statement addressed to the American people. Unlike many of his previous statements, this one made no overt threats and instead focused on what he called the failure of U.S. policy in Iraq, where he said events had gotten "out of control" for the United States. Several audiotapes of bin Laden also were broadcast in late 2007; some appeared to be aimed at influencing public opinion in the United States and other Western countries about what he claimed was the futility of Washington's support for the governments of Afghanistan and Iraq.

Perhaps the major development on the terrorism front during 2007 was a year-long series of suicide bombings and other attacks in North Africa, particularly Algeria, by a group apparently affiliated with al Qaeda. Algeria experienced several dozen terrorism-related incidents during the year, including two major suicide bombings: one in April killed thirty-three people, and another in December killed at least thirty-seven and was aimed at regional offices of the United Nations (UN).

New Warnings of al Qaeda Resurgence

The Bush administration during 2007 appeared to be engaged in a campaign to warn Americans that al Qaeda was planning future attacks inside the United States. Two major steps in this effort were a speech given by President Bush on May 23 to the U.S. Coast Guard Academy and the release in July of the unclassified portion of a new estimate by U.S. intelligence agencies reviewing terrorist threats to the "homeland."

Ever since the attacks of September 11, Bush had often warned that al Qaeda and other terrorist networks were intent on attacking the United States directly, not just U.S. soldiers and diplomats in Iraq and other faraway places. But with much of the public growing increasingly weary of the war in Iraq—and with more than five years having passed since the September 11 attacks—administration officials reportedly decided that the time had come to step up the reminders that al Qaeda remained a danger.

In his high-profile speech to the Coast Guard academy, Bush cited intelligence information—which he had declassified for the occasion—that he said showed links between anti-U.S. insurgents in Iraq and the worldwide al Qaeda network. "Victory in Iraq is important for Osama bin Laden, and victory in Iraq is vital for the United States of America," he said. Bush also expressed concerns that Americans had become complacent about the threat of terrorism. "Here in America, we are living in the eye of a storm," he said. "All around us, dangerous winds are swirling and these winds could reach our shores at any moment."

In subsequent weeks, administration officials offered numerous warnings that direct attacks against the United States were possible. The warnings suddenly appeared more realistic in late June, when British police found, and defused, two bombs in cars parked in London. Soon after, a sports utility vehicle loaded with bombs was driven into the entrance of Glasgow airport, where it burst into flames but failed to explode. British authorities attributed the failed attacks to groups associated with al Qaeda.

Less than two weeks after the incidents in England and Scotland, U.S. officials increased the intensity of their warnings. First, on July 10, Michael Chertoff, secretary of the Homeland Security Department, said he had a "gut feeling" that the summer of 2007 would be one of heightened risk to the United States. Two days later, Bush returned to the theme of relating al Qaeda to the conflict in Iraq, saying at a White House conference that "al Qaeda wants to hurt us here." Moreover, he said, "the same folks that are bombing innocent people in Iraq were the ones who attacked us in America on September the 11th, and that's why what happens in Iraq matters to the security here at home."

National Intelligence Estimate

The administration's most definitive statement on al Qaeda came with the July17 release of an unclassified summary of a National Intelligence Estimate (NIE) called "The Terrorist Threat to the U.S. Homeland." Said to be the product of nearly two years of research and analysis by the nation's sixteen intelligence agencies, the NIE was the second in two years dealing with terrorism issues. The estimate released in September 2006 was broad in scope, covering the implications for the United States of overall trends in global terrorism. The July 2007 estimate, by contrast, focused specifically on threats to the U.S. "homeland," addressing the capabilities and intentions of terrorist groups, such as al Qaeda, to attack the United States directly.

The heart of the estimate was that al Qaeda had been able to regenerate some of what the intelligence community called the "core operational capabilities" needed to attack the United States again. One of these was having a safe haven, a place where al Qaeda leaders could be relatively secure. Al Qaeda had established such a safe haven in the rugged mountainous area just inside Pakistan, along the border with Afghanistan. Known as the Federally Administered Tribal Areas, this was a lawless region controlled by local tribes with little interference from the central Pakistani government.

A second core capability, the NIE said, involved generating new second-tier leadership, in some cases to replace operatives who had been killed or arrested by the United States and its allies. According to the report, al Qaeda had been able to maintain a cadre of skilled "operational lieutenants" who directed specific attacks. News reports during 2007 also called into question previous claims by U.S. and Pakistani officials that many high-level al Qaeda members had been killed in various U.S. antiterrorism efforts.

A third core capability, the NIE said, was that al Qaeda's top leadership—meaning Osama bin Laden and his chief deputy, Ayman al-Zawahiri—had remained intact. The unclassified version of the assessment did not discuss bin Laden and Zawahiri in any detail, but in other public statements U.S. intelligence officials said the two al Qaeda leaders did not run their network on a day-to-day basis. Instead, the leaders appeared to retain importance for their ability to inspire followers with messages issued via the Internet and on audio and video tapes.

The one core capability that al Qaeda had not yet fully restored was the positioning of operatives in the United States. The NIE reported that the government had discovered "only a handful of individuals in the United States with ties to al Qaeda's senior leadership" since the September 11 attacks, adding that the network "will intensify its efforts to put operatives here."

In terms of potential targets for terrorism, the NIE indicated that al Qaeda likely would continue to focus on "prominent political, economic, and infrastructure targets with the goal of producing mass casualties, visually dramatic destruction, significant economic aftershocks, and/or fear among the U.S. population." In other words, al Qaeda would like to replicate the effect of the September 11 attacks against the World Trade Center in New York City and the Pentagon outside Washington, D.C. However, the estimate did not suggest what al Qaeda's precise targets might be.

According to the NIE, the weapons used by al Qaeda included conventional small arms and what the U.S. military called "improvised explosive devices" (IEDs)—the roadside bombs used in Afghanistan and Iraq that had become increasingly sophisticated in recent years. In an indirect reference to al Qaeda's use of commercial airplanes as flying bombs on September 11, the estimate noted that the network was "innovative in creating new capabilities and overcoming security obstacles." Again, the estimate did not specify what some of al Qaeda's new capabilities might be, but it did add that the network "would not hesitate to use" biological, chemical, or nuclear material in its attacks if it had "sufficient capability."

The estimate also highlighted the prospect of an attack inside the United States by the Lebanese group Hezbollah. Over the years, U.S. officials had accused Hezbollah of sponsoring, and in some cases carrying out, such major attacks as the bombings of the U.S. embassy and U.S. marine barracks in Lebanon in 1983, but the Iranian-backed group had never been accused of attempting to stage operations within the United States. The estimate said Hezbollah "may be more likely to consider attacking the Homeland over the next three years if it perceives the United States as posing a direct threat to the group or Iran."

Finally, the estimate suggested that, through use of the Internet, radical Islamist groups would continue to spread their influence in Western countries such as the United States and especially European nations. It was possible, the report warned, that some Muslims in the United States "may become sufficiently radicalized that they will view the use of violence here as legitimate." U.S. officials noted that bin Laden's September "message to the American people" appeared to be aimed specifically at aggrieved Muslims within the United States.

Reactions to the intelligence estimate broke largely along party lines, particularly on Capitol Hill, where all national security matters were viewed in the context of the political struggle over Iraq. Democrats in both the House of Representative and the Senate argued that the intelligence findings served as a reminder that the Bush administration had allowed Iraq to become a "distraction" from the original fight against al Qaeda in Afghanistan. Republicans maintained just the opposite, insisting that Iraq was the focal point for the U.S.-led war against terrorism.

CIA Report on Pre-September 11 Operations

Yet more evidence emerged in 2007 that the U.S. government had failed to do everything possible to prevent the September 11 attacks. At the insistence of Democrats in Congress, the administration on August 21 released the summary of a report by the CIA's inspector general, which indicated that the agency had failed to develop a "documented, comprehensive approach" to battling al Qaeda even after then CIA director George Tenet declared in December 1998 that "we are at war" with the terrorists.

The report by Inspector General John Helgerson was completed in June 2005 but kept secret until Congress passed legislation requiring it to be released. Helgerson's report suggested that the CIA's efforts against al Qaeda, even after the group launched major attacks against the United States in the 1990s, were fragmented and ad hoc. Tenet, who left the CIA in 2004, released a statement calling the inspector general's report "flat wrong" and insisting that the agency did have a "robust plan" to fight al Qaeda.

No Link between al Qaeda and Saddam Hussein

In another piece of old business, an internal investigation by the Pentagon's inspector general appeared to provide conclusive evidence refuting a claim made by President Bush, Vice President Dick Cheney, and other senior officials during 2002 and early 2003 that Iraqi leader Saddam Hussein was somehow connected to the al Qaeda network. This alleged connection, along with the allegation that Iraq was developing nuclear weapons, was one of Bush's principal justifications for the March 2003 invasion of Iraq.

In a report released on April 5, Inspector General Thomas Gimble said unreliable claims about the Iraq–al Qaeda link had been developed in 2002–2003 by a special intelligence-gathering arm established by Douglas J. Feith, the Pentagon's undersecretary for policy at the time. Gimble's report documented senior administration officials' use of faulty assessments developed by Feith's office, including Cheney's numerous public claims of a direct link between Hussein and al Qaeda. One official who did not seem impressed by Gimble's report was Cheney, who told conservative radio talk show host Rush Limbaugh on April 6 that al Qaeda was "present [in Iraq] before we invaded Iraq."

Attacks in Algeria

Apart from Afghanistan and Iraq—and a border area of Israel subjected to near-daily rocket attacks from Gaza—the place that endured the greatest level of terrorism during the year was Algeria. Starting in February, a series of at least three dozen suicide bombings and car bombings targeted police and army units, government officials, civilians, and even local representatives of the United Nations.

Algerian officials and international experts on terrorism said the attacks appeared to be carried out by groups associated with al Qaeda, and in some cases reportedly by militants who had served in Iraq or been trained by veterans of the Iraqi insurgency. Responsibility for several of the attacks was claimed by a group that called itself al Qaeda in the Islamic Maghreb, which until its incorporation into the broader al Qaeda network in 2006 was known as the Salafist Group for Call and Combat. The Salafist Group was itself a splinter faction of the older Armed Islamic Group, which had fought a long, bloody, and ultimately unsuccessful war against the Algerian government during the 1990s.

The first major attack in Algeria occurred on February 13, when four civilians and two police officers were killed in the simultaneous bombing of seven police stations. On April 11, the death toll was much higher: thirty-three people were killed, and more than 200 wounded, in two suicide bombings, one at the building housing the prime minister's office in Algiers and the other at a suburban police station. These were the deadliest terrorist attacks in Algeria since 2002, when a bombing of a market killed thirty-eight people.

Two apparently related attacks on September 6 and September 8 killed nearly fifty people. In the first case, a suicide bomber attacked a crowd of people waiting in the eastern city of Batna for a visit by President Abdelaziz Bouteflika; most of the twenty victims were civilians. A car bombing two days later killed twenty-eight Algerian coast guard officers at a barracks in the coastal town of Dellys.

Another high-profile attack came on December 11, when two car bombings outside UN and Algerian government offices killed more than thirty-seven people, including seventeen UN employees, and wounded dozens more. The attacks on the North African regional offices of the UN High Commissioner for Refugees and the UN Development Programme were the deadliest against the world body since a truck-bombing of the UN mission headquarters in Baghdad, Iraq, in August 2003; that attack killed twenty-three people, including nineteen UN staffers.

José Padilla Conviction

The Bush administration achieved a major legal victory on August 16 when a federal jury in Miami, Florida, convicted José Padilla on charges of conspiracy to commit terrorism. Padilla had been arrested in May 2002 at O'Hare International Airport in Chicago and taken into military custody. Attorney General John Ashcroft said at the time that Padilla had been at the center of a "terrorist plot to attack the United States by exploding a radioactive dirty bomb." The military held Padilla in isolation in a brig in South Carolina for more than three years but transferred him to the civilian court system in November 2002 after the Supreme Court indicated it was prepared to consider a habeas corpus petition brought by his lawyers. The government subsequently charged Padilla and two co-defendants with conspiring to kidnap, maim, and murder people outside the United States. Ashcroft's allegation that Padilla was plotting a radioactive bombing in the United States was never brought against him.

Padilla was convicted after a lengthy trial during which prosecutors presented thousands of telephone conversations and other evidence of the alleged conspiracy, while defense lawyers insisted Padilla and his co-defendants were religious Muslims who did not advocate or support terrorism. A sentencing hearing scheduled for early December was postponed until January 2008.

Following are the substantive sections from the unclassified summary of a National Intelligence Estimate, "The Terrorist Threat to the U.S. Homeland," released by the director of national intelligence on July 17, 2007.

DOCUMENT

National Intelligence Estimate on Terrorist Threats to the U.S. Homeland

July 17, 2007

The US Homeland Threat Estimate: How It Was Produced

The Estimate, *Terrorist Threats to the US Homeland,* followed the standard process for producing National Intelligence Estimates (NIEs), including a thorough review of sourcing, in-depth Community coordination, the use of alternative analysis, and review by outside experts. Starting in October 2006, the NIC [National Intelligence Council] organized a series of roundtables with IC [intelligence community] experts to scope out terms of reference (TOR) for the Estimate. Drafters from throughout the Community contributed to the draft. In May, a draft was submitted to IC officers in advance of a series of coordination meetings that spanned several days. The National Clandestine Service, FBI, and other IC collection officers reviewed the text for the reliability and proper use of the sourcing. As part of the normal coordination process, analysts had the opportunity—and were encouraged—to register "dissents" and provide alternative analysis. Reactions by the two outside experts who read the final product were highlighted in the text. The National Intelligence Board, composed of the heads of the 16 IC agencies and chaired by the ODNI [Office of the Director of National Intelligence], reviewed and approved the Estimate on 21 June. As with other NIEs, it is being distributed to senior Administration officials and Members of Congress. . . .

Key Judgments

We judge the US Homeland will face a persistent and evolving terrorist threat over the next three years. The main threat comes from Islamic terrorist groups and cells, especially al-Qa'ida, driven by their undiminished intent to attack the Homeland and a continued effort by these terrorist groups to adapt and improve their capabilities.

We assess that greatly increased worldwide counterterrorism efforts over the past five years have constrained the ability of al-Qa'ida to attack the US Homeland again and have led terrorist groups to perceive the Homeland as a harder target to strike than on 9/11. These measures have helped disrupt known plots against the United States since 9/11.

- We are concerned, however, that this level of international cooperation may wane as 9/11 becomes a more distant memory and perceptions of the threat diverge.

Al-Qa'ida is and will remain the most serious terrorist threat to the Homeland, as its central leadership continues to plan high-impact plots, while pushing others in extremist Sunni communities to mimic its efforts and to supplement its capabilities. We assess the group has protected or regenerated key elements of its Homeland attack capability, including: a safehaven in the Pakistan Federally Administered Tribal Areas (FATA), operational lieutenants, and its top leadership. Although we have discovered only a handful of individuals in the United States with ties to al-Qa'ida senior leadership since 9/11, we judge that al-Qa'ida will intensify its efforts to put operatives here.

- As a result, we judge that the United States currently is in a heightened threat environment.

We assess that al-Qa'ida will continue to enhance its capabilities to attack the Homeland through greater cooperation with regional terrorist groups. Of note, we assess that al-Qa'ida will probably seek to leverage the contacts and capabilities of al-Qa'ida in Iraq (AQI), its most visible and capable affiliate and the only one known to have expressed a desire to attack the Homeland. In addition, we assess that its association with AQI helps al-Qa'ida to energize the broader Sunni extremist community, raise resources, and to recruit and indoctrinate operatives, including for Homeland attacks.

We assess that al-Qa'ida's Homeland plotting is likely to continue to focus on prominent political, economic, and infrastructure targets with the goal of producing mass casualties, visually dramatic destruction, significant economic aftershocks, and/or fear among the US population. The group is proficient with conventional small arms and improvised explosive devices, and is innovative in creating new capabilities and overcoming security obstacles.

- We assess that al-Qa'ida will continue to try to acquire and employ chemical, biological, radiological, or nuclear material in attacks and would not hesitate to use them if it develops what it deems is sufficient capability.

We assess Lebanese Hizballah, which has conducted anti-US attacks outside the United States in the past, may be more likely to consider attacking the Homeland over the next three years if it perceives the United States as posing a direct threat to the group or Iran.

We assess that the spread of radical—especially Salafi—Internet sites, increasingly aggressive anti-US rhetoric and actions, and the growing number of radical, self-generating cells in Western countries indicate that the radical and violent segment of the West's Muslim population is expanding, including in the United States. The arrest and prosecution by US law enforcement of a small number of violent Islamic extremists inside the United States—who are becoming more connected ideologically, virtually, and/or in a physical sense to the global extremist movement—points to the possibility that others may become sufficiently radicalized that they will view the use of violence here as legitimate. We assess that this internal Muslim terrorist threat is not likely to be as severe as it is in Europe, however.

We assess that other, non-Muslim terrorist groups—often referred to as "single-issue" groups by the FBI—probably will conduct attacks over the next three years given their violent histories, but we assess this violence is likely to be on a small scale.

We assess that globalization trends and recent technological advances will continue to enable even small numbers of alienated people to find and connect with one another, justify and intensify their anger, and mobilize resources to attack—all without requiring a centralized terrorist organization, training camp, or leader.

- The ability to detect broader and more diverse terrorist plotting in this environment will challenge current US defensive efforts and the tools we use to detect and disrupt plots. It will also require greater understanding of how suspect activities at the local level relate to strategic threat information and how best to identify indicators of terrorist activity in the midst of legitimate interactions.

SOURCE: U.S. Director of National Intelligence. National Intelligence Council. "The Terrorist Threat to the US Homeland." National Intelligence Estimate. July 17, 2007. www.dni.gov/press_releases/20070717_release.pdf (accessed January 15, 2008).

OTHER HISTORIC DOCUMENTS OF INTEREST

FROM THIS VOLUME

- U.S. policy in Iraq, p. 9
- Afghanistan-Pakistan "peace jirga" communiqué, p. 442
- UN secretary-general on the situation in Afghanistan, p. 547
- Political crisis in Lebanon, p. 624
- Political situation in Pakistan, p. 647

FROM PREVIOUS *HISTORIC DOCUMENTS*

- U.S. National Intelligence Estimate on foreign terrorism, *2006*, p. 574

DOCUMENT IN CONTEXT

United Nations on the Political Situation and Violence in Somalia

JULY 18 AND NOVEMBER 7, 2007

Somalia descended into near total chaos in 2007, creating what United Nations officials called the most pressing humanitarian crisis in the world—worse even than the better publicized agony in the Darfur region of Sudan. Mogadishu, Somalia's seaside capital, was convulsed with violence for much of the year. Some 600,000 of the city's residents fled their homes, and a local human rights group counted 6,500 dead.

In theory, Somalia in 2007 was ruled by a fragile, internationally backed "transitional" government, which had been created by a regional diplomatic conference in 2004. In truth, that government controlled little territory and was heavily dependent on troops from Ethiopia. The government and the Ethiopians struggled all year to fend off an aggressive insurgency consisting of clan warlords and the remnants of an Islamist faction that had controlled most of Somalia for the last half of 2006. Although the Islamists had brought relative piece to Somalia for the first time in years, they posed a threat to neighboring Christian-majority Ethiopia, which sent in its army to oust them from power in December 2006. By the middle of 2007, some international observers said Ethiopia had gotten itself into a "quagmire" of occupation in Mogadishu and southern Somalia, with its only realistic hope being to win a prolonged war of attrition against the insurgents—similar in some respects to the war the United States was waging against insurgents in Iraq.

Security Situation

Somalia had rarely been a safe place since its long-time dictator, Mohamed Siad Barre, was ousted from power by warlords in 1991. But despite—or, more likely, because of—the presence of thousands of Ethiopian troops, who constituted the transitional government's only real security muscle, the violence that engulfed the capital and much of the country in 2007 was exceptionally intense even by Somali standards.

When Ethiopia invaded Somalia in December 2006, reportedly with the approval of and intelligence support from the United States, it succeeded in dispersing the recently installed Islamist regime known as the Islamic Courts Council. However, Ethiopia did not fully defeat the Islamists, whose fighters regrouped during 2007 and joined forces in Mogadishu with important clan warlords who were opposed to the transitional government that was beholden to Ethiopia's intervention.

The situation in Somalia was further complicated by another international component: the U.S. government in 2007 saw Somalia as the latest battleground in its "war" against the Islamist al Qaeda network. Washington had accused the Islamists who had briefly ruled Somalia in 2006 of harboring al Qaeda operatives, including several suspected of plotting the 1998 bombings of U.S. embassies in Kenya and Tanzania. Twice in January 2007, U.S. Special Forces attacked suspected al Qaeda fighters along the Somali coast; these air attacks represented the first known U.S. military intervention in Somalia since President Bill Clinton withdrew the American members of an international peacekeeping force there in 1994.

For much of 2007, Mogadishu was afflicted by nearly constant battles, rivaling Baghdad as the world's most violent city. On one side were Ethiopian troops and the militias of warlords who supported the transitional government. Opposed to them were various insurgent groups consisting of fighters from the former Islamic Courts regime and the militias of warlords who opposed the Ethiopian-backed government.

Early in the year, the insurgents conducted hit-and-run attacks in Mogadishu and surrounding areas. These attacks escalated into widespread violence in the capital starting on March 21, when Ethiopian troops entered an insurgent stronghold and were confronted by hundreds of fighters. In one incident, mobs killed two Ethiopian soldiers, mutilated their bodies, and dragged them through the streets—mimicking a famous incident in 1993 when mobs attacked U.S. peacekeepers.

Subsequent battles in the city featured Ethiopian tanks and helicopters blasting away at neighborhoods controlled by insurgents, and the insurgents returning fire with mortars and rocket-propelled grenades. In late March, the International Committee of the Red Cross described the fighting as the most intense Mogadishu had endured since 1991.

Civilians often were caught in the crossfire, and during the year's six weeks of heaviest combat in March and April, an estimated 1,000 people were killed. The UN said that at least 400,000 people fled from their Mogadishu homes between February and April—some simply sought to escape the fighting but many others were ordered to leave their neighborhoods by one side or the other. This round of fighting continued with only brief interruptions until April 27, when the Ethiopian army prevailed and forced the guerrillas to withdraw. But as had been the case at the end of 2006 when the Islamic Courts Council was ousted from power, the insurgents were dispersed, not defeated.

The fighting was so severe that Human Rights Watch, in an August 13 report, accused both sides of having committed war crimes. "The warring parties have all shown criminal disregard for the well-being of the civilian population of Mogadishu," said Human Rights Watch executive director Kenneth Roth.

Fighting escalated again in late October and continued through November, forcing another 200,000-some Mogadishu residents from their homes—many of whom had only recently returned home after the March–April violence. During this round of fighting, the UN ordered the withdrawal of all staff from Mogadishu; elsewhere in Somalia, UN staff were permitted to engage only in emergency operations. On December 31, a Somali group, the Elman Peace and Human Rights Organization, said it had counted 6,500 dead in the year's clashes in Mogadishu.

International Response

The violence in Somalia put new pressure on the international actors that normally would be expected to intervene in some way, notably the African Union and the UN. Both organizations, however, already were stretched by major peacekeeping missions in the Democratic Republic of the Congo and the Darfur region of Sudan. Humanitarian agencies also reported that donor nations were slow to respond to requests for money for

Somalia—an apparent manifestation of "donor fatigue" resulting from repeated appeals on behalf of chronically troubled countries.

On December 6, 2006, just before the Ethiopian invasion, the UN Security Council adopted Resolution 1725, authorizing a "peace support" mission of African troops to protect the Somali transitional government, which at the time was holed up in the town of Baidoa (south of Mogadishu) and under siege by militias from the Islamic Courts Council. Ethiopia's subsequent intervention to oust the Islamist regime changed the dynamic, leading UN officials to appeal for troops to fill an anticipated security vacuum when the Ethiopian troops were withdrawn. The Security Council on February 20, 2007, adopted Resolution 1744, again approving deployment of a peacekeeping force of as many as 8,000 troops organized by the African Union.

The first elements of the new peace mission—400 soldiers from Uganda—did not arrive in Somalia until March 6, 2007. By late March the African Union mission had reached just 1,600 soldiers, all from Uganda; other countries promised troops but did not deliver them, citing lack of money or equipment. Based at or near the Mogadishu airport, the Ugandans came under increasing attack from insurgents and, as a result, limited their operations to patrolling parts of Mogadishu and aiding the delivery of relief supplies.

The Security Council reauthorized the African force on August 20, in Resolution 1772. By this time, the African Union had asked the UN to bolster the mission, starting with equipment from other countries and possibly expanding to a full-scale UN peacekeeping mission. The Security Council asked UN secretary-general Ban Ki-moon to prepare contingency plans for such a mission, but in a November 7 report to the council, Ban said a UN force "cannot be considered a realistic and viable option" because of the anarchy in Somalia. "Given the complex security situation in Somalia, it may be advisable to look at additional security options, including the deployment of a robust multinational force or coalition of the willing," he said. The latter phrase echoed the Bush administration's description of countries that supported the U.S. invasion of Iraq in 2003.

While agreeing with Ban's assessment of Somalia's ills, the Security Council in late November and again in mid-December repeated its request for him to prepare a contingency plan for a UN peacekeeping force. Ban was working on his plan at year's end.

A Surplus of Weapons

As with many conflicts in developing countries, a significant factor in Somalia was the presence of enormous quantities of weapons, ranging from automatic rifles to mortars and rocket-propelled grenades. The UN Security Council in 1992 imposed an arms embargo against Somalia, but it apparently had little effect. The *New York Times* reported in January 2007 that the surplus of guns had reached the point that a Kalashnikov assault rifle—the main weapon used by militias and rebel forces worldwide—could be bought for $15. Weapons prices reportedly rose later in the year.

More formal confirmation of the weapons surplus came July 17 in a report from a committee appointed by the Security Council to monitor the arms embargo. The panel reported that Somalia "is literally awash in arms." In fact, the committee said, more weapons were pouring in than at any point since the early 1990s, when chaos first engulfed Somalia.

The panel traced several arms shipments from Eritrea to the Islamic Courts forces in Somalia during 2006 and 2007, constituting what the report called "a clear pattern of involvement by the government of Eritrea in arms embargo violations." Further, the panel said Eritrea had made "deliberate attempts to hide its activities and mislead the international community about its involvement." The Eritrean government denied the panel's charges. Eritrea's involvement in Somalia could be traced, at least in part, to the historic enmity between Eritrea and Ethiopia. Eritrea had been part of Ethiopia until it broke away in the 1980s, leading to a long and bloody war that ended only in 2000.

Political Situation

Until the Islamic Courts Council gained control over most of Somalia in mid-2006, the country essentially had been without a functioning government since the ousting of Siad Barre in 1991. A regional conference had appointed the "transitional government" in 2004—it was the fourteenth attempt to establish such a government since Barre's downfall—but many of the key leaders lived outside the country and the government exercised no real power until after Ethiopia's invasion in late 2006.

The government, most of which was based south of Mogadishu in the town of Baidoa, was headed by former warlord Abdullahi Yusuf Ahmed, who served as president. Hoping to unite many, if not all, of Somalia's numerous clans behind it, the transitional government early in 2007 turned for support to the major warlords, each of whom represented a clan or subclan grouping. As part of this effort, the government named several key warlords to high government posts, such as mayor of Mogadishu and chief of the national police. One important clan that appeared to be left out of this arrangement was the Ayr (itself a subclan of a larger clan). Many leaders of the Islamic Courts Council were Ayr, and transitional government officials charged that the clan was backing the Islamist guerrillas.

Although it exercised little real power, the government was subject to bitter internal disputes and underwent two major shakeups during the year that dumped the speaker of parliament and the prime minister. In a step that international officials said represented the government's best chance to broaden its support, President Yusuf on November 22 selected Nur Hassan Hussein, a respected law enforcement official and head of the country's Red Crescent Society, to be the new prime minister. The parliament quickly confirmed Hussein (also known as Nur Adde), who was sworn into office two days later. However, another political crisis erupted after Hussein ousted most clan leaders and warlords from the cabinet; the government remained essentially paralyzed at year's end.

Because the government appeared to have little popular support, international diplomats pressed for broad negotiations among all of Somalia's numerous factions—even including moderate representatives of the ousted Islamists. The government sponsored such "national reconciliation" talks during July and August, hoping to settle the longtime disputes among clans that had been at the heart of Somalia's anarchy. The six-week conference ended on August 31 with agreements on a cease-fire and other measures, including the holding of national elections in 2009. Even as the talks were concluding, however, insurgents mounted new attacks in Mogadishu. One week later, leaders of the Islamic Courts and other opposition groups met in the Eritrean capital, Asmara, to form the Alliance for the Liberation of Somalia with the goal of continuing the "military struggle" against the Ethiopian occupation.

Humanitarian Situation

More than fifteen years of war, anarchy, famine, drought, floods, and other disasters had turned most of Somalia into a perpetual zone of humanitarian crisis. Tens of thousands of Somalis had fled the country into neighboring Kenya and Ethiopia, and even across the Indian Ocean into Yemen. By 2007 more than 400,000 people were considered to be long-term "displaced" (and in permanent need of food and other aid) because their homes had been destroyed or their neighborhoods remained unsafe.

The violence of 2007 deepened the misery in Somalia, especially in Mogadishu, its capital and by far its largest city with an estimated population of about 1 million. The prolonged spasms of battle in 2007 drove about 600,000 people—more than half of the current residents—out of the city, bringing the country's total displaced population to about 1 million. At year's end the UN estimated that about 230,000 former city residents were now living along a fifteen-mile stretch of road between Mogadishu and the neighboring town of Afgooye. John Holmes, the UN under secretary general for humanitarian affairs, called it "probably the single largest gathering of internally displaced persons in the world." Most of these people lacked shelter and were using plastic bags and rags, lashed to sticks, to try to protect themselves from the elements.

In addition to conflict, about 1 million people in central and southern Somalia had been affected in late 2006 and early 2007 by severe flooding, said to be the worst in a half-century. Nearly half of those affected were forced from their homes at least temporarily, the UN said. The flood waters receded by February but left behind thousands of damaged homes and hundreds of people sickened by diarrhea and other diseases caused by contaminated water.

International agencies were facing their own difficulties getting food and other supplies to the neediest. Most food shipments arrived by sea, but piracy was so severe that shippers were reluctant to allow their vessels to travel to Somalia, reducing by half the number of available ships, according to the UN. Also, local warlords frequently demanded payment of "taxes" by aid agencies before allowing the delivery of food and other relief supplies.

Following are two documents. First, excerpts from a report, dated July 18, 2007, to the United Nations Security Council from the monitoring group established in 1992 to oversee an international arms embargo against all factions in Somalia. Second, excerpts from a quarterly report to the Security Council by Secretary-General Ban Ki-moon on the situation in Somalia.

DOCUMENT

Report to the Security Council by the Monitoring Group on Somalia

July 18, 2007

VII. Conclusions and Recommendations

A. Conclusions

Arms

109. It is the view of the Monitoring Group that the sheer numbers of arms currently in Somalia (central and southern Somalia, in particular) exceed those in the country since the early 1990s. There are at least three major groupings of arms: those in the possession of the Ethiopian forces, the Transitional Federal Government and the African Union-Ugandan military contingent in Mogadishu; those in the possession of the Shabaab [the armed faction of the Council of Islamic Courts, which briefly ruled Somalia in late 2006], including their weapons caches; and those in the possession of the warlords and clans.

110. Far from being stemmed by the presence and activities of the Ethiopians, the Transitional Federal Government and African Union military forces, arms continued to flow heavily into Somalia from November 2006 to mid-June 2007, prior to the preparation and submission to the Secretariat of the current report. Arms have been either openly brought into Somalia, as in the case of the Ethiopian and African Union forces, or brought in through clandestine channels, as in the cases of the Eritrean conduit and the Bakaraaha Arms Market [a large open-air market in central Mogadishu], for example, and variously distributed to the Shabaab, clans, warlords and others. While acknowledging the African Union exemption, it is the view of the Monitoring Group that the majority of the arms presently in Somalia were delivered or introduced into the Somali environment in violation of the arms embargo.

111. Moreover, importantly, no single actor or authority is in control of the majority of the arms in Somalia. That is noted in spite of Transitional Federal Government attempts to establish its authority and control over other key Somali actors.

112. In brief, Somalia is awash with arms. The current Somali environment contains more than at any time since the early 1990s. They were variously delivered and introduced into the Somali environment by different actors and continue to be held by a variety of important and potentially militarily powerful actors, and there is no clearly established authority that has the capability of exercising control over a majority of the arms. Furthermore, as of the submission of the present report, the foregoing is taking place in a context of persistent insecurity, a low-grade and deadly insurgency being waged by an undefeated militant group, the Shabaab, and increasingly disaffected clans and warlords, all of whom are trying to push back the clock and re-establish or continue their respective activities independent of the existence of the Transitional Federal Government and the presence of the Ethiopian and African Union military forces.

Finance

113. Owing to the division in the business community caused by the defeat of ICU [Islamic Courts Union], the military opposition to the Transitional Federal Government has experienced a decrease in funds from inside Somalia and depends more on finances from individuals in foreign countries.

114. Control over traditional revenue generators has shifted again to the Transitional Federal Government and the warlords, who benefit from those revenues. However, since the country is experiencing a war economy whereby financial efforts are concentrated in the military, very little space is left for businesses to flourish and pay taxes.

115. A combination of high tax rates, unsafe conditions, lack of confidence, fewer trade flows, fewer business transactions, high prices, inflation surge and decreased incomes have created a "fiscal reduction", resulting in less money to pay taxes.

Transport

116. The much heralded victory of the Transitional Federal Government forces, supported by the Ethiopians, over ICU has not yielded the stability that was anticipated. Instead, there is frustration and disappointment. General living conditions in Somalia have deteriorated and retaliation against the Government has increased greatly.

117. Whatever little confidence there was in the ability of the Transitional Federal Government to rule is fast eroding and antagonism against Ethiopia is at a crescendo—clearly not being helped by the Ethiopian Army's heavy-handed response to insurgent attacks, involving the use of disproportionate force to dislodge insurgents from their suspected hideouts.

118. The attacks on aircraft, acts of piracy and attempted hijackings of maritime vessels have a dampening effect on the confidence of trading companies to do business with Somalia, resulting in the scarcity of essential food items and medical supplies. That has also affected the delivery of much-needed humanitarian aid as ship owners are increasingly unwilling to venture into Somali waters.

119. There is no doubt that the increase in piracy attacks is caused by the climate of lawlessness that currently prevails on the mainland of Somalia, providing sanctuary and allowing the "lords of piracy" to carry out their operations unhindered.

B. Recommendations

Arms

120. The Transitional Federal Government, with regional and international assistance as noted in the present report, is in the process of attempting to gain control over Somalia and establish itself as a viable Government. The Monitoring Group recommends that the Transitional Federal Government consider taking the following actions with respect to gaining control over the rampant arms problem:

(a) Institute a formal programme, or continue with existing efforts to collect and destroy or register all weapons in areas under its control;

(b) Eliminate the Bakaraaha Arms Market;

(c) Establish a professional police force that, through community policing efforts and other organized information-gathering activities, makes it a priority to locate and appropriately deal with hidden arms caches.

Finance

121. The Transitional Federal Government should undertake combined efforts to allow economic growth and minimize threats, to include: improving overall security conditions; removing illegal tax collections at checkpoints and elsewhere; and promoting a climate of confidence within the business community. Adequate market policies, inflation reduction, fiscal consolidation and a regulated financial system are also priorities.

122. With regard to financial threats, it is essential to strengthen controls on inbound money flows physically or electronically entering the country, therefore curbing funds that may finance conflict escalation.

Transport

123. To curb the flow of unchecked imports into the country, consideration should be given to seeking the services of a reputable international inspection and verification company to monitor all imports into Somalia. The company would work in cooperation with the Somali Customs Authority and would be deployed at all harbours and airports. That system would also bring accountability to the revenues collected at all ports.

124. The Transitional Federal Government authorities should be encouraged to embrace the International Ship and Port Facility Security Code, developed by the International Maritime Organization, to enhance maritime security in ports, which would help in curbing the illegal arms trade.

125. Any peacekeeping mission to Somalia should include an element of maritime forces to enable it to effectively control and provide security for the long, remote Somali coastline.

126. To bring the dhow trade into some semblance of order, regional States should be encouraged to develop regulations for cargo ships and passenger-carrying vessels not covered by the provisions of international maritime conventions. That would serve to regulate trade by traditional vessels, including dhows, plying the waters off Somalia. It would be done under the auspices of the International Maritime Organization, which has already assisted other regions in developing model legislation along similar lines.

SOURCE: United Nations. Security Council. "Report of the Monitoring Group on Somalia Pursuant to Security Council Resolution 1724" (2006). S/2007/436. July 18, 2007. www.un.org/Docs/journal/asp/ws.asp?m=S/2007/436 (accessed December 12, 2007).

DOCUMENT

Report of the Secretary-General on the Situation in Somalia

November 7, 2007

VIII. OBSERVATIONS

81. As I mentioned in my letter to the Security Council of 20 September 2007 (S/2007/566), the United Nations is elaborating a two-track approach for Somalia based

on: (a) a political track to encourage dialogue within the Transitional Federal Government itself and with all opposition groups both inside and outside Somalia, in the hope of bringing about a cessation of hostilities and the establishment of broad-based and inclusive transitional institutions; and (b) a security track that would necessitate the strengthening of AMISOM [the African Union Mission in Somalia] to a level that would allow for the withdrawal of foreign forces and create the necessary conditions for stability. In addition, I also recommended the strengthening of the United Nations Political Office for Somalia by providing it with the necessary resources to implement the two-pronged approach.

82. In pursuit of the two-pronged approach, the United Nations system, including the United Nations Political Office for Somalia and the United Nations country team, has also embarked on developing a coherent peacebuilding strategy for Somalia. In support of these efforts, an interdepartmental and inter-agency integrated task force on Somalia is being established.

83. Following a series of consultations held by my Special Representative in Nairobi, I am pleased to note the support voiced by members of the international community for the Special Representative of the Secretary-General Ould Abdallah and the United Nations Political Office for Somalia to play a leadership role in coordinating the efforts of all stakeholders.

84. I welcome the amicable resolution of the divisions between President [Abdullahi] Yusuf [Ahmed] and Prime Minister [Ali Mohamed] Gedi and the conciliatory spirit of the statements issued by both on the latter's resignation. I call upon the Transitional Federal Government to continue to seek peaceful solutions to its internal differences so as to focus its efforts on national reconciliation.

85. Despite its shortcomings, the National Reconciliation Congress is an important development that has created opportunities for political and broad-based social reconciliation in Somalia. It should be considered as a step in the long process of national reconciliation. My Special Representative will intensify his efforts, in close cooperation with key Somali actors and international partners, to foster inter-Somali dialogue, including with opposition groups both inside and outside Somalia. To this end, I urge the Transitional Federal Government to demonstrate its political will by strengthening its unity and reaching out to the opposition groups. Similarly, I call upon all armed and unarmed opposition groups to renounce violence and accept the Transitional Federal Charter, and thus create a conducive environment for dialogue and genuine reconciliation. I commend all international partners for their tireless efforts and commitment to help the Somali parties reach a comprehensive agreement.

86. I also call upon the leaders of the transitional federal institutions to implement without delay the recommendations of the National Reconciliation Congress, including the development of a roadmap for the completion of the tasks provided for in the Transitional Federal Charter, particularly the constitutional process, preparation for the national population census and the holding of elections scheduled for 2009, as well as the National Security and Stabilization Plan.

87. I am concerned with the ongoing violence between "Puntland" and "Somaliland", and urge both to resort to peaceful means for resolving their dispute.

88. I condemn all acts of violence in Somalia and call upon all parties to cease hostilities and engage in the search for sustainable peace. I particularly call upon all parties to protect the civilian population and humanitarian workers. I also strongly condemn the attack on AMISOM premises in central Mogadishu, as well as the violation of the United

Nations premises in Mogadishu, and urge all parties to refrain from acts of violence, the arrest of or any form of harassment against humanitarian workers.

89. I commend the AMISOM troops for their professionalism in discharging their duties in a very difficult environment. The strengthening of AMISOM capabilities on the ground, including the completion of its full deployment, remains an urgent priority. As mentioned earlier, the United Nations remains committed to providing all possible support to the African Union in strengthening AMISOM and expediting the completion of its full deployment. I urge States members of the African Union that have pledged to send troops to Somalia to do so without further delay. I also appeal to the international community to assist the African Union with the necessary logistical and financial resources for the deployment of AMISOM.

90. As I have indicated in my letter dated 20 September 2007 to the President of the Security Council (see para. 81 above) the international community should also consider, in addition to AMISOM and a possible United Nations peacekeeping force, other options, including a multinational force or a coalition of willing partners.

91. In the meantime, the United Nations will continue its efforts to address the serious humanitarian needs in the country. I encourage the international community to continue to generously support humanitarian relief efforts in Somalia. I call upon all parties to provide unhindered access to relief efforts and to ensure strict compliance with international humanitarian law and human rights principles.

92. I reiterate the need to explore measures to deal with the regional dimensions of the Somali crisis, and to find ways to address the security concerns of Somalia and its neighbours, including respect of the sovereignty and territorial integrity of all States of the region.

93. I remain concerned by the continuing piracy off the coast of Somalia and its adverse impact on the safe delivery of humanitarian assistance and commercial shipping. Given the lack of capacity of the Transitional Federal Government to combat piracy, I reiterate the call to Member States of the United Nations with naval and military assets in the region to take action, in consultation with the Transitional Federal Government, to protect merchant shipping, with a particular focus on vessels transporting humanitarian aid.

94. Finally, I wish to reaffirm my deep appreciation to François Lonseny Fall, my former Special Representative, and his successor Ahmedou Ould Abdallah for their respective leadership and efforts to foster peace and reconciliation among the Somali people. I call upon all Somali parties and Member States to continue to give my new Special Representative their fullest support and cooperation in the pursuit of this goal.

SOURCE: United Nations. Security Council. "Report of the Secretary-General on the Situation in Somalia." S/2007/658. November 7, 2007. www.un.org/Docs/journal/asp/ws.asp?m=S/2007/658 (accessed December 12, 2007).

Other Historic Documents of Interest

From this volume

- Continuing conflict in the Democratic Republic of the Congo, p. 374
- Conflict and humanitarian crisis in the Darfur region of Sudan, p. 385

From previous *Historic Documents*

- Peacekeeping in Somalia, *2006,* p. 717
- European leaders on the prospect of war in Iraq, *2003,* p. 40
- UN on the war between Eritrea and Ethiopia, *2000,* p. 753
- U.S. embassy bombings in Africa, *1998,* p. 555

DOCUMENT IN CONTEXT

Executive Order and Statement on CIA Interrogation Techniques

JULY 20 AND DECEMBER 6, 2007

The Central Intelligence Agency (CIA), which had been buffeted for two years by reports about its "secret prisons" housing terrorism suspects in Europe and elsewhere, came in for more criticism in December 2007 following revelations that its agents had videotaped interrogations of some suspects years earlier, then destroyed the tapes against the advice of top officials and even some members of Congress. Two congressional committees announced investigations of the videotape destruction, and on January 2, 2008, the Justice Department appointed a prosecutor to conduct an independent criminal investigation into whether CIA officials had obstructed justice.

The videotape controversy capped a year in which previous controversies over the CIA's handling of terrorism suspects had begun to wind down. Early in the year, two European panels issued final reports on the agency's use of facilities in Poland and Romania for secret interrogations of suspected al Qaeda operatives. In July, President George W. Bush sought to end a related dispute over whether the CIA had used, and was authorized to use, abusive methods bordering on torture to extract information from terrorism suspects. The president issued an executive order authorizing the CIA to use methods harsher than those the U.S. military was allowed to employ—but he insisted that no U.S. official could engage in torture.

Controversies over CIA interrogations emerged again in October, however, when Bush's nominee for attorney general, Michael B. Mukasey, refused during Senate committee hearings to define as torture the practice known as "waterboarding." Under this procedure, interrogators tie a suspect to a board or other flat surface, place a cloth over his face, and pour water into his nose and mouth, making him feel that he is about to drown. Waterboarding is an ancient technique, dating at least to the Spanish Inquisition according to some sources, and some U.S. officials said it had proven effective in forcing prisoners to talk. The U.S. military and most European governments had barred its use for decades, but the CIA reportedly subjected two or three terrorism suspects to the technique in 2002 and 2003.

The CIA Tapes

The videotape controversy arose on December 6 when the *New York Times* disclosed that the CIA had destroyed at least two videotapes "documenting the interrogation of two al Qaeda suspects in the agency's custody." The *Times* had notified the CIA the day before that it planned to carry this story, so on December 6 CIA director Gen. Michael V. Hayden wrote to agency employees with his own version of events, saying he wanted to avoid "misinterpretations of the facts" in the matter.

Few people could have been surprised that the CIA had videotaped at least some of its interrogations, which reportedly began in 2002 when the Bush administration was stepping up its global war against terrorism in the wake of the September 11, 2001, attacks in New York and Washington. What was surprising was that the CIA had held on to some of its interrogation tapes for more than three years, only to destroy them because of fears they might become public. In essence, the CIA, which in the past had rarely done criminal investigative work, found itself deeply involved in the murky business of extracting information from alleged terrorists. Documenting some of its actions on videotape might have seemed an appropriate step, but if the tapes were to become public, the CIA would face yet another outcry over its secret methods.

According to the *Times* report, the CIA taped interrogations of at least two key suspects who had been captured early in 2002: Abu Zubaydah, said to be one of the most senior al Qaeda operatives, and Abd al-Rahim al-Nashiri, the alleged head of al Qaeda operations on the Arabian peninsula. Subsequent reports indicated that the CIA also had used waterboarding on Khalid Sheikh Mohammed, the alleged "mastermind" of al Qaeda's September 11 attacks.

The *Times* reported on December 30 that the CIA began taping its interrogations of Zubaydah during the spring of 2002, in large part to document his treatment so the agency could refute any subsequent allegations that he had been abused. Zubaydah had been seriously wounded at the time of his capture, and he reportedly survived because of extensive medical treatment arranged by the CIA. Another advantage of the taping, the newspaper said, was that language experts, psychologists, and CIA operatives being trained as interrogators could view the questioning afterward. The CIA reportedly videotaped Zubaydah for hundreds of hours—while he was sleeping and getting medical treatment, as well as during his interrogations.

The CIA reportedly stopped taping suspects in late 2002. It might have been coincidence, but this was about the same time, according to news reports, that the agency began subjecting a limited number of its prisoners, starting with Zubaydah, to waterboarding.

Destroying the Tapes

As of late 2007 the exact circumstances of the CIA's decision to destroy the tapes were still unclear. The *New York Times* reported that agency officials began worrying about the tapes in late 2002 and went to Capitol Hill in February 2003, asking for reaction to their possible destruction. The two top leaders of the House Intelligence Committee at the time, Porter J. Goss, R-Fla., and Jane Harman, D-Calif., reportedly urged the CIA not to destroy the tapes.

After the disclosure early in 2004 of abuses by army soldiers of Iraqi detainees at the Abu Ghraib prison near Baghdad, pressure again built within the CIA to destroy the tapes. Officials reportedly were concerned that, if the tapes became public, CIA officers shown on them might be subject to prosecution for actions that could be construed as abuse of prisoners, or even to revenge from al Qaeda.

In November 2005, the relatively new head of the CIA's clandestine service, Jose A. Rodriguez Jr., reportedly acting on his own authority, ordered the destruction of the tapes even though he had been warned against this step by Goss, who in 2004 had taken over as director of the CIA. News reports said Goss had given Rodriguez broad leeway to run

CIA covert operations, and Rodriguez obtained an opinion from CIA lawyers that he had authority to act on his own. As of late 2007, Rodriguez had not commented publicly on the matter.

In his letter to CIA employees, Hayden—who joined the CIA in May 2006—said he understood that the tapes had been destroyed because their existence "posed a serious security risk." The risk, he said, was that CIA officials shown on the tapes, and their families, could face "retaliation" from al Qaeda and its sympathizers if the tapes became public.

Hayden also said the tapes were destroyed "only after it was determined they were no longer of intelligence value and not relevant to any internal, legislative, or judicial inquiries—including the trial of Zacarias Moussaoui." A French citizen of Moroccan descent, Moussaoui was the only person to be convicted in the United States for playing a part in the September 11 attacks. He pleaded guilty in April 2006 to six conspiracy charges related to the attacks, although he never admitted having any direct role in those attacks. He was sentenced to life in prison. During his trial, his lawyers sought access to information, including videotapes, about CIA interrogations of other terrorism suspects but were told that no such tapes existed. The prosecutors in that case wrote to the trial judge in October 2005, informing her that the CIA had videotaped some terrorism suspects, but they insisted that the interrogations had no bearing on Moussaoui's case.

Reaction to the Tapes' Destruction

Hayden suggested in his message to CIA employees that the news about the interrogation tapes would provoke a firestorm of criticism—and that storm was not long in coming. The next day, December 7, Democrats in both the House of Representatives and the Senate made speeches and wrote letters to administration officials demanding more information about why the tapes were destroyed and calling for investigations into a possible obstruction of justice. Among those most concerned about the matter were past and current leaders of the congressional intelligence committees; of those members willing to talk about the matter, some said they never knew about the tapes in the first place, while others said they had known about the tapes and had urged that they not be destroyed. Sen. John D. Rockefeller IV, D-W.Va., chair of the Senate Intelligence Committee, released a statement in which he said that he had learned about the destruction of the tapes about one year after the fact.

On December 10, the leaders of the House Intelligence Committee—Reps. Silvestre Reyes, D-Texas, and Peter Hoekstra, R-Mich.—announced their plan to investigate "the issues surrounding the destruction" of the tapes. They noted that Hayden's letter to CIA employees had implied that the committee "had been properly notified" that the tapes were destroyed in 2005. "Based on our view of the record, this does not appear to be true," they said. Other members of Congress rejected Hayden's contention that the tapes had been destroyed solely to protect the identities of CIA employees. If that was the standard, the CIA would have to burn "millions" of documents containing names of agency employees, Sen. Carl Levin, D-Mich., said.

Hayden appeared before the Senate Intelligence Committee on December 11 and before the House panel the next day, in both cases behind closed doors. Hayden and members of the committees refused to give details of what was discussed, but it was clear that Hayden's presentation had satisfied few on the committees. After meeting with the House panel, Hayden did acknowledge that the CIA "could have done an awful lot better" job of informing Congress.

On December 14, the Justice Department and CIA inspector general John L. Helgerson asked the two congressional panels to postpone their investigations into the matter while the government's probes were under way. In a letter, officials said their investigations "would likely be jeopardized" if the CIA had to respond to multiple inquiries from Capitol Hill, as well as from the administration. Committee members rejected that request, saying they would proceed with their own investigations, and the CIA agreed on December 20 to open its files to the panels.

Under congressional pressure to ensure an impartial investigation, Attorney General Mukasey announced on January 2, 2008, that he had appointed John H. Durham, an assistant attorney general in Connecticut, to investigate any possible criminal misconduct in the CIA's handling of the tapes. News reports described Durham as a relentless, apolitical prosecutor who had handled several previous high-profile cases.

The existence of the CIA tapes came as a surprise to members and staffers of the high-level commission that had investigated the September 11 attacks and issued a comprehensive report in 2004. The chairmen of the National Commission on Terrorist Attacks upon the United States (known as the 9/11 Commission), Thomas H. Kean and Lee H. Hamilton, told reporters that the panel had asked the executive branch in 2003 and 2004 for all material related to the interrogation of terrorism suspects but had not been told about the videotapes. Philip D. Zelikow, who had been the panel's executive director, reviewed the commission's records and wrote a detailed memorandum on December 13, saying government officials had never disclosed the existence of the CIA videotapes to the commission. Kean and Hamilton said they viewed the CIA's withholding of information about the tapes as an "obstruction" of their inquiry.

Bush Executive Order on CIA Interrogations

The Bush administration tried early in 2007 to end the debate about CIA interrogations of suspected terrorists, only to watch as the debate grew even hotter once the videotaping was disclosed. After news reports in 2005 raised questions about the CIA's secret prisons and alleged use of torture against the people held in them, the Bush administration began drafting new guidelines for CIA interrogations of terrorism suspects. One spur to action was the Military Commissions Act of 2006 (PL 109-366), in which Congress authorized the president to allow the CIA to use interrogation techniques other than the nineteen methods cited in the U.S. Army field manual, which was revised in 2005 and explicitly barred harsh methods such as waterboarding.

Over the course of 2006 and into 2007, administration officials debated the exact language of the new CIA guidelines. According to news reports, State Department lawyers and officials argued for restrictive language clearly barring CIA interrogators from using harsh methods that could be construed as torture. Other officials, notably top aides to Vice President Dick Cheney, reportedly wanted to give CIA interrogators wide latitude to extract information from terrorism suspects.

The final result, an executive order signed by President Bush on July 20, apparently was a compromise. Key details of the order were kept secret, but the version made public by the White House barred "cruel, inhuman, or degrading treatment or punishment" of detainees; this was language sought by the State Department. News reports quoted administration officials as saying Bush's order, in effect, barred such techniques as waterboarding and exposure to extremes of cold and heat. However, the order did allow CIA interrogators to use some techniques that were harsher than those permitted by the U.S.

military, reportedly including sleep deprivation and loud noises. According to a White House statement, the order "clarified vague terms" of Common Article Three of the Geneva Conventions (barring cruel punishment) and was "consistent with" recent legal decisions, including those by the UN tribunal that was trying war crimes suspects from the former Yugoslavia.

Some administration critics said Bush's executive order appeared to be intended to evade the Geneva Conventions and other international treaties, rather than to comply with them. Human Rights Watch issued a statement on July 20 calling the CIA program "illegal to its core" because it still relied on secret detentions of suspects without due process of law. "Although the new executive order bars torture and other abuse, the order still can't purport to legalize a program that violates basic rights," said Joanne Mariner, terrorism and counterterrorism director for Human Rights Watch.

Another response to the president's executive order came from Democrats in Congress who tried, unsuccessfully, to adopt legislation barring the CIA from using harsh interrogation methods on terrorism suspects. On December 5, one day before the disclosure of the destroyed CIA interrogation tapes, House-Senate conferees on the fiscal 2008 authorization bill (HR 2082) for the intelligence agencies adopted language requiring the CIA to limit itself to interrogation methods allowed in the 2005 revision of the army field manual. The House approved the conference version on December 13 by a 222-199 vote, despite a veto threat from the White House, which opposed the CIA interrogation language and ten other provisions. The bill stalled in the Senate, however, when Sen. Lindsey Graham, R-S.C., and other Republicans blocked action because of the CIA language.

Mukasey on Waterboarding

The Senate's confirmation hearings for Mukasey became another venue for controversy over the interrogation of terrorist suspects. Bush selected Mukasey, a former federal judge in New York, as attorney general following the forced resignation in September of Alberto R. Gonzales, whose handling of the firing of federal prosecutors made him a topic of intense criticism in Washington.

Mukasey came up for confirmation in mid-October, shortly after the *New York Times* revealed the existence of secret legal opinions, drafted at the Justice Department in 2005, authorizing the use of harsh interrogation techniques. The *Times*' October 4 report implied that these opinions conflicted with a Justice Department document, released publicly in December 2004, that barred the use of torture as "abhorrent." The newspaper said the secret 2005 opinion had been opposed by then deputy attorney general James B. Comey, who reportedly said officials would be "ashamed" if it became public. White House officials rejected the *Times*' characterization, saying the 2005 memo did not authorize torture, but they refused to make the memo public.

In two days of confirmation hearings, on October 17–18, Mukasey refused to give the Senate Judiciary Committee a legal opinion on the interrogation methods used by the CIA, including waterboarding. Asked by Sen. Sheldon Whitehouse, D-R.I., if waterboarding was constitutional, Mukasey answered, "I don't know what is involved in the technique. If waterboarding is torture, torture is not constitutional." He refused to say whether the practice did amount to torture, angering Democrats and some Republicans.

Democrats on the committee wrote Mukasey a letter on October 23, expressing their disappointment with his refusal to take a stand on the issue. Mukasey responded with a letter on October 30, saying that the "coercive interrogation techniques" described by the senators—notably waterboarding—"seem over the line or, on a personal basis, repugnant to me." Even so, he said he still could not render a legal opinion on the matter because he had not been briefed on exactly what techniques U.S. officials had used. Mukasey also said he did not want to suggest that CIA officials who might have used these techniques could be in "personal legal jeopardy."

Mukasey's response still did not satisfy the majority of Democrats on the committee, who voted against him when the panel sent his nomination to the Senate floor on November 6. Two key Democrats did support Mukasey: Charles E. Schumer of New York (who had recommended Mukasey to Bush in the first place) and Dianne Feinstein of California. The full Senate confirmed Mukasey by a 53–40 vote on November 8.

Following are two documents. First, the text of an executive order, signed by President George W. Bush on July 20, 2007, laying out legal standards for the detention and interrogation of suspected terrorists by the Central Intelligence Agency. Second, the text of a December 6, 2007, statement by Gen. Mike Hayden, director of the Central Intelligence Agency, to CIA employees, explaining that the agency had videotaped the interrogations of some terrorism suspects in 2002 and 2003 but later destroyed the tapes.

DOCUMENT

Executive Order 13440: Interpretation of Geneva Conventions Common Article 3

July 20, 2007

By the authority vested in me as President and Commander in Chief of the Armed Forces by the Constitution and the laws of the United States of America, including the Authorization for Use of Military Force (Public Law 107–40), the Military Commissions Act of 2006 (Public Law 109–366), and section 301 of title 3, United States Code, it is hereby ordered as follows:

Section 1. *General Determinations.* (a) The United States is engaged in an armed conflict with al Qaeda, the Taliban, and associated forces. Members of al Qaeda were responsible for the attacks on the United States of September 11, 2001, and for many other terrorist attacks, including against the United States, its personnel, and its allies throughout the world. These forces continue to fight the United States and its allies in Afghanistan, Iraq, and elsewhere, and they continue to plan additional acts of terror throughout the world. On February 7, 2002, I determined for the United States that members of al Qaeda, the Taliban, and associated forces are unlawful enemy combatants who are not entitled to the protections that the Third Geneva Convention provides to prisoners of war. I hereby reaffirm that determination.

(b) The Military Commissions Act defines certain prohibitions of Common Article 3 for United States law, and it reaffirms and reinforces the authority of the President to interpret the meaning and application of the Geneva Conventions.

Sec. 2. *Definitions.* As used in this order:

(a) "Common Article 3" means Article 3 of the Geneva Conventions.

(b) "Geneva Conventions" means:

(i) the Convention for the Amelioration of the Condition of the Wounded and Sick in Armed Forces in the Field, done at Geneva August 12, 1949 (6 UST 3114);

(ii) the Convention for the Amelioration of the Condition of Wounded, Sick and Shipwrecked Members of Armed Forces at Sea, done at Geneva August 12, 1949 (6 UST 3217);

(iii) the Convention Relative to the Treatment of Prisoners of War, done at Geneva August 12, 1949 (6 UST 3316); and

(iv) the Convention Relative to the Protection of Civilian Persons in Time of War, done at Geneva August 12, 1949 (6 UST 3516).

(c) "Cruel, inhuman, or degrading treatment or punishment" means the cruel, unusual, and inhumane treatment or punishment prohibited by the Fifth, Eighth, and Fourteenth Amendments to the Constitution of the United States.

Sec. 3. *Compliance of a Central Intelligence Agency Detention and Interrogation Program with Common Article 3.* (a) Pursuant to the authority of the President under the Constitution and the laws of the United States, including the Military Commissions Act of 2006, this order interprets the meaning and application of the text of Common Article 3 with respect to certain detentions and interrogations, and shall be treated as authoritative for all purposes as a matter of United States law, including satisfaction of the international obligations of the United States. I hereby determine that Common Article 3 shall apply to a program of detention and interrogation operated by the Central Intelligence Agency as set forth in this section. The requirements set forth in this section shall be applied with respect to detainees in such program without adverse distinction as to their race, color, religion or faith, sex, birth, or wealth.

(b) I hereby determine that a program of detention and interrogation approved by the Director of the Central Intelligence Agency fully complies with the obligations of the United States under Common Article 3, provided that:

(i) the conditions of confinement and interrogation practices of the program do not include:

(A) torture, as defined in section 2340 of title 18, United States Code;

(B) any of the acts prohibited by section 2441(d) of title 18, United States Code, including murder, torture, cruel or inhuman treatment, mutilation or maiming, intentionally causing serious bodily injury, rape, sexual assault or abuse, taking of hostages, or performing of biological experiments;

(C) other acts of violence serious enough to be considered comparable to murder, torture, mutilation, and cruel or inhuman treatment, as defined in section 2441(d) of title 18, United States Code;

(D) any other acts of cruel, inhuman, or degrading treatment or punishment prohibited by the Military Commissions Act (subsection 6(c) of Public Law 109–366) and the Detainee Treatment Act of 2005 (section 1003 of Public Law 109–148 and section 1403 of Public Law 109–163);

(E) willful and outrageous acts of personal abuse done for the purpose of humiliating or degrading the individual in a manner so serious that any reasonable person, considering the circumstances, would deem the acts to be beyond the bounds of human decency, such as sexual or sexually indecent acts undertaken for the purpose of humiliation, forcing the individual to perform sexual acts or to pose sexually, threatening the individual with sexual mutilation, or using the individual as a human shield; or

(F) acts intended to denigrate the religion, religious practices, or religious objects of the individual;

(ii) the conditions of confinement and interrogation practices are to be used with an alien detainee who is determined by the Director of the Central Intelligence Agency:

(A) to be a member or part of or supporting al Qaeda, the Taliban, or associated organizations; and

(B) likely to be in possession of information that:

(1) could assist in detecting, mitigating, or preventing terrorist attacks, such as attacks within the United States or against its Armed Forces or other personnel, citizens, or facilities, or against allies or other countries cooperating in the war on terror with the United States, or their armed forces or other personnel, citizens, or facilities; or

(2) could assist in locating the senior leadership of al Qaeda, the Taliban, or associated forces;

(iii) the interrogation practices are determined by the Director of the Central Intelligence Agency, based upon professional advice, to be safe for use with each detainee with whom they are used; and

(iv) detainees in the program receive the basic necessities of life, including adequate food and water, shelter from the elements, necessary clothing, protection from extremes of heat and cold, and essential medical care.

(c) The Director of the Central Intelligence Agency shall issue written policies to govern the program, including guidelines for Central Intelligence Agency personnel that implement paragraphs (i)(C), (E), and (F) of subsection 3(b) of this order, and including requirements to ensure:

(i) safe and professional operation of the program;

(ii) the development of an approved plan of interrogation tailored for each detainee in the program to be interrogated, consistent with subsection 3(b)(iv) of this order;

(iii) appropriate training for interrogators and all personnel operating the program;

(iv) effective monitoring of the program, including with respect to medical matters, to ensure the safety of those in the program; and

(v) compliance with applicable law and this order.

Sec. 4. *Assignment of Function.* With respect to the program addressed in this order, the function of the President under section 6(c)(3) of the Military Commissions Act of 2006 is assigned to the Director of National Intelligence.

Sec. 5. *General Provisions.* (a) Subject to subsection (b) of this section, this order is not intended to, and does not, create any right or benefit, substantive or procedural, enforceable at law or in equity, against the United States, its departments, agencies, or other entities, its officers or employees, or any other person.

(b) Nothing in this order shall be construed to prevent or limit reliance upon this order in a civil, criminal, or administrative proceeding, or otherwise, by the Central Intelligence Agency or by any individual acting on behalf of the Central Intelligence Agency in connection with the program addressed in this order.

George W. Bush
The White House,
July 20, 2007.

SOURCE: U.S. Executive Office of the President. "Executive Order 13440—Interpretation of the Geneva Conventions Common Article 3 as Applied to a Program of Detention and Interrogation Operated by the Central Intelligence Agency." July 20, 2007. *Weekly Compilation of Presidential Documents* 43, no. 30 (July 30, 2007): 1000–1002. Washington, D.C.: National Archives and Records Administration. www.gpoaccess.gov/wcomp/v43no30.html (accessed February 6, 2008).

DOCUMENT

CIA Director on the Taping of Interrogations of Terrorism Suspects

December 6, 2007

The press has learned that back in 2002, during the initial stage of our terrorist detention program, CIA videotaped interrogations, and destroyed the tapes in 2005. I understand that the Agency did so only after it was determined they were no longer of intelligence value and not relevant to any internal, legislative, or judicial inquiries—including the trial of Zacarias Moussaoui. The decision to destroy the tapes was made within CIA itself. The leaders of our oversight committees in Congress were informed of the videos years ago and of the Agency's intention to dispose of the material. Our oversight committees also have been told that the videos were, in fact, destroyed.

If past public commentary on the Agency's detention program is any guide, we may see misinterpretations of the facts in the days ahead. With that in mind, I want you to have some background now.

CIA's terrorist detention and interrogation program began after the capture of Abu Zubaydah in March 2002. Zubaydah, who had extensive knowledge of al-Qa'ida personnel and operations, had been seriously wounded in a firefight. When President Bush officially acknowledged in September 2006 the existence of CIA's counter-terror initiative, he talked about Zubaydah, noting that this terrorist survived solely because of medical treatment arranged by CIA. Under normal questioning, Zubaydah became defiant and evasive. It was clear, in the President's words, that "Zubaydah had more information that could save innocent lives, but he stopped talking."

That made imperative the use of other means to obtain the information—means that were lawful, safe, and effective. To meet that need, CIA designed specific, appropriate interrogation procedures. Before they were used, they were reviewed and approved by the Department of Justice and by other elements of the Executive Branch. Even with the great care taken and detailed preparations made, the fact remains that this effort was new, and the Agency was determined that it proceed in accord with established legal and policy guidelines. So, on its own, CIA began to videotape interrogations.

The tapes were meant chiefly as an additional, internal check on the program in its early stages. At one point, it was thought the tapes could serve as a backstop to guarantee that other methods of documenting the interrogations—and the crucial information they produced—were accurate and complete. The Agency soon determined that its documentary reporting was full and exacting, removing any need for tapes. Indeed, videotaping stopped in 2002.

As part of the rigorous review that has defined the detention program, the Office of General Counsel examined the tapes and determined that they showed lawful methods of questioning. The Office of Inspector General also examined the tapes in 2003 as part of its look at the Agency's detention and interrogation practices. Beyond their lack of intelligence value—as the interrogation sessions had already been exhaustively detailed in written channels—and the absence of any legal or internal reason to keep them, the tapes posed a serious security risk. Were they ever to leak, they would permit identification of your CIA colleagues who had served in the program, exposing them and their families to retaliation from al-Qa'ida and its sympathizers.

These decisions were made years ago. But it is my responsibility, as Director today, to explain to you what was done, and why. What matters here is that it was done in line with the law. Over the course of its life, the Agency's interrogation program has been of great value to our country. It has helped disrupt terrorist operations and save lives. It was built on a solid foundation of legal review. It has been conducted with careful supervision. If the story of these tapes is told fairly, it will underscore those facts.

SOURCE: Central Intelligence Agency. "Statement to Employees by Director of the Central Intelligence Agency, General Mike Hayden on the Taping of Early Detainee Interrogations." December 6, 2007. www.cia.gov/news-information/press-releases-statements/press-release-archive-2007/taping-of-early-detainee-interrogations.html (accessed February 6, 2008).

OTHER HISTORIC DOCUMENTS OF INTEREST

FROM THIS VOLUME

- Council of Europe report on CIA "secret prisons," p. 275
- National Intelligence Estimate on terrorism, p. 335
- Resignation of Attorney General Alberto Gonzales, p. 457

FROM PREVIOUS *HISTORIC DOCUMENTS*

- Sentencing of Mousaoui for his role in the September 11 attacks, *2006,* p. 118
- President Bush on the secret detention of terrorism suspects, *2006,* p. 511
- Secretary of state on U.S. policy concerning torture of detainees, *2005,* p. 905
- U.S. Army report on abuses of Iraqi prisoners, *2004,* p. 207
- Justice Department on torture in terrorism interrogations, *2004,* p. 336

DOCUMENT IN CONTEXT

President's Commission on Caring for America's Wounded Soldiers

JULY 25, 2007

A two-part newspaper series describing shoddy living conditions and neglectful treatment of wounded outpatient soldiers at Walter Reed Hospital in Washington, D.C., touched off a host of investigations that called attention to serious problems in the way the government treated and cared for its injured soldiers and veterans. Together, the reporting and the investigations revealed a system in disarray, hampered by conflicting and outdated definitions of disability, incompatible records management systems, and a shortage of qualified medical and benefit claims personnel—all leading to lengthy delays in processing disability claims and providing adequate medical treatment. Many of the reports focused on the thousands of soldiers returning from Iraq and Afghanistan with conditions such as post-traumatic stress disorder (PTSD), traumatic brain injuries, and other mental and physical disorders that were difficult to diagnose and treat and that were often overlooked or ignored.

By the end of the year, the Defense Department, the Veterans Affairs (VA) Department, and Congress had all taken steps to start correcting some of the worst problems, and there were signs that improvements were under way. But it was likely to take years to integrate the multiple records managements systems, to train new personnel, and to significantly reduce the claims backlog. Meanwhile, the number of new claims was expected to continue to rise.

A Large and Growing Problem

One undisputed success in the wars in Iraq and Afghanistan was the outstanding medical attention wounded soldiers received on the battlefield. As a result of that care, the ratio of those wounded in combat to those killed was eight to one, the highest such ratio ever recorded. At the beginning of 2007, about 3,000 troops had been killed in the wars, while more than 22,000 had been injured in direct combat. If the troops injured in accidents or suffering from PTSD, traumatic brain injury, and other disorders sustained in the war zone were also counted, the ratio of total injured to total killed would rise to sixteen to one. Seriously injured soldiers were treated in military hospitals, where by virtually all accounts they received excellent care. They were eventually returned to active service or classified as disabled and turned over to the VA for further treatment and rehabilitation.

In addition, the VA was facing a growing caseload from veterans of the Iraq and Afghanistan wars who filed claims for disability benefits. According to the VA, of the nearly 690,000 veterans who had served in those wars, more than 180,000 had filed disability claims by the end of 2006. Linda Bilmes, a Harvard professor and expert in veterans' health care, estimated that number would climb to 731,000 by 2014, assuming no new troops were deployed. And that figure did not include claims still being filed by veterans of the 1990–1991 Gulf War or the Vietnam War. By the beginning of 2007, the VA had a backlog of about 400,000 disability claims to process. It took an average of 177 days to process a claim and an average of 657 days to process an appeal.

Legislators and others had been hearing complaints for years about slow claims processing and inadequate, and in some cases denial of, medical treatment. It was not until the *Washington Post* ran a series by reporters Dana Priest and Anne Hull on February 18 and 19, 2007, that the complaints received widespread national attention. The series focused not on Walter Reed Hospital itself, widely renowned for the advanced treatment it provided wounded soldiers, but on the "other Walter Reed," where some 700 outpatients lived in five buildings on campus and in nearby hotels and apartments leased by the U.S. Army. Many of the buildings were deteriorating. The reporters described Building 18 as moldy, smelly, and filthy, with rotting floors and battered walls. One room was missing part of its ceiling.

Occupying these rooms were Walter Reed outpatients—injured soldiers who had been released from the hospital but were still receiving medical treatment or awaiting a decision on a disability rating. Many of them told the reporters of having to deal with "disengaged clerks, unqualified platoon sergeants and overworked case managers," of heavily medicated patients who were left to make their own medical appointments and to work their way through a bureaucratic maze of forms and other necessary paperwork before their status could be finally determined. The *Post* reported that the typical soldier was required to file twenty-two documents with eight different military commands, which used sixteen different information systems to process the forms. According to the soldiers, lost paperwork was the most common bureaucratic problem they faced. The average stay for outpatients at Walter Reed was ten months, but some had been there as long as two years, according to the *Post*.

The series touched off a firestorm of outrage and sent Pentagon and VA officials scrambling to correct the situation. On February 23, Defense Secretary Robert M. Gates named an independent review panel to investigate what he described as an "unacceptable" situation at Walter Reed. These soldiers at Walter Reed "battled our foreign enemies," Gates said. "They should not have to battle an American bureaucracy." The army quickly repaired many of the worst housing problems at Walter Reed, and at least three top army officials lost their jobs as a result of the scandal: Maj. Gen. George W. Weightman, the commander at Walter Reed; Lt. Gen. Kevin C. Kiley, the surgeon general of the army and a former commander at Walter Reed; and Army Secretary Francis J. Harvey.

White House press secretary Tony Snow said on February 20 that President George W. Bush's reaction to the *Post* reports was "find out what the problem is and fix it." On March 6, Bush appointed a presidential commission to look into the problems uncovered by the news reports. "We have an obligation, we have a moral obligation to provide the best possible care and treatment to the men and women who have served our country. They deserve it. And they are going to get it," Bush said. He named former senator Robert Dole, a Republican from Kansas who was gravely wounded in World War II, and Donna Shalala, the secretary of health and human services during the Clinton administration, to chair the commission, which was to make its report by the end of July.

Fixing the Bureaucracy

The first investigative panel to make its report was the Pentagon Independent Review Group appointed by Gates, which was co-chaired by Togo D. West Jr., secretary of veterans affairs and secretary of the army in the Clinton administration, and John O. Marsh, Jr., secretary of the army in the Reagan administration and a former member of Congress from Virginia. Making its report on April 12, the panel sharply criticized the army's leadership for failing outpatients at Walter Reed. More broadly, the panel called for overhauling the disability rating system to give the Pentagon authority to determine whether an injured soldier is fit to return to duty and the VA Department authority to adjudicate disability ratings and eliminate inconsistencies. Currently, each department issued disability ratings. The panel also said the systems for managing clinical and bureaucratic records needed to be integrated.

A cabinet-level interagency task force, headed by Veterans Affairs Secretary Jim Nicholson, made similar recommendations to fix the bureaucratic inconsistencies and incompatibilities that were creating backlogs in providing treatment and processing disability benefit claims. That report, released on April 24, called for a joint Pentagon-VA electronic case management system that could follow each service member from military hospital to veteran facility. "If we can track a package in this country and know where it is at any given time, we certainly should be able to track a human being," Nicholson said when he announced the task force's findings. Later that day, President Bush ordered the two departments to work together to implement the task force's recommendations.

New Attention to Mental Health Problems

Both panels also urged the government to pay more attention to what they called the "signature injuries" of the Iraq and Afghanistan wars: PTSD and traumatic brain injury. The independent panel called for earlier and more comprehensive diagnoses and treatment for these and related conditions, as well as for a new "center of excellence" for treating brain injury cases. The interagency task force additionally urged that all Iraq and Afghanistan veterans being treated in VA facilities be screened for traumatic brain injury—neurological damage such as concussion, blindness, deafness, and mental retardation caused by blast waves from roadside bombs, mortars, and other weapons of war. Other symptoms included memory loss, lack of concentration, poor reasoning, headaches, confusion, depression, and irritability.

A Defense Department Task Force on Mental Health reported on May 3 that more than one-third of all troops in and veterans of the two wars were currently suffering from some mental health disorder, and that soldiers deployed for longer than six months or who had deployed multiple times were more likely to screen positive for such problems. The task force warned that the existing levels of funding and staffing available for treating these conditions would not be enough to meet anticipated needs. It also found that soldiers often failed to seek treatment for their conditions for fear they would be ridiculed as being "weak" or that the admission could damage their careers. Further, numerous veterans, particularly those who had served in the reserves or national guards, complained of being denied treatment even after they reported their condition.

Three days later, on May 6, a report prepared by a special panel of the Institute of Medicine at the request of the VA concluded that the method the government used for compensating veterans with PTSD and related emotional disorders was not based on science, was applied unevenly, and created little incentive for veterans to try to get better. "As the increasing number of claims to the VA shows, PTSD has become a very significant public health problem," said Nancy Andreasen, a mental health expert who led the year-long study. "Comprehensive revision is needed." The study found that most of the recent increase in claims for PTSD benefits came not from veterans of the current wars but from Vietnam veterans. The panel said it expected hundreds of thousands of veterans from the Gulf, Iraq, and Afghanistan wars to make claims over the coming decades.

Report of the President's Commission

The President's Commission on Care for America's Returning Wounded Warriors added its recommendations for change in a report released on July 25. The panel began by observing that a similar commission in 1956 had reported that there was "no clear national philosophy of veterans' benefits" and then had set out a proposed philosophy that neither Congress nor the executive branch ever acted on. "To this day," the panel said, "lack of a specific objective hinders the design, coordination, and evaluation of both individual veterans' programs and the disability system as a whole." Dole and Shalala said their commission's key proposal was to assign a "recovery coordinator" to help guide every injured service member through the military medical system. "Health care has to be customized and personalized" for each injured service member, Shalala said in a briefing on the report. The panel also said that every veteran deployed in Iraq and Afghanistan who needed care for PTSD should receive it from the VA, a not-so-subtle rebuke to the department, which had been accused of turning away veterans seeking help.

To eliminate the bureaucratic red tape that ensnarled so many injured soldiers, the commission recommended that the departments of Defense and Veterans Affairs create a single, comprehensive, and standardized medical examination that would be used to determine a soldier's status. A soldier determined to be unfit to return to service should be separated from the military and given a lifetime annuity based only on his or her military rank and years of service, the panel said. Any disability rating should be determined by the VA, and eligibility for continued payments should be assessed at least every three years.

Bush endorsed the commission's recommendations on July 25 and again on October 16, when he sent a formal set of proposals to Congress. Those formal recommendations came after members of both parties criticized the administration for being slow to correct the problems uncovered earlier in the year. On September 26 the Government Accountability Office (GAO) released a report saying that the Pentagon's promises to fix the system were falling victim to staff shortages and uncertainties about how to move forward. As a result, plans to provide recovery coordinators and streamline the disability rating system were behind schedule, the GAO said. "After so many promises but so little progress, we need to see more concrete results," said Rep. Thomas M. Davis III of Virginia, the ranking Republican on a House subcommittee on oversight of national security. "We're seven months into this process, and we're just getting off the ground," said subcommittee member John F. Tierney, D-Mass. "Why has it taken so long?"

One reason might have been the lengthy interlude between VA Secretary Nicholson's announcement on July 17 that he planned to retire and President Bush's appointment of a successor on October 30, nearly a month after Nicholson actually left his post. The new secretary was retired Army Lt. Gen. James Peake, a medical doctor. Peake had his work cut out for him; by the end of the year, the department had fallen even further behind in processing disability claims, taking an average of 183 days in 2007, compared with 177 days in 2006.

For its part Congress added several of the Dole-Shalala commission recommendations, including providing for a seamless transition between active duty and veteran status, to an authorization bill for the Defense Department. Bush pocket-vetoed that measure on December 28 for unrelated reasons, but he was expected to sign the legislation as soon as Congress removed the offending provisions after it convened in January 2008.

Following are excerpts from the final report of the President's Commission on Care for America's Returning Wounded Warriors, released on July 25, 2007.

DOCUMENT

Report of the President's Commission on Care for America's Returning Wounded Warriors

July 25, 2007

"It is almost cliché now to find examples of a wounded Marine having initially been treated by a Navy Corpsman find himself medevac'ed by an Army helicopter to undergo emergency surgery at an Air Force Theater Hospital."

LtCol [Andrew] Moore's testimony demonstrates how the skills and resources of the U.S. military can be brought together to aid an injured service member—without regard for traditional bureaucracies and hierarchies. Under the best circumstances, the entire system smoothly joins forces to provide exactly what is needed, precisely when it is needed. His example embodies the kind of efficient care, centered on the needs of the patient, that we envision for our injured service members throughout the process of treatment, rehabilitation, and return to their military unit or home community.

In our few months of operation, we nine Commissioners—health care, disability, and housing experts, injured service members, and family—have visited 23 Department of Defense (DoD), Department of Veterans Affairs (VA), and private-sector treatment facilities. We have heard first-hand from injured service members and their families, from health care professionals, and from the people who manage military and veterans' programs. More than 1,700 injured service members responded to a national survey we conducted, and we received more than 1,250 letters and emails from service members, veterans, family members, and health care personnel. We have analyzed the recommendations of past commissions and task forces, including several issued earlier in 2007. And, we have drawn on the extensive knowledge of our fellow Commissioners.

We want to emphasize that we've heard time and again about the overall high quality of our military's battlefield medicine and the care delivered by the staffs in our nation's military medical facilities and the VA health system. These clinical professionals' skill and intense commitment to the wounded is palpable. In the Vietnam era, five out of every eight seriously injured service members survived; today, seven out of eight survive, many with injuries that in previous wars would have been fatal. This is a remarkable record. The number of "seriously injured" service members on whom much of this report focuses is, without doubt, eminently manageable. . . .

Despite accomplishments in clinical care, problems do occur—particularly in handoffs between inpatient and outpatient care and between the two separate DoD and VA health care and disability systems. To resolve these problems, we have concentrated on ways to better:

- **Serve** the multiple needs of injured service members and their families
- **Support** them in their recovery and return to military duty or to their communities and
- **Simplify** the delivery of medical care and disability programs.

We believe our recommendations will produce a patient-centered system that fosters high-quality care, increases access to needed care and programs, promotes efficiency, supports families, and facilitates the work of the thousands of dedicated individuals who provide a gamut of health care and disability programs to injured service members and veterans. Our nation needs a system of care that enables injured service members to maximize their recovery and their opportunity to return to the mainstream of American life. Such a system not only should treat all service members—whether active duty or reserve component (that is, the National Guard and reserve)—evenhandedly, but it also must be perceived as doing so.

Our Commission was established at a time of great change in U.S. health care. Many of the statements—good and bad—that we have heard about care in the DoD and VA systems could apply to the nation's health care delivery system as a whole. While numerous aspects of U.S. medical care are excellent, problems in coordination and continuity of care are common; our nation's hospitals and health systems are struggling to develop effective information technology systems; the stigma associated with seeking mental health care is slowly diminishing, but far from gone; our overall health system is oriented to acute care, not long-term rehabilitation; and shortages in critical staff categories are felt nationwide.

In the past few months, the health care and disability systems for our service members and veterans have been under a media microscope and the subject of several reports cited earlier. Public concern arises because Americans recognize and respect the sacrifices of our young men and women fighting in Iraq and Afghanistan and the great debt we owe those injured and killed. Many of the concerns already are being addressed by Congress and in the two Departments.

The reports published earlier this year provided invaluable background information and analyses for our work. Because they are so recent, we did not need to reiterate their findings. Rather, we focused on ways to move forward. One other difference between our Commission and previous ones is that, while they addressed discrete pieces of the DoD and VA medical care and disability systems, President Bush charged us with looking at the whole continuum of care and programs for wounded service members, as well as what is needed to assure their successful return to military duty or civilian life.

We don't recommend merely patching the system, as has been done in the past. Instead, the experiences of these young men and women have highlighted the need for *fundamental changes* in care management and the disability system. Our recommendations address these fundamental changes. We believe they will help military service members and veterans of today and of tomorrow, as well.

Making the significant improvements we recommend requires *a sense of urgency* and *strong leadership.* The tendency to make systems too complex and rule-bound must be countered by a new perspective, grounded in an understanding of the importance of *patient-centeredness.* From the time injured service members are evacuated from the battlefield to the time they go back to active duty or are discharged home to complete their education, go to work, and be active family and community members, their needs and aspirations should inform the medical care and disability systems. . . .

Recommendations

Our recommendations will serve, support, and simplify health care and rehabilitation for injured service men and women, and return them as quickly as possible to their military duties or to civilian life. To make these recommendations a reality, the President, Congress, and Departments of Defense (DoD) and Veterans Affairs (VA) should initiate the steps described in this report.

1 Immediately Create Comprehensive Recovery Plans to Provide the Right Care and Support at the Right Time in the Right Place

Recommendation: Create a patient-centered Recovery Plan for every seriously injured service member that provides the right care and support at the right time in the right place. A corps of well-trained, highly-skilled Recovery Coordinators must be swiftly developed to ensure prompt development and execution of the Recovery Plan.

Goals: Ensure an efficient, effective and smooth rehabilitation and transition back to military duty or civilian life; establish a single point of contact for patients and families; and eliminate delays and gaps in treatment and services.

What it is:

The Recovery Plan should smoothly and seamlessly guide and support service members through medical care, rehabilitation, and disability programs.

The Recovery Plan will help service members obtain services *promptly* and *in the most appropriate care facilities*—whether DoD, VA, or civilian.

The Recovery Coordinator is the patient and family's single point of contact, who makes sure each service member receives the care specified for them in the plan when they need it, and that no one gets "lost in the system."

The Recovery Coordinator moves injured service members through the system in a timely way, because experience shows that people recover better when treatment and services are provided promptly.

Who oversees it:

A Recovery Coordinator would oversee implementation of the Recovery Plan. Recovery Coordinators would have the authority to coordinate medical care, rehabilitation, education, and employment-related programs, as well as disability benefits. This is a difficult and complex job, and both Departments must be committed to making it work.

Recovery Coordinators would ensure that patients and families understand the likely trajectory of the service member's recovery, the types of care and services that will be needed, and how much time recovery may take. . . .

2 Completely Restructure the Disability and Compensation Systems

Recommendation: DoD maintains authority to determine fitness to serve. For those found not fit for duty, DoD shall provide payment for years served. VA then establishes the disability rating, compensation and benefits.

Goals: Update and simplify the disability determination and compensation system; eliminate parallel activities; reduce inequities; and provide a solid base for the return of injured veterans to productive lives. . . .

Department of Defense Responsibilities

Each branch of the armed services would retain authority for determining whether a service member is fit for continued military service.

If not medically fit, the service member should receive DoD annuity payments, the dollar value of which would be *based solely on rank and length of service.*

Department of Veterans Affairs Responsibilities

The VA should assume all responsibility for establishing disability ratings and for all disability compensation and benefits programs.

The VA should initiate its education, training, and work-related benefits *early in the rehabilitation period.*

The Department's education, training, and employment programs should include incentives to encourage veterans to participate and stay enrolled. (Our survey found that 21 percent of demobilized reservists and 31 percent of retired/separated service members are enrolled in an educational program leading to a degree.)

Periodic Review

The disability status of veterans should be reevaluated every three years and compensation adjusted, if their condition has worsened or improved.

Vocational Rehabilitation & Education Program (VRE)

The effectiveness of various vocational rehabilitation programs is not well established, and the VA should undertake an effort to determine which have the greatest long-term success.

VA policies should encourage completion of effective programs by increasing the flexibility of scheduling for those whose disability does not permit taking a full course load. This can be done without increasing the dollar amount of the benefit. Also, the VA should develop financial incentives that would encourage completion. . . .

3 Aggressively Prevent and Treat Post-Traumatic Stress Disorder and Traumatic Brain Injury

Recommendation: VA should provide care for any veteran of the Afghanistan and Iraq conflicts who has post-traumatic stress disorder (PTSD). DoD and VA must rapidly improve prevention, diagnosis, and treatment of both PTSD and traumatic brain injury (TBI). At the same time, both Departments must work aggressively to reduce the stigma of PTSD.

Goals: Improve care of two common conditions of the current conflicts and reduce the stigma of PTSD; mentally and physically fit service members will strengthen our military into the future. . . .

Workforce Strategies

We recognize that augmenting DoD's mental health workforce will not be easy, because of national shortages in mental health professionals. DoD personnel requirements must take into account the expanding need for such personnel, due to the military's expanded prevention and education missions in behavioral health; and, both Departments should prepare for the expected long-term demand that may arise from chronic or delayed-onset symptoms of PTSD.

Reduce Stigma

DoD should intensify its efforts to reduce the stigma associated with PTSD. . . .

4 Significantly Strengthen Support for Families

Recommendation: Strengthen family support programs including expanding DoD respite care and extending the Family and Medical Leave Act for up to six months for spouses and parents of the seriously injured.

Goals: Strengthen family support systems and improve the quality of life for families. . . .

Many of the recommendations in this report serve and support families and simplify their lives. Prime examples are the Recovery Coordinator and increased availability of online resources that will be helpful to family caregivers.

DoD and VA should explore the applicability for service members and their families of innovative private-sector initiatives that have been developed and tested in the past few years.

DoD should establish a standby plan for family support of injured service members, drawing on the experiences and model programs developed during this conflict, to enable a quicker program ramp-up in any future large deployments. . . .

5 Rapidly Transfer Patient Information Between DoD and VA

Recommendation: DoD and VA must move quickly to get clinical and benefit data to users. In addition, DoD and VA should jointly develop an interactive "My eBenefits" website that provides a single information source for service members.

Goals: Support a patient-centered system of care and efficient practices.

Three Strong Caveats:

- Congress and the Departments should recognize that information technology is not the "silver bullet" that will solve various quality, coordination, and efficiency problems within the Departments' medical and benefits systems.
- Underlying organizational problems must be fixed first, or information technology merely perpetuates them.
- Every effort must be made not to make systems unnecessarily complex, difficult to use, or redundant.

DoD and VA should make information about benefits and services available online, via a password-protected site (which we call "My eBenefits"), in which service members and veterans can securely enter personal information. Based on this profile, they would receive tailored information about relevant programs and benefits in both the public and private sectors. . . .

6 Strongly Support Walter Reed By Recruiting and Retaining First-Rate Professionals Through 2011

Recommendation: Until the day it closes, Walter Reed must have the authority and responsibility to recruit and retain first-rate professionals to deliver first-rate care. Walter Reed Army Medical Center has a distinguished history and, with one in five injured service members going directly to Walter Reed, continues to play a unique and vital role in providing care for America's military.

Goals: Assure that this major military medical center has professional and administrative staff necessary for state-of-the-art medical care and scientific research through 2011.

Approximately one in five injured service members go directly to Walter Reed, and more than 700 outpatients remain on the campus.

Not only is it active today, but Walter Reed is scheduled to continue operation for at least four more years and must have the resources—professional and otherwise—to continue its historic role as a vital tertiary care and research center until the day it actually ceases operation. . . .

SOURCE: U.S. President's Commission on Care for America's Returning Wounded Warriors. "Serve, Support, Simplify." July 25, 2007. purl.access.gpo.gov/GPO/LPS84237 (accessed March 8, 2008).

OTHER HISTORIC DOCUMENTS OF INTEREST

FROM THIS VOLUME

- Ground war in Iraq, p. 407

FROM PREVIOUS *HISTORIC DOCUMENTS*

- Veterans affairs secretary on ALS in Gulf War veterans, *2001*, p. 907

DOCUMENT IN CONTEXT

United Nations on the Violence in the Democratic Republic of the Congo

JULY 27 AND NOVEMBER 14, 2007

Four years after the proclaimed end of a devastating war, and one year after generally successful elections, the Democratic Republic of the Congo in 2007 again faced violence in a flashpoint region. A rebellion by a renegade Congolese Tutsi general, who claimed to be protecting the Tutsi minority in eastern Congo, displaced more than 400,000 people from their homes and, even more significantly, raised the prospect of a return to the large-scale intercommunal violence that had plagued the country for most of the previous decade. At year's end, a United Nations (UN) peacekeeping mission in the Congo—the world body's largest ever—was struggling to prevent the violence from spreading.

One apparent outgrowth of the continuing conflict was what UN officials called an "epidemic" of sexual violence against women and girls. International experts said the extent and ferocity of rapes, maimings, and even murders of females in eastern Congo was the worst they had ever seen; to the experts, this suggested that armed and unemployed former fighters in the region were still venting anger after the end of the main part of the nation's war.

Background

The conflicts that plagued the Congo in 2007 had their immediate roots in the early 1990s, when ethnic violence between the Hutus and Tutsis in neighboring Burundi and Rwanda spilled across national borders in the Great Lakes region of central Africa. The conflicts became especially intense after the 1994 genocide in Rwanda, during which fanatical Hutus murdered an estimated 800,000 Tutsis and moderate Hutus. After being driven out of Rwanda by a Tutsi army, many of the Hutu fanatics (known as the Interahamwe) fled across the border into the Congo, then known as Zaire, where they continued their violent attack against the Tutsis and other Congolese ethnic groups. The Rwanda genocide set off a decade of war during which the Zairian dictator, Mobutu Sese Seko, was ousted from power. Rwanda and a half-dozen other countries intervened on various sides of the resulting civil war.

The massive conflict, often called "Africa's first world war" because so many nations were involved, lasted from 1998 through 2003 and resulted, according to some estimates, in the deaths of more than 4 million people—most of whom died from disease and starvation directly related to the conflict. The war was the largest and most complex anywhere in the world since World War II. Peace agreements in 2002 and 2003 ended most of the fighting and resulted in national elections in 2006, which were won by incumbent president Joseph Kabila, son of one of the central figures in the long conflict.

The Overall Political Situation

The peace agreements and subsequent elections ended the bulk of the conflict in most of the Congo but left some pockets of violent opposition to Kabila's government, which was supposed to be a government of national unity. At the center of one of these pockets was Jean-Pierre Bemba, a wealthy former rebel leader who had finished second to Kabila in the presidential race. Bemba insisted the election was fraudulent, and he refused to disarm the large militia that acted as his personal guard.

In January and February 2007, assemblies in each of the eleven provinces elected governors and members of the national senate. Bemba won a senate seat from the Bas-Congo province in southwestern Congo. The elections were marred by charges of vote-buying and other forms of corruption, and protests over the results in Bas-Congo province led to riots during which some 100 people died—many of them at the hands of the national army, which the UN said used "disproportionate force" against protesters.

Another episode of violence began on March 22, when intense clashes broke out in Kinshasa, the capital, between Bemba's private militia and the government's security forces. Explanations for the fighting varied, with some government officials charging that Bemba had attempted a coup against Kabila and Bemba's aides insisting that government forces had staged an unprovoked attack on his compound. Whatever the cause, the fighting lasted for two days and left about 300 people dead, according to a subsequent investigation by the UN high commissioner for human rights. A report summarizing the investigation's findings, published in January 2008, alleged serious human rights violations by both sides. In particular, the report cited "credible information" that at least forty civilians and members of Bemba's armed guard were summarily executed after they surrendered to the government.

During the fighting, Bemba took refuge in the South African embassy, and the Congolese government issued an arrest warrant charging him with treason. He left the Congo on April 12, escorted to Kinshasa's airport by UN troops, and eventually fled to Portugal.

Bemba's departure from the Congo left Kabila as the unchallenged leader of the national government but did not ensure that the government would be effective or operate harmoniously. One consequence of the peace agreements was that several former rebel leaders had entered the government, in some cases taking charge of ministries that became their personal fiefdoms. Numerous reports by news organizations and nongovernmental organizations said Kabila's government also struggled, with little success, to gain control over the vast natural resources—diamonds, gold, timber, and various minerals—that gave the Congo the potential to be one of the richest countries in Africa. These resources had been controlled by corrupt officials and businessmen during Mobutu's decades in power and had fallen into the hands of foreign armies and rebel leaders during the war following the overthrow of Mobutu. After the war, they were still being plundered for private gain by former rebel leaders, corrupt government officials, and reportedly even by foreign governments.

Rebellion in North Kivu

Many of the Congo's most valuable resources were located in the eastern provinces, which had been at the center of conflict since the early 1990s and remained the most volatile areas because of historical animosities among ethnic groups there and the proximity to countries with their own internal conflicts: Burundi, Rwanda, Sudan, and Uganda. The most serious flashpoint remained North Kivu province, in large part because of the continued presence there of thousands of Hutu fighters from Rwanda.

The Hutu fighters posed a constant threat to native Congolese Tutsis in North Kivu—or such was the argument of a renegade Congolese Tutsi general, Laurent Nkunda, who had his own militia, which he said was needed to protect the Tutsis against the ravages of the Hutu fighters. Ironically, the national elections—in which minority Tutsis lost the political power they had held in eastern Congo in previous years—added fuel to Nkunda's argument that Tutsis needed his protection.

An ethnic Tutsi who had lost family members in battles between Hutus and Tutsis following the 1994 genocide in Rwanda, Nkunda had fought as part of the rebel forces backed by Rwanda in the Congolese war that started with the overthrow of Mobutu in 1997. After the war ended, Nkunda kept his rebel army intact. Numerous UN reports documented Nkunda's practice of forcibly recruiting boys and girls, some of them under the age of ten.

After an outbreak of fighting in late 2006, Kabila's national government sought to gain control of the situation by agreeing, in January 2007, to combine its army units in North Kivu with Nkunda's forces. Called *mixage,* the process was supposed to pacify the area by putting Nkunda's armed men under government command. *Mixage* quickly proved to be a failure, however, because Nkunda retained effective control of his troops, who could roam even more widely under the guise of being aligned with the government. By mid-July, the United Nations called the step "disastrous" because it led to more fighting that displaced tens of thousands of additional people in an area were thousands previously had been forced from their homes.

Nkunda remained at large in North Kivu, commanding an estimated 8,000 fighters and even declaring himself the head of a political movement, the National Congress for the Defense of the People, with its own Web site and radio station. In late August, Nkunda launched new attacks against a government garrison in North Kivu, and the UN peacekeeping mission responded by airlifting army troops to combat the attacks. The UN negotiated a cease-fire but fighting continued anyway. The situation in North Kivu remained unstable, particularly in October when Nkunda's forces came under separate attacks from the army and a militia known as the Mai-Mai.

On November 11, diplomats from the Congo and Rwanda signed an agreement calling for a joint campaign against all rebel factions in eastern Congo, including Nkunda and the Hutu rebels whom Nkunda said he was fighting. The Congolese army launched an offensive against Nkunda in early December but lost ground when the rebel forces attacked the army. As in previous cases of fighting, UN peacekeepers intervened to separate the combatants and protect civilians, a move that shielded the undisciplined and demoralized government forces from an embarrassing defeat.

The government called a "peace summit" for December 27 in hopes of reaching broad agreement on ending the fighting. But that meeting was postponed until early January 2008, when negotiations collapsed because Nkunda's representatives withdrew.

According to UN figures, the various episodes of conflict in North Kivu between late 2006 and the end of 2007 forced more than 440,000 people from their homes—many of them permanently because their houses and crops were destroyed. This was in addition to some 600,000 people who remained displaced in eastern Congo as a result of earlier stages of fighting. Many of those people had lived in refugee camps for years, while others had crowded into cities, where the lack of clean water and sanitary facilities created epidemics of cholera and other diseases.

Violence against Civilians

The United Nations and several private international organizations sought during 2007 to focus attention on a dramatic explosion of violence against civilians, particularly women, in eastern Congo. Officials said the violence was committed by fighters on all sides of the region's conflicts: government troops, fighters for Nkunda and other rebel leaders, and former rebels who were no longer part of any organized fighting force. Many of the worse cases, UN officials said, appeared to be the work of Hutu extremists operating in the Congo, including the Interahamwe militia responsible for the Rwanda genocide.

The UN's most dramatic report on sexual violence in the Congo was made at a July 27 news conference in Kinshasa by Yakin Ertürk, special rapporteur on violence against women of the UN Human Rights Council. Summarizing an eleven-day visit to three areas of eastern Congo, Ertürk described the situation in South Kivu province as "the worst crisis I have encountered so far" in four years of work with the UN council. She said officials had tallied 4,500 cases of sexual violence in the first six months of the year—but the actual total was "many times higher."

Ertürk said rape was only part of the violence committed against women: "The atrocities are structured around rape and sexual slavery and aim at the complete and psychological destruction of women with implications for the entire society." This was a tactic used by the Hutu Interahamwe during the Rwanda genocide, she noted. On October 26 Ertürk told a UN General Assembly committee that the scale of violence was so acute that it amounted to war crimes.

The UN's chief humanitarian aid official, John Holmes, echoed Ertürk's concerns after his four-day visit to eastern Congo in early September. "Violence and rape at the hands of these armed groups has become all too common," he stated, according to a September 8 report by the *Washington Post*. "The intensity and frequency is worse than anywhere else in the world." Holmes said he had spoken with many women who recounted having been raped by members of armed groups, including the army. The rapes often were brutal and done publicly, as a form of humiliation; for example, women were gang-raped in public, sometimes while their husbands were forced to watch.

Human Rights Watch, a New York–based advocacy group, released a detailed report on October 23, detailing violence against civilians by both government and rebel forces in North Kivu. In particular, the report cited accounts by eyewitnesses and victims of attacks by Nkunda's forces on more than fifty villages—attacks that led to the deaths of dozens of people. Human Rights Watch said the UN peacekeeping mission had been "slow" to act against the violence.

International Criminal Court Cases

The massive war in the Congo led to two of the first cases to be handled by the UN-backed International Criminal Court (ICC), established in 2002 to bring justice in major war crimes cases. The ICC on January 29 ruled that former warlord Thomas Lubanga Dyilo should stand trial on charges of recruiting child soldiers and sending them into battles involving competing tribes in the Ituri region of eastern Congo during 2002 and 2003. Lubunga denied the charges. He was the first person charged with war crimes by the ICC, and his trial would be the court's first. He was arrested in Kinshasa in March 2005 and imprisoned in a high-security facility in the Netherlands.

Another former Congolese rebel leader, Germain Katanga, was taken into custody by the court in October. Katanga was charged with three counts of crimes against humanity and six war crimes, including the use of child soldiers and ordering the murder of some 200 civilians, during a violent conflict early in 2003 between rival ethnic groups in the Ituri region of Orientale province in northeastern Congo.

Following are two documents. First, excerpts from a statement, given at a news conference in Kinshasa on July 27, 2007, by Yakin Ertürk, special rapporteur on violence against women of the United Nations Human Rights Council, detailing her observations during a July 16–27 visit to the Democratic Republic of the Congo. Second, the "Observations and Recommendations" section from a November 14, 2007, report by Secretary-General Ban Ki-moon to the United Nations Security Council on the situation in the Congo.

UN Special Rapporteur on Violence against Women in the Congo

July 27, 2007

. . . In view of the seriousness and urgency of the situation in the DRC, my visit focused mainly on sexual violence, which is rampant and committed by non-state armed groups, the Armed Forces of the DRC, the National Congolese Police and increasingly also by civilians. However, I would like to caution against singling out sexual violence from the continuum of violence that Congolese women experience, which manifests itself in various forms in their homes and communities. Violence against women seems to be perceived by large sectors of society to be normal.

Tragically, in a resource rich country like DRC, poverty is all too striking and women disproportionately bear its hardships and burden. Empowerment and equality of women, socio-economic development and change of mentalities on gender must be prioritized as integral components of the reconstruction process if sustainable and just peace is to be achieved in the Democratic Republic of Congo. . . .

Sexualized Atrocities in South Kivu

I would like to start by drawing attention to the situation in South Kivu Province, which is alarming and requires immediate action. From the perspective of my mandate, which focuses on violence against women, the situation in the Kivus is the worst crisis I have encountered so far.

The South Kivu Provincial Synergie on Sexual Violence, a body bringing together representatives from Government, UN and civil society, has recorded 4500 sexual violence cases in the first six months of this year alone. The real number of cases is certainly many times higher as most victims live in inaccessible areas, are afraid to report or did not survive the violence.

Most of the sexual violence in South Kivu is reportedly perpetrated by foreign non-state armed groups. A number of their members appear to have been implicated in the

Rwandan genocide and subsequently fled to the DRC. Operating from forest areas, these armed groups raid local communities, pillage, rape, sexually enslave women and girls and subject them to forced labour.

The atrocities perpetrated by these armed groups are of an unimaginable brutality that goes far beyond rape. The atrocities are structured around rape and sexual slavery and aim at the complete physical and psychological destruction of women with implications for the entire society. They are in many ways reminiscent of those committed by the Interahamwe during the Rwandan genocide.

Women are brutally gang raped, often in front of their families and communities. In numerous cases, male relatives are forced at gun point to rape their own daughters, mothers or sisters. Frequently women are shot or stabbed in their genital organs, after they are raped. Women, who survived months of enslavement, told me that their tormentors had forced them to eat excrements or the human flesh of murdered relatives.

The Panzi Hospital, a specialized institution in Bukavu (South Kivu), receives annually about 3500 cases of women who suffer fistula and other severe genital injuries resulting from these sexualized atrocities. I spoke with a 10-year old girl at the hospital, who had been abducted together with her parents. She had to have an emergency operation, because the perpetrators had rammed a stick into her genital organs.

The FARDC [Armed Forces of the Democratic Republic of the Congo, the Congolese army] has so far proven to be unable to stop the atrocities in South Kivu, which have been raging for several years. The international community, in collaboration with the Government, must immediately exercise its responsibility to protect and become fully engaged at all levels to end these atrocities.

Sexual Violence Perpetrated by Army and Police

The FARDC, the National Congolese Police (PNC) as well as other State security forces continue to perpetrate sexual violence. In South Kivu and the Ituri, while non-state armed groups remain the main perpetrators of sexual violence, close to 20% of all cases of sexual violence are reportedly committed by the FARDC and the PNC.

FARDC units are said to be deliberately targeting civilian communities suspected of supporting militia groups and pillage, gang rape and, in some instances, murder civilians. Individual soldiers or police also commit such acts, considering themselves to be above the law. These acts amount to war crimes and, in some cases, crimes against humanity. International law requires the Government to bring all perpetrators, including the bearers of command responsibility, to justice.

Sexual violence committed by the State security forces is not restricted to areas of armed conflict in the East of the country. The problem is exacerbated by the fact that the process of integrating former militia in the regular armed forces does not entail any mechanism to exclude the major perpetrators of grave human rights violations from the armed forces. Consequently, a number of such men have assumed high ranks in the FARDC.

In Equateur Province, I was shocked to find out that the PNC and FARDC frequently respond to civilian unrest with organized armed reprisals that target the civilian population and involve indiscriminate pillaging, torture and mass rape. In December 2006, for instance, the PNC assembled about 70 police officers from duty stations across the region to take revenge for the burning of the police station in Karawa. The PNC officers pillaged the town, tortured civilians and raped at least 40 women, including an 11-year-old girl.

So far not a single police has been charged or arrested in relation to the Karawa atrocities. There are allegations that FARDC carried out similar punitive operations against the civilian population in Bonyanga (120 km south west of Gemena) in April 2007 and in Bongulu (90 km north of Bumba) in May 2007.

Sexual Violence Perpetrated by Civilians

Sexual violence is increasingly also committed by civilians. Some of these crimes are said to be committed by demobilized militiamen, who generally received no psycho-social rehabilitation in the reinsertion process.

More fundamentally, the widespread use of sexual violence in the armed conflict seems to have become a generalized aspect of the overall oppression of women in the DRC. Such behavioural norms will therefore remain a serious problem in the future—regardless of the security situation, unless Government and society are willing to make a serious effort to fundamentally change the prevailing gender relations that subordinate and devalue women.

As a starting point, the National Assembly can play a key role by undertaking basic legal reforms. The Family Code, for instance, effectively declares women to be minors under their husband's guardianship. While the new Constitution foresees gender parity, a Gender Parity Law to implement Article 14 of the Constitution has yet to be enacted.

Impunity

In July 2006, the Congolese Parliament passed the Law on the Suppression of Sexual Violence, which foresees strengthened penalties and more effective criminal procedures. In practice, however, little action is taken by the authorities to implement the law and perpetrators continue to enjoy impunity, especially if they wear the State's uniform.

The justice system is in a deplorable state. It is overwhelmed even by the limited number of cases, in which women brave all obstacles and dare to report sexual violence. Reports of corruption and political interference in the judicial process are widespread. In cases involving the FARDC or the Police, senior officers are said to shield the men under their command from prosecution and deliberately obstruct investigations. This results in impunity for perpetrators of mass rapes and other crimes against humanity.

I would like to acknowledge with admiration that there are some committed justice officials, who display a willingness to uphold the rule of law. In February 2007, for instance, the Military Tribunal of the Ituri Garrison, convicted 15 officers and soldiers in connection with a FARDC massacre in Bavi Village, Ituri. The commanding officer of the responsible unit was also convicted of rape and sentenced to life imprisonment. Such judgements, however, remain rare exceptions defying the rule of impunity, which prevails in the DRC.

State of Penitentiary System

Perpetrators of grave human rights violations, whether convicted or awaiting trial, seem to have no difficulty in escaping from prison. The penitentiary system is in a scandalous state and I was glad to hear from the Minister of Justice that the Government envisages a comprehensive reform of the penitentiary system.

Minimum standards for the humane treatment of prisoners were not even remotely upheld in any of the prisons I visited. Except for Makala Prison in Kinshasa, no prison receives funds from the Government to provide inmates with food. As a result, some prison directors have no choice but to allow inmates to leave the prison during the day to look for food and drinkable water.

In many cases of "escapes", there is also reason to believe that detainees simply walked out of prison with the complicity of those in charge. In April 2006, for instance, the Military Tribunal of Mbandaka Garrison convicted seven FARDC officers for crimes against humanity and sentenced them to life imprisonment. The officers were among a group of FARDC soldiers that raped at least 200 women and girls in Songo Mboyo (Equateur) in late December 2003. All seven perpetrators have since escaped from prison under most dubious circumstances.

Insufficient Support for Victims

Many rape victims are re-victimized because they are rejected by their own communities, families or husbands due to the stigma attached to rape, while the rapists enjoy impunity. Deprived of social support networks, these women are destitute and struggle for their mere physical survival. The problem is compounded by the fact that most survivors grapple with grave medical and psychological problems. The plight of the babies born of rapes is another serious concern yet to be addressed.

Provincial Synergies on Sexual Violence, bringing together dedicated local officials, civil society and the UN, have launched programmes that try to provide victims with the necessary medical, psycho-social, legal and economic support. However, these laudable initiatives must deal with a magnitude of needs with limited funds made available by international donors.

In a handful of cases, including the atrocities perpetrated by FARDC soldiers in Songo Mboyo and Bavi, courts have ordered individual perpetrators as well as the Congolese State to pay modest reparations to the victims. To this day, the Government has not paid reparations to a single victim who has suffered sexual violence at the hand of State agents.

Source: United Nations. MONUC: UN Mission in DR Congo. "South Kivu: 4,500 Sexual Violence Cases in the First Six Months of This Year Alone." July 27, 2007. www.monuc.org/news.aspx?newsID=15065 (accessed December 21, 2007).

DOCUMENT

UN Secretary-General on the Situation in the Congo

November 14, 2007

XIV. Observations and recommendations

65. Since my last report [on March 20, 2007], the Government has demonstrated a strong determination to exert its sovereign responsibilities throughout the country,

engaged with its neighbours, worked closely with the international community to advance planning for recovery and economic development, taken the first steps towards security sector reform, initiated key legislation and made a firm commitment to decentralization and the early holding of local elections. The Parliament is also to be commended for the assiduous manner in which it has collectively shouldered its responsibilities. The emergence of assertive provincial governments, as provided for in the Constitution, is also a welcome development, as effective decentralization is an essential element to good governance, stability and growth in a country as vast as the Democratic Republic of the Congo.

66. Notwithstanding these positive developments, prospects for the future continue to be overshadowed by long-standing security challenges in the eastern Democratic Republic of the Congo. The rule of law and respect for human rights, in particular by security services, must also be strengthened. These challenges require MONUC [the United Nations Organization Mission in the Democratic Republic of the Congo] to maintain a robust capacity in the eastern Democratic Republic of the Congo and a continued police, rule of law, human rights, political, and civil affairs presence throughout the country. I therefore recommend renewal of the Mission's mandate for a period of one year, and that it retain its current complement of military, police and civilian personnel at least until the end of the local elections. Gradual drawdown may then commence, subject to progress towards the broad benchmarks set out in the present report, including successful completion of the local elections, and, most importantly, towards ensuring the security of the population.

67. The crisis in the Kivus has many dimensions which call for a comprehensive solution, including an end to the fighting, the disarmament and demobilization or integration into the armed forces of militias and the re-establishment of State authority and rule of law. Steps are also urgently needed to resolve the problems presented by all foreign armed groups on Congolese soil, particularly FDLR [Democratic Forces for the Liberation of Rwanda, a Hutu militia], which constitutes the most destabilizing element. A purely military solution to this issue is neither desirable nor viable. In order to address the problem of foreign armed groups, a common approach and close cooperation will be needed between the Democratic Republic of the Congo and other States in the region, supported by the international community. In this regard, I am encouraged by the recent bilateral meetings between senior representatives of the Governments concerned and their contacts through multilateral and regional mechanisms.

68. Conflict in the eastern Democratic Republic of the Congo cannot be separated from its regional aspects. I welcome the emergence during the reporting period of increasing dialogue among Great Lakes countries, both bilaterally and through mechanisms such as the Great Lakes Pact and the Tripartite Plus Joint Commission. Improvement of regional relations, particularly but not exclusively related to security, should be sustained by concrete follow-up action by all Governments as a public sign of their determination to work together in resolving cross-border issues.

69. There is increasing recognition that the intertwined problems created by the activities of [dissident former general Laurent] Nkunda and his militia, FDLR and other foreign and Congolese armed groups must be addressed simultaneously through an approach that involves all major stakeholders, both within the Democratic Republic of the Congo and in the Great Lakes region. I therefore dispatched Haile Menkerios, Assistant Secretary-General for Political Affairs, on 1 November to undertake a special mission to the region to consult with the Government of the Democratic Republic of Congo and leaders in the region, as well as with the Government's bilateral and multilat-

eral partners, to find ways to resolve the immediate crisis and to address its underlying causes. Mr. Menkerios is coordinating closely with my Special Representative for the Democratic Republic of the Congo and with international partners currently engaged in initiatives to help resolve the crisis.

70. The protection of civilians of all communities is central to the resolution of the crisis in the eastern Democratic Republic of the Congo. MONUC will continue to focus on its mandate for the protection of civilians and to work in close cooperation with the Government and the Congolese security forces to this end.

71. In order to enable the Congolese military, police and judiciary and correctional services to assume their full responsibilities as MONUC begins its drawdown, a strong security sector reform programme is needed. Such a programme should reflect a unity of vision among the relevant Congolese actors and between the Democratic Republic of the Congo and its international partners. Such a programme should lead to an early operational capacity for essential military tasks, as well as steady progress in reform of the police and judiciary. Progress in security sector reform is essential for the Mission's exit strategy, and will require the committed effort of national and international actors. I urge the Government and its international partners to seize the opportunity presented by the planned round table on security sector reform to develop a detailed and coherent blueprint without delay.

72. Fighting impunity within the security services will further increase their effectiveness by allowing them to win the confidence and support of the people. I therefore call upon the Government to take advantage of the assistance and advice of the international community, including MONUC, and to further intensify its efforts to hold accountable members of the security services who are found guilty of human rights violations and other crimes.

73. While credible pressure is essential in dealing with the remaining armed groups in the east, experience in Ituri has shown that a gap in time between demobilization and reintegration, often caused by financial constraints, risks compromising the entire effort with serious consequences for security. Sufficient and timely donor funding is critical for channelling demobilized combatants to community reintegration. In this connection, I hope that the Government of the Democratic Republic of the Congo and the Multi-Country Demobilization and Reintegration Programme will reach early agreement on terms for the resumption of activities under the Programme in the Democratic Republic of the Congo. Inadequate reintegration programmes and the discontent they engender have also been major causes of unrest throughout the country. I would therefore propose a comprehensive review of disarmament, demobilization, and reintegration programmes in the Democratic Republic of the Congo with a view to capitalizing on lessons learned and determining how to ensure that former combatants can benefit from reintegration programmes without delay.

74. Sustainable peace and security will also require safer and more harmonious communities. In that connection, more attention to longer-term reintegration, including increased benefits for receiving communities, is required. Community disarmament programmes, which are about to be initiated by UNDP [United Nations Development Programme], must also be well supported in order to reduce the number of weapons in the hands of the population.

75. The decision of the Government to conduct local elections in 2008 represents a major step towards consolidating the considerable progress already achieved in the establishment of democracy. Encouraging progress has been made towards the adoption of

laws on the status of the opposition and the financing of political parties. I urge political leaders at all levels to respect the principles of transparency, inclusiveness and tolerance of dissent so as to create space for reform and build the credibility of the country's legitimate institutions. As requested by the Government, I recommend that the Council authorize MONUC to provide full support to the local elections on the same scale as for the national elections in 2006, subject to progress by the Government and relevant national institutions in putting in place the legal, institutional and financial frameworks needed to conduct credible polls.

76. I would like to thank my Special Representative, William Lacy Swing, and the women and men of MONUC and the United Nations country team for their determination and dedication, particularly over the past difficult months. Their service, often under difficult conditions, has provided invaluable support for the efforts of the people of the Democratic Republic of the Congo as they seek to realize their aspirations for peace, security and respect for human rights. My appreciation also goes to the countries contributing police and troops to MONUC, as well as to donor countries and the multilateral and non-governmental organizations that continue to provide the support needed to sustain progress in the Democratic Republic of the Congo.

SOURCE: United Nations. Security Council. "Twenty-Fourth Report of the Secretary-General on the United Nations Organization Mission in the Democratic Republic of the Congo." S/2007/671. November 14, 2007. www.un.org/Docs/sc/sgrep07.htm (accessed December 21, 2007).

OTHER HISTORIC DOCUMENTS OF INTEREST

FROM PREVIOUS *HISTORIC DOCUMENTS*

- Children at risk in the Democratic Republic of the Congo, *2006,* p. 414
- UN secretary-general on World Criminal Court, *2002,* p. 605
- Rwanda genocide, *1998,* p. 614

DOCUMENT IN CONTEXT

United Nations and Human Rights Coalition on the Violence in Darfur

JULY 31 AND DECEMBER 19, 2007

More than four years after fighting broke out in the region of western Sudan known as Darfur, the conflict had diminished in intensity but continued to resist international diplomacy and peacemaking efforts. The United Nations (UN) finally secured agreement on deployment to Darfur of a large peacekeeping mission, one that the Sudanese government had resisted. But that force began its work at year's end hobbled by a shortage of troops and equipment—making it appear little different from an ineffective African Union mission whose troops often had been forced to watch helplessly as combatants killed each other and thousands of civilians.

The latest round of peace talks between Sudan's government and rebel groups fighting for power, land, and resources in Darfur got under way in October but made no progress and offered little hope for success in the future. Previous rounds of diplomatic peacemaking had achieved limited agreements that collapsed.

In Darfur, the military situation became more complex than at any time since fighting erupted in 2003. Rebel groups splintered into numerous factions: one major rebel group aligned itself with the government while some militias that had been aligned with the government switched sides to support the rebels, with the result that outsiders often found it difficult to tell who was fighting whom and for what reason. Even so, the essence of the conflict remained the same: the ruling party in Khartoum, Sudan's capital, sought to consolidate its control of territory and of Sudan's limited political space in advance of elections scheduled for 2009, while numerous factions scrambled for their share of power and land.

Meanwhile, hundreds of thousands of civilians continued to suffer both from violence and from deprivation of food and shelter. Civilians remained at the mercy of all the various combatants; they were easy targets for rape and pillage, and they were too often seen as pawns to be manipulated. By late 2007 official UN figures showed that some 2.2 million people had been displaced from their homes but remained in Darfur; another 200,000 had fled across the border into neighboring Chad. About 4 million Darfurians remained in their home villages and towns, but at least half of them lacked adequate food or medical supplies. Altogether, the UN classified 4.2 million Darfurians—two-thirds of the region's total population—as dependent on international relief aid. The UN's estimate of the total number of people who had been killed or had died of disease or starvation since 2003 stood at "more than 200,000."

Since the early stages of the conflict in 2003, UN and private aid agencies had mobilized the largest relief effort in the world on behalf of Darfur's refugees and displaced people. This massive endeavor, involving nearly 15,000 Sudanese and international aid workers, encountered enormous logistical difficulties that were made worse by frequent cases of violence against aid agencies. Relief work often was suspended following attacks on aid convoys, and several agencies reluctantly withdrew from Darfur because of the danger to their staffs.

Combining the African Union and UN Peacekeepers

Until the UN began deploying a peacekeeping force in late 2007, the most significant international attempt to stop the killing in Darfur was made by the African Union Mission in Sudan (AMIS), an African Union (AU) force that was put in place as the result of a limited cease-fire agreement signed by the Sudanese government and two rebel groups in 2004. AMIS, which eventually grew to some 7,000 personnel, had only a limited mandate—basically, to monitor the situation but with little authority to prevent violence against civilians—and it had neither enough personnel nor adequate equipment to patrol an area comparable in size to all of France.

The international community in 2006 stepped up pressure on Sudan to accept a larger, better-equipped force commanded by the UN, but Sudanese president Omar al-Bashir resisted accepting non-African peacekeepers, saying they represented "colonial" forces. Bashir finally relented in November 2006, agreeing to allow a small number of troops, police, and civilian personnel from UN member states outside Africa. The AMIS force was to be expanded with what the UN called a "light support package" of additional personnel by March 2007 and a "heavy support package" consisting of 3,000 military police and military equipment, including attack helicopters to be delivered in stages during 2007.

The "light" expansion was completed early in 2007, but Bashir created obstacles to the more important "heavy" expansion. He relented in mid-April, having again come under strong international pressure—this time, China, a UN Security Council member that previously had been reluctant to push him to make concessions, joined in. China was thought to have considerable leverage over the Sudanese government because it was the biggest customer for Sudan's oil and the nation's most important source of foreign investment.

Numerous news reports in April suggested that China's sudden willingness to confront Sudan resulted from pressure against China itself—a publicity campaign by Hollywood stars and others linking Darfur to the 2008 Summer Olympic games in Beijing. One of the most active figures in this regard was actress Mia Farrow, who called the Beijing games the "Genocide Olympics"—a reference to China's supposed disregard for the killings in Darfur. Arab leaders also reportedly put some pressure on Bashir to make concessions on the UN force; Bashir in the past had gotten at least tacit support from colleagues in the Arab League, but by 2007 many were embarrassed by news reports holding them responsible for Darfur.

Once Bashir agreed to the "heavy" expansion, the next issue was a longstanding UN demand for Sudan to accept a much larger "hybrid" peacekeeping force, which would include thousands of troops and other personnel from countries outside Africa. After extensive diplomatic negotiations, the UN and AU on June 6 reached general agreement on how such a force would operate. Bashir on June 17 told a delegation of ambassadors from the UN Security Council that Sudan would accept this new force unconditionally—a step considered a major breakthrough because in the past he had always demanded restrictions on any form of international intervention in his country.

The Security Council formally endorsed creation of the hybrid force on July 31 with the unanimous adoption of its Resolution 1769, which authorized a force of up to 19,555 peacekeeping and 6,432 police personnel, to be called UNAMID (UN-AU Mission in Darfur). The headquarters for the new force was to be in place within three months, and by the end of 2007 the command of peacekeeping in Darfur was to pass from the AU force to the new hybrid mission. In perhaps its most important provision, the UN resolution gave the new force the authority "to take the necessary action" to protect not only its own personnel but also civilians. This was more explicit authority than the African force had received—and it offered the potential for a more effective peacekeeping force.

UN secretary-general Ban Ki-moon praised adoption of the resolution as a "historic and unprecedented" step by the Security Council. "You are sending a clear and powerful signal of your commitment to improve the lives of the people of the region and close this tragic chapter in Sudan's history," he told the council.

Despite the clear diplomatic achievement, the resolution was a watered-down version of an original proposal advanced by the United States and other countries. Reflecting Sudan's concerns, China blocked two provisions, one calling for sanctions against Khartoum if it obstructed the peacekeeping mission and another authorizing the peacekeepers to disarm rebels and progovernment militias known as the Janjaweed. Human rights activists warned that the Sudanese government would use various maneuvers to obstruct the UN mission, for example by delaying permission for peacekeepers and their equipment to enter the country.

From the outset, it was clear that another hurdle facing the new UN force would be getting member nations to contribute enough equipment. UN officials said on August 7 that they had already secured pledges of enough troops for UNAMID—most to come from Africa—but would need attack helicopters to help the troops protect both themselves and civilians. Even as the UN was gearing up, the weaknesses of the old African mission, and the difficulties facing its replacement, were brought into new focus in late September when an estimated 1,000 rebels attacked a small unit of AU peacekeepers in northern Darfur, killing ten of them.

The UNAMID force opened its headquarters in the town of El Fasher (the capital of Northern Darfur province) on October 31, an event hailed as a milestone by the force's political representative, Rodolphe Adada. One month later, however, the head of UN peacekeeping missions worldwide, Under-Secretary-General Jean-Marie Guéhenno, told the Security Council that Sudan's government was blocking key components of the UNAMID mission, including specialized troops, support staff, and military equipment. He said Sudan's obstruction might mean that the UN was deploying "a force that will not make a difference" in the conflict.

A similar warning was made in a December 19 report published by a coalition of thirty-five international nongovernmental organizations, spearheaded by Human Rights Watch. The bluntly worded report, "UNAMID Deployment on the Brink," said the Sudanese government was "actively obstructing and undermining" the peacekeeping force and argued that other countries were "wrangling over details and shirking their own responsibilities" rather than confronting Sudan on the matter.

The new UNAMID hybrid force officially took over security responsibilities in Darfur as scheduled on December 31. In effect, however, the handover of responsibility involved little more than 8,000-plus AU troops and police officers switching their green AU caps to blue UN berets. The bulk of the 26,000-person force was not expected to arrive in Darfur until well into 2008. It was more problematic that the helicopters, supply trucks, and other military gear that UN officials said would be essential for the new force still had not been pledged.

Peace Talks

The Darfur conflict had not lacked for plans to end it, but the participants lacked the incentive and the political will to embrace those plans. Several attempted cease-fires had failed, as had an ambitious peace accord signed, under intense international pressure, by the government and one major rebel group in May 2006. None of these efforts to stop the fighting lasted long because the warring parties all had their own reasons for pressing on with the bloodshed.

Yet another ambitious but unsuccessful peacemaking effort got under way on October 27, 2007, in an unlikely setting: a grand conference center in the Libyan coastal city of Sirte, a location chosen by the conference sponsor, Libya's mercurial leader, Muammar Qaddafi, apparently because it was his hometown. Sudanese leader Bashir appeared at the start of the talks but quickly left the scene. Most of the major Darfur rebel leaders never arrived, and those who did appear said they were not yet ready for a formal peace agreement. The peace conference lasted for just two days before it was adjourned, ostensibly to give the rebels more time to prepare a united position. Rebel groups in late November signed what they called a "united" platform, but this step did not herald the return of peace talks.

U.S. Policies

Ever since the violence escalated in Darfur in 2003, the administration of President George W. Bush had been among the world's most vocal critics of Sudan's government. In 2004 Secretary of State Colin Powell described the killings in Darfur as "genocide," making the United States the first—and only—country to use that term. But the United States had been no more willing than other non-African countries to back up words with deeds. Washington provided large quantities of food and medicine for Darfur's civilians, plus small amounts of military equipment for the AU peacekeeping force, but made it clear that American soldiers would never be sent to Darfur. The Bush administration also seemed reluctant to jeopardize its relationship with key Sudanese intelligence officials who reportedly had aided the U.S. fight against the al Qaeda network.

Under pressure from domestic U.S. groups that had campaigned for tougher action in Darfur, President Bush on April 18 signaled to Sudan that his patience had run out. He warned Bashir that the United States would tighten longstanding unilateral economic sanctions against Sudan unless the Sudanese president agreed to permit deployment of a UN peacekeeping force, drop barriers to aid agencies working in Darfur, and end support for the Janjaweed militias, which had done much of the killing.

When no further progress was made on any of these fronts, Bush acted on May 29, announcing that the United States would step up enforcement of existing economic sanctions against about 100 Sudanese companies; would add another 31 Sudanese companies (most controlled by the Khartoum government) to the list of those barred from doing business with the United States; and would prevent U.S. businesses and financial institutions from allowing any transactions with a junior Sudanese cabinet official and the leader of one rebel group, both of whom were accused of fomenting violence in Darfur. In a statement, Bush singled out Bashir for harsh criticism but took no direct action against him. "President Bashir's actions over the past few weeks follow a long pattern of promising cooperation while finding new methods for obstruction," Bush said. Representatives of human rights advocacy groups criticized Bush's steps as too little, too late, and said the sanctions would have virtually no impact on the most important sector of Sudan's economy: the oil industry.

Bush on December 31 signed legislation (PL 110-174), which had been adopted unanimously in both the House and Senate, intended to encourage private U.S. companies—as well as local and state governments—to withdraw investments from foreign companies doing business with Sudan's oil industry or military enterprises. Twenty-two states and more than fifty universities had withdrawn investments from companies doing business in Sudan, and the legislation signed by Bush sought to shield them from lawsuits filed by investors. Even while signing the bill, Bush objected to decisions by state and local governments to engage in foreign policy matters; he insisted the Constitution granted the president sole discretion to decide foreign policy for the United States.

Violence Continued in 2007

As the diplomacy intended to end the fighting in Darfur continued for yet another year, the fighting continued at a much reduced level but, in some respects, became even more complicated than before. The violence never had been a clear-cut case of an Arab-dominated government fighting black African rebels, as was widely portrayed in international news media. Instead, the Darfur conflict, since beginning in 2003, had witnessed shifting coalitions of rebel groups battling each other, the Sudanese government, and the Janjaweed militias that were aligned with (and in some cases sponsored and armed by) the government.

New features in 2007 included outbreaks of conflict among the Janjaweed militias and a decision by some Janjaweed leaders to side with the antigovernment rebels. This new type of factional fighting appeared to stem, in large part, from some Janjaweed militia leaders' resenting that they had not received a fair share of the land and other booty they said had been promised by the government as payment for fighting the rebels. UN and other international officials also said the leaders of various fighting groups appeared to be competing for control of territory in advance of deployment of the expanded UN-AU peacekeeping force. In a report issued on September 20, Human Rights Watch said the situation in Darfur had evolved "from an armed conflict between rebels and the government into a violent scramble for power and resources involving government forces, Janjaweed militia, rebels and former rebels, and bandits."

The year's fighting killed hundreds, perhaps even thousands, of people in Darfur, including civilians as well as combatants. Although the total was unclear, casualties were well below the height of the fighting in 2003–2004. Even so, the fighting in 2007 displaced more than 240,000 people and posed dangers to the aid workers providing food, shelter, and medicine for the more than 4 million Darfurians reliant on aid.

Regional Impact

Over the years the conflict in Darfur spilled over into two neighboring countries, Chad and the Central African Republic. More than 200,000 Darfurians had fled into Chad, seeking to escape the ravages of the Janjaweed. About 200,000 Chadians had also fled their homes or fled into Darfur to escape a conflict between rebels and the government there. Similarly, fighting among rebel groups had displaced tens of thousands of people in the northern regions of the Central African Republic, which bordered both Chad and Sudan.

In an effort to calm the fighting and protect civilians in these countries, the UN Security Council on September 25 authorized the European Union to create a small peacekeeping force of up to 300 police officers and 50 military liaison officers. This force, which had just begun to deploy by the end of 2007, had an initial assignment of training a Chadian police unit to provide security for refugees and displaced persons. In October the government of Chad and rebel groups from the east signed a power-sharing agreement, but this accord quickly fell apart.

International Criminal Court Action

The UN-backed International Criminal Court (ICC) on April 27 issued arrest warrants for two senior officials said to have been responsible for significant attacks against civilians in Darfur during 2003 and 2004: Ahmed Haroun (spelled as "Ahmad Harun" by the court), who was the Sudanese government's junior minister for humanitarian affairs and previously had been in charge of security in Darfur; and Ali Kosheib (spelled as "Ali Kushayb" by the court), the alleged leader of the Janjaweed militia. Haroun also was one of the officials targeted by U.S. economic sanctions approved by President Bush on May 29.

Sudan's government refused to surrender Haroun to the ICC and said it had no jurisdiction over Kosheib. A frustrated Luis Moreno-Ocampo, the ICC's chief prosecutor, told the Security Council on December 5 that he was broadening his investigation of Sudanese officials to determine who was responsible for harboring Haroun and who had ordered recent attacks against refugee camps. "When will be a better time to arrest Haroun?" he asked. "How many more women, girls, have to be raped? How many more persons have to be killed?" Sudan's ambassador to the UN, Abdalmahmood Abdalhaleem Mohamad, called Moreno-Ocampos's charges "fabrications."

"Peace" in the South

Another conflict in Sudan—one that was supposed to have ended with a peace agreement signed in 2005—appeared to be on the verge of resurrection at several points in 2007. This was the two-decades-long battle between the Khartoum government and a black rebel group, the Sudanese People's Liberation Movement, over control of the huge region known as South Sudan. Under the much-heralded 2005 peace accord, rebel leaders were to join the government and rebel fighters were to be integrated into the national army. The agreement also provided that southerners could vote in 2011 on whether to remain within Sudan or split off as an independent country.

That agreement encountered numerous difficulties, however, starting with the death later in 2005 of charismatic rebel leader John Garang, followed by complaints from each side that the other was not complying with the terms of peace. By 2007 the government and former rebel leaders were still disputing some of the central issues from the war, notably who would control (and therefore benefit from) oil fields located at the intersection between northern and southern Sudan. In addition to blocking progress in resolving political issues, the continuing disputes between the two sides had prevented a large UN peacekeeping mission from doing its job—a precedent UN officials feared might be duplicated once the new UN-AU mission arrived in Darfur.

On October 11 rebel leaders pulled out of the "unity" government that was created by the 2005 peace agreement, charging that Bashir had failed to keep his promises. The government and rebels again resolved their differences—at least for the time being—in November.

Following are two documents. First, the text of Resolution 1769, adopted by the United Nations Security Council on July 31, 2007, authorizing deployment of a 26,000-person peacekeeping force in the Darfur region of Sudan; this was to be a "hybrid" force commanded by the African Union and the United Nations. Second, the text of a December 19 report, "UNAMID Deployment on the Brink: The Road to Security in Darfur Blocked by Government Obstructions," by a coalition of thirty-five international human rights advocacy groups led by Human Rights Watch.

DOCUMENT

UN Security Council Resolution 1769

July 31, 2007

The Security Council,

Recalling all its previous resolutions and presidential statements concerning the situation in Sudan,

Reaffirming its strong commitment to the sovereignty, unity, independence and territorial integrity of Sudan, and to the cause of peace, and *expressing its determination* to work with the Government of Sudan, in full respect of its sovereignty, to assist in tackling the various problems in Darfur, Sudan,

Recalling the conclusions of the Addis Ababa high-level consultation on the situation in Darfur of 16 November 2006 as endorsed in the communiqué of the 66 meeting of the Peace and Security Council of the African Union held in Abuja on 30 November 2006 as well as the communiqué of [the] 79th meeting of the Peace and Security Council of the African Union on 22 June 2007, *recalling* the statement of its President of 19 December 2006 endorsing the Addis Ababa and Abuja agreements, *welcoming* the progress made so far and *calling* for them to be fully implemented by all parties without delay and for all parties to facilitate the immediate deployment of the United Nations Light and Heavy Support packages to the African Union Mission in the Sudan (AMIS) and a Hybrid operation in Darfur, for which back-stopping and command and control structures will be provided by the United Nations, and *recalling* that co-operation between the UN and the regional arrangements in matters relating to the maintenance of peace and security is an integral part of collective security as provided for in the Charter of the United Nations,

Re-affirming also its previous resolutions 1325 (2000) on women, peace and security, 1502 (2003) on the protection of humanitarian and United Nations personnel, 1612 (2005) on children and armed conflict and the subsequent conclusions of the Security Council Working Group on Children in Armed Conflict pertaining to parties to the armed conflict in Sudan (S/2006/971), and 1674 (2006) on the protection of civilians in armed conflict, as well as *recalling* the report of its Mission to Addis Ababa and Khartoum from 16 to 17 June 2007,

Welcoming the report of the Secretary-General and the Chairperson of the African Union Commission of 5 June 2007,

Commending in this regard the agreement of Sudan that the Hybrid operation shall be deployed in Darfur, as detailed in the conclusions of the high-level AU/UN consultations with the Government of Sudan in Addis Ababa on 12 June 2007 and confirmed in full during the Council's meeting with the President of Sudan on 17 June in Khartoum,

Recalling the Addis Ababa Agreement that the Hybrid operation should have a predominantly African character and the troops should, as far as possible, be sourced from African countries,

Commending the efforts of the African Union for the successful deployment of AMIS, as well as the efforts of member states and regional organisations that have assisted it in its deployment, *stressing* the need for AMIS, as supported by the United Nations Light and Heavy Support Packages, to assist implementation of the Darfur Peace Agreement until the end of its mandate, *calling upon* the Government of Sudan to assist in removing all obstacles to the proper discharge by AMIS of its mandate; and *recalling* the communiqué of the 79th meeting of the Peace and Security Council of the African Union of 22 June to extend the mandate of AMIS for an additional period not exceeding six months until 31 December 2007,

Stressing the urgent need to mobilise the financial, logistical and other support and assistance required for AMIS,

Welcoming the ongoing preparations for the Hybrid operation, including the putting in place of logistical arrangements in Darfur, at United Nations Headquarters and the African Union Commission Headquarters, force and police generation efforts and ongoing joint efforts by the Secretary General and the Chairperson of the African Union to finalise essential operational policies, and *further welcoming* action taken so that appropriate financial and administrative mechanisms are established to ensure the effective management of the Hybrid,

Re-iterating its belief in the basis provided by the Darfur Peace Agreement for a lasting political solution and sustained security in Darfur, *deploring* that the Agreement has not been fully implemented by the signatories and not signed by all parties to the conflict in Darfur, *calling* for an immediate cease-fire, *urging* all parties not to act in any way that would impede the implementation of the Agreement, and *recalling* the communiqué of the second international meeting on the situation in Darfur convened by the African Union and United Nations Special Envoys in Tripoli from 15–16 July 2007,

Noting with strong concern on-going attacks on the civilian population and humanitarian workers and continued and widespread sexual violence, including as outlined in the Report of the Secretary-General and the Chairperson of the African Union Commission on the Hybrid Operation in Darfur and the report of the Secretary-General of 23 February 2007, *emphasising* the need to bring to justice the perpetrators of such crimes and *urging* the Government of Sudan to do so, and *reiterating* in this regard its condemnation of all violations of human rights and international humanitarian law in Darfur,

Reiterating its deep concern for the security of humanitarian aid workers and their access to populations in need, *condemning* those parties to the conflict who have failed to ensure the full, safe and unhindered access of relief personnel to all those in need in Darfur as well as the delivery of humanitarian assistance, in particular to internally displaced persons [IDPs] and refugees, and *recognising* that, with many citizens in Darfur having been displaced, humanitarian efforts remain a priority until a sustained cease-fire and inclusive political process are achieved,

Demanding that there should be no aerial bombings and the use of United Nations markings on aircraft used in such attacks,

Reaffirming its concern that the ongoing violence in Darfur might further negatively affect the rest of Sudan as well as the region, *stressing* that regional security aspects must be addressed to achieve long-term peace in Darfur, and *calling* on the Governments of Sudan and Chad to abide by their obligations under the Tripoli Agreement of 8 February 2006 and subsequent bilateral agreements,

Determining that the situation in Darfur, Sudan continues to constitute a threat to international peace and security,

1. *Decides*, in support of the early and effective implementation of the Darfur Peace Agreement and the outcome of the negotiations foreseen in paragraph 18, to authorise and mandate the establishment, for an initial period of 12 months, of an AU/UN Hybrid operation in Darfur (UNAMID) as set out in this resolution and pursuant to the report of the Secretary-General and the Chairperson of the African Union Commission of 5 June 2007, and *further decides* that the mandate of UNAMID shall be as set out in paragraphs 54 and 55 of the report of the Secretary General and the Chairperson of the African Union Commission of 5 June 2007;

2. *Decides* that UNAMID, which shall incorporate AMIS personnel and the UN Heavy and Light Support Packages to AMIS, shall consist of up to 19,555 military personnel, including 360 military observers and liaison officers, and an appropriate civilian component including up to 3,772 police personnel and 19 formed police units comprising up to 140 personnel each;

3. *Welcomes* the appointment of the AU-UN Joint Special Representative for Darfur Rodolphe Adada and Force Commander Martin Agwai, and *calls* on the Secretary-General to immediately begin deployment of the command and control structures and systems necessary to ensure a seamless transfer of authority from AMIS to UNAMID;

4. *Calls* on all parties to urgently facilitate the full deployment of the UN Light and Heavy Support Packages to AMIS and preparations for UNAMID, and *further calls* on member states to finalise their contributions to UNAMID within 30 days of the adoption of this resolution and on the Secretary-General and the Chairperson of the African Union Commission to agree [on] the final composition of the military component of UNAMID within the same time period;

5. *Decides* that:

(a) no later than October 2007, UNAMID shall establish an initial operational capability for the headquarters, including the necessary management and command and control structures, through which operational directives will be implemented, and shall establish financial arrangements to cover troops costs for all personnel deployed to AMIS;

(b) as of October 2007, UNAMID shall complete preparations to assume operational command authority over the Light Support Package, personnel currently deployed to AMIS, and such Heavy Support Package and hybrid personnel as may be deployed by that date, in order that it shall perform such tasks under its mandate as its resources and capabilities permit immediately upon transfer of authority consistent with sub-paragraph (c) below;

(c) as soon as possible and no later than 31 December 2007, UNAMID having completed all remaining tasks necessary to permit it to implement all elements of its mandate, will assume authority from AMIS with a view to achieving full operational capability and force strength as soon as possible thereafter;

6. *Requests* the Secretary General to report to the Council within 30 days of the passage of this resolution and every 30 days thereafter, on the status of UNAMID's implementation of the steps specified in paragraph 5, including on the status of financial, logistical, and administrative arrangements for UNAMID and on the extent of UNAMID's progress toward achieving full operational capability;

7. *Decides* that there will be unity of command and control which, in accordance with basic principles of peacekeeping, means a single chain of command, *further decides* that command and control structures and backstopping will be provided by the United Nations, and, in this context, *recalls* the conclusions of the Addis Ababa high level consultation on the situation in Darfur of 16 November;

8. *Decides* that force and personnel generation and administration shall be conducted as set out in paragraphs 113–115 of the report of the Secretary-General and the Chairperson of the African Union Commission of 5 June 2007, and *requests* the Secretary-General to put in place without delay the practical arrangements for deploying UNAMID including submitting to the General Assembly recommendations on funding and effective financial management and oversight mechanisms;

9. *Decides* that UNAMID shall monitor whether any arms or related material are present in Darfur in violation of the Agreements and the measures imposed by paragraphs 7 and 8 of resolution 1556 (2004);

10. *Calls* on all Member States to facilitate the free, unhindered and expeditious movement to Sudan of all personnel, as well as equipment, provisions, supplies and other goods, including vehicles and spare parts, which are for the exclusive use of UNAMID in Darfur;

11. *Stresses* the urgent need to mobilise the financial, logistical and other support required for AMIS, and *calls* on member states and regional organisations to provide further assistance, in particular to permit the early deployment of two additional battalions during the transition to UNAMID;

12. *Decides* that the authorised strength of UNMIS [United Nations Mission in Sudan] shall revert to that specified in resolution 1590 (2005) upon the transfer of authority from AMIS to UNAMID pursuant to paragraph 5(c);

13. *Calls* on all the parties to the conflict in Darfur to immediately cease all hostilities and commit themselves to a sustained and permanent cease-fire;

14. *Demands* an immediate cessation of hostilities and attacks on AMIS, civilians and humanitarian agencies, their staff and assets and relief convoys, and *further demands* that all parties to the conflict in Darfur fully co-operate with AMIS, civilians and humanitarian agencies, their staff and assets and relief convoys, and give all necessary assistance to the deployment of the United Nations Light and Heavy Support Packages to AMIS, and to UNAMID;

15. Acting under Chapter VII of the Charter of the United Nations:

(a) *decides* that UNAMID is authorised to take the necessary action, in the areas of deployment of its forces and as it deems within its capabilities in order to:

(i) protect its personnel, facilities, installations and equipment, and to ensure the security and freedom of movement of its own personnel and humanitarian workers,

(ii) support early and effective implementation of the Darfur Peace Agreement, prevent the disruption of its implementation and armed attacks, and protect civilians, without prejudice to the responsibility of the Government of Sudan;

(b) *requests* that the Secretary-General, in consultation with the Chairperson of the African Union Commission, and the Government of Sudan conclude within 30 days a status-of-forces agreement with respect to UNAMID, taking into consideration General Assembly resolution 58/82 on the scope of legal protection under the Convention on the Safety of United Nations and Associated Personnel and General Assembly resolution 61/133 on the Safety and Security of Humanitarian Personnel and the Protection of United Nations Personnel, and *decides* that pending the conclusion of such an agreement the model status-of-forces agreement dated 9 October 1990 (A/45/594) shall provisionally apply with respect to UNAMID personnel operating in that country;

16. *Requests* the Secretary-General to take the necessary measures to achieve actual compliance in UNAMID with the United Nations zero-tolerance policy on sexual exploitation and abuse, including the development of strategies and appropriate mechanisms to prevent, identify and respond to all forms of misconduct, including sexual exploitation and abuse, and the enhancement of training for personnel to prevent misconduct and ensure full compliance with the United Nations code of conduct, and to further take all necessary action in accordance with the Secretary-General's Bulletin on special measures for protection from sexual exploitation and sexual abuse (ST/SGB/2003/13) and to keep the Council informed, and urges troop-contributing countries to take appropriate preventive action including the conduct of pre-deployment awareness training and, in the case of forces previously deployed under AU auspices, post-deployment awareness training, and to take disciplinary action and other action to ensure full accountability in cases of such conduct involving their personnel;

17. *Calls* on all concerned parties to ensure that the protection of children is addressed in the implementation of the Darfur Peace Agreement, and *requests* the Secretary-General to ensure continued monitoring and reporting of the situation of children and continued dialogue with parties to the conflict towards the preparations of time-bound action plans to end recruitment and use of child soldiers and other violations against children;

18. *Emphasises* there can be no military solution to the conflict in Darfur, *welcomes* the commitment expressed by the Government of Sudan and some other parties to the conflict to enter into talks and the political process under the mediation, and in line with the deadlines set out in the roadmap, of the United Nations Special Envoy for Darfur and the African Union Special Envoy for Darfur, who have its full support, *looks forward* to these parties doing so, *calls* on the other parties to the conflict to do likewise, and *urges* all the parties, in particular the non-signatory movements, to finalise their preparations for the talks;

19. *Welcomes* the signature of a Joint Communiqué between the Government of Sudan and the United Nations on Facilitation of Humanitarian Activities in Darfur, and *calls* for it to be fully implemented and on all parties to ensure, in accordance with relevant provisions of international law, the full, safe and unhindered access of relief personnel to all those in need and delivery of humanitarian assistance, in particular to internally displaced persons and refugees;

20. *Emphasises* the need to focus, as appropriate, on developmental initiatives that will bring peace dividends on the ground in Darfur, including in particular, finalising preparations for reconstruction and development, return of IDPs to their villages, compensation and appropriate security arrangements;

21. *Requests* the Secretary-General to report to the Council for its consideration no later than every 90 days after the adoption of this resolution on progress being made on, and immediately as necessary on any obstacles to:

(a) the implementation of the Light and Heavy Support Packages and UNAMID,

(b) the implementation of the Joint Communiqué between the Government of Sudan and the United Nations on Facilitation of Humanitarian Activities in Darfur,

(c) the political process,

(d) the implementation of the Darfur Peace Agreement and the parties' compliance with their international obligations and their commitments under relevant agreements, and

(e) the cease-fire and the situation on the ground in Darfur;

22. *Demands* that the parties to the conflict in Darfur fulfil their international obligations and their commitments under relevant agreements, this resolution and other relevant Council resolutions;

23. *Recalls* the reports of the Secretary-General of 22 December 2006 (S/2006/1019) and 23 February 2007 (S/2007/97) which detail the need to improve the security of civilians in the regions of eastern Chad and north-eastern Central African Republic [CAR], *expresses* its readiness to support this endeavour, and *looks forward* to the Secretary-General reporting on his recent consultations with the Governments of Chad and CAR;

24. *Emphasises* its determination that the situation in Darfur shall significantly improve so that the Council can consider, in due course and as appropriate, and taking into consideration recommendations of the Secretary-General and the Chairperson of the African Union, the drawing down and eventual termination of UNAMID;

25. *Decides* to remain seized of the matter.

SOURCE: United Nations. Security Council. "Resolution 1769 (2007)." S/RES/1769 (2007). July 31, 2007. www.un.org/Docs/sc/unsc_resolutions07.htm (accessed December 23, 2007).

DOCUMENT

Coalition of Human Rights Groups on UN Peacekeeping Force in Darfur

December 19, 2007

Ultimately, a strategic decision on the part of the government of the Sudan is necessary if we are to achieve our common goal: peace and security in Darfur and, indeed, in all of the Sudan. [. . .] The international community will be confronted with hard choices: do we move ahead with the deployment of a force that will not make a difference, that will not have the capability to defend itself and that carries the risk of humiliation of the Security Council and the United Nations and tragic failure for the people of Darfur?

—Jean-Marie Guéhenno,
Under-Secretary-General for Peacekeeping Operations, 27 November 2007

When Security Council Resolution 1769 was passed in July 2007, it was hoped that Darfur's struggling AU [African Union] peacekeepers would be rapidly reinforced by UN troops with the capacity to effectively protect civilians in Darfur. Five months later, on the brink of transfer of authority to UNAMID [United Nations/African Union peacekeeping operation], little if any additional capacity has been deployed or is on the horizon. After five years of suffering in Darfur, this is unacceptable.

The government of Sudan holds primary responsibility for the precarious state of the United Nations/African Union mission. They are obstructing deployment in five major ways, imposing conditions that would render the mission likely to fail. But rather than challenging Sudan to fulfil its commitment to Resolution 1769, politicians and members of the international community are instead wrangling over details and shirking their own responsibilities to support the force.

The credibility of the United Nations, especially the Security Council, is at stake. But far more urgently, the lives of millions of human beings continue to be put at risk by the failure to deploy effectively. Four million people in Darfur rely on humanitarian assistance in some form; more than 2 million men, women and children are displaced from their homes, and at least 500,000 of those in need of aid receive nothing because insecurity prevents aid agencies from reaching them. After years of horrific violence, insecurity, displacement and broken promises, the people of Darfur deserve protection now.

The Context: Five Months of Frustration; Five Years of Futile Efforts

On 31 July 2007 UN Security Council Resolution 1769 authorised a new "hybrid" United Nations/African Union peacekeeping operation (UNAMID) to take over from the under-resourced and ineffective African Union force (AMIS) deployed since 2004 in Darfur. The resolution called for UNAMID to consist of up to 26,000 military and police personnel, with an annual budget of $1.5 billion. If it were to reach full strength, UNAMID would be more than four times the size of AMIS, would have considerably better equipment and resources, and would be one of the largest peacekeeping missions ever deployed.

Nearly five months after the passage of Security Council Resolution 1769, however, it has become clear that the deployment of this force is in danger of failing. According to the Resolution, UNAMID is to assume control from AMIS on 31 December. Yet, mere weeks away from that deadline, only two battalions totalling 1,800 soldiers have deployed to bolster the AMIS force—one from Rwanda and one from Nigeria—along with important but insufficient support units. These include a 140-member police unit from Bangladesh and 200 members of a Chinese engineering unit. General Agwai, commander of the Darfur force, stated on 2 December that he expects no more than 6,500 troops and possibly 3,000 police to be on the ground in Darfur by the end-of-year deadline, the vast majority of whom will be the existing AMIS personnel. This is barely a third of the promised force.

Since 2003, the United Nations has passed 19 Resolutions on Darfur, to little effect so far. Of particular note, in 2006, Security Council Resolution 1706 authorised the existing United Nations Mission in Sudan (UNMIS)—deployed in support of the North-South peace agreement signed in 2005—to take over peacekeeping in Darfur, but subject to Khartoum's consent. Khartoum never approved the expanded mandate, and Resolution 1706 was never implemented—the first time that a UN peacekeeping force has been authorized and then failed to deploy. The international community must not permit this to be repeated with UNAMID.

Meanwhile AMIS, poorly resourced and with a weak (and weakly interpreted) mandate, is unable to protect civilians. The Sudanese government and some rebel movements have made AMIS's all-but-impossible job even harder. The government has caused severe delays in the deployment of vitally needed equipment and resources for the African Union mission. In 2005, it prevented the delivery of 105 Armoured Personnel Carriers (APCs) for three months, allowing them in only after an attack on AMIS, reportedly by government allied militia, left four troops dead. In 2007, Sudan refused to allow entry of six AMIS helicopters for five months until it received assurances that they would not be used for "offensive" purposes. The government has imposed curfews on AMIS, restrictions on patrolling, and a ban on night flights. Government officials have also prevented AMIS police from visiting people in detention facilities. Rebels and former rebels have also played their part, attacking and killing AMIS troops, and preventing them from entering their areas of control.

The Sudanese government's obstruction of UNAMID is consistent with its response to Resolution 1706 and its hindrance of AMIS. It appears to have no commitment to the deployment and operation of an effective peacekeeping force. Yet the international reaction to this has been muted and inadequate.

Five Obstructions to the Deployment of UNAMID

Under Resolution 1769, the government of Sudan is obliged to facilitate UNAMID deployment. However, it has instead adeptly constructed bureaucratic, administrative, and political hurdles to obstruct the mission, and potentially block it entirely.

1. Approval of troop contributions

***Specific Obstruction:* Failure to approve the list of UNAMID troop contributions**
After the passage of Resolution 1769, many countries responded quickly with troop pledges. On 18 September 2007 an initial list of pledges was transmitted to the government of Sudan, and a final list was sent on 2 October. As of 13 December 2007 the government of Sudan still has not formally approved the list or replied to the UN. Recent public statements by President [Omar] al-Bashir demonstrate Sudan's determination to reject certain units (see below), calling into question his willingness to comply with the Security Council Resolution.

Once the list has been approved, the AU and UN must complete pre-deployment assessments of each unit to establish if they are qualified and equipped according to UN standards. If the battalions do not meet UN standards, alternative forces or supplementary equipment must be found. Delaying approval of the list means delays in carrying out these assessments, and ultimately delay in deploying an effective force.

2. Refusal of 'non-African' contributions

***Specific Obstruction:* The government of Sudan has rejected troop units from Nepal, Thailand, and Nordic countries.**
Resolution 1769 notes that 'the Hybrid operation should have a predominantly African character and the troops should, as far as possible, be sourced from African countries'. This is consistent with similar language in the Report of the Secretary General and the President

of the African Union of 5 June 2007, describing proposed arrangements for the force. However, that same report goes on to state that if sufficient numbers of suitable troops and police cannot be obtained from Africa, offers from other countries will be considered.

Despite Khartoum's consent to these agreements, it is now challenging the inclusion of non-African units on specious grounds. The list of troop contributing countries transmitted to the government on 2 October 2007 was approximately 80 percent African, but on 14 November Jean-Marie Guéhenno, UN Under-Secretary-General for Peacekeeping Operations, announced that Sudan had specifically failed to approve certain non-African units—a Thai infantry battalion, a Nepalese force reserve and sector reserve unit, and an engineering unit made up of troops from Sweden, Norway and Denmark. If these units are refused it is not a simple matter of seeking units elsewhere. Guéhenno stated, "*There is no alternative to those units because to prepare for deployment takes time. We know that these units are ready; we know that they have made the preparation. They need now to deploy, and they are ready to deploy.*"

According to United Nations sources, the Nepalese Reserve Company is able to operate independently and on very short notice, and also have air capacity—all of which are critical for the force to be able to operate in this highly insecure environment. The Thai infantry battalion, unlike most of the other units pledged, is ready to deploy immediately and would arrive with all the capabilities and equipment required to meet UN standards.

The Nordic engineering unit provides technical expertise needed for the preparation of infrastructure to enable the rest of the force to function effectively. These units are all critical if UNAMID is to be more substantial than AMIS at the time of transfer of authority.

There is no valid justification for Khartoum's decision. Non-African peacekeepers are deployed as part of the UN Mission in Sudan (UNMIS). In a situation where African nations are unable to speedily offer forces with sufficient technical capacity and experience, if Khartoum continues to insist on UNAMID being all-African the result will be further delays in deployment while civilians continue to suffer and die.

3. Allocation of land for bases

Specific Obstruction: The government of Sudan has failed to provide land for bases.

After nearly five months, the government of Sudan has still failed to provide suitable land for a number of bases around Zalingei, one of the four areas of deployment in Darfur. Land was only agreed for bases in a second area (around El Geneina) in early December after more than two months of delay. UNAMID's success depends upon the provision of appropriate land for headquarters and forward operating bases near population centres and IDP [internally displaced person] camps. If the force is restricted to a few remote garrisons, it will be totally incapable of projecting force and deterring attacks on civilians. Furthermore, any additional troops will require additional bases and infrastructure. These will take some time to construct, but work cannot even begin until the government has formally allocated land for such purposes.

The Secretary General stated on 30 August that the UN and AU were working with the government towards "speedy conclusion of land-lease arrangements for all UNAMID locations." However, no agreement was concluded until late October, and then only for an initial parcel of sites that did not include land in El Geneina or Zalingei.

4. Status of Forces Agreement

***Specific Obstruction:* The government of Sudan has inserted unacceptable provisions into the Status of Forces Agreement.**

The relationship between UN peacekeeping forces and host governments is commonly governed by a—reasonably standard—'Status of Forces Agreement' (SOFA). UNAMID's SOFA was drawn up by the United Nations and African Union and transmitted to Sudan for its approval.

The Sudanese government immediately rejected the draft, and proposed unrealistic and unacceptable alternative provisions, such as a provision that would allow it to temporarily disable UNAMID's communications network when the government undertook "security operations", and another that would require UNAMID to give Sudanese authorities prior notification of all movements of troops and equipment. As Mr. Guéhenno reported, these egregious demands "would make it impossible for the mission to operate."

Edmund Mulet, Assistant Secretary General for Peacekeeping Operations, reported on 13 December that—following the dispatch by the Secretary General of two envoys to meet with the government of Sudan in Portugal—Khartoum had finally 'agreed to start talks' based on the UN/AU draft. As the government of Sudan has proven repeatedly, the start of talks is no guarantee of a resolution.

5. Night flights and Curfews

***Specific Obstruction:* The government of Sudan has refused to grant permission for UNAMID forces to fly at night, and continues to impose curfews on peacekeepers in certain areas.**

The government has consistently refused AMIS permission to fly at night and is threatening to do the same with UNAMID. After four months of negotiation, in early December the government finally conceded the right for UNAMID to use night flights for emergency medical evacuations only. This is a long way from the blanket authorisation for night flights that the force requires. There are also curfews on peacekeepers imposed in both government and rebel-controlled areas. Peacekeeping does not end at sunset and if the forces cannot operate at night they will be unable to protect civilians when they are most vulnerable.

The Pattern of Obstructions

These government imposed barriers cumulatively form a steadily-strengthening wall of resistance to UNAMID. The government of Sudan's obstruction, combined with the failure of the international community to contribute heavy equipment such as helicopters and trucks, means that by 31 December, the force on the ground will *not* be significantly different than it is today. There are high expectations for the UNAMID force among the people of Darfur, and transfer of authority to a force that is as weak as AMIS—or failure to handover authority at all—could provoke a serious backlash from both armed groups and civilians on the ground, seriously endangering UNAMID's future and the credibility of future UN peacekeeping operations in other parts of the world.

Sudan denies that it is being uncooperative and blames the slow progress on lack of funding for UNAMID, (even though the mission is to be funded through regular assessed contributions to the United Nations) and the failure of the 'West' to provide helicopters and transport. The pattern of obstructions suggests, however, that the government of Sudan aims to interminably delay the deployment of UNAMID, and is using these other issues to deflect attention from their primary culpability.

Five Reasons for Urgent Deployment

For more than four months, Sudan has blocked the timely and successful deployment of UNAMID. In that same period, the situation of civilians in Darfur has worsened. During October alone, an estimated 120 people were reportedly killed in Darfur, 30,000 were displaced, seven humanitarian workers were killed, and 10 humanitarian aid vehicles were hijacked.

1. Government of Sudan and Rebel Offensives

The government of Sudan and some rebel groups continue to launch attacks against civilians and civilian objects in Darfur. The government attacked the towns of Haskanita and Muhajariya during September and October 2007, killing dozens of people and displacing thousands more. Meanwhile, the rebel group JEM brazenly attacked an oil field in Western Kordofan, taking international oil-workers hostage for weeks before finally releasing them.

2. Continuing Displacement and Problems for Displaced

The UN Human Rights Council Group of Experts stated in their report of 28 November 2007 that between June and mid-November 2007 there were at least 15 air and ground attacks on civilians carried out by government, militia and rebel signatory groups, leading to approximately 170 civilian deaths and mass displacements. In October alone some 30,000 people were displaced in Darfur, bringing the total for the year to over 280,000. These individual human tragedies are overwhelming humanitarian aid agencies in the region and complicating peace negotiations. At the same time, the displaced persons camps themselves are becoming increasingly militarized and dangerous. Recent outbreaks of fighting at Kalma camp, one of the largest in Darfur, have highlighted the need for effective policing and protection of displaced persons camps, and UNAMID must have the resources to effectively provide such protection.

Additionally, UN humanitarian coordinator John Holmes recently reported that government security forces attempted to forcibly resettle residents of Kalma camp, in clear violation of existing agreements between the UN and the government. AMIS was unable to intervene and humanitarian agencies were unable to gain access to the camp to monitor the situation. When the humanitarian coordinator for South Darfur, Wael El-Haj Ibrahim, spoke out about these offences, Sudanese officials responded by expelling him from Darfur.

3. Endangered Humanitarian Response

Humanitarian operations cannot reach people in need without security. Twelve humanitarian workers have been killed in 2007. In the months since the passing of resolution 1769 at least five humanitarian workers have been shot and wounded, and 34 others temporarily abducted or physically or sexually assaulted. More than 60 UN or NGO [non-governmental organizational] vehicles and 18 trucks delivering humanitarian supplies have been hijacked or held up and looted. As a result, humanitarian operations are increasingly unable to travel or are forced to relocate staff leaving hundreds of thousands of people in need without access to humanitarian aid.

4. Rape and Gender-Based Violence

Throughout the conflict in Darfur, rape and other gender-based violence have been regularly used as weapons of war to systematically degrade, intimidate and destroy communities. Indictments issued by the International Criminal Court (ICC) in 2005 for crimes against humanity and war crimes cite incidents of targeted sexual violence and wide-scale rape. In 2007, the UN High Commissioner for Human Rights (OHCHR) released a comprehensive report on the December 2006 rebel attack on Deribat and eight other villages in East Jebel Marra, South Darfur, during which approximately 50 women were abducted, systematically raped, tortured and held as sex slaves for nearly one month before escaping. Tragically for Sudanese women, and evidence of failed international efforts, these are not isolated incidents. A strong and suitably trained and equipped UNAMID force could halt the epidemic of rape and other gender-based violence in Darfur.

5. Attacks on Peacekeepers

AMIS's under-resourced troops are currently facing difficulties defending themselves. In October 2007 an attack on the AMIS base outside of Haskanita by an unidentified armed group resulted in the death of ten African Union peacekeepers. Attacks against peacekeepers are unacceptable; in Darfur they threaten the very future of the mission and could discourage countries from volunteering troops for UNAMID.

While the government of Sudan pursues its calculated campaign of obstruction, the people of Darfur continue to suffer from rape and killings, displacement and critical obstacles to humanitarian relief. Without an effective peacekeeping force in Darfur with a mandate for civilian protection, the people of Darfur—along with the humanitarian workers who are delivering life-sustaining support—will be at the mercy of the Sudanese government and rebel forces, the very same players that are responsible for so much death and displacement over the last five years.

The government of Sudan should immediately cease its obstruction of AMIS and UNAMID and facilitate its deployment by:

- Immediately approving the list of troop contributions provided by the United Nations and African Union on 2 October 2007, including troops from non-African countries.
- Urgently allocating sufficient and appropriate land for the construction of bases in all areas of planned deployment in Darfur and ensuring the provision of other resources such as water and fuel. The government should further issue clear instructions to local authorities that they must provide all necessary support to the force.
- Providing a blanket authorisation for the use of night flights by UNAMID in Darfur, and lift all curfews imposed on peacekeepers.

- Immediately expediting the conclusion of the Status of Forces Agreement (SOFA), by disavowing provisions that would prevent the force from operating effectively, including provisions related to the disabling of communications equipment and the requirement for prior notification of troop movements.
- Expediting all procedures for the deployment of personnel and equipment associated with AMIS (including the Light Support Package and Heavy Support Package) and UNAMID.
- Taking any other action necessary to ensure that AMIS and UNAMID can carry out their mandates unhindered, including having freedom of movement and patrol throughout Darfur.

Recommendations

To the United Nations Security Council

- Convene an emergency session to address obstructions and delays to the UNAMID deployment.
- Issue a strong Presidential Statement condemning the government of Sudan's obstructions and requiring the government to take the explicit actions set out above.
- If the government of Sudan fails to take those actions within 30 days, the Council should immediately impose targeted sanctions on key government officials, including President Omar Al-Bashir, for non-compliance with its obligations under resolution 1769.
- The Council should continue to closely monitor the compliance of all parties with their obligations under Security Council Resolution 1769 to facilitate the expeditious deployment of UNAMID and ensure it can carry out its mandate unhindered, including having freedom of movement throughout Darfur.

To UN and AU member states

- Immediately fill the critical gaps in equipment and capabilities that have already been made public by the United Nations and African Union, especially air and ground transport and helicopter units.
- Respond rapidly and as fully as possible to all further requests for personnel, equipment, technical expertise, training and any other support required by UNAMID.
- Ensure that AMIS has sufficient funding to continue to operate effectively until the transfer of authority to UNAMID.
- Support multilateral targeted sanctions through the United Nations Security Council for Sudan's failure to take the required actions to facilitate the force.

Source: Human Rights Watch. "UNAMID Deployment on the Brink: The Road to Security in Darfur Blocked by Government Obstructions." Joint NGO Report. December 19, 2007. hrw.org/pub/2007/africa/unamid1207web.pdf (accessed December 23, 2007).

Other Historic Documents of Interest

From this volume

- Human rights abuses in Zimbabwe, p. 149
- Political situation in Somalia, p. 343
- Postwar situation in Liberia, p. 435

From previous *Historic Documents*

- Darfur crisis in Sudan, *2006,* p. 497

August

DOCUMENT IN CONTEXT

U.S. Intelligence Agencies and General Petraeus on Security in Iraq

AUGUST 2 AND SEPTEMBER 10, 2007

A "surge" of some 30,000 additional U.S. troops into Iraq helped reduce the violence that was tearing the country apart and provided the opportunity—not yet fully seized by year's end—for leaders of the embattled Iraqi government to reach the necessary political compromises to keep the country intact.

President George W. Bush in January ordered extra troops to Iraq because 2006 had proved to be the most violent year since the United States invaded the country in March 2003. Attacks by insurgents based among the Sunni minority and by a local branch of the al Qaeda network, combined with intense sectarian violence between Sunni and Shiite Muslims, had killed an estimated 34,000 Iraqis and 821 Americans during that year of unending violence.

Initially, the surge escalated violence even further as U.S. and Iraqi forces went on the offensive and insurgents responded with suicide bombs, truck bombs, and other attacks. Some of the deadliest bombings in the course of the war occurred in 2007, and the number of Iraqi civilians and U.S. troops killed ran at near-record rates in the first half of the year. But, with notable exceptions, the violence diminished markedly during most of the second half of 2007, enabling U.S. and Iraqi officials to declare the surge a qualified success to that point. By year's end, many people in Baghdad and Sunni-dominated Anbar province—the most violent places in Iraq in recent years—were able to enjoy something close to normal lives for the first time since the U.S.-led invasion.

Much of the initial success of the surge in 2007 could be attributed to two developments that coincided with the increased U.S. troop presence and were, at least in part, related to it. One factor was a decision by important Sunni tribes in Anbar province, west of Baghdad, to turn against the insurgents and align instead with the American forces. This so-called Sunni Awakening, which spread to other areas, stemmed from many people's disgust at the violence and destruction that had befallen their communities since the rise of the insurgency, particularly the bombings inflicted on fellow Iraqis by the Iraqi branch of the al Qaeda network. Another development was Shiite cleric Moqtada al-Sadr's sudden decision in late August to declare a six-month cease-fire in attacks by the militia loyal to him, the Mahdi Army, against U.S. forces, the Sunnis, and other Shiite militias. Sadr took this step after an especially bloody battle between his militia and Iraqi government forces dominated by a rival militia. His apparent intention was to regain control of renegade elements of his militia, but the resulting cease-fire helped reduce the overall level of violence in the central and southern parts of Iraq where Sadr's militia operated.

Yet another factor in the slowdown in violence was that the "ethnic cleansing" of formerly mixed Shiite and Sunni neighborhoods in Baghdad had largely succeeded. Many neighborhoods where Shiites and Sunnis once lived in relative harmony were now all-Shiite or all-Sunni—the result of prolonged sectarian violence since 2003 that forced tens of thousands of families to flee their homes, which were then taken over by families from the opposing sect. Baghdad had few mixed neighborhoods left by the middle of 2007, which reduced the motivation for some of the violence that had been associated with ethnic cleansing.

Very much in question at year's end was whether the decrease in violence could be sustained once—as U.S. officials said was inevitable—the number of American troops was reduced back to the pre-surge level or even lower. Because few Iraqi government security forces were fully capable of taking the place of the American troops, it was possible, if not likely, that some of the gains from the surge would prove temporary. Also worrisome was the fact that many insurgents had simply moved to remote parts of Iraq and perhaps planned to return to the central provinces once the security push ended. Over the longer term, humanitarian experts said, Iraqis had been so traumatized by the breadth and intensity of violence, following decades of a brutal dictatorship, that it would be years before they could rebuild Iraq as a normal nation.

Components of the Surge

As planned by the new U.S. commander in Iraq, Gen. David H. Petraeus, the surge had several components; the most widely known was the addition of five brigades of U.S. troops on the ground in Iraq. When Bush announced the surge in a speech to the American people on January 10, he used the figure 20,000 extra troops, but over the course of the year the actual increase averaged about 30,000 and, at its height in October, approached 40,000. At the start of the year, the United States had about 132,000 troops in Iraq, according to the Iraq Index compiled by the Brookings Institution. As five additional combat brigades reached Iraq, the total reached 160,000 in July, topped out at 171,000 in October, then fell back to about 158,000 at year's end.

The first drawdown of U.S. troops began in late November, when a combat brigade of the First Cavalry Division started withdrawing from Diyala province in preparation for a return to its home base in Texas. That brigade was being replaced in Diyala province by another unit already in Iraq, though, because Diyala had become dangerous in recent months.

In addition to the extra U.S. forces, Petraeus reported that Iraqi security forces added more than 100,000 soldiers and policemen during the year. That put the total number of trained Iraqi security forces at about 440,000.

The majority of additional U.S. troops were sent to the Baghdad area (home to nearly half of all Iraqis), and the rest to the violent provinces of Anbar (west of Baghdad) and Diyala (east of Baghdad). Most other areas received no additional U.S. troops, including Iraq's next-largest cities: Basra, in the south, and Mosul, in the north.

The surge was timed to coincide with a new security plan for Baghdad developed by Iraqi prime minister Nouri al-Maliki under intense U.S. pressure. The plan began to be implemented on February 14 with security sweeps across the city, including in some of the most lawless neighborhoods. This was the third such plan since early 2006. The previous two plans had relied primarily on increased levels of Iraqi security forces in the capital, something that did not happen in either case. By contrast, the 2007 plan depended on additional U.S. as well as Iraqi forces.

Because of the heightened U.S. involvement, the new plan came under strong criticism from some Iraqis, notably Shiite cleric Sadr, whose Mahdi Army had fought intense battles with U.S. forces in 2004. Despite his complaint about the U.S. role in the plan, Sadr generally cooperated during the year, ordering his men to avoid battles with Iraqi government or U.S. forces, even after March 4 when the Americans swept into the Shiite slum known as Sadr City for the first time in nearly three years. U.S. officials said many of Sadr's fighters had gone into hiding, some had simply put their guns away for the time being, and some had retreated to southern Iraq or even into Iran, possibly to wait out the surge.

Perhaps the most important aspect of Petreaus's counterinsurgency strategy was to get U.S. soldiers out of the massive bases that the Pentagon had built in Iraq and into the neighborhoods that had become Iraq's killing fields. The U.S. military had patrolled streets in the past—but only temporarily. Once the troops retreated to their bases, the insurgents, sectarian militias, and criminal gangs took over. To create a more lasting presence in danger zones, Petraeus stationed troops at about sixty small bases, called Combat Outposts or Joint Security Stations, in Iraqi neighborhoods that had been plagued by violence. Most of the posts were in Baghdad and many were at police stations or other government buildings; they were staffed by a few dozen U.S. and Iraqi soldiers. The theory was that the Iraqi troops would provide local knowledge while the Americans would train the Iraqis so they eventually could take over the responsibility themselves.

In some cases these neighborhood command posts were situated so that soldiers could view most nearby street intersections, enabling them to respond quickly when insurgents moved into the area. Troops patrolled the local neighborhoods during the day and used the bases as staging posts for nighttime raids of houses and buildings where insurgents were thought to be hiding.

One of the key facts about these forward operating posts was where they were not located: few of them were in Shiite-majority neighborhoods, such as Sadr City, which were controlled by militias, not by the government or U.S. forces. Also, due to limited military resources, only a few posts were located in some of the most recent zones of violence, including the city of Kirkuk (contested by Kurds, Shiites, and Sunnis) and Diyala province northeast of Baghdad.

In addition to stepping up patrols, the U.S. military expanded a previous tactic of building walls between the various sectarian neighborhoods of Baghdad in an effort to keep Shiites and Sunnis from killing one another. The creation of these "gated communities" brought protests from local leaders and Maliki's government, but by the end of 2007 much of Baghdad was a walled and segregated city.

The "Sunni Awakening"

Nothing in General Petreaus's plan for the surge had envisioned what turned out to be one of the most important developments of the year: the abandonment of al Qaeda in Iraq by many Sunni tribal leaders in Anbar province, some of whom probably had cooperated with al Qaeda in the past or even been active in the insurgency. The first evidence of this trend emerged in April with reports that several Sunni groups had broken with al Qaeda because of its killings of Sunnis. One of the Sunni groups, the Islamic Army, posted a letter on a Web site calling on al Qaeda leader Osama bin Laden to denounce those who killed Sunnis while invoking his name. The *Washington Post* on April 14 quoted an Islamic Army commander, Abu Mohammad al-Salmani, as saying that al Qaeda "has killed more Iraqi Sunnis in Anbar province during the past month than the soldiers of the American occupation have killed within three months. People are tired of the torture."

As an outgrowth of this new Sunni resistance to al Qaeda, tribal leaders in Anbar province formed what they called the Anbar Salvation Council and encouraged Sunni fighters to cooperate with the Americans. As a result, thousands of Sunnis volunteered to join government security forces or local Sunni volunteer forces that had the aim of fighting al Qaeda. The U.S. military called these local forces Concerned Local Citizens (Sunnis later changed the name to Sons of Iraq). Reversing their past aversion to accepting any help from the U.S. military, some of the tribal sheikhs asked U.S. commanders for weapons and other supplies for the volunteer forces.

By year's end, an estimated 72,000 Iraqis had signed up for about 300 Concerned Local Citizens groups, according to U.S. statistics; the United States was paying monthly stipends of $300 each to about 60,000 people, meaning that the rest were volunteers. The Pentagon said that about 80 percent of the members of these groups were Sunnis, and the rest were Shiites. Kurds did not participate because they had their own army, the *Peshmerga,* which was active in the northern provinces that Kurds called Kurdistan.

An unknown percentage of the Sunnis in these groups had been insurgents until they switched sides—a fact that unnerved American soldiers, who could not be sure of their new allies' motives, and angered Shiites, who wondered why the Americans were now trusting men who just weeks before might have been on the other side, firing weapons or planting bombs.

The main coalition of Shiite parties in parliament, the United Iraqi Alliance, issued a statement on October 2 demanding that the government and the U.S. military stop recruiting Sunnis with backgrounds in militant organizations. "We refuse and denounce giving protection to those terrorists who committed hideous crimes against the Iraqi people and allowing them to be responsible for security," the statement said. This statement appeared to be timed to thwart the recruitment of Sunnis in Baghdad, a step the government had begun just a few weeks earlier following the successful recruitment of Sunnis in Anbar province to the west. In response to this and similar complaints, U.S. officials insisted that the Concerned Local Citizens were "temporary" groups formed specifically to provide security to their neighborhoods.

Although U.S. officials said upward of 20 percent of the Sunni volunteers wanted to join the Iraqi security services, the government appeared to be reluctant to accept them. By year's end, only about 4,000 of the volunteers had been accepted by the government and fewer than half of them had been placed in the army or police forces.

At least a half-dozen tribal leaders who were active in collaborating with the Americans were killed during the year. Among them was the sheikh most identified with the Sunni Awakening, Abdul Sattar Buzaigh al-Rishawi, known as Abu Risha, who led the Anbar Salvation Council. He met with President Bush on September 3, during one of the president's in-and-out visits to Iraq, and was killed by a bomb ten days later near his home in Ramadi. His brother took his place. Another blow to U.S. reconciliation efforts came on September 24, when a suicide bomber attacked a U.S.-sponsored meeting of local Shiite and Sunni leaders in Baquba, the capital of Diyala province. At least twenty-four people were killed, including the local police chief. This bombing appeared to be part of a campaign of attacks against provincial leaders during the Islamic holy month of Ramadan.

The Shiite Militias

A major challenge for U.S. forces in Iraq, ever since late 2003, was the presence of large militias that were formed to advance sectarian, rather than national, interests. The most important of these were the Kurdish *Peshmerga* and two Shiite militias: Sadr's Mahdi Army and the Badr Organization, which was the armed wing of the Islamic Supreme Council of Iraq, the largest single Shiite political party and an important backer of Prime Minister Maliki. The Kurdish militia generally stayed within the three northern Kurdish provinces, but the two Shiite militias infiltrated thousands of their fighters into the Iraqi security forces (primarily the national and local police) while other fighters guarded the towns and neighborhoods where their leaders were strongest.

The most troublesome of these militias, from the U.S. viewpoint, was Sadr's Mahdi Army, which fought extensive battles with U.S. troops in 2004 and had been responsible since for numerous attacks against Americans, Sunnis, and opposing Shiites. The United States in 2006 began pressing the Iraqi government to move against the Mahdi Army, something Prime Minister Maliki was reluctant to do because Sadr had played a major role in his rise to power. Responding to the continuing U.S. pressure, Maliki said on January 17, 2007, that more than 400 members of the Mahdi Army had been arrested "within the last few days." U.S. officials said that figure appeared to be exaggerated but confirmed that some Mahdi fighters had been arrested and not released immediately, as had happened in the past. At about the same time, Sadr's aides announced the formation of a "golden unit" to purge the Mahdi Army of fighters who had engaged in violence for criminal reasons.

The key development of 2007 concerning the Mahdi Army began with a conflict on August 27–28 between members of that militia and government security forces dominated by the rival Badr Organization. The fighting, which was centered in the holy city of Karbala but spread to other areas including Baghdad, took place during a pilgrimage and appeared to be part of a power struggle for control of a region that was home to some of the most important Shiite religious shrines. The violence killed more than fifty people and resulted in damage to one of the shrines.

On August 29, Sadr's office in Karbala announced that he had ordered a suspension of all military activities by the Mahdi Army for six months. The step was widely seen as an admission by Sadr that the fighting in Karbala had undermined the public's support for him in an important constituency. In the months afterward, it appeared that Sadr was using the cease-fire to purge his Mahdi Army of commanders who had pursued their own agendas; many of these reportedly were gang leaders who used Sadr's militia to mask their personal criminal activities.

Assessing the Surge

Several attempts by Democrats in Congress to block, or at least pose limits on, the troop surge to Iraq failed during 2007. However, Congress did require the Bush administration to submit new reports on developments in Iraq, including assessments of how the Iraqi government was meeting numerous "benchmarks," such as adoption of legislation to reduce tension among the various sects.

The first of the administration's reports, in mid-July, said that progress was being made but it was too early to determine whether the surge ultimately would improve security in Iraq. At that time, violence had actually escalated as U.S. and Iraqi government forces stepped up their fight against insurgents.

A second significant assessment of the surge came in a National Intelligence Estimate (NIE)—the consensus view of the sixteen U.S. intelligence agencies—a declassified summary of which was released on August 2. This report was an update of a previous NIE that was released on February 2, shortly after the surge was announced. The February assessment had offered a gloomy look into Iraq's future, predicting that 2007 would be just as violent as the previous year unless the Iraqi government and security forces made rapid improvements. The August NIE found "measurable but uneven improvements" in the Iraqi security situation and suggested the situation would "continue to improve modestly" during the next six to twelve months if the United States kept up the military pressure. The estimate also identified numerous hurdles, though, most centering around the Iraqi government's inability to take advantage of the opportunity presented by the surge to make political compromises and the Iraqi security forces' continuing inability to conduct major operations unassisted.

The most anxiously awaited assessment of developments was due by September 15, and it was delivered in direct testimony to four congressional committees on September 10–11 by General Petraeus and Ryan C. Crocker, who had taken over as U.S. ambassador to Iraq in April. Petraeus focused on the security situation, while Crocker focused on Iraqi politics.

In his formal report, and in answers to questions from often-skeptical members of Congress, Petraeus said that the initial military objectives of the surge were being met, but neither Iraqis nor Americans should expect that Iraq would become an entirely peaceful place anytime soon. The general noted that violence was on the decline because of many Sunni groups' rejection of al Qaeda, the Iraqi security forces' gradual improvement, and Sadr's declaration of a cease-fire by his militia.

Petraeus suggested that, by the summer of 2008, the overall level of U.S. forces in Iraq could be reduced back to what it had been before the surge: from twenty to fifteen combat brigades, or back to about 130,000 troops. "Beyond that, while noting that the situation in Iraq remains complex, difficult, and sometimes downright frustrating, I also believe that it is possible to achieve our objectives in Iraq over time, though doing so will be neither quick nor easy," he said.

Petraeus stood by this assessment in comments he made later in the year. During a visit to Iraq by Defense Secretary Robert M. Gates, Petraeus told reporters on December 6 that bringing stability to Iraq would be a long-term process. "We see this as requiring a continued amount of very tough work," he said. "We see al Qaeda as a very, very dangerous adversary still able to carry out attacks and an adversary that we must continue to pursue." In contrast to this assessment, a Pentagon report issued at about the same time said al Qaeda "is mostly on the defensive and faces dwindling support within Iraq."

Although Petraeus was widely respected on Capitol Hill and in the broader community of U.S. foreign policy and defense experts, his generally upbeat assessment in September was greeted by a great deal of skepticism. Democrats and even some Republicans in Congress were particularly critical of Petreaus's suggestion that U.S. troop levels would be reduced only to the pre-surge levels by the middle of 2008. "Telling the Iraqis that the surge will end by the middle of next year and then we will make a decision as to whether to reduce our troop levels from the basic pre-surge level of 130,000 does not change our course in Iraq," Sen. Carl Levin, D-Mich., chair of the Senate Armed Services Committee, told Petraeus on September 11. "It presents an illusion of change to prevent a real change of course from occurring."

The Iraqi Army and Police

Since the United States abolished the existing Iraq security services in 2003 and began reassembling them into a new army and national police force, Washington's stated goal had been to turn responsibility for Iraq' security over to local forces as soon as possible. The timing of that goal kept receding into the distance, however. The principal reasons were that the training and equipping of Iraqi forces was taking longer than expected, Iraq had become a much more dangerous place than Washington's planners had anticipated, and the Iraqi forces—particularly the police—had been infiltrated by members of sectarian militias and criminal gangs that were pursuing narrow interests rather than national ones.

By mid-November 2007, according to a Pentagon report, the new Iraqi army had slightly more than 176,000 trained troops assigned to 117 battalions. The Pentagon said that about three-quarters of those units were able to plan, execute, and sustain operations with "minimal or no assistance" from U.S. forces. Even so, only a small portion of the units were left on their own for long or were given particularly dangerous assignments.

Nearly 85 percent of the soldiers were Shiites, suggesting that the army was heavily weighted toward that group as compared with the country's population (an estimated 60 percent of Iraqis were Shiites). Like other government agencies, the Iraqi army was plagued by desertions and absenteeism; more than 20,000 soldiers were dropped from the rolls because they had gone absent without leave (AWOL), according to the Pentagon report.

Iraq had three nationwide police forces, in addition to local police in the major cities. The largest was known as the Iraqi Police Service, with about 174,000 trained personnel as of November 2007. Another unit, the National Police (which despite its name was more of a paramilitary force than a police force in the conventional U.S. terminology) had about 36,000 trained personnel on its rosters, most of whom had undergone retraining since 2006 at a new U.S. facility near Baghdad. Another 31,000 trained personnel were assigned to a border patrol, and about 23,000 were in other security services.

A U.S.-sponsored investigation of the Iraqi police and the Interior Ministry that supervised them resulted in a blistering report in September. A panel headed by retired Marine Corps Gen. James L. Jones, the former commander of NATO, described the Interior Ministry as dysfunctional and the National Police force as "not viable in its current form." Jones's panel called for disbanding the national police (which was created by the United States in 2004) and shifting many of the policemen into the army, about 6,000 of them into specialized units. The Interior Ministry had been crippled by corruption and sectarianism, the report said, noting that top appointed positions were allocated to political parties. U.S. officials later expressed their belief that the National Police could be reformed, despite the findings of the Jones panel.

Top Iraqi officials routinely claimed that the army and police were making great strides, but these claims were undercut by numerous security lapses and by occasional admissions from mid-level officials that the picture was not so rosy. One revealing admission was made in early December by the Basra police chief, who told the Associated Press that he worried about the security situation once British forces withdrew entirely from the city and surrounding region. "Frankly speaking, we have rifles, machine guns and a few armored vehicles, which aren't as advanced as the British weaponry and are insufficient to maintain full control of the province," Maj. Gen. Jalil Khalaf said. Khalaf also acknowledged that many of the men in his force "don't act under my command and I don't know whom they're affiliated with or work for"—a reference to the fact that Shiite militias and criminal gangs had infiltrated all Iraqi police forces, particularly in Basra. The general's concerns were shared by the U.S. military because Basra province was home to Iraq's most important oil fields, and the major corridor for transporting supplies from Kuwait to Baghdad passed through the area.

A Mixed Security Picture for 2007

The U.S. military surge in Iraq unfortunately, but not unexpectedly, contributed to violence in the early months of 2007. By late in the year, the surge contributed to—but was not the sole cause of—a sharp reduction of violence in most areas. As a result, more U.S. soldiers and marines died in Iraq in 2007 than in any previous year, but other security indicators suggested an overall improvement by year's end.

The addition of more than 30,000 U.S. troops into Iraq, combined with General Petraeus's aggressive approach, exposed more American soldiers and marines to danger than ever before in the war. Total U.S. deaths for 2007 were 903, exceeding the previous record of 849 deaths in 2004, when the United States fought two major battles in the city of Fallujah. The months of April, May, and June, when the surge was getting under way, turned out to be the three deadliest months during the U.S. occupation of Iraq: 104 Americans died in April, 127 in May, and 100 in June. The May figure was exceeded only by the 137 Americans killed in October 2004. However, the monthly totals dropped below 40 for each of the last four months of 2007, dipping to a low of 23 U.S. deaths in December, the lowest of any month since the U.S. occupation began. At year's end, total U.S. deaths in Iraq since March 2003 stood at 3,903. The deadliest day of the year for U.S. troops—and one of the deadliest since the invasion—was January 20, when 25 service personnel were killed, including 13 who died in the crash of an army helicopter shot down north of Baghdad.

Counting Iraqi deaths and injuries due to war-related violence was never easy, and it became more difficult early in 2007 when the Iraqi government stopped supplying fatality figures to the United Nations and other agencies. Using reports from morgues, hospitals, and government offices, the UN had compiled the widely reported figure of more than 34,000 violent Iraqi deaths in 2006, a figure that proved deeply embarrassing to officials in Baghdad.

Most estimates of the number of Iraqis killed in 2007 were in the range of about 20,000. One of the highest estimates was by Iraq Body Count, an anti-war organization in Britain that compiled figures from media reports, morgues, and other sources. Its year-end report put the toll of Iraqi civilians killed violently at between 22,586 and 24,159. The organization's total count for the period since the U.S.-led invasion ranged between 81,174 and 88,585.

The Associated Press, which counted deaths reported by hospitals and other sources but acknowledged that its count likely was incomplete, tallied a total of 18,610 Iraqi civilians killed in 2007. The deadliest month in the AP's count was May, with 2,155 killed, and the low point was 710 deaths in December.

Another count, the Iraq Index compiled by the Brookings Institution in Washington, tallied 18,300 Iraqi civilian deaths during the year, basing its estimates on figures supplied by the Pentagon. According to those figures, January was the deadliest month for Iraqis (with 2,800 killed), followed by May (2,200 killed). Civilian deaths declined progressively to a low of 550 in December, according to this count.

The most widely cited count of deaths among Iraqi security forces was compiled by the Iraq Coalition Casualty Count, a U.S. group that gathered figures from government and media reports. That group counted 1,830 deaths of Iraq soldiers and policemen in 2007, putting the total since June 2003 at about 7,800.

Another important measure of security in Iraq was the number of attacks against civilians, Iraqi security forces, and troops from the United States and a handful of other countries. Whether measured on a daily, weekly, or monthly basis, the number of attacks soared to near-record levels through July then diminished sharply in the later months. Despite the overall number of attacks, the targets remained fairly constant over time: about two-thirds of all attacks were aimed at U.S. and other foreign forces, while the rest were directed at Iraqi civilians or Iraqi security forces (primarily the police).

According to Pentagon figures, the most violent months were May and June, when suicide bombings, car bombings, and other attacks were carried out at a rate of about 1,500 each week, or more than 200 per day. The average number of attacks each week fell below 1,000 by late August and was down to about 600 by the end of the year. In a year-end news conference on December 30, Petraeus told reporters that violent attacks had dropped, on average, by about 60 percent since June.

Baghdad and Anbar province remained by far the most violent places in Iraq, between them accounting for about one-half of all attacks in the country, the Pentagon said. Baghdad experienced a daily average of 58 attacks between early May and mid-July, the most in any period since the invasion, but the number of attacks fell by about one-half during the late July to early November period. Sunni-dominated Anbar province experienced its greatest violence in late 2006 and early 2007, but with the rise of the Sunni Awakening movement, violence there dropped sharply late in 2007. As had been the case since 2003, by far the safest areas of Iraq were the three Kurdish-controlled provinces in northern Iraq, which often went for weeks without any violent attacks.

Several of the deadliest days in Iraq's recent history took place in 2007, in each case the result of large-scale bombings, according to the database of the U.S. National Counterterrorism Center. Most of these bombings were aimed at Shiites, but the single deadliest attack since the invasion targeted Iraq's Yazidi community, a minority sect whose members speak Kurdish and live in the northern provinces. Simultaneous suicide truck bombings on August 14 killed 430 people and wounded about 500 others, nearly wiping out the Yazidi towns of Jazeera and Qahtaniya in the desert near the Syrian border.

Ethnic Turkmen, another of Iraq's small minorities, also were targeted in a truck bombing on July 7 in the village of Amurli in northern Iraq. Officials counted 156 people killed and 270 wounded.

Baghdad experienced more than a dozen attacks during the year, each of which killed 50 or more people. Several of the bombings took place in public markets, often on Fridays, the Muslim day of prayer, as people gathered to do their weekend shopping. The year's deadliest day in Baghdad was April 18, when more than 170 people were killed in a wave of car bombings and other attacks aimed primarily at Shiites. About 140 of these deaths resulted from a single car bombing at an intersection in a Shiite neighborhood. This was the same area where a bombing on February 3 killed 135 people and wounded 340 others.

Despite the continued violence in the capital, security conditions had noticeably improved in most of the city by year's end. A steady trickle of families who had fled Baghdad in previous years were returning to their old neighborhoods. Even so, most people in Baghdad were now living in neighborhoods with fellow Shiites or fellow Sunnis, not in mixed neighborhoods.

The city of Tal Afar, in northern Iraq, was the scene of intense violence on March 27–28, starting with a double suicide bombing attributed to al Qaeda, followed by revenge killings of Sunnis by Shiites. A total of 152 people were killed and 347 wounded in this violence. The killings came one year after President Bush had declared Tal Afar one of Iraq's success stories because U.S. military action there in late 2005 had succeeded in ousting insurgents. After U.S. troops left, however, insurgents returned and sectarian killings resumed.

Even the heavily fortified International Zone (formerly known as the Green Zone)—home to the Iraqi government, the U.S. embassy, and most other embassies and international offices—did not escape the violence. On April 12, a suicide bomber was able to penetrate deep into the Iraqi parliament building, where he set off his bomb, killing one parliamentarian and wounding twenty-two other people, half of them legislators. The International Zone often came under rocket and mortar attack, but this was the most serious attack there since the United States walled off the area after capturing Baghdad in April 2003. The Islamic State of Iraq, an umbrella group for various insurgents, claimed responsibility for the attack. After an upsurge in mortar attacks later in April, the U.S. embassy ordered all employees to wear flak jackets and helmets when they were outside in the Green Zone or in buildings that had not been "hardened" against explosives.

British Troops Leave Basra

One of the last remaining elements of the "coalition" that had supported the U.S. invasion and subsequent occupation of Iraq was fading away in late 2007. British forces who had taken control of Basra—Iraq's second-largest city—in 2003 withdrew from the city center in early September and in mid-December handed over full responsibility for security in the area to the Iraqi government.

A predominantly Shiite city with nearly 2 million residents, Basra was the center of the oil industry in southern Iraq. After the invasion, Britain controlled the city and the surrounding area with about 18,000 troops but gradually reduced its forces to about 5,500 by mid-2007. The British never were able to establish real security in Basra because of the prevalence of rival Shiite militias and hundreds of criminal gangs. Although Basra rarely endured the massive suicide and car bombs that afflicted Baghdad and other central Iraqi cities, thousands of people were killed there as the result of gang violence, intra-Shiite feuds, and common criminal activities. The British headquarters, at a palace in central Basra, also came under frequent attack from rockets and mortars, many of them reportedly launched by Sadr's Mahdi Army.

British forces relocated to the airport about twelve miles from the center of the city in the first week of September. On December 16 Britain handed formal responsibility for security in Basra province to the Iraqi government, which in effect meant security forces dominated by Shiite militias. At year's end the British contingent at the airport was down to 4,500, with plans for further reductions in 2008.

Following are two documents. First, the substantive sections from a National Intelligence Estimate issued by the National Intelligence Council on August 2, 2007, "Prospects for Iraq's Stability: Some Security Progress but Political Reconciliation Elusive." Second, excerpts from "Report to Congress on the Situation in Iraq," testimony by Gen. David H. Petraeus, commander of U.S. forces in Iraq, on September 10–11, 2007 (references to slides shown to members of Congress have been deleted).

DOCUMENT

U.S. National Intelligence Estimate on Stability in Iraq

August 2, 2007

Scope Note

This assessment updates the January 2007 National Intelligence Estimate (NIE) on Iraq entitled, *Prospects for Iraq's Stability: A Challenging Road Ahead;* it has been prepared at the direction of the Director of National Intelligence in response to a request from the National Security Council. It provides the Intelligence Community's analysis of the status of the critical factors identified in the January Estimate that are driving Iraq's security and political trajectory. Using the January Estimate as a baseline, this update examines the prospects for progress on the security and national reconciliation fronts over the next six to 12 months.

Analytic Caution: Driven largely by the accelerating pace of tribal engagement and the increasing tempo of Coalition operations, developments in Iraq are unfolding more rapidly and with greater complexity today than when we completed our January NIE. Regional variations in security and political circumstances are great and becoming increasingly more distinct—for example, intra-Shia violence in southern Iraq is very different from patterns of violence elsewhere. The intelligence assessments contained in this NIE largely focus on only a short period of the Iraqi conflict—the last six months—and in circumscribed areas—primarily the central provinces, which contain the center of gravity for Iraq's security prospects and in which we have a greater Coalition presence and therefore more information. The unfolding pace and scope of security and political realities in Iraq, combined with our necessarily limited focus of analysis, contain risks: our uncertainties are greater, and our future projections subject to greater chances of error. These issues, combined with the challenges of acquiring accurate data on trends in violence and continued gaps in our information about levels of violence and political trends in areas of Iraq without a substantial Coalition presence and where Intelligence Community collectors have difficulty operating, heighten our caution. Nonetheless, we stand by these judgments as our best collective assessment of security and political conditions in Iraq today and as likely to unfold during the next six to 12 months. . . .

Key Judgments

There have been measurable but uneven improvements in Iraq's security situation since our last National Intelligence Estimate on Iraq in January 2007. The steep escalation of rates of violence has been checked for now, and overall attack levels across Iraq have fallen during seven of the last nine weeks. Coalition forces, working with Iraqi forces, tribal elements, and some Sunni insurgents, have reduced al-Qa'ida in Iraq's (AQI) capabilities, restricted its freedom of movement, and denied it grassroots support in some areas. **However, the level of overall violence, including attacks on and casualties among civilians, remains high; Iraq's sectarian groups remain unreconciled; AQI retains the ability to conduct high-profile attacks; and to date, Iraqi political leaders remain unable to govern effectively.** **There have been modest improvements in economic output, budget**

execution, and government finances but fundamental structural problems continue to prevent sustained progress in economic growth and living conditions.

We assess, to the extent that Coalition forces continue to conduct robust counterinsurgency operations and mentor and support the Iraqi Security Forces (ISF), that Iraq's security will continue to improve modestly during the next six to 12 months but that levels of insurgent and sectarian violence will remain high and the Iraqi Government will continue to struggle to achieve national-level political reconciliation and improved governance. Broadly accepted political compromises required for sustained security, long-term political progress, and economic development are unlikely to emerge unless there is a fundamental shift in the factors driving Iraqi political and security developments.

Political and security trajectories in Iraq continue to be driven primarily by Shia insecurity about retaining political dominance, widespread Sunni unwillingness to accept a diminished political status, factional rivalries within the sectarian communities resulting in armed conflict, and the actions of extremists such as AQI and elements of the Sadrist Jaysh al-Mahdi (JAM) militia that try to fuel sectarian violence. Two new drivers have emerged since the January Estimate: expanded Sunni opposition to AQI and Iraqi expectation of a Coalition drawdown. Perceptions that the Coalition is withdrawing probably will encourage factions anticipating a power vacuum to seek local security solutions that could intensify sectarian violence and intra-sectarian competition. At the same time, fearing a Coalition withdrawal, some tribal elements and Sunni groups probably will continue to seek accommodation with the Coalition to strengthen themselves for a post-Coalition security environment.

- Sunni Arab resistance to AQI has expanded in the last six to nine months but has not yet translated into broad Sunni Arab support for the Iraqi Government or widespread willingness to work with the Shia. The Iraqi Government's Shia leaders fear these groups will ultimately side with armed opponents of the government, but the Iraqi Government has supported some initiatives to incorporate those rejecting AQI into Interior Ministry and Defense Ministry elements.

- Intra-Shia conflict involving factions competing for power and resources probably will intensify as Iraqis assume control of provincial security. In Basrah, violence has escalated with the drawdown of Coalition forces there. Local militias show few signs of reducing their competition for control of valuable oil resources and territory.

- The Sunni Arab community remains politically fragmented, and we see no prospective leaders that might engage in meaningful dialogue and deliver on national agreements.

- Kurdish leaders remain focused on protecting the autonomy of the Kurdish region and reluctant to compromise on key issues.

The IC [Intelligence Community] assesses that the emergence of "bottom-up" security initiatives, principally among Sunni Arabs and focused on combating AQI, represent the best prospect for improved security over the next six to 12 months, but we judge these initiatives will only translate into widespread political accommodation and enduring stability if the Iraqi Government accepts and supports them. A multi-stage process involving the Iraqi Government providing support and legitimacy for such initiatives could foster over the longer term political reconciliation between the participating Sunni Arabs and the national government. **We also assess that under some conditions "bottom-up initiatives" could pose risks to the Iraqi Government.**

- We judge such initiatives are most likely to succeed in predominantly Sunni Arab areas, where the presence of AQI elements has been significant, tribal networks and identities are strong, the local government is weak, sectarian conflict is low, and the ISF tolerate Sunni initiatives, as illustrated by Al Anbar Province.
- Sunni Arab resistance to AQI has expanded, and neighborhood security groups, occasionally consisting of mixed Shia-Sunni units, have proliferated in the past several months. These trends, combined with increased Coalition operations, have eroded AQI's operational presence and capabilities in some areas.
- Such initiatives, if not fully exploited by the Iraqi Government, could over time also shift greater power to the regions, undermine efforts to impose central authority, and reinvigorate armed opposition to the Baghdad government.
- Coalition military operations focused on improving population security, both in and outside of Baghdad, will remain critical to the success of local and regional efforts until sectarian fears are diminished enough to enable the Shia-led Iraqi Government to fully support the efforts of local Sunni groups.

Iraqi Security Forces involved in combined operations with Coalition forces have performed adequately, and some units have demonstrated increasing professional competence. However, we judge that the ISF have not improved enough to conduct major operations independent of the Coalition on a sustained basis in multiple locations and that the ISF remain reliant on the Coalition for important aspects of logistics and combat support.

- The deployment of ISF units from throughout Iraq to Baghdad in support of security operations known as Operation *Fardh al-Qanun* marks significant progress since last year when large groups of soldiers deserted rather than depart their home areas, but Coalition and Iraqi Government support remains critical.
- Recently, the Iraqi military planned and conducted two joint Army and police large-scale security operations in Baghdad, demonstrating an improving capacity for operational command and control.
- Militia and insurgent influences continue to undermine the reliability of some ISF units, and political interference in security operations continues to undermine Coalition and ISF efforts.
- The Maliki government is implementing plans to expand the Iraqi Army and to increase its overall personnel strength to address critical gaps, but we judge that significant security gains from those programs will take at least six to 12 months, and probably longer, to materialize.

The IC assesses that the Iraqi Government will become more precarious over the next six to 12 months because of criticism by other members of the major Shia coalition (the Unified Iraqi Alliance, UIA), Grand Ayatollah Sistani, and other Sunni and Kurdish parties. Divisions between Maliki and the Sadrists have increased, and Shia factions have explored alternative coalitions aimed at constraining Maliki.

- The strains of the security situation and absence of key leaders have stalled internal political debates, slowed national decisionmaking, and increased Maliki's vulnerability to alternative coalitions.

- We judge that Maliki will continue to benefit from recognition among Shia leaders that searching for a replacement could paralyze the government.

Population displacement resulting from sectarian violence continues, imposing burdens on provincial governments and some neighboring states and increasing the danger of destabilizing influences spreading across Iraq's borders over the next six to 12 months. The polarization of communities is most evident in Baghdad, where the Shia are a clear majority in more than half of all neighborhoods and Sunni areas have become surrounded by predominately Shia districts. Where population displacements have led to significant sectarian separation, conflict levels have diminished to some extent because warring communities find it more difficult to penetrate communal enclaves.

The IC assesses that Iraq's neighbors will continue to focus on improving their leverage in Iraq in anticipation of a Coalition drawdown. Assistance to armed groups, especially from Iran, exacerbates the violence inside Iraq, and the reluctance of the Sunni states that are generally supportive of US regional goals to offer support to the Iraqi Government probably bolsters Iraqi Sunni Arabs' rejection of the government's legitimacy.

- Over the next year Tehran, concerned about a Sunni reemergence in Iraq and US efforts to limit Iranian influence, will continue to provide funding, weaponry, and training to Iraqi Shia militants. Iran has been intensifying aspects of its lethal support for select groups of Iraqi Shia militants, particularly the JAM, since at least the beginning of 2006. Explosively formed penetrator (EFP) attacks have risen dramatically.

- Syria has cracked down on some Sunni extremist groups attempting to infiltrate fighters into Iraq through Syria because of threats they pose to Syrian stability, but the IC now assesses that Damascus is providing support to non-AQI groups inside Iraq in a bid to increase Syrian influence.

- Turkey probably would use a range of measures to protect what it perceives as its interests in Iraq. The risk of cross-border operations against the People's Congress of Kurdistan (KG) terrorist group based in northern Iraq remains.

We assess that changing the mission of Coalition forces from a primarily counterinsurgency and stabilization role to a primary combat support role for Iraqi forces and counterterrorist operations to prevent AQI from establishing a safehaven would erode security gains achieved thus far. The impact of a change in mission on Iraq's political and security environment and throughout the region probably would vary in intensity and suddenness of onset in relation to the rate and scale of a Coalition redeployment. Developments within the Iraqi communities themselves will be decisive in determining political and security trajectories.

- Recent security improvements in Iraq, including success against AQI, have depended significantly on the close synchronization of conventional counterinsurgency and counterterrorism operations. A change of mission that interrupts that synchronization would place security improvements at risk.

Source: U.S. Director of National Intelligence. National Intelligence Council. "Prospects for Iraq's Stability: Some Security Progress but Political Reconciliation Elusive." National Intelligence Estimate. August 2, 2007. odni.gov/press_releases/20070823_release.pdf (accessed March 3, 2008).

Gen. David H. Petraeus on the Security Situation in Iraq

September 10, 2007

. . . At the outset, I would like to note that this is my testimony. Although I have briefed my assessment and recommendations to my chain of command, I wrote this testimony myself. It has not been cleared by, nor shared with, anyone in the Pentagon, the White House, or Congress.

As a bottom line up front, the military objectives of the surge are, in large measure, being met. In recent months, in the face of tough enemies and the brutal summer heat of Iraq, Coalition and Iraqi Security Forces have achieved progress in the security arena. Though the improvements have been uneven across Iraq, the overall number of security incidents in Iraq has declined in 8 of the past 12 weeks, with the numbers of incidents in the last two weeks at the lowest levels seen since June 2006.

One reason for the decline in incidents is that Coalition and Iraqi forces have dealt significant blows to Al Qaeda-Iraq. Though Al Qaeda and its affiliates in Iraq remain dangerous, we have taken away a number of their sanctuaries and gained the initiative in many areas.

We have also disrupted Shia militia extremists, capturing the head and numerous other leaders of the Iranian-supported Special Groups, along with a senior Lebanese Hezbollah operative supporting Iran's activities in Iraq.

Coalition and Iraqi operations have helped reduce ethno-sectarian violence, as well, bringing down the number of ethno-sectarian deaths substantially in Baghdad and across Iraq since the height of the sectarian violence last December. The number of overall civilian deaths has also declined during this period, although the numbers in each area are still at troubling levels.

Iraqi Security Forces have also continued to grow and to shoulder more of the load, albeit slowly and amid continuing concerns about the sectarian tendencies of some elements in their ranks. In general, however, Iraqi elements have been standing and fighting and sustaining tough losses, and they have taken the lead in operations in many areas.

Additionally, in what may be the most significant development of the past 8 months, the tribal rejection of Al Qaeda that started in Anbar Province and helped produce such significant change there has now spread to a number of other locations as well.

Based on all this and on the further progress we believe we can achieve over the next few months, I believe that we will be able to reduce our forces to the pre-surge level of brigade combat teams by next summer without jeopardizing the security gains that we have fought so hard to achieve.

Beyond that, while noting that the situation in Iraq remains complex, difficult, and sometimes downright frustrating, I also believe that it is possible to achieve our objectives in Iraq over time, though doing so will be neither quick nor easy.

Having provided that summary, I would like to review the nature of the conflict in Iraq, recall the situation before the surge, describe the current situation, and explain the recommendations I have provided to my chain of command for the way ahead in Iraq.

The Nature of the Conflict

The fundamental source of the conflict in Iraq is competition among ethnic and sectarian communities for power and resources. This competition *will* take place, and its resolution is key to producing long-term stability in the new Iraq. The question is whether the competition takes place more—or less—violently. . . . Foreign and home-grown terrorists, insurgents, militia extremists, and criminals all push the ethno-sectarian competition toward violence. Malign actions by Syria and, especially, by Iran fuel that violence. Lack of adequate governmental capacity, lingering sectarian mistrust, and various forms of corruption add to Iraq's challenges.

The Situation in December 2006 and the Surge

In our recent efforts to look to the future, we found it useful to revisit the past. In December 2006, during the height of the ethno-sectarian violence that escalated in the wake of the bombing of the Golden Dome Mosque in Samarra, the [U.S.] leaders in Iraq at that time—General George Casey and Ambassador Zalmay Khalilzad—concluded that the coalition was failing to achieve its objectives. Their review underscored the need to protect the population and reduce sectarian violence, especially in Baghdad. As a result, General Casey requested additional forces to enable the Coalition to accomplish these tasks, and those forces began to flow in January.

In the ensuing months, our forces and our Iraqi counterparts have focused on improving security, especially in Baghdad and the areas around it, wresting sanctuaries from Al Qaeda control, and disrupting the efforts of the Iranian-supported militia extremists. We have employed counterinsurgency practices that underscore the importance of units living among the people they are securing, and accordingly, our forces have established dozens of joint security stations and patrol bases manned by Coalition and Iraqi forces in Baghdad and in other areas across Iraq.

In mid-June, with all the surge brigades in place, we launched a series of offensive operations focused on: expanding the gains achieved in the preceding months in Anbar Province; clearing Baqubah, several key Baghdad neighborhoods, the remaining sanctuaries in Anbar Province, and important areas in the so-called "belts" around Baghdad; and pursuing Al Qaeda in the Diyala River Valley and several other areas.

Throughout this period, as well, we engaged in dialogue with insurgent groups and tribes, and this led to additional elements standing up to oppose Al Qaeda and other extremists. We also continued to emphasize the development of the Iraqi Security Forces and we employed non-kinetic [non-combat] means to exploit the opportunities provided by the conduct of our kinetic operations—aided in this effort by the arrival of additional Provincial Reconstruction Teams.

Current Situation and Trends

The progress our forces have achieved with our Iraqi counterparts has, as I noted at the outset, been substantial. While there have been setbacks as well as successes and tough losses along the way, overall, our tactical commanders and I see improvements in the security environment. We do not, however, just rely on gut feel or personal observations; we also conduct considerable data collection and analysis to gauge progress and determine trends. We do this by gathering and refining data from coalition *and* Iraqi operations centers, using a methodology that has been in place for well over a year and that has benefited over the past seven months from the increased presence of our forces living among the Iraqi people. We endeavor to ensure our analysis of that data is conducted with rigor and consistency, as our ability to achieve a nuanced understanding of the security environment is dependent on collecting and analyzing data in a consistent way over time. Two US intelligence agencies recently reviewed our methodology, and they concluded that the data we produce is the most accurate and authoritative in Iraq.

As I mentioned up front . . . the level of security incidents has decreased significantly since the start of the surge of offensive operations in mid-June, declining in 8 of the past 12 weeks, with the level of incidents in the past two weeks the lowest since June 2006 and with the number of *attacks* this past week the lowest since April 2006.

Civilian deaths of *all* categories, less natural causes, have also declined considerably, by over 45% Iraq-wide since the height of the sectarian violence in December. . . . Periodic mass casualty attacks by Al Qaeda have tragically added to the numbers outside Baghdad, in particular. Even without the sensational attacks, however, the level of civilian deaths is clearly still too high and continues to be of serious concern.

. . . [T]he number of *ethno-sectarian* deaths, an important subset of the overall civilian casualty figures, has also declined significantly since the height of the sectarian violence in December. Iraq-wide . . . the number of ethno-sectarian deaths has come down by over 55%, and it would have come down much further were it not for the casualties inflicted by barbaric Al Qaeda bombings attempting to reignite sectarian violence. In Baghdad . . . the number of ethno-sectarian deaths has come down by some 80% since December. . . .

As we have gone on the offensive in former Al Qaeda and insurgent sanctuaries, and as locals have increasingly supported our efforts, we have found a substantially increased number of arms, ammunition, and explosives caches. . . . [W]e have, so far this year, already found and cleared over 4,400 caches, nearly 1,700 more than we discovered in all of last year. This may be a factor in the reduction in the number of overall improvised explosive device attacks in recent months . . . has declined sharply, by about one-third, since June.

The change in the security situation in Anbar Province has, of course, been particularly dramatic. . . . [M]onthly attack levels in Anbar have declined from some 1,350 in October 2006 to a bit over 200 in August of this year. This dramatic decrease reflects the significance of the local rejection of Al Qaeda and the newfound willingness of local Anbaris to volunteer to serve in the Iraqi Army and Iraqi Police Service. As I noted earlier, we are seeing similar actions in other locations, as well.

To be sure, trends have not been uniformly positive across Iraq. . . . The trend in Ninevah Province, for example, has been much more up and down, until a recent decline, and the same is true in Sala ad Din Province, though recent trends there and in Baghdad have been in the right direction. In any event, the overall trajectory in Iraq—a steady decline of incidents in the past three months—is still quite significant.

The number of car bombings and suicide attacks has also declined in each of the past 5 months, from a high of some 175 in March . . . to about 90 this past month. While this trend in recent months has been heartening, the number of high profile attacks is still too high, and we continue to work hard to destroy the networks that carry out these barbaric attacks.

Our operations have, in fact, produced substantial progress against Al Qaeda and its affiliates in Iraq. . . . [I]n the past eight months, we have considerably reduced the areas in which Al Qaeda enjoyed sanctuary. We have also neutralized 5 media cells, detained the senior Iraqi leader of Al Qaeda-Iraq, and killed or captured nearly 100 other key leaders and some 2,500 rank-and-file fighters. Al Qaeda is certainly not defeated; however, it is off balance and we are pursuing its leaders and operators aggressively. Of note, as the recent National Intelligence Estimate on Iraq explained, these gains against Al Qaeda are a result of the synergy of actions by: conventional forces to deny the terrorists sanctuary; intelligence, surveillance, and reconnaissance assets to find the enemy; and special operations elements to conduct targeted raids. A combination of these assets is necessary to prevent the creation of a terrorist safe haven in Iraq.

In the past six months we have also targeted Shia militia extremists, capturing a number of senior leaders and fighters, as well as the deputy commander of Lebanese Hezbollah Department 2800, the organization created to support the training, arming, funding, and, in some cases, direction of the militia extremists by the Iranian Republican Guard Corps' Qods [Jerusalem] Force. These elements have assassinated and kidnapped Iraqi governmental leaders, killed and wounded our soldiers with advanced explosive devices provided by Iran, and indiscriminately rocketed civilians in the International Zone and elsewhere. It is increasingly apparent to both Coalition and Iraqi leaders that Iran, through the use of the Qods Force, seeks to turn the Iraqi Special Groups into a Hezbollah-like force to serve its interests and fight a proxy war against the Iraqi state and coalition forces in Iraq.

The most significant development in the past six months likely has been the increasing emergence of tribes and local citizens rejecting Al Qaeda and other extremists. This has, of course, been most visible in Anbar Province. A year ago the province was assessed as "lost" politically. Today, it is a model of what happens when local leaders and citizens decide to oppose Al Qaeda and reject its Taliban-like ideology. While Anbar is unique and the model it provides cannot be replicated everywhere in Iraq, it does demonstrate the dramatic change in security that is possible with the support and participation of local citizens. . . . [O]ther tribes have been inspired by the actions of those in Anbar and have volunteered to fight extremists as well. We have, in coordination with the Iraqi government's National Reconciliation Committee, been engaging these tribes and groups of local citizens who want to oppose extremists and to contribute to local security. Some 20,000 such individuals are already being hired for the Iraqi Police, thousands of others are being assimilated into the Iraqi Army, and thousands more are vying for a spot in Iraq's Security Forces.

Iraqi Security Forces

As I noted earlier, Iraqi Security Forces have continued to grow, to develop their capabilities, and to shoulder more of the burden of providing security for their country. Despite concerns about sectarian influence, inadequate logistics and supporting institutions, and an insufficient number of qualified commissioned and non-commissioned officers, Iraqi units are engaged around the country.

. . . [T]here are now nearly 140 Iraqi Army, National Police, and Special Operations Forces Battalions in the fight, with about 95 of those capable of taking the lead in operations, albeit with some coalition support. Beyond that, all of Iraq's battalions have been heavily involved in combat operations that often result in the loss of leaders, soldiers, and equipment. These losses are among the shortcomings identified by operational readiness assessments, but we should not take from these assessments the impression that Iraqi forces are not in the fight and contributing. Indeed, despite their shortages, many Iraqi units across Iraq now operate with minimal coalition assistance.

As counterinsurgency operations require substantial numbers of boots on the ground, we are helping the Iraqis expand the size of their security forces. Currently, there are some 445,000 individuals on the payrolls of Iraq's Interior and Defense Ministries. Based on recent decisions by Prime Minister [Nouri al-] Maliki, the number of Iraq's security forces will grow further by the end of this year, possibly by as much as 40,000. Given the security challenges Iraq faces, we support this decision, and we will work with the two security ministries as they continue their efforts to expand their basic training capacity, leader development programs, logistical structures and elements, and various other institutional capabilities to support the substantial growth in Iraqi forces.

Significantly, in 2007, Iraq will, as in 2006, spend more on its security forces than it will receive in security assistance from the United States. In fact, Iraq is becoming one of the United States' larger foreign military sales [FMS] customers, committing some $1.6 billion to FMS already, with the possibility of up to $1.8 billion more being committed before the end of this year. And I appreciate the attention that some members of Congress have recently given to speeding up the FMS process for Iraq.

To summarize, the security situation in Iraq is improving, and Iraqi elements are slowly taking on more of the responsibility for protecting their citizens. Innumerable challenges lie ahead; however, Coalition and Iraqi Security Forces have made progress toward achieving sustainable security. As a result, the United States will be in a position to reduce its forces in Iraq in the months ahead.

Recommendations

Two weeks ago I provided recommendations for the way ahead in Iraq to the members of my chain of command and the Joint Chiefs of Staff. The essence of the approach I recommended is captured in its title: "Security While Transitioning: From Leading to Partnering to Overwatch." This approach seeks to build on the security improvements our troopers and our Iraqi counterparts have fought so hard to achieve in recent months. It reflects recognition of the importance of securing the population *and* the imperative of transitioning responsibilities to Iraqi institutions and Iraqi forces as quickly as possible, but without rushing to failure. It includes substantial support for the continuing development of Iraqi Security Forces. It also stresses the need to continue the counterinsurgency strategy that we have been employing, but with Iraqis gradually shouldering more of the load. And it highlights the importance of regional and global diplomatic approaches. Finally, in recognition of the fact that this war is not only being fought on the ground in Iraq but also in cyberspace, it also notes the need to contest the enemy's growing use of that important medium to spread extremism.

The recommendations I provided were informed by operational and strategic considerations. The *operational* considerations include recognition that:

- military aspects of the surge have achieved progress and generated momentum;
- Iraqi Security Forces have continued to grow and have slowly been shouldering more of the security burden in Iraq;
- a mission focus on either population security or transition alone will not be adequate to achieve our objectives;
- success against Al Qaeda-Iraq and Iranian-supported militia extremists requires conventional forces as well as special operations forces; and
- the security and local political situations will enable us to draw down the surge forces.

My recommendations also took into account a number of *strategic* considerations:

- political progress will take place only if sufficient security exists;
- long-term US ground force viability will benefit from force reductions as the surge runs its course;
- regional, global, and cyberspace initiatives are critical to success; and
- Iraqi leaders understandably want to assume greater sovereignty in their country, although, as they recently announced, they do desire continued presence of coalition forces in Iraq in 2008 under a new UN Security Council Resolution and, following that, they want to negotiate a long term security agreement with the United States and other nations.

Based on these considerations, and having worked the battlefield geometry with Lieutenant General Ray Odierno to ensure that we retain and build on the gains for which our troopers have fought, I have recommended a drawdown of the surge forces from Iraq. In fact, later this month, the Marine Expeditionary Unit deployed as part of the surge will depart Iraq. Beyond that, if my recommendations are approved, that unit's departure will be followed by the withdrawal of a brigade combat team without replacement in mid-December and the further redeployment without replacement of four other brigade combat teams and the two surge Marine battalions in the first 7 months of 2008, until we reach the pre-surge level of 15 brigade combat teams by mid-July 2008.

I would also like to discuss the period beyond next summer. Force reductions *will* continue beyond the pre-surge levels of brigade combat teams that we will reach by mid-July 2008; however, in my professional judgment, it would be premature to make recommendations on the pace of such reductions at this time. In fact, our experience in Iraq has repeatedly shown that projecting too far into the future is not just difficult, it can be misleading and even hazardous. The events of the past six months underscore that point. When I testified in January, for example, no one would have dared to forecast that Anbar Province would have been transformed the way it has in the past 6 months. Nor would anyone have predicted that volunteers in one-time Al Qaeda strongholds like Ghazaliyah in western Baghdad or in Adamiya in eastern Baghdad would seek to join the fight against Al Qaeda. Nor would we have anticipated that a Shia-led government would accept significant numbers of Sunni volunteers into the ranks of the local police force in Abu Ghraib. Beyond that, on a less encouraging note, none of us earlier this year appreciated the extent of Iranian involvement in Iraq, something about which we and Iraq's leaders all now have greater concern.

In view of this, I do not believe it is reasonable to have an adequate appreciation for the pace of further reductions and mission adjustments beyond the summer of 2008 until about mid-March of next year. We will, no later than that time, consider factors similar to those on which I based the current recommendations, having by then, of course, a better feel for the security situation, the improvements in the capabilities of our Iraqi counterparts, and the enemy situation. I will then, as I did in developing the recommendations I have explained here today, also take into consideration the demands on our Nation's ground forces, although I believe that that consideration should once again inform, not drive, the recommendations I make. . . .

One may argue that the best way to speed the process in Iraq is to change the MNF-I mission from one that emphasizes population security, counter-terrorism, and transition, to one that is strictly focused on transition and counter-terrorism. Making that change now would, in our view, be premature. We have learned before that there is a real danger in handing over tasks to the Iraqi Security Forces before their capacity and local conditions warrant. In fact, the drafters of the recently released National Intelligence Estimate on Iraq recognized this danger when they wrote, and I quote, "We assess that changing the mission of Coalition forces from a primarily counterinsurgency and stabilization role to a primary combat support role for Iraqi forces and counterterrorist operations to prevent AQI [Al Qaeda in Iraq] from establishing a safe haven would erode security gains achieved thus far."

In describing the recommendations I have made, I should note again that, like Ambassador [Ryan] Crocker, I believe Iraq's problems will require a long-term effort. There are no easy answers or quick solutions. And though we both believe this effort can succeed, it will take time. Our assessments underscore, in fact, the importance of recognizing that a premature drawdown of our forces would likely have devastating consequences.

That assessment is supported by the findings of a 16 August Defense Intelligence Agency report on the implications of a rapid withdrawal of US forces from Iraq. Summarizing it in an unclassified fashion, it concludes that a rapid withdrawal would result in the further release of the strong centrifugal forces in Iraq and produce a number of dangerous results, including a high risk of disintegration of the Iraqi Security Forces; rapid deterioration of local security initiatives; Al Qaeda-Iraq regaining lost ground and freedom of maneuver; a marked increase in violence and further ethno-sectarian displacement and refugee flows; alliances of convenience by Iraqi groups with internal and external forces to gain advantages over their rivals; and exacerbation of already challenging regional dynamics, especially with respect to Iran.

Lieutenant General Odierno and I share this assessment and believe that the best way to secure our national interests and avoid an unfavorable outcome in Iraq is to continue to focus our operations on securing the Iraqi people while targeting terrorist groups and militia extremists and, as quickly as conditions are met, transitioning security tasks to Iraqi elements.

Closing Comments

Before closing, I want to thank you and your colleagues for your support of our men and women in uniform in Iraq. The Soldiers, Sailors, Airmen, Marines, and Coast Guardsmen with whom I'm honored to serve are the best equipped and, very likely, the most professional force in our nation's history. Impressively, despite all that has been asked of them in recent years, they continue to raise their right hands and volunteer to stay in uniform.

With three weeks to go in this fiscal year, in fact, the Army elements in Iraq, for example, have achieved well over 130% of the reenlistment goals in the initial term and careerist categories and nearly 115% in the mid-career category. All of us appreciate what you have done to ensure that these great troopers have had what they've needed to accomplish their mission, just as we appreciate what you have done to take care of their families, as they, too, have made significant sacrifices in recent years.

The advances you have underwritten in weapons systems and individual equipment; in munitions; in command, control, and communications systems; in intelligence, surveillance, and reconnaissance capabilities; in vehicles and counter-IED [improved explosive devices, or roadside bombs] systems and programs; and in manned and unmanned aircraft have proven invaluable in Iraq. The capabilities that you have funded most recently—especially the vehicles that will provide greater protection against improvised explosive devices—are also of enormous importance. Additionally, your funding of the Commander's Emergency Response Program has given our leaders a critical tool with which to prosecute the counterinsurgency campaign. Finally, we appreciate as well your funding of our new detention programs and rule of law initiatives in Iraq.

In closing, it remains an enormous privilege to soldier again in Iraq with America's new "Greatest Generation." Our country's men and women in uniform have done a magnificent job in the most complex and challenging environment imaginable. All Americans should be very proud of their sons and daughters serving in Iraq today.

Thank you very much.

SOURCE: U.S. Department of Defense. "Report to Congress on the Situation in Iraq." Testimony by General David H. Petraeus at a joint hearing of the U.S. House of Representatives Committees on Foreign Affairs and Armed Services. September 10, 2007. www.defenselink.mil/pubs/pdfs/Petraeus-Testimony20070910.pdf (accessed March 3, 2008).

OTHER HISTORIC DOCUMENTS OF INTEREST

FROM THIS VOLUME

- President Bush on the "surge" of U.S. troops to Iraq, p. 9
- U.S. ambassador on U.S.-Iran diplomacy, p. 257
- GAO on Iraqi government issues, p. 482

FROM PREVIOUS *HISTORIC DOCUMENTS*

- Report of the Special Inspector General on Iraq Reconstruction, *2006,* p. 161
- President Bush and Prime Minister Maliki on the government of Iraq, *2006,* p. 171
- UN Secretary-General Annan on the situation in Iraq, *2006,* p. 725
- Iraq Study Group Report on Iraq, *2006,* p. 725

DOCUMENT IN CONTEXT

President Bush on the Passage of Electronic Surveillance Legislation

AUGUST 5, 2007

The Bush administration battled with congressional Democrats for much of the year regarding the permissible scope of the president's intelligence-gathering activities, including surveillance programs that eavesdropped on U.S. citizens. The heated debate pitted protection of civil liberties and privacy against national security in the face of international terrorist threats. But it also involved constitutional questions about the separation of powers among the branches of government, with President George W. Bush asserting he had constitutional authority to conduct the secret surveillance without first getting a judicial warrant, as required by law. Democratic legislators contended that the president had overstepped his constitutional boundaries, but they were largely stymied in their efforts to get the administration to disclose information about, or place limits on, the secret surveillance programs it put in place after the September 11, 2001, attacks.

Under intense White House pressure, which included raising the specter of a terrorist attack while lawmakers were on their August recess, Democratic leaders approved temporary legislation written by the administration that expanded the executive branch's authority to conduct warrantless surveillance on foreign targets, including any communications those targets might have with U.S. citizens. Once that legislation was in place, the administration began to push a measure that would make the expanded surveillance powers permanent and also protect private telecommunications companies against lawsuits for their participation in the surveillance program. Action on the legislation stalled in the Senate at the end of the year, however, and was put off until 2008.

Debate over the surveillance program was fought out in an atmosphere of increasing mutual distrust and animosity between congressional Democrats and the Republican administration. Consideration of the surveillance legislation occurred at the same time that House and Senate panels were investigating allegations that the administration had fired several federal prosecutors for partisan political purposes and acknowledgements that the FBI had misused its powers under the 2001 USA Patriot Act to illegally obtain telephone, e-mail, and financial records of thousands of U.S. citizens. Democratic and some Republican legislators expressed dismay and anger with administration witnesses' conflicting statements to congressional panels, the periodic emergence of new information suggesting that the administration had not been forthcoming in answering legislators' questions, and the White House's refusal to make key advisers available for public questioning and to provide all the documentation requested by congressional committees. Attorney General Alberto R. Gonzales, who as head of the Justice Department was integrally involved in all these matters, resigned in August, amid persistent questions about his competence and truthfulness.

A Secret Program Revealed

In the days following the September 11 al Qaeda attacks against the United States, the Bush administration made no secret of its intention to fight international terrorists with every tool, military and nonmilitary, at its disposal. What officials did not make public were the tools that they thought they had a right to use. Some of those came to light in December 2005, when the *New York Times* reported that the top-secret National Security Agency (NSA) had been monitoring, without court warrants, thousands of telephone conversations and e-mail messages between people in the United States and people overseas believed to be connected to terrorist groups. Bush acknowledged that he had ordered the warrantless surveillance and defended the program as necessary to intercept communications that might enable the government to thwart a potential terrorist attack.

The news report set off a legal and political furor that lasted for months. Much of the controversy concerned whether Bush had overstepped his authority by ordering the eavesdropping without seeking warrants from a special court established in 1978 under the Foreign Intelligence Surveillance Act (FISA, PL 95-511). Consisting of federal judges appointed by the chief justice, the court's sole function was to review, in secret, the administration's requests for warrants to monitor telephone calls of people suspected of posing national security threats to the United States. The court rarely turned down such requests. The Bush administration said its surveillance of terrorism suspects did not need to be reviewed by the FISA court because the president had the constitutional power, as commander in chief of the armed forces, to order the surveillance. Administration officials also said Congress had given Bush broad authority in September 2001 when it passed a joint resolution authorizing him to take "necessary" action to combat terrorism. Bush's assertions were greeted with skepticism by many members of Congress. Patrick J. Leahy of Vermont, then the ranking Democrat on the Senate Judiciary Committee, chided Bush for conducting the wiretapping "illegally," for never seeking congressional approval, and for thwarting congressional efforts to learn the details.

Amid increasingly loud calls from legislators in both parties for measures putting the program on firm legal ground and imposing some limits on it, the White House endorsed a Republican bill, passed by the House in September 2006, giving the president authority to conduct warrantless wiretaps under certain conditions. The Senate did not act, however, and after Democrats won control of both chambers of Congress in the November 2006 elections, they said they would conduct new hearings on the administration's antiterrorism programs early in 2007.

Temporary Authorization

Soon after Congress reconvened in 2007, with the Democrats at the helm, the Bush administration announced that it would allow the FISA court to oversee the wiretapping program. Although legislators welcomed the news, many continued to question why the administration had taken so long to turn to the FISA court; they also pressed for more details of the surveillance program itself. On January 31, the administration agreed to turn over documents on the program to the top members of both chambers' judiciary committees as well as to their intelligence committees and the senior congressional leadership, which had already been briefed on the program.

In April the administration began pressing for legislation to modernize the FISA law. Michael McConnell, the director of national intelligence, said that technological developments in telecommunications and electronics had made some aspects of the law obsolete and that requirements to seek warrants through the secret FISA court were hampering the government's ability to spy on communications between overseas targets that were routed through the U.S. telecommunications network. Testifying before the Senate Select Intelligence Committee on May 1, McConnell also refused to guarantee that the administration would seek warrants from the FISA court before eavesdropping on U.S. citizens, as it had promised in January.

Democratic legislators immediately pushed back, saying that they needed more information about the program before they would agree to expand the president's powers and that any FISA overhaul should maintain the court's oversight of the government's surveillance of U.S. citizens. "The trick is, we want to go after the bad guys. We want to get the information that we need," Florida senator Bill Nelson said at the May 1 hearing. "But we're a nation of laws and we want to prevent the buildup of a dictator who takes the law into his own hands. . . . So now we have to find the balance."

Over the next few months, McConnell and Republican legislators continued to press the Democrats to act on the proposed legislation. Then on July 31, House minority leader John A. Boehner, R-Ohio, revealed during an interview on Fox News that passage of the legislation was crucial because a FISA court judge had secretly ruled earlier in the year that the administration could not legally collect information from foreign calls and e-mails routed through the U.S. telecommunications system to another foreign country. "There's been a ruling, over the last four or five months, that prohibits the ability of our intelligence services and our counterintelligence people from listening in to two terrorists in other parts of the world where the communication could come through the United States," Boehner said in the interview. "This means that our intelligence agencies are missing a wide swath of potential information that could help protect the American people."

With Boehner's revelation, the administration and its Republican allies in Congress stepped up pressure on the Democrats to act before leaving for their August recess. Intelligence officials talked of heightened "chatter" among terrorist suspects, pointed to two bomb scares in London and Glasgow in June, and cited a new national intelligence estimate concluding that direct attacks against the United States were possible. "Protecting America is our most solemn obligation," Bush said in a statement issued August 4. Senate Republican leader Mitch McConnell of Kentucky said he could not "imagine [Democrats] would take a month-long vacation without fulfilling their obligation to keep America safe."

Unable to pass their own legislation and fearing the political repercussions of not acting, Democrats reluctantly approved legislation (S 1927 – PL 110-55), written by the administration, that would expire in six months. The measure expanded the executive branch's authority to conduct surveillance on foreign targets without a warrant, whether or not the surveillance included communications with U.S. citizens. It also allowed the government to monitor, without a warrant, foreign calls and e-mails being routed over the U.S. telecommunications system.

Signing the bill on August 5, President Bush said that it "gives our intelligence professionals . . . greater flexibility while closing a dangerous gap in our intelligence gathering activities that threatened to weaken our defenses." He called on Congress to enact permanent legislation as soon as they returned to work in September. He also pressed for "meaningful liability protection to those who are alleged to have assisted our Nation" after the September 11 attacks. For their part, several Democrats said they regretted their vote in favor of the temporary legislation and promised to scale back the expanded authority granted the executive branch in any permanent legislation that was passed.

Immunity for Telecommunications Companies

In an interview with the *El Paso Times,* a transcript of which was posted on the paper's Web site on August 22, National Intelligence Director McConnell acknowledged for the first time that U.S. telecommunications companies had helped NSA conduct its warrantless eavesdropping program and said it was vital for Congress to give the companies, which he would not name, retroactive immunity from lawsuits. By year's end, forty lawsuits had been filed against AT&T, Verizon, and several other major carriers by individuals and organizations such as the American Civil Liberties Union alleging that the companies had violated privacy rights by participating in government surveillance programs. "Under the president's program . . . the private sector has assisted us because if you're going to get access, you've got to have a partner," McConnell said, adding that "if you play out the [law]suits at the value they're claimed, it would bankrupt these companies."

By the end of the year, it was not clear whether Congress would grant the administration's request for immunity for these companies. In November the House passed a measure (HR 3773) cutting back on the administration's authority to conduct surveillance without a warrant on foreign targets who were communicating with U.S. citizens. It did not contain retroactive immunity for companies that allegedly participated in the NSA program. Democrats in the Senate were divided on the issue, however, and floor action on the legislation was postponed until 2008. The Intelligence Committee reported a bill granting the retroactive immunity, but the Judiciary Committee version withheld it. Judiciary Committee chair Leahy said in October that the reason the White House was lobbying so intensely for the immunity was "because they know that it was illegal conduct and that there is no saving grace for the president to say, 'Well, I was acting with authority.' " But Intelligence Committee chair John D. Rockefeller IV of West Virginia said the companies had "made a very strong case" and that if it were "not for these companies, there is no way we could conduct surveillance."

Following is the text of remarks made by President George W. Bush on August 5, 2007, as he signed legislation that temporarily expanded executive branch authority to conduct warrantless surveillance of foreign targets.

DOCUMENT

President Bush on Signing the Protect America Act

August 5, 2007

When our intelligence professionals have the legal tools to gather information about the intentions of our enemies, America is safer. And when these same legal tools also protect the civil liberties of Americans, then we can have the confidence to know that we can preserve our freedoms while making America safer.

The Protect America Act, passed with bipartisan support in the House and Senate, achieves both of these goals by modernizing the Foreign Intelligence Surveillance Act. Over the past three decades, this law has not kept pace with revolutionary changes in technology. As a result, our intelligence professionals have told us that they are missing significant intelligence information that they need to protect the country.

S. 1927 reforms FISA by accounting for changes in technology and restoring the statute to its original focus on appropriate protections for the rights of persons in the United States and not foreign targets located in foreign lands.

Today, we face a dynamic threat from enemies who understand how to use modern technology against us. Whether foreign terrorists, hostile nations, or other actors, they change their tactics frequently and seek to exploit the very openness and freedoms we hold dear. Our tools to deter them must also be dynamic and flexible enough to meet the challenges they pose. This law gives our intelligence professionals this greater flexibility while closing a dangerous gap in our intelligence gathering activities that threatened to weaken our defenses.

We know that information we have been able to acquire about foreign threats will help us detect and prevent attacks on our homeland. Mike McConnell, the Director of National Intelligence, has assured me that this bill gives him the most immediate tools he needs to defeat the intentions of our enemies. And so in signing this legislation today, I am heartened to know that his critical work will be strengthened, and we will be better armed to prevent attacks in the future.

I commend Members of Congress who supported these important reforms and also for acting before adjourning for recess. In particular, I want to thank Mitch McConnell and John Boehner for their strong leadership on this issue and Senators Kit Bond and Dianne Feinstein for coming together in the Senate on an effective bipartisan solution. In the House of Representatives, Pete Hoekstra and Heather Wilson were instrumental in securing enactment of this vital piece of legislation before the August recess, and I thank them for their leadership.

While I appreciate the leadership it took to pass this bill, we must remember that our work is not done. This bill is a temporary, narrowly focused statute to deal with the most immediate shortcomings in the law.

When Congress returns in September, the intelligence committees and leaders in both parties will need to complete work on the comprehensive reforms requested by Director McConnell, including the important issue of providing meaningful liability protection to those who are alleged to have assisted our Nation following the attacks of September 11, 2001.

SOURCE: U.S. Executive Office of the President. "Statement on Congressional Passage of Intelligence Reform Legislation." August 5, 2007. *Weekly Compilation of Presidential Documents* 43, no. 32 (August 13, 2007): 1048-1049. Washington, D.C.: National Archives and Records Administration. www.gpoaccess.gov/wcomp/v43no32.html (accessed April 3, 2008).

OTHER HISTORIC DOCUMENTS OF INTEREST

FROM THIS VOLUME

- National Intelligence Estimate on terrorism, p. 335
- Gonzales resignation as attorney general, p. 457

FROM PREVIOUS *HISTORIC DOCUMENTS*

- Statements on executive power and NSA surveillance authority, *2006*, p. 61
- USA Patriot Act, *2006*, p. 449

DOCUMENT IN CONTEXT

UN Secretary-General on the Postwar Situation in Liberia

AUGUST 8, 2007

Liberia made significant progress during 2007 on the road to recovery from two decades of political upheaval and civil war. An important sign of the progress was that the government of President Ellen Johnson-Sirleaf—elected at the end of 2005—had faced no violent opposition for two full years and continued to win international backing for the rebuilding of a nation largely destroyed by the war.

Liberia had experienced coups and other upheavals all during the 1980s, then more than a decade of bloody civil conflict that killed, by most estimates, at least 200,000 of its approximately 3 million citizens. The violence ended in 2003 when President Charles Taylor, a former warlord who had won an election but fostered wars in neighboring Guinea and Sierra Leone, was forced to flee. An interim government ruled for two years until the internationally sponsored elections that were won by Johnson-Sirleaf, a U.S.-educated economist who had been both a government official and a political prisoner in Liberia and was known as the "Iron Lady" because of her toughness.

Paying tribute to Johnson-Sirleaf and other Liberians who were trying to reassemble their country, UN secretary-general Ban Ki-moon nevertheless warned, in an August 8 report to the Security Council, that violence remained a threat to Liberia. The "prevailing peace is very fragile and Liberia is still susceptible to lawlessness," Ban reported.

Many Liberians said they still had faith in their new, elected government to make things better. "The changes in Liberia are taking place a little slow," a former boy soldier named Norman told an Associated Press reporter in December. "But we expect this government to do something. We have faith it is doing its best to improve our lives."

Taylor Goes on Trial

One of the most important events in Liberia's recent sad history took place several thousand miles away, at a courtroom in The Hague, Netherlands. On June 4, Charles Taylor went on trial on eleven charges of war crimes, all concerning his support for antigovernment rebels in the civil war in Sierra Leone. The case was being heard by a UN tribunal dealing with crimes from the war in Sierra Leone, which at times exceeded even Liberia's war in brutality before its end in 2002. According to the charges, the former Liberian president had provided weapons, money, and other support for rebels in Sierra Leone in exchange for access to diamonds and other natural resources in the rebel-controlled areas of that country.

Taylor in 2003 had fled into exile in Nigeria, which protected him from the UN tribunal until 2006, when he disappeared. He was taken into custody when he tried to cross into Cameroon, then turned over to the UN tribunal, which later in 2006 transferred him to The Hague to await his trial. Taylor pleaded not guilty to all charges, and he fired his lawyer and refused to appear in court on the opening day of his trial. He finally did appear in court on July 3, but six weeks later the tribunal postponed further sessions until early January 2008 to give Taylor's new lawyers time to prepare his defense.

Taylor's trial brought mixed reactions in his own country. Despite the devastation he had wrought, Taylor remained popular among many Liberians, some of whom erected billboards in Monrovia, the capital city, proclaiming his innocence. Many others were relieved to see Taylor in court, but some said he was just one of many Liberians responsible for the civil war. "If you start prosecuting war crimes in Liberia, you'll prosecute every Liberian," a former child soldier, Paul Tolbert, told the Associated Press.

Diamond Ban Lifted

The UN Security Council had imposed sanctions against Taylor's government in 2001 and 2002, including bans on international trading in diamonds and timber from Liberia, on exports of weapons to Liberia, and on international travel and financial transactions by Taylor and the government officials and rebel leaders associated with him. Shortly after taking office in 2006, President Johnson-Sirleaf asked the council to lift the bans on diamond and timber exports, saying the country desperately needed the money from exports of its most valuable natural resources. The council lifted the timber ban in June 2006 but retained the diamond ban (as well as the arms and travel bans) into 2007. In the case of diamonds, the country had been unable to meet the requirements of an international program, the Kimberley Process, intended to prevent international trading in so-called blood diamonds from conflict zones such as Liberia.

The Security Council voted on April 27 to lift the ban on diamond exports, pending confirmation that Liberia was indeed complying with the Kimberley Process regulations. Diamond exports resumed on a limited scale in September, and in December Liberia signed an agreement with Israel, which had one of the world's largest diamond-trading and -processing industries. Under the agreement, the Israeli Diamond Institute was to provide technical services to help Liberia revive its diamond industry.

The Security Council on December 19 extended for one more year the prohibitions on arms sales to Liberia and travel by Taylor's associates. The council acted to keep the weapons ban after receiving a report from a committee of experts decrying a surge of armed robberies and other violent crimes involving firearms in Liberia.

Anticorruption Campaign

The decades of civil war and upheaval in Liberia had featured not only death and destruction but also the theft of the country's resources, in many cases by government officials. After taking office, Johnson-Sirleaf launched an anticorruption campaign that included the dismissal of dozens of officials accused of having engaged in corrupt actions. The government also sought and won indictments in 2007 of two senior officials. Gyude Bryant, who had led the transitional government in the period between the ousting of Taylor in 2003 and the 2005 elections, was charged with economic sabotage in February, then arrested in December after he failed to appear at court hearing. Edwin Snowe, a former speaker of the House of Representatives, was indicted on charges of stealing $1 million from the Liberia Petroleum Refining Corporation when he had been its managing director in the 1990s. The allegations against Snowe precipitated a political crisis that included rump sessions by competing factions of the House of Representatives; Snowe resigned on February 15, and the crisis was resolved when a successor was selected in April.

The new government came under suspicion in mid-June, however, when the country's auditor general, John Morlu, charged that Johnson-Sirleaf's government was "three times more corrupt" than its predecessor. He alleged that "millions of dollars are unaccounted for" in the draft budget, then under review, for fiscal year 2007–2008. Morlu provided no specifics for his charges, which led to an extended dispute between him and the government, still unresolved at year's end.

UN Peacekeeping

Although the various phases of Liberia's civil war ended in 2003, security in the country remained fragile because an estimated 100,000 former fighters lacked jobs, some militia leaders remained at large in Liberia or elsewhere in West Africa, and many of the underlying tribal tensions that had contributed to the war had not been resolved. The continuing risk of instability was highlighted by the arrests on July 17 of two former senior officials who were accused of attempting to destabilize the government. George Koukou, speaker of parliament during the transitional period after Taylor's downfall, and Charles Julu, a general who was army chief of staff under former president Samuel Doe, were charged with treason. The government alleged that they had conspired with a third man who was arrested in the Ivory Coast on charges of attempting to buy and transport weapons to Liberia. This incident, and continuing violence in Liberia, suggested that "there is still a risk of the possible resurgence of armed groups, which may be easily organized to cause political instability," Secretary-General Ban said in his August report.

The main source of security in the country was a 15,000-troop UN peacekeeping force known as the United Nations Mission in Liberia (UNMIL). Ban in March asked the Security Council to reauthorize the force for another twelve months, but the council—concerned about demand for peacekeepers in the Darfur region of Sudan and other current zones of conflict—instead on March 30 approved a six-month extension and asked Ban to submit a plan for a gradual withdrawal of the Liberia mission. Ban submitted that plan in his August report, suggesting that the peacekeeping force (then down to 14,123 members) be reduced by about 2,450 troops between October 2007 and September 2008, after which further troop withdrawals would depend on events on the ground. Ban gave no estimate of when the UN could close down its mission, noting that national elections scheduled for October 2011 would need to be taken into account in such planning. The Security Council on September 21 endorsed Ban's recommendation for the gradual cutback in the force during 2007 and 2008. The council also extended the force's mandate for one year.

Rebuilding Liberia

The war left much of Liberia in ruins: farms and factories were destroyed, and nearly all public buildings and services in the capital, Monrovia, were severely damaged. Unemployment was estimated to be about 80 percent. Rebuilding efforts were under way, but by 2007 change was still modest. Street lights and electrical service were installed in parts of Monrovia (most of which still lacked running water), and many primary schools had been reopened, although in many cases without enough textbooks or classroom space.

Liberia received substantial aid from donor countries to help it rebuild and, just as important, relief from much of the approximately $4 billion in debt that previous governments had accumulated. The administration of President George W. Bush on February 13 announced plans to forgive $391 million in past loans for development projects and military aid; similar pledges came from China, Germany, Britain, and other countries. The International Monetary Fund and the World Bank, which together held about $1.5 billion of Liberia's debt, also announced plans to forgive most or even all of that debt once they received financing from donor members.

Following are excerpts from the "Findings" and "Observations" sections of the "Fifteenth Progress Report of the Secretary-General on the United Nations Mission to Liberia," submitted by Secretary-General Ban Ki-moon to the Security Council on August 8, 2007.

DOCUMENT

United Nations Progress Report on Its Mission in Liberia

August 8, 2007

FINDINGS

A. Security situation and threat assessment

15. Liberia has become a generally stable country in a volatile subregion. However, the prevailing peace is very fragile and Liberia is still susceptible to lawlessness. The most immediate threats to sustained peace and stability in Liberia at this stage include increasing violent criminal activities, especially armed robbery and rape; the limited capacity of the security sector to curb violent crime; the weak justice system; the limited capacity of key national institutions to deliver on the promised peace dividend; the proliferation of disaffected groups such as unemployed ex-combatants, deactivated soldiers and police personnel, and elements from the dismantled irregular militias; economic insecurity, in particular youth unemployment; resurfacing ethnic and social cleavages; and the perception by some opposition political parties that the Government is not genuinely pursuing national reconciliation.

16. There is still a risk of the possible resurgence of armed groups, which may be easily organized to cause political instability. Some Liberian stakeholders expressed concern that elements whose interests are threatened by the Government's reform agenda, as well as individuals loyal to former President Charles Taylor, could attempt to instigate political instability. These factors, as well as the risk of a possible spillover from the unstable situation in Guinea, Sierra Leone and Côte d'Ivoire, underscore the country's continued vulnerability. Nonetheless, Liberia's relations with its neighbours have improved to the extent that potential insurgents would find it difficult to use a neighbouring country's territory to destabilize the country.

17. Prior to the arrival of the technical assessment mission, four joint security assessment teams, co-led by UNMIL [United Nations Mission in Liberia] and the Government, conducted a comprehensive evaluation of the security situation in the country. The teams, which included representatives of UNMIL, the United Nations country team and the Ministries of Internal Affairs, Defence and National Security, as well as the National Security Agency, the National Bureau of Investigation, the Bureau of Immigration and Naturalization and the Liberian National Police, concluded that security, State administration and the rule of law in the counties were extremely limited. While the security situation in Bomi, Grand Cape Mount and Margibi Counties was assessed as relatively benign, Sinoe and Lofa Counties were assessed as relatively high risk areas.

18. The joint security assessment teams also determined that the weak presence of State authority in some counties, aggravated by poor communications and the lack of livelihood opportunities, have contributed to the movement of ex-combatants, youth and foreigners into areas that are rich in natural and mineral resources. In some of these areas, the tension between the local residents and the "enclaves" who are engaged in these illegal activities is very high. In addition, larger communities of ex-combatants tend to maintain liaison with their former factional commanders. The joint security assessment teams concluded that the security cover provided by UNMIL remains a critical guarantee of peace and stability in Liberia. . . .

XII. Observations

77. President [Shirley] Johnson-Sirleaf's Government has made great strides in consolidating peace and promoting economic recovery in the country. The main achievements of the Government include completing the implementation of the measures required to lift timber and diamond sanctions; increasing public revenues by 48 per cent; completing the staff-monitored programme of the International Monetary Fund; preparing and implementing an interim poverty reduction strategy; restoring electricity and water supply to some parts of Monrovia for the first time in 15 years; increasing school enrolment by 40 per cent; improving the human rights situation and cultivating mutually beneficial relations with Liberia's neighbours. In addition, the Government has remained focused on the important priorities of consolidating its authority, fighting corruption, implementing the Governance and Economic Management Assistance Programme, reforming the security sector, regaining control and regulation of its natural resources and strengthening the capacity of its institutions.

78. These are remarkable achievements for a country that is emerging from a situation of complete lawlessness and whose State institutions and infrastructure had crumbled. However, the remaining challenges are formidable. The process of consolidating peace and rebuilding State institutions is still in its formative stage. Until the army and police can stand on their own and the justice system is rehabilitated and accessible to all Liberians, the country will remain vulnerable to the risk of a return to lawlessness. Moreover, providing alternative livelihoods for ex-combatants and deactivated security sector personnel, creating employment opportunities, ensuring genuine national reconciliation, addressing the needs of the victims of the conflict, alleviating poverty and delivering basic social services to the population are critical tasks that must be accomplished to ensure durable stability in Liberia. In order to meet these challenges, appropriate measures to promote economic growth must be implemented in order to generate the requisite public revenues.

79. The slow progress in strengthening the security sector is a source of great concern. The training of the Armed Forces of Liberia has faced considerable delays, which has resulted in the postponement of the operationalization date of its units. Meanwhile, although significant progress has been made in meeting the training benchmarks for the Liberian National Police, its operational effectiveness is constrained by the lack of adequate funding, vehicles, communication equipment and accommodation. These deficiencies are a major obstacle to the full deployment of the police throughout the country. The management of the Liberian National Police will also need to be strengthened. I appeal to the international community to generously support the equipping and deployment of the Liberian National Police and to assist in completing the training of the new Armed Forces of Liberia expeditiously. I also call on the Government of Liberia to finalize its national security strategy and architecture in the coming months.

80. Although illegal diamond mining continues to pose serious challenges and remains a potential source of instability, it is gratifying to note the commendable efforts of the Government to curb these activities, which have resulted in the lifting of the sanctions on diamonds and Liberia's admission into the Kimberley Process Certification Scheme. I encourage the Government to further strengthen its regulation of this important sector and to ensure, in particular, that Liberia becomes fully compliant with the Kimberley Process.

81. President Johnson-Sirleaf is to be commended for the positive steps that she has taken to foster national reconciliation and political inclusiveness in the country. However, the ethnic and social cleavages that have plagued the country in the past could still resurface. I therefore call on the Government of Liberia to intensify its efforts to promote national and local reconciliation in the interest of sustaining peace and stability in the country.

82. The uncertain situation in Côte d'Ivoire and Guinea poses additional challenges to stability in Liberia. I am encouraged however by the continuing efforts to strengthen cordial relations among the countries of the Mano River Basin.

83. Considering the many challenges that still face Liberia, in particular its complex and fragile security environment, I recommend that the Security Council approve the drawdown concept set out in section XI above [providing for a gradual reduction of UNMIL forces through 2008], as well as the force levels described in paragraph 73 and the plan for the adjustment of the UNMIL police component set out in paragraph 75. The proposed plan allows UNMIL to adapt to evolving priorities and to conduct a gradual, phased and deliberate transfer of responsibility for the security of Liberia to the Government in a manner that gives the Government the opportunity to build its capacity, while the Mission continues to help to maintain the prevailing stability. The plan emerged from a careful analysis of the existing security situation, as well as a thorough threat assessment and evaluation of the unfinished tasks under the UNMIL mandate. The plan is also the product of close consultations with the Government of Liberia, national stakeholders and Liberia's partners.

84. The drawdown process will need to be continually assessed in the context of the specific benchmarks identified in the present report. The Government and the international community are therefore strongly encouraged to make a timely and effective investment in the accomplishment of these benchmarks. Should the Council approve these proposals, I intend to submit regular updates on the implementation of the drawdown process. This would provide the Security Council, the Government and UNMIL with the opportunity to take stock of the security situation and to evaluate the progress made in

accomplishing the benchmarks. In the meantime, I recommend that the Mission's mandate be extended for a further period of one year, until September 2008. . . .

SOURCE: United Nations. Security Council. "Fifteenth Progress Report of the Secretary-General on the United Nations Mission in Liberia." S/2007/479. August 8, 2007. http://daccess-ods.un.org/access.nsf/Get?Open&DS=S/2007/479&Lang=E&Area=UNDOC (accessed March 20, 2008).

OTHER HISTORIC DOCUMENTS OF INTEREST

FROM THIS VOLUME

- Bishops' letter on Zimbabwe, p. 145
- United Nations reports on the Democratic Republic of the Congo, p. 374
- Report on World Bank programs in Africa, p. 592

FROM PREVIOUS *HISTORIC DOCUMENTS*

- Johnson-Sirleaf on her inauguration as president of Liberia, *2006,* p. 3
- European Union observation mission on elections in Liberia, *2005,* p. 800
- President of Sierra Leone on end of the war, *2002,* p. 247

DOCUMENT IN CONTEXT

Afghan and Pakistani Presidents Sign a "Peace Jirga" Communiqué

AUGUST 12, 2007

After sniping at each other for several years, the leaders of Afghanistan and Pakistan finally agreed in August on the need to control the flow of men and weapons that threatened the stability of Afghanistan's fragile Western-backed government. Afghanistan's president Hamid Karzai and Pakistan's president Pervez Musharraf met in Kabul, Afghanistan's capital city, on August 12 as part of the first-ever "peace jirga" (assembly) between the two countries and issued a declaration calling terrorism "a common threat" to both nations.

The meeting between the two leaders was long on symbolism and short on substance. However, it did hold out modest promise that their countries could set aside historic differences long enough to battle an insurgency that already was a serious threat to Karzai's government and was looming as a potential threat to Musharraf's as well.

The risk of instability in Pakistan grew even greater at year's end following the assassination of opposition leader Benazir Bhutto, who had returned from a long exile to compete in parliamentary elections in 2008. Musharraf blamed the killing of Bhutto on Islamist extremists in the lawless North-West Frontier province of Pakistan—an area that also was a base for insurgents battling Karzai's government and the U.S. and North Atlantic Treaty Organization (NATO) troops who were supporting it.

A Cross-Border Insurgency

Afghanistan in 2007 faced what international leaders called an increasingly dangerous insurgency. The key elements of this insurgency were the former Taliban regime—an extreme Islamist faction that the United States had ousted from power in Kabul in the wake of the September 11, 2001, attacks because it had harbored al Qaeda—and several groups composed of other extremist factions, local warlords, and narcotics traffickers. Most of the actual fighters in these insurgent groups reportedly were Afghan citizens, but Karzai's government and its Western backers insisted that many of the leaders, as well as sources of weapons and money for the insurgency, were located in Pakistan. The most prominent of the insurgent leaders was Mullah Omar, who had headed the Taliban regime during its years in power and was said to be based in Quetta, the capital of Baluchistan province in Pakistan. Al Qaeda leaders were also believed to live along the mountainous border between Afghanistan and Pakistan.

The U.S. and NATO forces that were battling the Afghan insurgency, on behalf of Karzai's government, found it impossible to control the long, mountainous border between the two countries, even with the use of spy satellites and unmanned aerial drones. Western commanders also could not enter Pakistan in hot pursuit of insurgents because Musharraf refused to allow Western troops to operate in his country, insisting his army was capable of controlling its own territory. The falseness of this claim was evident nearly every day, as the Taliban and other insurgent groups ferried men and supplies from Pakistan into Afghanistan. An October report by a London think tank, the Royal Institute of International Affairs (also known as Chatham House), put the problem in stark terms: "As long as parts of Pakistan serve as a safe haven for the Taliban and Al-Qaeda, coalition forces will not be able to control Afghanistan."

Relations between Afghanistan and Pakistan

Against this backdrop, relations between Karzai and Musharraf had been frosty at best. Karzai, a leader of the Pashtun ethnic group that straddled the Afghan-Pakistani border, had come to power in 2002, with strong Western backing, as head of a transitional government and won the presidential election in 2004. Musharraf, the head of Pakistan's army, had seized power from an elected government in 1999 and continued to hold on to office despite growing opposition to his military rule.

Karzai complained repeatedly that Pakistan was not doing enough to prevent the use of Pakistani territory by the Afghan insurgents. Even though the Pakistani military long had been the chief international sponsor of the Taliban, Musharraf insisted that the insurgents were indigenous to Afghanistan and that his government did not allow them to operate on his side of the border. The dispute between the two leaders erupted on several occasions in 2006 and seemed likely to continue through 2007, particularly after Karzai told the *New York Times,* in an interview published on April 1, that his government had "solid, clear information" that Pakistani security forces were protecting Mullah Omar.

Karzai and his Western backers also said Musharraf had virtually turned a particularly troublesome part of the border area over to tribal leaders who supported, or at least tolerated, the Afghan insurgents. Musharraf in 2006 had signed agreements with leaders in the Wajiristan districts of Pakistan's North-West Frontier province; these gave control of the land to the leaders, who pledged to prevent insurgents from crossing over into Afghanistan but failed to do so.

Pakistan did take a significant step in February 2007, when its agents arrested the former Taliban defense minister, Mullah Obaidullah. News agencies reported that Obaidullah was arrested around the time of a visit to Pakistan by Vice President Dick Cheney, apparently in an effort by Pakistani authorities to demonstrate to the United States that they were acting against the Taliban.

The year's first effort toward a reconciliation between Karzai and Musharraf came in late April, when Turkey hosted a tripartite summit that ended with a pledge of cooperation. A potentially more significant event was Karzai's convening of the peace jirga in Kabul on August 9, which brought together, for the first time, 700 parliamentarians, religious figures, and tribal leaders—most of them Pashtun—from both sides of the border, with the stated purpose of developing a common strategy to combat the insurgents. At the outset, there were two signs of trouble for this assembly: Musharraf initially refused to attend, sending instead his prime minister, Shaukat Aziz; and tribal elders from the Wajiristan districts in Pakistan, where many fighters for the al Qaeda network and other insurgent groups were based, also refused to come. Another indication of the difficulties facing any move toward peace came on August 10, when forty-five people reportedly were killed as the result of various Taliban attacks across Afghanistan.

It later emerged that Musharraf had stayed away from the event because he was contemplating imposing emergency rule to quell civil unrest in Pakistan during the period leading up to presidential and parliamentary elections set for later in 2007. U.S. secretary of state Condoleezza Rice pressured him, by telephone, not to impose emergency rule and, instead, to attend the meeting in Kabul.

Musharraf finally showed up for the last scheduled day of the peace jirga, August 12, and delivered a remarkably conciliatory speech that included a rare admission that the insurgency was not just an Afghan problem. "I realize this problem goes deeper, there is support from these areas," he said of the border regions in his own country. "There is no doubt Afghan militants are supported from Pakistan soil. The problem that you have in your region is because support is provided from our side."

Karzai and Musharraf then signed the six-point declaration recognizing the common threat of terrorism and pledging "noninterference" in one another's affairs. Each leader also agreed to appoint twenty-five representatives to a smaller assembly that would meet every two months to work on specific ways of cooperating and would plan follow-up peace assemblies. Reporting to the UN Security Council five weeks later, on September 21, UN secretary-general Ban Ki-moon said the declaration presented "a unique opportunity for their respective countries to pursue a joint strategy for cross-border peace and security, aimed at defeating extremism and terrorism in both countries."

Iran's Role in Afghanistan

Just three days before the peace jirga convened, Karzai met with President George W. Bush at Camp David, the presidential retreat near Washington, D.C. The two men met frequently and appeared to share similar views on many issues, but this meeting brought out an important difference between them on the role of another of Afghanistan's neighbors: Iran.

The Bush administration had been focusing increasing attention on what it called Iran's "meddling" in Afghanistan. Its prime evidence was the discovery of several shipments of weapons used to create roadside bombs, which U.S. officials said came from Iran and were intended for Afghan insurgents.

Just before his trip to the United States, Karzai told an interviewer from CNN that Iran had been "a helper" in Afghanistan, citing that country's aid for numerous reconstruction projects. After Karzai and Bush met at Camp David, the U.S. president made clear that he did not share that view. "I would be very cautious about whether the Iranian influence in Afghanistan is a positive force," Bush said.

Karzai's Government Still Weak

In 2007 it became increasingly obvious that Karzai's government still had very little authority and was fast losing popular support, primarily because many of its officials—particularly at the provincial level—were abusive and corrupt. Karzai continued to enjoy warm support from Bush and other Western leaders, but in private they reportedly increased pressure on him to exercise greater control over his government, including former warlords and others who were using newfound political power for their own benefit.

The UN secretary-general offered an unusually blunt assessment in a March 15 report to the Security Council, citing concerns of "popular alienation" from the government in Kabul. "The central government's frequent tolerance of weak governance has diminished public confidence in its responsiveness and its readiness to hold officials accountable for their transgressions," Ban wrote. Even when the government appointed "capable governors" (specifically those in Paktika, Uruzgan, and Zabul provinces), "it has failed to provide them with the resources necessary to maintain the goodwill that they have generated."

Ban followed up with similarly blunt language, in his September 21 report to the Security Council, on the Karzai government's inability to curb corruption: "The resulting sense of impunity has encouraged a culture of patronage and direct involvement in illegal activities, including the drug trade, especially within the police force." Ban noted that the population had greater contact with the police than with any other level of government, and the rampant corruption of police personnel undermined public support for the government generally.

A reminder that Afghanistan once had enjoyed stability came with the death of former king Zahir Shah on July 23 at the age of ninety-two. Zahir Shah had ruled from 1933 until he was deposed in 1973; it was a period that many people in Afghanistan remembered as one of relative harmony and prosperity. Six years after the king's ouster, the Soviet Union invaded Afghanistan, launching a quarter-century of war that devastated the country and led more than 15 percent of the population to flee into Iran and Pakistan.

Economic and Humanitarian Indicators

Afghanistan remained one of the poorest countries in the world. With its infrastructure and public services yet to recover from the war years, Afghanistan ranked at or near the bottom of countries in just about every category of development, including education, health care, housing, jobs, and access to such public services as electricity and clean water.

The United States, European countries, Japan, and other donors had poured hundreds of millions of dollars into various aid projects by 2007. The results could be seen in new schools and housing, particularly in rural areas. A tiny elite of business executives and government officials in Kabul and a few other cities enjoyed some elements of an international lifestyle, but the vast majority of Afghanistan's approximately 30 million people were still mired in deep poverty, earning less than the equivalent of $1 per person each day.

Kabul had taken on the look of a giant, semipermanent refugee camp. Its population had more than tripled, since the fall of the Taliban, to 4.5 million people, a large proportion of whom lived in makeshift structures, were unemployed, and had little access to social services. Many of these new residents in the capital were among the 4 million refugees who had returned to Afghanistan from Pakistan and Iran since the ousting of the Taliban.

Afghanistan remained heavily dependent on outside aid, both for economic development and to meet emergency humanitarian needs. At a meeting in London in January 2006, donor nations and Karzai's government agreed on a joint plan, called the London Compact, under which rich countries would provide more than $10 billion in additional development aid over five years. In return, the government pledged sweeping reforms intended to improve its capacity to use the aid. A UN coordinating committee monitored Kabul's steps to keep its promises and issued periodic reports, starting in 2006, saying progress was being made, although much more slowly than was needed.

Providing humanitarian aid to the hundreds of thousands of Afghan citizens who were still homeless after the decades of war remained difficult, particularly in remote mountainous regions and in the most heavily conflicted zones in the east, south, and west of the country. UN spokesman Charlie Higgins told reporters on December 17, 2007, that aid agencies continued to face "an unacceptably high risk" in delivering food, medicine, and other supplies to the zones of heaviest fighting. Even so, he said, agencies had been able to deliver about 90 percent of the 23,000 tons of food needed for an estimated 326,000 beneficiaries during the winter of 2007–2008. That winter turned out to be one of the most severe in recent history, forcing UN and other agencies to ship additional emergency supplies into rugged areas.

One bit of good news was on the education front; in 2007 some 6 million Afghan children were enrolled in school, a record for the country, and about one-third of the students were girls. Even so, Islamists opposed to the education of girls continued to attack schools in the east and south of the country. More than one-half of all schools in several of the most insecure provinces were forced to close at least temporarily. In June insurgents shot schoolgirls in Logar province, killing three and wounding eight; this was the first time girls had been specifically targeted.

Several other social indicators, such as the rates of infant and maternal mortality, improved during the year. Afghanistan also had one of its best wheat harvests in many years, meaning that requirements for outside food aid were somewhat less than had been projected.

Opium Production Soars, Again

The U.S. government had warned in March 2005 that Afghanistan was in danger of becoming a "narcotics state" because of the rapidly expanding production of opium poppies, used to make heroin. Little that happened during the next two-plus years appeared likely to change that assessment.

According to figures from the UN and the United States, the area dedicated to opium poppy cultivation expanded by 59 percent in 2006 to 407,000 acres, and by another 17 percent in 2007 to more than 470,000 acres. This level of production solidified Afghanistan's status as the world's dominant source of opium and heroin. (Most of Afghanistan's heroin went elsewhere in Asia and to Europe and the Middle East; most heroin entering the United States came from Colombia.)

The soaring production came despite U.S.-led efforts that focused on helping Afghan farmers plant and market alternative crops, along with a small-scale opium poppy–eradication program. These efforts were failing because farmers were paid higher prices for opium poppies than for other crops and because of what the UN called "alliances of convenience between narco interests and insurgents." An estimated two to three dozen narco-bosses, some associated with key figures in Karzai's government, controlled overall opium production in Afghanistan. Most of the opium poppy production also took place in the southern parts of Afghanistan, where the Taliban and other insurgents were strongest; insurgent groups protected farmers who grew opium poppies in return for a cut of the profits.

Numerous reports suggested that high-level Afghan government officials also benefited from the drug trade, including several cabinet or subcabinet officers. It was widely believed in the country that Karzai's own brother, Ahmed Wali Karzai, had a role in the narcotics trade, either directly or indirectly, but no formal charges had been brought against him as of the end of 2007.

Following is the text of a "declaration" issued at the end of a "peace jirga" (assembly) of some 700 representatives from Afghanistan and Pakistan, held in Kabul August 9–12, 2007. The declaration was signed by Afghan president Hamid Karzai and Pakistani president Pervez Musharraf.

DOCUMENT

Pakistan and Afghanistan Hold First "Peace Jirga"

August 12, 2007

To reaffirm and further strengthen the resolve of two brotherly countries to bring sustainable peace in the region, Afghan-Pak Joint Peace Jirga was convened in Kabul, Afghanistan from August 09 to August 12, 2007 as a result of initiative taken by the presidents of the Islamic Republic of Afghanistan and the Islamic republic of Pakistan on September 27, 2006.

This was the first historic event of its kind that opened a channel of people to people dialogue I which around 700 people including members or the parliaments, political parties, religious scholars, tribal elders, provincial councils, civil society and business community of both countries participated.

The inaugural session was addressed by HE Hamid Karzai, President of the Islamic republic of Afghanistan and HE Shaukat Aziz, Prime Minster of the Islamic republic of Pakistan. The concluding session of the Joint Peace Jirga was addressed by HE Hamid Karzai, President of the Islamic Republic of Afghanistan and HE General Pervez Musharraf, President of the Islamic Republic of Pakistan.

The main recommendations made by the first Joint Peace Jirga are summarised as follows:

1. Joint Peace Jirga strongly recognises the fact that terrorism is a common threat to both countries and the war on terror should continue to be an integral part of the national policies and security strategies of both countries. The participants of this Jirga unanimously declare to an extended, tireless and persistent campaign against terrorism and further pledge that government and people of Afghanistan and Pakistan will not allow sanctuaries/training centres for terrorists in their respective countries.

2. The Joint Peace Jirga resolved to constitute a smaller Jirga consisting of 25 prominent members from each side that is mandated to strive to achieve the following objectives:

   a) Expedite the ongoing process of dialogue for peace and reconciliation with opposition.

   b) Holding of regular meetings in order to monitor and oversee the implementation of the decisions/ recommendations of the Joint Peace Jirga.

c) Plan and facilitate convening of the next Joint Peace Jirgas.

d) Both countries will appoint 25 members each in the committee.

3. The Joint Peace Jirga once again emphasises the vital importance of brotherly relations in pursuance of policies of mutual respect, non-interference and peaceful coexistence and recommends further expansion of economic, social, and cultural relations between the two countries.

4. Members of the Joint Peace Jirga in taking cognisance of the nexus between narcotics and terrorism condemn the cultivation, processing and trafficking of poppy and other illicit substances and call upon the two governments to wage an all out war against this menace. The Jirga takes note of the responsibilities of the international community in enabling Afghanistan to provide alternative livelihood to the farmers.

5. The governments of Islamic Republic of Afghanistan and Islamic Republic of Pakistan, with the support of the international community, should implement infrastructure, economic and social sector projects in the affected areas.

SOURCE: The High Commission for Pakistan in United Kingdom. "Pakistan and Afghanistan Hold the First Ever Joint Peace Jirga." August 12, 2007. www.phclondon.org/News/NewsItem209.asp (accessed February 8, 2008).

OTHER HISTORIC DOCUMENTS OF INTEREST

FROM THIS VOLUME

- UN secretary-general on Afghanistan, p. 547
- Political situation in Pakistan, p. 647

FROM PREVIOUS *HISTORIC DOCUMENTS*

- London Conference "Compact" on Afghanistan, *2006*, p. 48
- UN secretary-general Annan on the situation in Afghanistan, *2006*, p. 525
- President Karzai addresses the new parliament of Afghanistan, *2005*, p. 970
- President Karzai's inauguration, *2004*, p. 912

DOCUMENT IN CONTEXT

Federal Reserve Board on the State of the U.S. Economy

AUGUST 17, SEPTEMBER 18, OCTOBER 31, AND DECEMBER 11, 2007

Many words were used to define the U.S. economy in 2007, among them "crisis" and "recession," but the most apt description was probably "uncertainty." Which was the bigger problem, inflation or faltering growth? Had the sinking housing markets struck bottom yet, or did they have further to go? Had the damage to financial markets from the subprime mortgage debacle been fully disclosed, or would more problems and losses be revealed? At year's end, the answers to these and many other questions about the stability of the economy remained to be answered. But one thing did seem certain: economic growth slowed markedly in the final quarter of the year, even as inflation ticked upward, ensuring that uncertainty would continue at least in the first few months of 2008.

International Credit Crunch

The major economic crisis of the year was an international credit crunch triggered by the collapse of the U.S. housing market and the related subprime mortgage market. In recent years, investment banks, hedge funds, and other institutions around the world had invested billions of dollars in complicated securities packages, known as collateralized debt obligations (CDOs), which bundled subprime mortgages—mortgages given to the riskiest borrowers—with other debt obligations. The CDOs were rated by credit rating agencies, and investors felt safe buying them without paying much attention to the risks involved. CDOs of all types were popular with investors because they appeared to spread risk widely and they were highly profitable—as long as the markets were rising. When the pace of foreclosures on subprime mortgages began to quicken toward the middle of the year, however, investors began to question the value of the CDOs they held and to withdraw from them as fast as they could. Uncertainty about the worth of securities backed by subprime mortgages and other debt fed into uncertainties about the size of potential losses and made banks extremely reluctant to extend short-term loans to creditworthy customers and even other banks.

Although a few analysts had been warning of an impending meltdown in subprime mortgage securities for months if not years, Wall Street did not appear to pay much attention until mid-June, when the brokerage firm Bear Stearns announced that investors were withdrawing from two of its hedge funds that specialized in subprime debt. "That was the match that lit a very dry field of risk on fire," Jack Malvey, a chief strategist for Lehman Brothers, told the *New York Times*. Two weeks later, Standard & Poor's, the rating agency, downgraded $7.3 billion worth of securities tied to subprime mortgages. By the end of

July, stock markets around the world began to fall, concerned that the subprime mortgage mess would spill over into other forms of credit. Over the next few days, the stock markets seesawed up and down in response to one development after another. Matters came to a head on August 9, when BNP Paribas, one of Europe's largest banks, froze deposits in three of its hedge funds, citing the "evaporation of liquidity" in the U.S. securitization market. Goldman Sachs and several other U.S. institutions cited similar concerns the same day, sending stock markets around the world tumbling and the interest rates banks charged each other on overnight loans soaring. Led by the European Central Bank, the world's central banks, including the Federal Reserve (the Fed), acted quickly to inject some liquidity into the money markets.

The financial markets had been anxiously awaiting the Fed's evaluation of the growing credit crisis and of its potential to slow the "real" economy. Under the leadership of its relatively new chair, Ben S. Bernanke, the Fed had held its key federal funds rate (the benchmark rate for overnight lending between banks) steady, at 5.25 percent, since June 2006 in an effort to discourage rising inflation without slowing economic growth. The Fed's discount rate, the rate at which the Fed lent money to creditworthy banks, was also held steady at a slightly higher percentage. Meeting on August 7, the Fed again left both rates unchanged. Acknowledging that the financial markets were volatile and credit conditions tighter, the central bankers said that the economy nevertheless "seems likely to continue to expand at a moderate pace over coming quarters, supported by solid growth in employment and incomes and a robust global economy."

With the financial markets still in turmoil, the Fed pumped $24 billion into the financial system on August 9, followed by another $38 billion on August 10. On August 11, the Fed issued a brief statement that sought to assure jittery markets by affirming that it would provide "reserves as necessary" to "facilitate the orderly functioning of financial markets." Then on August 17, the Fed partly reversed its decision of August 7 and announced that it was lowering its discount rate by half a percentage point, to 5.75 percent. "The downside risks to growth have increased appreciably," the central bankers said.

On September 18, at a regularly scheduled meeting of its Federal Open Market Committee, the Fed moved more aggressively, cutting its federal funds rate by half a percentage point, to 4.75 percent, on concerns that "the tightening of credit conditions has the potential to intensify the housing correction and to restrain economic growth more generally." The committee said its action was "intended to forestall" any adverse effects on the broader economy from the disruptions in financial markets. In two more steps, on October 31 and December 11, the Fed lowered its federal funds rate to 4.25 percent. In both cases, the central bank said it was hoping to forestall at least some of the anticipated slowdown in the broader economy. On December 12, the Fed announced a series of four auctions to infuse at least $64 billion more into the banking system. The announcement was coordinated with the Bank of Canada, the European Central Bank, the Bank of England, and the Swiss National Bank, which also infused new money into the international banking system in an effort to ease liquidity. Credit markets had eased a bit in September and October but had begun to tighten again in mid-November amid continuing fallout from the subprime mortgage crisis.

Already that crisis had cost several bank executives their jobs. Merrill Lynch's Stanley O'Neal and Citigroup's Charles O. Prince III both stepped down in late October and early November after the two companies each announced losses connected to subprime mortgages of more than $8 billion. Morgan Stanley on December 19 reported a fourth quarter loss of $3.59 billion—the first quarterly loss in its seventy-two-year history. The company took a $9.4 billion charge on investments linked to subprime mortgages and said its remaining subprime exposure was $1.8 billion.

By the end of the year, the subprime crisis was threatening another aspect of the financial industry—bond insurers, the companies that traditionally insured municipal bonds against default but that had in recent years begun to insure bonds backed by mortgages and CDOs. On December 19, MBIA Inc., the largest bond insurer in the United States, announced that its total exposure to such bonds was nearly $31 billion and that slightly more than $8 billion of that involved what some analysts considered to be the riskiest CDOs. MBIA was just one of the several bond insurers that were facing a potential calamity if they were not able to make good on their insurance commitments.

Meanwhile, the subprime crisis showed little sign of easing. As many as half a million homeowners were considered likely to find themselves in foreclosure in 2008. In December the Bush administration announced that a group of lenders, mortgage counselors, and others had agreed to offer some relief to some homeowners, but most analysts thought the voluntary relief program would be able to help only a fraction of those facing default.

Stock Market Seesaw

"Volatile" was the only word to describe stock market activity in 2007. Markets swung widely, and wildly, posting record highs and then plunging steeply. Not all of the swings were attributable to the credit crisis. On February 28, the Dow Jones industrial average and Standard & Poor's (S&P) 500 stock index both lost more than 3 percent of their value after China's stock market plummeted. Two weeks later, the markets sagged again, apparently over concerns about slowing consumer spending and rising home foreclosures. By the end of April, some measure of confidence had returned, and the Dow rose above 13,000 for the first time in its history. On July 20, it crossed the 14,000 mark for the first time, closing at 14,000.41. That was a remarkable feat, given that it had taken nearly nine months for the Dow to move from 12,000 to 13,000. Strong corporate profits and corporate buyouts apparently overcame concerns about foreclosures, inflation, and interest rates.

A week later, however, as the subprime mortgage crisis began to unwind, the Dow lost more than 500 points in two days, and the S&P 500 index had its worst showing in five years. On August 15, the Dow closed below 13,000 for the first time since it had risen above that marker in April. The next day it lost 340 points before turning around and closing the day with a slight gain. The markets bounced around for a few more weeks before beginning another upward climb. On October 9, both the Dow and the S&P 500 index closed at record levels, the Dow at 14,164.53 and the S&P index at 1,565.15. In the next week, third quarter corporate earnings reports showing a spate of poor profits precipitated another steep selloff.

The Dow closed out the year at 13,264.82, up 6.4 percent for the year overall. The S&P 500 index finished at 1,468.36, for a gain of 3.5 percent for the year. Not surprisingly, financial stocks were hit hard, falling 20.8 percent in value over the year. Stocks in the consumer discretionary spending sector, which included home builders and retailers, also declined significantly, but many stocks not directly tied to the financial markets or the housing collapse held up well. "If you were very credit sensitive, you got hammered, and if you were not, you did not," Tobias Lekovich, chief U.S. equity strategist at Citigroup, told the *New York Times*. "Equity markets did not have the same kind of bubble conditions that the credit markets had."

The Real Economy

For all the turmoil in the financial markets, the standard indicators showed that the real economy, where people buy and sell goods and services, held up fairly well for much of the year before slowing markedly in the last quarter. The economy grew at an inflation-adjusted rate of 3.8 percent and 4.9. percent in the middle two quarters of the year before slipping to a preliminary estimate of 0.6 percent in the final quarter. The jobless rate remained stable, hovering around 4.7 percent before closing the year at 5 percent. Consumer spending was still a main driver of the economy, although spending growth slowed to 1.9 percent in the last quarter, heightening concerns about possible recession in 2008. And while core inflation remained relatively stable during the year, at 2.4 percent, increases in food and energy prices brought total inflation for the year to 4.1 percent, compared with 2.5 percent in 2006.

Despite the numbers, many Americans were feeling the pinch of higher food and gas prices, lower home values, job layoffs, and tighter credit. Preliminary figures showed that energy costs rose by 17.4 percent in 2007, while food prices went up 4.9 percent. Both were the largest price increases since 1990. Gasoline prices were up 29.6 percent, according to the Commerce Department. Workers at the bottom end of the wage scale saw some help midway through 2007, after Congress raised the minimum wage for the first time in a decade, from $5.15 to $7.25 per hour in two steps over two years. Overall, though, wages of rank-and-file workers, who made up about four-fifths of the workforce, did not keep up with inflation; average weekly wages adjusted for inflation fell nearly a percentage point during the year. Although the economy was creating new jobs overall, workers in the financial services and construction industries were hard hit by the housing slowdown, bankruptcies among builders and mortgage brokers, and layoffs in financial institutions suffering losses from the subprime crisis. Unionized auto workers were accepting buyouts from General Motors, Ford, and Chrysler, and under the terms of new union contracts, many of those departing workers would be replaced by workers making half as much.

Even many people who were not directly affected by the subprime mortgage crisis may have felt poorer as the decline in home values around the country reduced their equity in their homes—in some cases below the balance of their mortgages. The volatility of the stock market also affected consumer confidence, which by the end of the year had sunk to its lowest level in two years. Meanwhile, even businesses and consumers with good credit ratings were finding it increasingly difficult to obtain short-term credit, a situation that if prolonged could affect future business investment and hiring as well as consumer spending.

Toward the end of the year, President George W. Bush tried to shake off talk of recession, assuring Americans that the economy was still sound. "There's definitely some storm clouds and concerns, but the underpinning is good," he told a Rotary Club in Fredericksburg, Virginia, on December 17. "We'll work our way through this period." Others were less sanguine. One of these was Alan Greenspan, the former Fed chair and economic guru whom many blamed for fueling the housing boom by keeping interest rates too low in the early 2000s and then contributing to the credit crisis by not clamping down on abusive subprime lending practices sooner. Speaking on National Public Radio's *Morning Edition* on December 13, Greenspan said the odds of falling into a recession were "clearly rising" as the economy was "getting close to stall speed."

Following are the texts of four statements from the Federal Reserve's Federal Open Market Committee on August 17, September 18, October 31, and December 11, 2007, announcing its decisions on key interest rate levels.

DOCUMENT

The Federal Reserve Holds Key Interest Rate Steady

August 17, 2007

Financial market conditions have deteriorated, and tighter credit conditions and increased uncertainty have the potential to restrain economic growth going forward. In these circumstances, although recent data suggest that the economy has continued to expand at a moderate pace, the Federal Open Market Committee judges that the downside risks to growth have increased appreciably. The Committee is monitoring the situation and is prepared to act as needed to mitigate the adverse effects on the economy arising from the disruptions in financial markets.

Voting in favor of the policy announcement were: Ben S. Bernanke, Chairman; Timothy F. Geithner, Vice Chairman; Richard W. Fisher; Thomas M. Hoenig; Donald L. Kohn; Randall S. Kroszner; Frederic S. Mishkin; Michael H. Moskow; Eric Rosengren; and Kevin M. Warsh.

SOURCE: U.S. Federal Reserve System. "Federal Open Market Committee Statement." Press release. August 17, 2007. www.federalreserve.gov/newsevents/press/monetary/20070817b.htm (accessed February 10, 2008).

DOCUMENT

The Federal Reserve Lowers Key Interest Rate

September 18, 2007

The Federal Open Market Committee [FOMC] decided today to lower its target for the federal funds rate 50 basis points to 4-3/4 percent.

Economic growth was moderate during the first half of the year, but the tightening of credit conditions has the potential to intensify the housing correction and to restrain economic growth more generally. Today's action is intended to help forestall some of the adverse effects on the broader economy that might otherwise arise from the disruptions in financial markets and to promote moderate growth over time.

Readings on core inflation have improved modestly this year. However, the Committee judges that some inflation risks remain, and it will continue to monitor inflation developments carefully.

Developments in financial markets since the Committee's last regular meeting have increased the uncertainty surrounding the economic outlook. The Committee will continue to assess the effects of these and other developments on economic prospects and will act as needed to foster price stability and sustainable economic growth.

Voting for the FOMC monetary policy action were: Ben S. Bernanke, Chairman; Timothy F. Geithner, Vice Chairman; Charles L. Evans; Thomas M. Hoenig; Donald L. Kohn; Randall S. Kroszner; Frederic S. Mishkin; William Poole; Eric Rosengren; and Kevin M. Warsh.

In a related action, the Board of Governors unanimously approved a 50-basis-point decrease in the discount rate to 5-1/4 percent. In taking this action, the Board approved the requests submitted by the Boards of Directors of the Federal Reserve Banks of Boston, New York, Cleveland, St. Louis, Minneapolis, Kansas City, and San Francisco.

SOURCE: U.S. Federal Reserve System. "Federal Open Market Committee Statement and Board Approval of Discount Rate Requests of the Federal Reserve Banks of Boston, New York, Cleveland, St. Louis, Minneapolis, Kansas City, and San Francisco." Press release. September 18, 2007. www.federalreserve.gov/newsevents/press/monetary/20070918a.htm (accessed February 10, 2008).

DOCUMENT

The Federal Reserve Lowers Key Interest Rate a Second Time in 2007

October 31, 2007

The Federal Open Market Committee [FOMC] decided today to lower its target for the federal funds rate 25 basis points to 4-1/2 percent.

Economic growth was solid in the third quarter, and strains in financial markets have eased somewhat on balance. However, the pace of economic expansion will likely slow in the near term, partly reflecting the intensification of the housing correction. Today's action, combined with the policy action taken in September, should help forestall some of the adverse effects on the broader economy that might otherwise arise from the disruptions in financial markets and promote moderate growth over time.

Readings on core inflation have improved modestly this year, but recent increases in energy and commodity prices, among other factors, may put renewed upward pressure on inflation. In this context, the Committee judges that some inflation risks remain, and it will continue to monitor inflation developments carefully.

The Committee judges that, after this action, the upside risks to inflation roughly balance the downside risks to growth. The Committee will continue to assess the effects of financial and other developments on economic prospects and will act as needed to foster price stability and sustainable economic growth.

Voting for the FOMC monetary policy action were: Ben S. Bernanke, Chairman; Timothy F. Geithner, Vice Chairman; Charles L. Evans; Donald L. Kohn; Randall S. Kroszner; Frederic S. Mishkin; William Poole; Eric S. Rosengren; and Kevin M. Warsh. Voting against was Thomas M. Hoenig, who preferred no change in the federal funds rate at this meeting.

In a related action, the Board of Governors unanimously approved a 25-basis-point decrease in the discount rate to 5 percent. In taking this action, the Board approved the requests submitted by the Boards of Directors of the Federal Reserve Banks of New York, Richmond, Atlanta, Chicago, St. Louis, and San Francisco.

SOURCE: U.S. Federal Reserve System. "Federal Open Market Committee Statement and Board Approval of Discount Rate Requests of the Federal Reserve Banks of New York, Richmond, Atlanta, Chicago, St. Louis, and San Francisco." Press release. October 31, 2007. www.federalreserve.gov/newsevents/press/monetary/20071031a.htm (accessed February 10, 2008).

DOCUMENT

The Federal Reserve Lowers Key Interest Rate a Third Time

December 11, 2007

The Federal Open Market Committee [FOMC] decided today to lower its target for the federal funds rate 25 basis points to 4-1/4 percent.

Incoming information suggests that economic growth is slowing, reflecting the intensification of the housing correction and some softening in business and consumer spending. Moreover, strains in financial markets have increased in recent weeks. Today's action, combined with the policy actions taken earlier, should help promote moderate growth over time.

Readings on core inflation have improved modestly this year, but elevated energy and commodity prices, among other factors, may put upward pressure on inflation. In this context, the Committee judges that some inflation risks remain, and it will continue to monitor inflation developments carefully.

Recent developments, including the deterioration in financial market conditions, have increased the uncertainty surrounding the outlook for economic growth and inflation. The Committee will continue to assess the effects of financial and other developments on economic prospects and will act as needed to foster price stability and sustainable economic growth.

Voting for the FOMC monetary policy action were: Ben S. Bernanke, Chairman; Timothy F. Geithner, Vice Chairman; Charles L. Evans; Thomas M. Hoenig; Donald L. Kohn; Randall S. Kroszner; Frederic S. Mishkin; William Poole; and Kevin M. Warsh. Voting against was Eric S. Rosengren, who preferred to lower the target for the federal funds rate by 50 basis points at this meeting.

In a related action, the Board of Governors unanimously approved a 25-basis-point decrease in the discount rate to 4-3/4 percent. In taking this action, the Board approved the requests submitted by the Boards of Directors of the Federal Reserve Banks of New York, Philadelphia, Cleveland, Richmond, Atlanta, Chicago, and St. Louis.

SOURCE: U.S. Federal Reserve System. "Federal Open Market Committee Statement and Board Approval of Discount Rate Requests of the Federal Reserve Banks of New York, Philadelphia, Cleveland, Richmond, Atlanta, Chicago, and St. Louis." Press release. December 11, 2007. www.federalreserve.gov/newsevents/press/monetary/20071211a.htm (accessed February 10, 2008).

Other Historic Documents of Interest

From this volume

- The struggling auto industry, p. 232
- The housing crisis, p. 514
- World economic outlook, p. 603
- Long-term deficit problems, p. 728

From previous *Historic Documents*

- Outlook for the U.S. economy, *2006*, p. 151

DOCUMENT IN CONTEXT

President Bush and Attorney General Gonzales on His Resignation

AUGUST 27, 2007

After months of fending off allegations ranging from incompetence, to possible perjury, to sacrificing the judicial independence of the Justice Department for partisan political purposes, Alberto R. Gonzales announced on August 27 that he was resigning as attorney general. President George W. Bush staunchly defended his longtime friend and ally even as he accepted Gonzales's resignation, claiming the attorney general was a casualty of a partisan political climate engendered by Democrats. The president nominated Michael Mukasey, a former federal judge, to succeed Gonzales as attorney general. Mukasey was narrowly confirmed in December after frustrating many legislators by refusing to offer a legal opinion on certain forms of torture.

Gonzales had been besieged by calls for his resignation from Democrats and Republicans alike as Congress pursued investigations into the firings of nine U.S. attorneys in 2006 and Gonzales's conflicting comments on a secret domestic wiretapping program that President Bush had ordered in the wake of the September 11, 2001, attacks. Gonzales's inept handling of the incidents, particularly of the attorneys' firings, led to broader Democratic charges that the Bush administration had politicized the quasi-independent department to serve its own partisan ends.

Gonzales's announcement that he was stepping down came two weeks after White House political adviser Karl Rove announced that he was giving up his job and returning to Texas. A lightning rod for controversy, Rove was generally regarded as the mastermind who engineered Bush's rise to the presidency and built the religious conservative GOP base that propelled him into a second term. The departures signaled White House acknowledgement that Rove and Gonzales had become political liabilities whose continuing presence was likely to involve the administration in partisan wrangling with Democrats at the expense of advancing the president's legislative agenda during his last two years in office.

Gonzales's Career

The grandson of Mexican immigrants, Gonzales was the first Hispanic attorney general in the nation's history. A graduate of Harvard Law School, Gonzales had been part of Bush's inner circle of advisers since the president served as governor of Texas in the 1990s. He was Bush's gubernatorial counsel for three years before Bush tapped him to be Texas secretary of state and later a Texas Supreme Court justice.

Gonzales served as White House counsel during Bush's first term, then replaced retiring John D. Ashcroft as attorney general at the beginning of Bush's second term in 2005. Later that year, Bush reportedly considered nominating Gonzales to a vacant seat on the U.S. Supreme Court, which would have made him the first Hispanic to serve on the Court. But Bush backed away from the nomination after conservatives, who disapproved of Gonzales's relatively moderate stance on most social issues, threatened to oppose the nomination.

Gonzales's uneasy relationship with Democrats began after the September 11 attacks, when he helped develop the administration's controversial positions on the legal treatment of prisoners and enemy combatants in the war on terrorism. Democratic legislators said his positions were in clear opposition to core civil liberties protected by the Constitution. Among other things, Gonzales helped draft an order allowing terrorism suspects captured on the battlefield to be tried by military tribunals rather than in civilian courts. The Supreme Court later overturned the order, ruling that only Congress could authorize such tribunals (which it subsequently did). In a January 2002 memo, Gonzales successfully argued to Bush that the president had the legal authority to exempt these detainees from the human rights protections of the Geneva Conventions, justifying his position by claiming that the need to obtain information quickly from these terrorist suspects made the conventions' rule on treatment of detainees "obsolete." The Supreme Court also overruled that argument, but not before a Defense Department panel said that the administration's acceptance of Gonzales's position had contributed to the inhumane and abusive treatment of prisoners at Abu Ghraib and other Iraqi prisons in 2003–2004. As White House counsel, Gonzales also endorsed a legal opinion written for the CIA by Justice Department lawyers that loosened restrictions on acts that would be considered torture. When the memo's release to the public created a wave of criticism, Gonzales repudiated the document as an "unnecessary, overbroad" discussion of abstract theories, and the administration withdrew key parts of it in December 2004.

Gonzales's tenure as attorney general had been equally tumultuous, starting with a contentious confirmation hearing. Democratic legislators were especially scornful of his defense of the National Security Agency's (NSA) authority to eavesdrop on U.S. citizens without first obtaining a warrant. Later, in testimony before Congress in February 2006, Gonzales argued that Congress had implicitly authorized the warrantless surveillance when it authorized the war in Afghanistan after the September 11 attacks. Democrats and some Republicans rejected this rationale, saying Congress had authorized no such thing. In early 2007, Gonzales agreed to put the surveillance program under the authority of the Foreign Intelligence Surveillance Court. By that time, Democrats were beginning to investigate allegations that the Justice Department had fired several U.S. attorneys so that they could be replaced with attorneys more in tune with the administration's political agenda.

Firings of Nine U.S. Attorneys

Speculation about the administration's motives began shortly after the Justice Department notified seven U.S. attorneys on December 7, 2006, that their services would no longer be needed. U.S. attorneys, who serve as chief prosecutors in each of the ninety-three federal judicial districts, are typically appointed by the president at the beginning of his term and remain on the job until the president leaves office or they choose to move on. Resignations are not particularly unusual, but it was unusual for so many to leave at the same time. Among the seven was Carol Lam of San Diego, who oversaw the successful prosecution of former Republican representative Randy "Duke" Cunningham on bribery charges and who was continuing to direct investigations growing out of that prosecution. Another was Kevin Ryan in San Francisco, who oversaw the BALCO steroid case involving baseball slugger Barry Bonds and was currently handling an investigation involving the backdating of stock options.

Various rumors circulated about the firings, with some critics suggesting the attorneys had been fired because they were not pursuing the types of cases, such as vote fraud, that the administration wanted them to; because they were prosecuting politically sensitive cases that could affect Republicans and not pursuing cases that might have similarly affected Democrats; or because the administration wanted to fill the slots with people to whom it owed favors. That last claim was given some credibility when it was learned that an eighth U.S. attorney, H. E. "Bud" Cummins of Little Rock, Arkansas, had been dismissed in June 2006 so that he could be replaced by J. Timothy Griffin, a military prosecutor who had been a deputy to Rove in the White House and who had also worked at the Republican National Committee. (A ninth attorney was added to the list of fired prosecutors later in the congressional investigation.)

Testifying before the Senate Judiciary Committee on January 18, Gonzales flatly denied any politics was involved in the dismissals. "What we're trying to do here is ensure that for the people in each of these respective districts, we have the very best possible representatives for the Department of Justice," he said. "I would never, ever make a change in a United States attorney for political reasons or if it would in any way jeopardize an ongoing serious investigation. I just would not do it."

Gonzales's remarks were backed up on February 7 by Deputy Attorney General Paul J. McNulty, who told the Senate Judiciary Committee that the department had fired six attorneys for performance-related issues. Two of the attorneys, John McKay of Seattle and Daniel G. Bogden of Nevada, contradicted McNulty publicly, saying that they had not been told of any performance problems and were given no explanation for being asked to leave.

Then, on February 28, another of the fired attorneys, David C. Iglesias of New Mexico, told reporters he had been pressured by two members of Congress, whom he declined to name, to bring charges before the November 2006 elections against a state Democratic legislator involved in a criminal investigation. In an interview with the *Washington Post,* Iglesias said, "I didn't give them what they wanted. That was probably a political problem that caused them to go to the White House or whomever and complain that I wasn't a team player." Sen. Pete V. Domenici and Rep. Heather Wilson, both Republicans, acknowledged that they had contacted Iglesias about the investigation but denied that they had pressured him to speed it up. Iglesias told the Senate Judiciary Committee on March 6 that he saw it differently: "I felt leaned on; I felt pressure to get these matters moving."

Justice Department officials acknowledged on March 2 that the firings were based not on poor performance but on the attorneys' failure to go along with administration policies on immigration, firearms, and other issues. The officials said the firings were appropriate because the attorneys served at the pleasure of the president and were dismissed for policy, rather than political, reasons. The officials also admitted that the White House had approved the firings. After Iglesias and other dismissed attorneys testified, several Republicans on the Senate Judiciary Committee voiced their skepticism of the Justice Department's explanation as well as their displeasure with Gonzales. Two of the Justice Department's most ardent defenders on the committee, Republicans Jon Kyl of Arizona and Jeff Sessions of Alabama, also objected to the way the firings had been handled, saying the reputations of the fired attorneys might have been damaged needlessly.

Democrats began to call for Gonzales's resignation shortly after he and FBI director Robert S. Mueller III acknowledged on March 9 that the FBI had improperly used secret warrants, known as national security letters, to obtain telephone records and financial information about American citizens and legal aliens. Senate Judiciary Committee member Charles E. Schumer, D-N.Y., said on March 11 on CBS's *Face the Nation* that Gonzales was "a nice man, but he either doesn't accept or doesn't understand that he is no longer the president's lawyer, but has a higher obligation to the rule of law and the Constitution even when the president should not want it to be so."

Speaking at a March 13 news conference, Gonzales said he accepted that "mistakes were made" in the attorney firings but added that he was not involved "in seeing any memos, was not involved in any discussions about what was going on." Gonzales also placed much of the blame for the inept handling of the firings on his chief of staff, D. Kyle Sampson, who had resigned the day before. He said Sampson was in charge of identifying underperforming U.S. attorneys after Gonzales rejected a proposal from the White House in January 2006 that the department remove all ninety-three U.S. attorneys. Gonzales also implied that he had fired Sampson because Sampson had not told him of the extent of the White House's involvement in the firings of the attorneys.

But on March 29, Sampson refuted Gonzales's assertions. "I don't think the attorney general's [March 13] statement that he was not involved in any discussions about U.S. attorneys removals is accurate," Sampson told the Senate Judiciary Committee. Sampson said he had talked to the attorney general about the U.S. attorney issue several times over a two-year period, including at a November 27, 2006, meeting in Gonzales's conference room with other senior officials. "Ultimately," Sampson said, Gonzales "approved both the list and the notion of going forward and asking for those resignations." The next day, at a news conference, Gonzales insisted that he had not been involved in choosing which attorneys should be fired and that his role was more one of process than of substance. "At the end of the day, I know what I did, and I know that the motivations for the decisions that I made were not based upon improper reasons," he said.

By the time Gonzales appeared before the Senate Judiciary Committee on April 19, most committee Democrats had already turned against him and his support among Republicans on the panel was in doubt. GOP senator Sessions began his remarks by advising Gonzales: "Be alert and direct and honest with this committee. Give it your best shot." In view of many, his best shot fell short of the mark. Many legislators just shook their heads as the attorney general repeated his inability to recall several details about the firings. Republicans on the panel described Gonzales's explanations of the firings as "a stretch," his handling of the affair as "incompetent" and "deplorable," and his credibility as "significantly impaired."

A Bedside Visit

Ironically, Gonzales's fate was probably sealed in May with testimony about an incident that had nothing to do with the attorney firings. In dramatic testimony before the Senate Judiciary Committee on May 15, former deputy attorney general James B. Comey vividly described a March 2004 showdown over the NSA's warrantless surveillance program in the hospital room of then attorney general Ashcroft, who was suffering from acute pancreatitis. Ashcroft had formally turned over the running of the Justice Department to

Comey while he was hospitalized. On the evening of March 10, after Comey had refused a White House request to authorize a renewal of the surveillance program because he believed it had no legal basis, he received a call from Ashcroft's chief of staff saying that Gonzales (then White House counsel) and Andrew H. Card (then the president's chief of staff) were on their way to Ashcroft's room in the intensive care unit of a local Washington hospital to get him to override Comey's decision. Comey said he ordered his security detail to use sirens, which allowed him to arrive in Ashcroft's room shortly before Gonzales and Card. Once the two men had made their request, Comey testified, the ailing Ashcroft raised his head long enough to express his refusal to override his deputy's decision "in very strong terms."

Comey's testimony appeared to contradict earlier testimony from Gonzales, who had said there had been no disagreement over the program. But beyond that, Comey's testimony raised new and troubling questions about Gonzales's temperament and judgment. Democrats said the incident demonstrated that even before becoming attorney general, Gonzales was willing to try to steamroll Justice officials, including a bedridden cabinet secretary, to please his White House patrons. Republicans were also appalled. Gonzales "has failed this country," said Sen. Chuck Hagel of Nebraska. "He has lost the moral authority to lead."

Gonzales did not appear before the committee again until July 24, when he said that he had gone to Ashcroft's hospital room after an "emergency meeting" in the White House involving senior administration officials, congressional leaders, and top members of the House and Senate intelligence panels. He denied, however, that the meeting was about what the administration called the "terrorist surveillance program." Two days later, FBI director Mueller came before the same committee and appeared to contradict Gonzales's version of events. The same day, four Democratic members of the panel asked Solicitor General Paul D. Clement to name a special prosecutor to investigate whether Gonzales committed perjury.

Disarray at Justice

Meanwhile the congressional investigation into the attorney firings continued, with more testimony damaging to Gonzales. Appearing before the House Judiciary Committee on May 23, Monica Goodling, senior aide to Gonzales and the department's White House liaison until she retired earlier in the year, denied that she had discussed the firings with Rove or Harriet E. Miers, who had succeeded Gonzales as White House counsel but then resigned in January 2007. Goodling acknowledged, however, that she had "crossed the line of the civil service rules" in taking political considerations into account in hiring other Justice Department officials. Goodling also said that Gonzales had tried to discuss the firings with her after the congressional investigations began early in the year. She quoted him as saying "let me tell you what I can remember" about the firings and then gave his "general recollection" of events. Goodling's statements flatly contradicted those of her former boss, who repeatedly asserted in his testimony on May 10 that he had refrained from talking with other department officials about the firings because he was a "fact witness" in the ongoing investigations. Some legislators suggested that the competing statements raised questions about whether Gonzales's actions amounted to illegal witness tampering.

Although Senate Democrats failed in their attempt to take a "no-confidence" vote on Gonzales in June, both the House and Senate Judiciary Committees were pursuing several lines of investigation, including more details about the bedside visit to Ashcroft and the department's involvement in the NSA warrantless surveillance program and about the FBI's misuse of national security letters. Investigations were also continuing into the U.S. attorney firings, as well as into allegations that the administration had politicized the hiring of other Justice Department officials. Other still simmering issues were the release of White House e-mails and other documents that the judiciary committees wanted to review, as well as whether the Bush administration would make top senior advisers, including Rove and Miers, available for public questioning. The White House had agreed to make some top officials available to the committees for questioning in private, not under oath, and with no transcript, conditions that Democrats on the judiciary committees rejected. By the end of the year, the committees had cited Rove, Miers, and White House Chief of Staff Joshua Bolten for contempt of Congress for citing executive privilege in refusing to testify and turn over documents requested in connection with the attorney firings. The contempt citations required approval by Congress to go forward.

The Justice Department's inspector general, Glen Fine, was also investigating many of the same issues, including whether partisan politics influenced hiring and firing decisions at the department and whether Gonzales had made "intentionally false, misleading, or inappropriate" statements in his testimony to Congress about both the prosecutor firings and the warrantless surveillance program.

Although President Bush loyally stuck by his longtime ally, it was clear by the time Congress left for its August break that Gonzales's continuing presence as the head of the Justice Department would only create more problems for the administration in the waning months of Bush's presidency. In his resignation letter, Gonzales said he was "profoundly grateful" to Bush for the "many opportunities" the president had given him "to serve the American people." Accepting the resignation, Bush referred to the "months of unfair treatment that has created a harmful distraction at the Justice Department" and said it was "sad" when the name of "a talented and honorable" person like Gonzales was "dragged through the mud for political reasons."

Mukasey Nomination

If his confirmation hearings were any indication, Gonzales's replacement, former federal judge Michael Mukasey, was off to a difficult start. Mukasey, who served eighteen years as a U.S. district judge in New York before joining a private law firm in 2006, was hailed by Republicans and some Democrats as one of the most qualified candidates to be named to the post. But opposition to his September 17 nomination grew after his response to questions during his confirmation hearing about "waterboarding," an interrogation technique that simulated drowning and was considered torture by human rights groups and others. Although Mukasey indicated that he considered such tactics "repugnant," he stopped short of saying they were illegal under U.S. law. The Senate confirmed Mukasey's nomination on November 8 by a vote of 53-40, the lowest level of support for an attorney general in fifty-five years. Even Gonzales won seven more votes for his confirmation.

Tensions eased somewhat when, in one of his first actions, Attorney General Mukasey on December 19 limited the people in the Justice Department who could initiate conversations about civil and most criminal cases with the White House to himself and the deputy attorney general. Such conversations would be restricted to the president's counsel and deputy counsel and would involve only cases necessary for the president to fulfill his constitutional duties. "This limitation recognizes the president's ability to perform his constitutional obligation 'to take care that the laws be faithfully executed' while ensuring that there is public confidence that the laws of the United States are administered and enforced in an impartial manner," Mukasey wrote in a memo to department employees. Sen. Sheldon Whitehouse, D-R.I., called the new policy "a clear, unmistakable, and welcome repudiation of the Gonzales era."

Following are the texts of two statements dated August 27, 2007: the first from Alberto R. Gonzales, announcing his resignation as U.S. attorney general effective September 17, 2007; the second from President George W. Bush, accepting the resignation.

DOCUMENT

Attorney General Gonzales Announces His Resignation

August 27, 2007

Thirteen years ago, I entered public service to make a positive difference in the lives of others. During this time, I have traveled a remarkable journey from my home state of Texas to Washington, D.C., supported by the unwavering love and encouragement of my wife Rebecca and our sons Jared, Graham, and Gabriel. Yesterday, I met with President Bush and informed him of my decision to conclude my government service as Attorney General of the United States, effective as of September 17, 2007.

Let me say that it has been one of my greatest privileges to lead the Department of Justice. I have great admiration and respect for the men and women who work here. I have made a point as Attorney General to personally meet as many of them as possible and today I want to again thank them for their service to our nation. It is through their continued work that our country and our communities remain safe, that the rights and civil liberties of our citizens are protected and the hopes and dreams of all of our children are secured.

I often remind our fellow citizens that we live in the greatest country in the world and that I have lived the American dream. Even my worst days as Attorney General have been better than my father's best days. Public service is honorable and noble, and I am profoundly grateful to President Bush for his friendship and for the many opportunities he has given me to serve the American people.

Thank you and God bless America.

SOURCE: U.S. Department of Justice. "Remarks of Attorney General Alberto R. Gonzales Announcing His Resignation." August 27, 2007. www.usdoj.gov/ag/speeches/2007/ag_speech_070827.html (accessed August 27, 2007).

DOCUMENT

President Bush Accepts Attorney General Gonzales's Resignation

August 27, 2007

This morning Attorney General Alberto Gonzales announced that he will leave the Department of Justice after 2 1/2 years of service to the Department. Al Gonzales is a man of integrity, decency, and principle. And I have reluctantly accepted his resignation, with great appreciation for the service that he has provided for our country.

As Attorney General and before that as White House Counsel, Al Gonzales has played a critical role in shaping our policies in the war on terror and has worked tirelessly to make this country safer. The PATRIOT Act, the Military Commissions Act, and other important laws bear his imprint. Under his leadership, the Justice Department has made a priority of protecting children from Internet predators and made enforcement of civil rights laws a top priority. He aggressively and successfully pursued public corruption and effectively combated gang violence.

As Attorney General, he played an important role in helping to confirm two fine jurists in Chief Justice John Roberts and Justice Samuel Alito. He did an outstanding job as White House Counsel, identifying and recommending the best nominees to fill critically important Federal court vacancies.

Alberto Gonzales's tenure as Attorney General and White House Counsel is only part of a long history of distinguished public service that began as a young man when, after high school, he enlisted in the United States Air Force. When I became Governor of Texas in 1995, I recruited him from one of Texas's most prestigious law firms to be my general counsel. He went on to become Texas's 100th secretary of state and to serve on our State's supreme court. In the long course of our work together, this trusted adviser became a close friend.

These various positions have required sacrifice from Al, his wife, Becky, their sons, Jared, Graham, and Gabriel. And I thank them for their service to the country.

After months of unfair treatment that has created a harmful distraction at the Justice Department, Judge Gonzales decided to resign his position, and I accept his decision. It's sad that we live in a time when a talented and honorable person like Alberto Gonzales is impeded from doing important work because his good name was dragged through the mud for political reasons.

I've asked Solicitor General Paul Clement to serve as Acting Attorney General upon Alberto Gonzales's departure and until a nominee has been confirmed by the Senate. He's agreed to do so. Paul is one of the finest lawyers in America. As Solicitor General, Paul has developed a reputation for excellence and fairness and earned the respect and confidence of the entire Justice Department.

Thank you.

SOURCE: U.S. Executive Office of the President. "Remarks on the Resignation of Attorney General Alberto R. Gonzales in Waco, Texas." August 27, 2007. *Weekly Compilation of Presidential Documents* 43, no. 35 (September 3, 2007): 1118. Washington, D.C.: National Archives and Records Administration. www.gpoaccess.gov/wcomp/v43no35.html (accessed February 5, 2008).

Other Historic Documents of Interest

From this volume

- Standards for CIA interrogations, p. 354

From previous *Historic Documents*

- Executive power and NSA surveillance authority, *2006*, p. 61.
- Justice Department memo on terrorist interrogation guidelines, *2004*, p. 337

DOCUMENT IN CONTEXT

Senator Craig on His Arrest for Solicitation and Possible Resignation

AUGUST 28, SEPTEMBER 1, AND OCTOBER 4, 2007

Despite actions by both political parties to tighten rules governing congressional ethics and lobbying, allegations of individual misconduct continued to dog Congress in 2007. Perhaps the most embarrassing incident involved Sen. Larry Craig, R-Idaho, who was arrested in a men's restroom at a Minneapolis airport for allegedly soliciting sex from an undercover policeman. Craig announced he would resign from the Senate but then reneged, deciding instead to try to clear his name. Rep. William J. Jefferson, D-La., was indicted for bribery and other criminal conduct in connection with business deals he tried to arrange in Africa. Ethics problems plagued all three members of Alaska's delegation, while other legislators were drawn into the controversy over the firings of several U.S. attorneys who said they were let go for political reasons rather than for cause. Investigations stemming from the activities of disgraced Republican lobbyist Jack Abramoff also continued during the year. Abramoff, who was serving at least six years in prison, was at the helm of influence peddling schemes that also figured in the resignation, in June 2006, of House majority leader Tom DeLay, R-Tex.

The 2007 investigations and revelations followed two years in which personal scandal and allegations of corruption drove several Republican members of the House from office. They included DeLay, who was under indictment in Texas on campaign finance money-laundering charges and who had been closely associated with Abramoff when both men were at the heights of their careers in Washington. Randy "Duke" Cunningham of California and Bob Ney of Ohio both resigned after pleading guilty to bribery charges in separate incidents. And Mark Foley of Florida resigned after it was revealed that he had written suggestive e-mails to underage male House pages. These and other scandals contributed to poor public approval ratings of Congress and the Democratic takeover of Congress in the November 2006 elections.

New Ethics Rules

Democratic leaders in both the House and Senate made an overhaul of ethics and lobbying rules a top priority during the 2006 midterm elections, assailing what they described as a "culture of corruption" on Capitol Hill during twelve years of Republican control. Under new Democratic leadership, both chambers quickly passed legislation that restricted members' ties to lobbyists. A dispute over disclosure of all earmarks—projects inserted by a single lawmaker into legislation on behalf of local interests—held up final passage of the legislation until August.

As passed, the legislation (PL 100-81) sought to give the public more information about the work of lobbyists and their political fundraising by requiring more reporting about how lobbyists bundle contributions to congressional and presidential campaigns. It also lengthened to two years the time period before senators leaving office could join lobbying firms and placed new rules on contact between a legislator's staff and their boss's spouse if he or she were a lobbyist. In addition, the measure tightened rules on accepting trips on private planes. "Congress deserves credit for finally changing the rules of the game," said Mary G. Wilson, president of the League of Women Voters. "Now starts the hard part, actually playing by the rules they've created."

Craig's Airport Scandal

The scandal involving Larry Craig began on June 11, when the sixty-two-year-old senator was arrested at the Minneapolis-St. Paul airport in a sting operation by a police officer investigating complaints of men soliciting sexual activity in a men's restroom. Initially accused of lewd behavior, Craig pleaded guilty to a lesser misdemeanor charge of disorderly conduct and paid a $500 fine. He signed the guilty plea on August 1, but the episode was not made public until *Roll Call,* a Capitol Hill newspaper, broke the news on August 27. In a public statement the next day from his home in Idaho, Craig said his guilty plea had been a mistake, that he had hoped the plea would end the case quickly and quietly, and that he had consulted no one—not his wife or other members of his family, attorneys, or political aides—before signing the plea agreement. Craig, who had fought allegations of homosexuality since 1982, also denied repeatedly that he was gay.

The news created a furor in Idaho and Washington, D.C., which was exacerbated with the release of details about the exchange in the men's room. Craig's hand and foot signals, interpreted by the arresting officer as an invitation to sex, quickly became a primary source of late-night television comedy. Several Republicans called for Craig's resignation, and the GOP Senate leadership, headed by Minority Leader Mitch McConnell of Kentucky, demanded that he give up the committee leadership positions he held and called for a Senate Ethics Committee investigation of the episode. On September 1, Craig announced that he would resign his seat by the end of the month. Four days later, however, he told Republican leaders he was leaving open the possibility of staying, pending a court decision on his request to withdraw his guilty plea.

On October 4, Craig announced that his request to withdraw the plea had been denied but that he had decided to remain in his Senate seat through the end of his term, which expired at the end of 2008. Craig said that while waiting for the court's decision, he had "seen that it was possible for me to work here [in the Senate] effectively." He also said he wanted to continue his effort to clear his name in the Senate Ethics Committee, "something that is not possible if I am not serving in the Senate."

Craig's presence in the Senate continued to be an embarrassment to Republicans, particularly after the *Idaho Statesman* in December published detailed accounts from two men who claimed to have had sex with Craig and two others who said he had made passes at them. As he had when the newspaper made similar allegations in the past, Craig dismissed the claims. "Despite the fact that the *Idaho Statesman* has decided to pursue its own agenda and print these falsehoods without any facts to back them up, I won't let this paper's attempt to malign my name stop me from continuing my work to serve the people of Idaho," he said. The Senate Ethics Committee had taken no formal action against Craig as of the end of the year.

Another Republican senator, David Vitter of Louisiana, was also caught in a sex scandal. In July Vitter, who was married, was forced to admit that he had "a very serious sin in my past" after his Washington phone number was found in the phone records of an escort service that federal prosecutors said was a prostitution ring. The senator, however, vigorously denied allegations that he had used at least one prostitute in New Orleans in 1999 before winning a seat in the House to replace Rep. Robert L. Livingston, who had resigned due to a sex scandal in 1999. Vitter, a conservative, had built his political identity around a message of family values, morality, and ethics.

The Alaska Delegation

The Craig scandal came on the heels of a July 30 raid on the Girdwood, Alaska, home of another Republican, Sen. Ted Stevens, the ranking minority member on the Senate Appropriations Committee. A federal grand jury in Washington, D.C., was investigating issues related to work done on the senator's house by Veco Corp., an oil services company, and contractors the company hired, according to the *Anchorage Daily News.* A Stevens neighbor, the newspaper reported, was ordered by the grand jury to produce documents involving Stevens, his wife, and his son Ben, a former state senate president whose own office was raided in 2006. Two Veco executives pleaded guilty in May 2007 to bribing state legislators. The state's only House member, Republican Don Young, was also drawn into the Veco inquiry. According to a report in the *Wall Street Journal* in July, federal investigators were trying to determine whether Stevens or Young received bribes or gifts from Veco, which had been a major recipient of federal contracts.

The state's junior senator, Lisa Murkowski, raised eyebrows when she bought land from a political supporter at a price that appeared to be well below market value. Murkowski subsequently sold the land back to the supporter.

Indictment of Rep. Jefferson

In an embarrassment for Democrats, a federal grand jury sitting in Alexandria, Virginia, indicted Rep. William J. Jefferson on June 4 on sixteen counts that included racketeering, money-laundering, wire fraud, and conspiracy to solicit bribes by a public official. The charges came out of a longstanding FBI investigation of allegations that Jefferson took as much as $500,000 in bribes to promote high-technology business ventures in Nigeria and other African countries for a Kentucky-based telecommunications firm, iGate. If convicted on all counts, Jefferson could receive up to 235 years in prison. Jefferson pleaded not guilty to all counts on June 8. "I am absolutely innocent of the charges against me," he said at a news conference. Later in the year, federal prosecutors filed papers accusing Jefferson of soliciting bribes in two other schemes. Prosecutors said they would not file additional charges against Jefferson but would use the new allegations to establish a pattern of intentional wrongdoing. Jefferson's trial was scheduled to begin in January 2008.

The investigation first came to public attention in 2005, when the FBI raided Jefferson's Washington home and found $90,000 stashed in a freezer. In May 2006, FBI agents touched off a lengthy legal battle when they raided Jefferson's Capitol Hill office and seized numerous documents. Lawmakers, including Jefferson, said the search was an abuse of the separation of powers and demanded that the FBI return the documents. The federal district judge who issued the search warrant, Thomas F. Hogan, ruled in July 2006 that the raid was constitutional, but the U.S. Court of Appeals for the District of

Columbia Circuit disagreed in a decision handed down on August 3, 2007. The appeals court said that "the congressman is entitled to the return of all legislative materials (originals and copies) that are protected by the Speech or Debate Clause." That clause protects members of Congress from being questioned "in any other place" for what they have said in the House or Senate. In late December, the Justice Department appealed the ruling to the U.S. Supreme Court.

House Democrats had forced Jefferson to give up his seat on the Ways and Means Committee in 2006, and he faced strong opposition in his reelection race in November, which he nevertheless won. After the indictment, he voluntarily gave up his seat on the House Small Business Committee. At the time, House Speaker Nancy Pelosi said that Jefferson, like any other citizen, "must be considered innocent until proven guilty," but she added that if the charges "are proven true, they constitute an egregious and unacceptable abuse of public trust and power." On June 5, the House voted, 373-26, to call for an ethics panel inquiry aimed at determining whether Jefferson should be expelled from the House. The FBI investigation had already led to a guilty plea from the former head of iGate and to an admission by a former congressional aide that he had solicited bribes on Jefferson's behalf.

Fallout from Other Scandals

At least a dozen people pleaded guilty or were convicted in 2007 in connection with the Abramoff lobbying scandal. Among them were J. Steven Griles, deputy secretary of the interior during President George W. Bush's first term. Griles pleaded guilty in March to lying to the Senate Indian Affairs Committee about his relationship with Abramoff. On June 26, he was sentenced to ten months in prison, twice the length recommended by the Justice Department. Former Ohio Republican representative Bob Ney also received a longer sentence than had been recommended: he was sentenced on January 19 to thirty months in prison for accepting illegal gifts from Abramoff in return for legislative favors. Ney pleaded guilty in August 2006 to two counts of conspiracy and making false statements and resigned his seat in November 2006. Prosecutors had asked for a sentence of twenty-seven months, but a federal district judge in Washington said Ney deserved a longer sentence because of his "significant and serious abuse of the public trust."

In addition, the Justice Department was still investigating Abramoff's dealings with two California legislators, John T. Doolittle and Jerry Lewis. Doolittle temporarily gave up his seat on the House Appropriations Committee on April 19, a week after the FBI raided his Northern Virginia home. Doolittle and Lewis were also tied to San Diego military contractor Brent Wilkes, who in November was convicted of bribing former representative Randy "Duke" Cunningham in return for Cunningham's help in obtaining nearly $90 million worth of defense contracts. Cunningham pleaded guilty to bribery and other charges in 2005 and was serving an eight-year prison sentence. Wilkes had reportedly given Doolittle and Lewis together about $110,000 in campaign contributions, and Doolittle had acknowledged helping Wilkes obtain about $37 million in defense contracts. Wilkes was also facing trial in another federal bribery case involving his childhood friend, former CIA executive director Kyle "Dusty" Foggo.

One former legislator who had been under an ethical cloud was cleared in April. The Justice Department and the Securities and Exchange Commission announced they would not file insider-trading charges against Republican Bill Frist of Tennessee, the former Senate majority leader. Frist had come under suspicion in 2005 after selling holdings in his family's giant hospital conglomerate, HCA Inc., just weeks before the company issued a gloomy earnings forecast and its stock price tanked. Frist said he sold the stock "to eliminate the appearance of a conflict of interest." At the time, he was considering making a run for the Republican presidential nomination in 2008. Frist did not seek reelection to the Senate when his term expired at the end of 2006.

Following are excerpts of the texts of three press releases issued on August 28, September 1, and October 4, 2007, by the office of Sen. Larry Craig, R-Idaho, in connection with his arrest for alleged solicitation of sex in a men's restroom at the Minneapolis-St. Paul Airport.

DOCUMENT

Craig Apologizes, Calls His Guilty Plea a Mistake

August 28, 2007

Senator Craig made the following statement to Idaho at 2:30 pm:

"First, please let me apologize to my family, friends, staff, and fellow Idahoans for the cloud placed over Idaho. I did nothing wrong at the Minneapolis airport. I regret my decision to plead guilty and the sadness that decision has brought to my wife, family, friends, staff, and fellow Idahoans. For that I apologize.

"In June, I overreacted and made a poor decision. While I was not involved in any inappropriate conduct at the Minneapolis airport or anywhere else, I chose to plead guilty to a lesser charge in the hope of making it go away. I did not seek any counsel, either from an attorney, staff, friends, or family. That was a mistake, and I deeply regret it. Because of that, I have now retained counsel and I am asking my counsel to review this matter and to advise me on how to proceed.

"For a moment, I want to put my state of mind into context on June 11. For 8 months leading up to June, my family and I had been relentlessly and viciously harassed by the *Idaho Statesman*. If you've seen today's paper, you know why. Let me be clear: I am not gay and never have been.

"Still, without a shred of truth or evidence to the contrary, the *Statesman* has engaged in this witch hunt. In pleading guilty, I overreacted in Minneapolis, because of the stress of the *Idaho Statesman*'s investigation and the rumors it has fueled around Idaho. Again, that overreaction was a mistake, and I apologize for my misjudgment. Furthermore, I should not have kept this arrest to myself, and should have told my family and friends about it. I wasn't eager to share this failure, but I should have done so anyway.

"I love my wife, family, friends, staff, and Idaho. I love serving Idaho in Congress. Over the years, I have accomplished a lot for Idaho, and I hope Idahoans will allow me to continue to do that. There are still goals I would like to accomplish, and I believe I can still be an effective leader for Idaho. Next month, I will announce, as planned, whether or not I will seek reelection.

"As an elected official, I fully realize that my life is open for public criticism and scrutiny, and I take full responsibility for the mistake in judgment I made in attempting to handle this matter myself.

"It is clear, though, that through my actions I have brought a cloud over Idaho. For that, I ask the people of Idaho for their forgiveness.

"As I mentioned earlier, I have now retained counsel to examine this matter and I will make no further comment."

SOURCE: U.S. Senate. Office of Senator Larry Craig. "Statement of Senator Craig." August 28, 2007, Boise, Idaho. craig.senate.gov/keyportal.cfm (accessed August 29, 2007).

DOCUMENT

Craig Announces That He Will Resign from the Senate

September 1, 2007

Senator Craig made the following statement to Idaho:

"First and foremost this morning, let me thank my family for being with me. We're missing a son who's working in McCall, and simply couldn't make it down. But for my wife Suzanne and our daughter Shae, and Mike to be with me is very humbling.

"To have the governor standing behind me, as he always has, is a tremendous strength for me. To have Bill Sali who has never wavered, and who has been there by phone call and by prayer, and his wife, is tremendously humbling.

"For the leader of our party, Kirk Sullivan, to be standing here, who sought immediate counsel with me in this, is humbling. For Tom Luna—for any public official at this moment in time—to be standing with Larry Craig is a humbling experience. . . .

"For most of my adult life, I had the privilege of serving the people of Idaho. I'm grateful for the opportunity they have given me. It has been a blessing. I am proud of my record and accomplishments, and equally proud of the wonderful and talented people with whom I have had the honor and the privilege to work and to serve.

"I choose to serve because I love Idaho. What is best for Idaho has always been the focus of my efforts, and it is no different today. To Idahoans I represent, to my staff, my Senate colleagues, but most importantly, to my wife and my family, I apologize for what I have caused. I am deeply sorry.

"I have little control over what people choose to believe, but clearly my name is important to me and my family is so very important also. Having said that, to pursue my legal options, as I continue to serve Idaho, would be an unwanted and unfair distraction of my job and for my Senate colleagues. These are serious times of war and of conflict—times that deserve the Senate's and the full nation's attention.

"There are many challenges facing Idaho that I am currently involved in. And the people of Idaho deserve a senator who can devote 100 percent of his time and effort to the critical issues of our state and of our nation.

"Therefore it is with sadness and deep regret that I announce that it is my intent to resign from the Senate, effective September 30. In doing so, I hope to allow a smooth and orderly transition of my loyal staff and for the person appointed to take my place at William E. Borah's desk. I have full confidence that Governor Otter will appoint a successor who will serve Idaho with distinction.

"I apologize to the people of our great state for being unable to serve out a term to which I have been elected. Few people have had the privilege and the pleasure to represent Idaho for as many years as I have. Each day, each week, each year brought new challenges and opportunities to create a better life for Idahoans. I have enjoyed every moment and cannot adequately put into words how much I appreciate what you have given me: the chance to work for this great state. I hope you do not regret the confidence you have placed in me over all of these years. I hope I have served you and our state to the best of my ability.

"Lastly, Suzanne and I have been humbled beyond words by the tremendous outpouring of support we have received from our friends, our family, our staff and fellow Idahoans. We are profoundly and forever grateful. Thank you all very much."

SOURCE: U.S. Senate. Office of Senator Larry Craig. "Senator Craig Announces Intent to Resign From the Senate." September 1, 2007, Boise, Idaho. craig.senate.gov/releases/pr090107a.cfm (accessed December 23, 2007).

DOCUMENT

Craig Announces Decision to Remain in the Senate

October 4, 2007

Idaho Senator Larry Craig issued the following statement in reaction to today's ruling by the State of Minnesota District Court Fourth Judicial District allowing the guilty plea to stand:

”I am extremely disappointed with the ruling issued today. I am innocent of the charges against me. I continue to work with my legal team to explore my additional legal options.

"I will continue to serve Idaho in the United States Senate, and there are several reasons for that. As I continued to work for Idaho over the past three weeks here in the Senate, I have seen that it is possible for me to work here effectively.

"Over the course of my three terms in the Senate and five terms in the House, I have accumulated seniority and important committee assignments that are valuable to Idaho, not the least of which are my seats on the Appropriations Committee, the Energy and Natural Resources Committee and the Veterans' Affairs Committee. A replacement would be highly unlikely to obtain these posts.

"In addition, I will continue my effort to clear my name in the Senate Ethics Committee—something that is not possible if I am not serving in the Senate.

"When my term has expired, I will retire and not seek reelection. I hope this provides the certainty Idaho needs and deserves."

SOURCE: U.S. Senate. Office of Senator Larry Craig. "Craig Reaction to Court Ruling." October 4, 2007, Washington, D.C. craig.senate.gov/releases/pr100407b.cfm (accessed October 9. 2007).

OTHER HISTORIC DOCUMENTS OF INTEREST

FROM THIS VOLUME

- Resignation of Attorney General Gonzales, p. 457

FROM PREVIOUS *HISTORIC DOCUMENTS*

- Rep. Randy "Duke" Cunningham case, *2006,* p. 103
- Rep. William Jefferson investigation, *2006,* p. 105
- Rep. Tom DeLay resignation, Jack Abramoff scandal, *2006,* p. 264
- Rep. Ney plea bargain and resignation, *2006,* p. 266
- Rep. Foley scandal, *2006,* p. 595
- Rep. Tom DeLay indictment, *2005,* p. 631
- Sen. Frist stock sale, *2005,* p. 635

September

DOCUMENT IN CONTEXT

Government Accountability Office on Iraqi Actions to Meet "Benchmarks"

SEPTEMBER 4, 2007

Iraq's faltering and deeply divided government was unable to take advantage of an opportunity provided it in 2007 to make the political compromises necessary for peace and long-term stability. A U.S. "surge" of some 30,000 additional troops—combined with internal Iraqi developments—helped reduce the level of sectarian violence that was threatening to tear Iraq apart. President George W. Bush said in January that he was sending the extra troops to give the Iraqi government time to resolve sectarian differences, and he warned that the patience of both the Iraqi and American people was limited.

Despite the somewhat calmer security situation late in the year, the sectarian divisions in Iraq remained acute, and the government in Baghdad had taken only halting steps toward settling them. Shiites and Sunnis remained deeply suspicious of one another and were reluctant to compromise on matters they saw as fundamentally affecting their sectarian interests. Shiites were determined to hold onto the power they had gained after centuries of Sunni domination, and Sunnis were reluctant to admit that they were a minority in Iraq. Disagreements within each sect further reduced possibilities for compromise; at least two major factions were vying for primacy among the Shiites, while Sunnis were divided into numerous factions, some of which supported various elements of the antigovernment insurgency that was the cause of much of the violence. Kurds, meanwhile, were primarily interested in protecting the semi-independent region of Kurdistan, which was the most peaceful and stable part of Iraq. Attempts to establish cross-sectarian alliances emerged in 2007, but no significant progress had been made by year's end.

The Iraqi parliament was unable to adopt any major legislative proposals to settle disputes between the Shiites and Sunnis—although one important proposal to allow members of former Iraqi leader Saddam Hussein's dissolved Baath Party back into the government was on the table at year's end and won approval in January 2008. A review of the constitution, adopted in 2005 but fiercely criticized by Sunnis, failed to produce any changes. Particularly vexing was the future of the city of Kirkuk, which was disputed among all Iraqi factions; a referendum planned for 2007 among the city's residents was postponed because the matter was so divisive.

By late 2007 many U.S.-based experts on Iraq were arguing that Iraq's leaders needed international help to bridge their differences, perhaps through mediation by the United Nations or a group of countries in the region. Iraqi officials were understandably resistant to even more international intrusion into their affairs, and the Bush administration was reluctant to expose Iraq's fragile political process to additional intervention by Iran.

In a report for the U.S. Institute of Peace, Rend al-Rahim Francke, Iraq's representative to the United States from 2003 to 2005, said ordinary Iraqis were feeling helpless and victimized by the squabbling among their political leaders. "The reactions of Iraqis from all walks of life to the political stalemate range from bewilderment and incomprehension to anger, condemnation, and suspicions of deep conspiracy," she wrote in the report, "Political Progress in Iraq during the Surge," published in December.

Continuing Quarrels in Baghdad

Iraqis in December 2005 had elected members to a 275-seat parliament, the Council of Representatives, which in turn selected an executive branch headed by Nouri al-Maliki, a representative of the conservative Shiite Dawa Party. Prime Minister Maliki succeeded a fellow Dawa leader, Ibrahim al-Jaafari, who had led a transitional government during 2005 but had proven ineffectual. Maliki himself often appeared weak and indecisive—so much so that President Bush's national security adviser, Stephen J. Hadley, sent Bush a memo in November 2006 (later leaked to the press) questioning whether Maliki was "both willing and able to rise above the sectarian agendas being promoted by others." Among the "others" referred to in Hadley's memo were the two main Shiite political groupings that had provided the votes for Maliki's rise to power: the Supreme Islamic Iraqi Council (formerly known as the Supreme Council for the Islamic Revolution in Iraq), headed by Abdul Aziz al-Hakim; and a party aligned with radical Shiite cleric Moqtada al-Sadr and therefore known as the Sadrists. Both of these parties had their own militias that provided much of the muscle for their political influence, and both had ties to the Shiite-majority government of Iran.

One of Maliki's main initiatives early in his tenure was what he called a "National Reconciliation Plan," published in June 2006, that offered amnesty and other incentives for insurgents to stop their attacks against the government, something he said would hasten the insurgents' stated goal of a U.S. withdrawal from Iraq. Little of substance came from that reconciliation plan, however. Rather than turning themselves in and accepting amnesty, insurgents stepped up their attacks throughout the rest of 2006 and into the first half of 2007. The insurgency slowed after mid-2007, in large part because Sunni tribal chiefs in Anbar province, west of Baghdad, turned against the local branch of the al Qaeda terrorist network, which carried out the bloodiest attacks.

One of Maliki's political problems was a testy relationship with Sadr, a mercurial figure with little apparent understanding of the give-and-take of democratic politics. Sadr's party had won thirty seats in parliament; although small in number, this bloc gave Sadr a powerful voice in Maliki's government, plus the right to name several cabinet ministers. In November 2006 Sadr temporarily suspended the participation of his cabinet appointees and members of parliament in the Maliki government; this was widely seen as an attempt to force Maliki to resist pressure from the United States to act against Sadr's militia, the Mahdi Army. Sadr ultimately withdrew his cabinet appointees from the government and his party members from parliament in several stages during 2007, signifying a final break with Maliki. The withdrawal did not leave Sadr without influence, however. His Mahdi Army provided security for large portions of Baghdad and other cities, and a civilian arm controlled a large business empire that owned gas stations and real estate and dispensed food and clothing from a network of storefront offices to needy Shiites.

The break between Maliki and Sadr was just one of several quarrels that plagued the Iraqi government during 2007, all of which reflected underlying disputes left hanging in the debate over the new constitution in 2005. Maliki had running quarrels with Sunni leaders and factions, most importantly with the Iraqi Accordance Front, a coalition of Sunni parties that twice during the summer withdrew its forty-four members from parliament and six ministers from the cabinet. The Sunnis cited several grievances, including a charge that the government was protecting Shiite militias that received money and weapons from Iran.

The difficulty of tackling contentious issues was magnified by ongoing violence and by the fact that none of Iraq's political leaders had any experience in democratic governance. The thirty-eight-member cabinet did not function as a cohesive body, partly a result of the boycotts but also because some cabinet ministers rarely showed up for work and several ministries (notably the Health Ministry) were widely believed to be extremely corrupt and incompetent. Parliament often had trouble assembling a quorum, and even when it did, it rarely acted as a decisive body. Under strong U.S. pressure, parliament cut back its planned summer recess from two months to the month of August—a crucial time because the U.S. Congress was expecting reports in mid-September on Iraq's political progress.

Personality clashes also interfered with government decision making. The most notable during 2007 was the unseating on June 11 of parliament speaker Mahmoud al-Mashhadani, a Sunni who many members found to be excessively abrasive (he once slapped a member in the face and called him "scum"). Parliament returned Mashhadani to office five weeks later. By most accounts, nearly all important decisions were made by the prime minister and his advisers or by an informal caucus of Maliki, President Jalal Talabani, and two deputy presidents. Parliament thus had a choice of rubber-stamping decisions made elsewhere or holding decisions hostage to narrow interests, and it did both of these.

Another consequence of the squabbling was that some of the alliances-of-convenience that had been formed to contest the December 2005 parliamentary elections were beginning to fracture. This was particularly true of the broad Shiite coalition known as the United Iraqi Alliance, which won the largest bloc of seats in parliament and selected Maliki as prime minister. Sadr's party had been in this alliance but effectively split from it in late 2006. A small but influential moderate party, the Fadhila, also withdrew from the alliance in March 2007.

August 26 Communiqué

The central question hanging over Iraq's political leaders was posed by President Bush in January: If the surge of U.S. troops into Iraq helped quell the violence, could the leaders take advantage of the relative calm to reach a consensus on the vexing issues that threatened to split the country into sectarian camps? Maliki's reconciliation plan of June 2006 had been a modest step toward such a consensus, but its failure highlighted the difficulties of addressing underlying questions of how money and power would be distributed among Shiites, Sunnis, and Kurds. Another faint glimmer of compromise came early in 2007 when the leaders of several moderate parties attempted to put together a cross-sectarian alliance to replace the Maliki government. Those parties lacked the votes to affect real change, however, and their coalition-building efforts went nowhere.

Facing increasing pressure from Washington, Maliki and other leaders met all through August in a new search for consensus. On August 16 Maliki, President Talabani, and other leaders announced a new parliamentary coalition to replace the Kurdish-Shiite governing alliance, formed in 2006, which had crumbled into ineffectiveness with the withdrawal of Sadr's party and other factions. The coalition gave Maliki a strong plurality, but not a majority, in parliament, and it lacked Sunni representatives. Vice President Tariq al-Hashemi, a Sunni and head of the moderate Iraqi Islamic Party, refused to join the new coalition when the other parties rejected his demands for top posts for his party's officials.

Formation of a new coalition spurred further attempts to resolve some of the main issues facing the government and broaden the coalition's support. During this process U.S. officials stepped up the pressure. Ambassador Ryan C. Crocker openly speculated about the pros and cons of replacing Maliki. The top leaders of the Senate Armed Services Committee—Chair Carl Levin and ranking Republican John W. Warner—warned after visiting Iraq on August 20 that the Iraqi government was facing its "last chance" to end the political paralysis. Levin even called on the parliament to dump Maliki. President Bush, who held monthly videoconferences with Maliki and had repeatedly defended him, on August 21 acknowledged that he shared "a certain level of frustration with the leadership" in Iraq. Apparently annoyed by such criticism, Maliki said during a visit to Syria on August 22 that Iraq "can find friends elsewhere" if Washington did not approve of its government.

The search for a broad governing consensus finally yielded some progress on August 26 when Hashemi, the Sunni vice president, joined Maliki and three other top leaders in issuing a statement saying they had reached tentative agreement on some of the most contentious matters. These included allowing some members of the Baath Party to regain government jobs, setting procedures for provincial elections, and releasing prisoners held without charges. Also present at a news conference where the agreement was announced were President Talabani; Masoud Barzani, president of the Kurdish Regional Government; and Vice President Adel Abdul-Mahdi, a Shiite leader.

The leaders offered no details of their agreement, but just the appearance of a consensus was greeted with relief by the U.S. government. Ambassador Crocker issued a statement calling it "an important step forward for political progress, national conciliation, and development," and Bush called the five leaders the next day to offer his congratulations. U.S. officials also announced that they would step up the release of some of the thousands of Iraqis detainees, long a central demand of Sunnis. Even so, several Sunni political leaders said they were unimpressed by the August 26 agreement, and even Vice President Hashemi said the next day that his Iraqi Islamic Party and other Sunni parties would not yet return to parliament.

It soon became apparent that the August 26 consensus had not eliminated sectarian suspicions or resolved the most contentious issues. One of the first matters to fall victim to continuing dispute was a proposed law governing oil production. Key leaders had agreed on a tentative compromise on some oil-related issues in February but had been unable to push legislation through parliament. By mid-September it appeared that even the earlier agreement was unraveling, this time because of an insistence by Kurds on controlling oil production in their northern provinces.

Prospects for final parliamentary approval of any legislation to settle the controversial issues by the end of the year disappeared on December 6 when the parliament adjourned for the rest of the month, ahead of the Islamic holiday of Eid al-Adha. Parliament had been working on legislation to allow many former Baath Party members back into the government but set it aside amidst a new round of sectarian bickering. Parliament ultimately adopted the legislation—one of its most important accomplishments ever—on January 12, 2008. Even so, that action brought more complaints from some Sunnis, who said it imposed too many restrictions and might even force some Sunnis currently employed by the government to leave their jobs.

Yet another "reconciliation" agreement was signed on December 24. In this case, President Talabani and fellow Kurdish leader Barzani (each acting in their capacity as head of a major Kurdish political party) agreed with Vice President Hashemi to push for national consensus.

U.S. Assessments of Iraq's Government

As part of its effort to convince Congress that it was pressing for political compromise in Iraq, the Bush administration in 2006 set out a series of "benchmarks" for specific steps toward national reconciliation. Maliki's government published two sets of its own benchmarks in 2006, one as part of the June "national reconciliation plan" and the other announced by Maliki in September 2006 and ratified in October by the Presidency Council, which consisted of President Talabani and his two deputies.

Maliki and President Bush added several additional benchmarks in various statements in January and February 2007. The White House included key benchmarks in "The Way Forward in Iraq," a report issued at the time of President Bush's January 10 speech announcing the military surge in Iraq. Members of Congress, regardless of their positions on the surge or the underlying nature of U.S. involvement in Iraq, embraced the idea of holding the Iraq government accountable for the benchmarks. Congress in May laid out eighteen of the benchmarks in its supplemental appropriations bill (PL 110–28) providing money for the surge and other purposes; that bill instructed the administration and the Government Accountability Office (GAO) to report by September on the status of those benchmarks.

Most of the benchmarks also were incorporated in the International Compact with Iraq, a United Nations–sponsored document endorsed by the Iraqi government, other regional governments, the United States, and other interested parties at a conference in Sharm al-Shaykh, Egypt, on May 3–4. Public attention at that conference centered not on the Iraqi benchmarks but on the fact that U.S. secretary of state Condoleezza Rice was sitting in the same room with Iranian foreign minister Manuchehr Mottaki despite the longtime break in relations between Washington and Tehran.

Some of the benchmarks were general in nature, such as reducing the level of sectarian violence in Iraq. Other benchmarks embodied specific steps that the Iraqi government promised to take by certain dates. Prominent among those steps was the adoption of laws and other changes intended to ease sectarian strains and resolve questions about the federal nature of Iraq's government: a law governing the distribution of revenue from Iraq's exports of oil and natural gas, a law allowing former Baathists back into the government, revisions to the 2005 constitution to ameliorate Sunni complaints that it discriminated against them, and laws laying out the duties of provincial governments and providing for the election of provincial councils and governors. Maliki had pledged to win parliamentary approval of legislation on all these contentious matters by the end of 2007.

The Bush administration reported to Congress on July 12 that Iraq had made "satisfactory" progress on nine of the eighteen benchmarks and "unsatisfactory" or mixed progress on the others. The administration also said it was too early to make a definitive judgment about the matter.

Congress received a more critical assessment on the status of the benchmarks on September 4 from David M. Walker, the comptroller general of the United States and head of the GAO. Walker said in testimony to congressional committees and in a detailed report that Iraq had met three of the eighteen benchmarks, had partially met four, and had not met eleven. "Overall, key legislation has not been passed, violence remains high, and it is unclear whether the Iraqi government will spend $10 billion in reconstruction funds," Walker told the committees. The $10 billion reference concerned the Maliki government's pledge to spend that amount of Iraq's own money to deliver essential services to its citizens "on an equitable basis." Only about one-fourth of the money had been spent by late August, Walker reported. This money was in addition to about $40 billion in U.S. taxpayer funds that Congress had allocated since 2003 for reconstruction and related expenses, such as the training of Iraqi security forces.

Congress received a slightly more upbeat assessment on September 10–11, when Ambassador Crocker and Gen. David H. Petraeus, the U.S. commander in Iraq, testified to four committees. Crocker reviewed the many challenges facing Iraqi leaders and concluded that they were trying, albeit slowly and painfully, to overcome them. "I do believe that Iraq's leaders have the will to tackle the country's pressing problems, although it will take longer than we originally anticipated because of the environment and the gravity of the issues before them," he said. Crocker cited, in particular, the August 26 agreement among top leaders on moving important pieces of legislation in the parliament. "This agreement by no means solves all of Iraq's problems," he said. "But the commitment of its leaders to work together on hard issues is encouraging."

In a follow-up report to Congress on September 14, Bush rejected the GAO's assessment and said Iraq had made "satisfactory" progress on eight of the eighteen benchmarks, had partly met three others, and had not met seven. Even so, Bush said he was releasing $15 billion in economic aid to Iraq that Congress had made contingent on the Iraqi government meeting the benchmarks. Administration officials also played down the importance of the benchmarks, saying some of them were no longer relevant to the political situation in Iraq. Vice President Dick Cheney, in a speech in Michigan on September 14, said the president would make his decisions "based on the national interest and nothing else—not by artificial measures, not by political calculations, certainly not by poll numbers."

Following are excerpts from "Securing, Stabilizing, and Rebuilding Iraq: Iraqi Government Has Not Met Most Legislative, Security, and Economic Benchmarks," a report presented to congressional committees on September 4, 2007, by the U.S. Government Accountability Office.

DOCUMENT

GAO Assessment of Progress by Iraqi Government in Meeting Benchmarks

September 4, 2007

Congressional Committees:

Over the last 4 years, the United States has provided thousands of troops and obligated nearly $370 billion to help achieve the strategic goal of creating a democratic Iraq that can govern and defend itself and be an ally in the War on Terror. These troops have performed courageously under dangerous and difficult circumstances. The U.S. Troop Readiness, Veterans' Care, Katrina Recovery, and Iraq Accountability Appropriations Act of 2007 (the Act) requires GAO to submit to Congress by September 1, 2007, an independent assessment of whether or not the government of Iraq has met 18 benchmarks contained in the Act, and the status of the achievement of the benchmarks. The benchmarks cover Iraqi government actions needed to advance reconciliation within Iraqi society, improve the security of the Iraqi population, provide essential services to the population, and promote economic well-being. The benchmarks contained in the Act were derived from benchmarks and commitments articulated by the Iraqi government beginning in June 2006. . . .

The January 2007 U.S. strategy, *The New Way Forward in Iraq*, is designed to support the Iraqi efforts to quell sectarian violence and foster conditions for national reconciliation. The U.S. strategy recognizes that the levels of violence seen in 2006 undermined efforts to achieve political reconciliation by fueling sectarian tensions, emboldening extremists, and discrediting the Coalition and Iraqi government. Amid such violence, it became increasingly difficult for Iraqi leaders to make the compromises necessary to foster reconciliation through the passage of legislation aimed at reintegrating former Ba'athists and sharing hydrocarbon revenues more equitably, among other steps. Thus, the new strategy was aimed at providing the Iraqi government with the time and space needed to help address reconciliation among the various segments of Iraqi society.

As required by the Act, this report provides 1) an assessment of whether or not the Iraqi government has met 18 key legislative, security, and economic benchmarks, and, 2) provides information on the status of the achievement of each benchmark. Among these 18 benchmarks, eight address legislative actions, nine address security actions, and one is economic-related. In comparison, the Act requires the administration to report in July and September 2007 on the status of each benchmark, and to provide an assessment on whether satisfactory progress is being made toward meeting the benchmarks, not whether the benchmarks have been met. In order to meet our statutory responsibilities in a manner consistent with GAO's core values, we decided to use "partially met" criteria for selected benchmarks. . . .

Legislative Benchmarks

Our analysis shows that the Iraqi government has met one of the eight legislative benchmarks and partially met another. Specifically, the rights of minority political parties in the Iraqi legislature are protected through existing provisions in the Iraqi Constitution and Council of Representatives' by-laws; however, minorities among the Iraqi population are vulnerable and their rights are often violated. In addition, the Iraqi government partially met the benchmark to enact and implement legislation on the formation of regions; this law was enacted in October 2006 but will not be implemented until April 2008.

Six other legislative benchmarks have not been met. The benchmark requiring a review of the Iraqi Constitution has not been met. Fundamental issues remain unresolved as part of the constitutional review process, such as expanded powers for the presidency, the resolution of disputed areas (such as Kirkuk), and power sharing between federal and regional governments over issues such as distribution of oil revenue. In addition, five other legislative benchmarks have not been met. . . .

. . . [L]legislation on de-Ba'athification reform has been drafted but has yet to be enacted. Hydrocarbon legislation is in the early stages of legislative action; although three key components have been drafted, none are under active consideration by the Council of Representatives. Although the government of Iraq has established an independent electoral commission and appointed commissioners, the government has not implemented legislation to establish provincial council authorities, provincial elections law, or a date for provincial elections. No legislation on amnesty or militia disarmament is being considered because the conditions for a successful program, particularly the need for a secure environment, are not present, according to U.S. and Iraqi officials.

Prospects for additional progress in enacting legislative benchmarks have been complicated by the withdrawal of 15 of 37 members of the Iraqi cabinet. According to an August 2007 U.S. interagency report, this boycott ends any claim by the Shi'ite-dominated coalition to be a government of national unity and further undermines Iraq's already faltering program of national reconciliation. In late August, Iraq's senior Shi'a, and Sunni Arab and Kurdish political leaders signed a Unity Accord signaling efforts to foster greater national reconciliation. The Accord covered draft legislation on de-Ba'athification reform and provincial powers laws, as well as setting up a mechanism to release some Sunni detainees being held without charges. However, these laws need to be passed by the Council of Representatives. . . .

The Administration's July 2007 report cited progress in achieving some of these legislative benchmarks but provided little information on what step in the legislative process each benchmark had reached. . . .

Security Benchmarks

Our analysis shows that the Iraqi government has met two of the nine security benchmarks. Specifically, it has established political, communications, economic, and services committees in support of the Baghdad security plan and, with substantial coalition assistance, 32 of the planned 34 Joint Security Stations across Baghdad. Of the remaining 7 benchmarks, the Iraqi government partially met 2 and did not meet five. . . . The Iraqi government partially met the benchmark of providing three trained and ready brigades to support Baghdad operations. Since February 2007, the Iraqi government deployed nine Iraqi army battalions equaling three brigades for 90-day rotations to support the Baghdad Security Plan. The administration's July 2007 report to Congress noted problems in manning the Iraqi brigades, but stated that the three brigades were operating in support of Baghdad operations. Our classified report provides additional information on the readiness levels and performance of these units, which supports our assessment of this benchmark.

The Iraqi government also partially met the benchmark of ensuring that the Baghdad security plan will not provide a safe haven for any outlaws regardless of their sectarian or political affiliation. Even though the Baghdad Security Plan is aimed at eliminating safe havens, and U.S. commanders report satisfaction with the coalition's ability to target extremist groups, opportunities for creating temporary safe havens exist due to the political intervention of Iraqi government officials . . . and the strong sectarian loyalties and militia infiltration of security forces.

The Iraqi government has not met the benchmark to reduce sectarian violence and eliminate militia control of local security. . . . [M]ilitia control of local security forces remains a problem. Several U.S. and UN reports have found that militias still retain significant control or influence over local security in parts of Baghdad and other areas of Iraq.

On trends in sectarian violence, we could not determine if sectarian violence had declined since the start of the Baghdad Security Plan. The administration's July 2007 report stated that MNF-I trend data demonstrated a decrease in sectarian violence since the start of the Baghdad Security Plan in mid-February 2007. The report acknowledged that precise measurements vary, and that it was too early to determine if the decrease would be sustainable. Measuring sectarian violence is difficult since the perpetrator's intent is not always clearly known. Given this difficulty, broader measures of population security should be used in judging these trends. The number of attacks targeting civilians and population displacement resulting from sectarian violence may serve as additional indicators. For example . . . the average number of daily attacks against civilians remained about the same over the last six months. The decrease in total average daily attacks in July is largely due to a decrease in attacks on coalition forces rather than civilians.

While overall attacks declined in July compared to June, levels of violence remain high. Enemy initiated attacks have increased around major religious and political events, including Ramadan and elections. For 2007, Ramadan is scheduled to begin in mid-September. Our classified report provides further information on measurement issues and trends in violence in Iraq obtained from other U.S. agencies. The unclassified August 2007 National Intelligence Estimate (NIE) on Iraq reported that Coalition forces, working with Iraqi forces, tribal elements, and some Sunni insurgents, have reduced al Qaeda in Iraq's (AQI) capabilities and restricted its freedom of movement. However, the NIE further noted that the level of overall violence, including attacks on and casualties among civilians remains high and AQI retains the ability to conduct high-profile attacks.

For the remaining four unmet security benchmarks, we found that:

- The Iraqi government has not always allowed Iraqi commanders to make tactical and operational decisions without political intervention, resulting in some operational decisions being based on sectarian interests.
- The government had not always ensured that Iraqi security forces were providing even-handed enforcement of the law, since U.S. reports have cited continuing sectarian-based abuses on the part of Iraqi security forces.
- Instead of increasing, the number of Iraqi army units capable of independent operations had decreased from March 2007 to July 2007.
- Iraqi political authorities continue to undermine and make false accusations against Iraqi security force personnel. According to U.S. government officials, little has changed since the administration's July 2007 report. . . .

Economic Benchmark

The Iraqi government partially met the benchmark to allocate and spend $10 billion because it allocated $10 billion in reconstruction funds when it passed its 2007 budget in February, 2007. *The New Way Forward in Iraq* cited Iraq's inability to spend its own resources to rebuild critical infrastructure and deliver essential services as an economic challenge to Iraq's self-reliance. Iraqi government funds represent an important source of financing for rebuilding Iraq since the United States has obligated most of the $40 billion provided to Iraq for reconstruction and stabilization activities since 2003.

However, it is unclear whether the $10 billion allocated by the Iraqi government will be spent by the end of Iraq's fiscal year, December 31, 2007. Preliminary Ministry of

Finance data reports that Iraq's central ministries spent about $1.5 billion, or 24 percent, of the approximately $6.5 billion in capital project funds allocated to them through July 15, 2007. The remaining funds from the $10 billion were allocated to the provinces and the Kurdish region. . . .

Conclusions

As of August 30, 2007, the Iraqi government met 3, partially met 4, and did not meet 11 of its 18 benchmarks. The Iraqi government has not fulfilled commitments it first made in June 2006 to advance legislative, security, and economic measures that would promote national reconciliation among Iraq's warring factions. Of particular concern is the lack of progress on the constitutional review that could promote greater Sunni participation in the national government and comprehensive hydrocarbon legislation that would distribute Iraq's vast oil wealth. Despite Iraqi leaders recently signing a unity accord, the polarization of Iraq's major sects and ethnic groups and fighting among Shi'a factions diminishes the stability of Iraq's governing coalition and its potential to enact legislation needed for sectarian reconciliation.

Reconciliation was also premised on a reduction in violence. While the Baghdad security plan was intended to reduce sectarian violence, measuring such violence may be difficult since the perpetrator's intent is not clearly known. Other measures of violence, such as the number of enemy-initiated attacks, show that violence has remained high through July 2007.

As the Congress considers the way forward in Iraq, it must balance the achievement of the 18 Iraqi benchmarks with the military progress, homeland security, foreign policy and other goals of the United States. Future administration reports on the benchmarks would be more useful to the Congress if they clearly depicted the status of each legislative benchmark, provided additional quantitative and qualitative information on violence from all relevant U.S. agencies, and specified the performance and loyalties of Iraqi security forces supporting coalition operations.

Recommendations

In preparing future reports to Congress and to help increase transparency on progress made toward achieving the benchmarks, we recommend that:

1. The Secretary of State provide information to the President that clearly specifies the status in drafting, enacting, and implementing Iraqi legislation;
2. The Secretary of Defense, and the heads of other appropriate agencies, provide information to the President on trends in sectarian violence with appropriate caveats, as well as broader quantitative and qualitative measures of population security, and
3. The Secretary of Defense, and the heads of other appropriate agencies, provides additional information on the operational readiness of Iraqi security forces supporting the Baghdad security plan, particularly information on their loyalty and willingness to help secure Baghdad.

As discussed below, State and DOD concurred with these recommendations.

Agency Comments

We provided a draft of this report to the Departments of State and Defense, the National Intelligence Council, and the Central Intelligence Agency. The National Intelligence Council and the Central Intelligence Agency provided technical comments, which we incorporated as appropriate.

The Department of State provided written comments. . . . State also provided us with technical comments and suggested wording changes that we incorporated as appropriate. State agreed with our recommendation to provide the President with additional information on the specific status of key Iraqi legislation in preparing future reports to Congress. State suggested that we note the standards we used in assessing the 18 benchmarks differ from the administration's standards. The highlights page and introduction of our report discuss these differing standards. State also suggested that we take into consideration recent political developments in Iraq, such as the communiqué released by Iraqi political leaders on August 26, 2007. We added additional information to the report about this communiqué and related developments.

The Department of Defense also provided written comments. . . . DOD also provided us with technical comments and suggested wording changes that we incorporated as appropriate. Defense agreed with our recommendations to provide, in concert with other relevant agencies, information to the President on trends in sectarian violence with appropriate caveats, as well as broader quantitative and qualitative measures of security. Defense also agreed to provide the President with additional information on the operational readiness of Iraqi security forces supporting the Baghdad security plan.

DOD also provided additional oral comments. DOD disagreed with our conclusion in the draft report that trends in sectarian violence are unclear. Further information on DOD's views, and our response, are contained in our classified report. However, the additional information that DOD provided did not warrant a change in our assessment of "not met." We note that the unclassified August 2007 NIE stated that the overall violence in Iraq, including attacks on and casualties among civilians, remains high, Iraq's major sectarian groups remain unreconciled, and levels of insurgent and sectarian violence will remain high over the next six to twelve months.

DOD disagreed with our initial assessment of "not met" for the training and readiness of the Iraqi brigades supporting operations in Baghdad and provided additional information on this issue. While acknowledging that some of these Iraqi units lacked personnel, fighting equipment, and vehicles, the U.S. commander embedded with the units attested to their fighting capabilities. Based on this additional information, and our classified and unclassified information, we changed our rating from "not met" to "partially met."

DOD did not agree with our initial assessment that the benchmark related to safe havens was not met. DOD provided additional information describing MNF-I efforts to conduct targeted operations in Sadr City. For example, from January to August 2007, Coalition forces and Iraqi security forces conducted over eighty operations that span each sector of Sadr City. However, due to sectarian influence and infiltration of Iraqi security forces, and support from the local population, anti-coalition forces retain the freedom to organize and conduct operations against coalition forces. Based on this additional information, we changed this assessment to "partially met."

Source: U.S. Congress. Government Accountability Office. "Securing, Stabilizing, and Rebuilding Iraq: Iraqi Government Has Not Met Most Legislative, Security, and Economic Benchmarks." GAO-07-1195. September 4, 2007. purl.access.gpo.gov/GPO/LPS85646 (accessed March 11, 2008).

Other Historic Documents of Interest

From this volume

From previous *Historic Documents*

DOCUMENT IN CONTEXT

Congressional Testimony on the Minneapolis Bridge Collapse

SEPTEMBER 5, 2007

It was a typical evening rush hour on the eight-lane Interstate 35 West (I-35W) bridge in downtown Minneapolis on August 1, when suddenly the bridge shuddered, swayed, and then collapsed. Several cars, trucks, and their drivers and passengers were plunged into the Mississippi River sixty feet below, while other vehicles came to rest on broken pavement, crashed with other cars, or slammed into broken concrete and twisted steel girders and guard rails. "Boom, boom, boom, and we were just dropping, dropping, dropping," one witness told the Associated Press. The bridge collapse killed thirteen people and injured another 123. It also called dramatic attention to a nationwide problem that most drivers and government officials would just as soon not have to think about—the safety of the nation's 600,000 bridges.

Within hours, the news media were reporting that the bridge, built in 1967, had been considered "structurally deficient"—as were 70,000 other bridges across the country. Even as federal officials assured the driving public that a structurally deficient classification did not mean a bridge was unsafe, they were warning states to immediately inspect steel-deck truss bridges similar to the span that had collapsed. "Even though we don't know what caused this collapse, we want states to immediately and thoroughly examine all similar spans out of an abundance of caution," Transportation Secretary Mary E. Peters said in a statement issued late on August 1. According to federal highway officials there were 756 such spans scattered across the country. The I-35W collapse also prompted calls for revamping the nation's bridge inspection program and for putting more money into the country's aging transportation infrastructure.

The I-35W Bridge Collapse

Eyewitnesses to the collapse described a loud boom, clouds of dust and smoke, shooting sprays of water, screams, and cries for help as people climbed from their cars and struggled to find safety. One eyewitness told the *Minneapolis Tribune* that he was stuck in a traffic jam on the bridge when he saw the roadway ahead of him rolling down into river. "It kept collapsing, down, down, down, until it got to me," he said, explaining that his car rolled forward, stopping just before he thought it would fall into the water. He climbed out and over the wreckage and jumped to safety. "I thought I was dead," he said. "I honestly did. I thought it was over."

Police, fire, and rescue crews were on the scene almost immediately, as were transportation and other local, state, and federal officials. FBI officials quickly ruled out terrorism as a cause of the collapse. "This is a catastrophe of historic proportions for Minnesota," Governor Tim Pawlenty said at a news conference later that evening, when there was still great uncertainty about the number of people killed or injured in the accident. The bridge was also part of a major commuter artery, and its collapse was expected to cause significant upheaval in the city over the coming days and weeks as commuters sorted out new routes to get to work. In addition, it was doubtful that a new bridge could be constructed before the Republican National Convention, scheduled to be held in Minneapolis in September 2008.

On August 5, President George W. Bush visited the site of the collapse, offering condolences to the families of the victims and pledging to help the city rebuild the bridge. The next day he signed an emergency bill (HR 3311—PL 110-56) authorizing $250 million in emergency funds for the reconstruction project.

A month later, Mark V. Rosenker, chairman of the National Transportation Safety Board (NTSB), described what investigators had so far determined about the accident. Testifying to the House Transportation and Infrastructure Committee on September 5, Rosenker said about 1,000 feet of the 1,900-foot-long bridge collapsed, with the 456-foot-long center span falling into the river below. A total of 110 vehicles were on the portion of the bridge that collapsed; of those, Rosenker said, investigators thought 17 had fallen into the river. It took twenty days to complete victim recovery. The toll might have been worse if all eight lanes of the bridge had been open to traffic, but four were shut down while construction crews repaved the decking.

Finding the Cause of the Collapse

Federal and state investigators began searching for the cause or causes of the collapse within hours of the event. Several factors entered into the equation. One factor was the age of the bridge. As steel ages, it grows brittle, especially in cold climates such as Minnesota's, which could make it more vulnerable to stress caused by traffic vibration. Metal fatigue could result in fractures and larger cracks, while weathering, standing water, and debris could lead to poor drainage and corrosion, affecting bearings and joints in particular and lowering the integrity of a bridge. The I-35W bridge was classified as "structurally deficient" in 1990 and had been subjected to annual inspections since 1993. That classification means that the load that a bridge could safely carry should be recalculated to determine the maximum weight a bridge could carry safely. In some cases, the recalculation would require that maximum load limits be posted. In extreme cases a bridge might have to be closed. But most structurally deficient bridges are safe to keep open provided that they are properly inspected and maintained.

The I-35W bridge had long suffered from corrosion around its joints and cracking in its superstructure; many of those problems had been fixed over the years, and an inspection in June 2006 found no evidence of growth in preexisting cracks, according to Dan Dorgan, Minnesota's chief bridge engineer. Later in 2006, a consulting firm recommended that the state either reinforce the superstructure with steel plates or do a close inspection and put steel plating only where there were signs of metal fatigue. Fearing that drilling thousands of small holes into the superstructure to fasten the steel plates could weaken the bridge, the state "chose the inspection route," Dorgan said. "In May [2007] we began inspections. We thought we had done all we could, but obviously something went terribly wrong."

Another factor in many bridge collapses is soil erosion in the river bed around the bridge's piers. According to Jean-Louis Briaud, a civil engineer with the Texas Transportation Institute, underwater erosion caused almost 60 percent of the 1,502 recorded bridge failures between 1966 and 2005. Erosion is "the number one killer of bridges," Briaud told the *Washington Post.* "If you create a hole around the bridge support, then the foundation cannot carry the load of the deck."

Initial attention at the I-35W collapse, however, appeared to focus on metal fatigue. One question was whether the repaving might have contributed to the accident. In his September 5 congressional testimony, Rosenker said that about 383,000 pounds of construction materials had been delivered to the work site the day of the accident, and that the combined weight of these materials and construction vehicles was about 287 tons. NTSB investigators were also looking into a possible design flaw in the bridge gusset plates—steel plates that tie steel girders together. Rosenker would say only that his agency was "in the process of verifying the loads and stresses on the gusset plates" at various locations on the bridge. A typical gusset plate measured 5 feet by 5 feet and was a half-inch thick. A determination by NTSB that the design of the gusset plates was flawed would indicate that the bridge had been structurally unsound since the day it was opened. Such a finding could have implications for the thousands of other bridges throughout the country that used similar plates, but it would not answer the immediate question of what caused the I-35W span to collapse when it did. NTSB was not expected to have a complete answer to that question until well into 2008.

Although it was not a cause of the collapse, the possibility that inspectors had missed serious structural flaws or had not made the right decisions about how to deal with the ones they found was also a concern. In 2007 states were responsible for inspecting bridges and hiring inspectors under guidelines issued by the Federal Highway Administration (FHWA). Bridge inspectors were required to meet minimum standards, but checks by the FHWA found that training and inspection skills varied considerably and in some cases were inadequate. Several experts said visual bridge inspections should be supplemented with new technology, such as permanent sensors that could monitor stress and strain on the bridge superstructure. "Visual instruction is just not enough to be absolutely certain you have no cracks," Steven Fenves, a researcher at the National Institute of Standards and Technology, told the Associated Press. "Much of it involves bridge inspection by a very tired bridge inspector who has just climbed up a bridge and is dodging pigeons. . . . It's a very inadequate process."

On August 2, Transportation Secretary Peters announced that she had directed the Transportation Department's inspector general to review whether the FHWA was monitoring bridge inspections adequately. "What happened in Minnesota is simply unacceptable," she said. Speaking at the House hearing on September 5, the inspector general, Calvin L. Scovel III, said his inquiry would focus on how well the FHWA was following recommendations his office had made in March 2006 for improving monitoring of bridge inspections. At that time, Scovel's office found that the agency could do a better job of ensuring that structurally deficient bridges most in need of weight limits were identified and corrective actions taken. Scovel also said his office would study whether states were using federal funding meant for bridge maintenance and repairs effectively or possibly were using the funding for other purposes. States received more than $5 billion in federal funding in fiscal year 2007 for bridge construction and repair.

A Backlog of Repairs

The Minneapolis bridge collapse called attention to a nationwide problem—a steadily deteriorating highway system in need of steadily more costly maintenance and repairs.

"We're falling further and further behind," Robert Poole, director of transportation studies at the Reason Foundation, told the *Washington Post.* "We're prospering as a nation, driving more as commuters, and shipping more goods, and that's pounding the highways and wearing them out," he said.

According to the FHWA, about 72,500 of the nation's 600,000 highway bridges were considered structurally deficient and nearly another 80,000 were considered "functionally obsolete." A study done in 2005 by the American Society of Civil Engineers estimated that the country would need to spend $188 billion over the next twenty years just to repair deteriorating bridges. In January 2007, the FHWA said that about $65 billion could be used immediately to correct deficiencies just on the 6,149 bridges in the national highway system that were categorized as structurally deficient.

The main source of revenue for construction and repair of highways and bridges was the federal highway trust fund, which was funded through a federal tax on gasoline. The transportation spending bill for fiscal 2008 (PL 110–161) that Congress passed in December boosted the amount to be drawn from the trust fund to $1 billion above the level set in 2005 to finance bridge inspection and repair activities. But many legislators said that was not enough. Since 1993 the fuel tax had remained at 18.3 cents per gallon, despite the ever-growing backlog in bridge repair and maintenance needs. Rep. James Oberstar, a Minnesota Democrat and chairman of the House Transportation and Infrastructure Committee, proposed raising that tax by 5 cents per gallon to pay for bridge repairs. Other members of the committee, including some Republicans, called for indexing the fuel tax to inflation. Both proposals had been made in previous years and been rejected.

Oberstar and other Democrats blamed the Bush administration for not supporting greater investment in the nation's infrastructure. The administration countered that Congress had to share some of the blame for doling out federal highway money based on individual legislators' pet projects—a practice called earmarking—rather than on maintenance and repair priorities set by the individual states. Testifying before the House Transportation Committee on September 5, Transportation Secretary Peters said the cost of the earmarks in the 2005 legislation authorizing funding for highways and bridges amounted to $23 billion. Moreover, she said, the department had found that on average federal earmark amounts covered only about 10 percent of the total cost of a project, forcing states either to delay the project indefinitely or to reallocate funding from higher-priority projects.

Following are excerpts from statements delivered September 5, 2007, to the House Committee on Transportation and Infrastructure on the federal government's response to the August 1 collapse of the I-35W bridge in Minneapolis, Minnesota. The first statement is from Transportation Secretary Mary E. Peters; the second is from the department's inspector general, Calvin E. Scovel III.

DOCUMENT

Secretary Peters's Statement to House Transportation Committee

September 5, 2007

. . . America was stunned on the evening of August 1, 2007, when the Interstate 35 West (I-35W) bridge over the Mississippi River in Minneapolis, Minnesota, collapsed. Numerous vehicles were on the bridge at the time and there were 13 fatalities and 123 people injured. We extend our deepest sympathy to the loved ones of those who died and to the injured.

We do not yet know why the I-35W Bridge failed, and our Department is working closely with the National Transportation Safety Board (NTSB) as it continues its investigation to determine the cause or causes. In the interim, we are taking every step to ensure that America's infrastructure is safe. I have issued two advisories to States in response to what we have learned so far, asking that States re-inspect their steel deck truss bridges and that they be mindful of the added weight construction projects may bring to bear on bridges. . . .

FHWA [Federal Highway Administration] is assisting the NTSB as they conduct a thorough investigation, which includes a structural analysis of the bridge. Within days of the collapse, development of a computer model based upon the original design drawings for the bridge began at FHWA's Turner Fairbank Highway Research Center in McLean, Virginia. This model can run simulations to determine the effect on the bridge of removing or weakening certain elements to recreate, virtually, the actual condition of the bridge just prior to and during its collapse.

By finding elements that, if weakened or removed, result in a bridge failure similar to the actual bridge failure, the investigators' work is considerably shortened. While examination of the physical members of the bridge being recovered from the site provide the best evidence of why the bridge collapsed, the analytical model allows the evaluation of multiple scenarios which can then be validated against the physical evidence. This work is expected to take several months and my experts will be there, on the ground, to provide assistance. We need to fully understand what happened so we can take every possible step to ensure that such a tragedy does not happen again. Data collected at the scene, with the help of the Federal Bureau of Investigation's 3-D laser scanning device, are being used to assist in the investigation.

On August 2, the day after the collapse, I requested that the DOT Inspector General conduct a rigorous assessment of the Federal-aid bridge program and the National Bridge Inspection Standards (NBIS). The NBIS, in place since the early 1970s, generally requires safety inspections for all highway bridges in excess of 20 feet in total length on public roads at least every two years. Safety is ensured through hands-on inspections and rating of components, such as the deck, superstructure, and substructure, and the use of non-destructive evaluation methods, and other advanced technologies. The composition and condition information is collected in the national bridge inventory (NBI) database, maintained by FHWA.

The I-35W bridge has been inspected annually by the Minnesota Department of Transportation (MNDOT). The most recent inspection was begun by MNDOT on May 2, 2007. No imminent dangers were observed and MNDOT planned to continue inspecting the bridge in the fall following completion of construction work on the bridge.

Federal, State, and local transportation agencies consider the inspection of our nearly 600,000 bridges to be of vital importance and invest significant funds in bridge inspection activities each year. We strive to ensure that the quality of our bridge inspection program is maintained at the highest level and that our funds are utilized as effectively as possible. The Inspector General will be monitoring all of the investigations into the collapse and reviewing our inspection program to decide and advise us what short- and long-term actions we may need to take to improve the program. Though we will have to wait for the NTSB's report before we can conclude if the inspection program played any role in this collapse, we must have a top-to-bottom review to make sure that everything is being done to keep this kind of tragedy from occurring again.

In the aftermath of this tragedy, a necessary national conversation has begun concerning the state of the Nation's bridges and highways and the financial model used to build, maintain and operate them. It is important to understand that, while we must do a better job of improving the Nation's transportation systems, we do not have a broad transportation infrastructure "safety" crisis.

Since 1994, the percentage of the Nation's bridges that are classified as "structurally deficient" has declined from 18.7% to 12.0%. The term "structurally deficient" is a technical engineering term used to classify bridges according to serviceability, safety, and essentiality for public use. The fact that a bridge is classified as "structurally deficient" does not mean that it is unsafe for use by the public. Since 1995 the percentage of travel taking place on roads that are considered "good" has increased from 39.8% to 44.2%. Overall, approximately 85% of travel takes place on pavement that is considered "acceptable." FHWA estimates that it will cost approximately $40 billion a year to maintain the physical condition of our Nation's highways and bridges and approximately $60 billion a year to substantially improve the quality of current roads and bridges. In 2005, Federal, State, and local governments together made over $75 billion in capital investment to rehabilitate highways and bridges in the U.S. and improve their operational performance. If we include operational, administrative, and debt service costs in addition to capital investments, the U.S. spent nearly $153 billion on highways and bridges in 2005.

These infrastructure quality numbers should and can be improved with more targeted investment strategies, but it is inaccurate to conclude that the Nation's transportation infrastructure is subject to catastrophic failure. We have quality control systems that provide surveillance over the design and construction of bridges. We have quality control systems that oversee the operations and use of our bridges. And we have quality control over inspections of bridges to keep track of the attention that a bridge will require to stay in safe operation. These systems have been developed over the course of many decades and are the products of the best professional judgment of many experts. We will ensure that any findings and lessons that come out of the investigation into the I-35W bridge collapse are quickly learned and appropriate corrective actions are institutionalized to prevent any future occurrence.

A more accurate description of our current and broader problem is that we have an increasingly flawed investment model and a system performance crisis. Many are calling for a renewed national focus on our Nation's highway infrastructure. I applaud Ranking Member Mica for starting the conversation about a multimodal National Strategic Transportation Plan. And while I agree that our infrastructure models need to be reexamined, it is imperative that we actually focus on the right problem.

When faced with an underperforming division, the response of any credible business organization is to assess the cause of underperformance and to implement policies and practices intended to reverse performance declines. In my assessment, the underperformance in the highway sector is fundamental, not incremental. In other words, increases in Federal taxes and spending would likely do little, if anything, without a more basic change in how we analyze competing spending options and manage existing systems more efficiently.

Because tax revenues are deposited into a centralized Federal trust fund and re-allocated on the basis of political compromise, major decisions on how to prioritize investments—and thus, spend money—are made without consideration of underlying economic or safety merits. The degree to which one capital investment generates more returns than a competing investment is the most basic question asked in virtually every other capital intensive sector of the economy. Yet, when it comes to some of our largest and most critical investments we make as a Nation—highways and bridges—there is virtually no analysis of this question. There is no clearer evidence of this failure to prioritize spending than the disturbing evolution of the Federal highway program. This program has seen politically-designated projects grow from a handful in the surface transportation bill enacted in the early 1980s to more than 6,000 enacted in SAFETEA-LU. The cost of these earmarks totaled $23 billion—a truly staggering figure.

The real cost of these earmarks is much higher. Looking at a sample of various recent earmarks, we found that the Federal earmark amounts themselves comprised on average only 10% of the total project cost. Because of this, State departments of transportation will typically either delay the earmarked project indefinitely or re-allocate resources from higher priorities to fill the funding gap. In addition, earmarks present large administrative burdens for States that must dedicate scarce personnel resources to managing lower priority projects that are subject to earmarking. In short, earmarks ripple through the entire Federal-aid program structure.

In addition to earmarks, there are more than 40 special interest programs that have been created to provide funding for projects that may or may not be a State and local priority. As a former State DOT director, I have had first-hand experience with the difficulties created when Washington mandates override State priorities.

While it is true that not all of these investments are wasteful, it is also true that virtually no comparative economic analysis is conducted to support these spending decisions. No business could survive for any meaningful period of time utilizing a similar investment strategy. Not surprisingly, new economic literature reveals that the returns on our highway investments have plummeted into the low single digits in recent years.

The Department is working with States to encourage them to regularly use benefit cost analysis (BCA) when making project selection decisions. Currently, approximately 20 States make some use of BCA, while 6 States use the technique regularly. . . .

Moreover, since Federal transportation funding levels are not linked to specific performance-related goals and outcomes, the public has rightfully lost confidence in the ability of traditional approaches to deliver. Performance-based management can help establish and maintain accountability. As former Washington State DOT Secretary Doug MacDonald noted, "transportation agencies need to demonstrate to taxpayers that they get a dollar's worth of value for a dollar's worth of tax." The use of performance measures, by helping to identify weaknesses as well as strengths, can improve the transportation project selection process and the delivery of transportation services.

In addition to an insufficient performance and cost-benefit focus, the current gas tax-dependent model does virtually nothing to directly address the growing costs of congestion and system unreliability. Indirect taxes on gasoline, diesel fuel, motor vehicles, tires, property and consumer products—the dominant means of raising revenues for transportation—are levied regardless of when and where a driver uses a highway. This leads to a misperception that highways are "free," which in turns encourages overuse and gridlock at precisely the times we need highways the most. Consistent with the views of almost every expert that has looked at the issue, GAO recently released a report arguing that gas taxes are fundamentally incapable of balancing supply and demand for roads during heavily congested periods.

The data simply do not lie in this case. Relying extensively on gas and motor vehicle taxes, virtually every metropolitan area in the U.S. has witnessed an explosion in traffic delays over the last 25 years. Meanwhile, in recent years, the increase in surface transportation funding has significantly outpaced the overall growth of non-defense, non-homeland security Federal discretionary spending. And, since 1991, capital outlays at all levels of government have nearly doubled. Economists have long understood the connection between payment mechanisms and system performance, but technology and administrative complexities limited the ability of policymakers to explore alternatives. Today, those barriers no longer exist.

This is one of the main reasons that our Department has been strongly supporting States that wish to experiment with electronic tolling and congestion pricing. Nationwide, the majority of projects in excess of $500 million currently in development are projected to be financed at least in part with electronic tolls. In the middle of August, we announced Federal grants in excess of $800 million to some of the country's largest cities to fully explore the concept of electronic tolling combined with expanded commuter transit options and deployment of new operational technologies. Nationwide, the trends are inescapable and encouraging.

We believe that to the extent feasible, users should finance the costs of building, maintaining and operating our country's highways and bridges. What is increasingly clear is that directly charging for road use (similar to the way we charge for electricity, water, and telecommunications services) holds enormous promise to generate large amounts of revenues for re-investment and to cut congestion. Equally important, however, prices send better signals to State DOTs, planners, and system users as to where capacity expansion is most critical. Prices are not simply about demand management, they are about adding the right supply.

The current financial model is also contradictory to other critical national policy objectives. As a country, we are rightfully exploring every conceivable mechanism to increase energy independence, promote fuel economy in automobiles, stimulate alternative fuel development, and reduce emissions. President Bush has urged Congress to pass laws that will substantially expand our alternative energy capabilities and increase Corporate Average Fuel Economy requirements for automobiles and light trucks. The Federal Government should be strongly encouraging States to explore alternatives to petroleum-based taxes, not expanding the country's reliance upon them.

Finally, the current highway and bridge financial model fails to provide strong incentives for technology development and deployment, particularly when contrasted to other sectors of the economy. It is imperative that we find more effective means to ensure that the rewards of a given advancement—for example, in extended life pavements or more sophisticated traveler information systems—can accrue in part to those firms or individuals that come forward with creative ideas. It is no coincidence that we are seeing a technology boom in markets that have pricing structures that reward innovation. Pricing infrastructure usage more closely to its true costs will not only reduce congestion and more appropriately target resources, it will also provide new incentives for innovation.

The I-35W bridge collapse was both a tragedy and a wake-up call to the country. We have a duty to ensure a safe transportation system for all who use it. Moreover, our country's economic future is tied in large part to the safety and reliability of our transportation infrastructure. Before reaching the conclusion that additional Federal spending and taxes is the right path, we should critically examine how we establish spending priorities today. We need a data-driven, performance based approach to building and maintaining our Nation's infrastructure assets—a process where we are making decisions based on safety first, economics second, and politics not at all. And we need an underlying framework that is responsive to today's and tomorrow's challenges, not those of the 1950s. . . .

SOURCE: U.S. Department of Transportation. "Statement of the Honorable Mary E. Peters, Secretary of Transportation, before the Committee on Transportation and Infrastructure, U.S. House of Representatives." September 5, 2007. testimony.ost.dot.gov/test/peters6.htm (accessed January 13, 2008).

DOCUMENT

Federal Highway Administration's Oversight of Deficient Bridges

September 5, 2007

. . . Thank you for the opportunity to testify today on the National Bridge Inspection Program, particularly the Federal Highway Administration's (FHWA) oversight of structurally deficient bridges within the National Highway System. . . . As you know, under the current National Bridge Inspection Program, the states, with oversight by FHWA, are responsible for inspecting bridges on public roads. The primary purpose is to identify and evaluate bridge deficiencies in order to ensure public safety. . . .

. . . Today, I will discuss our previous work dealing with structurally deficient bridges and make several observations regarding FHWA's actions to address our prior recommendations to improve its oversight of bridges. Specifically:

- Federal oversight of bridge inspections and funding for bridge rehabilitation and replacement constitute significant issues for the U.S. Department of Transportation (DOT).
- FHWA needs to develop a data-driven, risk-based approach to bridge oversight to better identify and target those structurally deficient bridges most in need of attention.
- Action can be taken now to strengthen the National Bridge Inspection Program and FHWA's oversight.

Federal Oversight of Bridge Inspections and Funding for Bridge Rehabilitation and Replacement Constitute Significant Issues for DOT

Federal oversight of bridge inspections and funding of bridge rehabilitation and replacement have been significant issues for DOT for years. The safety of the Nation's bridges depends upon a complex web of Federal, state, and local activities, including such items as maintenance and rehabilitation, inspections and reviews, and load ratings and postings. While states are ultimately responsible for ensuring that bridges within their jurisdictions are safe, FHWA is responsible for overseeing the states in this effort, and for providing technical expertise and guidance in the execution of bridge inspection, repair and maintenance, and remediation activities.

The National Bridge Inventory comprises data on 599,976 bridges, including 116,086 bridges on the National Highway System, as well as bridges maintained and operated by various state and local entities. Many bridges require enhanced attention: nationwide, almost 80,000 bridges are considered functionally obsolete and nearly 72,500 are structurally deficient. In five states, more than 20 percent of the bridges are considered structurally deficient. The term "structurally deficient" refers to bridges that have major deterioration, cracks, or other deficiencies in their structural components, including decks, girders, or foundations. Regular inspections that check for corrosion, decay, and other signs of deterioration are important tools for ensuring that bridges are safe. In some cases, structurally deficient bridges require repair of structural components, or even closure. But most bridges that are classified as structurally deficient can continue to serve traffic safely if they are properly inspected, the bridges' maximum load ratings are properly calculated, and, when necessary, the proper maximum weight limits are posted.

Of the National Highway System's bridges, 6,149, or 5.3 percent, are categorized as structurally deficient. National Highway System bridges carry over 70 percent of all bridge traffic. The price of repair or remediation of these bridges is high. An FHWA report issued in January of this year estimated that about $65 billion could be invested immediately to address current bridge deficiencies.

Bridge safety first emerged as a high-priority issue in the United States in the 1960s. In 1967, corrosion caused the Silver Bridge on the Ohio River between Ohio and West Virginia to collapse, killing 46 people. In 1968, in hopes of avoiding further catastrophes, Congress responded by holding hearings on bridge design, inspection, and maintenance, determining that serious safety concerns and problems of lost investment and replacement costs "elevate bridge inspection and maintenance problems to national priority." In 1971, FHWA issued standards for identifying, inspecting, evaluating, and acting upon bridge deficiencies to ensure that bridges are safe for the traveling public. However, disaster struck again with further bridge collapses, including those of the Mianus River Bridge in Connecticut in 1983 (with 3 deaths), the Schoharie Creek Bridge in New York in 1987 (10 deaths), the Hatchie River Bridge in Tennessee in 1989 (8 deaths), and the Arroyo Pasajero Bridge (sometimes called Twin Bridges) in California in 1995 (7 deaths). Investigations showed that these collapses were caused at least in part by structural deficiencies created by the elements. The loss of lives, injuries, and significant economic impact resulting from these collapses, as well as the recent Minneapolis bridge collapse, underscore the significance of bridge safety as a major issue for DOT.

National Bridge Inspection Standards. According to current inspection standards, when bridge inspectors identify deficiencies that pose safety problems, a bridge should either be repaired to correct the deficiencies, posted with signs to restrict the size and weight of vehicles allowed, or, if the deficiencies are serious enough, closed to vehicular traffic.

While FHWA provides the oversight of state bridge inspections and programs, the states themselves are responsible for performing actual bridge inspections on public roads. The inspection standards provide a definition of bridges (greater than 20 feet long) and outline requirements regarding the frequency of inspections, qualifications of inspection personnel, and data to be collected. According to the standards:

- Most bridges are to be inspected at 2-year intervals.
- Each state is required to have a bridge inspection organization capable of performing inspections, preparing reports, and determining bridge ratings in accordance with the American Association of State Highway and Transportation Officials (AASHTO) standards and provisions in the Code of Federal Regulations.
- Each bridge shall be rated as to its safe load-carrying capacity. If the calculated load rating is less than the state's maximum legal load, the bridge must have signs posted as to the maximum permitted load, or be closed.
- The findings and results of bridge inspections, including safe load ratings, shall be recorded by state inspectors on standard paper or electronic forms, and submitted to the National Bridge Inventory.

Each year, FHWA's Office of Bridge Technology collects bridge inventory data from the states for use in updating its inventory. Along with maintaining the inventory of public highway bridges, FHWA is responsible for submitting a biennial report to Congress on the conditions of all bridges in that inventory. FHWA also performs an annual review of each state's bridge inspection program and compliance with inspection standards. Bridge inventory data provide important information on bridge location, age, ownership, and condition.

Structurally Deficient Bridges, Load Ratings, and Postings. A total of 6,149 National Highway System bridges (of the 116,086 National Highway System bridges in the inventory) were classified as structurally deficient as of last month. . . .

Proper reviews of the calculations of a bridge's maximum safe load ratings are important because as a bridge ages, corrosion and decay can decrease its capacity to support vehicles.

The practice of calculating the load rating of structurally deficient bridges and, if necessary, posting signs to keep heavier vehicles from crossing them, serves to protect structurally deficient bridges from powerful stresses caused by loads that exceed a bridge's capacity. The load rating is a calculation of the weight-carrying capacity of the bridge and is critical to its safety. A load rating is performed separately from the bridge inspection, but is based upon design capacities supplemented with data and observations of the bridge's physical condition provided by a bridge inspector. The load rating, expressed in tons, serves as the basis for posting signs noting the vehicle weight limit restriction, which can be referred to more simply as the bridge's maximum weight limit. Some bridges are weakened to the point that signs must be posted to bar vehicles heavier than the calculated maximum load.

Federal Funding for the Nation's Bridges. Congress has long recognized the vital national interest of assisting states in improving the condition of bridges. In 1978, Congress passed legislation authorizing the Highway Bridge Replacement and Rehabilitation Program and the Discretionary Bridge Program to provide states with funds needed to correct structural deficiencies. In 2005, Congress replaced the Highway Bridge Replacement and Rehabilitation Program and the Discretionary Bridge Program with the Highway Bridge Program, and broadened the scope to include systematic preventive maintenance. Overall, a total of $21.6 billion was authorized for the Highway Bridge Program through 2009.

For fiscal year 2007, states were allocated more than $5 billion to be used for bridge construction, repair, and remediation under the Highway Bridge Program. According to FHWA officials, while the agency tracks all Federal bridge funding, its financial management system does not differentiate between spending on structurally deficient bridges and other bridge-related expenditures. As a result, FHWA is unable to tell how much of the funding it provides to the states is actually spent on structurally deficient bridges. As part of our comprehensive audit of FHWA's oversight of the bridge program, we will be evaluating this issue and will report back to the Secretary of Transportation.

FHWA Needs to Develop a Data-Driven, Risk-Based Approach to Bridge Oversight to Better Identify and Target Those Structurally Deficient Bridges Most in Need of Attention

Our March 2006 report found that FHWA could improve its oversight of the states to ensure that maximum weight limit calculations and postings are accurate. The need for improved oversight was evidenced by our finding that, based on a statistical projection, the load ratings for as many as 10.5 percent of the structurally deficient bridges on the National Highway System are inaccurate.

To address deficiencies in its oversight, we recommended that FHWA develop a risk-based, data-driven approach with metrics to target the bridge problems most in need of attention. Since last year, FHWA has taken steps to address these deficiencies. In April 2006, for example, FHWA convened a working group to evaluate options and make recommendations for action. Based on the work of this group, FHWA has initiated several specific efforts to improve oversight of structurally deficient bridges, including load ratings and posting. However, more action is needed. In the coming months, we plan to continue our evaluation of these initiatives.

FHWA did not require its Division Offices to analyze bridge inspection data to better identify and target those structurally deficient bridges most in need of load limit recalculation and posting. FHWA's Division Offices in the three states we reviewed in depth—Massachusetts, New York, and Texas—did not ensure that the states' bridge load ratings were properly calculated and corresponding postings performed. Our statistical sample showed similar problems nationwide. . . .

Going forward, FHWA needs to ensure the effectiveness of these new risk management initiatives:

- As part of FHWA's risk management process, Division Offices are given the latitude to analyze, prioritize, and manage identified risks across their program areas. FHWA needs to take aggressive action to ensure that the Division Offices are conducting a rigorous and thorough assessment of potential risks associated with load rating and posting practices of structurally deficient bridges as part of the risk assessment process. FHWA should also ensure that these evaluations are completed by Division Offices and done in a rigorous and thorough manner.
- Further, FHWA needs to ensure that, if a high-risk area is identified, the Division Office follows up with an in-depth review and conducts it in a timely and rigorous manner. The recent bridge collapse in Minneapolis has increased the

urgency of making sure that any potential risks are identified and corrective actions taken expeditiously.

The time that FHWA engineers have available for bridge oversight is limited. An FHWA Division Office exists in every state as well as the District of Columbia and Puerto Rico. Each FHWA Division Office has a bridge engineer, in some cases assisted by additional engineering staff, designated to handle Federal bridge program oversight responsibilities. In addition, FHWA bridge engineers perform other activities. We found that time constraints restricted bridge engineers' reviews to only a small percentage of the total number of bridges in the state. For example, one FHWA engineer in a large state informed us that he spent only about 15 percent of his time on oversight of the bridge inspection program. The majority of his time was spent providing technical assistance, construction inspection, and in committee meetings, among other tasks. FHWA needs to examine whether bridge engineers are devoting sufficient time and effort to examining the structurally deficient bridges most in need of attention, including those requiring load rating recalculations and postings. Based on the results of this assessment, FHWA should make the necessary resource decisions to strengthen oversight in this area.

FHWA would benefit from an oversight program that makes substantially greater use of data and metrics to target bridge inspections for its compliance reviews. Given the thousands of bridges that FHWA oversees and the limited time its engineers have available, a data-driven approach would help FHWA bridge engineers focus on inspections and compliance reviews. That is, they could address the bridge problems most in need of attention. FHWA has undertaken several initiatives to make greater use of such an approach, although more aggressive action must be taken going forward. Specifically, FHWA has:

- *Modified the Bridge Program Manual to provide better guidance to Division Office bridge engineers conducting the annual compliance reviews. . . .*
- *Implemented new National Bridge Inventory reports that are intended to identify problem areas in load rating data. . . .*
- *Agreed to promote greater use of computerized bridge management systems. . . .*

Action Can Be Taken Now to Strengthen the National Bridge Inspection Program and FHWA's Oversight

The bridge collapse in Minneapolis has focused attention on FHWA's oversight of the Nation's bridges and underscores the importance of vigilant oversight of states' efforts to inspect and repair structurally deficient bridges. FHWA must be more aggressive in implementing the initiatives it has already identified as being critical to improving its oversight of structurally deficient bridges, as well as identifying any other needed changes. As we evaluate the National Bridge Inspection Program, we will make recommendations where appropriate to improve the program and how it is implemented by FHWA.

FHWA Needs to Take Aggressive Action Going Forward. The implementation of FHWA's recent initiatives to improve oversight of structurally deficient bridges is the responsibility of its 52 Division Offices. It is too early to tell the extent to which each Division Office has started to implement these new initiatives, or whether they are working effectively. FHWA needs to ensure that it carefully monitors the progress of implementing these initiatives in its Division Offices, systematically evaluates their effectiveness, and shares lessons learned about what is working well or not working well in each state. The Minneapolis bridge collapse increases the urgency of making sure that these new initiatives are being fully implemented in a timely manner and working as intended.

FHWA can take action immediately to improve oversight of the nation's bridges. Specifically, FHWA should:

- Identify and target those structurally deficient bridges most in need of recalculation of load ratings and postings, using a data-driven, risk-based approach.
- Finalize and distribute the revised Bridge Program Manual to the Division Offices as quickly as possible and ensure that FHWA engineers make greater use of existing bridge data as part of the annual compliance review process.
- Ensure that each of the 52 Division Offices conducts rigorous and thorough assessments of any potential risks associated with structurally deficient bridges, as directed in February 2007, and define how it will respond to any specific high-priority risks that Division Offices have identified.

We Are Undertaking a Comprehensive Audit of the National Bridge Inspection Program. Shortly after the Minneapolis bridge collapse, the Secretary of Transportation asked us to undertake an audit of the National Bridge Inspection Program. Our work will be separate and distinct from the National Transportation Safety Board's investigation, which will focus specifically on the events and conditions that led to the Minneapolis bridge collapse.

Our audit work will proceed in three concurrent phases, with sequential reporting dates. Specifically, our audit work will focus on the following efforts.

- An assessment of the corrective actions that FHWA has taken to address the recommendations we made in our March 2006 report on structurally deficient bridges. We have already initiated this effort and plan to issue a report later this year.
- A study of Federal funding provided to states for bridge rehabilitation and repair. We will assess FHWA's management and tracking of such funding, the extent to which states effectively and efficiently use these funds to repair or replace structurally deficient bridges, and whether states are using bridge funding for other purposes.
- A comprehensive review of FHWA's oversight activities to ensure the safety of National Highway System bridges across the country.

Going forward, our overall objective is to evaluate FHWA's implementation of the National Bridge Inspection Program and make recommendations for improvement in order to provide assurance that FHWA is doing everything that should be done to ensure bridge safety. We will report back to the Committee and the Secretary of Transportation as we identify additional steps that could be taken to improve the National Bridge Inspection Program. . . .

SOURCE: U.S. Department of Transportation. Office of Inspector General. "Federal Highway Administration's Oversight of Structurally Deficient Bridges." Testimony by Calvin L. Scovel III, Inspector General, U.S. Department of Transportation, before the House Committee on Transportation and Infrastructure. September 5, 2007. CC-2007-095. www.oig.dot.gov/item.jsp?id=2115 (accessed January 13, 2008).

OTHER HISTORIC DOCUMENTS OF INTEREST

FROM PREVIOUS *HISTORIC DOCUMENTS*

- Protecting U.S. infrastructure, *1997,* p. 711
- Democratic platform on infrastructure needs, *1984,* p. 576
- Highway repair programs, *1981,* p. 73

DOCUMENT IN CONTEXT

Japanese Prime Ministers Abe and Fukuda on Government Changes

SEPTEMBER 12 AND OCTOBER 1, 2007

Japan's creaky political system—and the party that had dominated it for more than a half-century—underwent extraordinary stress in 2007. The central events were the ruling party's loss of control of the upper chamber of parliament and the subsequent resignation of Prime Minister Shinzo Abe just one year after he took over from his popular predecessor, Junichiro Koizumi. Moreover, late in the year the political system was stymied for weeks in a controversy over Japanese logistical support for U.S.-led military operations in Afghanistan.

The political dissension mirrored continuing unease in Japanese society about slow economic growth (particularly in rural areas), declining population, and the challenges posed to Japan by the rapid rise of China. Six decades after it began a remarkable recovery from the ravages of World War II, Japan in 2007 was confronting serious questions about how long its success could be sustained and whether its political system was up to the challenges the nation faced.

Abe's Fall

Abe's misfortune was to follow in the footsteps of Koizumi, Japan's most popular and successful prime minister in decades. Koizumi had been unconventional in the traditionally dour world of Japanese politics: he was charismatic and challenged the "old boy" network of his own party, the Liberal Democratic Party (LDP), which had ruled Japan, with only one brief exception, for a half-century.

Abe, who was fifty-one when he succeeded Koizumi in September 2006, represented the younger generation and was Japan's first prime minister born after World War II. But in other respects, he was a throwback to the more conventional political system that Koizumi had sought to break: a system dominated by a conservative (despite its name) probusiness party that really was a coalition of nine factions, each headed by an aging leader whose influence was based largely on patronage. Abe's family had played a significant role in that system: his grandfather and an uncle each served as prime minister, and his father, a foreign minister, was deprived of a shot at the top job only by his early death from cancer. Abe himself had followed the conventional political path of serving in various government posts before he won the nod from party leaders, and the party's 1 million members, to follow Koizumi once he voluntarily stepped aside.

Abe's one year in office had two general themes, one of his choice and the other a result of his government's actions: patriotism and incompetence. Whereas Koizumi had sought to shake up Japan by breaking the government's hold over much of the economy through patronage and businesses that were owned or controlled by the government, Abe used nationalistic rhetoric to try to revive Japan's sense of greatness. He pledged to instill pride in Japan as a "beautiful country," a reference to its culture and history, not just its landscape. More specifically, he promised to revise the country's constitution to reflect changes in Japanese society since U.S. occupation authorities drafted it following World War II.

Once in office, Abe followed through on much of his nationalist rhetoric. He ordered the revision of school textbooks to soften statements about Japan's violations of human rights in China and Korea during the war, and schools revised their curricula to promote patriotic themes. In one of his most controversial acts, Abe on March 1 denied that there was any proof that the Japanese army forced Asian women into sexual slavery during the war—a practice that had been extensively documented by historians and even by Japan's government. Abe later expressed sorrow that the so-called comfort women had been abused, but he insisted that private companies, not the army, were at fault.

The prime minister's tenure also was marred by a string of scandals, all of them resulting from what critics, and even many colleagues, said was his flawed selection of members for his cabinet. The most serious scandal erupted when opposition politicians exposed a years-old problem of government mishandling of more than 50 million pension payments. Abe failed to respond aggressively, making him appear unconcerned about pensioners who were deprived of their full benefits.

Three cabinet members resigned, and the agriculture minister committed suicide in May, as a result of other scandals. Public approval for Abe's performance, which had soared over 70 percent after he took office, dropped steadily during the first half of 2007, reaching just 26 percent in mid-July after yet another scandal symbolized the government's bumbling: the defense minister, Fumio Kyuma, made statements appearing to justify the U.S. use of atomic bombs against Hiroshima and Nagasaki in the closing stages of World War II. Kyuma resigned, but not before more political damage had been done to Abe's government.

Abe also undermined much of the support for his party, the LDP, that Koizumi had won in his campaigns to reform the government, notably the extensive patronage system and the awarding of contracts to businesses friendly to the ruling party. Abe welcomed back into the LDP several politicians whom Koizumi had ousted, suggesting that he favored returning to the old ways in which government business was done behind closed doors.

A Split Parliament

The prime minister might have been able to survive all these political troubles if the electoral calendar did not require elections for half of the 242 seats in the upper chamber of parliament, the House of Councilors. To maintain control of that chamber, the ruling coalition (consisting of the LDP and its partner, the New Komei Party) needed to win 64 of the 121 seats up for election. The election thus was seen as a referendum on the performance of the LDP in general and Abe in particular.

Voting was scheduled for July 22, but Abe pushed through a one-week postponement to give his party's candidates more time to campaign. The extra week did not help. The election turned into a rout, with the LDP losing 21 seats, including those of several of the party's most senior and best-known politicians.

The victor was the relatively new opposition party, the Democratic Party of Japan (DPJ), which won 39.5 percent of the vote—enough to give it control of the upper chamber. The party was led by Ichiro Ozawa, one of Japan's most famous political mavericks. Ozawa had once been a rising star in the LDP but had quit the party in 1993; he subsequently formed two other parties before creating the DPJ. With its control of the upper chamber, the DPJ could block legislation, up to a point. The LDP coalition could use its super-majority in the lower house to override the upper chamber, but such a step had never been taken in Japan's modern history.

In years past, any poor showing for the ruling party in elections usually meant an immediate resignation by the prime minister. Abe refused to step aside, however. While acknowledging that he had to "repent" for mistakes, he added: "To pursue reforms, to build a new country, I have to fulfill my duties as prime minister from now on as well."

Abe reshuffled his cabinet on August 27, bringing back into office several senior LDP politicians who had served in previous governments, including Koizumi's. The step was seen as one of desperation because Abe's personal popularity continued to shrink. Yet more trouble for Abe came early in September, when three more senior officials—two cabinet members and one subcabinet member—were embroiled in scandals, raising new questions about the prime minister's judgment in picking top aides.

The final straw for Abe was a decision by Ozawa's party to seize on a foreign policy matter as a means of demonstrating its new clout. Ozawa said his party would block approval by the upper house of legislation extending, past a November 1 deadline, the legal mandate for Japanese ships to provide fuel, water, and other nonmilitary supplies to ships from the United States and other countries that were operating in the Indian Ocean as part of the U.S.-led campaign to bolster the beleaguered government in Afghanistan. Although this support role did not involve direct military action on Japan's part, it had created domestic controversy because of the country's pacifist constitution. Moreover, some critics said the Japanese military was more heavily engaged than the government admitted, for example by transporting U.S. troops into Iraq.

Abe pledged to fight on behalf of the refueling operation, but he quickly gave up just three days into the new legislative session when Ozawa appeared to be adamant. On September 12, Abe announced that he was resigning despite his previous insistence that he would stay in office. "However, under the current situation it has become very difficult to advance government policy vigorously with the trust and support of the people," he said at a news conference.

Public reaction to Abe's resignation was muted. Many people told pollsters and reporters that the prime minister had finally acknowledged his failures. "I think he was cornered, and he is just not the leader type," Takako Katayama, an employee of a Tokyo securities firm, told the *Washington Post.* Abe entered a hospital the next day, complaining of a stress-related stomach ailment, and stayed there for two weeks. He eventually apologized for "causing trouble" for the public and regretted that he "couldn't meet people's expectations."

Fukuda Takes Office

Abe's resignation left the LDP without an obvious successor because none of the party's leaders appeared to have enough personal popularity to reclaim the ground Abe had lost. However, party leaders quickly selected Yasuo Fukuda, a former oil company official who for a few weeks in mid-2006 had been Abe's main challenger in the intraparty elections for a successor to Koizumi. Like Abe, Fukuda came from a political family: his father, Takeo Fukuda, had been prime minister from 1976 to 1978. But at age seventy-one, Fukuda was a member of the generation of Japanese leaders who were leaving the stage of power in business and politics. Facing one of the party's most serious crises ever, leaders turned to Fukuda as a calming figure who could keep things going and head off likely pressures for early elections; the next elections were not scheduled to be held until September 2009.

Most political analysts said Fukuda's big challenge would be to regain the LDP's traditional bedrock support in rural areas. The party had built that support over the decades through patronage and large-scale public works projects, many of which had no apparent purpose except to create jobs and stimulate the economy. Koizumi had curtailed this type of spending to cut back Japan's massive deficit but in so doing had weakened the LDP's standing in its key constituencies. It was widely expected that Fukuda would restore the old emphasis on public works, even at the risk of expanding the deficit.

The party elected Fukuda as its new chairman on September 23, automatically making him the prime minister because the LDP still had two-thirds of the seats in the lower house of parliament. Fukuda acknowledged that the party had lost public trust and said it could be won back only slowly, "one block at a time." He took office on September 26 and appointed a cabinet consisting primarily of holdovers from Abe's tenure. In his first speech to parliament on October 1, the new prime minister pledged "sincere consultations" with the opposition DPJ, particularly over the still-pending matter of Japan's participation in the Afghanistan mission.

The November 1 deadline for extending that mission came and went without any action in parliament, suggesting that Fukuda's effort to break the impasse over the issue through dialogue had fallen short. The last Japanese refueling of a coalition ship took place in late October, and Fukuda was forced to order Japanese ships to return home.

The early days of November brought a series of events that were exceptionally bizarre by Japanese standards. On the evening of November 2, Fukuda met with opposition leader Ozawa and offered an unprecedented power-sharing arrangement under which the ruling LDP would broaden its coalition to include Ozawa's DPJ. "We must do something to break through this situation," Fukuda later told reporters.

Although Ozawa apparently was willing to consider Fukuda's proposal, fellow members of his party were not. He announced two days later that he was resigning as party leader because he had caused "political confusion" by meeting with Fukuda. In another two days, Ozawa changed his mind and announced that he was staying on.

Fukuda met with President George W. Bush in Washington on November 16 and promised to try again to revive the Indian Ocean refueling mission. He had not succeeded by year's end, however, because Ozawa's party refused to relent. The new prime minister's next major challenge was likely to come in the spring of 2008, when parliament would have to pass a new budget and Ozawa would have a chance to press for early elections.

Following are two documents. First, excerpts from statements by Japanese prime minister Shinzo Abe at a September 12, 2007, news conference during which he announced his decision to resign. Second, excerpts from a speech by the new prime minister, Yasuo Fukuda, to the Japanese Diet (parliament) on October 1, 2007, announcing his government's legislative program.

DOCUMENT

Press Conference by Prime Minister Abe Announcing His Resignation

September 12, 2007

Today I decided that I should resign from my position as Prime Minister of Japan. The House of Councillors election on July 29 delivered a very harsh result. In light of that severe result I resolved to remain in my position out of the conviction that the momentum of reform must not be halted, and that the move to break away from the post-war regimes must not alter its course. I thus continued to make my very best efforts for the country as Prime Minister, right up until the present.

Furthermore, I announced in Sydney that the fight against terrorism, for which the international community has great expectations of the role played by Japan and for which Japan has gained strong appreciation from the international community, must not be suspended. At that time I stressed the need to continue our activity, whatever it takes.

Contributing to the international community lies at the very core of my "Proactive Diplomacy," and it is my responsibility to carry through this policy at all costs. I therefore announced that I would stake my position on doing everything to ensure that there is no suspension of Japan's activity. I stated that I will not cling to my job as Prime Minister. I have always believed that I have to make every possible effort to continue the fight against terrorism, and likewise that I need to work to the fullest to create the appropriate environment for this. I have been fully prepared to relinquish my post if need be to advance these aims.

I wanted to convey my frank thoughts and feelings to President of the Democratic Party of Japan (DPJ) Ichiro Ozawa, and as such, I proposed to him that we have a political party leaders' meeting today. However, it was unfortunate that my proposal was in fact refused. Earlier, President Ozawa voiced his criticism that I failed to receive a public mandate. It was indeed unfortunate.

As such I asked myself what I should do in order to maintain the ongoing fight against terrorism. I decided that the time has come to change the current situation; that my aim should be for the fight against terrorism to continue under the leadership of a right, new prime minister; and that in order to bring change to the current state of affairs it would be better for a new prime minister to attend the upcoming General Assembly of the United Nations. My determination was to continue to advance reform, and it was from this determination that I remained at my post and reshuffled the Cabinet. However, under the current situation it has become very difficult to advance Government policy vigorously with the trust and support of the people. I therefore reached the conclusion that I had to make a breakthrough in the situation by acting to take responsibility.

Just prior to this press conference I addressed the top five executives of the Liberal Democratic Party (LDP) and explained my thoughts and my decision. I gave them instructions to start work today toward the election of a new LDP president, which should take place at the soonest date possible in order to avoid a political vacuum now that I have decided to resign.

I believed that were I to postpone my decision, it would lead to larger disorder in the Diet proceedings. Recognizing that, I concluded that I had to make a decision as soon as possible.

That concludes my comments.

SOURCE: Japan. Prime Minister of Japan and His Cabinet. "Press Conference by Prime Minister Shinzo Abe." September 12, 2007. www.kantei.go.jp/foreign/abespeech/2007/09/12press_e.html (accessed September 16, 2007).

DOCUMENT

Policy Speech to the Diet by Japan's New Prime Minister

October 1, 2007

Introduction

I have been recently appointed as Prime Minister of Japan. I am bracing myself up, in keen awareness of the gravity of taking charge of the government, at a time when we face a period of great change. Putting first and foremost Japan's future development and the stability of people's lives, I will do my utmost to discharge my duties under the coalition government of the Liberal Democratic Party and the New Komeito Party.

In delivering my policy speech, I should like to apologize to all members of the Diet and the entire nation for having inconvenienced the conduct of Diet affairs due to the holding of the LDP presidential election, and I am resolved to make sincere efforts to respond to the Diet.

Conduct of Diet Affairs

The results of the recent ordinary elections for the House of Councillors, which gave the majority to the opposition, were extremely severe for the ruling parties. Under such circumstances, it becomes difficult for the country to advance new policies when the House of Representatives and the House of Councillors take different decisions. It is the mission of politics to safeguard people's lives and uphold the national interests. As the leader in charge of the government, I would like to engage in sincere consultations with opposition parties on important policy issues in conducting the affairs of state.

Recovering Trust in Politics and the Administration

I accept forthright the people's distrust in politics and in the administration. Without the trust of the people, it is impossible to realize any policies or achieve any necessary reforms. Recovering the trust in politics and in the administration is an urgent task.

On the issue of political funding, on which we have received severe criticisms from the people, the ruling parties have just compiled their views on what must be done to improve the situation and enhance transparency in political funding. We would like to have full discussions with the opposition parties. I have given thorough instructions on political funding to the Cabinet members, in recognition of the need for them first to act upright, directing them to conduct strict management of political funds based on law to enable themselves to demonstrate full accountability when questions are raised, and also to fully abide by all of the matters stipulated in the Norms of Ministers, and not to contravene political ethics by abiding by laws and regulations and to observe morality as politicians. I intend to hold myself to a particularly high level of standards in this regard.

Similarly, for civil servants, who serve the greater good, it is essential to maintain the self-awareness that they are public servants, faithfully carry out their duties and avoid any actions that could bring shame upon themselves. In order to recover the trust in the administration, we will ensure that senior officials in particular of each ministry and agency will take charge of their respective duties in full and conduct administration responsibly from the standpoint of the public. At the same time, we will advance reform of the public service system to create a comprehensive system that will enable each and every civil servant to maintain high morals, enhance their capacities, and focus on their work with pride.

If we allow administrative waste and inefficiency to continue without redress, we not only pass the burden on to the next generation, but we will never be able to regain the trust of the people. We will further advance integrated reform of expenditures and revenues, including definitely achieving a surplus in the primary balance of the central and local governments combined in FY [fiscal year] 2011, by promoting stable growth and through measures such as reducing administrative costs. We will continue to advance administrative reforms vigorously in order to create a simple yet efficient government that befits the 21st Century.

Even after implementing expenditure reforms and administrative reforms, for any possible increases in burden caused by social security services and the declining birthrate, we must secure a stable supply of revenue sources, and avoid any shift of the burden onto the shoulders of future generations. We will proceed with full-fledged discussions without delay aiming at a national consensus, and endeavor to realize fundamental reform of the taxation system, including the consumption tax.

Developing a Trustworthy Social Security System

We must ensure that all elements of our social security system, such as pensions, medical care, nursing care and welfare, must serve the interest of the people. Although we are facing an extremely severe fiscal situation, we need to develop a social security system based on the principle of self-reliance and mutual cooperation that remains sustainable into the future and that gives everyone, including the elderly and the young, a sense of reassurance.

The pension issue that has recently surfaced stems greatly from the fact that the perspective of the people was taken lightly. It is important that the pension records of each and every person be reviewed and that pension payments be correctly made. We will work from the perspective of those who receive pensions, and faithfully resolve the various issues that relate to the pension system by conducting a review of the organization and its management.

Since the pension system concerns the entire nation and serves as the foundation for the lives of the elderly, it is essential that we design the system from a long-term perspective to ensure that pension payments will be made stably into the future. I should like to call on the Diet members to resume discussions in the Diet that transcend partisan positions, and engage in transparent and constructive consultations. . . .

Continuing to Advance Reform and Achieving Stable Growth

Japan has been making efforts at structural reform in all areas of its economy and society. A certain degree of success has been achieved, and we have seen a recovery in our economy and an expansion in employment. Still, Japan is facing difficult issues such as the arrival of a society with full-fledged population decline; increasing expenses for social security services caused by falling birth rates and aging society; structural changes in domestic and overseas economy; and global environmental issues. In order to overcome these challenges and create a more mature society, we must advance reforms with our eyes fixed on the future of Japan and revise systems and organizations that are no longer befitting the present times.

Reform and stable economic growth are two wheels on the same axle, and we will promote both of them. In response to changes in the domestic economic environment, the level of interdependence with overseas economies will continue to grow. We will work to promote domestic and foreign investment, and take steps to materialize the Asian Gateway Initiative, which utilizes the strengths of being inside the remarkably growing Asia, and advance measures to make Japan a tourist destination and strengthen our competitiveness in the financial sector. With a view to further development of science and technology, we will promote concentrated investment in strategic areas and enhance measures for human resource development. At the same time, we will advance an intellectual property strategy that will aim to place Japan at the cutting-edge in the world.

Response to the Issues of "Disparities"

In the process of forging ahead with the structural reforms, various problems have arisen that are broadly called the issues of "disparities." Without changing the direction of the reforms, I will never close my eyes to the reality of the situation, and do my utmost to provide solutions to each and every problem in a proper way.

The local regions have fallen into a negative spiral in which the population declines, and as a result, the facilities that support people's daily lives, including schools and hospitals, becomes less convenient, thereby reducing the attraction of the local regions, and contributing to further population decline. In order to get out of this spiral, we have to think about what is necessary to maintain daily life and vitalize industry in accordance with various situations faced by each local region, and pave the way for what is needed.

We will carry out our policies in a coordinated and comprehensive manner, by integrating the implementation systems established under the Cabinet, including the one for regional revitalization, and creating a system to plan and implement strategy for regional revitalization in an integrated way. We will sincerely listen to the voices of people in the local regions through regular exchange of views between the national and local governments, and we will advance structural reforms for regional revitalization, not through pork-barrel spending, but by adding ingenuity to our policies and making attentive responses, including by establishing a Regional Vitality Restoration Organization.

People's lives are not conducted in cities alone. Under the concept of "mutual cooperation," in which the local regions and the cities support each other, we will further transfer power to local governments so that a system will be developed that enables local regions to explore ideas on their own and implement them, and we will strive for local tax and financial reforms so that the local regions can become financially independent as well. Moreover, we will accelerate our deliberations toward realizing the system of a broader regional government (doshu-sei), the final completion of the decentralization of power from the central government to local governments.

For the cities, we will aim at safe and secure urban development, including ensuring safety in the event of major disasters.

Today marks the start of the privatization of the postal service. We will advance this endeavor steadily so as not to cause inconvenience to users. . . .

Diplomacy which Contributes to Peace

Maintaining the solid Japan-U.S. alliance and promoting international cooperation are the foundation of Japan's diplomacy. World peace cannot be realized unless the international community works together in solidarity. Fixing my eyes on the future of the world in a drastically changing international environment, I will carry forward a diplomacy which contributes to world peace, so that Japan will realize its responsibilities commensurate with its national strength in the international community, and become a country which is relied upon internationally. The most pressing issues we are facing are the continuation of the Maritime Self-Defense Force's support activities in the Indian Ocean and the early resolution of issues related to North Korea.

The support activities based on the Anti-Terrorism Special Measures Law are part of the international community's joint efforts to prevent the proliferation of terrorists. They serve the national interests of Japan which depends on maritime transportation for much of its natural resources, and also constitute the responsibilities that Japan should fulfill in the international community. They are highly appreciated by the international community including the U.N., and we have received specific requests to continue these activities from various countries. We will continue to make our utmost efforts to explain in detail the necessity of continuing these activities to the people and the Diet, so as to gain their kind understanding.

The resolution of issues related to the Korean Peninsula is indispensable for peace and stability in Asia. For the denuclearization of North Korea, we will further strengthen coordination with the international community, through fora such as the Six-Party Talks. The abduction issue is a serious human rights issue. We will exert our maximum efforts to realize the earliest return of all the abductees, settle the "unfortunate past," and normalize the relations between Japan and North Korea.

The Japan-U.S. alliance is the cornerstone of Japan's diplomacy, and we will work to further consolidate our relationship of trust. We will steadily implement the realignment of U.S. Forces in Japan, based on the idea of maintaining deterrence and reducing burdens, while listening closely to the earnest voices of local communities including Okinawa, and exerting our best efforts to promote the development of these communities.

It is truly regrettable that a Japanese citizen was killed in Myanmar where the situation became deteriorated. Though Asia is achieving remarkable growth, it also contains such a vulnerability. We will promote active diplomacy toward Asia, so that the consolidation of the Japan-U.S. alliance and the promotion of diplomacy toward Asia will make a resonance and stability and growth will take root in all Asian countries.

With China, we will establish a mutually beneficial relationship based on common strategic interests, and work together to contribute to the peace and stability in Asia. With South Korea, we will further strengthen a future-oriented relationship of trust. With the ASEAN [Association of Southeast Asian Nations] and other countries, we will promote our endeavors, such as economic partnerships, toward further strengthening of our relationships. With Russia, we will endeavor patiently toward resolution of the Northern

Territories issue, while striving to promote exchanges between the two countries.

In order to be able to make further contribution to the international community, we will pursue U.N. Security Council reform and permanent membership on the Security Council, and strive for an early conclusion of the WTO [World Trade Organization] Doha Round negotiations. Based on the principle of "self-reliance and mutual cooperation," we will actively promote assistance on issues such as global environment and poverty by utilizing Official Development Assistance, etc., while maintaining the principle of self-help.

Conclusion—Toward a Society of Self-Reliance and Mutual Cooperation

Even though Japan has come out of the recent stagnant economic situation, it is still undergoing great changes of the times, and faces unclear prospects and uncertainties in various areas spanning from economic, social, international situations to the natural environment. I am sure that not few people are having various anxieties about themselves, their families and the future of their children.

Under such an insecure situation, it is my duty to take the helm of the country and present a broad vision by turning our thoughts to future generations and safeguarding what should be safeguarded, nurturing what should be nurtured, and inheriting what should be inherited.

I will continue to advance reforms from the people's perspective, by fixing my eyes on the vision of Japan's future, and always considering how best to approach that vision.

In continuing to advance reforms, I will implement policies based on the principle of "self-reliance and mutual cooperation." I will conduct politics with warm compassion, based on the concept that it is necessary for the young and elderly people, big companies and small and medium enterprises, and cities and rural areas, to respect each other, and support and help each other, while maintaining the principle of self-help. I am convinced that this will lead us to a country of "hope and reassurance," where the young people have hope for the future and the elderly people have a sense of reassurance. I will give all that I possess and make endeavors, so that we can overcome the fierce currents of the times with the entire nation, and feel for ourselves that we are making steady steps along the path to the future.

I would like to ask from the bottom of my heart for the understanding and cooperation of the entire nation as well as all the members of the Diet.

Source: Japan. Prime Minister of Japan and His Cabinet. "Policy Speech by Prime Minister Yasuo Fukuda to the 168th Session of the Diet." Provisional translation. October 1, 2007. www.kantei.go.jp/foreign/hukudaspeech/2007/10/01syosin_e.html (accessed November 17, 2007).

Other Historic Documents of Interest

From this volume

- The situation in Afghanistan, p. 547

From previous *Historic Documents*

- Abe's inauguration as prime minister, *2006*, p. 582

DOCUMENT IN CONTEXT

CBO Director and President Bush on the Housing Crisis

SEPTEMBER 19 AND DECEMBER 6, 2007

The U.S. housing boom that began in 1995 finally went bust in 2007, and with a bigger crash than almost anyone had expected. A rash of delinquencies on subprime mortgages—those given to the riskiest borrowers—exacerbated a natural correction in an overheated housing market, throwing families out of their homes and triggering a worldwide financial crisis that many analysts feared would tip the U.S. economy into recession in 2008. By the end of 2007, sales of new and existing homes had dropped to their lowest levels in decades and median home prices had declined for the first time since the government began keeping such statistics in the 1950s. By some estimates, 500,000 families were expected to face foreclosure in 2008.

That picture was gloomy enough, but the housing crisis also touched off an unanticipated crisis in worldwide financial markets, where many of the world's largest banks and investment houses had bundled subprime mortgages into complicated securities packages that were then sold to investors on Wall Street and elsewhere. When large numbers of homeowners began to default on their mortgage payments in late 2006 and early 2007, the value of mortgage-backed securities began to plummet. By early August, some of the world's largest banks had been forced to write down billions of dollars in losses, and the availability of credit not only for mortgages but also for other types of loans had constricted. Central banks around the world injected cash into the financial system to reassure jittery investors. In the United States, the Federal Reserve also took steps to loosen credit by lowering key interest rates. Although markets seemed to stabilize in September and early October, dismal earnings reports from major banks, concerns that the subprime problem might have spilled over into bond insurance markets, and suggestions that the "real" economy was slowing renewed concerns that the United States was headed into a recession.

The Bush administration and Congress were slow to respond to the mortgage crisis. In December, President George W. Bush announced a voluntary plan that would provide some relief to some distressed homeowners, and the Federal Reserve announced a new set of regulations for mortgage lenders, which included requiring them to do a better of job of assessing borrowers' ability to repay the loans. But critics noted that while these and other actions might help avert future crises, they were unlikely to stem the tide of expected foreclosures.

The Making of a Crisis

From the 1960s to the mid-1990s, the rate of home ownership in the United States remained roughly stable, at about 65 percent, but from 1995 to 2005 it climbed to 69 percent. That increase translated to an additional 4.5 million homeowners who otherwise would have been renters. The number of investors and second-home owners buying properties also increased. Low interest rates, especially after 2000, were a major factor in the housing boom, making housing more affordable and increasing demand, which led to higher housing prices. During that period, expectations that home prices would continue to rise became something of a self-fulfilling prophecy. As Peter R. Orszag, director of the Congressional Budget Office (CBO), explained to the Joint Economic Committee on September 19, "if people believe that prices will rise, demand for homes increases, which puts upward pressure on prices."

Another major factor in the housing boom was the development of the subprime mortgage. Designed for borrowers who did not qualify for regular, or prime, mortgages because they had poor credit ratings, could not fully document their incomes, or were able to make only small down payments, these subprime mortgages typically had higher interest rates than prime mortgages. Interest on the mortgages could be fixed over the life of the loan or adjusted. Many of the subprime mortgages that were likely to end up in default began with low, or "teaser," rates that ballooned by as much as 30 percent after the first or second year. By one estimate, 70 percent of subprime mortgages carried prepayment penalties, making it more expensive for borrowers to refinance their mortgages with more affordable loans. Only 2 percent of regular mortgages carried such penalties.

According to the Center for Responsible Lending, a nonprofit group working to end abusive lending practices, by the middle of 2007, 7.2 million homeowners had subprime mortgages (or about 14.4 percent of total mortgages), with a total outstanding value of $1.3 trillion. Three factors propelled the growth of the subprime mortgage market. Legislative and regulatory changes made in the 1980s expanded the mortgage brokerage business to new types of companies and loosened restrictions on the interest rates that could be charged. Computerized credit-scoring technologies made it easier for lenders to assess and price risks associated with lending to subprime borrowers. Finally, financial innovations that pooled these mortgages into securities that were then sold to investors spread the risks more widely and encouraged further lending into the subprime market. The subprime market grew quickly, accounting for more than one-fifth of all mortgages (measured in dollar value) originated in 2005 and 2006.

The growth of the subprime market allowed many young and low-income people, who might otherwise have remained renters, to own their own homes. Subprime lenders were particularly active in minority communities. According to the Center for Responsible Lending, more than half of all home loans given to African Americans and more than 40 percent of those given to Hispanics in 2006 were in the subprime category.

But lax—some critics said fraudulent—underwriting practices by some mortgage originators, and a lack of understanding by some borrowers of exactly what they were committing to, meant that many subprime borrowers could probably not afford their monthly payments—especially once the higher interest rates on adjustable rate mortgages (ARMs) kicked in. Advocacy groups faulted loan originators for failing to fully document a borrower's ability to repay; the Center for Responsible Lending estimated that as many as half of all subprime loans had been made without such documentation.

Delinquencies on subprime mortgages began to rise in 2004, and by mid-2007, 17 percent of the holders of subprime ARMs were behind on their payments or in foreclosure. An estimated 1.8 million ARMs (prime and subprime) were scheduled to reset to higher rates in 2007 and 2008, and as many as half a million homeowners were expected to default. Foreclosures could have a devastating effect not only on borrowers but on neighborhoods, lowering the value of surrounding homes and making it more difficult to sell houses in the area. By the end of the year, there were numerous reports of empty foreclosed houses that had been looted and stripped of anything of value.

Meanwhile, in the broader housing market, rising interest rates and the record-high price of housing were beginning to slow home sales and home construction. According to the National Association of Realtors, sales of existing single-family homes were down 13 percent compared with 2006, while new homes sales were projected to fall 23 percent. The Labor Department reported that construction employment had fallen by 257,000 jobs between September 2006 and December 2007. Some areas were affected much more than others, notably states such as California, Florida, and Nevada, where the housing boom had been strongest, as well as states like Michigan and Ohio, which had been hard-hit by problems in the auto and other manufacturing industries. Homeowners everywhere, however, were feeling less wealthy as falling house prices reduced the equity they had in their homes and rising interest rates and a tighter credit market made it more difficult for homeowners to access that equity, either by selling or refinancing their homes or by taking out home equity loans.

Falling home prices could also have repercussions for local property taxes and ultimately the amount of money available in communities for education and other services that rely heavily on property taxes. In communities across the nation, homeowners who had seen the market values of their properties fall below the assessed values were asking to have their homes reassessed and consequently their taxes lowered. "Government has been the beneficiary of increasing home prices," Relmond Van Daniker, the executive director of the Association of Government Accountants, told the *New York Times* in December. "And now they are on the other side of that, and they will have to reduce expenses."

An Industry Relief Plan

Predisposed to market solutions and wanting to avoid any action that might be construed as a bailout of irresponsible lenders and borrowers, the Bush administration did not take any significant steps to ease homeowners' distress until the second half of the year. On August 31, President Bush announced that he was asking Treasury Secretary Henry M. Paulson Jr. and Housing and Urban Development Secretary Alphonso R. Jackson to persuade the mortgage industry to voluntarily help at least some homeowners avoid foreclosure by adjusting the terms of their mortgages.

On December 6, Bush announced that a group of lenders and loan servicers, mortgage counselors, and investors known as the HOPE NOW Alliance had agreed to freeze, for five years, the current interest rates for homeowners who could afford those rates but not the higher payments they would owe as a result of the reset on their adjustable rate mortgages. The rate freeze was part of a set of industrywide standards that could be applied to anyone requesting relief. Under the standards, lenders could also refinance borrowers into new mortgages or move them into what the administration called FHASecure loans, through which the Federal Housing Administration (FHA) would help stressed homeowners with good credit histories refinance their loans. In remarks at the White House, Bush said that the voluntary industry plan could help as many as 1.2 million responsible homeowners avoid foreclosure, but it would "not bail out lenders, real estate speculators, or those who made the reckless decision to buy a home they knew they could not afford."

Several critics of the plan questioned the president's assessment of how many people might be helped, arguing that many of those who would be eligible for the five-year interest rate freeze were already qualified financially to receive cheaper conventional mortgages. The Greenlining Institute, a housing advocacy group in California, estimated that only 12 percent of all subprime borrowers and only 5 percent of minority borrowers were likely to benefit from the rate freeze. "This grossly inadequate plan is likely to harm the president's desire to close the minority homeownership gap and create an ownership society," Robert Gnaizda, the organization's general counsel, told the *New York Times.*

The leading Democratic presidential candidates also took issue with the plan, saying it left too many people without relief. But some liberal Democratic legislators praised the relief effort. "I welcome it," Rep. Barney Frank of Massachusetts, chairman of the House Financial Services Committee, said at a committee hearing. "It is a recognition that the increase in the [foreclosure] rates would cause serious problems and that some public-sector concern with that is appropriate, that the market cannot be left entirely on its own." In November Frank had won House passage of a measure to tighten regulation of mortgage brokers. The legislation would establish a nationwide licensing registry for mortgage brokers, who were currently regulated state by state, and establish minimum standards for making home loans. Among other things, it would prohibit brokers from steering consumers into mortgages that they were unlikely to be able to repay. Under the measure, companies that packaged home loans into securities to be sold on the secondary markets would have greater liability if they bought and sold mortgages that failed to meet the minimum standards set out in the legislation. Similar legislation was pending in the Senate.

Also pending in Congress was legislation requested by the president that would modernize the FHA by lowering requirements for down payments and allowing the agency to insure bigger mortgages in states with high housing costs and to charge risk-based insurance premiums. The House passed the legislation, but the Senate had not acted by the end of 2007. The administration said quick action on the Senate's part would allow the FHA to help an additional 250,000 homeowners in 2008. Just before recessing for the year, Congress did approve legislation that would give temporary tax relief to certain homeowners who refinanced their homes at a lower value.

On December 18, the Federal Reserve Board proposed new regulations to curb unfair and deceptive lending practices. Under the proposed rule, subprime mortgage lenders would have to verify that borrowers had the income and assets they said they had, prepayment penalties would be permitted only in certain circumstances, and creditors would have to establish escrow accounts for taxes and insurance payments. (A significant number of subprime borrowers had defaulted when they found they could not afford to pay the property taxes or insurance policies on their homes.) The proposed rules, which were expected to take effect in the first half of 2008, also sought to curb several abuses common in advertising for subprime mortgages. "Unfair and deceptive practices have harmed consumers and the integrity of the home mortgage market," said Randall S. Krozner, a Federal Reserve governor. "We have listened closely and developed a response to abuses that we believe will facilitate responsible lending."

Following are excerpts from two documents. First, testimony before the Joint Economic Committee on the housing crisis on September 19, 2007, by Peter R. Orszag, director of the Congressional Budget Office. Second, remarks on December 6, 2007, by President George W. Bush, outlining a voluntary industry plan for helping certain homeowners avoid foreclosure.

DOCUMENT

CBO Director Orszag on Turbulence in the Mortgage Markets

September 19, 2007

. . . Housing markets entered a period of sustained growth in the mid-1990s—the rate of home ownership expanded rapidly, and in the early 2000s, housing prices increased dramatically. Since 2005, however, the markets have softened substantially, and in many areas of the country, housing has now entered a deep slump. Sales of new and existing homes have dropped, and many forecasters expect further declines in coming months. The construction of new single-family homes has contracted sharply. The inventory of unsold existing homes has climbed to record levels. At today's sales rates, it will take about nine and a half months to clear the current inventory of existing homes on the market. Home prices have stopped climbing in many areas of the country and have begun to fall in some. Many forecasters now believe that the national average home price could decline significantly before housing markets stabilize.

Those developments have raised a number of important questions. What factors account for the recent slump in housing markets? How will developments in housing affect the rest of the economy? To what extent will consumers retrench in the face of declining home values, and to what extent will turmoil in certain parts of the mortgage markets spill over into other credit markets and affect the intermediation of funds between borrowers and lenders? And how should policymakers respond to the situation?

My testimony reviews issues raised by those questions and comes to the following conclusions:

- Innovations in mortgage markets, including the development of subprime mortgages, permitted many more people to become homeowners by reducing credit restraints. The home ownership rate had varied within a narrow range from the 1960s to the mid-1990s but then increased from about 65 percent in 1995 to about 69 percent in 2006.

- The boom in housing prices between 1995 and 2005 was caused by several factors, including low interest rates, buyers' expectations of price increases, and easier availability of credit, especially through subprime mortgages (which played a particularly prominent role over the past few years).

- Over the past two years, prices have softened, and problems in the subprime market in particular have become apparent. To date, the problems with subprime mortgages are disproportionately concentrated in California, Nevada,

Arizona, and Florida. Other areas of the country, however, have also been significantly affected.

- The turbulence in housing markets could affect the broader macroeconomy through four channels: reduced investment in housing; a reduction in consumer spending because household wealth declines; contagion in financial markets, which can impede business investment and some household spending, especially for consumer durables; and a lessening of consumers' and businesses' confidence about the future, which can constrain economic activity.
  - The available data and evidence suggest that the first two channels (reduced investment in housing and reduced consumer spending because of a decline in wealth) will impose a significant drag but are unlikely, by themselves, to tip the economy into recession. The other two channels—contagion in financial markets and weakened confidence—are more difficult to predict but could pose serious economic risks.
  - The economic outlook is thus particularly uncertain right now. Analysts have lowered their economic forecasts as a consequence of this summer's turmoil in financial markets, and the risk of a recession is heightened. But the most likely scenario involves continued (albeit more sluggish) economic growth, and few analysts expect an outright recession next year. Even the average for the bottom 10 forecasts included in the *Blue Chip* survey (an average of about 50 private sector forecasts) released in early September suggested 2.0 percent real growth in 2008, and not a single forecaster projected negative growth in 2008.
- Policy proposals for addressing the financial difficulties originating in the subprime market could be classified into three categories: sustaining the overall economy, helping homeowners facing foreclosures, and preventing future crises by protecting homeowners and reducing the chances of a recurrence of financial instability.
  - In evaluating policies to achieve those goals, it is important to recognize that although significant problems have arisen, not all current housing and credit policies are broken and that the seeds of future crises are often sown by the reaction to current crises.
  - Policy interventions need to reach an appropriate balance between assisting people at risk from events beyond their reasonable control and allowing people to assume responsibility for the consequences of their own decisions.
  - The challenge is to find ways of correcting the abuses and instability that are now becoming apparent while strengthening successful institutions and continuing the benefits of market innovation.

Background

The current contraction of housing markets comes after several years of extraordinary growth in the residential sector, and the recent slump in housing partly reflects an inevitable correction to more normal levels after that remarkable growth. By 2005, home sales had climbed to record levels. The residential construction industry boomed, and home prices soared in many areas of the country.

Many people who had previously been renters became homeowners. As a result, the rate of home ownership, which had varied within a narrow range from the 1960s to the mid-1990s, increased from about 65 percent in 1995 to about 69 percent in 2006. . . . That rise meant that approximately 4-1/2 million more families that otherwise would have been renters owned their homes. Investors and second-home buyers also purchased a growing number of properties, accounting for more than one-sixth of all first-lien loans to purchase one-to-four-family site-built homes in 2005 and 2006.

The housing boom stemmed from many factors. Low interest rates, both short- and long-term, in the early 2000s spurred demand for houses. The Federal Reserve kept short-term rates low through mid-2004 in an effort to promote growth, as the growth of gross domestic product (GDP) was slow to recover from the recession of 2001 and as some analysts expressed concerns in 2003 about the possibility of deflation. The housing sector is generally more sensitive to interest rates than most other sectors, so the effect of monetary policy is often channeled to the economy through housing markets. Rates for 30-year conventional mortgages, which had averaged 7.6 percent from 1995 through 2000, dropped to 5.8 percent in 2003 and generally remained below 6 percent until the fourth quarter of 2005. The low rates increased the affordability of homes, increased demand, and ultimately caused housing prices to be bid up. More people decided to live in separate households than would have occurred in the absence of the housing boom; that phenomenon both reflects and partially caused that boom.

Homebuyers' expectations of continued and rapid home price inflation also appear to have played a central role in propelling prices upward. If people believe that prices will rise, demand for homes increases, which puts upward pressure on prices. Thus, the expectation of higher prices can become a self-fulfilling prophecy in the short run. But that temporary cycle may not be tied to underlying fundamentals (such as demographic forces, construction costs, and the growth of household income), and in the long run, prices will ultimately evolve back toward becoming aligned with those fundamentals. To the extent that the underlying fundamentals are reflected in rental prices, the ratio of housing prices to rents may provide insight into the degree to which prices are deviating from the fundamentals. The ratio tended to vary within a relatively narrow range between 1975 and 1995 before climbing steeply between 1995 and 2005. . . . To be sure, homebuyers' expectations of home prices may deviate from long-term fundamentals for extended periods of time, as shown by evidence that Professor Robert Shiller of Yale University and others have developed, and the prolonged rise in the ratio of house prices to rents between 1995 and 2005 is consistent with the possibility of such extended deviations of prices from underlying fundamentals.

Another major factor in the housing boom was the plentiful supply of credit, which manifested itself most dramatically in the expansion of the subprime mortgage industry. Subprime mortgages are extended to borrowers who for one reason or another—a low credit rating, insufficient documentation of income, or the capacity to make only a low down payment—do not qualify as prime borrowers. The share of subprime mortgages rose rapidly after 2002, and more than 20 percent of all home mortgage originations (in dollar terms) in the past two years were for subprime loans. By the end of 2006, the outstanding value of subprime mortgages totaled an estimated $1.2 trillion and accounted for about 13 percent of all home mortgages.

Subprime mortgages include fixed-rate mortgages, adjustable-rate mortgages (ARMs), and combinations of the two, such as the 2/28 mortgage, in which the interest rate is fixed for two years and then varies for the 28 years remaining on the life of the loan. Many adjustable-rate loans have so-called "teaser" rates, which offer lower-than-market rates during the loans' early years. Subprime mortgages may be interest-only loans and negative amortization loans, in which the principal can actually grow during the initial years of the loans. A common characteristic of many subprime loans is that they offer borrowers low monthly payments in the loans' early years but higher ones in later years. Prepayment penalties (which impose fees on borrowers who want to pay off the remaining balance on a mortgage early) are common on subprime mortgages that have teaser rates but relatively uncommon on prime mortgages.

Subprime mortgages have provided significant benefits to many borrowers. The availability of subprime mortgages has expanded home ownership, especially in minority and low-income communities. Many borrowers in such communities have low income, have less than stellar credit histories, or can only make down payments that are smaller than prime lenders require. Subprime loans may be particularly appropriate for people whose income is expected to rise—for instance, if they are in the early stages of a career. The number of borrowers with first-lien subprime mortgages has climbed to about 7-1/2 million, and many of them would not have been eligible for a prime mortgage and might not become homeowners in the absence of subprime mortgages. Although the foreclosure rates on subprime mortgages have received a great deal of attention and are higher than those on prime mortgages, over 85 percent of the borrowers who currently hold subprime mortgages (including both fixed-rate and adjustable-rate ones) are still making their payments on time.

The growth of the subprime mortgage industry stemmed from three factors. First, legislative and regulatory changes made in the 1980s lifted constraints on the types of institutions that could offer mortgages and the rates that could be charged. Second, the development of new credit-scoring technology in the 1990s made it easier for lenders to evaluate and price the risks of subprime borrowers. Third, the expansion of the securitization of subprime mortgages allowed the market to bear the risks of those mortgages more efficiently and at lower costs.

As has become apparent, the underwriting standards of some originators in the subprime mortgage market slipped. Some made loans to borrowers who put little money down—and had little to lose if they defaulted—and to borrowers with particularly weak credit histories. Some subprime lenders also required little or no documentation of borrowers' income and assets, and determined borrowers' qualification for mortgages on the basis of initial teaser rates. That approach created opportunities for both borrowers and originators to exaggerate borrowers' ability to repay the loans. Those problems fundamentally stemmed from a failure of lenders to provide the right incentives to and oversight of originating brokers. In the traditional form of mortgage financing, the originator of the loan also holds the loan in its portfolio and therefore has a strong incentive to learn about the borrower's ability to repay. By contrast, in the securitized form of mortgage financing, the originator sells the mortgage to a third party and earns a fee for origination but receives little immediate reward for discovering relevant information about the borrower. As a result, the originator may not have adequate incentives to exercise care and discretion in its underwriting unless the ultimate purchaser carefully structures such incentives.

Some borrowers may also have not understood the complex terms of their mortgages, and some mortgage originators may also have taken advantage of unsophisticated borrowers. Certain adjustable-rate mortgages may have been among the more difficult mortgages for first-time borrowers to understand. Many of those mortgages made in recent years included teaser rates, which may have confused some borrowers about the eventual size of their mortgage payments when their mortgage rates were reset. Most of those mortgages also included prepayment penalties, which protected lenders from the potential churning of mortgages with very low initial rates but also made it more expensive for borrowers to refinance their loans when their monthly payments rose. As Edward Gramlich asked in a speech that was delivered on his behalf just before he died, "Why are the most risky loan products sold to the least sophisticated borrowers?"

The subprime market began to experience growing problems after 2004, when delinquencies on subprime ARMs began to rise. By the second quarter of 2007, almost 17 percent of subprime ARMs were delinquent, up from a recent low of 10 percent in the second quarter of 2005. . . . In addition, the share of subprime ARMs entering foreclosure increased from an average of 1.5 percent in 2004 and 2005 to 3.8 percent in the second quarter of 2007. Although delinquencies have also risen for fixed-rate subprime loans, the level of delinquencies for fixed-rate loans has been lower and its increase has been slower.

Housing markets have weakened throughout the country, but a only few states have had significant increases in foreclosure rates. . . .

Several factors seem to have contributed to the growing delinquencies of subprime mortgages. Mortgage rates moved upward during the period as monetary policy tightened, and some ARM borrowers may have been surprised at how high their mortgage rate became. Many ARM borrowers appear to have defaulted after the initial period of low rates expired and their monthly payments were reset at significantly higher levels. Such ARM borrowers often found it difficult to refinance their mortgages to avoid increasing payments. In addition, some borrowers who had purchased their home with little money down may have seen their equity vanish as home prices began to decline in some areas. In the industrial Midwest, especially in Michigan, those problems were aggravated by the slowdown of the regional economy as the automotive industry retrenched.

The problems have undermined investors' confidence in the securities backed by subprime mortgages. During the boom years, investors may not have fully appreciated the risks of subprime loans and seem to have underpriced them. Investment managers around the globe were seeking securities that offered higher yields but apparently did not fully appreciate the risks that they were taking on. The price that investors charged for taking on risk in the subprime mortgage market, as well as other financial markets, plummeted to abnormally low levels. The rating agencies, too, appear to have not kept up with some fast-emerging problems in the quality of securities backed by subprime loans, and they may have placed undue emphasis on the unusual period of substantial price appreciation in evaluating the risks of mortgage-related securities. This year, when the risks of subprime mortgages were recognized, the prices for securities backed by them dropped sharply. Liquidity in both the primary and secondary markets for subprime mortgage-backed securities has also declined, as some of the country's largest originators of such loans collapsed.

Risks to Individuals and the Broader Economy

The shakeout in housing markets has already affected both individuals and the overall economy. House prices have declined in some areas of the country, mortgage delinquencies and foreclosures have risen, and housing investment has fallen dramatically. The effects that have occurred to date, however, may only be the beginning. Even if the economy manages to maintain a fairly steady pattern of growth, many homeowners will face dramatically higher mortgage payments, which will probably lead to additional foreclosures, and some mortgage investors will experience further losses. Moreover, the problems in the subprime mortgage market have spilled over into the broader financial markets, raising borrowing costs for other mortgage and nonmortgage borrowers and threatening to further depress economic activity. Although the consensus forecast for the economy still indicates real growth of about 2-1/2 percent next year, economists generally agree that the probability of a recession next year has risen and is now quite elevated relative to normal conditions.

Individuals

Mortgage payments, delinquencies, and foreclosures will be a problem for many years as interest rates are reset on prime and subprime ARMs that were originated during the 2004–2006 period. Rates have already been reset for some of those ARMs, and the remaining instances (most of which will occur before the end of 2010) will eventually add about $30 billion to annual payments. Although that increase is not large relative to total household income of $10 trillion, many households will be hard pressed to make the higher payments, and some will become delinquent on their mortgages.

New foreclosures on ARMs have risen over the past year and are likely to remain high for some time. About 1.65 percent of the 8.7 million ARMs (both prime and subprime) included in data tabulated by the Mortgage Bankers Association (MBA) went into foreclosure in the second quarter of this year, about twice the rate during the second quarter of last year. Extending that percentage to all 12.4 million ARMs that were outstanding during the second quarter of 2007 suggests that about 200,000 may have gone into foreclosure.

The rate of new foreclosures in the future depends upon a wide variety of factors, particularly the overall state of the economy and housing prices, so forecasts vary widely—from an additional 1 million over the next few years to more than 2 million. The lower estimates suggest that the pace of foreclosures may slow next year, reflecting the fact that many of the recent foreclosures stem from the expiration of extremely low and very short-term (one- to six-month) teaser rates on some ARMs. Such mortgages will not have as large an effect on the overall foreclosure rates in the future as they have had recently. The higher estimates, however, reflect a concern about the outlook for the overall economy and the possibility that a negative cycle may develop—higher rates of foreclosure may depress housing prices, undermining efforts to refinance mortgages, pushing more homes into foreclosure, and lowering prices further.

Individuals who owned assets that were affected by the recent turmoil in financial markets have experienced losses as a result of the problems in mortgage markets. No data are available about how the losses were distributed among various categories of investors—domestic or foreign, individuals or institutions—nor about how pension funds may have been affected.

The Broader Economy

The problems created by mortgage markets threaten to slow economic activity, possibly by a substantial amount. Four channels exist through which the turbulence in housing markets could affect the broader economy:

- **Reduced Housing Investment.** Between 1995 and 2005, investment in residential housing directly contributed an average of 0.3 percentage points per year to economic growth. The slump in residential housing has already weakened the economy, and more weakness in the housing market could constrain growth further by reducing that source of investment.

- **Less Consumer Spending Based on Housing Wealth**. Lower house prices also are likely to weaken economic activity through the housing wealth effect: Reduced housing wealth causes a decline in consumer spending. The effect could be somewhat larger than expected if households have increased difficulty withdrawing equity from their homes.
- **Contagion in Mortgage and Financial Markets**. Higher mortgage rates and weaker house prices, contributing to higher foreclosure rates and losses for mortgage lenders, threaten to precipitate a spiral of tighter mortgage standards, lower house prices, and more foreclosures. The broader spillover, or contagion, of the subprime mortgage problems into other credit markets, causing stricter standards and terms for other types of borrowing, could reduce economic activity by weakening business investment.
- **A Decline in Consumers' and Businesses' Confidence**. A slowdown in economic activity and employment growth triggered by the problems in mortgage markets, especially if associated with spillover effects in financial markets, could weaken consumers' and businesses' confidence about income growth in the future. Such a reaction could then constrain economic activity further.

Those various channels through which the problems in mortgage markets could spread to the broader economy make the current situation particularly uncertain; the potential effects involving contagion and confidence are especially difficult to evaluate because they depend in part on how financial market participants, consumers, and business executives perceive the situation. . . .

SOURCE: U.S. Congress. Congressional Budget Office. "Turbulence in Mortgage Markets: Implications for the Economy and Policy Options." Prepared testimony by Director Peter R. Orszag at a hearing by the Joint Economic Committee, U.S. Congress. September 19, 2007. www.cbo.gov (accessed January 29, 2008).

DOCUMENT

President Bush on the Housing Market and the Government's Response

December 6, 2007

Good afternoon. . . .

I just had an important discussion on the housing market with [Treasury] Secretary [Henry M.] Paulson [Jr.], [Housing and Urban Development] Secretary [Alphonso R.] Jackson, and members of the mortgage industry.

The housing market is moving through a period of change. In recent years, innovative mortgage products have helped millions of Americans afford their own homes, and that's good. Unfortunately, some of these products were used irresponsibly. Some lenders made loans that borrowers did not understand, especially in the subprime sector. Some borrowers took out loans they knew they could not afford. And to compound the problem, many mortgages are packaged into securities and sold to investors around the world. So when concerns about subprime loans begin to mount—began to mount, uncertainty spread to the broader financial markets.

Secretary Paulson and Secretary Jackson and [Federal Reserve] Chairman [Ben S.] Bernanke are monitoring developments in the housing market and working to limit the disruption to our overall economy. Data released this morning confirmed the difficulties facing the housing market. Yet one reason for confidence is that the downturn in housing comes against a backdrop of solid fundamentals in other areas, including low inflation, a healthy job market, record-high exports. America's economy has proved itself highly resilient. And it is strong, and it is flexible, and it is dynamic enough to weather this storm.

For individual homeowners, the problem is more difficult. Many of those feeling financial stress have an adjustable rate mortgage, which typically starts with a lower interest rate and then resets to a higher rate after a few years. Many of those borrowers cannot afford the higher payments. And now some are fearing foreclosure, which is a terrible burden for hard-working families and a source of concern for communities and neighborhoods across our country.

The rise in foreclosures would have negative consequences for our economy. Lenders and investors would face enormous losses, so they have an interest in supporting mortgage counseling and working with homeowners to prevent foreclosure.

The Government has a role to play as well. We should not bail out lenders, real estate speculators, or those who made the reckless decision to buy a home they knew they could never afford. Yet there are some responsible homeowners who could avoid foreclosure with some assistance. And in August, I announced a series of targeted actions to help them. My administration has moved forward in three key areas.

First, we've launched a new initiative at the Federal Housing Administration called FHASecure. This program gives the FHA greater flexibility to offset refinancing to homeowners—to offer refinancing to homeowners who have good credit histories but cannot afford their current payments. In just 3 months, the FHA has helped more than 35,000 people refinance. And in the coming year, the FHA expects this program to help more than 300,000 families.

Second, in August, I asked Secretaries Paulson and Jackson to work with lenders and loan servicers and mortgage counselors and investors on an initiative to help struggling homeowners find a way to refinance. They assembled a private sector group called HOPE NOW Alliance. Their leaders are with us today. HOPE NOW is an example of Government bringing together members of the private sector to voluntarily address a national challenge, without taxpayer subsidies or without Government mandates. I'm pleased to announce that our efforts have yielded a promising new source of relief for American homeowners.

Representatives of HOPE NOW just briefed me on their plan to help homeowners who will not be able to make the higher payments on their subprime loan once the interest rates go up, but who can at least afford the current starter rate. HOPE NOW members have agreed on a set of industry-wide standards to provide relief to these borrowers in one of three ways: by refinancing an existing loan into a new private mortgage; by moving them into an FHASecure loan; or by freezing their current interest rate for 5 years.

Lenders are already refinancing and modifying mortgages on a case-by-case basis. With this systematic approach, HOPE NOW will be able to help large groups of homeowners all at once. This will bring more relief to more homeowners more quickly. HOPE NOW estimates there are up to 1.2 million American homeowners who could be eligible for this assistance.

Public awareness is critical to this effort because the group can only help homeowners who ask for it. So HOPE NOW recently mailed hundreds of thousands of letters to borrowers falling behind on their payments, and they have set up a counseling hotline that Americans can call 24 hours a day. I've directed Secretaries Paulson and Jackson to expand the public awareness campaign. And I have a message for every homeowner worried about rising mortgage payments: The best you can do for your family is to call 1-800-995-HOPE [1-888-995-HOPE—White House correction]. . . .

Third, the Federal Government is taking several regulatory actions to make the mortgage industry more transparent, reliable, and fair. Later this month, the Federal Reserve intends to announce stronger lending standards that will help protect borrowers. At the same time, HUD and the Federal banking regulators are taking steps to improve disclosure requirements so that homeowners can be confident that they are receiving complete, accurate, and understandable information about their mortgages.

As we take these steps, the Department of Justice will continue to pursue wrongdoing in the banking and housing industries so we can help ensure that those who defraud American consumers face justice.

These measures will help many struggling homeowners, and the United States Congress has the potential to help even more. Yet in 3 months since I made my proposals, the Congress has not sent me a single bill to help homeowners. If Members are serious about responding to the challenges in the housing market, they can start with the following steps.

First, Congress needs to pass legislation to modernize the FHA. In April 2006, I sent Congress an FHA modernization bill. This bill would increase access to FHA-insured loans by lowering down payment requirements, allowing the FHA to insure bigger mortgages in high-cost States, and expanding FHA's authority to price insurance fairly with risk-based premiums. This bill could allow the FHA to reach an additional 250,000 families who could not otherwise qualify for prime-rate financing. Last year, the House passed the bill with more than 400 votes, and this year, the House passed it again. Yet the Senate has not acted. The liquidity and stability that FHA provides the market are needed more than ever, and I urge the United States Senate to move as quickly as possible on this important piece of legislation.

Second, Congress needs to temporarily reform the Tax Code to help homeowners refinance during this time of housing market stress. Under current law, if the value of your house declines and your bank forgives a portion of your mortgage, the Tax Code treats the amount forgiven as taxable income. When you're worried about making your payments, higher taxes are the last thing you need. The House agrees and recently passed this relief with bipartisan support. Yet the Senate has not responded. This simple reform could help many American homeowners in an hour of need, and the Senate should pass it as soon as possible.

Changing the Tax Code can also help State and local governments do their part to help homeowners. Under current law, cities and States can issue tax-exempt bonds to finance new mortgages for first-time home buyers. My administration has proposed allowing cities and States to issue these tax-exempt mortgage bonds for an additional purpose: to refinance existing loans. This temporary measure would make it easier for State housing authorities to help troubled borrowers, and Congress should approve it quickly.

Third, Congress needs to pass funding to support mortgage counseling. Nonprofit groups like NeighborWorks provides essential services to—by helping homeowners find affordable mortgage solutions and prevent foreclosures. My budget requests nearly $120 million for NeighborWorks and another 50 million for HUD's mortgage counseling programs. Congress has had these requests since February, yet it has not sent me a bill, and they need to get the funding to my desk.

Fourth, Congress needs to pass legislation to reform Government-sponsored enterprises [GSEs] like Freddie Mac and Fannie Mae. These institutions provide liquidity in the mortgage market that benefits millions of homeowners, and it is vital they operate safely and operate soundly. So I've called on Congress to pass legislation that strengthens independent regulation of the GSEs and ensures they focus on their important housing mission. The GSE reform bill passed by the House earlier this year is a good start, but the Senate has not acted, and the United States Senate needs to pass this legislation soon.

The holidays are fast approaching, and unfortunately, this will be a time of anxiety for Americans worried about their mortgages and their homes. There's no perfect solution, but the homeowners deserve our help. And the steps I've outlined today are a sensible response to a serious challenge. I call on Congress to move forward quickly and join with me in delivering relief to homeowners in need so we can keep our economy healthy and the American Dream alive.

God bless.

SOURCE: U.S. Executive Office of the President. "Remarks Following a Meeting with Secretary of the Treasury Henry M. Paulson, Jr., and Secretary of Housing and Urban Development Alphonso R. Jackson." December 6, 2007. *Weekly Compilation of Presidential Documents* 43, no. 49 (December 10, 2007): 1567–1569. Washington, D.C.: National Archives and Records Administration. www.gpoaccess.gov/wcomp/v43no49.html (accessed January 29, 2008).

OTHER HISTORIC DOCUMENTS OF INTEREST

FROM THIS VOLUME

- Struggling auto industry, p. 232
- U.S. economic situation, p. 449
- World economic outlook, p. 603

FROM PREVIOUS *HISTORIC DOCUMENTS*

- Outlook for the U.S. economy, *2006*, p. 151

DOCUMENT IN CONTEXT

Reports on Violence and Unrest in Myanmar

SEPTEMBER 19, NOVEMBER 7, AND DECEMBER 7, 2007

For a few days in late September, Myanmar (also known as Burma), one of the most closed societies on Earth, captured the world's attention as its military government brutally suppressed public protests led by Buddhist monks. The protests were the largest and most important in Myanmar since 1988, when students led similar antigovernment demonstrations.

The Burmese generals' harsh crackdown on dissent was widely condemned around the world, even bringing a rare rebuke by the country's key ally, China. At least in the short term, however, the generals who ran Myanmar succeeded in crushing opposition and blocking any prospects for a democratic opening.

Background

Myanmar, a Southeast Asian nation of about 45–50 million people, had been under military rule since 1962, when generals led by Ne Win seized power. For more than twenty-five years, General Win ruled with an iron hand and imposed his own version of socialism based on isolating Burma from the rest of the world. But in 1988, a sharp increase in the price of rice and a devaluation of the currency sparked student-led, large-scale protests around the country. As many as 3,000 people reportedly were killed by the military during the protests, which led a group of generals to oust Win. In subsequent years, a junta of generals ruled Burma in almost total secrecy, first as the State Law and Order Restoration Council (SLORC) and later under the title of State Peace and Development Council, headed by General Than Shwe starting in 1992. The military changed the name of the country from Burma to Myanmar in 1989.

The student leaders of the 1988 protests formed the 88 Generation Students Group, which became one of Burma's most important opposition groups, along with a political party, the National League for Democracy. This party was led by Aung San Suu Kyi, who was under house arrest at the time. Suu Kyi was the daughter of the guerrilla leader Aung San, who won independence for Burma after World War II.

The ruling generals agreed to allow elections in May 1990. When the results showed the opposition National League for Democracy winning by a landslide, the government annulled the elections. Suu Kyi was awarded the Nobel Peace Prize in 1991 in an apparent effort by the Nobel committee to focus world attention on repression in Myanmar. The government freed her in 1995, but sent her to prison in 2000 for two years, then sent her back into house arrest in 2003 after she resumed political activity. For much of the time, the government prevented her from having any visitors or any form of contact with her political allies. One other Burmese political prisoner had been held even longer: poet and journalist U Win Tin had been imprisoned since July 1989. Other opposition figures also spent years in prison.

Ever since its suppression of the 1990 election, the government had claimed to be working on a new constitution that it said would lead to new elections and restore democracy. Opposition figures boycotted the process starting in 1996, and work was suspended until 2004. As the 2007 protests were getting under way, the government on September 3 unveiled a list of "principles" for a new constitution that appeared aimed at solidifying the military's monopoly of political power.

Another illustration of the military's determination to retain power at any cost was its secret decision to move the government from Yangon (or Rangoon) to a remote mountain town, Naypyitaw ("abode of kings"). The relocation was announced in November 2005, after the government had been moved. One apparent reason for the move was to make it difficult for civil servants to participate in protests, as many had in 1988.

Before the events of 2007, United Nations (UN) officials and human rights organizations said the Burmese government was holding more than 1,100 political prisoners, most of them student leaders from the 1988 demonstrations and members of Suu Kyi's party. Like North Korea, Myanmar was closed to nearly all outside contact and influences. With few exceptions the government banned foreign journalists, and it prohibited the domestic circulation of foreign newspapers and magazines and reception of foreign broadcast media, including through the Internet.

Buildup of Protests

Although the nation had abundant natural resources, including one of Asia's largest reserves of natural gas, and once was known as Asia's "rice bowl," Myanmar under the generals' rule had declined economically—in contrast to Malaysia, Thailand, and other countries in the region whose economies had exploded in recent decades. By 2007 the annual per capita income in Burma was less than $200, one of the lowest in Asia, and an estimated one-third to one-half of the population lived in what the UN categorized as severe poverty.

On August 15, without public announcement, the government removed subsidies of gas used for cooking, diesel fuel, and gasoline. This resulted in price increases of up to 500 percent; bus fares and some prices for food and other commodities more than doubled.

Several small protests took place in subsequent days, led by the 88 Generation Students Group. Overnight on August 22, the government arrested thirteen protest leaders, including Min Ko Naing, the most prominent opposition leader not then in prison. The government later announced that the thirteen had been charged with terrorism and would face long prison terms.

The following day, several dozen people marched again in Yangon before police intervened to break up the demonstration. More demonstrations took place in the following weeks, notably after August 28, when Buddhist monks joined a protest in the city of Sittwe, about 300 miles northwest of Yangon. The government sought to suppress these demonstrations by arresting dozens of protesters and their leaders. The government also employed gangs of men, called the Swan-ar Shin (or Masters of Force), to pose as civilian onlookers who would break up demonstrations.

Although the government tightly controlled domestic reporting by newspapers and broadcast media, opposition leaders were able to circulate descriptions and photographs of the demonstrations nationally and internationally via the Internet. One of the first international condemnations of the crackdown came from the White House, which issued a statement on August 30 by President George W. Bush denouncing the government for "arresting, harassing, and assaulting prodemocracy activists for organizing and participating in peaceful demonstrations." U.S. officials later said they had received reports that arrested demonstrators were beaten and interrogated harshly.

The situation escalated after September 5, when the government used violence to quell a protest by monks in Pakokku, a religious center about 400 miles northwest of Yangon. Several monks reportedly were beaten, and unconfirmed rumors spread that one monk was killed. The next day, a group of monks briefly took hostage a group of government officials who came to a Pakokku monastery.

On September 9, the All Burma Monks Alliance, a new group, issued a statement demanding that the government apologize for the violence in Pakokku, reduce fuel prices, release all political prisoners, and enter into a dialogue with opposition groups. The government refused to respond to these demands by the September 17 deadline set by the monks, so hundreds of monks in Yangon and the nearby city of Bago began large demonstrations the next day—the anniversary of the current regime's coming to power in 1988. These and subsequent demonstrations over the next eight days in Yangon, Mandalay, Pakokku, and Sittwe were peaceful, with the monks chanting and marching while police stood by.

At a protest in Yangon on September 18, a leader of the monks told his colleagues, according to a recording transcribed by the Asian Human Rights Commission in Hong Kong: "The clergy boycotts the violent, mean, cruel, ruthless, pitiless kings, the great thieves who live by stealing from the national treasury. The clergy hereby also refuses donations and preaching." At this and other demonstrations, the monks carried their begging bowls upside down, a symbolic rebuke to the government signaling that they refused to accept donations.

A key event took place in Yangon on September 22 when, in a driving rain, a group of several hundred protesters were allowed to pass through a police checkpoint and walk by the house of Suu Kyi, who greeted them at her gate. This was the first time she had been seen in public in four years, and the sight of her energized the protesters. A photograph of Suu Kyi, her hands clasped in prayer, barely visible behind policemen in their riot gear, was transmitted around the world via the Internet and became an iconic image of the protests.

The demonstrations swelled into a giant crowd of an estimated 20,000 demonstrators in Yangon on September 23 and nearly 100,000 the following day. Led by monks and nuns, these were by far the biggest protests in the country since the 1988 upheaval. However, police prevented protesters from repeating the march past Suu Kyi's house.

The protests on September 24 brought a warning from the generals that their patience was wearing thin. In a televised statement, the general who served as religious affairs minister warned senior monks that the government would take action "according to the law" if the protests were not curtailed. This warning reminded protesters that public meetings of more than five people were against the law. The government followed this warning the next day by placing security forces around temples in Yangon and other cities and by imposing all-day curfews in Yangon and Mandalay. Even so, thousands of people again marched in the streets on September 25.

A Violent Crackdown

A violent crackdown—something that had seemed increasingly likely as the protests built momentum—came on September 26. With an estimated 5,000 monks in the lead, tens of thousands of protesters gathered at or near the two main pagodas in Yangon. Security forces attacked the protesters with tear gas and smoke grenades, rubber bullets and live ammunition, and they beat them with rubber batons and wooden and bamboo sticks. Police arrested several hundred people, many of them monks. Eyewitness reports said students and other protesters tried to come to the aid of monks who were being attacked by the police, only to be attacked themselves.

The government acknowledged that one person died and three people were wounded but otherwise played down the extent of the violence. Antigovernment groups claimed that more people had been killed, perhaps as many as eight, and they provided details of the crackdown to foreign news agencies. After these initial reports, the government disconnected mobile telephone lines and closed down Internet providers in an effort to keep photographs, videos, and written reports about the crackdown from reaching the rest of the world. Among the dead was Japanese news photographer Kenji Nagai, the Japanese embassy later reported.

The crackdown continued the next day as security forces raided monasteries, arresting many monks and forcing others to leave, and attacked a crowd of protesters in Yangon—a much smaller crowd than had gathered on the previous day. Several people were reportedly killed.

The government's use of force succeeded in quelling the protests. By September 28, the demonstrations had dwindled and in the following days stopped altogether as thousands of police were deployed in Yangon and other cities. Even so, the government continued to arrest political opponents well into October and did not lift its night-time curfew until October 20.

Myanmar's minister of home affairs told the UN special rapporteur on human rights, Paulo Sérgio Pinheiro, in mid-November that 2,927 people had been arrested, all but 91 of whom had been released. In a report to the UN General Assembly on December 7, however, Pinheiro estimated that 3,000–4,000 people actually had been arrested, with 500–1,000 of them still in detention. The most important opposition leaders were housed at Insein prison in Yangon, which was known for its poor conditions and abusive treatment of detainees, he said.

Pinheiro also reported that the government had detained family members or friends of protesters who had gone into hiding, a practice he said amounted to "hostage-taking" intended to pressure the protesters to turn themselves in. In addition, security forces reportedly raided fifty-two monasteries around the country, beating monks and arresting large numbers of them. The government announced that it had detained 533 monks as of October 5, of whom 398 had been released. However, the government said it continued to hold "bogus monks" who had infiltrated the ranks of "peaceful monks." It defined the bogus monks as "violent persons and political opportunists instigated by foreign media."

In the weeks that followed, a major point of contention between the government and outside observers concerned how many people had been killed in the protests. Shortly after the crackdown, the government said no more than ten people had died but later revised that figure upward to twelve and then to fifteen. Opposition figures, foreign diplomats, and Burmese exiles insisted the real total was many times higher. In his

December 7 report, the UN special rapporteur said he had received information on another sixteen deaths, and he noted allegations that an unknown number of bodies were cremated secretly immediately after the September 26 crackdown at a crematorium under police control.

Myanmar's government made only a few statements about the protests, one of which came in a November 5 letter to the UN from the country's ambassador, Kyaw Tint Swe. He diminished the importance of the protests, saying they had been "sullied by political opportunists, who sought . . . a political showdown with the aim of derailing the government's seven-step road map to democracy."

UN Intervention

The events in Myanmar created a difficult situation for the UN, various components of which had long pressed the generals to ease their repression. But UN action was limited by the reluctance of some key member states—notably China and Russia, veto-bearing members of the Security Council—to intervene in what they viewed as Myanmar's internal affairs. This had been the case in January when the United States and several European countries offered a resolution at the Security Council condemning Myanmar's human rights record, only to have it scuttled by a joint veto from China and Russia.

Secretary-General Ban Ki-moon, a former foreign minister of South Korea, responded to the crackdown in late September with strong denunciations, such as in an October 5 statement calling it "abhorrent and unacceptable." Ban also sent his special envoy on the country, former Nigerian foreign minister Ibrahim Gambari, to Burma immediately after the crackdown. Gambari arrived on September 29 and during a four-day stay was able to meet with Gen. Than Shwe and other top officials. He also met twice with Suu Kyi, the second time after his meeting with Than Shwe.

Reporting to the Security Council on October 5, Gambari sought to rebuff contentions by the Burmese generals—implicitly supported by China—that political events in Myanmar were of no concern to other countries. "As I have said before, the world is not what it was twenty years ago, and no country can afford to act in isolation from the standards by which all members of the international community are held," he told the council. "It is therefore essential for Myanmar's leadership to recognize that what happens inside the country can have serious international repercussions." Six days later, on October 11, China for the first time joined a Security Council statement—not a formal resolution—deploring the use of violence against demonstrators and calling for the release of all political prisoners.

One apparent result of Gambari's visit was an agreement by the government to meet with Suu Kyi. The deputy labor minister, retired general Aung Kyi (no relation), met with her three times later in the year but made no apparent concessions except to allow a brief meeting between Suu Kyi and some members of her party on November 9. Suu Kyi was allowed to issue just one public statement: on November 8, she told Gambari, during his follow-up visit to Myanmar, that she was willing to cooperate with the government "in the interest of the nation."

U.S. Sanctions

The United States since 1997 had maintained a series of sanctions against Myanmar, including an arms embargo that blocked U.S. arms sales to the country but did not stop extensive equipping of Myanmar's army by China and India, among other countries. The *Washington Post* on December 31 reported that India had halted its arms sales to Myanmar, a step that some analysts said could put pressure both on the Myanmar generals and on China.

President Bush on September 24 imposed new sanctions targeting the junta's top fourteen leaders, including Gen. Than Shwe. The Bush administration on October 19 added eleven other names to the list of Myanmar officials prohibited from doing business with U.S. banks and other entities. These measures were expected to have only limited impact, however; most experts believed Myanmar's top generals had enriched themselves through corruption and kept their fortunes in secret accounts in Hong Kong and Singapore.

The Karen Separatist Movement

Deep in the jungles of southeast Burma, near the border of Thailand but far from the cities of either country, the longest-running civil war in the world continued to play out in 2007. For more than sixty years, a succession of Burmese governments had fought a secessionist movement known as the Karen National Union, which was fighting for the independence of Karen state. The government and the guerrillas agreed in January 2004 to end the fighting, but the two sides did not sign a cease-fire and the conflict continued at a low level.

By 2007 the fighting had displaced an estimated 500,000 people, tens of thousands of whom had fled across the border into Thailand, where they lived in refugee camps. Fighting between the separatists and militias allied to the junta heated up in April, displacing several hundred more Burmese. The area also was the site for major hydroelectric projects being developed jointly by Burma and Thailand. Plans called for three new dams that would displace more than 80,000 people, most of them in Burma.

Following are three documents. First, an English-language transcript of a statement made by Buddhist monks during a demonstration in Yangon on September 19, 2007, calling on all monks to boycott the government. Second, excerpts from a Memorandum on the Situation of Human Rights in the Union of Myanmar, sent to the secretary-general of the United Nations on November 7, 2007, by Kyaw Tint Swe, the ambassador to the UN from Myanmar. Third, excerpts from a report to the UN Human Rights Council, dated December 7, 2007, from Paulo Sérgio Pinheiro, the UN special rapporteur on human rights in Myanmar.

DOCUMENT

Statement by Burmese Monks at a Protest in Yangon

September 19, 2007

Reverend clergy, may you listen to my words. The violent, mean, cruel, ruthless, pitiless kings [military leaders]—the great thieves who live by stealing from the national treasury—have killed a monk at Pakokku, and also arrested reverend clergymen by trussing them up with rope. They beat and tortured, verbally abused and threatened them. The clergy who are replete with the Four Attributes [worthy of offerings, hospitality, gifts and salutation] must boycott the violent, mean, cruel, ruthless, pitiless soldier kings, the great thieves who live by stealing from the national treasury. The clergy also must refuse donations (of four types) and preaching. This is to inform, advise and propose.

Reverend clergy, may you listen to my words. The violent, mean, cruel, ruthless, pitiless soldier kings—the great thieves who live by stealing from the national treasury—have killed a monk at Pakokku, and also arrested reverend clergymen by trussing them up with rope. They beat and tortured, verbally abused and threatened them. Clergy replete with the Four Attributes—boycott the violent, mean, cruel, ruthless, pitiless kings, the great thieves who live by stealing from the national treasury. Clergy—also refuse donations and preaching. If the reverends consent and are pleased at the boycott and refusal of donations and preaching, please stay silent; if not in consent and displeased, please voice objections.

[Silence.]

The clergy boycotts the violent, mean, cruel, ruthless, pitiless kings, the great thieves who live by stealing from the national treasury. The clergy hereby also refuses donations and preaching."

SOURCE: Asian Human Rights Commission. "Burma: The Alms Bowl and the Duty to Defy." Press release, AS-228-207. September 19, 2007. www.ahrchk.net/statements/mainfile.php/2007statements/1203 (accessed January 17, 2008).

DOCUMENT

Letter to the United Nations from Myanmar's UN Ambassador

November 7, 2007

I. Introduction

1. The General Assembly on 22 December 2006 adopted resolution 61/232, "Situation of human rights in Myanmar". The country-specific resolution initiated by the European Union was rejected by Myanmar.

2. In the debate in the Third Committee [of the General Assembly] on the draft resolution, several delegations voiced their opposition to country-specific resolutions on human rights. It was pointed out that such resolutions not only contravene the letter and spirit of General Assembly resolution 60/251, establishing the Human Rights Council, but also run counter to General Assembly resolution 61/166 on the promotion of equitable and mutually respectful dialogue on human rights. It was stressed that, as the Human Rights Council had already been instituted to address human rights issues in the global context, it would be redundant for the Third Committee to consider country-specific resolutions.

3. On 15 September 2006, the United States took further measures and placed Myanmar on the agenda of the Security Council, alleging that the situation in Myanmar posed a potential threat to international peace and stability. That allegation was not

accepted by a sizable number of Security Council members. As a result, when the United States and the United Kingdom co-sponsored in the Security Council a draft resolution on Myanmar in January 2007, the draft failed to carry owing to the negative votes cast by two permanent members of the Security Council.

4. The United States called for a closed meeting of the Security Council on 5 October 2007 purportedly to be briefed by the Special Adviser to the United Nations Secretary-General, Ibrahim Gambari, on his visit to Myanmar from 29 September to 2 October 2007, following the demonstrations there in September. Subsequently, on 11 October the United Nations Security Council issued a presidential statement on the situation in Myanmar.

5. It is evident that the intention of the countries that initiated the draft resolution on the situation of human rights in Myanmar did so only to channel the domestic political process in the direction of their choosing and not to promote human rights per se.

6. Human rights in a given country can best be promoted through understanding and cooperation, rather than through country-specific resolutions that are confrontational and prescriptive.

7. This memorandum is being circulated to provide factual information on the situation in Myanmar, particularly with regard to the progress made in promoting peace and stability and the transition to democracy.

II. Brief political background of Myanmar

8. Myanmar is a multi-ethnic and multi-religious society. It is home to 8 major ethnic groups comprising over 100 ethnic nationalities. They have always lived in weal and woe throughout history. It was only with the advent of the "divide and rule" policy of the colonialists that seeds of discord were sown among them. This resulted in insurgency which took a heavy toll on the country. The challenges facing Myanmar are delicate and complex, and it would be a grave mistake to conclude that they can be overcome overnight.

9. In 1988 the Government replaced the socialist system and the centrally planned economy with a multiparty system and a market-oriented economy. It initiated reforms and has been making untiring efforts to maintain peace and stability and promote economic and social development.

10. National unity is vital for political, economic and social progress in Myanmar. Accordingly, the Government has been promoting national unity. Peace overtures to armed groups have resulted in the return to the legal fold of 17 out of 18 armed insurgent groups. They are now cooperating with the Government in regional development programmes. The Government has not shut the door on peace talks with the remaining armed group, the Kayin National Union (KNU).

11. With the return of peace and stability the Government is now focusing on the political, economic and social development of the country, particularly in the farflung border areas, which lagged behind the heartland. Most importantly, the Government is focusing on a seven-step road map for transition to democracy.

III. Recent developments in Myanmar

12. There have been notable political developments in Myanmar in recent months. On 3 September 2007 the National Convention successfully concluded its work and adopted basic principles for drafting a new State Constitution.

13. The National Convention was an inclusive forum. It was attended by 1,088 delegates representing political parties, ethnic nationalities, peasants, workers, intellectuals and civil servants. Representatives of the 17 insurgent groups that returned to the legal fold also took part in the process. The basic principles adopted by the National Convention will ensure that the rights of all nationalities will be guaranteed.

14. At the same time, there have also been positive developments in Myanmar's cooperation with the United Nations. The Secretary-General's Special Adviser, Under-Secretary-General Ibrahim Gambari, visited Myanmar from 29 September to 2 October 2007, at the invitation of the Government of Myanmar. He was accorded the opportunity to call on the Chairman of the State Peace and Development Council, Senior General Than Shwe. He also met twice with Daw Aung San Suu Kyi. Following the visit, the Government appointed U Aung Kyi as Minister for Relations to liaise with Daw Aung San Suu Kyi. On 25 October U Aung Kyi met for the first time with Daw Aung San Suu Kyi. A 54-member committee was established to draft a new constitution. The Government has invited both Special Adviser Ambassador Gambari and Special Rapporteur on the situation of human rights in Myanmar Professor Pinheiro to visit Myanmar again in November 2007. A total of 2,677 demonstrators who were briefly detained for questioning have been released. Another 80 more detainees had also been released most recently.

15. Due to the relentless negative media campaign, Myanmar has become an emotive issue. It would however be more constructive to view the situation in a wider perspective rather than through tinted lens. It is no coincidence that the demonstrations took place soon after the successful conclusion of the National Convention, which laid down the basic principles for a new constitution. It is also no secret that the destructive elements both inside and outside the country are strongly opposed to the seven-step road map.

16. The recent demonstrations started with a small group expressing their concern over the rise in fuel price. The situation was sullied by political opportunists, who sought to turn it into a political showdown with the aim of derailing the Government's seven-step road map to democracy. They also took advantage of protests staged initially by a small group of Buddhist clergy demanding apology for maltreatment of fellow monks by local authorities. The destructive elements instigated the march of the few hundred monks chanting prayers and turned it into a political rally. It is completely against the precepts of the Buddhist religion and the code of conduct for the monks to engage in mundane matters, let alone politics.

17. The Government exercised restraint and did not intervene for nearly a month. The security forces were called in to restore law and order only when the mob became unruly and provocative and the situation got out of hand, posing a challenge to the peace and stability of the nation. The situation would hardly have deteriorated had it not been for the subversive acts carried out by political opportunists aided and abetted by their foreign supporters. Citizens who had no wish to disturb the peace and stability stayed away from the demonstrations. They even prevented the demonstrators from entering their townships. A certain political party in collusion with certain western embassies disseminated malicious news. The authorities have since discovered a conspiracy to commit terrorist acts and that some terrorists were involved in a bombing attempt. In this connection, the Government, on 18 October 2007, released information regarding the plots and the seizure of high explosive cartridges from the perpetrators, which include bogus monks.

18. The country has weathered the recent storm and normalcy has been restored. The curfew placed on a few urban centres has been completely lifted. The rule of law is a fundamental principle on which nations are established. Without it there can be neither

the orderly conduct of the day-to-day affairs of state nor the enjoyment of human rights and democracy by the people. Peace and stability are fundamental prerequisites for democracy and economic development. . . .

[Sections IV through XIII deal with Myanmar's relations with the United Nations and specific allegations about human rights violations in Myanmar, all of which are denied.]

XIV. Conclusion

57. It is generally held worldwide that serious violations of human rights occur in situations of armed conflict. In Myanmar, the national reconciliation policy of the Government has resulted in the return to the legal fold of 17 out of 18 insurgent groups. The Government has effectively put an end to the 40 years of insurgency. The prevailing conditions of peace and stability provide better opportunities for the people of Myanmar to enjoy human rights. The sustained endeavours by the Government have also led to significant progress in the economic and social sectors. When the first Human Development Report by UNDP came out in 1990, Myanmar was placed in the third category: country enjoying low human development. The country's steady progress in development was reflected when, starting eight years ago, we were elevated to the second category: country enjoying medium human development. The report also showed that the percentage of the undernourished population has declined from 10 per cent to 6 per cent. The percentage of the population with sustained access to improved sanitation increased from 21 per cent to 73 per cent. Myanmar's adult literacy rate is 93.3 per cent, youth literacy rate is 96.5 per cent and the net primary school enrolment is 84.5 per cent.

58. We have also been making strides in the implementation of our seven-step political road map. The first and crucial step of the road map, the National Convention, has successfully completed the task of laying down the basic principles to be enshrined in the new constitution. Myanmar today is steadfastly proceeding on its chosen path for democracy. The challenges faced by Myanmar are complex and multifaceted. Undue pressure from the outside without fully comprehending the challenges faced by Myanmar can in no way facilitate the country's home-grown political process. It cannot be stressed enough that no one can address the complex challenges facing Myanmar better than its Government and people. The international community can best help Myanmar by lending its understanding, encouragement and cooperation.

Source: United Nations. General Assembly. Sixty-Second Session. Third Committee. "Annex to the Letter Dated 5 November 2007 from the Permanent Representative of Myanmar to the United Nations Addressed to the Secretary-General. Memorandum on the Situation of Human Rights in the Union of Myanmar." November 7, 2007. A/C.3/62/7. www.un.org/Docs/journal/asp/ws.asp?m=A/C.3/62/7 (accessed January 10, 2008).

DOCUMENT

Report on Protests in Myanmar by the Special Rapporteur on Human Rights

December 7, 2007

V. PRELIMINARY FINDINGS

29. As a result of his investigations to date, the Special Rapporteur would like to present the following preliminary findings to the Human Rights Council, recalling that his visit cannot be considered as a full-fledged fact-finding mission, which would require a number of conditions, such as independent access to all places and people, to verify the information collected.

A. Excessive use of force against civilians, including use of unnecessary and disproportionate lethal force

30. The Special Rapporteur found that security forces, including the army and riot police, used excessive force against civilians from 26 to 29 September 2007, in spite of several international appeals calling upon the Government of Myanmar to show restraint in policing the demonstrations. This included the use of live ammunition, rubber bullets, tear gas and smoke grenades, bamboo and wooden sticks, rubber batons and catapults (slingshots). This largely explains the killings and severe injuries that have been reported. Victims included monks, as well as men, women, and children who were either directly participating in the protests or were onlookers in the vicinity. In some cases these beatings were administered indiscriminately, while in other cases the authorities deliberately targeted individuals, chasing them down to beat them. At least one demonstrator, Ko Ko Win, an NLD [National League for Democracy] member, died as a result of injuries sustained when he was beaten near Sule Pagoda in Yangon on 27 September. Allegations of targeted killings and the use of snipers were also received but not yet verified.

31. In a letter dated 1 November 2007, the Special Rapporteur requested from the Government of Myanmar a list of the people who died. The Government has acknowledged the death of 15 people during the demonstrations and provided full details as to the causes of death. However several reports of killings indicate that the figure provided by the authorities may greatly underestimate the reality. To date the Special Rapporteur has received information regarding the killing of 16 additional persons as a result of the crackdown on the demonstrations in September and October, in addition to the 15 individuals included in the information provided by the Government. The Special Rapporteur has transmitted this information to the Government for clarification.

32. According to information received and based on credible eyewitness reports, there were more than 30 fatalities in Yangon associated with the protests on 26-27 September 2007, primarily on 27 September and in the vicinity of Sule Pagoda. No deaths were reported during the demonstrations outside Yangon. According to diplomats more than 500 protesters remain in detention in Yangon, Mandalay, Sittwe, Mytkyina, and Mawlamyine.

33. Among those killed by the security forces during the demonstrations was the Japanese photojournalist, Kenji Nagai. The TV footage of the killing of Mr. Nagai raises

the possibility that he may have been deliberately targeted from a short distance rather than caught in crossfire between the security forces. While the Tokyo Metropolitan Police Agency conducted an autopsy on Mr. Nagai's body on 4 October at Kyorin University (Mitaka City, Tokyo), his post-mortem certificate was also provided to the Special Rapporteur by the Htain Bin crematorium.

34. During his visit to the Htain Bin crematorium, the Special Rapporteur was informed by the authorities that during the disturbances in September, the Yangon General Hospital transferred 14 dead corpses, with the relevant burial certificates, to the crematorium. These were consequently registered and cremated accordingly. The hospital certified 11 deaths due to injuries (mostly firearms), 2 deaths due to illness and 1 death due to drowning. The Crematorium was not able to identify three corpses. The families and relatives of the identified bodies were reportedly able to participate in the cremations. The non-identified corpses were cremated on 1 October. It was noted that 25 persons are cremated on a daily basis at this crematorium and that corpses were only received from the General Hospital. While the Special Rapporteur was informed that there were no monks among the 14 corpses, the pictures did not provide sufficient indications to confirm this. Pictures and burial certificates from the register were shared with the Special Rapporteur.

35. Despite his request, the Special Rapporteur, was not given access to the second crematorium in Yangon, the Ye Way crematorium under the control of the Police Controller and Central Department, where credible sources report a large number of bodies (wrapped in plastic and rice bags) were burned during the night, between 4 a.m. and 8 a.m., on 27-30 September. Sources indicate that it was not usual practice for the crematorium to operate during the hours in question, that normal employees were instructed to keep away, and that the facility was operated on those nights by State security personnel or State-supported groups. At least one report indicates that some of the deceased being cremated had shaved heads and some had signs of serious injuries. The Special Rapporteur has expressed his concerns to the Government regarding these allegations and hopes that future investigations will shed light on these alleged cremations during the nights of the incidents in Yangon. The remains of the deceased should be returned to families or relatives in order to enable them to give their dead proper funerals in accordance with their religion and belief.

36. The Special Rapporteur asked officials from the General Hospital how many demonstrators were wounded, following allegations that they were only treated in the public hospital. The General Hospital recorded 30 admissions in Yangon, of which 23 were accidents and emergencies. According to the list, provided to the Special Rapporteur following clearance from the capital, the patients suffered injuries due to gunshots and assaults, among others. The Special Rapporteur enquired whether the wounded were detained. Once received in the emergency ward and after being sent to the general surgery wards, some were discharged. The information was also provided to the security forces who interviewed the patients at the hospital.

37. The use of lethal force by law enforcement officials from 26 to 29 September 2007 in Myanmar was inconsistent with the fundamental principles reflected in the basic international norms deriving from international customary law. They ignored the principles of necessity and proportionality which are included in article 3 of the Code of Conduct for Law Enforcement Officials and its commentary. Article 3 states that: "Law enforcement officials may use force only when strictly necessary and to the extent required for the performance of their duty." The commentary appended to this provision

explains that "in no case should this provision be interpreted to authorize the use of force which is disproportionate to the legitimate objective to be achieved". Similarly, in the Basic Principles on the Use of Force and Firearms by Law Enforcement Officials, the most general statement on the use of lethal force, principle 9, provides that: "In any event, intentional lethal use of firearms may only be made when strictly unavoidable in order to protect life." Whereas the Myanmar Code of Criminal Procedure provides for the use of civil force (art. 128) and military force (art. 129) to disperse an assembly, it also provides for the use of as little force as is consistent with dispersing an assembly, in order to avoid "injury to person and property" (art. 130). From 26 to 29 September, the security forces without doubt exceeded the limits of the power conferred on them by the law.

38. The Special Rapporteur found that whereas the Government and its agents showed some diligence in preventing a massacre, the decision by the security forces to shoot to kill and to severely beat protesters causing death constitutes an arbitrary deprivation of life and violates the right to life, as the lethal force used was unnecessary and disproportionate.

B. The use of non law enforcement officials

39. The Special Rapporteur considers that the participation of USDA members and SAS militia largely contributed to the excessive use of force against the peaceful protesters. It is unfortunate that the Myanmar Code of Criminal Procedure provides for the use by the authorities of civil forces to disperse assemblies (art. 128). In addition to Government soldiers and riot police, members of the Government-backed USDA and Swan Ah Shin (SAS) militia took violent action against the protesters with Government acquiescence or approval. Whether this group acted on direct Government orders is not clear. There is evidence that the Myanmar authorities have been complicit in the abuses perpetrated by these groups, or negligent in failing to intervene, punish or prevent them.

40. The USDA was established by the State Peace and Development Council (SPDC) in 1993 and in 2006 announced its intention to become a political party and field candidates in the next election. The Special Rapporteur expressed concerns in his previous reports over various allegations of involvement by members of USDA in acts of political and criminal violence. The existence of the SAS was first reported in 2003 when they were allegedly involved in the tragic incident of Depayin. According to sources, the SAS was reportedly already involved in incidents in 1997. The SAS, which has no legal status, is a grassroots force composed of civilians who reportedly assist the authorities in providing law enforcement, paramilitary services and military intelligence without being on the payroll of the Government. It includes members of the fire brigades, first aid organizations, women's organizations and USDA, as well as criminals/convicts released from jails, members of local gangs and the very poor and unemployed.

C. Arbitrary arrest and detention

41. From 18 September to the end of the curfew on 20 October, people were arrested on a daily basis with massive numbers of arrests on 26, 27, 28 and 29 September. It should be stressed that since the lifting of the curfew on 20 October, the Special Rapporteur continues to receive reports alleging the arrests of people, as well as further releases. After reviewing various reports and testimonies, it is estimated that between 3,000 and 4,000 people were arrested in September and October, and between 500 and

1,000 are still detained at the time of writing. In addition, 1,150 political prisoners held prior to the protests have not been released. Most of the arrests took place during the crackdown on the demonstrations and the night raids carried out by the security forces and non law enforcement officials (USDA and SAS). The analysis of several credible reports has strengthened the Special Rapporteur's view that relatives of people in hiding have also been taken as hostages during the raids. In the context of the preparation of his visit, in a letter dated 1 November 2007 to the Government, the Special Rapporteur requested the lists of people arrested, those released and the persons who are still detained, including information on their whereabouts, their detention conditions and the charges for their detention. He further asked under which law they were kept in custody.

42. The Minister of Home Affairs informed the Special Rapporteur that 2,927 persons have been arrested for investigation since the start of the crackdown in September 2007, with 2,836 having been released, and 91 remaining in detention. Most of them are detained on charges under the criminal code for terrorism while others are still under investigation. At least 15 individuals arrested in relation to the peaceful protests since August have been sentenced to prison terms of up to 9.5 years. Five of these individuals were reportedly tried in proceedings likely to have been closed and grossly flawed, in a court inside Thayet prison, Magway division on 24 and 26 September according to reliable sources. It should be noted that the Special Rapporteur has not been able to verify the figures collected.

43. The Special Rapporteur is particularly concerned about the numerous accounts of the use of large capacity informal detention centres, unacknowledged by State authorities, which are regarded as "secret" facilities. Detainees have included children and pregnant women. According to various reports, people have been held in six places of detention, including Government Technology Institute (GTI) in Insein Township, Police Centre No. 7 in Thanyin Township, Aung Tha Paye in Mayangone Township, Riot Police No. 5 in Hmawbe Township, Plate Myot Police Centre in Mandalay and Kyaik Ka San Interrogation Centre in Tamwe Township. Since many people have been released, it is believed that the remaining detainees are kept in custody in a few places of detention, including GTI and Police Centre No. 7, locations that the Special Rapporteur visited during his official mission.

44. During his visit to GTI the Special Rapporteur was informed by the police that from 27 September to 15 October, security forces took 1,930 demonstrators there (under responsibility of the Yangon Community since July 2007) out of which 80 persons were sent to Insein prison as violators of the security laws. The others were reportedly immediately released. He was presented with a detailed map indicating the detention rooms (women and men were separated) which he visited. He was informed that GTI, which is no longer a technical college, was planned as a shelter in case of emergency (in coordination with the Red Cross of Myanmar). While GTI could only host 1,500 persons at a time, Government officials informed him that 2,500 blankets were made available; 488 persons had reportedly been sick under the responsibility of 5 doctors and 15 nurses; and 5 persons were transferred to the General Hospital for urgent treatment. The Special Rapporteur visited the rooms where 153 women and 140 men had been detained (70 per room). One hundred police officers had ensured security. The Special Rapporteur was told information on the injuries and investigations of the detainees was classified.

45. The Special Rapporteur also welcomed the access provided to No. 7 Police Battalion Control Command Headquarters in Kyauktan, Thanlyin, located around 60 km from Yangon. It was reported that those brought here were being moved in and out, as it

had a maximum capacity of 30 at a time. The facility is under the control of the Security Force Battalion of Southern District Township, their main activity being VIP escort for embassies, security in Nay Pyi Taw and working along the border areas. The Special Rapporteur asked about their participation in law enforcement activities, to which they noted that they were responsible for receiving those detainees sent by other security forces. He further asked why the suspects were brought to such an isolated and remote area, in response to which he was told that the facility covers Yangon downtown area. The authorities noted that those involved in the demonstrations were to be separated, interrogated and investigated. When asked by whom the detainees were interrogated the authorities noted that this was not a place for interrogations, but only investigations. There were reportedly no wounded and all the people brought to Kyauktan had been transferred back, although it was not specified where. Despite his request, the Special Rapporteur was not granted access to the records, which were to be cleared by the Minister for Home Affairs and Police Chief.

46. The Special Rapporteur was informed that 10,000 prisoners are detained in Insein prison, managed by 500 guards, with 70 detainees reportedly placed in a separate building. Prisoners do receive visits from friends and family members, medicine, parcels and newspapers but are only allowed to write letters. Most prisoners need medical care and are in poor health due to the prison environment. Many of the 88 Generation Students are weak and can barely walk. The Special Rapporteur noted that most political prisoners from the NLD and the 88 Generation Group, as well as the monks, are labelled as terrorists by the authorities and had been prosecuted on the basis of the security law. Many political prisoners are in the so-called Insein Annex Dormitory 5 Building where not even prison guards are allegedly allowed access (70 detainees are in cell No. 8). The Special Rapporteur was provided with commercial satellite pictures of the place. Min Ko Naing was reportedly placed in the Annex a day before the Special Rapporteur's arrival at the Insein prison. Others in this dormitory are Htay Kywe, Min Zeya, Mie Mie, Mya Aye, Aung Thu, Ko Ko Gyi, Aung Naing, U Pyi Kyaw and U Zin Payit.

47. Credible sources report that detainees were held in degrading conditions in a special punishment area of Insein prison, commonly known as the "military dog cells", a compound of 9 tiny isolation cells measuring 2 meters by 2 meters constantly guarded by a troop of 30 dogs. The cells lack ventilation or toilets, and the detainees (mostly political prisoners) have to sleep on a thin mat on the concrete floor and are only allowed to bathe with cold water once every three days for five minutes. A recently released detainee testified that he was made to kneel bare-legged on broken bricks and also made to stand on tiptoe for long periods. Further reports confirm that monks held in detention were disrobed and intentionally fed in the afternoon, a time during which they are religiously forbidden from eating.

48. State security groups have continued to search for and detain specific individuals suspected of involvement in the anti-Government protests primarily through night raids on homes. It has also been confirmed that the authorities have resorted to arbitrary and unlawful detention of family members or close friends and suspected sympathizers of protesters currently in hiding. This constitutes hostage taking—explicit or implicit pressure on the suspected protesters to come forward as a condition for releasing or not harming the hostage. It is a violation of fundamental rules of international law. For example, before Thet Thet Aung was detained on 19 October, her mother and mother-in-law, otherwise unwanted by the authorities, were arbitrarily detained by Myanmar authorities seemingly to intimidate and pressure Thet Thet Aung to come forward. Both have since been released, though

her mother was kept in detention until 2 November. Similarly, before poet Ko Nyein Thit was detained by Myanmar authorities, his wife, Khin Mar Lar, was taken into custody on 1 October and not released until 21 October. When Di Nyein Lin evaded arrest on 12 October, the owner of the house in which he was hiding, Thein Aye, was arbitrarily arrested. Di Nyein Lin was arrested on 23 October, and Thein Aye remains in custody.

49. The Special Rapporteur received allegations indicating that 106 women, including 6 nuns, are being held in custody in Yangon after being arrested in connection with September's demonstrations and would like to praise the more than 25 women activists who paraded through downtown Yangon on 26 November in the first public display of opposition to the military regime since the September crackdown, in commemoration of the International Day for the Elimination of Violence Against Women. The group, which included housewives and students, marched from the Sule Pagoda to the Botataung pagoda, where they prayed for the monks and other protesters who died in the September demonstrations and for the release of detainees. The women were shadowed by members of the Government-backed USDA and the paramilitary SAS, but they did not intervene.

50. On 20 November, a week after the Special Rapporteur's visit and call for the release of all political prisoners in accordance with his proposed plan of action in his last report to the General Assembly (A/62/223), 58 prisoners had been released on humanitarian grounds, according to a statement by the Government. It said that 9 men over the age of 65, and 49 women, either pregnant or with children, were set free. It did not say if they were political prisoners and made no mention of pro-democracy leader Daw Aung San Suu Kyi. "The Government will continue to release those that will cause no harm to the community nor threaten the existing peace, stability and the unity of the nation as the country goes through a steady evolution towards a democracy", the statement said.

51. The Special Rapporteur condemns, however, the new arrests of political activists, despite the commitment by Prime Minister Thein Sein to the Special Adviser to the Secretary-General on Myanmar, Ibrahim Gambari, in early November that no more arrests would be carried out. Credible reports confirm that the following arrests have occurred since early November: U Gambira, head of the All-Burma Monks Alliance and a leader of the September protests, his father, Min Lwin and brother, Aung Kyaw Kyaw who were previously detained as hostages in an attempt to force him out of hiding; Su Su Nway, a member of the youth wing of the NLD and fellow youth activist Bo Bo Win Hlaing. Authorities raided a monastery in western Rakhine State, and arrested monk U Than Rama, wanted for his involvement in the September protests, whose whereabouts remain unknown. Myint Naing, a senior member of the NLD was detained. Ethnic Arakanese leader U Tin Ohn was detained and his whereabouts remain unknown. Other ethnic leaders, including Arakanese Cin Sian Thang and U Aye Thar Aung, Naing Ngwe Thein from the Mon National Democracy Front, and Kachin political leader U Hkun Htoo were rounded up but released after questioning. Aung Zaw Oo, a member of the Human Rights Defenders and Promoters group, was arrested in Yangon, likely on account of his involvement in planning events for International Human Rights Day on 10 December. Three further persons were arrested, Win Maw, lead guitarist in the popular Shwe Thansin band, Myat San, a member of the Tri-Colour Students Group and Aung Aung, a friend of the two above. Moreover, eight members of the Kachin Independence Organization (KIO) were arrested in Daw Hpum Yang, Momauk Township, Bamaw District. It is believed that this was on account of the KIO's refusal to accede to the SPDC's demand that they publicly renounce the recent statements by Daw Aung San Suu Kyi, made public by the Special Adviser, Mr. Gambari.

52. The Special Rapporteur is therefore urgently calling on the Government of Myanmar to release all those detained or imprisoned merely for the peaceful exercise of their right to freedom of expression, assembly and association, including both long-term and recent prisoners of conscience, as well as in the context of the peaceful demonstrations, and to stop making further arrests. He notes with grave concern the long-standing use of arbitrary detention by the authorities against prisoners of conscience including Daw Aung San Suu Kyi, U Win Tin, and senior opposition figures from ethnic minority groups, such as U Khun Htun Oo. It has been confirmed that the release of many detainees to date has been conditional on their signing an agreement to refrain from further political activity.

D. Disappearances

53. In the course of his investigation to date, the Special Rapporteur is aware of at least 74 cases of enforced disappearance, where the Myanmar authorities are either unable or unwilling to account for the whereabouts of individuals where there are reasonable grounds to believe that they have been taken into custody by State agents. The figures provided by different sources may underestimate the reality, as not all family members reported missing persons, fearing reprisals and severe punishment. The Special Rapporteur engaged in a dialogue with the authorities during his mission, requesting them to disclose information about the fate and whereabouts of the persons concerned. The authorities only partially met with his requirements.

54. The allegation of the burning of a large amount of bodies documented earlier is very disturbing. Without expressing at this stage an opinion on the accuracy of these reports, careful attention should be given to this allegation as it may explain why the Government has not been able, so far, to provide information on the whereabouts of a number of detainees and missing persons. It may also explain the numerous reports received about the removal of dead bodies by the security forces during the crackdowns and night raids on some monasteries.

E. Death in custody

55. According to credible reports received from an independent source, 1 monk who was in the GTI detention centre from 27 September to 5 October reported that around 14 individuals died during that period in custody, including 8 monks and 1 young boy who died on the first day. According to the monk, who was held in one cell with hundreds of people, the deaths were due more to the poor conditions of detention than injuries sustained during the crackdown. The NLD member Win Shwe, who was arrested on 26 September near Mandalay reportedly died during questioning in Plate Myot Police Centre on 9 October. His body was not returned to his family. Likewise, Venerable U Thilavantha, Deputy Abbot of the Yuzana Kyaungthai monastery in Myitkyina, was allegedly beaten to death in detention on 26 September, having also been beaten the night before when his monastery was raided.

F. Cruel, inhuman and degrading treatment and torture

56. Increasing reports from people who have been released describe degrading conditions of detention and the practice of torture. The Special Rapporteur's general impres-

sion is that the detainees are undergoing harsh conditions during the interrogation phase, lasting from four to eight days, undertaken at separate locations from the places of detention (such as the Tax Commission Office and the Ministry of Home Affairs in Yangon). Many interrogations are conducted with the detainees handcuffed, and they sleep on cold and wet floors. Food and drink are provided depending upon the answers given by the detainees. Some prisoners are kept in isolation, with only one hour for exercise in each of the morning and the afternoon (during the Special Rapporteur's visit these times were extended by half an hour).

57. The practice of torture in Myanmar has been documented by various observers, including by the Special Rapporteur for the last seven years. Experience shows that political activists and human rights defenders have been particularly targeted during their arrest, interrogation and detention. Reports have confirmed appalling detention conditions which fail to meet international standards on the treatment of prisoners and in fact constitute cruel, inhuman and degrading treatment prohibited under international law. Since the crackdown there have been an increasing number of reports of death in custody as well as beatings, ill-treatment, lack of food, water or medical treatment in overcrowded unsanitary detention facilities across the country. Provision of basic necessities, including food, water, blankets, and access to sleeping space and sanitary facilities has been lacking.

G. Severe reprisals against peaceful protesters

58. In his last report to the General Assembly (A/62/223), the Special Rapporteur gave special attention to sustained practices of restriction on the right to freedom of expression, the right to peaceful assembly and the right to freedom of movement. The events of September and October 2007 represent another manifestation of the severe methods of persecution and harassment that prevail in Myanmar. From 26 September to 20 October, the ban on gatherings (five people or more) enshrined in the Myanmar law was strictly applied and a curfew severely restricted the freedom of movement of people, lending a hand to the security forces for the conduct of night raids.

59. Night raids have been reportedly committed during curfew hours. On 26 September, overnight, the security forces arrested Myint Thein, the spokesman for opposition leader Daw Aung San Suu Kyi's political party. Relatives of people in hiding are reportedly taken hostage during these raids. The reduced curfew hours decided on 2 October have had no impact on the incidents which are reportedly committed between 11 p.m. and 3 a.m.

60. From 26 September to 6 October, the security forces reportedly raided 52 monasteries across the country, looting the possessions of monks and beating and arresting them in large numbers. Allegations of killings were also received. Early on Thursday 27 September at 12.30 a.m., security forces raided the Ngwe Kyar Yan monastery, a famous Buddhist teaching centre in Yangon (South Okkalapa Township), where they allegedly opened fire, physically assaulted and arrested an estimated 70 monks. Pictures taken at the scene after the curfew show blood spattering at different locations in the monastery and destruction of property, including gates, windows and other furniture. The pictures also suggest looting, which has been alleged by various sources, including direct testimonies. According to unconfirmed reports, some of the monks left after the violent raid, reported several arrests and the removal of dead bodies of several monks allegedly beaten to death by the security forces. Ngwe Kyar Yan was the site later this same day of a huge confrontation between security forces and civilians. There were rings of

soldiers and civilians around the monastery from late afternoon until the evening, with shots heard.

61. The Special Rapporteur was taken to the empty Ngwe Kyar Yan monastery, without being able to enter. The authorities showed him pictures of items (weapons, defamatory signs, gambling and pornographic images) reportedly found in the monastery. The total number of monks initially staying at the monastery was between 180 and 200. He was informed that 92 monks were moved on 27 September to another monastery under the State's responsibility, though not detained. He was able to engage in a closed meeting with 10 of these 92 remaining monks on the last day of his visit. The Special Rapporteur is concerned regarding the whereabouts of the remaining monks, who according to the authorities had absconded and returned to their families (allegedly dismissed for their conduct, according to the monks' disciplinary rules requiring permission from the head monk to leave the monastery). The Special Rapporteur noted that he will return to visit the monks on his follow-up mission.

62. The authorities announced that, as of 5 October, it had detained 533 monks, of whom 398 were released after sorting out what they called real monks from bogus ones. Twenty-one monks are reportedly detained in Insein prison. Reliable sources believe, however, that many more were detained or disappeared. Many young monks who used to study Buddhist literature have not dared to come back to Yangon, as the monasteries are still under surveillance by the authorities and vacant ones have been occupied by USDA members who immediately became trustees after the crackdown. There have been surprise checks in monasteries subjected to scrutiny by local authorities. On 29 November, monks assisting HIV/AIDS patients were forced by the military to leave the Maggin Monastery which was sealed off by the authorities. . . .

SOURCE: United Nations. General Assembly. Human Rights Council. "Report of the Special Rapporteur on the Situation of Human Rights in Myanmar, Paulo Sérgio Pinheiro, Mandated by Resolution S-5/1 Adopted by the Human Rights Council at its Fifth Special Session." December 7, 2007. A/HRC/6/14. www.unhcr.org/cgi-bin/texis/vtx/refworld/rwmain?docid=475fc7e52 (accessed January 10, 2008).

OTHER HISTORIC DOCUMENTS OF INTEREST

FROM PREVIOUS *HISTORIC DOCUMENTS*

- Human trafficking in Myanmar, *2005,* p. 343
- Tsunami relief effort, *2005,* p. 1005

DOCUMENT IN CONTEXT

UN Secretary-General on the Situation in Afghanistan

SEPTEMBER 21, 2007

Afghanistan's sixth postwar year saw limited progress, but continued insecurity, the increasingly evident weakness of the new government in Kabul, faltering economic development, and exploding narcotics production raised serious concerns internationally about the prospect of the county falling into a new version of the chaos that had reined there throughout the last decades of the twentieth century. At year's end the Bush administration and U.S. allies in Europe were conducting policy reviews in the hope of salvaging a broad international effort to bring stability and a measure of prosperity to Afghanistan.

There continued to be widespread agreement—both in foreign capitals and among Afghanistan's small community of leaders—that the country could not be allowed to become a failed state once more, as it had been after the withdrawal of the Soviet Union in the early 1990s. The results of that failure had included years of civil war, the Taliban's rise to power, and the creation of a sanctuary within Afghanistan's borders for the al Qaeda terrorist network headed by Osama bin Laden. But recognizing the dangers of failure in Afghanistan proved to be much easier than creating the conditions that would prevent it. In a report to the United Nations Security Council on September, Secretary-General Ban Ki-moon offered a generally downbeat assessment, acknowledging that the transition in Afghanistan had come under "increasing strain owing to insurgency, weak governance and the narco-economy."

As if the trends in Afghanistan were not worrisome enough, at year's end turmoil in neighboring Pakistan threatened to destabilize the entire region. Pakistan's president, Pervez Musharraf, who had been an uncertain partner in the fight against extremist forces along the border between the two countries, faced the greatest challenge yet to his leadership, particularly after the assassination on December 27 of opposition leader Benazir Bhutto. At year's end both Afghanistan and Pakistan were facing increasingly dangerous situations.

A Year of Frustration

In many ways 2007 was a year of extreme frustration for those who were trying to transform Afghanistan into a stable and relatively prosperous country. Leaders of the international community—the United Nations, the United States, European countries, humanitarian groups, and other donors—were becoming increasingly frustrated with the inadequacies of the government they had promoted in Afghanistan, headed by President Hamid Karzai. Western officials became more open in expressing concerns about Karzai's reluctance to exercise power over local warlords and his willingness to tolerate corruption and incompetence among government officials, even at the highest levels. Diplomats also expressed concern that Karzai had reacted to criticism by clamping down on the country's newly independent news media, the creation of which had been one of the government's principal achievements since the ousting of the Taliban.

Karzai, some of his colleagues in government, and an unknown portion of the Afghan population were becoming increasingly frustrated by the slow pace of the Western-led efforts to bring security to the country and to rebuild its infrastructure after decades of war. Perhaps the most damaging irritant, from an Afghan perspective, was the perceived lack of Western concern for innocent civilians who were killed or whose lives were disrupted by military operations. Air strikes and raids on villages killed several hundred civilians, leading Karzai to complain that Westerners considered Afghan lives to be "cheap."

Frustration within the North Atlantic Treaty Organization (NATO) alliance also became strikingly evident during the year. NATO since 2003 had gradually taken over more responsibility, from the United States, for security in Afghanistan and in 2006 had assumed full command of the United Nations-mandated International Security Assistance Force (ISAF). But the burden of actually fighting insurgents fell heavily on just a few countries—notably the United States, Great Britain, Canada, and the Netherlands—while other European countries contributed troops to help with reconstruction projects but would not let them fight. As the year wore on, U.S. defense secretary Robert Gates stepped up calls on the European allies to do more in Afghanistan. In response, some of them suggested that Washington was focusing too much on fighting insurgents and not enough on rebuilding the country.

For a while in the latter part of the year, it appeared that Afghanistan's international patrons were looking for a strong man as the solution—replacing the generally powerless top UN representative in Afghanistan with a higher-profile figure who had more political clout to make sure security and reconstruction efforts were more closely coordinated. U.S. and British officials suggested Lord Paddy Ashdown, a former British politician who had served a few years earlier as the UN's chief representative in Bosnia, where, by nearly all accounts, he had effectively wielded near-dictatorial powers. Ashdown's appointment seemed so certain that the top UN official in Afghanistan, Thomas Koenings, stepped down as planned on December 31. Early in 2008, however, Karzai objected to the idea of having such a powerful international diplomat looking over his shoulder, and Ashdown's appointment fell by the wayside.

Another Violent Year

Aside from the presence of substantially more Western troops, the overall security situation in Afghanistan was only somewhat different at the end of 2007 from what it had been at the beginning of the year. The NATO-led ISAF had many more troops on hand—41,700 at year's end, compared to about 32,000 twelve months earlier. Most of the soldiers were from Europe, but about 15,000 of the total were U.S. troops assigned to the NATO command. Another 11,000 U.S. troops—many of them special operations forces—remained under a separate American command known as Operation Enduring Freedom; their responsibility was to target remnants of the al Qaeda network based along the Afghanistan-Pakistan border and train Afghan soldiers.

The ISAF, working with the still-developing Afghan army and national police, had general control over the northern and western provinces, as well as most of the central areas around Kabul, the capital. Eastern and southern provinces continued to be more problematic, however.

Throughout the year NATO units launched offensives against a variety of antigovernment groups that were based along the border between Afghanistan and Pakistan. The most prominent of these groups was the former Taliban regime, now reconstituted as a guerrilla force whose leaders were said to be based in Quetta, Pakistan. Fighters for bin Laden's al Qaeda network also conducted attacks in Afghanistan from bases in Pakistan, as did sizable groups of guerrillas loyal to two different Afghan warlords, Jalaluddin Haqqani and Gulbuddin Hekmatyar (the latter of whom had been one of the prime recipients of U.S. aid in the 1980s, when he and other Afghan militants fought the Soviet Union's occupation of their country).

The year turned out to be the most violent in Afghanistan since a U.S.-led invasion ousted the Taliban from power in late 2001. According to an unofficial count by the Associated Press (AP), more than 6,500 people were killed during 2007, most of whom (about 4,500) were alleged militants. The total was an increase of more than 50 percent from 2006, when about 4,000 died, and more than four times the reported death toll of 1,600 in 2005.

According to the AP, the United States lost 110 service personnel, the largest number since the invasion; previous figures were 93 in 2005 and 87 in 2006. Great Britain lost 41 troops, Canada 30, and other NATO nations a total of 40. More than 900 Afghan soldiers and policemen died, many of the deaths the result of suicide bombs.

The rapidly rising death toll was due to two major factors. First, NATO and U.S. military operations against the insurgents escalated dramatically in 2007, thereby leading to greater casualties both of combatants and civilians caught in the cross-fire. Second, the Taliban and other insurgents also stepped up their attacks. According to UN figures, the rate of insurgent ambushes, suicide bombings, and other attacks was at least 20 percent higher in 2007 than in 2006, reaching a monthly average of nearly 550 attacks.

Suicide bombings, which had been rare in Afghanistan, suddenly emerged as a favorite tactic of the Taliban and related insurgents starting in mid-2005 and continuing through 2006 and 2007. The Pentagon's count of such attacks rose from 27 in 2005 to 139 in 2006 and stabilized at slightly more than 140 in 2007, although the attacks in 2007 killed many more people than previous attacks.

UN and U.S. officials said most of the suicide bombers appeared to be trained and directed by extremists from elsewhere, notably Chechnya (the embattled Muslim-majority province of Russia), neighboring countries in Central Asia, and the Arabian peninsula. The UN mission in Afghanistan on September 10 released a detailed study of suicide bombers. It said the typical bomber "is not crazed, fanatical, or brainwashed. Some are recruited in madrassas [Islamic schools], but many are not. Of those we've seen, most are young, poor, uneducated, and easily influenced."

Several major suicide bombings shook Afghanistan during the year. On February 27, during a visit by Vice President Dick Cheney to the massive U.S. military base at Bagram, north of Kabul, a suicide bomber struck just outside the base, killing 23 people. On June 17 a suicide bomber killed 35 police recruits on a bus in central Kabul, and on July 10 a bombing at a marketplace in Kabul killed 13 schoolchildren. On September 10 a suicide bomber on a motorized rickshaw killed 28 people (most of them civilians but about a dozen policemen) in the town of Gereshik in Helmand province. The year's single deadliest incident—indeed the deadliest since the Taliban were ousted—occurred on November 6 when a suicide bomber detonated his explosives during a parade, in the northern province of

Baghlan, welcoming a visiting delegation of parliamentarians. More than 70 people—most of them schoolchildren, but also including six legislators—died as a result of the bombing and indiscriminate firing by security guards in the chaos that ensued.

The Taliban and other insurgents had made a practice over the years of killing local, regional, and national leaders, including clergymen, school principals, and government officials. Several hundred such people had been assassinated, often in gruesome killings clearly intended to warn others against cooperating with the government. The most prominent government leader killed in 2007 was Mohammad Islam Mohammedi, a former Taliban governor who had won election to the new parliament in 2005. He was gunned down while walking to Friday prayers in Kabul on January 26.

Insurgents also turned increasingly to kidnappings, both of Afghan civilians and of foreigners, including aid workers. The most prominent case during the year was a kidnapping on July 19 of twenty-three South Korean missionaries as they traveled by bus from Kabul along the dangerous road to Kandahar. The Taliban demanded that eight Taliban prisoners held by the Afghan government be freed, as the price for releasing the Koreans. This kidnapping drew special attention within Afghanistan because two of the missionaries were women. The Taliban killed two of the male hostages when the demand was not met but released the others gradually over the course of the next month as the result of direct negotiations with South Korean diplomats. South Korea denied reports that it had paid ransom for the hostages, but it did promise to prevent any more Korean Christian missionaries from going to Afghanistan, and it reaffirmed a previously announced decision to withdraw, by the end of the year, all 200 of its troops who had been aiding the NATO mission in Afghanistan.

Civilian Deaths

Various estimates by UN officials, aid workers, and the Afghan government put the total number of civilians killed during the year at more than 1,000—the greatest number for any year since the Taliban were ousted. Western officials and nongovernmental agencies, such as Human Rights Watch, said most civilian deaths were caused by the Taliban and other insurgents. However, several hundred civilians died as the result of fighting between Western forces and insurgents, and locals routinely blamed NATO and U.S. troops rather than the insurgents.

One of the most controversial incidents came on March 4, when an estimated twelve civilians were killed and about three dozen were wounded when U.S. marines were surrounded by thousands of people following a suicide bombing in Nangahar province east of the city of Jalalabad. U.S. officials at first said they believed most of the civilians were killed by insurgents, but witnesses said at least some of the dead had been shot by American troops. In a separate incident the next day, nine members of one family northeast of Kabul died as the result of a U.S. air strike during a battle between insurgents and U.S. forces.

The March 4 incident apparently troubled U.S. commanders, who three weeks later ordered the unit of 120 marines involved in it to leave Afghanistan. A subsequent investigation determined that the marines used excessive force, and commanders ordered a criminal inquiry. The military in May officially apologized and paid compensation to families of the victims. The *New York Times* on September 5 quoted lawyers for two of the marines as disputing the findings of the investigation and defending their clients' actions. Legal proceedings in the case were still under way at year's end.

The issue of civilian deaths caused by NATO and U.S. military operations became troublesome during the year, particularly at the height of fighting midyear. Karzai had raised the matter repeatedly in previous years, and he became more vocal in 2007. "We can no longer accept civilian casualties the way they occur," he said on May 2, complaining that 51 civilians had been killed in a NATO offensive in Herat province. "It is not understandable anymore." Karzai complained specifically that international commanders were failing to coordinate their combat operations with his government—something he insisted would help reduce civilian casualties.

Western and Afghan aid agencies said much the same thing as Karzai. On June 19 the coordinating body for agencies working in Afghanistan filed a formal protest with military authorities, saying more than 230 civilians had been killed in recent months by U.S. and NATO air strikes and raids into villages. The "excessive use of force" against civilians was "undermining support" among Afghans both for Western military operations and for international aid efforts, the Agency Coordinating Body for Afghan Relief said in a statement.

In response, NATO and U.S. officials insisted that civilians were never targeted intentionally and that commanders did their best to avoid civilian casualties when attacking the Taliban and other insurgents. They noted, however, that Taliban fighters often hid in villages—even in homes of civilians who did not necessarily support them. Soldiers from Canada, the United States, and Western Europe had trouble distinguishing farmers from insurgents, and in too many cases made the wrong choice. Also, some of the largest-scale killings of civilians came when NATO and U.S. troops called in air strikes against suspected insurgents, only to discover that the insurgents were hiding among civilians or had never been in the area. Western commanders said they had evidence that insurgents often pressured village elders into claiming that civilian deaths were caused by NATO or U.S. troops rather than by the insurgents themselves. In late July NATO commanders announced new efforts to reduce the problem. These steps included using smaller bomb loads in air strikes against suspected insurgents and turning the job of conducting house-to-house searches over to the Afghan army.

Splinters in the U.S.-Led Coalition

The NATO-led ISAF force originally was intended as a small peacekeeping unit to protect Karzai's government in Kabul. As violence escalated over the years, the ISAF gradually expanded to a nationwide, multinational army that battled insurgents and helped rebuild the country. NATO gradually assumed control over security for most of Afghanistan, taking formal command from the United States in 2006. Even so, U.S. soldiers and marines continued to represent more than one-third of the ISAF's troop strength, and in 2007 the ISAF was commanded by a U.S. Army general, Dan K. McNeill.

For several years the Bush administration had pressed its European allies to step up their contributions to the NATO force, but Washington's requests generally produced only modest results. European leaders agreed in February to provide an additional 3,800 troops; the largest additions came from Britain (1,400 troops, bringing its total commitment to about 7,700 soldiers) and Poland (1,000 troops total).

Governments in several NATO-member countries faced mounting public opposition to the Afghan war, but parliaments nevertheless agreed to continue previous commitments to the ISAF, albeit in some cases by very narrow margins. Among them were Canada (which had some 2,500 troops on the front line in southern Afghanistan),

Germany, and the Netherlands. One exception was Japan, which sent no troops because of its pacifist constitution but for several years had provided refueling for ships from the United States and other countries that supported NATO operations in Afghanistan. A domestic political dispute halted that aid in November, but the Japanese government succeeded in restoring it in January 2008.

Several European countries placed restrictions on how their troops could be used, thus making it difficult for commanders on the ground to shift soldiers from one front to another in response to events. According to one estimate late in the year, only about 7,000 of the 41,700 NATO troops were available for offensive combat operations at any given time. Among the most important of these restrictions, called "national caveats," were that German, Italian, and Turkish troops were effectively barred from engaging in active combat.

Washington's frustration with its allies spilled out into the open in October, when Defense Secretary Gates used a speech at a military conference in Germany to warn of the dangers of inadequate foreign support for Afghanistan. "The failure to meet commitments puts the Afghan mission—and with it the credibility of NATO—at real risk," he said on October 25. Gates later toned down his criticism, and in mid-December he and his counterparts from seven other NATO countries agreed to work on new ways to get enough troops and equipment to Afghanistan.

Adding to the complexity of international operations in Afghanistan was the role of two dozen provincial reconstruction teams (PRTs) that had both military and nonmilitary functions. First launched by the United States in 2002, the PRTs were intended to provide both security and economic reconstruction in specific locations. By 2007, however, some of the teams—notably those managed by Germany and several other European countries—focused primarily on reconstruction in the relatively safe provinces in the north and west of the country. The PRTs managed by Britain, Canada, the Netherlands, and the United States worked in the more dangerous eastern and southern provinces and were more heavily focused on security matters.

NATO and U.S. Offensives

In the two previous years, NATO and U.S. troops found themselves reacting to an annual "spring offensive" by the Taliban. Bolstered with additional troops from Western Europe, NATO conducted its own spring offensive in 2007 to catch the Taliban off-guard. NATO in early March launched its largest-ever offensive, called Operation Achilles, in Helmand province. In addition to attacking the Taliban, a major goal of this offensive was to provide enough security to enable resumption of work on a hydroelectric project known as the Kajaki dam. The offensive succeeded in thwarting the Taliban's annual offensive, and it achieved some of its goals. However, NATO failed to create a safe zone for work on the dam, a project that had been one of the Afghan government's and international community's highest priorities for the year.

Another major NATO offensive, Operation Silicon, successfully targeted a key Taliban leader, known as Mullah Dadullah, who was killed in May. In these and smaller operations, NATO commanders claimed to kill hundreds of insurgents—in some cases on the order of 100 or more a day. Even so, it was not clear that the overall number of insurgents declined; in their few contacts with Western reporters, insurgent leaders claimed to have no trouble recruiting new fighters to replace those who were killed.

Perhaps because of the NATO offensives, the Taliban avoided the type of large-scale attacks it had mounted in 2006 and instead concentrated on suicide bombings, ambushes, kidnappings or beheadings of prominent local leaders, and hit-and-run attacks against civilian targets and the Afghan police. The apparent goal was to make life so intolerable for civilians that they would abandon their support for Karzai's government, which, after all, was supposed to be protecting them.

One of the most embarrassing setbacks for NATO forces in 2007 was the failure of an experiment in Helmand province that was supposed to put local leaders in charge of their region's security. With the blessing of British commanders in the area, the provincial governor and local elders in September 2006 negotiated an agreement calling for tribal leaders to provide security around the town of Musa Qala. NATO commanders agreed to respect the agreement by keeping their troops out of the district, thus freeing them for work elsewhere. This arrangement worked for several months, but on February 2 the Taliban seized Musa Qala, the largest town the insurgents had captured in more than five years of fighting. For most of the rest of 2007, the Taliban used the town as a base to conduct operations in a broad area of Helmand province. Afghan and NATO troops retook the town on December 10 as part of a major offensive.

Building up the Afghan Army and Police

A central element of the international community's long-term strategy for Afghanistan was to train and equip national army and police forces, which eventually could take over all responsibility for the country's security. Building national security forces in a country wracked by three decades of war and long dominated by private militias turned out to be an agonizingly slow process.

By 2007 most reports suggested that the Afghan army was growing in effectiveness, while the police force was unpopular and plagued by corruption at all levels. "The police are thieves and drug mafia, but the soldiers treat us with respect and try to provide security," Alaji Akhter, a tribal leader in the southeastern province of Zabul told the Associated Press in late April. The AP and other news organizations also quoted ISAF soldiers as praising the fighting spirit of Afghan soldiers, especially when compared with the revamped but ineffective Iraqi army.

The army in midyear reached an official strength of 40,360 troops, of which "approximately 22,000 are consistently present for combat duty," according to UN Secretary-General Ban's September 21 report to the Security Council. These figures were in line with other estimates that approximately one-half of Afghan army soldiers were absent at any given time, often for trips home to visit family members. Under U.S. pressure, the Afghan government agreed in 2007 to speed up the expansion of the army to a total strength of 70,000 by mid-2008—three years earlier than had been planned—and to set an eventual goal of 82,000 soldiers.

Afghan army units operated alongside Western troops, and in late June–early July the army for the first time took the lead in a major operation. Called Operation Maiwind, the movement into Ghazni province, in the south, resulted in little actual combat and was intended more as a public relations gesture to highlight the capabilities of the Afghan army.

About 70,000 police officers were listed on official billets as of November, of whom some 50,000 had been trained and equipped. As with the army, a substantial portion of Afghan policemen did not show up for duty on a regular basis. The national police also was considered one of the most corrupt institutions in Karzai's government; from the national command down to the local level, many police officers used their positions to extract bribes from the people they were supposed to protect. Germany was in charge of training the Afghan police for nearly four years after the ousting of the Taliban but failed to provide more than a token number of trainers. The United States took over the police-training mission in 2005, pumping in considerably more money and effort. By 2007 U.S. officials said they were retraining thousands of Afghan police officers in hopes of increasing their effectiveness and reducing corruption.

Following are excerpts from "The Situation in Afghanistan and Its Implications for International Peace and Security," a report released to the United Nations Security Council on September 21, 2007, by Secretary-General Ban Ki-moon.

DOCUMENT

Report on Afghanistan by United Nations Secretary-General

September 21, 2007

II. Overview

A. Political developments

2. The political transition that began with the signing of the Bonn Agreement [resulting from an international conference in Bonn, Germany, in December 2001 that created a transitional government for Afghanistan] nearly six years ago has come under growing internal and external pressure. While certain institutions and ministries continue to mature, public confidence in the Government and its leaders is wavering owing to increasing corruption and weak governance, particularly at the subnational level. An intensifying Taliban-led insurgency that increasingly relies on suicide bombing and other terrorist tactics is undermining confidence in the future and denying access of the Government and international aid organizations to a growing number of districts. Despite these pressures, there has been progress in terms of economic growth, education, health, road building and rural development. Furthermore, international support for Afghanistan has intensified rather than wavered. Progress, at this key moment, will depend on the international community and the Government of Afghanistan better coordinating their efforts to defeat the insurgency, promote good governance and provide tangible improvements to the lives of Afghans.

3. All of this must be achieved at a time when strains have emerged in the Afghan coalition that supported the Bonn Agreement and the Afghanistan Compact. The National Front of Afghanistan, a new alliance of former members of the Northern Alliance with representatives of the former Communist Government and the royal family, has become a prominent voice critical of Government shortcomings. While the emergence of an institutionalized political opposition will be a healthy development in the long term, it remains vital for Afghanistan to avoid a fragmentation of power in the short term, especially if that fragmentation weakens the political consensus that has underpinned the successful implementation of the Bonn Agreement.

4. On a more positive note, at the regional level, a collaborative atmosphere has begun to prevail in Afghan-Pakistani relations as the common challenge of terrorism has come into focus on both sides of the border. With regard to the international level, since March my Special Representative has travelled to capitals to promote the importance of a coherent approach to Afghanistan, amid a groundswell of political, financial and diplomatic support for the country. The increased support for Afghanistan underscores the urgent need for an integrated political and military strategy that complements the Afghanistan National Development Strategy, but also encompasses wider issues and provides a sharper focus on the achievement of national reconciliation and regional stability.

B. Security situation

5. Although the expanded International Security Assistance Force (ISAF) and the increasingly capable Afghan National Army have accrued multiple military successes during the reporting period, the Taliban and affiliated insurgent groups continue to prevent the attainment of full security in a number of areas. Access to rural areas of south and south-eastern Afghanistan for official and civil society actors has continued to decline. The boldness and frequency of suicide bombings, ambushes and direct fire attacks have increased.

6. Following counter-insurgency operations in the south and east, the Taliban have lost a significant number of senior and mid-level commanders. In Hilmand, Kunar, Paktya and Uruzgan Provinces, insurgent leaders have been forced to put foreigners in command positions, further undermining the limited local bases of support. This has heightened the importance to the Taliban of the support it receives from the border regions of Pakistan.

7. Rates of insurgent and terrorist violence are at least 20 per cent higher than in 2006; an average of 548 incidents per month were recorded in 2007, compared to an average of 425 per month in 2006. There have been over 100 suicide attacks to date in 2007, compared to 123 in all of 2006. While 76 per cent of all suicide missions target international military and Afghan security forces, their victims have been largely civilian bystanders: 143 civilians lost their lives to suicide attacks between 1 January and 31 August 2007. Suicide attacks have been accompanied by attacks against students and schools, assassinations of officials, elders and mullahs, and the targeting of police, in a deliberate and calculated effort to impede the establishment of legitimate Government institutions and to undermine popular confidence in the authority and capability of the Government of Afghanistan.

8. Defeating the insurgency has been complicated by the growth of criminal and drug gangs, which enjoy a symbiotic relationship with anti-Government armed groups. While these groups may not share the political goals of the Taliban, they do have a common interest in preventing the imposition of State authority in certain areas or corrupting what State authority exists. In the poppy-cultivating Provinces of Badakhshan, Hilmand and Kandahar, the State is extremely weak or non-existent throughout much of the countryside, while corruption is endemic in provincial centres.

9. The successes of the counter-insurgency in conventional battles and in eliminating Taliban and other insurgent leaders are undeniable. If the trends of the past two years are to be reversed, however, a more comprehensive counterinsurgency strategy will be needed to reinforce political outreach to disaffected groups and address the security gaps that allow insurgents to recover from their losses and, with very few resources, still manage to terrorize local populations or enlist criminal gangs to further their goals.

C. Institutional challenges

10. While some institutions within the judicial, executive and legislative branches of Government continue to gain capacity and effectiveness, internal disputes and institutional corruption threaten efforts to consolidate and legitimize these institutions. The Anti-Corruption Commission established by President [Hamid] Karzai has not yet delivered results and faces an uncertain future. The resulting sense of impunity has encouraged a culture of patronage and direct involvement in illegal activities, including the drug trade, especially within the police force.

11. Relentless pressure from the international community has resulted in the removal of some prominent human rights offenders from positions in the Ministry of the Interior and their replacement by more professional officers. The lack of alignment among the international partners involved in reforming the Ministry, however, has contributed to a notable failure to prevent or curb the use of parochial connections and bribes to determine appointments. Tolerance of corruption has had a particularly corrosive effect on the police. Since many communities' only contact with the national Government is through the police, poor police behaviour often translates into a negative perception of the Government and, to some degree, the international community that supports it.

12. The Office of the Attorney General has attempted a proactive strategy to combat corruption by arresting, investigating and prosecuting medium- and high-level Government officials, as well as some of its own prosecutors. These efforts are frustrated, however, by the fact that those targeted are often able to defend themselves through their personal relationships with powerful figures. Police or justice officials are commonly bribed to prevent arrests, arrange releases from detention or dismiss charges. Sentences, when they are imposed, tend to be lenient. Furthermore, there have been no substantial successes in the restitution or forfeiture of the proceeds from corruption-related crimes.

13. Another measure to combat corruption is the creation of a professional and adequately paid civil service. The Action Plan on Peace, Reconciliation and Justice, adopted as part of the Afghanistan Compact, calls for the Government of Afghanistan to establish a clear and transparent national appointments mechanism for all senior-level appointments. Progress on this front has been inadequate. The senior appointments panel created for this purpose still has no approved rules of procedure that guarantee transparency and impartiality. In addition, it has not been provided with premises and is, in general, under-resourced and underperforming.

14. For lower level appointments, the Independent Administrative Reform and Civil Service Commission has finalized a revised public administration reform framework and implementation programme. This effort has been underpinned by a more coordinated approach among donors. The Cabinet, after much deliberation, approved a pay and grading reform, with a salary range of $80 to $650 per month for civil servants. Implementation, however, will be incremental over a four-year period, meaning that it is unlikely to have a significant effect on reducing corruption in the immediate term.

15. Finally, in August 2007, the Wolesi Jirga (Lower House) approved the United Nations Convention against Corruption. United Nations agencies, including the United Nations Development Programme (UNDP) and the United Nations Office on Drugs and Crime (UNODC), are providing technical assistance to develop legislation, strategies and capacity to implement the provisions of the Convention. . . .

X. Observations

74. As the transition in Afghanistan comes under increasing strain owing to insurgency, weak governance and the narco-economy, the Government of Afghanistan, supported by the international community, will need to demonstrate political will by taking the bold steps necessary to recapture the initiative in each of these fields and restore confidence to the population in tangible ways. Without stronger leadership from the Government, greater donor coherence—including improved coordination between the military and civilian international engagement in Afghanistan—and a strong commitment from neighbouring countries, many of the security, institution-building and development gains made since the Bonn Conference may yet stall or even be reversed.

75. The most urgent priority must be an effective, integrated civilian-military strategy and security plan for Afghanistan. A coordinated military response is still needed to defeat insurgent and terrorist groups, but success in the medium term requires the engagement of communities and the provision of lasting security in which development can take place. To achieve that end, Afghan civilian and military leaders need to play a greater role in planning security operations and ensuring that military gains are consolidated with the provision of basic security by State institutions. At the same time, the different goals and movements within the insurgency present opportunities for political outreach and inclusion that must be seized.

76. A key to sustaining security gains in the long term is increasing the capability, autonomy and integrity of the Afghan National Security Forces, especially the Afghan National Police. The Government and its partners should develop, through the International Police Coordination Board, a unified vision for police reform and definitive structure for the national police that addresses the requirements of both law enforcement and counter-insurgency. It must also tighten financial and administrative accountability to end corruption and absenteeism in police ranks.

77. An effective, integrated and coherent Government-led subnational governance programme should be developed in partnership with the international community. The Government must be prepared to take painful decisions now to bring credibility to emerging institutions. It should avoid rotating underperforming officials into new positions, especially in the provinces, and replace them instead with effective administrators who both enjoy the confidence of the population, including tribal and religious leaders, and display a capacity to manage security, development and reconstruction processes in their provinces and districts. The extension of central authority and the stabilization of the country will be possible only if the Ministry of the Interior resolutely tackles corruption and improves popular perceptions of the police.

78. Building on the successes at the Conference on the Rule of Law in Afghanistan, the Government should finalize the justice sector strategy and begin implementation of the emerging national justice programme funded through the Afghanistan Reconstruction Trust Fund. The Government should simultaneously address the apparent impunity enjoyed by those Government officials perceived to be abusing their offices. ISAF should support Government efforts to enforce the law and to implement the project for the disbandment of illegal armed groups in areas less affected by the insurgency.

79. The continued increase in opium production poses an increasingly grave threat to reconstruction and nation-building in Afghanistan. The Government must prioritize interdiction and bring drug traffickers to justice. The international community, supported by a strengthened UNODC [the United Nations Office on Drugs and Crime], should

unite behind a truly Afghan-led plan that moves beyond eradication efforts, which have proved ineffective in isolation.

80. The finalization and the future funding of the Afghanistan National Development Strategy must remain the overriding focus of donor engagement. Following its expected launch in March 2008, careful management of public expectations, follow-up at the provincial and district levels and further outreach that builds on the initial consultations will be crucial to the credibility of the Strategy. If the Strategy is to become an enduring vehicle for partnership between the Government and people of Afghanistan, it must be seen to deliver genuine results in response to priorities defined by the communities themselves.

81. The Government of Afghanistan has demonstrated its determination to contribute to regional security and prosperity by means of a series of major regional foreign policy achievements, including the Afghanistan-Pakistan peace jirga, the visit of the President of the Islamic Republic of Iran to Kabul and President Karzai's participation in the summit of the Shanghai Cooperation Organization in Bishkek. The Government must retain the trust of its neighbours by engaging constructively in bilateral and multilateral initiatives, including on the counter-narcotics and migration issues, and by building its capacity to manage and deepen complex bilateral relationships.

82. The recognition by Presidents Karzai and [Pervez] Musharraf at the peace jirga in Kabul of the cross-border nature of the insurgency provides a unique opportunity for their respective countries to pursue a joint strategy for crossborder peace and security, aimed at defeating extremism and terrorism in both countries.

83. National reconciliation will require agreement as to which insurgent leaders ought to be subject to military operations or law enforcement. It will also require both Afghanistan and Pakistan to undertake outreach and dialogue with those political forces capable of contributing to a peace process. It will be vital for all Member States to ensure implementation of the sanctions provided for under resolution 1267 (1999) and to include new insurgent and terrorist leaders on the consolidated list or remove them after reconciliation, as appropriate.

84. The Government of Afghanistan must investigate allegations of arbitrary detentions, inhumane treatment and torture of detainees by the authorities, and in particular by the National Directorate for Security. The Government should invite the Special Rapporteur on torture and other cruel, inhuman or degrading treatment or punishment and the Working Group on Arbitrary Detention to visit Afghanistan as part of a cooperative process to combat arbitrary detention, torture and ill-treatment. The Government should renew its political commitment towards the full implementation of the Action Plan on Peace, Reconciliation and Justice.

85. Presidential elections are due to be held in 2009. The upcoming cycle of elections will require urgent attention by both the Government and the National Assembly in order to ensure the adoption of the electoral law by the end of 2007. I reiterate my appeal to donors to meet the remaining shortfall from past Afghan elections and provide the resources necessary to support a new voter registry, capacity-building for the Independent Electoral Commission and planning and preparations for the elections themselves.

86. The United Nations will remain fully engaged in Afghanistan and continue to play its central and impartial coordinating role. I personally visited Afghanistan in July and shortly thereafter co-chaired the Conference on the Rule of Law in Afghanistan in Rome. I will, in addition, co-chair a high-level conference on Afghanistan in September

with President Karzai, in the margins of the General Assembly. I shall continue to remain personally engaged in working with President Karzai and other partners to ensure success in Afghanistan.

87. Finally, I would also like to pay tribute to the dedicated efforts of my Special Representative and the staff of the United Nations in Afghanistan, who continue to carry out their mission under difficult and increasingly dangerous circumstances, and whose courage and commitment to Afghanistan have been essential to the progress achieved so far.

SOURCE: United Nations. General Assembly and Security Council. "The Situation in Afghanistan and Its Implications for International Peace and Security: Report of the Secretary-General." A/62/345-S/2007/555. September 21, 2007. daccess-ods.un.org/TMP/1359383.html (accessed February 1, 2008).

OTHER HISTORIC DOCUMENTS OF INTEREST

FROM THIS VOLUME

- Security developments in Iraq, p. 9
- Declaration of Afghanistan-Pakistan "peace jirga," p. 442
- Political developments in Japan, p. 504
- Political developments in Pakistan, p. 647

FROM PREVIOUS *HISTORIC DOCUMENTS*

- London conference "compact" on Afghanistan, *2006,* p. 48
- UN Secretary-General Annan on the situation in Afghanistan, *2006,* p. 525
- President Karzai addresses the new parliament of Afghanistan, *2005,* p. 970
- Karzai on his inauguration as president of Afghanistan, *2004,* p. 912

DOCUMENT IN CONTEXT

FBI Report on Crime in the United States

SEPTEMBER 24, 2007

Violent crime in the United States edged upward in 2006, led by a sharp increase in the estimated number of robberies. It was the second year in a row that robbery, murder, and other violent crime had increased, indicating that the slight decline in crime seen from 2002 to 2004 might be ending. Although the volume of violent crime in 2006 was nearly the same as it had been in 2002, it was still 13.3 percent lower than it had been ten years earlier, in 1997. Property crimes continued to show a downward trend in 2006, according to the Federal Bureau of Investigation.

The number of people in prison or jail continued to increase, rising 2.8 percent in 2006 over 2005. Overall, the Justice Department reported, 2.38 million adults—nearly one of every thirty-one adults in the United States—were behind bars or on parole or probation at the end of 2006. According to the advocacy group Human Rights Watch, the United States had the highest incarceration rate of any country in the world. There was some chance that those numbers might begin to decline in coming years. Decisions by the Supreme Court and the U.S. Sentencing Commission late in the year made it almost certain that thousands of inmates jailed on crack cocaine offenses would have their sentences reduced in the coming years.

Crime Rates

The increase in violent crime was the second in two years and represented the first two-year increase in violent crime since 1993, according to the FBI's annual report on crime statistics, *Crime in the United States, 2006,* released September 24. The 1.9 percent jump in 2006 was also higher than the 1.3 percent increase the FBI had projected earlier in the year. The biggest jump came in the number of robberies: an estimated 447,403 in 2006, for an increase of 7.2 percent over 2005. The number of murders also rose 1.8 percent in 2006, to 17,034. In the other two categories of violent crime, forcible rapes were down 2 percent from 2005, while aggravated assaults were down just a fraction. Firearms were used in 68 percent of the murders, 42 percent of the robberies, and 22 percent of the aggravated assaults.

Property crime—burglary, larceny-theft, motor vehicle theft, and arson—continued to decline in 2006. The number of property crimes, an estimated 9.98 million, fell 1.9 percent from its level in 2005, while the property crime rate was down 2.8 percent from the previous year. That, in turn, lowered the overall crime rate.

In a September 24 statement, Justice Department spokesman Brian Roehrkasse said that the rate of crime in 2006—the number of crimes per 100,000 people—had fallen to its lowest level in more than thirty years. Nonetheless, Roehrkasse acknowledged, "violent crime remains a challenge for some communities."

The truth of Roehrkasse's latter statement was reflected in the number of murders in 2007, which varied substantially from city to city. Preliminary counts showed that New York City, which had 2,245 homicides in 1990, had fewer than 500 in 2007—the lowest number since the city began keeping reliable statistics in 1963. The murder rates in two other big cities, Los Angeles and Chicago, also declined, but the rates increased in Atlanta, Baltimore, Detroit, and Oakland. The number of murders in New Orleans rose nearly 30 percent between 2006 and 2007.

The Police Executive Research Forum, an association of chief executives at all levels of law enforcement, said it found a similar volatility in violent crime from city to city. Its survey of fifty-six jurisdictions, released in December 2007, showed that the number of violent crimes declined overall in the first six months of 2007, but several cities experienced significant increases over 2006. The forum said it viewed the differences in crime levels as evidence that police departments in some cities, working closely with their communities, had been able to implement new policing initiatives that successfully reduced violent crime. The report also asked participating police forces to list what they saw as the chief factors contributing to the rise in violent crime. The top five were: gangs (cited by 77 percent); juvenile crime (74 percent); impulsive violence and "disrespect" issues (66 percent); the economy, poverty, and unemployment (63 percent); and the release of offenders back into communities (63 percent).

Prison Trends

The 2.8 percent increase in the prison population was the fastest growth in that population in the last five years, according to the Bureau of Justice Statistics. In addition to the 2.38 million people serving time in federal or state prison or local jails, a record 5 million people were on parole or probation at the end of 2006, according to the bureau's annual report on incarceration, released on December 5. The number of people detained by immigration authorities grew the fastest in 2006, rising 43 percent to 14,482, from 10,104 in 2005. The number of women in state and federal prison was at a record high of 112,498, an increase of 4.6 percent over 2005. The rate of female incarceration had been growing at double the rate for men since 1980.

The number of African American inmates also reached a new high of 905,600, reflecting continuing, if somewhat lower, racial disparities in the nation's prisons and jails. Blacks made up 38 percent of the male prison population in 2006, down from 43 percent in 2000, according to the bureau. White males accounted for 34 percent, while Hispanics made up 21 percent. The rate of incarceration for black women was also declining, while that for white and Hispanic women was increasing.

The national rate of incarceration for black males was 6.2 times higher than the incarceration rate for whites, according to the bureau. A separate analysis conducted by the Sentencing Project, a group advocating fairness in sentencing, found significantly higher incarceration rates for blacks compared with whites in some states. In Iowa blacks were jailed at 13.6 times the rate of whites; other states where black incarceration rates were at least 10 times higher than those of whites included Vermont, New Jersey, Connecticut, Wisconsin, and North and South Dakota. In some cases the disparity occurred because the rate of black incarceration was exceptionally high; in others, the rate of black incarceration was about average, while the rate of white incarceration was substantially below average. The lowest ratios were in Hawaii (where black incarceration rates were 1.9 times those of whites), Georgia, Mississippi, Alabama, and Arkansas. The Sentencing Project noted that the racial disparity was low in Georgia, Mississippi, and Alabama largely because the white incarceration rate was higher than average.

Reduced Sentences for Crack Cocaine

One reason for the racial disparity in incarceration rates was a controversial federal law that set much stiffer penalties for crack cocaine offenders, who tended to be black, than for powder cocaine offenders, who tended to be white. Enacted in 1986, when crack and related street violence appeared to reach epidemic proportions in some cities, the federal law set a mandatory minimum sentence of five years for crimes involving five grams of crack (about one-fifth of an ounce); the five-year minimum trigger for powder cocaine was five hundred grams (about seventeen ounces). A ten-year sentence was mandatory for possession of fifty grams of crack, as opposed to five kilograms of powder. Those sentencing disparities were then built into the federal sentencing guidelines used to determine each offender's actual sentence. At the time the differences were justified on grounds that crack was a more dangerous drug than powder cocaine.

The disparity in sentencing not only meant that low-level crack drug users and pushers were treated at least as harshly as mid- and high-level drug traffickers, but that black drug users were disproportionately affected. Although the two forms of cocaine were chemically similar, crack was much cheaper to produce and was most popular among young, disadvantaged drug users, often from inner cities. The U.S. Sentencing Commission, which wrote the sentencing guidelines for those convicted of federal crimes, estimated that in 2006 more than 80 percent of those convicted on crack cocaine charges were African Americans; they comprised just over 25 percent of those convicted on powder cocaine charges.

Civil rights groups and others lobbied to lower the minimum sentences for crack cocaine, arguing that they were discriminatory and unfair. During the 1990s the sentencing commission, backed by the Clinton administration, also recommended that the disparity be reduced, but Congress rejected those calls. Over the next few years, a growing number of federal judges also joined the call for reducing the disparity. In April 2007 the commission tried again, sending a recommendation to Congress that would lower sentencing ranges for crack cocaine but not the underlying minimum sentences. The recommendation took effect on November 1, when Congress took no action to block it.

Then on December 10 the Supreme Court added its voice to those approving a reduction in the sentencing disparity. By a 7–2 vote, the Court ruled that a federal judge in Virginia had acted reasonably when he imposed a lower sentence than the one called for in the sentencing guidelines for a former crack dealer. "The cocaine guidelines, like all other guidelines, are advisory only," Justice Ruth Bader Ginsburg wrote for the majority, adding that "the court of appeals erred in holding the crack/powder disparity effectively mandatory." The ruling further clarified a decision the Court issued in 2005 that declared the federal sentencing guidelines unconstitutional but said that judges should nonetheless treat them as advisory.

On December 11 the sentencing commission voted unanimously to make its new, lighter sentencing ranges for crack cocaine retroactive. The move, which was scheduled to take effect on March 3, 2008, could result in reduced sentences for an estimated 19,500 federal prisoners. The commission said the average sentence could be lowered by about twenty-seven months, and that about 2,500 prisoners could be released within a year. Those serving time on crack charges would have to apply to have their sentences reduced.

The Bush administration opposed making the changes retroactive. "We are going to see an influx of the very people who are most likely to reoffend and are most likely to upset . . . fragile neighborhoods," Gretchen C. F. Shappert, a U.S. attorney in Charlotte, North Carolina, told the commission in November. Civil rights activists said they would now turn their attention to Congress and push to abolish or greatly reduce the mandatory minimum sentences for crack. "Where we go from here is Capitol Hill," said Julie Stewart, the founder of Families Against Mandatory Minimums.

Following are excerpts from the Federal Bureau of Investigation's annual report, Crime in the United States, 2006, *released on September 24, 2007.*

DOCUMENT

Statistics on Crime in the United States

September 24, 2007

Violent Crime

. . .

- An estimated 1,417,745 violent crimes occurred nationwide in 2006.
- There were an estimated 473.5 violent crimes per 100,000 inhabitants.
- When data for 2006 to 2005 were compared, the estimated volume of violent crime increased 1.9 percent. The 5-year trend (2006 compared with 2002) indicated that violent crime decreased 0.4 percent. For the 10-year trend (2006 compared with 1997) violent crime fell 13.3 percent.
- Aggravated assault accounted for the majority of violent crimes, 60.7 percent. Robbery accounted for 31.6 percent and forcible rape accounted for 6.5 percent. Murder, the least committed violent offense, made up 1.2 percent of violent crimes in 2006.
- In 2006, firearms were used in 67.9 percent of the Nation's murders, in 42.2 percent of the robbery offenses, and in 21.9 percent of the aggravated assaults.

. . .

Murder

. . .

- An estimated 17,034 persons were murdered nationwide in 2006, an increase of 1.8 percent from the 2005 estimate.
- Murder comprised 1.2 percent of the overall estimated number of violent crimes in 2006.
- There were an estimated 5.7 murders per 100,000 inhabitants.
- In 2006, an estimated 90.6 percent of the murders occurring in the Nation were within Metropolitan Statistical Areas. . . .

Forcible Rape

. . .

- In 2006, there were an estimated 92,455 forcible rapes reported to law enforcement, a 2.0-percent decrease from the 2005 estimate.
- When compared with 2002 data, the estimated number of forcible rapes decreased 2.9 percent; when compared with 1997 data, the number of forcible rape offenses declined 3.8 percent.
- The rate of forcible rapes in 2006 was estimated at 60.9 offenses per 100,000 female inhabitants, a 2.9-percent decrease when compared with the 2005 estimate of 62.7 forcible rapes per 100,000 female inhabitants.
- Based on the number of rape offenses reported to the UCR Program in 2006, rapes by force comprised 91.9 percent of reported rape offenses, and assaults to rape attempts accounted for 8.1 percent of reported rape offenses. This equated to 56.0 rapes by force per 100,000 female inhabitants and 4.9 assaults to rape attempts per 100,000 females in 2006. . . .

Robbery

. . .

- Nationwide in 2006, there were an estimated 447,403 robbery offenses.
- The estimated number of robbery offenses increased 7.2 percent from the 2005 estimate. However, the estimated number of offenses declined 10.3 percent in a comparison with the data from 10 years earlier (1997 and 2006).
- By location type, most robberies (44.5 percent) were committed on streets or highways.
- The average dollar value of property stolen per robbery offense was $1,268. By location type, bank robbery had the highest average dollar value taken—$4,330 per offense.
- Losses estimated at $567 million were attributed to robberies during 2006.

- Firearms were used in 42.2 percent of robberies for which the UCR Program received supplementary data. . . .

Aggravated Assault

. . .

- Nationwide, there were an estimated 860,853 aggravated assaults during 2006.
- The 2- and 10-year trends indicate that the estimated number of aggravated assaults in 2006 declined less than 1 percent (-0.2) when compared with the offense estimates from 2005 and declined 15.9 percent from the 1997 estimate.
- In 2006, the rate of aggravated assaults in the Nation was estimated at 287.5 offenses per 100,000 inhabitants.
- An examination of the 10-year trend data for the rate of aggravated assaults revealed that the rate in 2006 declined 24.8 percent when compared with the rate for 1997.
- In 2006, 21.9 percent of the aggravated assaults for which law enforcement agencies provided expanded data involved a firearm.
- The use of firearms during aggravated assaults increased 2.8 percent when 2006 data were compared with data for 2005. . . .

Property Crime

. . .

- There were an estimated 9,983,568 property crimes in the Nation in 2006.
- The 2- and 10-year trends showed that the number of property crimes in 2006 decreased 1.9 percent when compared with the 2005 estimate and declined 13.6 percent when compared with the 1997 estimate.
- In 2006, the rate of property crime offenses was estimated at 3,334.5 property crimes per 100,000 inhabitants.
- Two-year and 10-year trends indicated that the rate of property crimes in 2006 decreased 2.8 percent when compared with the 2005 data and declined 22.7 percent when compared with the 1997 data.
- Two-thirds of all property crimes were larceny-thefts.
- Property crimes accounted for an estimated $17.6 billion dollars in losses. . . .

Burglary

. . .

- In 2006, there were an estimated 2,183,746 burglary offenses—an increase of 1.3 percent when compared with 2005 data.

- An examination of 5- and 10-year trends revealed an increase of 1.5 percent in the number of burglaries when compared with the 2002 estimate and a decline of 11.2 percent when compared with the 1997 estimate.
- Burglary accounted for 21.9 percent of the estimated number of property crimes committed in 2006.
- In 2006, burglary offenses cost victims an estimated $4 billion in lost property.
- The average dollar loss per burglary offense in 2006 was $1,834.
- Of the burglary offenses in 2006, 66.2 percent were of residential structures.
- Of the burglaries for which the time of occurrence was known, 63.1 percent of residential burglaries took place during the day. . . .

Larceny-theft

. . .

- There were an estimated 6.6 million (6,607,013) larceny-theft offenses nationwide during 2006.
- An examination of 2- and 10-year trends revealed a 2.6-percent decrease in the estimated number of larceny-thefts compared with the 2005 figure, and a 14.7-percent decline from the 1997 estimate.
- Two-thirds of all property crimes in 2006 were larceny-thefts.
- During 2006, there was an estimated rate of 2,206.8 larceny-theft offenses per 100,000 inhabitants.
- From 2005 to 2006, the rate of larceny-thefts declined 3.5 percent, and from 1997 to 2006, the rate declined 23.7 percent.
- The average value for property stolen during the commission of a larceny-theft was $855 per offense.
- There were an estimated $5.6 billion dollars in lost property in 2006 as a result of larceny-theft offenses. . . .

Motor Vehicle Theft

. . .

- Nationwide in 2006, there were an estimated 1.2 million motor vehicle thefts, or a rate of approximately 398.4 motor vehicles stolen for every 100,000 inhabitants.
- The estimated number and rate of motor vehicle thefts in 2006 decreased 3.5 percent and 4.4 percent, respectively, when compared with data for 2005.
- When considering data from 10 years earlier, the estimated number of motor vehicle thefts in 2006 decreased 11.9 percent. The estimated rate of motor vehicle thefts decreased 21.2 percent when compared with estimates for 1997.

- An estimated 93.5 percent of the Nation's motor vehicle thefts occurred in Metropolitan Statistical Areas in 2006.
- Property losses due to motor vehicle theft in 2006 were estimated at $7.9 billion, averaging $6,649 per stolen vehicle.
- Among vehicle types, automobiles comprised 73.5 percent of the motor vehicles reported stolen in 2006.

Arson

. . .

- Nationally, 69,055 arson offenses were reported by 13,943 agencies that submitted arson data in 2006 to the UCR Program. . . .
- Arsons involving structures (residential, storage, public, etc.) accounted for 42.3 percent of the total number of arson offenses. Mobile property was involved in 28.2 percent of arsons. The rest were arsons of other types of property.
- The average value loss per arson offense was $13,325.
- Arsons of industrial and manufacturing structures resulted in the highest average dollar losses (an average of $66,856 per arson).
- In 2006, arson offenses increased 2.1 percent when compared with arson data from the previous year.
- The rate of arson was 26.8 offenses for every 100,000 inhabitants of the United States in 2006. . . .

Clearances

. . .

- Nationwide in 2006, 44.3 percent of violent crimes and 15.8 percent of property crimes were cleared by arrest or exceptional means.
- Of the violent crimes (murder and nonnegligent manslaughter, forcible rape, robbery, and aggravated assault), murder had the highest percentage of offenses cleared at 60.7 percent.
- Of the property crimes (burglary, larceny-theft, and motor vehicle theft), larceny theft had the highest percentage of offenses cleared at 17.4 percent.
- Eighteen percent of arson offenses were cleared by arrest or exceptional means.
- Nationwide in 2006, 40.2 percent of arson offenses cleared by arrest or exceptional means involved only juveniles (individuals under age 18), the highest percentage of all offense clearances involving only juveniles. . . .

Arrests

. . .

- In 2006, the FBI estimated that 14,380,370 arrests occurred nationwide for all offenses (except traffic violations), of which 611,523 were for violent crimes, and 1,540,297 were for property crimes.
- Law enforcement made more arrests for drug abuse violations in 2006 (an estimated 1.9 million arrests, or 13.1 percent of the total number of arrests) than for any other offense.
- Nationwide, the 2006 rate of arrests was estimated at 4,832.5 arrests per 100,000 inhabitants; for violent crime, the estimate was 207.0 per 100,000; and for property crime, the estimate was 524.5 per 100,000.
- Although the number of arrests for violent crimes increased less than 1 percent (0.3) when compared with arrest data from 2005, arrests for robbery offenses rose 8.6 percent.
- Arrests of juveniles (under 18 years of age) for murder rose 3.4 percent in 2006 compared with 2005 arrest data; for robbery, arrests of juveniles increased 18.9 percent over the same 2-year period.
- In 2006, 76.3 percent of all persons arrested were male, 82.2 percent of persons arrested for violent crime were male, and 68.8 percent of persons arrested for property crime were male.
- Among the four categories of race reflected in UCR arrest data, 69.7 percent of persons arrested were white, 58.5 percent of persons arrested for violent crime were white, and 68.2 percent of persons arrested for property crime were white.
- White juveniles comprised 67.1 percent of the juveniles arrested in 2006.
- Black juveniles accounted for 51.0 percent of the juveniles arrested for violent crime, and white juveniles comprised 66.3 percent of juveniles arrested for property crime. . . .

Source: U.S. Department of Justice. Federal Bureau of Investigation. *Crime in the United States, 2006.* September 24, 2007. www.fbi.gov/ucr/cius2006/index.html (accessed October 20, 2007).

Other Historic Documents of Interest

From previous *Historic Documents*

- FBI annual crime report for 2006, *2006,* p. 547
- FBI annual crime report for 2005, *2005,* p. 677
- Court on federal sentencing guidelines, *2005,* p. 680
- Sentencing commission on minimum cocaine penalties, *1997,* p. 245

DOCUMENT IN CONTEXT

Supreme Court on Constitutionality of Lethal Injections

SEPTEMBER 25, 2007

The number of death row inmates executed in the United States dropped dramatically in 2007, the result largely of the Supreme Court's decision on September 25 to hear a case challenging the constitutionality of the most common method used to put convicts to death. The Supreme Court was expected to hear arguments in January 2008 on whether the three-drug "cocktail" used by most capital punishment states inflicted an unnecessary risk of pain and suffering in violation of the Eighth Amendment. Recent botched executions in Florida and elsewhere that appeared to cause the inmate substantial and perhaps unnecessary pain had already led several states to stop executions temporarily while the protocols and procedures for administering the lethal injections were reviewed and litigation was resolved.

All told, forty-two people were executed during the year, the lowest number since 1994, when thirty-one death row inmates were executed. In 2006, fifty-three people were executed. The executions in 2007 took place in just ten states; the majority, twenty-six, occurred in Texas. South Dakota conducted its first execution in sixty years; the inmate had pleaded guilty and asked to be put to death even before he had exhausted all possible appeals. The number of death sentences also continued to decline, dropping to an estimated 110 in 2007.

Continuing Concerns about Capital Punishment

No one expected the Court to overturn capital punishment altogether, but there were increasing signs that concerns about the fairness, accuracy, and cost of imposing the death penalty were affecting public attitudes. Depending on the poll, roughly 60–69 percent of Americans said they supported the death penalty, and that percentage had remained stable for several years. When they were given a choice between sentencing someone to the death penalty and life in prison without possibility of parole, their support for the death penalty dropped below 50 percent.

In December, New Jersey became the first state to abolish its death penalty since the Supreme Court reinstated capital punishment in 1976. Governor Jon Corzine signed an abolition measure into law on December 17. The day before, he had commuted the sentences of eight death row inmates to life in prison without possibility of parole. That brought to thirteen the number of states that did not have the death penalty. In addition, the Death Penalty Information Center, an advocacy group opposed to capital punishment, said it no longer considered New York a death penalty state. The state's capital punishment law had been declared unconstitutional in 2004, the legislature had taken no action to correct the flaw, and no one had been sentenced to death or executed there since before 1976.

Concerns about racial disparities in imposing the death sentence coupled with fears that innocent people had been put to death because of shoddy police and legal work were among the factors leading more Americans to prefer life imprisonment without parole as an alternative to the death penalty. Three men sentenced to death—one each in North Carolina, Oklahoma, and Tennessee—were exonerated in 2007 when new evidence was uncovered, bringing to 126 the number of death row exonerations since 1976. On October 28, an American Bar Association committee that reviewed the way the death penalty operated in eight states recommended a moratorium on further executions until states had reviewed their procedures for fairness and accuracy. The ABA Death Penalty Moratorium Implementation Project said its review had uncovered a general failure to preserve DNA material until after a prisoner was either executed or released from prison, as well as incidences of misidentification by eyewitnesses, false confessions, and prosecutorial misconduct among other shortcomings. "After carefully studying the way states across the spectrum handle executions, it has become crystal clear that the process is deeply flawed. The death penalty system is rife with irregularity," said Stephen F. Hanlon, who chaired the moratorium project. The states studied were Alabama, Arizona, Georgia, Florida, Indiana, Ohio, Pennsylvania, and Tennessee.

The cost to the state of defending indigent defendants charged with capital crimes was leading some states to seek a life sentence without the possibility of parole rather than the death penalty. In many death penalty cases, states had to provide defense attorneys not only for the trial but also for the lengthy course of appeals that often followed. And in recent years, judicial rulings and new guidelines intended to ensure that defendants received full and adequate representation had pushed the costs higher. According to one study done in 2005 by a liberal research group, the New Jersey Policy Perspective, the state of New Jersey spent an estimated $256 million since 1983 on death penalty cases, including $60 million for defense. Not a single person was sentenced to death or executed in the state during that time period.

International pressure to halt executions worldwide also continued in 2007. On December 18, the United Nations General Assembly passed a nonbinding resolution calling on member countries "to progressively restrict" the use of the death penalty, for example, by reducing the number of capital offenses. The vote on the resolution, which stopped short of asking for an outright moratorium, was 104 to 54, with 29 abstentions. In previous years, resolutions calling for moratoriums were rejected.

Earlier in the year, Amnesty International reported that the number of executions worldwide fell by 25 percent to 1,591 people in 2006, from 2,148 in 2005. China had by far the highest number at 1,010, followed by Iran (177), Pakistan (82), Iraq (65), Sudan (65), and the United States (53). The human rights organization cautioned that the number of executions was likely higher than shown by official statistics in some countries. A law passed in China at the end of 2006 requiring all executions to be approved by the country's highest court was expected to reduce the number of executions there. But a spokesperson for Amnesty International said the organization was "particularly concerned about a disturbing 'revival' of executions in countries like Iraq, Sudan, and Pakistan." The report noted that 128 countries had abolished the death penalty altogether and that fewer than half of the 69 countries that still allowed capital punishment were "currently carrying out executions."

A Constitutional Standard for Lethal Injection

Since the Supreme Court reinstated capital punishment in 1976, all but one of the death penalty states used lethal injection (the exception was Nebraska, which still used the electric chair). Prisoners in most states received three drugs: first, an injection of sodium pentothal, a powerful barbiturate intended to make the prisoner unconscious; second, an injection of pancuronium bromide, a paralytic to prevent involuntary motions that might be disturbing to witnesses; and finally, a shot of potassium chloride, to stop the heart. The combination, which had been in use since the early 1980s, was intended to avoid the "unnecessary and wanton infliction of pain" that the Supreme Court had said would amount to cruel and unusual punishment in violation of the Eighth Amendment.

Problems occurred if the drugs were not administered properly or in the proper dosage. If the barbiturate injected first was not strong enough, for example, the paralytic given next would make the prisoner feel as if he or she were suffocating. Further, an inmate who had not been made unconscious by the barbiturate but was too paralyzed to move might suffer excruciating pain, described as feeling like one's blood is on fire, when administered the potassium chloride. Moreover, professional medical associations had advised doctors and other medical personnel not to participate in executions, so the task of administering the drugs usually fell to prison personnel, who were often poorly trained in the procedures. Inserting IV lines, through which the drugs would be injected, could be difficult in an inmate whose veins had been damaged by past drug abuse or who was anxious about his or her impending death.

In 2006, several botched executions brought these problems to the attention of the news media and the public. In May 2006, an Ohio inmate told his executioners that the injections were not working. In December 2006, Florida inmate Angel Nieves Diaz took thirty-four minutes to die, about twice as long as normal, and required two rounds of the lethal chemicals. Florida governor Jeb Bush imposed a moratorium on further executions until the state had reviewed its protocols for administering the drugs. (That moratorium was lifted in 2007, but no executions were conducted pending the Supreme Court's decision on lethal injection.) Executions were also suspended at least temporarily in California, Maryland, and South Dakota.

The picture grew more confused in the first months of 2007 as state and federal courts issued conflicting rulings on whether lethal injection procedures ran the risk of inflicting "unnecessary and wanton" pain. Several more states temporarily halted executions while their protocols were reviewed. Some states went ahead with planned executions, using essentially the same procedures that courts later found objectionable. In Tennessee, for example, the governor halted executions for three months while the state revised its guidelines governing lethal injections. Soon after the moratorium was lifted, an inmate was executed in early May. Four months later, in a separate case, a federal judge ruled that Tennessee's procedure was unconstitutional because it presented a "substantial risk" of unnecessary pain.

On September 25, the Supreme Court announced that it would hear arguments in a case from Kentucky, *Baze v. Rees,* challenging whether the three-drug lethal injection procedure was constitutional. The two inmates argued that the risk that the three-drug protocol would produce needless pain was unnecessary because there were alternative methods of lethal injection virtually guaranteed to be painless. One alternative was the standard veterinarians had adopted for euthanizing animals—a massive dose of barbiturate that, if administered properly, was believed to cause death painlessly. They asked that the standard for determining whether a method of execution was unconstitutional be broadened to include "unnecessary risk." Their argument was rejected by the Kentucky Supreme Court, which ruled that the three-drug protocol did not present a "substantial risk" of pain and suffering. "The prohibition is against cruel and unusual punishment," the state court wrote, "and does not require a complete absence of pain." The U.S. Supreme Court was scheduled to hear arguments in the case in January 2008 and to issue its decision before the term ended near the end of June. According to papers filed in the case, the Supreme Court has not reviewed the constitutionality of a method of execution since 1878, when it upheld Utah's use of a firing squad.

In agreeing to hear arguments in the Kentucky case, the Court did not issue a blanket moratorium on executions. Later that day, Texas went ahead with a scheduled execution, after the presiding judge of the Texas Court of Criminal Appeals declined to delay the court's closing time for twenty minutes so that the inmate's attorneys could file a rush appeal. It was the state's twenty-sixth execution of the year, and the last for the year anywhere in the nation. Three days later, on September 28, the Supreme Court stayed the execution of another Texas inmate slated to be killed later that day. In mid-October and again on October 30, the Court granted two more stays of execution, actions that taken together indicated the Court would block all executions until it announced its decision in the Kentucky case.

Following is the text of the petition for a writ of certiorari (review) filed with the Supreme Court in the case of Baze v. Rees, *asking the Court to "provide guidance to the lower counts on the applicable legal standard for method of execution cases." The Court on September 25, 2007, agreed to hear arguments on the first three questions set out in the petition.*

DOCUMENT

Supreme Court Order Granting Certiorari for Baze v. Rees

September 25, 2007

QUESTIONS PRESENTED:

Although the Court has authorized civil actions challenging portions of a method of execution, it has not addressed the constitutionality of a method of execution or the legal standard for determining whether a method of execution violates the Eighth Amendment in over 100 years—leaving lower courts with no guidance on the law to apply to the many lethal injection challenges filed since the Court's rulings allowing the claim in a civil action. Lower courts have been left to look to cursory language in the Court's opinions dealing with the the [sic] death penalty on its face and prison conditions. As a result, the law applied by lower courts is a haphazard flux ranging from requiring "wanton infliction of pain," "excessive pain," "unnecessary pain," "substantial risk", "unnecessary risk," "substantial risk of wanton and unnecessary pain," and numerous other ways of describing when a method of execution is cruel and unusual.

Considering that at least half the death row inmates facing an imminent execution in the last two years have filed suit challenging the chemicals used in lethal injections, certiorari petitions and stay motions on the issue are arriving before the Court so often that this issue is one of the most common issues. Thus, it is important for the Court to determine the appropriate legal standard, particularly because the difference between the standards being used is the difference between prevailing and not.

This case presents the Court with the clearest opportunity to provide guidance to the lower courts on the applicable legal standard for method of execution cases. This case arrives at the Court without the constraints of an impending execution and with a fully developed record stemming from a 20-witness trial. The record contains undisputed evidence that any and all of the current lethal injection chemicals could be replaced with other chemicals that would pose less risk of pain while causing death than the tri-chemical cocktail currently used. Although this automatically makes the risk of pain associated with the use of sodium thiopental, pancuronium bromide, and potassium chloride unnecessary, relief was denied on the basis that a "substantial risk of wanton and unnecessary pain" had not been established. This squarely places the issue of whether "unnecessary risk" is part of the cruel and unusual punishment equation and whether an "unnecessary risk" exists upon a showing that readily available alternatives are known.

The Kentucky Supreme Court's decision gives rise to the following important questions:

I. Does the Eighth Amendment to the United States Constitution prohibit means for carrying out a method of execution that create an unnecessary risk of pain and suffering as opposed to only a substantial risk of the wanton infliction of pain?

II. Do the means for carrying out an execution cause an unnecessary risk of pain and suffering in violation of the Eighth Amendment upon a showing that readily available alternatives that pose less risk of pain and suffering could be used?

III. Does the continued use of sodium thiopental, pancuronium bromide, and potassium chloride, individually or together, violate the cruel and unusual punishment clause of the Eighth Amendment because lethal injections can be carried out by using other chemicals that pose less risk of pain and suffering?

IV. When it is known that the effects of the chemicals could be reversed if the proper actions are taken, does substantive due process require a state to be prepared to maintain life in case a stay of execution is granted after the lethal injection chemicals are injected?

CERT. GRANTED 9/25/2007

. . . THE PETITION FOR A WRIT OF CERTIORARI IS GRANTED LIMITED TO QUESTIONS 1, 2, AND 3 PRESENTED BY THE PETITION.

SOURCE: Supreme Court of the United States. *Baze v. Rees.* Docket 07-5439. Order granting petition for a writ of certiorari. September 25, 2007. www.supremecourtus.gov/qp/07-05439qp.pdf (accessed October 21, 2007).

Other Historic Documents of Interest

From previous *Historic Documents*

- Questions about lethal injections, *2006*, p. 339
- Statistics on capital punishment, *2004*, p. 794
- Supreme Court on fairness of the death penalty, *2001*, p. 387
- Restoraton of the death penalty, *1973*, p. 305

DOCUMENT IN CONTEXT

Prime Minister Surayud on a "Sustainable" Democracy in Thailand

SEPTEMBER 26, 2007

Fourteen months after Thailand's military staged a coup that ousted the elected government and created a new government in its place, Thai voters essentially reversed the coup in the December 23 election by giving broad support to a reincarnation of the political party the generals had ousted. The new People Power Party (PPP) had most of the characteristics of the ousted Thai Rak Thai (TRT) party. Missing, however, was the party's controversial founder: former prime minister Thaksin Shinawatra, who during his five years in office had manipulated the democratic process to accumulate exceptional power. Thaksin was out of the country at the time of the coup in September 2006 and remained abroad all through 2007. At year's end, he said he was considering returning home, but not to politics.

The generals had ousted the controversial Thaksin, a billionaire businessman, following several months of political turmoil resulting from protests over a decision by Thaksin's family to sell a minority stake in his telecommunications company to an investment firm owned by the government of Singapore. The generals installed what they described as an interim government, headed by retired army general Surayud Chulanont, and promised to restore democratic rule as soon as possible. Most Thais believed the real power in the new government was Gen. Sonthi Boonyaratglin, who was army commander at the time of the coup but later stepped down from that post to become deputy prime minister.

Late in the year, Thais set aside their political differences for at least one day: December 5, the eightieth birthday of King Bhumibol Adulyadej, who had been on the throne since 1946 and thus was the world's longest serving monarch. Many Thais revered King Bhumibol as a divine figure and source of national stability. Military officials close to him had been involved in the 2006 coup, but this did not appear to diminish his popularity, as demonstrated by the tens of thousands of people who gathered in Bangkok for the birthday celebration—many of them wearing yellow T-shirts in honor of the royal color.

Banning Thaksin's Party

After the 2006 coup, the military government brought charges against the country's two largest political parties—Thaksin's TRT party and the long-established Democrat Party—for alleged violations of election laws during the failed elections in April 2006. The country's Constitutional Tribunal ruled in the cases on May 30, 2007, clearing the Democrat Party of six counts of cheating but convicting Thaksin's party of bribing several smaller parties to help it skirt rules on minimum turnout in legislative districts. As punishment, the court disbanded the party and barred Thaksin and 110 other senior party officials from engaging in politics for five years.

Party leaders counseled their followers to accept the ruling because the party would register under a new name and its leaders would remain active politically, though not as candidates. "This is not the last day of Thai Rak Thai, brothers and sisters," a party spokesman told about 1,000 supporters who gathered in a protest the day after the ruling was released.

The party later emerged in its new guise of the People Power Party, headed by Samak Sundaravej, a conservative politician and former governor of Bangkok known for his pugnacious style. Samak made it clear he was standing in for Thaksin, who he said had asked him to take the job. Many of the party's candidates were new to national politics, however, because more than 100 of the top leaders remained barred from seeking office.

Charges against Thaksin

A new line of struggle between the military government and Thaksin began on June 11, when a government anticorruption commission announced plans to seize about $1.5 billion in bank accounts held by Thaksin and his wife. The panel said "Thaksin and his cronies had been corrupt and committed wrongdoings." The next day, Prime Minister Surayud told reporters he would personally guarantee Thaksin's safety if the deposed prime minister wanted to return to Thailand to contest the seizure of his assets—but Thaksin would not be permitted to engage in politics.

The government stepped up its pressure on Thaksin a week later, ordering him to return to Thailand by the end of June or face an arrest warrant on corruption charges. Thaksin's lawyer said that despite the prime minister's reassurances, the former prime minister was worried about his safety should he return home. Thaksin did not return, and on August 14, the Supreme Court in Bangkok issued arrest warrants for him and his wife on charges involving a disputed land transaction. However, the court on September 25 suspended Thaksin and his wife's trial on the charges when they failed to appear at a hearing.

A New Constitution

One of the generals' key achievements of the year was to draft, and win public support for, a new constitution—the eighteenth since the end of absolute rule by the monarchy in 1932. This constitution replaced a 1997 version that had been intended to stabilize Thai politics by eliminating political procedures that had led to the several short-lived, ineffective coalition governments during the 1980s and 1990s.

Drafted early in the year by a committee appointed by the generals, the new constitution was widely viewed as weakening the role of political parties and increasing the influence of unelected officials. One of the most important clauses abolished an elected upper house of parliament and replaced it with a senate, nearly half of whose members would be appointed by a seven-member committee of judges and government officials. The constitution also reduced the lower house from 500 to 480 seats. Other provisions appeared intended to reduce the influence of major parties and increase the role played by the country's numerous smaller parties—in effect, setting the stage for the type of unsteady coalition governments that were in power before Thaksin took office and consolidated power in his own hands.

Despite these controversial provisions, just under 58 percent of voters approved the constitution in an August 19 referendum. The "no" vote of about 42 percent was widely characterized as a protest vote by supporters of Thaksin's party. Tulsathit Taptim, editor of *The Nation* newspaper, editorialized prior to the referendum that the vote would not be a decisive step in Thailand's political history; rather, it would only "confirm that we are a divided nation blindly struggling for the true meaning of democracy."

Approval of the constitution set in motion a series of events leading up to elections, starting with an August 27 announcement that the voting would be held on December 23. Previously, the government had said only that elections would be scheduled by the end of the year.

As the political campaign got under way, General Surayud, the prime minister, visited New York City in late September in what seemed to be an effort to improve the government's international image. In a September 26 address to the Asia Society—a group of business leaders, academics, and others promoting U.S. interests in Asia—and in a September 27 address to the United Nations General Assembly, Surayud said Thailand had made "great reforms" and was moving rapidly to restore democracy. "We know we must do away with money politics," he told the Asia Society. "We know we must promote good governance. We know we must create a political culture driven by public service rather than greed. These are challenges that cannot be resolved overnight, or even in one year. What my government has done is try to lay a foundation upon which future governments can build."

Despite these words, it soon emerged that one aspect of the government's democratic foundation was the drafting of legislation giving the military broad powers to intervene again in politics. An internal security act, drafted in September, proposed to allow the military to respond as it saw fit in response to security threats—without the necessity of declaring a state of emergency or martial law. The government withdrew the draft law in October in the face of widespread criticism that the military had no intention of restoring democracy. The revised version of the law adopted by the government in December omitted the clauses giving the generals such sweeping powers.

The military also stirred up controversy with plans—elaborated in a memorandum that was made public in October—to discredit the new PPP through a covert propaganda campaign. PPP officials complained about this memorandum to the government's electoral commission, which ruled that the plan did not violate the law because it had not been carried out.

The Election

Thais went to the polls on December 23 and handed a strong—but not overwhelming—victory to the PPP. The new party won 233 seats in parliament, just eight votes short of a majority of the 480 total seats. These results were better than had been expected based on opinion polls taken earlier in December. The PPP drew much of its support, as had Thaksin's predecessor party, from poor and middle-class areas in northern Thailand. Thaksin had built a deep base of support in these areas with social programs, such as low-cost health care and loans for small businesses. The Democrat Party, which had been openly backed by the military and much of Thailand's political establishment, won 165 seats; this was a stronger showing than it made in the previous elections in 2006 but not enough to form a government without getting support from nearly all of the five smaller parties that won seats. Under the new constitution, the upper chamber of parliament, the

Senate, was to include some members chosen by judges and government officials, with the rest selected in a subsequent election, likely to be held in early 2008.

PPP leader Samak said on election night that he had already begun discussing formation of a coalition government, which he would head as prime minister. And he had this warning for the military, should it be considering another coup: "Please think carefully."

A jubilant Thaksin said at a news conference in Hong Kong on December 25 that he would return to Thailand sometime early in 2008—but not as a political figure. "I will not take any political position," he said.

PPP leaders reported on December 26 that they had secured enough support from smaller parties to take control of the new government, with a comfortable parliamentary majority of 254 seats. One possible obstacle to the party's plans was an investigation by the national electoral commission into allegations that numerous candidates had violated stringent campaign regulations. The commission in the past had used even minor infractions of the regulations to disqualify large numbers of winning candidates. At year's end, however, it was unclear whether the commission would intervene again in a manner that might overturn the results of an election just one year after a military coup.

Following are excerpts from a September 26, 2007, address by retired general Surayud Chulanont, then prime minister of Thailand, to a meeting of the Asia Society in New York City.

DOCUMENT

Prime Minister Surayud on Political Developments in Thailand

September 26, 2007

. . . It is a great honour to address the distinguished members and guests of the Asia Society today. As you are all friends of Thailand, or at least friends of Asia, I must say I feel quite at home. At the very least, I know that here, if I say I am Thai, no one will ask what I think about reunification with the mainland! This confusion is perhaps not too surprising. After all, these days most people learn about world affairs through bite-sized video chunks on cable news. The need for speed in this day and age often means we sacrifice depth for instant analysis, and understanding for easy formulas. Over the past year, much has been written in the Western press about Thailand. Most of it has been based [on] one simple premise: military coup overthrows popularly elected prime minister. And that premise conjures up all sorts of stereotypes—about the military, elected political leaders, and democracy. Stereotypes are handy things.

They allow us to judge based on very little information. But as those of you who are area specialists know, if you want to really understand something, you need to go beyond generalities. Today I would like to share with you what Thailand has been doing this past year to make our democracy and our development more sustainable. Let me start with where we stand today. Thailand has a new Constitution that was approved by a majority of voters on 19 August, in the first-ever national referendum. This gave the green light for election preparations to proceed. My Government has set the election date for 23

December. As we speak, political parties are gearing up and preparing their election campaigns. The economy is picking up; business confidence is also up. The public, in general, is eager to move on.

The key question, of course, is this: will the elections usher in a period of sustained democracy, or will it lead to another political cycle of corruption, culminating in another military intervention? As a retired professional soldier, I can say to you in all frankness that I have no taste for politics, and the Thai military as a whole shares this sentiment. The past year has thus been a time-out in Thailand's democratic evolution. It has been a time to take stock, review what went wrong, adjust our strategy and resume our efforts. The Thai people have always desired democracy. They have fought for it and they have died for it. The desire for democracy has taken deep root in Thailand. But we have not paid enough attention to the conditions for democracy to be sustainable. Thailand's constitutional rule began 75 years ago. Yet our democratic development has been uneven. When the previous government was elected to power with a commanding majority, many Thais had high hopes. Unfortunately, the axiom that absolute power corrupts absolutely proved correct once again.

Still, despite all the missteps, all the stops and starts, Thailand has always returned to the path of democracy. And after every ordeal, democracy has always emerged stronger. That is what we have to look forward to. To be sure, much needs to be done to further strengthen democracy and make it sustainable. And I think we have made a good start. The new Constitution aims to strengthen democracy where it counts most—accountability, rule of law, respect for civil and human rights, and public participation. It encourages ethics, transparency and predictability in policy processes. Some have criticized the new charter as less perfect than the previous one, the so-called "People's Charter" of 1997. But the future will decide. To avoid the mistakes of the past, to make democracy sustainable, we must put what we have learned into practice. We know we must do away with money politics. We know we must promote good governance. We know we must create a political culture driven by public service rather than greed. These are challenges that cannot be resolved overnight, or even in one year. What my Government has done is try to lay a foundation upon which future governments can build.

Distinguished Guests, Ladies and Gentlemen,

Another aspect of Thailand that must be made sustainable is our economic development. For the past several decades, Thailand has been one of the fastest growing economies in the world. That growth, however, came at a high cost. As the income gap between rich and poor widened, as the environment declined, as families and communities broke up, we realized that what we needed was quality growth, growth that is sustainable, inclusive and human-centred. For guidance, we have looked to the Sufficiency Economy philosophy of His Majesty the King. Since ascending the throne over 60 years ago, His Majesty has travelled to every corner of the country, learning about his people and the problems they faced daily. He distilled what he learned into a philosophy that, if practiced, would foster inner peace and sustainable development. But the country only took notice when the Asian economic crisis struck in 1997.

Since then, Sufficiency Economy virtues such as moderation, rationality, mindfulness and strengthening one's inner resilience have come to be recognized as compatible with such post-crisis concerns as good governance, sustainable development and risk management. Indeed, the solid business performance of many of our blue-chip companies and SMEs [small and medium-size enterprises] attest to the benefits of this philosophy.

Some have mistaken the Sufficiency Economy approach as an inward-looking one. Far from it. The importance given to the development of inner strength is exactly so that one may engage fruitfully with the outside world. Thailand is not just Bangkok. Overall living standards have improved substantially, but much of the countryside remains poor and undereducated. Through His Majesty's teachings, our poor are learning to lift themselves out of poverty in a way that is sustainable. The philosophy can also be usefully applied to families, communities, organizations, the country itself and even well beyond.

Distinguished Guests, Ladies and Gentlemen,

Sustainable democracy and sustainable economic development, important as they are, would be of less value if our society is not at peace. Thailand faces a delicate and volatile situation in our southern border provinces. Militants who hide behind religion and culture have killed thousands of innocents, including women, teachers and monks. Borrowing techniques of terror from elsewhere, they have apparently sought to be associated in the public's mind with international terrorism. However, from all intelligence available, we have found no such links.

To build a sustainable, harmonious society, we have tackled the problem at its roots. We recognize that the situation in the southern border provinces is not one of religious conflict or discrimination. Thais of all faiths have lived side by side for generations. Rather, the situation involves political, economic, social and cultural issues. The overall situation has improved significantly over recent months, as we have gained more trust and cooperation from the local people. We realize that we're facing an extremely sensitive issue. And so we work closely with the Organization of the Islamic Conference (OIC), which understands the situation and appreciates our peaceful approach that relies on perseverance, rule of law, and tolerance. The Malaysian Government has also pitched in to help train and educate our young Muslims on the so-called 3 Es—education, employment, and entrepreneurship. The situation has taken decades to reach the current crisis point. We hope it will take much less time than that to resolve.

Distinguished Guests, Ladies and Gentlemen,

Compounding the many challenges confronting us is the way our friends perceive us. I think by now most of them realize that my Government is earnest about creating a more sustainable, more democratic future for our country. However, there have been fears that Thailand might turn to economic nationalism and shut its door on foreign investors. That perception couldn't be farther from the truth. The proposed amendment of the Foreign Business Act, for example, was an attempt to realign the legal and practical aspects of doing business in Thailand, not to raise the entry bar for foreign business. In the long run, it should even help make doing business in Thailand a lot easier.

The fact is that the Thai economy is so deeply integrated into the international economy that it would be unimaginable for Thailand to turn inward or try to do without foreign investment. In fact, efforts to modernize many aspects of our economy are well under way. For example, the Second Financial Master Plan is expected early next year, while several mega-projects are getting the go-ahead to upgrade our infrastructure. Indeed, our efforts need the support of our friends to fully bear fruit. And few of our friends are as close to our hearts as the United States. Thailand and the United States go back a long way as allies and partners in so many ways. Next year we will celebrate 175 years of our relations. Thai soldiers fought alongside American comrades in all the major regional wars—Korea, Vietnam. More recently, in Iraq and Afghanistan, Thailand has sent technical and medical support teams. We appreciate the US's understanding of our political situation. I hope the US will continue to support Thailand along the reform path.

And increasingly these days, when the US deals with Thailand, you are also dealing with us as part of ASEAN [Association of Southeast Asian Nations]. With a combined population of over 566 million and a gross GDP of about 1,173 billion US dollars, ASEAN is the most populous emerging free trade area. In many ways ASEAN also embodies the hopes and dreams of our region. Now 40 years old, the grouping is regarded with hope as a vehicle toward sustainable development and democracy. It has embraced closer integration and community-building in all aspects. Always outward-looking, ASEAN is becoming more cohesive and rules-based with the drawing up of the ASEAN Charter. The progressiveness of the Charter has surprised some observers, but it shows ASEAN's growing maturity, which includes a newfound willingness to address issues such as human rights. And as this year sees the United States and ASEAN celebrate 30 years of relations, we must build on this solid past to forge a partnership to address the opportunities and challenges of the future.

Distinguished Guests, Ladies and Gentlemen,

Thailand and the United States share more than a long history of friendship. They also enjoy a special partnership that has survived the test of time. The past year has been an important though trying year for Thailand and for Thai-US relations. But I am optimistic about the future. Before long, Thailand will again be back on the path of full democracy. And we shall continue to look to US support as we attempt to strengthen the substance of democracy in such areas as good governance, transparency and respect for human rights. By enhancing our resilience and strengthening relations with our most important partners, such as the United States and ASEAN, we can look forward to a future where sustainable democracy and development bring peace and hope to generations to come.

Thank you.

SOURCE: Kingdom of Thailand. Royal Thai Government. " 'Thailand: Moving Toward a More Sustainable and Democratic Future.' Keynote Address by His Excellency General Surayud Chulanont (Ret.) Prime Minister of the Kingdom of Thailand at the Asia Society." New York, NY. September 26, 2007. www.thaigov.go.th/eng/index.aspx?PageNo=1&parent=467&directory=1941&pageid=467&pagename=content20&directory2=1942 (accessed March 30, 2008).

Other Historic Documents of Interest

From this volume

- Chinese president and rights activists on preparation for the Olympic games, p. 284
- Political crisis in Pakistan, p. 647

From previous *Historic Documents*

- Military leaders on coup in Thailand, *2006*, p. 555

October

DOCUMENT IN CONTEXT

President Bush on His Vetoes of Child Health Insurance Legislation

OCTOBER 3 AND DECEMBER 12, 2007

With the cost of health care and the number of uninsured Americans climbing steadily, an increasing number of government officials, businesses, insurers, and health care professionals explored options for helping a greater percentage of the population get health insurance coverage. Massachusetts became the largest state so far to mandate that its citizens have health insurance, and similar plans were being debated in California and other states in 2007. A broad coalition of consumer and business groups, doctors, and drug companies as well as a group of insurers offered two separate plans during the year aimed at sharply reducing the number of uninsured people. And health care reform was a hot topic among the candidates on the 2008 presidential campaign trail.

If the debate in Congress over expanding a federal program that provided health insurance to low-income children was any indication, reaching a consensus on any sort of universal health care plan was likely to be long and difficult. President George W. Bush twice vetoed an ambitious Democratic plan to expand the program to cover many middle-income families. Bush said the Democratic plan went in "the wrong direction" because it would encourage millions of families currently covered by private insurance to sign up for the government program. Democrats were unable to override and settled for a short-term extension of the existing program.

Costs of Health Care and Insurance

According to the federal Centers for Medicare and Medicaid Services, total spending on health care in the United States was projected to reach $2.2 trillion in 2007, or a little less than $7,500 for each U.S. resident. Slightly more than 50 percent of that total was spent on care provided by hospitals and doctors; prescription drug costs accounted for another 10 percent. About 45 percent of the total spending was covered by government programs such as Medicare and Medicaid. The remaining 55 percent was paid for through private spending, with private insurers picking up about 64 percent of the tab and individuals paying about 23 percent in out-of-pocket costs. The remainder was covered by other private sources such as charities.

About 60 percent of Americans received health insurance coverage through their employers. In September, the Kaiser Family Foundation, a nonprofit organization that studied health care costs, reported that premiums for employer-sponsored health benefits rose by an average of 6.1 percent in 2007, the smallest annual increase since 1999. Nonetheless, health care premiums rose 78 percent between 2002 and 2007, four times faster than either the rate of inflation or cumulative wage growth. In 2007 the average annual cost for family coverage premiums was $12,106. Because doctor and hospital costs were rising at a faster pace, the foundation said the slower increase in premium rates reflected cutbacks in coverage, with insurers requiring employees to pick up a larger share of their health care costs.

Rising insurance premiums were a major factor behind an increase in the number of Americans without either private or government-funded health insurance. According to the Census Bureau, that number rose to 47 million in 2006, from 44.8 million in 2005. (A separate survey by the federal Centers for Disease Control and Prevention found that about 43.6 million people did not have health insurance at the time of the survey. An estimated 54.5 million had no insurance for at least part of the year.) Even for many people who did have private insurance, public opinion polls showed that "affordable" health insurance was one of their chief worries.

State Universal Care Experiments

Ever since President Bill Clinton's health care reform proposal was declared dead in 1994, the victim of overambition and intense opposition from insurers and voters, politicians in Washington had shied away from talk about universal health care coverage. That reluctance began to change in recent years as the number of uninsured Americans grew larger and the economic costs of failing to provide adequate health care to millions of uninsured people became more apparent. In January, a coalition of sixteen business and consumer groups, insurers, health care organizations, and others put forth a plan that would provide insurance coverage for roughly half the uninsured population through a combination of expanded federal benefits programs and tax credits. Known as the Health Coverage Coalition for the Uninsured, the group included the AARP (formerly the American Association of Retired Persons), the American Hospital Association, the American Medical Association, the Blue Cross and Blue Shield Association, Kaiser Permanente, Pfizer, and the Chamber of Commerce of the United States of America.

In September, the Mayo Clinic proposed a plan that would guarantee all individuals basic insurance coverage by requiring private insurance companies to offer standard plans containing many different options. Individuals would pay for the plans, with some assistance from employers; low-income persons would receive help from the federal government on a sliding scale. A somewhat similar proposal was offered in December by the insurance industry's main trade group, America's Health Insurance Plans. Under this proposal, states would offer affordable coverage to people who were likely to incur very high medical bills and the insurers would agree to offer insurance to other high-risk people not eligible for the state programs. Premiums for that insurance would be capped at 150 percent of the market rate. Insurers would also have to submit any cancellations of coverage to a third party for review. "The health care industry is coming to grips with the fact that practices that are clearly driven by market forces are giving the industry a black eye," Paul B. Ginsburg, the president of the Center for Studying Health System Change, a Washington-based research group, told the *New York Times.*

Massachusetts became the first large state to require all of its residents to have health insurance or face a fine. Under the law, which was enacted in 2006, health insurers in the state were required to offer a range of affordable insurance options; low-income residents were eligible for subsidized plans; and employers with ten or more workers were required to offer insurance or pay a fee for each worker to help finance the program. Although it was too soon to know how well the law was working, several concerns about the program had already surfaced in 2007, including whether insurance companies could continue to afford the low premiums they were offering in the face of rising health care costs and whether taxes would have to increase to cover the costs of subsidized insurance.

Other states, most notably California, were considering similar plans. California's Republican governor, Arnold Schwarzenegger, proposed a mandatory health insurance plan, similar to the Massachusetts plan, which won approval from the Democratic-controlled state assembly in December. Passage by the Democratic-controlled Senate was uncertain, however; several senators were concerned about expanding the health care program in the face of an already large budget deficit.

The major Democratic presidential candidates were also supporting universal health care coverage. New York senator Hillary Clinton and former North Carolina senator John Edwards offered plans that would require individuals to have insurance; Illinois senator Barack Obama's plan would not. The major Republican candidates were also offering plans, including tax deductions and tax credits that they said would help currently uninsured people buy private insurance, but none of the plans was as inclusive as those proposed by the Democrats.

Expanded Children's Health Plan Vetoed

An effort by congressional Democrats to expand a popular health insurance program for children in low-income families quickly turned into a political dogfight over the future of the nation's health care system. Enacted by a Republican-controlled Congress in 1997, the State Children's Health Insurance Program (SCHIP) was designed to extend health insurance to children of low-income families who were not poor enough to qualify for Medicaid. By 2007, the program served 6.6 million children at an annual cost of about $5 billion. Most states covered children in families earning 200 percent or less of the poverty level ($20,650 for a family of four in 2007), but sixteen states had used a waiver procedure to extend coverage to families earning as much as 350 percent of the poverty level.

The battle was joined early in the year, when Democrats announced they wanted to expand the program by $50 billion or more over five years to help ease financial strains on middle-class families and to extend health insurance coverage to a large proportion of the 8 million children who were not currently insured. Democrats argued it was unconscionable for a country as wealthy as the United States to leave so many children without insurance. They were supported by many moderate Republicans in Congress and governors of both parties. Leading the opposition, President Bush said he favored taking care of poor children but objected to the expansion, viewing it as the first step in a Democratic plan to create a government-run health care system that he said would lead to huge increases in government spending. He and his supporters particularly objected to extending coverage to people making three times the poverty level, arguing that millions of families who currently had private insurance would abandon that coverage to rely on taxpayer-funded care. The president called for a $5 billion increase over five years and said Congress would just be wasting time if it sent him a broader expansion.

Gambling that enough Republicans would be unwilling to vote against expanded health care for children just a year before an election, Democrats decided to forge ahead despite the administration's objections. On August 1, the House approved a five-year, $50 billion expansion of the program. The next day, the Senate approved a less expensive version designed to win the support of more Republicans; that bill called for a $35 billion expansion that would cover an additional 4 million children from families earning up to three times the poverty level (about $62,000 for a family of four). The cost would be offset by an increase in tobacco taxes, including a 61-cent boost in the federal cigarette tax to $1 a pack. The House, in turn, agreed to that version on September 25 by a vote of 265–159. As expected, and with little fanfare, Bush vetoed the measure on October 3, saying that bill went "in the wrong direction," moving one of every three new subscribers from private to government insurance. Bush also objected to provisions of the bill that he said would expand coverage for adults and make families earning nearly $83,000 a year eligible for coverage, claims the bill's supporters said were wildly exaggerated.

House Democratic leaders delayed an override vote, hoping to build enough pressure on Republicans to support the bill, but that effort failed. The House on October 18 supported the motion to override by a vote of 273–156, falling thirteen votes short of the necessary two-thirds majority of those present and voting. In the next few days, Democrats sought to win over more House Republicans by tightening language related to eligibility, but the major outlines remained the same and the new version won no new Republican support. Bush vetoed the second bill on December 12. "Like its predecessor this bill does not put poor children first and it moves our country's health care system in the wrong direction," Bush said in his veto message. "Ultimately, our nation's goal should be to move children who have no health insurance to private coverage—not to move children who already have private health insurance to government coverage."

With an override attempt, scheduled for January 23, 2008, all but certain to fail in the House, Congress quickly passed a short-term extension of SCHIP that would provide enough funding to cover children already in the program though March 31, 2009, signaling that Democrats had little hope of enacting an expansion while Bush remained in the White House. Bush signed that legislation (S 2499–PL 110-173) on December 30.

Prescription Drug Costs and Retiree Benefits

Democrats were also unable to come through on another of their long-held priorities—giving Medicare administrators the power to negotiate prices for drugs purchased through the new Medicare Part D drug benefit. Democrats said such authority would lead to lower costs for seniors and for the government. The Republican-controlled Congress had expressly prohibited giving the government such authority when it created the benefit in 2003. The House easily passed legislation granting the negotiation authority in the first days of the 2007 session. But in the Senate the measure ran into problems, including a veto threat and strong opposition from the pharmaceutical industry as well as from conservative senators. A compromise that would allow but not require government negotiation could not win enough support to cut off a filibuster, and the measure died.

Meanwhile, prices for prescription drugs continued to climb. IMS Health, which gathered data for the pharmaceutical and health care industries, reported in March that U.S. prescription drug sales rose 8.3 percent in 2006, to $274.9 billion; in 2005 the increase was 5.8 percent. Prescriptions provided under the Medicare drug benefit accounted for 17 percent of retail prescription sales. Sales of unbranded generics also increased 22 percent during the year, for a total of $27.4 billion in sales.

In another development affecting seniors, the Equal Employment Opportunity Commission issued a final rule on December 26 allowing employers to reduce or eliminate health benefits for retirees when they turned sixty-five and became eligible for Medicare. Employers said the rising cost of health insurance premiums, combined with longer lives, made insurance coverage for their retirees over age sixty-five increasingly unaffordable. According to a lawyer for the commission, many employers and unions said they would drop retiree health benefits altogether if they had to provide the same benefits to retirees over sixty-five as they did to younger retirees. A spokesperson for the AFL-CIO told the *New York Times* that "given the enormous cost pressures on employer-sponsored health benefits, we support the flexibility reflected in the rule as a way to maximize our ability to maintain comprehensive coverage for active and retired workers."

Following are the texts of two messages from President George W. Bush, dated October 3, 2007, and December 12, 2007, vetoing legislation that would have significantly expanded funding for the State Children's Health Insurance Program.

DOCUMENT

President Bush's First Veto of Child Health Insurance Legislation

October 3, 2007

To the House of Representatives:

I am returning herewith without my approval H.R. 976, the "Children's Health Insurance Program Reauthorization Act of 2007," because this legislation would move health care in this country in the wrong direction.

The original purpose of the State Children's Health Insurance Program (SCHIP) was to help children whose families cannot afford private health insurance, but do not qualify for Medicaid, to get the coverage they need. My Administration strongly supports reauthorization of SCHIP. That is why I proposed last February a 20 percent increase in funding for the program over 5 years.

This bill would shift SCHIP away from its original purpose and turn it into a program that would cover children from some families of four earning almost $83,000 a year. In addition, under this bill, government coverage would displace private health insurance for many children. If this bill were enacted, one out of every three children moving onto government coverage would be moving from private coverage. The bill also does not fully fund all its new spending, obscuring the true cost of the bill's expansion of SCHIP, and it raises taxes on working Americans.

Because the Congress has chosen to send me a bill that moves our health care system in the wrong direction, I must veto it. I hope we can now work together to produce a good bill that puts poorer children first, that moves adults out of a program meant for children, and that does not abandon the bipartisan tradition that marked the enactment of SCHIP. Our goal should be to move children who have no health insurance to private coverage, not to move children who already have private health insurance to government coverage.

George W. Bush
The White House,
October 3, 2007.

SOURCE: U.S. Executive Office of the President. "Message to the House of Representatives Returning Without Approval the 'Children's Health Insurance Program Reauthorization Act of 2007.' " October 3, 2007. *Weekly Compilation of Presidential Documents* 43, no. 40 (October 8, 2007): 1298. Washington, D.C.: National Archives and Records Administration. www.gpoaccess.gov/wcomp/v43no40.html (accessed December 19, 2007).

DOCUMENT

President Bush's Second Veto of Child Health Insurance Legislation

December 12, 2007

To the House of Representatives:

I am returning herewith without my approval H.R. 3963, the "Children's Health Insurance Program Reauthorization Act of 2007." Like its predecessor, H.R. 976, this bill does not put poor children first and it moves our country's health care system in the wrong direction. Ultimately, our Nation's goal should be to move children who have no health insurance to private coverage—not to move children who already have private health insurance to government coverage. As a result, I cannot sign this legislation.

The purpose of the State Children's Health Insurance Program (SCHIP) was to help low-income children whose families were struggling, but did not qualify for Medicaid, to get the health care coverage that they needed. My Administration strongly supports reauthorization of SCHIP. That is why in February of this year I proposed a 5-year reauthorization of SCHIP and a 20 percent increase in funding for the program.

Some in the Congress have sought to spend more on SCHIP than my budget proposal. In response, I told the Congress that I was willing to work with its leadership to find any additional funds necessary to put poor children first, without raising taxes.

The leadership in the Congress has refused to meet with my Administration's representatives. Although they claim to have made "substantial changes" to the legislation, H.R. 3963 is essentially identical to the legislation that I vetoed in October. The legislation would still shift SCHIP away from its original purpose by covering adults. It would still include coverage of many individuals with incomes higher than the median income in the United States. It would still result in government health care for approximately 2 million children who already have private health care coverage. The new bill, like the old bill, does not responsibly offset its new and unnecessary spending, and it still raises taxes on working Americans.

Because the Congress has chosen to send me an essentially identical bill that has the same problems as the flawed bill I previously vetoed, I must veto this legislation, too. I continue to stand ready to work with the leaders of the Congress, on a bipartisan basis, to reauthorize the SCHIP program in a way that puts poor children first; moves adults out of a program meant for children; and does not abandon the bipartisan tradition that marked the original enactment of the SCHIP program. In the interim, I call on the

Congress to extend funding under the current program to ensure no disruption of services to needy children.

George W. Bush
The White House,
December 12, 2007.

Source: U.S. Executive Office of the President. Office of the Press Secretary. "Message to the House of Representatives." December 12, 2007. www.whitehouse.gov/news/releases/2007/12/20071212-10.html (accessed December 19, 2007).

Other Historic Documents of Interest

From this volume

- President Bush on housing crisis, p. 524

From previous *Historic Documents*

- Medicare drug benefit reform, *2003,* p. 1119; *2004,* p. 577
- Census Bureau on health insurance coverage, *2002,* p. 666; *2003,* p. 846

DOCUMENT IN CONTEXT

World Bank on Its Agriculture Programs in Africa

OCTOBER 12, 2007

Misguided policies, missed opportunities, and armed conflicts were major explanations for Africa's failure to keep up with most of the rest of the developing world in recent decades. These were the conclusions of major reports, issued in 2007, that sought to explain why sub-Saharan Africa continued to fall behind other regions by most measures of economic development and societal advancement.

An unusually frank internal assessment by the World Bank—by far the largest source of outside development aid for Africa—found in October that the bank itself had used its money and influence to impose inappropriate policies for agriculture, which was still the main source of income for most Africans. The study said the bank, and by extension Africa, failed to take advantage of recent advances in agricultural techniques that would have helped the continent's people feed themselves. The bank announced major policy changes, which had been in the works, later in the month.

Similarly, a landmark study by three major nongovernmental organizations estimated that civil wars and other conflicts since 1990 in twenty-three of Africa's fifty-three countries had cost the continent at least $284 billion. That amount was equal to what the World Bank and other donors had spent in Africa during the same period. If the money spent on weapons had instead been spent on education or health programs, some of Africa's most persistent problems could have been eliminated or sharply reduced, the study concluded.

The year 2007 marked the fiftieth anniversary since the beginning of the postcolonial period in Africa. In March 1957, Ghana became the first sub-Saharan African country to achieve independence (from Britain); by the early 1960s, nearly every other country in the region had gained independence from European colonial powers. Although the early decades of independence were full of ambition, the period was also marked by failure and strife, particularly during a region-wide economic downturn during most of the 1980s.

Much of Africa had made some progress since the mid-1990s. Most countries experienced regular economic growth, the overall level of poverty fell, and nearly all the wars ended eventually. Another World Bank report, issued on November 14, found that Africa had posted average economic growth gains of 5.4 percent during 2005 and 2006—on par with most other regions. A significant factor in that growth, however, was the rapid rise in world oil prices starting in 2004. This pumped billions of dollars into the seven African countries that produced oil but drained billions of dollars away from most of the other countries, according to the bank's African Development Indicators study. Zimbabwe was the only African country regressing in 2007; the failed policies of long-time president Robert Mugabe had sapped the life out of what had been one of Africa's most vibrant economies.

Agriculture in Africa

The study of the World Bank's lending to support agriculture in Africa was conducted by the bank's Independent Evaluation Group, its in-house watchdog office that reported to the bank's board of directors rather than its management. The group regularly conducted frank assessments of the bank's activities, but its report released on October 12, entitled "World Bank Assistance to Agriculture in Sub-Saharan Africa," was unusually blunt. Perhaps just as important as the study itself was the wide attention it received as the result of a report about it in the *New York Times* on October 15, just a few days before world finance ministers were to convene in Washington for the bank's annual meeting.

The general thrust of the evaluation was that the World Bank—along with other international donors and many African nations themselves—had neglected agriculture during previous decades, even though farming was the principal form of livelihood for about two-thirds of the people in sub-Saharan Africa, generated about one-third of the income, and, of course, provided most of the region's food. The bank had been successful in urging many African countries to adopt free-market economic policies, the report said, but had failed to provide adequate support to enable ordinary Africans to take advantage of the free markets. These failures included lack of investment in such essentials as irrigation, better roads so farmers could get their crops to market, advice to farmers on sustainable land-use policies, and low-cost credit for farmers to buy the equipment and seeds needed for improved yields.

The report noted that despite the region's dependence on agriculture, sub-Saharan Africa had been a net importer of food since 1973. "Since that time, food production has not kept pace with the rapidly growing population, and food imports have grown rapidly," the report said. Because of inadequate investments, it added, cereal yields were less than half of those in South Asia and one-third of those in Latin America. The production gains Africa did experience resulted from cultivating more land rather than from increasing per-acre yields.

The Independent Evaluation Group offered an extensive series of recommendations for increasing the bank's focus on, and assistance for, agricultural development in Africa. Many of the recommendations dealt with improving the bank's own technical capabilities so it could provide better advice for African farmers and governments. The study noted that the bank had only seventeen technical experts assigned to Africa in 2006, compared with forty experts a decade earlier.

In its formal response, the bank's management team generally agreed with the analysis in the study and disputed only some of the points. The management response said the bank had "already moved" to restore a greater emphasis on agriculture "and strengthen the sector's contribution to the reduction of poverty" in the region.

Development Report

The World Bank's more important response to criticism may have come in the 2007 edition of its annual development report, which for the first time in a quarter-century was focused on agriculture. Released on October 19, "Agriculture for Development" called for major policy changes by the bank and other donor institutions as well as by African governments on the full range of issues affecting agriculture. In particular, the report focused on the need for policies to help ordinary farmers get access to credit and markets, which would help them increase their production and sell the commodities at prices that would give them adequate returns on their investments and labor.

"While the worlds of agriculture are vast, varied, and rapidly changing, with the right policies and supportive investments at local, national and global levels, today's agriculture offers new opportunities to hundred of millions of rural poor to move out of poverty," the report said.

The World Bank's new president, Robert B. Zoellick, had signaled his interest in promoting agricultural development, particularly in Africa, during a high-profile speech to the National Press Club on October 10 concerning globalization. Citing the "green revolution" that used new growing techniques to power food production in Asia during the second half of the twentieth century, Zoellick said: "We need a twenty-first century green revolution designed for the special and diverse needs of Africa, sparked by greater investments in technological research and dissemination, sustainable land management, agricultural supply chains, irrigation, rural micro credit, and policies that strengthen market opportunities while assisting with rural vulnerabilities and insecurities. More countries need to open their markets to farm exports, too."

Several other reports made public during the year identified one of Africa's most critical needs as improving soil- and water-conservation methods so the region could reduce the expected impact of climate change. Some scientists had suggested that agricultural production in sub-Saharan Africa could decline by a third or more by the middle of the twenty-first century due to drought and other factors brought on by climate change. Agriculture experts said climate change could be especially damaging to Africa's topsoil, which in many regions was perilously thin and already was being stripped by wind and rain. "We're losing 400 million tons of soil every year," said James Breen, an emergency agronomist for the Food and Agriculture Organization, said in a September 25 report on "cutting edge" farming techniques in Africa by the United Nations Office for the Coordination of Humanitarian Affairs. "The production of this year's food crop is shockingly low and it's going to get worse with global warming. We really are facing a meltdown."

The Cost of Conflicts

Another report issued in 2007 suggested that much of the work to promote economic development in Africa had been undone by the many conflicts that had plagued the region, particularly those since 1990. The study, "Africa's Missing Billions: International Arms Flows and the Cost of Conflict," was produced by three British-based nongovernmental groups: Oxfam International, the International Action Network on Small Arms (IANSA), and Saferworld. The latter two organizations lobbied international governments and at the UN in favor of controls on international arms sales.

Using economic indicators, the study concluded that civil wars in twenty-three African countries had cost at least $284 billion in lost economic production. The authors said this was a conservative figure because it included only production that was lost during periods of actual conflict and therefore did not count societal costs once the fighting had stopped. The study also did not estimate the cost of civil conflict to neighboring countries in terms of lost trade or dealing with refugees.

The direct costs of conflict were obvious, the study noted, and included military and medical expenses and the destruction of infrastructure. Indirect costs were higher and included damage to the economy and the diversion of natural resources by self-interested individuals or groups. "More people, especially women and children, die from the fall-out of conflict than die in conflict itself," the study noted.

The three groups sponsoring the study used it to promote a proposed Arms Trade Treaty, which was being negotiated in the UN. The proposed treaty would ban global arms transfers if the weapons were likely to be used to commit serious violations of international human rights and humanitarian laws.

Following is the executive summary from "World Bank Assistance to Agriculture in Sub-Saharan Africa: An IEG Review," a report prepared by the Independent Evaluation Group of the World Bank and made public on October 12, 2007.

DOCUMENT

Evaluation of World Bank Agriculture Programs in Africa

October 12, 2007

Sub-Saharan Africa is a highly complex Region of 47 countries with 7 distinctly different colonial histories. It is also highly diverse, with more than 700 million people and at least 1,000 different ethnic groups. The Region is a critical development priority. It includes some of the world's poorest countries, and during the past two decades the number of poor in the Region has doubled to 300 million—more than 40 percent of the Region's population. Africa remains behind on most of the Millennium Development Goals (MDGs) and is unlikely to reach them by 2015.

A major drag on Africa's development is the underperformance of the agriculture sector. This is a critical sector in the Region, because it accounts for a large share of gross domestic product (GDP) and employment. The weak performance of the sector stems from a variety of constraints that are particular to agriculture in Africa and make its development a complex challenge. Poor governance and conflict in several of the countries further complicate matters. IEG [Independent Evaluation Group] has assessed the development effectiveness of World Bank assistance in addressing constraints to agricultural development in Africa over the period of fiscal years 1991–2006 in a pilot for a wider assessment of the Bank's assistance to agriculture worldwide.

The central finding of the study is that the agriculture sector has been neglected by both governments and the donor community, including the World Bank. The Bank's strategy for agriculture has been increasingly subsumed within a broader rural focus, which has diminished its importance. Both arising from and contributing to this, the technical skills needed to support agricultural development adequately have also declined over time.

The Bank's limited—and, until recently, declining—support for addressing the constraints on agriculture has not been used strategically to meet the diverse needs of a sector that requires coordinated intervention across a range of activities. The lending support from the Bank has been "sprinkled" across various agricultural activities such as research, extension, credit, seeds, and policy reforms in rural space, but with little recognition of the potential synergy among them to effectively contribute to agricultural development. As a result, though there have been areas of comparatively greater success—research, for example—results have been limited because of weak linkage with extension and limited availability of such complementary and critical inputs as fertilizers and water. Hence the Bank has had limited success in contributing to the development of African agriculture.

The Challenges of African Agriculture

Agricultural output has grown in Africa, but it is difficult to calculate a reliable growth rate for the Region over the study period because of wide variations across countries and over time. Some countries, such as Gabon, moved from poor performance in 1990–2000 to better performance in 2000–04; others, such as Malawi, moved in the opposite direction. The change has often been dramatic, which makes aggregate growth rates misleading. For example, agriculture in Angola grew at 13.7 percent a year during 2000–04, although growth had retreated by 1.4 percent yearly during 1990–2000. Only about a quarter of the countries in the Region, among them Benin, Burkina Faso, Ghana, Nigeria, and Tanzania, show consistent agricultural growth of over 3 percent in the 1990–2004 period.

Total agricultural output in Africa consists primarily of food crops. Agricultural export crops account for less than 10 percent of total production. While some export crops, including cotton, have contributed to poverty alleviation in countries such as Burkina Faso, food crops have performed poorly in most countries. Cereal yields in Africa, even in 2003–05, were less than half those in South Asia and one-third those in Latin America. Africa also lags behind other Regions in the percentage of cropland irrigated, fertilizer use, and labor and land productivity per worker. While the great strides in South Asia's agricultural production from 1961 to 2001 were mainly the result of increased yields, gains in food production in Africa were produced primarily through the expansion of cultivated land. Meanwhile, crop yields stagnated.

Beginning in 1973, Africa became a net food importer. Since that time, food production has not kept pace with the rapidly growing population, and food imports have grown rapidly. Meanwhile, Africa's exports, which are primarily agriculture-based, declined; for several commodities, including coffee, the Region's share of the world market evaporated. Agricultural subsidies in Organisation for Economic Co-operation and Development (OECD) countries have played a major role in keeping world prices low for several of these crops. This, among other factors, has impacted the adequacy of returns to farmers.

Agriculture in Africa is primarily a family activity, and the majority of farmers are smallholders who own between 0.5 and 2.0 hectares of land, as determined by socio-cultural factors. Women provide about half of the labor force and produce most of the food crops consumed by the family.

Agricultural land in Africa falls into several agroecological zones that run across countries. It is largely characterized by poor soils, highly variable rainfall, and frequent droughts. Transport infrastructure is poor, access to irrigation is limited, and under rain-fed conditions, chronic food insecurity is a reality for millions of small farmers. To survive in this harsh environment, most farmers rely on diversified coping strategies. To ensure at least some produce from their land, African farmers normally plant several varieties of crops (typically 10 or more) with different maturation periods, together with trees. Livestock is also an important source of security for farmers in Africa, particularly in lean years. The average smallholder's access to credit is also extremely limited. Hardy crops such as millet, sorghum, cassava, and other root crops are more important than cereals such as rice and wheat, which were the mainstay of the Asian Green Revolution.

In this environment, for farmers to have an incentive to practice intensive agriculture and take risks with new crop varieties, a number of factors need to come together at the same time, or at least appear in an optimal sequence, including improved seeds, water, credit, and access to markets; good extension advice; and adequate returns through undistorted prices for inputs and outputs. A strategy for development of agriculture in Africa must consider each of these factors in the context of Africa's unique characteristics and specific local conditions.

Past Approaches to African Agriculture

Until very recently, agricultural development in Africa was neglected by both governments and donors. During the 1960s, immediately following independence, governments in several African countries considered agriculture primarily a source of resources for industrialization. Then, in the 1970s, the World Bank led the shift toward a broader development model in Africa that was consistent with a more general shift in the understanding of development. This committed the institution to integrated rural development to directly attack Africa's rural poverty and underdevelopment. In the mid-1980s, when African countries faced severe fiscal crises, donors prioritized improvements in the efficiency of resource allocation and pressed agriculture marketing reforms. But structural reforms also fell short of producing the desired growth effects.

The Role of Aid

Bilateral and multilateral donor aid for development of African agriculture declined from \$1,921 million in 1981 to \$997 million in 2001 (in 2001 dollars). Lending from both sources has since rebounded with the increasing focus on African development. OECD data show that although bilateral donors as a group have played a comparatively larger role, the World Bank was the single largest donor to African agriculture between 1990 and 2005. The largest bilateral donors were the United States and Japan.

Foreign private sector flows into Africa are modest in comparison with bilateral and multilateral aid. . . . Private commercial investment in African agriculture has been largely limited to export crops and higher potential zones. A number of international seed companies have invested in maize seed multiplication, and in September 2006 the Rockefeller and Bill and Melinda Gates Foundations together launched a new partnership to help Africa develop its agriculture.

Agriculture's Potential and the Bank's Strategy

For Africa to meet the MDGs, it will be necessary to realize the potential of the agriculture sector, to provide the support needed for it to contribute to growth and poverty reduction. Research by Dorosh and Haggblade . . . and IFPRI . . . found that investments in agriculture generally favor Africa's poor more than similar investments in manufacturing.

The World Bank has not had a separate strategy for agriculture in Africa except as part of its wider rural development strategies, and over time the agriculture strategy was subsumed in a broader rural focus. More recently, however, the Africa Action Plan has recognized the agriculture sector as a potential driver of growth.

The Bank's Overall Assistance and Its Assessment

Over fiscal years 1991–2006, the Bank provided the countries of the Africa Region with $2.8 billion in investment lending (as distinct from adjustment lending) in agriculture, constituting 8 percent of total Bank investment lending to the Region. A large part of this lending has been in the form of agriculture components in rural projects. In addition, there have been 77 Development Policy Loans with agriculture components, and in 18 of these, agriculture was a significant dimension.

This limited investment lending has performed below par. IEG data show that the percentage of satisfactory outcome ratings for largely agricultural investment projects during 1991–2006 is lower than that for non-agriculture investments in the Region (60 against 65 percent satisfactory). It is also lower than the percentage for similar investment projects in other Bank Regions (73 percent satisfactory). Sustainability ratings are also below average. Although further analysis is needed, the study found that largely agricultural projects in countries with less favorable agricultural conditions have done better than similar projects in countries with more favorable conditions.

The Bank's activities in support of agricultural development in Africa have comprised lending, analytical work, and policy advice. Until very recently the analytical work—necessary for the diagnosis of issues and actions and to help shape the policy advice and lending—has been limited, scattered, of variable quality, and not easily available. In addition, IEG found that there are no specific procedures in place to ensure that the findings of analytical work are systematically reflected in lending and policy dialogue.

IEG found that the lending support provided by the Bank has not reflected the interconnected nature of agriculture activities. Rather, the lending has been "sprinkled" across an array of activities in rural space, including research, extension, marketing reform, drought relief, seed development, and transport, but with little recognition of the relationships among them and the need for all of these areas to be developed at the same time, or at least in an optimal sequence, to effectively contribute to agricultural development. While the Bank's broader rural focus from the mid-1980s was justified, an unintended result was that it led to less focused attention on the need for various activities that are critical for agricultural development in rural space to come together at the same time or to take place in some optimal sequence.

This review found that none of the top 10 borrowers, among them Côte d'Ivoire, Ethiopia, Tanzania, and Uganda, had received *consistent* and *simultaneous* support across all critical subsectors. That is not to suggest that the Bank should do this alone—it might well be done better in partnership—but the Bank could reasonably be expected to take the lead in fostering such a multifaceted approach, based on its comparative advantage as a multisector lending institution.

Thematic Performance

An assessment of the achievements and shortcomings in the Bank's support by main theme reveals a mixed record:

Agro-ecological diversity. Bank support has helped build the capacity of national research systems and develop zonal stations to give an agro-ecological focus to research.

However, there is little indication that Bank projects other than research interventions have systematically adapted activities to diverse country agro-ecological conditions. The ability to respond to local conditions has been the primary appeal of projects that use community-based approaches, but there is little evidence that these approaches, as used in projects in Ghana and Tanzania, for example, are able to respond to agro-ecological diversity.

Fluctuating rainfall and droughts. Bank projects completed through fiscal 2006 have been responsive to drought emergencies and have helped governments set up drought management systems. But they have not been able to help countries such as Malawi, for example, develop a long-term strategic approach to address the basic factors that create food insecurity—that is, to help countries increase agricultural productivity sufficiently to arrest declining per capita food availability. In this connection, while the Bank has contributed to development of improved millet and cassava varieties through support to research, it has missed the opportunity to recognize the important role that cassava can play in promoting food security in most countries.

Poor soil fertility. The Bank has been party to several international and regional initiatives on this issue, including the *Terr Africa Regional Initiative,* launched in 2005. This multidimensional partnership is expected to promote a collective approach to sustainable land management in the Region. But Bank lending appears to have addressed soil fertility more as an environmental than as an agricultural productivity issue.

Access to water. Though the Bank has identified the need for investment in irrigation, it has done very limited lending for that purpose. The Bank interventions that support water management in rain-fed areas have achieved physical targets, but because of poor monitoring and evaluation (M&E), it is difficult to tell what has worked and what has not.

Improved seeds. The Bank has contributed to the Consultative Group on International Agricultural Research (CGIAR), which has made significant contributions in this area, and Bank projects have also provided opportunity for testing and scaling up technologies, as in Ethiopia and Togo. Nonetheless, seed-related activities have so far made only a modest contribution to increases in crop production. Bank projects have also not been able to address the issue of limited use of seeds by farmers because of inadequate access to complementary inputs.

Farmers' access to credit and rural finance. Overall support from the Bank in this critical area has been limited. Aside from institutional capacity weaknesses in client countries, one reason for this low level of support has been weak project performance in this area, brought about by, among other things, weak implementation of Bank guidelines, particularly regarding eligibility and performance of financial intermediaries. There is need for the Bank to take greater care in designing and supervising these operations, and all options should continue to be explored for the most appropriate way to provide farmers with the means necessary to increase productivity and incomes.

Poor transport infrastructure. Bank-supported agriculture interventions have made only a limited contribution to improving transport infrastructure to promote market access for agricultural development.

Weak extension. The Bank has helped raise client awareness about the importance of extension to agricultural development. It currently supports a range of partnership approaches (public-private, demand-driven, nongovernmental organizations, and so on), as in Uganda. But the cost, effectiveness, and sustainability of these approaches need to be systematically evaluated.

Price and marketing reform. Though results have been variable across countries, the Bank's effort has contributed to improving the macroeconomic environment and fiscal discipline in several countries. However, these changes were not enough to stimulate private sector investments in several critical areas from which the public sector withdrew. Consequently, most countries in Africa face exorbitant fertilizer prices, inadequate seed production, poor transport, and limited credit access. While the reform process had limited positive impact on food production, it nevertheless boosted production of nontraditional export crops such as mangoes from Mali and flowers from Kenya. Beyond individual countries, the Bank lobbied for a genuinely pro-development Doha Round and for elimination of OECD agricultural subsidies in international forums, but with limited success to date.

Insecurity of tenure. Analytical work has contributed to a better understanding of property rights regimes. But the Bank has found it difficult to provide effective support in this area because of its political, social, and cultural sensitivity. The Millennium Development Project Hunger Task Force concluded in 2005 that the world could meet the MDG of halving hunger by 2015. Development of African agriculture is critical to achieving this goal, and the World Bank can make a major contribution because it is one of the largest sources of development finance for agriculture and can provide policy advice to governments.

Key Findings on Bank and Country Factors of Performance

Bank factors

- The institution's strategy for the development of the agriculture sector has been part of its rural strategy, and over time the importance of agriculture in the Bank's rural strategy has declined. Both arising from and contributing to this, technical skills to support agricultural development adequately have also declined over time. Data from the Human Resources Department of the World Bank show that there were 17 technical experts mapped to the Agriculture and Rural Development Department in Sub-Saharan Africa in 2006, compared with 40 in 1997.
- The Bank's diagnosis of a country's development status and priorities in the agriculture sector is carried out primarily through analytical work. Until very recently this work has been limited and not readily available. Nor have the findings from analytical work strategically informed Bank–client policy dialogue and lending program design.
- Bank policy advice appears to have had far-reaching implications for the direction of agricultural development in African countries, in particular its policy advice associated with the adjustment agenda. However, results have fallen short of expectations because of weak political support and insufficient appreciation of reality on the ground, among other things.
- The Bank's data systems and support for M&E have been insufficient to adequately inform the institution's effort to develop agriculture in Africa across a broad front. Current data systems do not allow the institution to track in

enough detail how much is being provided for development of specific activities such as seed development and credit. M&E at the project level has been of limited value in answering fundamental questions about outcome, impact, and efficiency, such as who benefited, which crops received support and how, what has been the comparative cost effectiveness, and to what can one attribute gains.

Country factors

- Although the governance environment in several African countries continues to be weak, political commitment for the development of agriculture in client countries appears stronger than in the past. African governments, many of which were allocating less than 1 percent of their budget to agriculture, agreed in July 2003 at the African Union Summit to allocate at least 10 percent of national budgetary resources for programs to support agricultural growth in the next five years.
- Considerable agricultural research capacity exists, although the sustainability of the activities supported remains uncertain. Overall, government capacity in several countries remains weak, and local agriculture ministries are still relatively ineffective partners in promoting development of the agriculture sector. Though further analysis is needed, the study finding that largely agricultural projects in countries with less favorable agricultural conditions have done better than similar projects in countries with more favorable conditions suggests that other factors—such as political economy and country capacity—are also a challenge for agricultural development in Africa.

Recommendations

To effectively support the implementation of the Africa Action Plan and its appropriate focus on agricultural development as a key priority, IEG recommends that the Bank:

1. Focus attention to achieve improvements in agricultural productivity:
    - Establish realistic goals for expansion of irrigation and recognize the need to increase productivity of rain-fed agriculture through improvements in land quality, as well as water and drought management.
    - Help design efficient mechanisms, including public-private partnerships, to provide farmers with critical inputs, including fertilizers, water, credit, and seeds.
    - Support the development of marketing and transport infrastructure.

2. Improve its work on agriculture:
    - Increase the quantity and quality of analytical work on agriculture and ensure that policy advice and lending are grounded in its findings.

- Support public expenditure analyses to assess resource availability for agriculture and to help set Bank priorities.
- Rebuild its technical skills, based on a comprehensive assessment of current gaps.

3. Establish benchmarks for measuring progress:
   - Improve data systems to better track activities supported by the Bank.
   - Strengthen M&E to report on project activities in various agro-ecological zones and for different crops and farmer categories, including women.
   - Develop a system to coordinate agricultural activities in a country with road access, market proximity, and soil conditions.

SOURCE: World Bank. Independent Evaluation Group. "World Bank Assistance to Agriculture in Sub-Saharan Africa." October 2007. web.worldbank.org/WBSITE/EXTERNAL/EXTOED/EXTASSAGRSUBSAHAFR/0,,menuPK:4174793~pagePK:64168427~piPK:64168435~theSitePK:4174768,00.html (accessed February 12, 2008).

Other Historic Documents of Interest

From this volume

- United Nations reports on Somalia, p. 343
- UN report on conflict in the Democratic Republic of the Congo, p. 374
- UN report on the Darfur region of Sudan, p. 385

From previous *Historic Documents*

- Group of Eight leaders on aid to Africa, *2006*, p. 405
- Chinese president Hu on relations between China and Africa, *2006*, p. 629
- Group of Seven aid program for Africa, *2002*, p. 446

DOCUMENT IN CONTEXT

International Monetary Fund on the World Economic Outlook

OCTOBER 17, 2007

Mimicking the occasional March in temperate climates, the world economy came into 2007 like a lamb and went out like a lion—or, more accurately in economic terms, a bear. A relatively gentle and prosperous start to the year turned violent and uncertain in August, when world financial markets were shaken by a credit crunch resulting from huge losses in risky lending for subprime mortgages in the United States. The downturn forced economists to reduce their projections for global economic growth in 2007 and led some of the more pessimistic experts to suggest there could be a lengthy U.S. recession that could drag down some other economies in 2008 and perhaps beyond.

Although the suddenly exposed credit crisis in the United States was the proximate cause of economic worries, longer-term factors also contributed to widespread uncertainty that, in some quarters, darkened into gloom at year's end. These included worries that the global credit system could be buffeted by other underlying problems that had not emerged or had not yet been fully accounted for in the financial markets; the continued rise of oil prices, which nudged the $100 per barrel mark late in the year and transferred hundreds of billions of additional dollars from oil-importing to oil-exporting countries; concerns that inflation could be on the rise because of oil prices, sharply increased food costs worldwide, and the growing demands for consumption in China and India; a steady decline since 2002 in the value of the U.S. dollar, which boosted sales of American goods and services abroad and narrowed the country's trade gap but made it more expensive for the United States to keep financing its enormous deficits by borrowing from foreign creditors; and, because of the continued failure of global trade negotiations, the prospect that economic upheaval would magnify increased protectionist sentiment in many countries, thus stifling some of the positive aspects of globalization.

Surveying this scene, the retiring managing director of the International Monetary Fund (IMF), Rodrigo de Rato, said on October 22 that the credit crisis had been an "earthquake" and the world should be prepared for long-term consequences. "Like most earthquakes, it has been something distant for most people, something they read about in the newspapers," he said. "But there is still a risk of aftershocks, and the full effects of the disruption we have already had will be felt over time."

The International Credit Crisis

As in the United States, the credit crunch in summer 2007 revealed a dirty little secret of international finance: in pushing for higher rates of return than standard debt and equity markets were producing, investment banks, hedge funds, and other institutions created a series of complex securities that were much more highly leveraged, and therefore riskier, than most people wanted to admit. Indeed, it emerged that even experienced bankers and managers of multibillion-dollar funds had bought these securities without looking very closely into what the risks might be.

These securities, known generically as collateralized debt obligations (CDOs), bundled together different kinds of debt, such as bonds and mortgages, in creative ways that reduced the cost of buying them. The CDOs produced strong returns for investors—but only so long as markets were rising and nearly everyone was buying. Once the subprime mortgage crisis in the United States exposed the risky nature of that one class of CDOs, investors panicked, banks refused to extend credit not just to hedge funds and private investors but to other banks as well, and the artificial house of cards began falling in on itself.

U.S. financial markets began tumbling in late July, with revelations that banks were suffering because of a high rate of defaults on subprime mortgages. The global impact came into sharp focus on August 9, when BNP Paribas, one of Europe's biggest banks, froze deposits in three of its hedge funds with a combined value of $2.2 billion. The bank cited a "complete evaporation of liquidity in certain market segments of the U.S. securitization market"—an explicit reference to the subprime mortgage mess. Goldman Sachs and several other U.S. financial institutions cited similar concerns on the same day. Stock markets around the world fell sharply, with losses in European and Asian markets averaging above 2 percent.

In response, the European Central Bank injected $131 billion into European money markets. This was to be the first of several steps by that bank, and other central banks, to attempt to restore some of the liquidity that even major banks such as BNP Paribus had found lacking. The injection of $131 billion failed to turn the tide, however. Less than two weeks later, on August 22, the European Central Bank set up a short-term emergency fund to promote interbank lending, and the Bank of England on September 5 made its first direct injection of emergency money into the credit markets.

In Europe, perhaps the most dramatic manifestation of the credit crunch came in Britain. Early in September, a large mortgage bank, Northern Rock, fell into danger because its main source of finance, the money markets, was quickly drying up due to the worldwide lack of liquidity. The Bank of England, which in previous weeks had tried to stay out of the credit crisis, suddenly intervened on September 14 by saying it would stand behind the bank's loans as the lender of last resort. Rather than reassuring depositors, this move spooked them, sparking the first mass run on a British bank since 1866. Northern Rock was rescued three days later with a promise by the government to guarantee 100 percent of deposits in the bank for an indefinite period. By year's end, it appeared likely that the government eventually would have to take over control of the bank.

As the experience with Northern Rock demonstrated, the credit crunch, coupled with other economic concerns, left central bankers and government policymakers with few good options for effective action. The one immediate reaction in Europe was to halt what had been a process of raising interest rates in hopes of heading off inflation. The European Central Bank (which set monetary policy for the thirteen countries using the euro as their currency) had raised its rates by a half-point, in two steps in March and June, to 4.00 percent but stopped there. The Bank of England had taken rates up in three steps (in January, May, and July) to 5.75 percent but went no further during 2007. However, European banks did not follow the lead of the U.S. Federal Reserve Bank, which cut interest rates by a half-point on September 18. In Japan, the central bank called a pause in its policy of recent years of increasing interest rates gradually from the near-zero basis of the 1990s to a more normal level.

High Growth, Low Growth

As 2007 got under way, the IMF and most economists were projecting another rather quiet year, the fourth in a row, of solid growth in nearly every part of the world. The global economy had expanded by 5.0 percent in 2006, and most forecasts put the growth for 2007 at about the same level, or even slightly higher. Rato, the IMF's managing director, on January 16 expressed pleasure at what he called a "benign global economic environment."

The IMF's overall growth projections masked sharp regional differences. China was still growing at a blistering pace of about 11 percent annually; most other developing countries, including those that exported oil, were growing in the range of 6–10 percent; but the United States, Japan, and major European countries were all stuck in the range of 2–3 percent growth.

Although there were warning signs early in the year that the long-expanding bubble in the U.S. housing market might break, very little happened to change the economists' forecasts until the credit crisis emerged in August. All of a sudden, the warning signs became more obvious, and economists started marking down their forecasts, if not for 2007 then for 2008, when the full impact of the credit crunch was expected to be evident in the United States.

In its annual "World Economic Outlook," published on October 17, the IMF kept its projection for 2007 at 5.2 percent growth but shaved 0.4 percent off its previous projection (made in July) for the same rate of growth in 2008. The biggest change was for the United States, where IMF economists projected growth to fall to 1.9 percent in both 2007 and 2008; the latter figure was nearly a full percentage point below the IMF's earlier estimates. (The IMF estimates were to fall even further in projections published in late January 2008, to global growth of 4.1 percent for 2008, with the United States again leading the downhill plunge.)

Hedging their bets, the IMF economists said in October that risks to the global economy were "firmly on the downside, centered around the concern that financial market strains could deepen and trigger a more pronounced global slowdown." The fund said it assumed that the credit crisis gradually would ease, thereby making it easier for businesses and individuals to borrow again—although in many cases at higher rates to offset higher risks. Even so, the report said, "there remains a distinct possibility that turbulent financial market conditions could continue for some time. An extended period of tight credit conditions could have a significant dampening impact on growth, particularly through the effect on housing markets in the United States and some European countries." The IMF also highlighted the economic effects of rising oil prices and said "a further spike in prices cannot be ruled out," in part because of a lack of spare production capacity.

Even before the slowdown, 2007 was on course to mark a major turning point in economic history: it was the first year that China contributed more to global economic growth than did the United States. The U.S. economy was still about four times bigger than China's ($13.2 trillion in 2006 versus $2.8 trillion), but China's growth was so fast, and the U.S. growth so slow in comparison, that China surpassed the United States as an economic engine. China achieved another landmark in 2007, when it surpassed Germany as the world's third-biggest national economy. According to most projections, if China continued its economic expansion in at least the high single digits each year, it would be on course to surpass the United States as number one by about 2020.

Another trend that became more apparent in 2007 was that the rise in oil prices had produced one of the most dramatic transfers of wealth in world history, in this case primarily from the United States and other industrialized countries to the oil-producing countries. By 2007, members of the Organization of the Petroleum Exporting Countries (OPEC) were flush with cash. Russia spent billions of its petro-dollars paying off its foreign debt (some of it dating from the Russian financial collapse in 1998) and buying gold and foreign currencies for a reserve fund. Some of the Persian Gulf states were building giant tourism complexes to underpin their economies when the oil runs out. These same states also created "sovereign wealth funds" that shopped for distressed assets—real estate, hedge funds, even entire companies—in the newly cash-poor industrialized countries. This latter trend heightened protectionist sentiment in the United States and Europe, where politicians and others worried about allowing the Gulf sheikhdoms to own companies considered important to domestic economies or national security.

One curious aspect of the rise in oil prices was that it did not, by itself, lead to a sudden surge in inflation. In two previous periods of dramatic rises in oil prices—after the 1973 Arab-Israeli war and after the 1979 Iranian revolution—the world experienced soaring inflation; in some places, including the United States, this was accompanied by an economic downturn, creating an unpleasant combination called stagflation. As of the end of 2007, inflation was becoming a concern but had not yet developed into a serious problem in most economies. One reason offered by economists was that the current round of oil price rises, which began in 2004, was more gradual than the two spikes in the 1970s. Another explanation was that in recent decades the United States and most European nations had become much more efficient in using energy, so oil consumption represented a smaller share of their economies than it had in the 1970s. Even so, many economists and central bankers worried that inflation could still pose a problem by 2008. One concern was that food prices were rising as a result of higher energy costs and the diversion of large swaths of the American Corn Belt into the production of corn for ethanol, a fuel additive.

Trade Talks Still in Limbo

At the annual World Economic Forum in Davos, Switzerland, in January 2007, international negotiators decided to try to revive global trade talks that had been on hold since July 2006. The negotiations were called the Doha Round, after the capital of Qatar where the first talks took place in 2001, and their original intent had been to make it easier for developing countries to compete in international markets. An optimistic Tony Blair, the British prime minister, said at Davos on January 27 that it was "now more likely than not, though by no means certain, that we will reach a deal within the next few months." He cited "a re-ignition of political energy and drive, and an increased recognition of the dire consequences of failure." By the time Blair's prediction proved wrong later in the year, he was out of office, having resigned after serving as prime minister for more than ten years.

As in the past, agricultural trade, particularly by Europe and the United States, continued to be a major stumbling block. Developing countries, led by Brazil and India, demanded greater access to agricultural markets on both sides of the Atlantic. But the United States and Europe insisted on protecting their farmers, generally by providing them with subsidies. Brazil and India also were part of the problem, however, because they insisted on protecting their own markets from foreign competition—in their case, competition from even poorer countries.

In one attempt to bridge the gaps, negotiators from the four major players—Brazil, the European Union, India, and the United States—met in Potsdam, Germany, in June. That effort collapsed in discord when the United States accused Brazil and India of refusing to open their own markets. Another attempt took place in July, when negotiating committees released draft proposals covering the gamut of subjects still in dispute. These proposals did little to narrow longstanding differences, however, and it seemed no more likely at the end of the year that any kind of a global agreement was possible.

Instead, the United States, the European Union, and many other countries were turning to regional or bilateral trade deals that gave them specific advantages and that, over the long haul, could make a global agreement even more difficult to achieve. Congress in 2007 approved a bilateral trade deal with Peru but left on hold other agreements the Bush administration had negotiated with Colombia, Panama, and South Korea. The last deal, which was concluded in April, was described as the largest bilateral agreement the United States had ever negotiated, and it would eliminate tariffs on more than 90 percent of the products traded by the United States and South Korea. Congress also gave no serious consideration to a request by President George W. Bush for renewal of his authority to negotiate trade agreements that would then be submitted to Congress for straightforward yes-or-no votes. Congress allowed this so-called fast track authority to lapse on July 1, thus making future trade deals more problematic because they would be subject to numerous amendments once they reached Capitol Hill.

While international negotiators continued to flail around on the agriculture trade issues, Congress moved closer during 2007 to passage of a five-year, $286 billion extension of farm legislation. The bill (HR 2419) was passed by both chambers but had not been through conference negotiations as of year's end. As always, it came complete with billions of dollars worth of subsidies, which the Doha trade negotiations were trying to reduce or even eliminate.

Following are excerpts from the executive summary of the "World Economic Outlook," published by the International Monetary Fund on October 17, 2007.

DOCUMENT

World Economic Outlook by the International Monetary Fund

October 17, 2007

Executive Summary

The global economy grew strongly in the first half of 2007, although turbulence in financial markets has clouded prospects. While the 2007 forecast has been little affected, the baseline projection for 2008 global growth has been reduced by almost ½ percentage point relative to the July 2007 *World Economic Outlook Update*. This would still leave global growth at a solid 4¾ percent, supported by generally sound fundamentals and strong momentum in emerging market economies. Risks to the outlook, however, are firmly on the downside, centered around the concern that financial market strains could deepen and trigger a more pronounced global slowdown. Thus, the immediate focus of policymakers is to restore more normal financial market conditions and safeguard the expansion. Additional risks to the outlook include potential inflation pressures, volatile oil markets, and the impact on emerging markets of strong foreign exchange inflows. At the same time, longer-term issues such as population aging, increasing resistance to globalization, and global warming are a source of concern.

Global Economic Environment

The global economy continued to expand vigorously in the first half of 2007, with growth running above 5 percent. . . . China's economy gained further momentum, growing by 11½ percent, while India and Russia continued to grow very strongly. These three countries alone have accounted for one-half of global growth over the past year. Robust expansions also continued in other emerging market and developing countries, including low-income countries in Africa. Among the advanced economies, growth in the euro area and Japan slowed in the second quarter of 2007 after two quarters of strong gains. In the United States, growth averaged 2¼ percent in the first half of 2007 as the housing downturn continued to apply considerable drag.

Inflation has been contained in the advanced economies, but it has risen in many emerging market and developing countries, reflecting higher energy and food prices. In the United States, core inflation has gradually eased to below 2 percent. In the euro area, inflation has generally remained below 2 percent this year, but energy and food price increases contributed to an uptick in September; while in Japan, prices have essentially been flat. Some emerging market and developing countries have seen more inflation pressures, reflecting strong growth and the greater weight of rising food prices in their consumer price indices. The acceleration in food prices has reflected pressure from the rising use of corn and other food items for biofuel production and poor weather conditions in some countries. . . . Strong demand has kept oil and other commodity prices high.

Financial market conditions have become more volatile. As discussed in the October 2007 *Global Financial Stability Report* (GFSR), credit conditions have tightened as increasing concerns about the fallout from strains in the U.S. subprime mortgage market led to a spike in yields on securities collateralized with such loans as well as other higher-risk securities. Uncertainty about the distribution of losses and rising concerns about counter-party risk saw liquidity dry up in segments of the financial markets. Equity markets initially retreated, led by falling valuations of financial institutions, although prices have since recovered, and long-term government bond yields declined as investors looked for safe havens. Emerging markets have also been affected, although by relatively less than in previous episodes of global financial market turbulence, and asset prices remain high by historical standards.

Prior to the recent turbulence, central banks around the world were generally tightening monetary policy to head off nascent inflation pressures. In August, however, faced by mounting market disruptions, major central banks injected liquidity into money markets to stabilize short-term interest rates. In September, the Federal Reserve cut the federal funds rate by 50 basis points, and financial markets expect further reductions in the coming months. Expectations of policy tightening by the Bank of England, Bank of Japan, and European Central Bank have been rolled back since the onset of the financial market turmoil. Among emerging markets, some central banks also provided liquidity to ease strains in interbank markets, but for others the principal challenge remains to address inflation concerns.

The major currencies have largely continued trends observed since early 2006. The U.S. dollar has continued to weaken, although its real effective value is still estimated to be above its medium-term fundamental level. The euro has appreciated but continues to trade in a range broadly consistent with fundamentals. The Japanese yen has rebounded strongly in recent months but remains undervalued relative to medium-term fundamentals. The renminbi has continued to appreciate gradually against the U.S. dollar and on a real effective basis, but China's current account surplus has widened further and its international reserves have soared.

Outlook and Risks

In the face of turbulent conditions in financial markets, the baseline projections for global growth have been marked down moderately since the July *World Economic Outlook Update*, although growth is still expected to continue at a solid pace. The global economy is projected to grow by 5.2 percent in 2007 and 4.8 percent in 2008—the latter forecast is 0.4 percentage point lower than previously expected. The largest downward revisions to growth are in the United States, which is now expected to grow at 1.9 percent in 2008; in countries where spillovers from the United States are likely to be largest; and in countries where the impact of continuing financial market turmoil is likely to be more acute. . . .

The balance of risks to the baseline growth outlook is clearly on the downside. While the underlying fundamentals supporting growth are sound and the strong momentum in increasingly important emerging market economies is intact, downside risks emanating from the financial markets and domestic demand in the United States and western Europe have increased. While the recent repricing of risk and increased discipline in credit markets could strengthen the foundations for future expansion, it raises the near-term risks to growth. The extent of the impact on growth will depend on how quickly more normal market liquidity returns and on the extent of the retrenchment in credit markets. The IMF staff's baseline forecast is based on the assumption that market liquidity is gradually restored in the coming months and that the interbank market reverts to more normal conditions, although wider credit spreads are expected to persist. Nonetheless, there remains a distinct possibility that turbulent financial market conditions could continue for some time. An extended period of tight credit conditions could have a significant dampening impact on growth, particularly through the effect on housing markets in the United States and some European countries. Countries in emerging Europe and the Commonwealth of Independent States region with large current account deficits and substantial external financing inflows would also be adversely affected if capital inflows were to weaken.

Several other risks could also have an impact on the global outlook. While downside risks to the outlook from inflation concerns have generally been somewhat reduced by recent developments, oil prices have risen to new highs and a further spike in prices cannot be ruled out—reflecting limited spare production capacity. Risks related to persistent global imbalances still remain a concern.

Policy Issues

Policymakers around the world continue to face the immediate challenge of maintaining strong noninflationary growth, a challenge heightened by recent turbulent global financial conditions. In the advanced economies, after a period of tightening that has brought monetary stances close to or above neutral, central banks have addressed the recent drying up of market liquidity and associated financial sector risks while continuing to base monetary policy decisions on judgments about the economic fundamentals. In the United States, signs that growth was likely to continue below trend would justify further interest rate reductions, provided that inflation risks remain contained. In the euro area, monetary policy can stay on hold over the near term, reflecting the downside risks to growth and inflation from financial market turmoil. However, as these risks dissipate, further tightening eventually may be required. In the event of a more protracted slowdown, an easing of monetary policy would need to be considered. In Japan, while interest rates will eventually need to return to more normal levels, such increases should await clear signs that prospective inflation is moving decisively higher and that concerns over recent market volatility have waned.

In due course, lessons will need to be drawn from the current episode of turbulent global financial market conditions. One set of issues concerns the various approaches that central banks have used to provide liquidity to relieve financial strains and the linkage of this liquidity support with financial safety nets. A series of regulatory issues will need to be addressed, as discussed in the October 2007 GFSR. Greater attention will need to be given to ensuring adequate transparency and disclosure by systemically important institutions. It will also be relevant to examine the regulatory approach to treating liquidity risk, the relevant perimeter around financial institutions for risk consolidation, the approach to rating complex financial products, and whether the existing incentive structure ensures adequate risk assessment throughout the supply chain of structured products.

Substantial progress has been made toward fiscal consolidation during the present expansion in advanced economies, but more needs to be done to ensure fiscal sustainability in the face of population aging. Much of the recent improvement in fiscal positions has reflected rapid revenue growth driven by strong growth in profits and high-end incomes, and it is not clear to what extent these revenue gains will be sustained. Further, current budgetary plans envisage limited additional progress in reducing debt ratios from current levels over the next few years. Governments should adopt more ambitious medium-term consolidation plans, together with reforms to tackle the rising pressures on health and social security spending, although in most countries there is scope to let the automatic fiscal stabilizers operate in the event of a downturn.

A number of emerging markets still face over-heating pressures and rising food prices, and further monetary tightening may be required. Moreover, notwithstanding recent financial market developments, strong foreign exchange inflows are likely to continue to complicate the task of policymakers. . . . [T]here is no simple formula for dealing with these foreign exchange inflows. Countries need to take a pragmatic approach, finding an appropriate blend of measures suited to their particular circumstances and longer-term goals. Fiscal policy is likely to play a key role. While fiscal positions have improved, this reflects strong revenue growth generated by high commodity prices that may not be sustained. At the same time, government spending in many countries has accelerated, which has added to the difficulties of managing strong foreign exchange inflows. The avoidance of public spending booms, particularly in emerging Europe but also in Latin

America, would help both in managing inflows and in continuing to reduce public debt levels. In fuel-exporting countries, however, there is scope to further increase spending, subject to absorptive capacities and the cyclical position of the economy. A tightening of prudential standards in financial systems and steps to liberalize controls on capital outflows can all play useful roles. In some cases, greater exchange rate flexibility would provide more room for better monetary control. Specifically for China, further upward flexibility of the renminbi, along with measures to reform the exchange rate regime and boost consumption, would also contribute to a necessary rebalancing of demand and to an orderly unwinding of global imbalances.

Across all countries, a common theme is the need to take advantage of the opportunities created by globalization and technological advances, while doing more to ensure that the benefits of these ongoing changes are well distributed across the broad population. A key part of this agenda is to make sure that markets work well, with priorities being to boost productivity in the financial and service sectors in Europe and Japan; resist protectionist pressures in the United States and Europe; and improve infrastructure, develop financial systems, and strengthen the business environment in emerging market and developing countries.

Globalization is often blamed for the rising inequality observed in most countries and regions. . . . [T]his report finds that technological advances have contributed the most to the recent rise in inequality, but increased financial globalization—and foreign direct investment in particular—has also played a role. Contrary to popular belief, increased trade globalization is actually associated with a decline in inequality. It is important that policies help ensure that the gains from globalization and technological change are more broadly shared across the population. Reforms to strengthen education and training would help to ensure that workers have the appropriate skills for the emerging "knowledge-based" global economy. Policies that increase the availability of finance to the poor would also help, as would further trade liberalization that boosts agricultural exports from developing countries.

. . . [T]his report examines the current global expansion from a historical perspective. It finds that not only has growth been stronger than in other recent cycles, but also the benefits are being more widely shared across the world and economic volatility has been lower. Indeed, better monetary and fiscal policies, improved institutions, and increased financial development mean that it is likely that business cycles will be of longer duration and lesser magnitude than in the past. Nevertheless, the prospects for future stability should not be overstated, and recent increased financial market volatility has underlined concerns that favorable conditions may not continue. The abrupt end to the period of strong and sustained growth in the 1960s and early 1970s provides a useful cautionary lesson of what can happen if policies do not adjust to tackle emerging risks in a timely manner.

In some key areas, joint actions across countries will be crucial. The recent slow progress with the Doha Trade Round is deeply disappointing, and major countries should demonstrate leadership to re-energize the process of multilateral trade liberalization. Concerns about climate change and energy security also clearly require a multilateral approach. . . . [G]lobal warming may be the world's largest collective action problem where the negative consequences of individual activities are felt largely by others. It will be important that countries come together to develop a market-based framework that balances the long-term costs of carbon emissions against the immediate economic costs of mitigation. Energy policy should focus less on trying to secure national sources of energy and more on ensuring the smooth operation of oil and other energy markets, encouraging diversification of energy sources (for example, by reducing barriers to trade in biofuels), and paying greater attention to price-based incentives to curb the growth of energy consumption. . . .

SOURCE: International Monetary Fund. "World Economic and Financial Surveys, World Economic Outlook: Globalization and Inequality." October 17, 2007. www.imf.org/external/pubs/ft/weo/2007/02/index.htm (accessed February 26, 2008).

OTHER HISTORIC DOCUMENTS OF INTEREST

FROM THIS VOLUME

- Resignation of British prime minister Blair, p. 221
- Federal Reserve Board on the state of the U.S. economy, p. 449
- Congressional Budget Office, President Bush on U.S. housing crisis, p. 514
- Comptroller general on the federal deficit, p. 728

FROM PREVIOUS *HISTORIC DOCUMENTS*

- WTO director general on suspension of trade talks, *2006,* p. 424
- Argentine financial crisis, *2002,* p. 80
- Russian financial crisis, *1998,* p. 601
- Asian financial crisis, *1998,* 722

DOCUMENT IN CONTEXT

President Bush and Cuban Foreign Minister on U.S. Policy toward Cuba

OCTOBER 24 AND 30, 2007

President George W. Bush signaled in October that he was not prepared to do business with the government of Cuba even after Fidel Castro, the island nation's longtime leader, departed from power. Bush denounced the transition of power that had taken place since July 2006, when Castro was hospitalized and handed over his chief duties to his brother, Raul, and other top aides. Bush called this a "succession from one dictator to another," insisted the United States would continue its economic embargo against Cuba, and urged the Cuban people to resist the continuation of the Castro regime in any form.

The hard-line rhetoric from Washington—mirrored six days later by Cuba's foreign minister—seemed likely to lock in place a status quo that had existed for nearly fifty years. Located just ninety miles from the Florida Keys, Cuba would remain economically and politically isolated from the United States until one side or the other changed its government and its policies.

Perhaps the most remarkable aspect of the period after Fidel Castro gave up his power was what did not happen. For years, Cuban exiles had predicted a great uprising of antigovernment protest—possibly even a popular revolution—once Castro was no longer in power. However, there were no substantial protests in Cuba of any kind during the nearly eighteen months after Castro stepped down. The gradual transition from one Castro to another might have weakened the incentive for Cubans to take to the streets, Cubans might have still feared the wrath of the Castro regime, or popular opposition to the regime might not have been as intense as the exiles in Miami assumed.

Yet another surprise, at least for some observers, was that Castro—aged eighty and reportedly suffering from a serious stomach ailment—did not die during 2007. In January, John Negroponte, the U.S. director of national intelligence, said Castro's days "seem to be numbered." Other U.S. officials said they expected Castro to die within a few months or by year's end. As he had so many times in the past, Castro once again foiled American expectations.

The Transition Continues in Cuba

The workings of the upper echelons of the Cuban government under Castro were never transparent. Ordinary Cubans, and foreign observers who studied the country, knew little about how or why decisions were made. The one obvious fact was that Castro was the ultimate authority by virtue of his numerous government posts and his iconic status as leader of the 1959 revolution that overthrew Gen. Fulgencio Batista's authoritarian regime. After falling ill from an undisclosed illness, Castro on July 31, 2006, handed his top government posts—"temporarily," it was said—to his brother Raul, who was five years younger. Castro gave his posts dealing with the economy, education, health, and other matters to other top aides. Day-to-day government affairs were said to be under the supervision of Vice President Carlos Lage, who, at age fifty-five, was among the younger leaders at the top of the Castro regime.

Castro did not appear in public at any time after ceding power, although the government-controlled news media occasionally carried statements attributed to him and photographs of him in the hospital or greeting visiting dignitaries. He was listed as a candidate for reelection to the National Assembly in January 2008, ensuring him at least a formal role in government. Raul Castro shunned the limelight that his brother had so clearly enjoyed; he made few speeches and rarely appeared in public.

By most accounts, the transition went smoothly, with no known interruptions of government services and, just as important, no public protests. Most experts on Cuba, both in the United States and elsewhere in Latin America, said it appeared that the Castro brothers and other officials had carefully planned out such a transition and simply were putting their plans into effect. "Power has been successfully transferred to a new set of leaders, whose priority is to preserve the system while permitting only very gradual reform," Julia E. Sweig, director of Latin America Studies at the Council of Foreign Relations (and a frequent visitor to Cuba), wrote in the January–February 2007 edition of *Foreign Affairs* magazine.

A similar assessment came in January from the head of the U.S. Defense Intelligence Agency, Army Lt. Gen. Michael D. Maples, who said Raul Castro appeared to be in charge of Cuba and "will likely maintain power and stability after Fidel Castro dies, at least for the short term."

Such assessments ran directly counter to long-time expectations by Castro's most visible and vocal opponents: leaders of the Cuban American community, based in Miami. The Cuban American National Foundation and other groups for years had lobbied in favor of tough anti-Castro U.S. policies, saying such a stance ultimately would weaken Castro's grip on power and help foster a rapid democratic transition once he passed from the scene.

The major problem faced by the Castro government was the economy, which had suffered for a half century, in part because of tight state controls and in part because of the unilateral U.S. embargo. The economy had gone into a tailspin after the collapse of the Soviet Union in late 1991. Moscow had heavily subsidized Cuba as an outpost of communism in the Western hemisphere, but once Soviet communism collapsed, the flow of Soviet money and cheap oil stopped. Cuba had survived in subsequent years with the help of limited amounts of foreign investment and private enterprise. Even so, most families were barely able to meet their basic needs on government-controlled salaries averaging about $15 a month; the result was widespread theft and corruption. A new savior, at least for the time being, was Venezuela's leftist president, Hugo Chavez, who provided subsidized oil in exchange for the services of Cuban doctors and nurses in Venezuelan slums.

News reports in late 2006 and early 2007 suggested that Raul Castro was studying ways of boosting the economy, perhaps by copying some of the methods used by China's communist leaders during the previous two decades. It was widely assumed that any changes would be modest and that Cuba would not suddenly embrace the central tenets of capitalism.

The new leader offered his first real suggestion of change in a speech during the annual July 26 ceremonies marking the beginning in 1953 of the Cuban revolution. While vowing to retain the economy's socialistic underpinnings, Raul Castro said the country could benefit from access to foreign "capital, technology, or markets." The simple use of the words "capital" and "markets" was a change from his brother's rhetoric. During the year, the government also quietly released several dozen of the estimated 300 political prisoners, including some who had served terms of more than a dozen years for such infractions as criticizing Fidel Castro. The government did not lift the restrictions on free speech or assembly that had helped stifle the emergence of a broad-based political opposition.

Bush's Speech on Policy toward Cuba

Fidel Castro's departure from an active role in government gave new life to a long-simmering debate in the United States about the best ways of encouraging change in Cuba. The politically powerful Cuban American lobby, which had become deeply embedded in the Republican Party, argued that the United States needed to continue pressuring the Castro regime in every way possible, starting with the economic embargo. First imposed by President Dwight D. Eisenhower in 1960, the embargo had been alternately tightened and relaxed over the years but still barred nearly all forms of U.S. trade with and economic support for Cuba.

In recent years a coalition of farm-state politicians, Democrats, and others had argued that the embargo clearly had failed to achieve its purpose and should be modified or even eliminated. Congress voted on several occasions to ease the embargo, notably in 2000 when it authorized cash-only sales to Cuba of food and agricultural products—a provision Cuba began using in 2001. President Bush, however, adamantly opposed any weakening of the embargo and, in 2004, imposed new restrictions intended to reduce the flow of money that Cuban Americans sent to their relatives in Cuba.

Any question that the Bush administration was reconsidering its position was answered in February, when Commerce Secretary Carlos Gutierrez, its highest ranking Cuban American, said the embargo should remain in place. "The question is not when will the U.S. change its policy," he told a Council of the Americas forum on February 21. "The question is when will the Cuban regime change its policy."

Bush adopted a similar hard-line stance in his speech on October 24—his first major speech on policy toward Cuba in four years. Speaking at the State Department before a select audience of Cuban American leaders and conservative Republican politicians, Bush also aimed his address at the Cuban people, including government officials, and members of the security services. "You may have once believed in the revolution," he said. "Now you can see its failure. When Cubans rise up to demand their liberty . . . the liberty they deserve, you've got to make a choice. Will you defend a disgraced and dying order by using force against your own people? Or will you embrace your people's desire for change?" Bush also made clear that he did not see the transition from Fidel to Raul Castro as a genuine change of power. "Life will not improve for Cubans under their current system of government," he said. "It will not improve by exchanging one dictator for another." The United States, he said, "will have no part in giving oxygen to a criminal regime victimizing its own people."

Bush insisted that the embargo remain in place, but he offered one new policy element: offers to buy computers for Cubans to gain access to the Internet, and to provide scholarships for Cuban students to attend U.S. universities. However, the president suggested that he did not expect Cuba's leaders to accept these offers.

U.S. government services that broadcast into Cuba, including Radio Marti, carried Bush's speech live. In an unusual move, the Cuban government allowed local television to carry about half of the speech, and local newspapers printed extended excerpts from it. What appeared to be an official Cuban response to Bush's speech came on October 30 from Foreign Minister Felipe Pérez Roque, in a speech to the United Nations General Assembly during its annual debate on the U.S. embargo. "Cuba, delegates, will not surrender," he said. "It fights and it will fight with the conviction that defending our rights today is tantamount to defending the rights of all the peoples represented in this Assembly." As it had for the previous fifteen years, the General Assembly overwhelmingly adopted a resolution demanding an end to the embargo.

Following are excerpts of two documents. First, President George W. Bush's speech at the State Department on October 24, 2007, restating U.S. policy toward Cuba. Second, Foreign Minister Felipe Pérez Roque's speech to the United Nations General Assembly on October 30, 2007, during the debate on a resolution (subsequently adopted) calling for an end to the U.S. economic embargo against Cuba.

DOCUMENT

President Bush on U.S. Policy toward Cuba

October 24, 2007

. . . Few issues have challenged this Department—and our Nation—longer than the situation in Cuba. Nearly half a century has passed since Cuba's regime ordered American diplomats to evacuate our Embassy in Havana. This was the decisive break of our diplomatic relations with the island, a troubling signal for the future of the Cuban people and the dawn of an unhappy era between our two countries. In this building, President John F. Kennedy spoke about the U.S. economic embargo against Cuba's dictatorship. And it was here where he announced the end of the missile crisis that almost plunged the world into nuclear war.

Today another President comes with hope to discuss a new era for the United States and Cuba. The day is coming when the Cuban people will chart their own course for a better life. The day is coming when the Cuban people have the freedom they have awaited for so long. . . .

One country in our region still isolates its people from the hope that freedom brings and traps them in a system that has failed them. Forty-eight years ago, in the early moments of Cuba's revolution, its leaders offered a prediction. He said—and I quote—"The worst enemies which the Cuban revolution can face are the revolutionaries themselves." One of history's great tragedies is that he made that dark prophecy come true.

Cuba's rulers promised individual liberty. Instead, they denied their citizens basic rights that the free world takes for granted. In Cuba, it is illegal to change jobs, to change houses, to travel abroad, to read books or magazines without the expressed approval of the state. It is against the law for more than three Cubans to meet without permission. Neighborhood watch programs do not look out for criminals. Instead, they monitor their fellow citizens, keeping track of neighbors' comings and goings, who visits them, and what radio stations they listen to. The sense of community and the simple trust between human beings is gone.

Cuba's rulers promised an era of economic advancement. Instead, they brought generations of economic misery. Many of the cars on the street predate the revolution, and some of—Cubans rely on horse carts for transportation. Housing for many ordinary Cubans is in very poor condition, while the ruling class lives in mansions. Clinics for ordinary Cubans suffer from chronic shortages in medicine and equipment. Many Cubans are forced to turn to the black market to feed their families. There are long lines for basic necessities—reminiscent of the Soviet bread lines of the last century. Meanwhile, the regime offers fully stocked foodstores for foreign tourists, diplomats, and businessmen in communism's version of apartheid.

Cuba's rulers promised freedom of the press. Instead, they closed down private newspapers and radio and television stations. They've jailed and beaten journalists, raided their homes, and seized their paper, ink, and fax machines. One Cuban journalist asked foreigners who visited him for one thing: a pen. Another uses shoe polish as ink—as a typewriter ribbon.

Cuba's rulers promised, quote, "absolute respect for human rights." Instead, they offered Cubans rat-infested prisons and a police state. Hundreds are serving long prison sentences for political offenses such as the crime of "dangerousness"—as defined by the regime. Others have been jailed for the crime of "peaceful sedition," which means whatever Cuban authorities decide it means.

Joining us here are family members of political prisoners in Cuba. I've asked them to come because I want our fellow citizens to see the faces of those who suffer as a result of the human rights abuses on the island some 90 miles from our shore. . . .

These are just a few of the examples of the terror and trauma that is Cuba today. The "socialist paradise" is a tropical gulag. The quest for justice that once inspired the Cuban people has now become a grab for power. And as with all totalitarian systems, Cuba's regime no doubt has other horrors still unknown to the rest of the world. Once revealed, they will shock the conscience of humanity, and they will shame the regime's defenders and all those democracies that have been silent. One former Cuban political prisoner, Armando Valladares, puts it this way: It will be a time when "mankind will feel the revulsion it felt when the crimes of Stalin were brought to light." And that time is coming.

As we speak, calls for fundamental change are growing across the island. Peaceful demonstrations are spreading. Earlier this year, leading Cuban dissidents came together for the first time to issue the Unity of Freedom, a declaration for democratic change. They hear the dying gasps of a failed regime. They know that even history's cruelest nightmares cannot last forever. A restive people who long to rejoin the world at last have hope, and they will bring to Cuba a real revolution, a revolution of freedom, democracy, and justice.

Now is the time to support the democratic movements growing on the island. Now is the time to stand with the Cuban people as they stand up for their liberty. And now is the time for the world to put aside its differences and prepare for Cuban's transition to a future of freedom and progress and promise. The dissidents of today will be the nation's leaders tomorrow. And when freedom finally comes, they will surely remember who stood with them.

The Czech Republic and Hungary and Poland have been vital sources of support and encouragement to Cuba's brave democratic opposition. I ask other countries to follow suit. All nations can make tangible efforts to show public support for those who love freedom on the island. They can open up their Embassies in Havana to prodemocracy leaders and invite them to different events. They can use their lobbies of the Embassies to give Cubans access to the Internet and to books and to magazines. They can encourage their country's nongovernmental organizations to reach out directly to Cuba's independent civil society.

Here at home we can do more as well. The United States Congress has recently voted for additional funding to support Cuban democracy efforts. I thank you all for your good work on this measure, and I urge you to get the bill to my desk as soon as we possibly can. I also urge our Congress to show our support and solidarity for fundamental change in Cuba by maintaining our embargo on the dictatorship until it changes.

Cuba's regime uses the U.S. embargo as a scapegoat for Cuba's miseries. Yet Presidents of both our political parties have long understood that the source of Cuba's suffering is not the embargo but the Communist system. They know that trade with the Cuban Government would not help the Cuban people until there are major changes to Cuba's political and economic system. Instead, trade with Cuba would merely enrich the elites in power and strengthen their grip. As long as the regime maintains its monopoly over the political and economic life of the Cuban people, the United States will keep the embargo in place.

The United States knows how much the Cuban people are suffering, and we have not stood idle. Over the years, we've granted asylum to hundreds of thousands who have fled the repression and misery imposed by the regime. We've rallied nations to take up the banner of Cuban liberty, and we will continue to do so. We've authorized private citizens and organizations to provide food and medicine and other aid—amounting to more than $270 million last year alone. The American people, the people of this generous land, are the largest providers of humanitarian aid to the Cuban people in the entire world.

The aid we provide goes directly into the hands of the Cuban people, rather than into the coffers of the Cuban leaders. And that's really the heart of our policy: to break the absolute control that the regime holds over the material resources that the Cuban people need to live and to prosper and to have hope.

To further that effort, the United States is prepared to take new measures right now to help the Cuban people directly, but only if the Cuban regime, the ruling class, gets out of the way.

For example—here's an interesting idea to help the Cuban people—the United States Government is prepared to license nongovernmental organizations and faith-based groups to provide computers and Internet access to Cuban people, if Cuba's rulers will end their restrictions on Internet access for all the people.

Or the United States is prepared to invite Cuban young people whose families suffer oppression into the Partnership for Latin American Youth Scholarship Programs to help them have equal access to greater educational opportunities, if the Cuban rulers will allow them to freely participate.

We make these offers to the people of Cuba, and we hope their rulers will allow them to accept. You know, we've made similar offers before, but they've been rejected out of hand by the regime. It's a sad lesson, and it should be a vivid lesson for all: For Cuba's ruling class, its grip on power is more important than the welfare of its people.

Life will not improve for Cubans under their current system of government. It will not improve by exchanging one dictator for another. It will not improve if we seek accommodation with a new tyranny in the interests of "stability." America will have no part in giving oxygen to a criminal regime victimizing its own people. We will not support the old way with new faces, the old system held together by new chains. The operative word in our future dealings with Cuba is not "stability." The operative word is "freedom."

In that spirit, today I'm also announcing a new initiative to develop an international multibillion dollar Freedom Fund for Cuba. This fund would help the Cuban people rebuild their economy and make the transition to democracy. I have asked two members of my Cabinet to lead the effort, Secretary [of State Condoleezza] Rice and Secretary [of Commerce Carlos] Gutierrez. They will enlist foreign governments and international organizations to contribute to this initiative.

And here's how the fund will work. The Cuban Government must demonstrate that it has adopted, in word and deed, fundamental freedoms. These include the freedom of speech, freedom of association, freedom of press, freedom to form political parties, and the freedom to change the Government through periodic, multiparty elections. And once these freedoms are in place, the fund will be able to give Cubans—especially Cuban entrepreneurs—access to grants and loans and debt relief to help rebuild their country.

The restoration of these basic freedoms is the foundation of fair, free, and competitive elections. Without these fundamental protections in place, elections are only cynical exercises that give dictatorships a legitimacy they do not deserve.

We will know there is a new Cuba when opposition parties have the freedom to organize, assemble, and speak with equal access to the airwaves. We will know there is a new Cuba when a free and independent press has the power to operate without censors. We will know there is a new Cuba when the Cuban Government removes its stranglehold on private economic activity.

And above all, we will know there is a new Cuba when authorities go to the prisons, walk to the cells where people are being held for their beliefs, and set them free. It will be a time when the families here are reunited with their loved ones and when the names of free people—including dissidents such as Oscar Elias Biscet, Normando Hernandez Gonzalez, and Omar Rodriguez Saludes—are free. It will be a moment when Cubans of conscience are released from their shackles—not as a gesture or a tactic, but because the Government no longer puts people in prison because of what they think or what they say or what they believe.

Cuba's transition from a shattered society to a free country may be long and difficult. Things will not always go as hoped. There will be difficult adjustments to make. One of the curses of totalitarianism is that it affects everyone. Good people make moral compromises to feed their families, avoid the whispers of neighbors, and escape a visit from the secret police. If Cuba is to enter a new era, it must find a way to reconcile and forgive those who have been part of the system but who do not have blood on their hands. They're victims as well.

At this moment, my words are being transmitted into—live into Cuba by media outlets in the free world, including Radio and TV Marti. To those Cubans who are listening—perhaps at great risk—I would like to speak to you directly.

Some of you are members of the Cuban military or the police or officials in the Government. You may have once believed in the revolution. Now you can see its failure. When Cubans rise up to demand their liberty, they deserve—they—the liberty they deserve, you've got to make a choice. Will you defend a disgraced and dying order by using force against your own people? Or will you embrace your people's desire for change? There is a place for you in the free Cuba. You can share the hope found in the song that has become a rallying cry for freedom-loving Cubans on and off the island: "*Nuestro Dia Ya Viene Llegando*"—our day is coming soon.

To the ordinary Cubans who are listening: You have the power to shape your own destiny. You can bring about a future where your leaders answer to you, where you can freely express your beliefs, and where your children can grow up in peace. Many experts once said that that day could never come to Eastern Europe or Spain or Chile. Those experts were wrong. When the Holy Father came to Cuba and offered God's blessings, he reminded you that you hold your country's future in your hands. And you can carry this refrain in your heart: *Su dia ya viene llegando*—your day is coming soon.

To the schoolchildren of Cuba: You have a lot in common with young people in the United States. You both dream of hopeful futures, and you both have the optimism to make those dreams come true. Do not believe the tired lies you are told about America. We want nothing from you except to welcome you to the hope and joy of freedom. Do not fear the future. *Su dia ya viene llegando*—your day is coming soon.

Until that day, you and your suffering are never far from our hearts and prayers. The American people care about you. And until we stand together as free men and women, I leave you with a hope, a dream, and a mission: *Viva Cuba Libre.*

SOURCE: U.S. Executive Office of the President. "Remarks at the Department of State." October 24, 2007. *Weekly Compilation of Presidential Documents*, no. 43 (October 29, 2007): 1406–1410. Washington, D.C.: National Archives and Records Administration. www.gpoaccess.gov/wcomp/v43no43.html (accessed February 7, 2008).

DOCUMENT

Cuban Foreign Minister on the U.S. Embargo against Cuba

October 30, 2007

. . . The economic, commercial and financial blockade imposed by the United States of America against Cuba, and also against the rights of the peoples that you represent in this Assembly, has already lasted for nearly half a century.

According to conservative estimates, it has caused losses to Cuba in the order of over US$ 89 billion. At the dollar's current value, that accounts for no less than US$ 222 billion. Anyone can understand the level of socio-economic development that Cuba would have attained had it not been subjected to this unrelenting and obsessive economic war.

The blockade is today the main obstacle to the development and well-being of the Cubans, and a blatant, massive and systematic violation of the rights of our people.

The blockade attempts to subdue the Cuban people through starvation and disease. . . .

Seven in every ten Cubans, distinguished delegates, have only known the perennial threat of aggression against our Homeland and the economic hardships caused by the relentless persecution of the blockade.

The United States has ignored, with both arrogance and political blindness, the fifteen resolutions adopted by this General Assembly calling for the lifting of the blockade against Cuba. What is more, over the last year they have adopted new measures, bordering on madness and fanaticism, which further tighten the sanctions and the extraterritorial persecution of our relations with the countries that you represent.

The blockade had never been enforced with such viciousness as over the last year.

On 14 August 2006, the US Government went as far as penalizing the Alliance of Baptist Churches, claiming that some of its faithful "did tourism" during a visit to Cuba with religious purposes.

In December 2006, the US Government prevented American companies from providing Internet services to Cuba. Then, if you try to access the services of Google Earth, as done by millions of users around the world every day, you get the response that: "This service is not available in your country."

Cuban children have been particularly harmed by the blockade that President Bush has promised to strengthen.

Cuban children cannot receive Sevorane, an inhalation anesthetic manufactured by the American company Abbott, which is the best for children's general anesthesia. We have to use lower-quality substitutes. President Bush will certainly explain it by saying that those Cuban children are "collateral victims" of his war against Cuba.

The Cuban children suffering from arrhythmias can no longer receive the pacemakers that the American company Saint-Jude used to sell to us. There was extreme pressure from OFAC, the Office for Foreign Assets Control, and Saint-Jude was forced to part with Cuba.

The US delegation should explain to this Assembly why the Cuban children suffering from cardiac arrhythmias are enemies of the US Government.

The Cuban delegation cannot explain—perhaps the US can—why culture has been one of the main targets in the persecution of the blockade.

The US Government prevents Cuba from participating in the Puerto Rico Book Fair. Blocking the participation of Cuban writers and publishers in a Book Fair is a barbaric deed. . . .

The blockade persecutes the human exchanges and relations between the peoples of Cuba and the United States. It also prevents normal relations between the Cuban families on both sides of the Florida Straits. Fines of up to a million dollars for companies and US$250,000 for individuals and prison penalties of up to 10 years for the offenders is the price to be risked by an American visiting our country as a tourist or by a Cuban residing in the United States who wants to visit a sick relative in Cuba.

Delegates:

More than once, this Assembly has heard the US representatives say that the issue that we are now discussing is a bilateral matter, which should not be dealt with by this forum. They will probably repeat this false argument during their explanation of vote.

However, as you are very well aware, the ruthless economic war imposed on Cuba not only affects the Cubans. If that were just the case, it would be extremely serious. But it is even worse. It is an effrontery to International Law, to the purposes and principles enshrined in the Charter of the United Nations and to the right of any country to engage in free and sovereign trade with whom it chooses to.

The extraterritorial enforcement of American laws, scorning the legitimate interests of third countries—the countries that you represent, distinguished delegates, in this Assembly—in investing and developing normal economic and trading relations with Cuba, is an issue concerning all the States gathered here. . . .

Over the last year, more than a score of banks from various countries have been grossly threatened in order to disrupt any kind of relation or transaction with Cuba. For logical reasons, I cannot give more information to this Assembly on such a sensitive issue, for that would facilitate the obsessive persecution of the American agencies fully entrusted with this ignoble task.

Mr. President:
Delegates:

A few days ago, the President of the United States said that "Cuba's regime uses the US embargo as a scapegoat for Cuba's miseries."

However, the Secretary-General's Report contained in document A/62/92, with the information provided by 118 countries and 21 international agencies, clearly and thoroughly proves the actions undertaken by the US Administration in the course of the last year to reinforce the blockade and its serious consequences to Cuba.

Today, this General Assembly is provided with the opportunity to freely and openly voice the opinion of the international community on the policy of blockade and aggressions that the United States has imposed on the Cubans for nearly 50 years.

As we speak, back in Cuba our people are following with both intent and hope the decision that you will make. They do so recalling Fidel's remarks: "Never had a nation such sacred things to defend or such profound convictions for which to fight."

Cuba, delegates, will not surrender. It fights and it will fight with the conviction that defending our rights today is tantamount to defending the right of all the peoples represented in this Assembly.

On behalf of Cuba, I ask you to vote in favor of the draft resolution entitled "Necessity of Ending the Economic, Commercial and Financial Embargo Imposed by the United States of America against Cuba."

I ask you, distinguished delegates, to vote in favor of the draft resolution presented by Cuba, despite the lies that have been uttered by the US delegation and the threats that have been made in previous days.

We ask you to vote in favor of Cuba's draft resolution, which is also to vote in favor of the rights of all the peoples on the planet.

I will now conclude recalling the words by José Martí, Apostle of Cuba's Independence: "He who rises with Cuba today will be rising for all time to come."

Freedom to the Five Cuban Heroes, fighters against terrorism and political prisoners in US jails!

Freedom to the Five Cuban Heroes!

I do have the legitimate right, distinguished delegates, to say:

¡Viva Cuba Libre! (Long live Cuba!)
¡Viva Cuba Libre! (Long live Cuba!)
¡Viva Cuba Libre! (Long live Cuba!)

SOURCE: Republic of Cuba. Permanent Mission to the United Nations. "Statement by Felipe Pérez Roque, Minister of Foreign Affairs of the Republic of Cuba, Under Agenda Item 'Necessity of Ending the Economic, Commercial and Financial Embargo Imposed by the United States of America Against Cuba.'" October 30, 2007. http://embacuba.cubaminrex.cu/Default.aspx?tabid=6594 (accessed February 7, 2008).

Other Historic Documents of Interest

From previous *Historic Documents*

- President Castro on the temporary transfer of power in Cuba, *2006,* p. 439.
- Bush administration commission on policies toward Cuba, *2004,* p. 246
- Jimmy Carter's address to the Cuban people, *2002,* p. 232

DOCUMENT IN CONTEXT

United Nations on the Political Situation in Lebanon

OCTOBER 24 AND DECEMBER 11, 2007

Lebanon faced a stark choice during 2007 between a renewed civil war or a prolonged political stalemate. It was a testament to the still-fresh memories of the country's bloody civil war from 1975 to 1990 that a political stalemate appeared to be almost everyone's preferred alternative. At year's end, Lebanon was in the grip of a standoff over electing a new president but had avoided, at least for the time being, another descent into civil war.

One year after suffering through a punishing month-long war between Israel and Hezbollah, the Lebanese Shiite militia, Lebanon remained a microcosm of most of the problems afflicting the Middle East. The country faced sectarian tensions involving Shiite Muslims, Sunni Muslims, Christians, Druze, other minorities, and a large population of disenfranchised and impoverished Palestinian refugees. These divisions prevented the parliament from meeting at all through 2007, even to elect a new president after the incumbent's term expired in November. Syria, which had never reconciled itself to Lebanese independence and had dominated the country for decades, continued to intervene, including through its support (along with that of Iran) for Hezbollah. Israel, Lebanon's neighbor to the south, no longer occupied large sections of Lebanese territory (as it had from 1982 to 2000) but remained a constant presence, not least with its daily flights over Lebanese territory by pilotless drones collecting intelligence information.

One victim of the political stalemate was a United Nations plan for a joint panel with Lebanon to try suspects, likely including senior Syrian intelligence officials, in the February 2005 assassination of former Lebanese prime minister Rafiq Hariri. The UN Security Council, in November 2006 and again in February 2007, authorized such a tribunal. The Lebanese government headed by Prime Minister Fuad Siniora had agreed, but the parliament needed to give its approval as well. Because parliament could not meet, the Security Council instead voted on May 30 to establish an international tribunal on the killing of Hariri and other Lebanese leaders; it was to be based in The Hague, Netherlands.

Rearming of Hezbollah

Hezbollah remained the dominant military force in Lebanon in 2007 despite two UN Security Council resolutions—numbers 1559 (passed in 2004) and 1701 (which ended the 2006 Israel-Hezbollah war)—that had demanded the disarming of all of Lebanon's numerous sectarian militias. Even after being pounded by Israeli bombs in 2006, Hezbollah reportedly had several thousand armed fighters and retained a stock of thousands of missiles. Hezbollah officials made it clear that they would not disarm their fighters. Neither the Lebanese army nor a UN peacekeeping force in southern Lebanon (which was expanded after the 2006 war) was capable of forcing Hezbollah to disarm.

The one strategic setback Hezbollah suffered as a consequence of the 2006 war was that it was forced to withdraw from its military positions south of the Litani River, which generally runs on an east-west axis about ten to fifteen miles north of the Israeli border. The Lebanese army and the 13,000-troop UN force moved into the area south of the river, establishing dozens of military posts and observation towers. This area had been a military zone for a quarter-century; Hezbollah had controlled it most recently, following Israel's withdrawal from Lebanon in 2000.

Numerous reports by UN officials and journalists said that Hezbollah had created new military posts, including underground bunkers and rocket-launching stations, immediately north of the Litani River and in the Bekaa Valley of southeastern Lebanon. In some cases, according to these reports, Hezbollah and its agents bought land long occupied by Christian and Druze farmers, apparently using money from Iran, which had bankrolled Hezbollah ever since its founding in 1982.

The UN in June warned that Lebanon's historic inability to control its borders with Syria had enabled Hezbollah to restock its missiles and other military supplies after the 2006 war. The Security Council on June 12 issued a statement expressing "deep concern" about the rearming of Hezbollah and other militias, which directly violated the council's resolutions demanding their disarmament.

On June 26, a team of experts appointed by UN secretary-general Ban Ki-moon reported that Lebanon's border with Syria was wide open to arms smuggling. Lebanon's army and other security forces had no capacity to patrol the border, particularly in mountainous areas, and Syria had made no effort to stem the flow of weapons across that border, the Lebanon Independent Border Assessment Team said. "There is no cooperation between the Lebanese agencies on the operational level and their Syrian counterparts." The team made several recommendations for improved border security, starting with Syrian cooperation to patrol the border. Ban endorsed the recommendations, which the UN had no power to enforce, and little action to strengthen Lebanon's border was taken during the year.

Continued Political Violence

Lebanon's political scene was deeply polarized in 2007. On one side were several parties that wanted a complete end to the long domination of Lebanon by Syria, which, under international pressure, had withdrawn its military in 2005 but remained a potent political force in Lebanon. These anti-Syrian parties controlled the Lebanese government, which was headed by Prime Minister Siniora, a former finance minister and strong critic of Syria. In opposition was a coalition consisting of Hezbollah; an older Shiite faction known as Amal; and a Christian party headed by former general Michel Aoun, who in the 1980s had been violently anti-Syrian but later joined the pro-Syrian factions, apparently in hopes of winning the presidency.

After the 2006 war, Hezbollah began demanding Siniora's resignation or at least the formation of a "national unity" government—in essence, giving Hezbollah a de facto veto power over all major government decisions. Hezbollah supporters, starting in December 2006, occupied a Beirut square adjacent to the parliament building, blocking further sessions by the parliament and paralyzing the city center.

The antigovernment protests turned violent on several occasions and included bombings that appeared to be aimed at intimidating progovernment forces. Most of

Beirut was brought to a standstill on January 23 by blocked roads and numerous clashes between progovernment and pro-Hezbollah groups.

Even more dangerous, from the government's viewpoint, was what appeared to be a campaign of assassinations of progovernment politicians—as well as journalists and others who opposed a Syrian role in Lebanese politics. One of the most significant such killings had taken place in November 2006, when Pierre Gemayel, a scion of a leading Maronite Christian family, was killed in Beirut. The political killings continued in 2007, starting with a bombing on June 13 that killed prominent anti-Syrian lawmaker Walid Eido and nine other people.

Even in this tense atmosphere, Lebanon was able to hold elections on August 5 to fill the seats of two legislators who had been assassinated. Eido's seat was won by a fellow anti-Syrian politician. The other election—for the seat that had been held by Pierre Gemayel in a Maronite Christian district north of Beirut—produced an unexpected victory for the antigovernment coalition. Seeking to succeed Gemayel were Amin Gemayel, his father and a former Lebanese president during the civil war who was running as a supporter of the government, and Kamil Khoury, a little-known politician backed by former general Aoun of the antigovernment coalition. Khoury won a narrow victory, which dealt a major political blow to the government and represented a similarly important gain for the opposition, particularly Aoun, who was maneuvering to win election as president.

The government lost another supporter on September 19 when a car bomb killed parliamentarian Antoine Ghanem and six other people. This killing reduced the government's already slim majority in parliament to 68 of 128 seats. The next day about three dozen progovernment legislators moved into the luxury Phoenicia Hotel, near the parliament building, where they were guarded by government security forces. In a report to the UN Security Council on October 24, Secretary-General Ban said the pattern of political assassinations in Lebanon "strongly suggests a concerted effort aimed at undermining the democratic institutions of Lebanon and the continued exercise of the political functions of the democratically elected representatives of the sovereign people of Lebanon."

A Political Stalemate

The killing of Ghanem also reduced the chances that Lebanon's opposing political forces could agree on a new president. Emile Lahoud, a pro-Syrian former general whose term as president had been extended in 2004 for three years under pressure from Damascus, was due to step down on November 24. By tradition, Lebanon's president was a Maronite Christian (whereas the prime minister was a Sunni Muslim and the speaker of parliament a Shiite Muslim). Several candidates to succeed Lahoud had emerged by September 25, when parliament was due to start acting to elect a new president, but none of them had strong enough support from both the progovernment and the opposition factions. One potential compromise candidate whose name emerged in September was Gen. Michel Suleiman, the head of the army. Speaker Nabih Berri postponed an opening vote on the matter when his fellow opposition members boycotted the session, thus denying parliament a required two-thirds quorum. It was to be the first of eleven such postponements by year's end.

As the November 24 date approached for Lahoud to step down, Lebanon was awash with rumors that he would refuse to quit or would establish what some opposition figures called a "parallel government" in competition with Prime Minister Siniora. This would not have been unprecedented. After parliament failed to select a new president in

1988, General Aoun, then head of the army, declared himself president and set up a competing government, sparking the last gasp of Lebanon's long civil war.

The prospect of another stalemate, or even a renewed civil war, increasingly worried diplomats and political leaders around the world, who stepped up their calls on both sides to reach a compromise. Secretary-General Ban traveled to Beirut for meetings on November 15–16 with both pro- and antigovernment leaders. He warned that the country was approaching "the brink of the abyss." Also attempting to arrange a compromise was Bernard Kouchner, the foreign minister of France, which had been the colonial power in Lebanon and Syria during the first half of the twentieth century and retained some influence there. Kouchner appeared unable to use that influence to bring about an agreement, however.

Lahoud stepped down, as required, when his term ended, but he did not follow through on threats to establish a separate government. He did declare a "state of emergency" and order the army to take control of security, a step that appeared to have more political than military significance because the army already controlled security where it could; much of Lebanon was under the de facto control of Hezbollah and other militias.

The departure of the president changed little else in Lebanon. Various factions continued to meet in search of an elusive compromise, and Speaker Berri continued to call, then postpone, parliamentary votes. For a moment in late November it appeared that a compromise was possible when presidential aspirant Aoun said he would support General Suleiman for the job. However, Suleiman's candidacy rapidly became mired in new bickering over a renewed opposition demand for a veto over major government decisions, and more votes were postponed.

On December 11, the UN Security Council put more international pressure on both sides by issuing a statement expressing "deep concern" at the political paralysis. The council called on all Lebanese parties "to continue to exercise restraint, and to show responsibility with a view to preventing, through dialogue, further deterioration of the situation in Lebanon."

If anyone needed a reminder of the prospects for deterioration, it came on December 13 when a car bomb killed one of General Suleiman's top aides, Brig. Gen. François al-Hajj, and his bodyguard as they drove to work. No one claimed responsibility for the killing, and under the circumstances, it was unclear who benefited by his death.

The final attempt to elect a new president in 2007 was scheduled for December 29, but it, too, was postponed at the last minute by Speaker Berri because the parties still had not agreed on whether and how to share power. It was the eleventh postponement of a vote. Berri called another vote for January 12, 2008, but that vote was also postponed, and the dispute lingered into another year.

Battle at a Palestinian Refugee Camp

For three months in mid-2007, Lebanon faced a new version of one of its oldest challenges: a threat from an extremist group, this time one operating out of a Palestinian refugee camp. The threat came from a pan-Arab, Sunni Muslim militant group, Fatah al-Islam, which late in 2006 had crossed into Lebanon from Syria and taken over part of the Nahr al-Barad Palestinian camp, located outside Tripoli on the northern coast. Many of the group's estimated 200 fighters reportedly had participated in the anti-U.S. insurgency in Iraq.

Blaming the group for several attacks, the Lebanese army began shelling the Palestinian camp on May 20; this was the army's first major offensive operation in nearly three decades. The assault drove out many of the camp's 40,000-some Palestinian refugees, but thousands remained trapped in the camp for weeks. The army kept up its barrage during the summer, gradually taking control of the camp as the Palestinians left and the Fatah al-Islam fighters defended their positions. A final assault on September 2 killed many of the remaining fighters, including their leader, Shakir al-Abssi, a Palestinian with links to al Qaeda in Iraq. According to official figures, the three months of fighting killed 163 soldiers and around 120 militants, as well as about 40 civilians, most of them Palestinian refugees.

Despite the heavy toll, the army—and its commander, General Suleiman—emerged as heroes among the Lebanese, who had feared the takeover of the camp might be the opening shot in a campaign against the country by Islamist extremists. Prime Minister Siniora was among those sharing this fear. In an October 8 letter to UN secretary-general Ban, Siniora charged that the occupation of the camp by Fatah al-Islam had been part of a "carefully drawn plot, of very serious and dangerous proportions," to seize much of northern Lebanon, destabilize the government, and mount attacks against the UN peacekeepers to force their withdrawal. Ban said the UN could not independently verify these claims, and he quoted a letter from the Syrian foreign ministry, dated October 19, denying that Damascus was trying to destabilize its neighbor. Ban's report appeared to give some weight to Siniora's charges, however, by saying that Lebanon's "sovereignty, territorial integrity, unity and political independence continue to suffer further infringements."

Following are two documents. First, excerpts from a report to the UN Security Council by Secretary-General Ban Ki-moon, dated October 24, 2007, on the implementation of Security Council Resolution 1559 (adopted in 2004), which required the withdrawal of all foreign military forces from Lebanon. Second, the text of a presidential statement by the Security Council on December 11, 2007, calling for political reconciliation in Lebanon.

DOCUMENT

Report of UN Secretary-General on Resolution 1559 Concerning Lebanon

October 24, 2007

III. Observations

73. Since the adoption of resolution 1559 (2004), Lebanon has continued to suffer setbacks in its struggle to reassert, beyond dispute, its sovereignty, territorial integrity, unity and political independence. Once again I salute the brave Lebanese people and their political leaders, who stand firm in that struggle. The United Nations remains as committed as ever to helping them complete the historical transition that has been under way in Lebanon since September 2004. For this purpose, I have remained in close contact with all relevant parties in the region and beyond.

74. Over the past six months, Lebanon has lived through yet another difficult chapter in its efforts to assert its sovereignty, territorial integrity, unity and political independence, extend governmental control over all Lebanese territory and ensure that there are no weapons outside the Government's control. In combination, the explosions, assassinations, incidents in south Lebanon and the prolonged fighting between the Lebanese Armed Forces

and Fatah al-Islam have manifested the precarious state of security in Lebanon. Security conditions have combined with the political stalemate to create a climate of enduring crisis, with adverse effects more widely on Lebanese society and the economy. Many Members of Parliament are spending most of their time abroad. Member of Parliament Antoine Ghanem, for example, returned from a prolonged sojourn abroad just two days before his assassination. Overall, the conditions prevailing in Lebanon are not conducive to the reassertion of the country's sovereignty, territorial integrity, unity and political independence.

75. In the context of prolonged political crisis, the challenge from militias and allegations of widespread rearming and paramilitary training, the authority of the Government of Lebanon has remained constrained and contested, as has its monopoly on the legitimate use of violence. The most notable challenge during this period has come from Fatah al-Islam. I commend and congratulate the people and Government of Lebanon and the Lebanese Armed Forces for successfully weathering a critical test on the road to a truly free and sovereign Lebanon.

76. Yet, many challenges remain if Lebanon is to free itself from the vestiges of a captive past in a sustainable manner. First and foremost, I call for political dialogue in Lebanon to resume on all relevant matters, most notably the issue of the Lebanese presidency and the disarming and disbanding of Lebanese and non-Lebanese militias.

77. In the aftermath of the victory over Fatah al-Islam, it is essential now that the Government of Lebanon and the Lebanese Armed Forces maintain their vigilance and efforts, for the welfare and security of all people living in Lebanon. It is also paramount that political discussions resume among all Lebanese parties, leading them to reaffirm their commitment to the disarmament of Palestinian militias in Lebanon, in fulfilment of the terms of resolution 1559 (2004).

78. I am also deeply conscious of the conditions in Palestinian refugee camps in Lebanon and the challenges that arise from them. It is imperative that the close cooperation that has been established between PLO [Palestine Liberation Organization] and the Lebanese authorities continue, for the welfare of the Palestinian refugees who already have too often paid the price for the misdeeds of others. I commend both the Government of Lebanon and PLO for their role in re-establishing security in the camps, but call on them to undertake tangible measures now to improve significantly the conditions in which the refugee population lives, without prejudice to the settlement of the Palestinian refugee question in the context of an eventual Israeli-Palestinian peace agreement. The United Nations family stands ready to work with our Lebanese and Palestinian partners towards this goal, while we also exert all efforts to help bring about an Israeli-Palestinian peace agreement at the earliest time possible.

79. The information that I continue to receive, suggesting that Hizbullah has rebuilt and increased its military capacity compared to the period prior to the war of July and August 2006, is deeply disconcerting and stands in stark contradiction to the terms of resolution 1559 (2004). I restate my conviction that the eventual disarmament of Hizbullah in the sense of the completion of its transformation into a solely political party, consistent with the requirements of the Taif Agreement [the 1989 agreement that ended Lebanon's civil war], is an element of critical importance for the future of a fully sovereign, united and politically independent Lebanon. I urge renewed political dialogue in Lebanon to affirm the commitment of all parties to the disarmament of Lebanese militias in Lebanon, including Hizbullah, in fulfilment of the terms of resolution 1559 (2004). I also expect the unequivocal cooperation of all relevant regional parties who have the ability to support such a process, most notably the Syrian Arab Republic and the Islamic

Republic of Iran, which maintain close ties with the party, for the sake of the security, stability and welfare of both Lebanon and the wider region.

80. Not once since the end of the civil war has there been a presidential election in Lebanon conducted according to constitutional rules, without any constitutional amendments and without foreign interference. In 1989, President Elias Hrawi was elected in Shtoura, far away from the parliamentary chamber, to replace the assassinated René Mouawwad, who had been appointed at a military airport in the north of Lebanon. President Hrawi's term of office was extended in 1995 for an additional three years beyond the constitutionally prescribed regular term of six years. President Émile Lahoud's election in 1998 was made possible by a constitutional amendment allowing the former chief of the Lebanese Armed Forces to run in the elections. President Lahoud's term of office was extended for three additional years in 2004 by another constitutional amendment. Therefore, this time, after the withdrawal of Syrian troops, military assets and military intelligence apparatus, the Lebanese have the opportunity to conduct a free and fair presidential election process, according to Lebanese constitutional rules and without any foreign interference, for the first time since the end of the civil war. Such an election would signify a major milestone on the road towards the full re-assertion by Lebanon of its sovereignty, territorial integrity, unity and political independence, as is the goal of resolution 1559 (2004).

81. With the upcoming presidential election, it is my strong belief that the Lebanese people and their political representatives must rise to the occasion and turn a new page in their difficult history. There must not be a constitutional void at the level of the presidency, nor two rival governments. Constitutional provisions should be fully respected. In consequence, political dialogue must make it possible for a new president to be elected before the constitutional deadline of 24 November. I urge the Lebanese political parties to engage in a constructive dialogue and aim for conciliation, in full respect of the Taif Agreement. The president should enjoy the broadest possible acceptance.

82. I continue to be deeply concerned at the prevailing security conditions in Lebanon. The fact that these have forced many Members of Parliament to reside permanently abroad or in seclusion, under extremely tight security, in their own country, is unacceptable. It can also not go unnoticed that the most recent assassination of Member of Parliament Antoine Ghanem, in combination with previous assassinations of members of the ruling coalition, reduced its majority to 68 out of the now 127 Members of Parliament and raised the spectre not only of further deterioration of the situation, but also of an upset of the political balance that has existed since the parliamentary elections in the spring of 2005. The pattern of political assassinations in Lebanon strongly suggests a concerted effort aimed at undermining the democratic institutions of Lebanon and the continued exercise of the political functions of the democratically elected representatives of the sovereign people of Lebanon.

83. It is equally disconcerting to observe that most political parties in Lebanon are apparently preparing for the possible further deterioration of the situation. Rearmament and military training directly contravene the call contained in resolution 1559 (2004) for the disarming and disbanding of all Lebanese and non-Lebanese militias. I commend the Government of Lebanon and the Lebanese security services for their continued vigilance in this regard and their efforts aimed at calming the situation. I also repeat my urgent call on all Lebanese parties to immediately halt all efforts to re-arm and engage in weapons training, and to return instead to dialogue and conciliation as the only viable method of settling issues and resolving the ongoing political crisis.

84. A return to political dialogue among the Lebanese parties is absolutely imperative under the current conditions, and the only way to resolve all relevant issues. Lebanon must preserve its comprehensive and, most importantly, conciliatory political framework, as manifested in the Taif Agreement.

85. I am of course acutely aware that such a framework also necessitates the renewed support and engagement of all relevant external parties and supporters of Lebanon. Without it, Lebanon will not be able to take further steps towards reasserting its sovereignty, territorial integrity and political independence, or to sustain such progress in the long term. But I am equally convinced that the deep foreign involvement in Lebanon have done little to decrease tension in that country. Instead, the foreign penetration and interference in Lebanon have only worsened the crisis. It is time that foreign interference stop and that the Lebanese people and their political representatives alone determine the fate of Lebanon.

86. In this context, I reiterate my expectation vis-à-vis the Syrian Arab Republic, in particular, that it cooperate on all relevant issues related to the full implementation of all provisions of resolutions 1559 (2004), 1680 (2006) and 1701 (2006). I welcome the assertions and pledges in the Syrian Arab Republic's recent letter to me and expect to see the Syrian Arab Republic's commitment to the sovereignty, territorial integrity, unity and political independence of Lebanon reflected in further tangible steps in the coming period.

87. I remain keenly aware of the interlinkages between the various conflicts in the region. It is my most profound belief that all possible efforts must be exerted to attain a just, comprehensive and lasting peace for all peoples in the region. The achievement of such peace throughout the entire Middle Eastern region, consistent with all relevant Security Council resolutions, especially resolutions 242 (1967) and 338 (1973), and the full restoration of the territorial integrity, full sovereignty and political independence of Lebanon will remain contingent upon each other.

88. I will continue my efforts to assist all parties in the quest for peace and stability in the region, and in the full implementation of resolutions 1559 (2004), 1680 (2006) and 1701 (2006). I also reiterate my call on all parties and actors to support the reconstruction and political transformation of Lebanon, and to urgently take all enabling measures to this end, as outlined in the Taif Agreement and in resolutions 1559 (2004), 1680 (2006) and 1701 (2006).

SOURCE: United Nations. Security Council. "Sixth Semi-Annual Report of the Secretary-General on the Implementation of Security Council Resolution 1559 (2004)." S/2007/629. October 24, 2007. Released October 31, 2007. www.un.org/Docs/journal/asp/ws.asp?m=s/2007/629 (accessed March 19, 2008).

DOCUMENT

United Nations Security Council on the Political Situation in Lebanon

December 11, 2007

At the 5799th meeting of the Security Council, held on 11 December 2007, in connection with the Council's consideration of the item entitled "The situation in the Middle East", the President of the Security Council made the following statement on behalf of the Council:

"The Security Council stresses its deep concern at the repeated postponements of the presidential election in Lebanon.

"It emphasizes that the ongoing political impasse does not serve the interest of the Lebanese people and may lead to further deterioration of the situation in Lebanon.

"The Security Council reiterates its call for the holding, without delay, of a free and fair presidential election in conformity with Lebanese constitutional rules, without any foreign interference or influence, and with full respect for democratic institutions.

"It emphasizes the importance of Lebanese constitutional institutions, including the Government of Lebanon, as well as the importance of the unity of the Lebanese people, in particular on the basis of reconciliation and political dialogue.

"It calls upon all Lebanese political parties to continue to exercise restraint and to show responsibility with a view to preventing, through dialogue, further deterioration of the situation in Lebanon.

"The Security Council therefore commends the course adopted by the democratically-elected Government of Lebanon and the Lebanese Armed Forces in carrying out their respective responsibilities in the period until the presidential election occurs.

"The Security Council reiterates its call for the full implementation of all its resolutions on Lebanon."

SOURCE: United Nations. Security Council. "Statement by the President of the Security Council." SPRST/2007/46. December 11, 2007. http://daccessdds.un.org/doc/UNDOC/GEN/N07/638/84/PDF/N0763884.pdf? (accessed March 19, 2008).

Other Historic Documents of Interest

From this volume

- Upheavals in the Palestinian territories, p. 39
- Israel's handling of the 2006 war in Lebanon, p. 200
- U.S. Ambassador Crocker on U.S.-Iranian diplomacy, p. 257

From previous *Historic Documents*

- UN Security Council resolution on war between Israel and Hezbollah, *2006*, p. 454
- UN commission on Hariri assassination in Lebanon, *2005*, p. 686
- UN Security Council resolution demanding Syrian withdrawal from Lebanon, *2004*, p. 557

November

DOCUMENT IN CONTEXT

Government Accountability Office on the Safety of Imported Drugs

NOVEMBER 1, 2007

Even as Congress passed legislation in 2007 to improve the safety of domestic drugs, new concerns arose about the safety of imported drugs and the ability of the Food and Drug Administration (FDA) to keep unsafe or counterfeit drug imports out of the U.S. market. Reports and testimony before Congress showed that the FDA did not have the technological capacity or the necessary personnel to keep track of and inspect foreign manufacturers exporting drugs and drug ingredients to the United States. A hard-hitting report by a subcommittee of the FDA's Science Advisory Board also said the agency did not have the scientific knowledge and capacity to keep up with new developments in the drug industry, including genetics and nanotechnologies that showed promise for significant breakthroughs in drug treatments and medications.

DRUG REGULATION OVERHAUL

Congressional action to overhaul the FDA's regulatory authority over domestic drug manufacturers came after several years in which new drugs were pulled off the market after their safety was questioned. The most-publicized withdrawal involved the best-selling painkiller Vioxx, which the pharmaceutical giant Merck removed from drugstore shelves in September 2004 following studies that showed it increased risk of heart attack and stroke. The abrupt move raised questions about Merck's slowness in responding to earlier warnings about the drug's safety. The withdrawal also focused attention on whether the FDA should have approved the drug in the first place and whether it was adequately monitoring safety once new drugs came on the market. A claim by one of the FDA's own researchers that the agency had tried to suppress his findings about the dangers of Vioxx further inflamed the controversy. Altogether, drug makers had pulled at least ten major drugs from the market since 2000 after deaths and other injuries called their safety into question.

Safety issues for drugs already on the market could arise for several reasons. The FDA was under intense pressure from drug manufacturers and patient advocacy groups to get potentially beneficial and life-saving drugs to the market as fast as possible. Critics of the industry and the FDA's approval process also argued that it was too easy for drug companies to cover up a new drug's potential safety problems. Two significant studies published in 2006, one by the Government Accountability Office (GAO) and one by the Institute of Medicine, attributed many of the FDA's problems to poor management, bureaucratic squabbling, lack of authority, and underfunding. Both reports recommended that Congress give the FDA express authority to require drug manufacturers to do follow-up safety studies on drugs already on the market.

That authority was a key provision of legislation passed by Congress and signed into law on September 27 (HR 3580—PL110–85). The legislation renewed for five years funding for the 1992 Prescription Drug User Fee Act, under which drug makers paid the FDA about $393 million annually in fees for expedited review of new products. A parallel program for medical devices was also renewed. Under its new authority, the FDA could require risk-management plans for new drugs suspected to have potentially dangerous side effects. It could also require clinical studies of new drugs after they were approved. In the past, the FDA had often asked drug makers to study specific drugs already on the market for adverse side affects, but the agency reported that nearly two-thirds of such requests were never fulfilled. Under the new law, companies that did not comply with FDA requests could be fined up to $250,000 per violation and $1 million for several violations. Fines for continued violations could go as high as $10 million.

To make more information available to doctors and consumers, the new law called for the creation of a public database, accessible over the Internet, where companies would be required to post the results of their clinical trials for approved drugs. Drug makers would also have to post information about, but not the results of, ongoing clinical trials for products in development. The FDA could require additional data posting, and companies that did not comply would be fined. The new law also gave the FDA the authority to require drug companies to make changes to product labels; under the previous law the FDA had to negotiate warning labels with drug makers. In addition, the law granted the agency new authority to review advertisements aimed directly at consumers before they were aired and to fine companies for any false or misleading ads, but the FDA would not be allowed to ban the ads, even temporarily. Further, under the new law, experts serving on FDA advisory panels would be required to disclose all financial conflicts of interest, but many of them would be allowed to continue to serve under a waiver system. Critics had complained that the integrity of the review system was undermined when panel members who had conducted research with grants from a drug company whose product was under review did not recuse themselves.

The final measure was a compromise between legislators who thought the versions passed by the House and Senate gave the FDA too much authority and those who thought stronger protections were needed. "This is not a perfect bill, and compromises were made to assure its passage," said Sen. Edward M. Kennedy, D-Mass., whose amendment to allow the importation of cheaper prescription drugs was ultimately defeated. "But after so many recent instances in which Americans have been harmed by unsafe prescription drugs and contaminated food, American cannot afford inaction on this important measure."

Weaknesses in the Drug Import Safety Program

The rise of pharmaceutical industries in China and India, coupled with high-profile incidents of contaminated products imported from China, focused heightened attention during the year on the FDA's system for inspecting drug exports from those and other foreign countries. In the late 1990s, India exported only a handful of generic drugs to the United States; by 2007, Americans were buying nearly 350 varieties and strengths of generics made in India, according to a report in the *Washington Post.* Chinese drug makers, which were rapidly expanding their share of the market for active ingredients, sold $675 million worth of drug products in the United States in 2006, more than double the value sold five years earlier. Industry analysts estimated that India and China made one-fifth of all finished generic and over-the-counter drugs and more than two-fifths of all active ingredients used in drugs manufactured in the United States. Some experts predicted that within fifteen years the two countries would be the source of 80 percent of all active ingredients used in medications sold in the United States.

While drug experts affirmed that many Indian and Chinese firms made high-quality products, they also noted that both countries had experienced problems related to counterfeit drugs, poor quality control, and very little regulation. Most recently, the former head of China's food and drug safety agency was executed after being convicted of taking bribes from companies he regulated. American consumers were shaken in May when the FDA warned them not to buy toothpaste from China, because it might contain diethylene glycol, a common component of antifreeze. Unscrupulous Chinese and Indian manufacturers were also suspected of being a major source of counterfeit drugs that included everything from fake malaria pills to best-selling drugs for treating high blood pressure, high cholesterol, osteoporosis, and other conditions. Many of these counterfeit drugs were sold over the Internet to consumers who had no way of knowing where or how the drugs were manufactured.

Under U.S. law in 2007, a company that wanted to import drug products into the United States had to register with the FDA and comply with quality and labeling requirements. The FDA maintained border controls meant to catch drugs that were not registered or made at an approved facility, and it was also authorized to inspect foreign manufacturing facilities. In addition, U.S. drug makers routinely tested the drugs and active ingredients they bought from companies overseas, if for no other reason than to protect their own reputations. But investigations by the Government Accountability Office, congressional investigators, and others found the FDA's management of its drug imports program to be woefully inadequate.

Testifying before the House Energy and Commerce Subcommittee on Oversight and Investigations on November 1, Marcia Crosse, GAO's director of health care, said the FDA did not know how many foreign drug makers were subject to inspection. It did not know either the number of registered firms whose products were currently imported into the United States or the number of firms that were not required to register but whose products were ultimately used in drugs that were sold in the United States. The FDA used three separate databases, none of which was designed expressly for the purpose, to try to keep track of foreign firms subject to inspection. One of the databases, according to GAO's investigation, indicated that about 3,000 firms were subject to inspection; another database put that number at about 6,800. Moreover, registration information was not up to date, and the FDA did not routinely verify the registration information it did receive. Errors in data entry, such as misspelled company names, meant that some companies were included in a database more than once.

GAO reported that the FDA conducted "relatively few inspections" of foreign drug establishments; the precise percentage could not be determined because the FDA did not have an accurate count of the firms actually subject to inspection. GAO found that the FDA conducted an average of 241 inspections per year during the last five years. Using the FDA's own planning figure of 3,249 firms needing inspection, GAO calculated that the agency inspected only about 7 percent of foreign drug-making establishments in a given year and would need thirteen years to inspect every firm on the list. In the previous five years, FDA had conducted 200 inspections in India compared with 88 in China, even though by fiscal 2007 China had nearly twice as many registered firms (566) as India

(299). Moreover, the great majority of the inspections were associated with new drug approval applications; a far smaller number of inspections monitored the quality of the drugs and of the manufacturing process. The FDA told GAO that it had conducted only 30 such inspections in fiscal 2007, although it did perform some quality monitoring along with its new drug inspections.

Staff investigators for the committee reported on October 30 on several other shortcomings in the inspection program. They found that inspectors often had to rely on translators provided by the company being inspected, that companies were notified of the inspection well in advance of the inspectors' arrival, and that inspectors were generally not given enough time to do adequate investigations. The investigators said establishing permanent FDA offices in China and India could "greatly facilitate" the inspection process. Senior officials from India's government and drug firms supported the idea, the investigators said, but China's position was unclear. But even adopting all of their recommendations for improvements, the investigators said, "would not compensate for the severe limitations in resources that appear to plague the foreign inspection process."

Responding to these reports in testimony on November 1, FDA commissioner Andrew C. von Eschenbach described numerous steps the agency was taking to improve its surveillance of foreign drugs, including the negotiation of an agreement with the Chinese government for monitoring the manufacture and export of certain Chinese drugs. That agreement, which was signed on December 11, covered ten specific drug products that had been counterfeited or otherwise misused in the past, including several antibiotics, the cholesterol-lowering drug atorvastatin, dietary supplements intended to treat erectile dysfunction, human growth hormone, glycerin, and condoms. Chinese makers that wanted to export those products to the United States would be required to register with the Chinese government, and the two governments would collaborate to develop a process for certifying that the products met U.S. safety standards.

As with food safety, however, these improvements were likely to have little real effect until the FDA was given sufficient resources to meet its responsibilities. A subcommittee of the FDA's Science Advisory Board took a yearlong look at FDA's food and drug safety programs and concluded that the agency had been loaded with new responsibilities over the past few years but had not received any additional funding to meet those mandates. As a result, the Subcommittee on Science and Technology said in a report released on December 1, the agency was unable to recruit and retain top scientists and experts, maintain up-to-date and integrated information technology programs, and keep current with emerging scientific developments in the areas of food and drugs. "In contrast to previous reviews that warned crises would arise if funding issues were not addressed, recent events and our findings indicate that some of those crises are now realities and American lives are at risk," the panel concluded.

Following are excerpts from testimony by Marcia Crosse, director of health care for the Government Accountability Office, to the House Energy and Commerce Subcommittee on Oversight and Inspections on November 1, 2007.

DOCUMENT

GAO on the Safety of the FDA Drug Import Program

November 1, 2007

Mr. Chairman and Members of the Subcommittee:

I am pleased to be here today as you examine the Food and Drug Administration's (FDA) inspections of foreign drug manufacturers whose products are imported into the United States. In 1998, we reported that FDA needed to improve its foreign drug inspection program. Among other things, we noted that FDA had serious problems managing its foreign inspection data and that it lacked a comprehensive automated system for tracking this important information. We were also critical of the number of inspections FDA conducted at foreign manufacturers. At that time, FDA reported on our growing dependence on imported pharmaceutical products, noting that as much as 80 percent of the bulk drug substances used by manufacturers in the United States to produce prescription drugs was imported and that the number of finished drug products manufactured abroad for the U.S. market was increasing. Today, we are still dependent on foreign establishments manufacturing drugs for the U.S. market as the value of pharmaceutical products coming into the United States from abroad continues to increase.

Given the importance of FDA's foreign drug inspection program, you expressed concern about FDA's ability to oversee foreign establishments manufacturing drugs and asked whether FDA has improved its management of the foreign drug inspection program since our previous report was issued. My testimony today will summarize preliminary findings from our ongoing work to update our 1998 report. . . .

In summary, our preliminary results indicate that more than 9 years after we issued our last report on this topic, FDA's effectiveness in managing the foreign drug inspection program continues to be hindered by weaknesses in its data systems. FDA does not know how many foreign establishments are subject to inspection. FDA relies on information from several databases that were not designed for this purpose. One of these databases contains information on foreign establishments that have registered to market drugs in the United States, while another contains information on drugs imported into the United States. One database indicates about 3,000 foreign establishments could have been subject to inspection in fiscal year 2007, while another indicates that about 6,800 foreign establishments could have been subject to inspection in that year. Despite the divergent estimates of foreign establishments subject to inspection generated by these two databases, FDA does not verify the data within each database. For example, the agency does not routinely confirm that a registered establishment actually manufactures a drug for the U.S. market. However, FDA used these data to generate a list of 3,249 establishments from which it prioritized establishments for inspection.

Because FDA is not certain how many foreign establishments are actually subject to inspection, the percentage of foreign establishments that have been inspected cannot be calculated with certainty. We found that FDA inspects relatively few foreign establishments. Using the list of 3,249 establishments from which FDA prioritized establishments for inspection, we found that the agency may inspect about 7 percent of foreign establishments in a given year. At this rate, it would take FDA more than 13 years to inspect each foreign establishment on this list once, assuming that no additional establishments are subject to inspection. FDA cannot provide the exact number of foreign establishments that have never been inspected. Most of the foreign inspections are conducted as part of processing a new drug application (NDA) or an abbreviated new drug application (ANDA), rather than as GMP [Good Manufacturing Practice] surveillance inspections, which are used to monitor the quality of marketed drugs. Although FDA used a risk-based process to develop a prioritized list of foreign establishments for GMP surveillance inspections in fiscal year 2007, few such inspections are completed in a given year. According to FDA, about 30 such inspections were completed in fiscal year 2007 and at least 50 are targeted for inspection in fiscal year 2008. Further, the data on which this risk-based process depends limits its effectiveness.

Finally, the very nature of the foreign inspection process involves unique circumstances that are not encountered domestically. For example, FDA does not have a dedicated staff to conduct foreign inspections and relies on those inspecting domestic establishments to volunteer. While FDA may conduct unannounced GMP surveillance inspections of domestic establishments, it does not arrive unannounced at foreign establishments. It also lacks the flexibility to easily extend foreign inspections if problems are encountered, due to the need to adhere to an itinerary that typically involves multiple inspections in the same country. Finally, language barriers can make foreign inspections more difficult to conduct than domestic ones. FDA does not generally provide translators to its inspection teams. Instead, they may have to rely on an English-speaking representative of the foreign establishment being inspected, rather than an independent translator.

Because of the preliminary nature of our work, we are not making recommendations at this time.

Background

FDA is responsible for overseeing the safety and effectiveness of human drugs that are marketed in the United States, whether they are manufactured in foreign or domestic establishments. Foreign establishments that market their drugs in the United States must register with FDA. As part of its efforts to ensure the safety and quality of imported drugs, FDA is responsible for inspecting foreign establishments whose products are imported into the United States. The purpose of these inspections is to ensure that foreign establishments meet the same manufacturing standards for quality, purity, potency, safety, and efficacy as required of domestic establishments.

Requirements governing foreign and domestic inspections differ. Specifically, FDA is required to inspect registered domestic establishments that have been previously approved to market their drugs in the United States every 2 years, but there is no comparable requirement for inspecting foreign establishments. FDA does not have authority to require foreign establishments to allow the agency to inspect their facilities. However, FDA has the authority to conduct physical inspections of the imported product or prevent its entry at the border.

Within FDA, CDER [Center for Drug Evaluation and Research] sets standards for and evaluates the safety and effectiveness of prescription drugs and over-the-counter drugs. . . . CDER requests that ORA [Office of Regulatory Affairs] inspect both foreign and domestic establishments to ensure that drugs are produced in conformance with federal statutes and regulations, including current GMPs. CDER requests that ORA conduct inspections of establishments that produce finished drug products. CDER also requests inspections of those that produce bulk drug substances, including the active pharmaceutical ingredients (API) used in finished drug products. These inspections are performed by investigators and laboratory analysts. ORA conducts two primary types of inspections:

- Preapproval inspections of domestic and foreign establishments are conducted before FDA will approve a new drug to be marketed in the United States. These inspections occur following FDA's receipt of an NDA or ANDA and focus on the manufacture of a specific drug product. Preapproval inspections are designed to verify the accuracy and authenticity of the data contained in these applications and ensures that the manufacturer of the finished drug product, as well as each manufacturer supplying a bulk drug substance used in the finished product, manufactures, processes, and packs the drug adequately to preserve its identity, strength, quality, and purity.
- Postapproval GMP surveillance inspections are conducted to ensure compliance with applicable laws and regulations pertaining to the manufacturing processes used by domestic and foreign establishments in the manufacture of finished drug products marketed in the United States and bulk drug substances used in the manufacture of those products. These inspections focus on a manufacturer's systemwide controls for ensuring that drug products are high in quality. Systems examined during these inspections include those related to quality control, production, and packaging and labeling. These systems may be involved in the manufacture of multiple drug products.

. . .Typically, ORA investigators and laboratory analysts travel abroad for about 3 weeks at a time, during which they inspect approximately three establishments. Each establishment inspection typically lasts a week, with 1 day of each week set aside for documenting the inspection or for extending the inspection, if necessary.

CDER uses a risk-based process to select some domestic and foreign establishments for postapproval GMP surveillance inspections. According to an FDA report, the agency developed the process after recognizing that it did not have the resources to meet the requirement for inspecting domestic establishments every 2 years. The process uses a risk model to identify those establishments that, based on characteristics of the establishment and of the product being manufactured, have the greatest public health risk potential should they experience a manufacturing defect. . . . For example, FDA considers the risk to public health from poor quality over-the-counter drugs to be lower than for prescription drugs, and consequently establishments manufacturing only over-the-counter drugs receive a lower score on this factor than other manufacturers. Through this process, CDER annually prepares a prioritized list of domestic establishments and a separate, prioritized list of foreign establishments. CDER began applying this risk-based process to domestic establishments in fiscal year 2006 and expanded it to foreign establishments in fiscal year 2007.

FDA relies on multiple databases to manage the foreign drug inspection program. FDA assigns unique numeric identifiers to establishments, known as the FDA establishment identifier (FEI) number. An FEI number could be assigned at the time of registration, importation, or inspection.

- DRLS [Drug Regulation and Listing System] contains information on foreign and domestic drug establishments that have registered with FDA. Establishments that market their drugs in the United States must register with FDA. These establishments provide information, such as company name and address and the drug products they manufacture for commercial distribution in the United States, on paper forms that are entered into DRLS by FDA.

- OASIS [Operational and Administrative System for Import Support] contains information on drugs and other FDA-regulated products imported into the United States, including information on the establishment that manufactured the drug. The information in OASIS is automatically generated from data managed by U.S. Customs and Border Protection, which are originally entered by customs brokers based on the information available from the importer. Each establishment is assigned a manufacturer identification number that is generated from key information entered about an establishment's name, address, and location.
- FACTS [Field Accomplishments and Compliance Tracking System] contains information on FDA's inspections of domestic and foreign drug establishments. FDA investigators and laboratory analysts enter information into FACTS, following completion of an inspection.

According to DRLS, in fiscal year 2007, China and India had more establishments registered to manufacture drugs for the U.S. market than any other country. Other countries that had a large number of establishments registered to manufacture drugs for the U.S. market in this year were Canada, France, Germany, Italy, Japan, and the United Kingdom. . . . These countries are also listed in OASIS as having the largest number of manufacturers importing drugs into the United States.

FDA Lacks Accurate Information to Effectively Manage the Foreign Drug Inspection Program

FDA does not know how many foreign establishments are subject to inspection; including the number of establishments that are registered and whose products are currently imported into the United States and establishments that are not required to register but whose products are ultimately used in drugs that are marketed here. Instead of maintaining a list of such establishments, FDA relies on information from several databases that were not designed for this purpose. . . .

FDA's data suggest that between 3,000 and 6,760 establishments could be subject to FDA inspection. However, FDA officials told us that the two databases—DRLS and OASIS—cannot be electronically integrated or interact with one another, so any comparisons are done manually for each individual establishment. Because comparisons of the data and error identification are done manually, the databases are not conducive to routine data analysis. FDA officials told us that they have not generated an accurate count of the establishments whose drugs are imported into the United States. . . .

FDA Conducts Relatively Few Foreign Establishment Inspections and Relies on the NDA and ANDA Review Process as the Primary Selection Factor

FDA conducts relatively few inspections of foreign drug establishments. However, because FDA is not certain how many foreign establishments are actually subject to inspection, the percentage of foreign establishments that have been inspected cannot be calculated with certainty. Most foreign establishments are selected for inspection as part of the agency's review process associated with an NDA or ANDA. Therefore, the vast majority of foreign inspections include a preapproval inspection. In addition, although FDA has implemented a risk-based process in selecting foreign establishments for GMP surveillance inspections, relatively few such inspections are conducted. FDA tries to make efficient use of its resources by selecting establishments for these inspections that allow it to coordinate travel with preapproval inspections.

Relatively Few Foreign Establishments Are Inspected by FDA Each Year

In each year we examined, FDA inspected a small portion of foreign establishments through either preapproval or GMP surveillance inspections. However, its lack of a list of foreign establishments subject to inspection makes it difficult to determine an exact percentage. Based on our review of data on inspections, FDA conducted an average of 241 foreign establishment inspections per year from fiscal year 2002 through fiscal year 2007. Comparing this average number of inspections with FDA's count of 3,249 foreign establishments it used to plan its fiscal year 2007 prioritized GMP surveillance inspections suggests that the agency inspects about 7 percent of foreign establishments in a given year. At this rate it would take FDA more than 13 years to inspect this group of establishments once, assuming that no additional establishments are subject to inspection.

FDA's data indicate that some foreign drug manufacturers have not received an inspection, but the exact number of establishments not inspected was unclear. Of the list of 3,249 foreign establishments, there were 2,133 foreign establishments for which the agency could not identify a previous inspection. Agency officials told us that this count included registered establishments whose drugs are being imported into the United States that have never been inspected but also included other types of establishments, such as those whose products were never imported into the United States or those who have stopped importing drugs into the United States without notifying FDA. FDA was unable to provide us with counts of how many establishments fall into each of these subcategories. Of the remaining 1,116 establishments on FDA's list, 242 had received at least one inspection, but had not received a GMP surveillance inspection since fiscal year 2000, and the remaining 874 establishments had received at least one GMP inspection since fiscal year 2000. Of these 874 establishments, 326 had last been inspected in fiscal years 2005 or 2006, 292 were last inspected in fiscal years 2003 or 2004, and the remaining 256 received their last inspection from fiscal year 2000 through fiscal year 2002.

FDA has increased the number of foreign establishments it inspects, most of which are concentrated in a small number of countries. From fiscal year 2002 through fiscal year 2007, the number of foreign establishment inspections FDA conducted annually varied from year to year, but increased overall from 222 in fiscal year 2002 to 295 in fiscal year 2007. During this period, FDA inspected establishments in a total of 51 countries. More than three quarters of the 1,445 foreign inspections the agency conducted during this period were of establishments in ten countries. . . . The country with the most inspections during this period was India, which had 200 inspections. Inspections of establishments located in India increased from 11 in fiscal year 2002 to 65 in fiscal year 2007. . . .

Challenges Unique to Foreign Inspections Influence the Manner in Which FDA Conducts Such Inspections

Inspections of foreign drug establishments pose unique challenges to FDA—in both human resources and logistics. For example, unlike domestic inspections, FDA does not have a dedicated staff devoted to conducting foreign inspections and relies on volunteers. In addition, unlike domestic GMP surveillance inspections, foreign establishment GMP surveillance inspections are announced in advance and inspections cannot be easily extended due to travel itineraries that involve more than one establishment. Other factors, such as language barriers, can also add complexity to the challenge of completing foreign establishment inspections.

According to FDA officials, the agency does not have a dedicated staff to conduct foreign inspections. They explained that the same investigators and laboratory analysts are responsible for conducting both foreign and domestic inspections. These staff members must meet certain criteria in terms of their experience and training to conduct inspections of foreign establishments. For example, they are required to take certain training courses and have at least 3 years of experience conducting domestic inspections before they can be considered to conduct a foreign inspection. FDA reported that it currently has approximately 335 employees who are qualified to conduct foreign inspections of drug manufacturers. Approximately 250 of these employees are investigators and 85 are laboratory analysts. These counts do not represent the number of individuals that actually conduct foreign inspections in a given year. Not all investigators and laboratory analysts who are qualified to conduct a foreign inspection do so in a given year, while others may perform multiple inspections during the same period. Using data from FACTS, we found that the total number of employees conducting pre-approval and GMP surveillance inspections of drug manufacturing establishments, either foreign or domestic, decreased from 587 in fiscal year 2002 to 446 in fiscal year 2007. . . . However, of these, the number of employees who conducted foreign inspections of drug manufacturers increased from 100 to 141 during that same period. While an investigator and analyst team may participate in foreign inspections, FDA officials stated that in certain circumstances, such as inspections that do not involve the review of laboratory facilities, only an investigator is sent. . . .

FDA relies on investigators and laboratory analysts to volunteer to conduct foreign inspections. FDA officials told us that it is difficult to recruit investigators and laboratory analysts to voluntarily travel to certain countries. However, officials noted that the agency provides various incentives to recruit employees for foreign inspection assignments. For example, employees receive a $300 bonus for each three week trip completed. FDA indicated that if the agency could not find an individual to volunteer for a foreign inspection trip, it would mandate the travel. However, FDA does not typically send investigators and laboratory analysts to countries for which the U.S. Department of State has issued a travel warning nor would it mandate travel to such a country. We found that 49 foreign establishments registered as manufacturers of drugs for the U.S. market were located in 10 countries that had travel warnings posted as of October 2007. However, FDA officials told us that in the past they have conducted inspections in countries with travel warnings. They also provided us with one example in which an establishment in a country with a travel warning hired security through the U.S. Department of State to protect the inspection team.

FDA also faces several logistical challenges in conducting inspections of foreign drug manufacturing establishments. FDA guidance states that inspections at foreign facilities are to be approached in the same manner as domestic inspections. However, the guidance notes that one main difference posing a significant challenge to the inspection team abroad is the logistics borne by the program itself. For example, FDA is unable to conduct unannounced inspections of foreign drug manufacturers, as it sometimes does with domestic manufacturers. FDA policy states that the agency, with few exceptions, initiates inspections of establishments without prior notification to the specific establishment or its management so that the inspection team can observe the establishment under conditions that represent normal day-to-day activities. However, prior notification is routinely provided to foreign establishments. FDA recognizes that the time and expense associated with foreign travel requires them to ensure that the foreign establishment's managers are available and that the production line being inspected is operational during the inspection. In addition, FDA does not have explicit authority to inspect establishments in foreign countries, and it therefore may have to obtain permission from the government and company prior to the inspection. FDA officials explained that, in some cases, investigators and laboratory analysts may need to obtain a visa or letters of invitation to enter the country in which the establishment is located. In addition, FDA does not have the same flexibility to extend the length of foreign inspection trips if problems are encountered as it does with domestic inspections because of the need to maintain the inspection schedule, which FDA officials told us typically involves inspections of multiple establishments in the same country.

FDA officials also told us that language barriers can make foreign inspections more difficult to conduct than domestic inspections. The agency does not generally provide translators in foreign countries, nor does it require that foreign establishments provide independent interpreters. Instead, they may have to rely on an English-speaking representative of the foreign establishment being inspected, who may not be a translator by training, rather than rely on an independent translator.

Concluding Observations

Millions of Americans depend on the safety and effectiveness of the drugs they take. More than nine years ago we reported that FDA needed to make improvements in its foreign drug inspection program. Yet, our preliminary work indicates that fundamental flaws that we identified in the management of this program in 1998, continue to persist. FDA still does not have a reliable list of foreign establishments that are subject to inspection. As more imported drugs enter the United States, it becomes increasingly important that foreign establishments receive appropriate scrutiny. We understand that FDA currently cannot inspect all foreign establishments every few years. We also recognize that FDA has taken steps to improve its management of the foreign drug inspection program by enhancing the risk-based process it uses to select establishments for GMP surveillance inspections. In addition, FDA is pursuing an initiative that is intended to improve its identification of foreign drug establishments. However, until FDA responds to systemic weaknesses in the management of this important program, it cannot provide the needed assurance that the drug supply reaching our citizens is appropriately scrutinized, and safe. . . .

Source: U.S. Congress. Government Accountability Office. "Drug Safety: Preliminary Findings Suggest Weaknesses in FDA's Program for Inspecting Foreign Drug Manufacturers." Testimony by Marcia Crosse, Director of Health Care, before the Subcommittee on Oversight and Investigations, Committee on Energy and Commerce. November 1, 2007. GAO-08-224T. purl.access.gpo.gov/GPO/LPS88548 (accessed January 13, 2008).

Other Historic Documents of Interest

From this volume

- Safety of imported foods, p. 51

From previous *Historic Documents*

- GAO on prescription drug safety, *2006,* p. 129
- FDA advisory on antidepressants and suicide, *2004,* p. 746
- FDA officials on Vioxx withdrawal, *2004,* p. 850
- Task force report on prescription drug importation, *2004,* p. 981
- GAO on prescription drug advertising, *2002,* p. 945

DOCUMENT IN CONTEXT

Pakistani Leaders on the Political Situation in Pakistan

NOVEMBER 3 AND 10, 2007

Pakistan suffered through a tumultuous series of political and social upheavals during 2007 that raised serious questions about its near-term, let alone long-term, stability. President Pervez Musharraf took several risky gambles to preserve his power, including twice sacking the chief justice of the Supreme Court of Pakistan, imposing a state of emergency, and even—under intense pressure—giving up his post as head of the army. Another gamble was an attempt to negotiate a political settlement with exiled opposition leader and two-time prime minister Benazir Bhutto, one of Pakistan's most popular politicians, who returned to the country in October. Bhutto survived one assassination attempt on the day of her return but succumbed to another on December 27.

Bhutto's death capped a year of violence and unrest that left Pakistan facing perhaps the most dangerous prospects it had confronted since 1971, when, during a war with India, its eastern half was severed from it and became Bangladesh. Musharraf promised to hold parliamentary elections in February 2008, but even successful elections could not guarantee an end to turmoil in a country that has possessed nuclear weapons for nearly a decade.

Pakistan's neighbors had obvious concerns about instability there, as did the United States. President George W. Bush and his administration depended heavily on Musharraf for help in fighting the al Qaeda network and the Islamist militant group known as the Taliban, both of which had been rooted out of Afghanistan by the United States in 2001, only to relocate to Pakistan. Just how effective Musharraf's help was in this regard always had been open to question, but never more so than in 2007, when the United States and its NATO allies had to step up military operations to protect the government in Afghanistan from attacks by insurgents based in Pakistan.

Increasing Unrest, Opposition to Musharraf

Musharraf came to power as the result of a military coup in 1999, ousting an elected government headed by Prime Minister Nawaz Sharif. As often is the case in such events, the general promised to restore democratic rule once he had guaranteed stability. Musharraf's was the latest in a series of coups in Pakistan, which had been run by the military for more than half of the time since the country gained independence in 1947. Musharraf assumed the presidency in 2001, and in 2003 he promised to step down from his army post as part of a deal to win parliamentary approval for him to continue as president. One year after parliament gave its approval, he reneged on his promise, saying he needed to remain head of the army to defeat terrorists who had twice sought to kill him.

It was clear from the beginning that 2007 could be a pivotal year for Pakistan, if only because Musharraf's term as president expired in October and parliamentary elections were due by the end of the year, or early 2008 at the latest. A series of bombings and political assassinations, starting in mid-January, served notice that the political calendar would be marked by violence.

The first of several political crises during 2007 started on March 9, when Musharraf suddenly suspended the chief justice of the Supreme Court, Iftikhar Mohammad Chaudhry, and placed him under de facto house arrest. The government announced that Chaudhry was under investigation by a judicial council on allegations of abuse of power and nepotism.

The suspension of Chaudhry, who had ruled against the government in high-profile cases, sparked widespread protests by lawyers, opposition political parties, and others with grievances against Musharraf. For weeks, police and angry lawyers, clad in their black suits and ties, clashed in the streets of the capital, Islamabad, and several other cities. The most violent day was May 12, when more than forty people were killed in the clashes. Musharraf harshly denounced the protests on May 21, saying his political opponents were conspiring against him "and want to incite the people." He took strong action to stop further incitement, issuing a decree on June 4 authorizing the government to shut down independent television stations. By June, the battle over Chaudhry had shifted primarily to the Supreme Court, which had taken up the case against him and his appeal of his suspension.

While the dispute over Chaudhry was still pending, another major controversy came to a head, this one involving a months-long standoff between the government and Islamist militants. Students at one of Islamabad's most prominent mosques, the Lal Masjid, or Red Mosque, had confronted the government by holding several hostages, including police officers and alleged prostitutes, since January. The situation deeply embarrassed the government because it highlighted the ambiguous relationship between Musharraf and radical Islamists, some of whom had close ties to the army and some of whom had tried to kill him. The standoff came to a fiery end during a week-long siege early in July, when security forces stormed the mosque and killed many of the students and a leading cleric. After the fighting ended on July 11, most reports put the death toll at about 100 people, most of them student militants. The siege of the mosque was followed by yet more violence, including suicide bombings in Islamabad by Islamist groups during subsequent days that killed some 50 people.

Any respite Musharraf might have thought his due from the resolution of the Red Mosque crisis ended on July 20, when the Supreme Court reinstated Chaudhry by ruling that the president's suspension of him had been illegal. The ruling, which the government accepted, amounted to a stunning rebuke of the president's actions and raised questions about the legality of his seeking reelection while still in the army.

Bhutto Returns

The daughter of a Pakistani civilian leader hanged by the military in 1979, Benazir Bhutto served twice as prime minister during the 1980s and early 1990s, representing the country's largest political party, the Pakistan People's Party, which was based in her home province of Sind. Bhutto fled Pakistan in 1999 to avoid corruption charges pressed against her and her husband, Asif Ali Zardari, but she remained head of the party and a figure beloved by millions of Pakistan's poor. Reports surfaced early in 2007 that she was hoping to return to Pakistan to lead her party in parliamentary elections. Prospects for her return heightened in April, when the government dropped some, but not all, of the charges against her. This led to speculation that she and Musharraf were negotiating a deal under which she would run the government as prime minister, if she won the parliamentary elections, and he would remain as president. Left out of such a deal would have been Bhutto's longtime political rival, Nawaz Sharif, the prime minister whom Musharraf had ousted in 1999. Sharif led the original wing of the Pakistan Muslim League (a different faction was aligned with Musharraf) and also was living in exile.

Musharraf and Bhutto held a closed-door meeting on July 27 in Abu Dhabi, the capital of the United Arab Emirates, sparking more speculation that the two were working on a political agreement. A sign that the increasingly unpopular Musharraf felt cornered by the scarcity of his political options came on August 9 when he backed down from a reported plan to impose a state of emergency, apparently acting under intense pressure from U.S. secretary of state Condoleezza Rice. Subsequent news reports said the Bush administration also was pressing Musharraf to agree to a power-sharing arrangement with Bhutto as the best way to shore up his own position.

The prospect of a direct challenge to Musharraf from a different front arose on August 23, when the reinvigorated Supreme Court ruled that Sharif could return to Pakistan. After Musharraf ousted him, Sharif had gone into exile on the condition that he stay abroad for ten years; in return, the government had dropped several criminal charges against him. Sharif acted quickly to take advantage of the opportunity afforded him by the court. He flew from London to Pakistan on September 10 but was detained briefly at the airport, charged with corruption and money laundering, then hustled onto an airplane and sent to Saudi Arabia.

Still unclear in the legal and political maneuvering over Bhutto and Sharif was whether Musharraf intended to give in to demands by his opponents that he resign as army chief of staff before seeking reelection as president. Some clarification came on September 18, when his senior legal adviser said Musharraf would remain army chief until after his election but would then resign from the army before taking the oath of office for a new term as president. This statement failed to satisfy critics, who noted that Musharraf had pledged before to quit the army, only to backtrack on the promise.

The pledge did clear the way for Musharraf to win reelection handily on October 6 after other candidates withdrew and opposition parties boycotted the vote. Under Pakistan's system, the president was elected by a combination of the national parliament and provincial assemblies. This was one of the last major acts of the outgoing national parliament before its term ended on November 15. It was not known at this point, however, whether the Supreme Court would uphold the validity of the election, which was challenged by opponents as unconstitutional on several grounds, including that Musharraf was still in the army.

The event that millions of Pakistanis had been awaiting occurred on October 18, when Bhutto returned home after eight years in exile. She was greeted by cheering throngs upon her midday arrival in Karachi, the capital of her home province. Hours later, during a procession through crowds estimated at some 200,000 people, the celebration turned to tragedy as two bombs exploded near the armored truck in which Bhutto was riding. Bhutto was unhurt but an estimated 140 people were killed and about 400 wounded. It was the worst terrorist attack in recent Pakistani history, and it signaled that any revitalization of democracy would come at a high price. Bhutto blamed the government for failing to heed her warnings that extremist elements of the military were trying to kill her. Government officials blamed Islamist militants, but the leader of one party supporting Musharraf blamed Bhutto herself, suggesting she had organized the bombings to generate sympathy for her cause.

Declaration of Emergency

In this highly charged atmosphere, Musharraf intervened by taking the action he reportedly had contemplated three months earlier—the declaration of a state of emergency. On November 3, he suspended the constitution, fired Chief Justice Chaudhry and six of his colleagues, ordered the opposition news media closed, and declared his own actions beyond the review of courts. In a lengthy televised speech, Musharraf told Pakistanis he had acted because the country faced "an internal crisis" and he had no other choice. Speaking in English at the close of his address, Musharraf appealed to the rest of the world to "please give us time" to finish developing democracy.

Although government officials insisted Musharraf's emergency order did not represent the imposition of martial law, the resulting crackdown bore strong resemblance to full rule by the military. Thousands of troops took control of the streets in major cities, an estimated 2,500 opposition leaders and other dissidents were detained for at least the following two days, and security forces harshly suppressed protests by thousands of lawyers on November 5.

Bhutto, who briefly had been out of the country visiting her children at the time of Musharraf's emergency decree, quickly returned and announced plans for a major protest on November 9 in Rawalpindi, the headquarters city of the army. The police locked down that city, however, and briefly confined Bhutto in what amounted to house arrest at her home in Islamabad. In an apparent move to quell protests, Musharraf announced that parliamentary elections would be held by February 15, 2008, and that he, in fact, would give up his army post before beginning his new term as president.

The emergency decree prevented Bhutto and other opposition leaders from addressing their supporters in public, but Bhutto was able to reach an international audience with a speech to foreign diplomats in Islamabad on November 10. Bhutto acknowledged that, before her return to Pakistan on October 18, she and Musharraf had negotiated a "road map for fair elections and a transition to democracy." She said the president had upended this agreement with his emergency decree. Bhutto demanded that Musharraf revive the constitution, retire as army chief, and hold elections by January 15, 2008. In subsequent days, Bhutto and Musharraf gave dueling interviews to the international news media, each apparently seeking to gain support of foreign governments—particularly the United States—to put pressure on the other. Bhutto also said she was no longer willing to negotiate the terms of a new government with Musharraf.

Musharraf's main motivation in imposing emergency rule had been to prevent the Supreme Court from overturning his recent election as president. A new court, which he had appointed, on November 22 cleared the way for him to retain the presidency by dismissing all challenges to that election. On November 28, Musharraf reluctantly did what his critics had been demanding he do for many years: he resigned from the army, turning over his post as chief of staff to Gen. Ashfaq Parvez Kayani, a former head of the military's Inter-Services Intelligence agency. The next day, now a civilian, Musharraf was sworn in for a second five-year term as president, and he said that he would lift the state of emergency on December 16 and restore the constitution before parliamentary elections. In an address to foreign diplomats, Musharraf defended his actions against international criticism. "We want democracy, I am for democracy," he said. "We want human rights, we want civil liberties, but we will do it our way, as we understand our society, our environment, better than anyone in the West."

During the first two weeks of December, the government slowly relaxed some of the restrictions of Musharraf's emergency rule, for example by allowing private television channels back on the air—but only after they signed a "code of conduct" that amounted to self-censorship of news displeasing to the government. Musharraf lifted the state of emergency and restored the constitution on December 16. Before he did so, however, he permanently replaced Chaudhry and the other Supreme Court justices he had dismissed, and he issued several decrees legalizing the steps he had taken under emergency rule.

Bhutto Assassination

The easing of the state of emergency intensified the parliamentary election campaign, which had been under way since late November, when Sharif was allowed to return to Pakistan (but not to run as a political candidate) and Bhutto was allowed to meet with her supporters. These two leaders, who had been bitter enemies for many years, promised to cooperate—in a manner left unclear—to defeat Musharraf's supporters in the elections.

The climactic event of Pakistan's bloody and tumultuous year took place on the evening of December 27. As Bhutto was in her car leaving a large outdoor rally in Rawalpindi, one or two shots rang out, a bomb went off, and Bhutto collapsed. After being taken to a hospital, she was declared dead, at age fifty-four. More than twenty other people also died as a result of the explosion, which appeared to be caused by a suicide bomb.

Almost immediately, the cause of Bhutto's death became yet another source of contention. Bhutto's supporters said she was killed by the gunfire, which they darkly suggested came from someone associated with the government or the military. Government officials said Bhutto had died from head wounds suffered when she ducked from the gunfire and struck a lever on the sunroof opening in her car. The bombing had been carried out by Islamists, government officials said, citing in particular a commander of the Taliban in the tribal region of South Waziristan. The scene of the crime was swept clean, limiting the usefulness of any subsequent police investigations. The next day, Bhutto's representative in Washington, D.C., Mark Siegel, released an e-mail he had received from her on October 26 saying she "wld [would] hold Musharraf responsible" if anything happened to her. "I have been made to feel insecure by his minions," she added.

Bhutto's death sparked a wave of anguish among her supporters, many of whom had looked to her—as they had to her father before her—for relief from their grinding poverty. The Bhutto clan and its Pakistan People's Party had delivered little of this relief, but Benazir Bhutto's powerful charisma had kept the hope alive even during her many years of exile.

Protests raged for several days, mostly in Karachi and other parts of Sind province, with grieving Bhutto supporters venting most of their anger at Musharraf. Bhutto was buried on December 28 at an elaborate family mausoleum in Larkana, near Karachi.

One of the many consequences of Bhutto's death was that it left her party without a well-known, dynamic leader. To fill the void, party leaders announced on December 30 that Bhutto's nineteen-year-old son, Bilawal, would take over the party chairmanship. He was still an undergraduate student at Oxford University, however, so his place would be taken for the time being by his father and Bhutto's husband, Asif Ali Zardari. This move put the party in the hands of someone who had been even more controversial than

Bhutto. Zardari had been jailed in 1990 on corruption charges and later developed a reputation as "Mr. Ten Percent" for the commissions he allegedly earned as a cabinet minister during Bhutto's second term in office. Zardari said he would not be a candidate for parliament in the elections, which were rescheduled to February 18, 2008.

Following are two documents. First, the text of a "Proclamation of Emergency" issued on November 3, 2007, by Pakistani president Pervez Musharraf, acting in his capacity as chief of staff of the army. Second, the text of a November 10, 2007, address by Benazir Bhutto, leader of the Pakistan People's Party, to foreign diplomats at a meeting in Islamabad.

DOCUMENT

General Musharraf's Proclamation of Emergency

November 3, 2007

WHEREAS there is visible ascendancy in the activities of extremists and incidents of terrorist attacks, including suicide bombings, IED [improvised explosive devices] explosions, rocket firing and bomb explosions and the banding together of some militant groups have taken such activities to an unprecedented level of violent intensity posing a grave threat to the life and property of the citizens of Pakistan;

WHEREAS there has also been a spate of attacks on State infrastructure and on law enforcement agencies;

WHEREAS some members of the judiciary are working at cross purposes with the executive and legislature in the fight against terrorism and extremism thereby weakening the Government and the nation's resolve and diluting the efficacy of its actions to control this menace;

WHEREAS there has been increasing interference by some members of the judiciary in government policy, adversely affecting economic growth, in particular;

WHEREAS constant interference in executive functions, including but not limited to the control of terrorist activity, economic policy, price controls, downsizing of corporations and urban planning, has weakened the writ of the government; the police force has been completely demoralized and is fast losing its efficacy to fight terrorism and Intelligence Agencies have been thwarted in their activities and prevented from pursuing terrorists;

WHEREAS some hard core militants, extremists, terrorists and suicide bombers, who were arrested and being investigated were ordered to be released. The persons so released have subsequently been involved in heinous terrorist activities, resulting in loss of human life and property. Militants across the country have, thus, been encouraged while law enforcement agencies subdued;

WHEREAS some judges by overstepping the limits of judicial authority have taken over the executive and legislative functions;

WHEREAS the Government is committed to the independence of the judiciary and the rule of law and holds the superior judiciary in high esteem, it is nonetheless of paramount importance that the Honourable Judges confine the scope of their activity to the judicial function and not assume charge of administration;

WHEREAS an important Constitutional institution, the Supreme Judicial Council, has been made entirely irrelevant and non est by a recent order and judges have, thus, made themselves immune from inquiry into their conduct and put themselves beyond accountability;

WHEREAS the humiliating treatment meted to government officials by some members of the judiciary on a routine basis during court proceedings has demoralized the civil bureaucracy and senior government functionaries, to avoid being harassed, prefer inaction;

WHEREAS the law and order situation in the country as well as the economy have been adversely affected and trichotomy of powers eroded;

WHEREAS a situation has thus arisen where the Government of the country cannot be carried on in accordance with the Constitution and as the Constitution provides no solution for this situation, there is no way out except through emergent and extraordinary measures;

AND WHEREAS the situation has been reviewed in meetings with the Prime Minister, Governors of all four Provinces, and with Chairman Joint Chiefs of Staff Committee, Chiefs of the Armed Forces, Vice-Chief of Army Staff and Corps Commanders of the Pakistan Army;

NOW, THEREFORE, in pursuance of the deliberations and decisions of the said meetings, I General Pervez Musharraf, Chief of the Army Staff, proclaim Emergency throughout Pakistan.

2. I hereby order and proclaim that the Constitution of the Islamic Republic of Pakistan shall remain in abeyance.

3. This Proclamation shall come into force at once.

SOURCE: Islamic Republic of Pakistan. Associated Press of Pakistan. "Text of 'Proclamation of Emergency.' " November 3, 2007. www.app.com.pk/en_/index.php?option=com_content&task=archivecategory&id=0&year=2007&month=11&module=1&limit=9&limitstart=2124 (accessed March 17, 2008).

DOCUMENT

Benazir Bhutto's Address to Foreign Diplomats

November 10, 2007

I appreciate the opportunity to have this conversation with you, especially at this most critical time in the history of my nation.

Obviously Pakistan is at a turning point, and the direction we follow will not only impact the future of my nation, but I strongly believe will have a direct and immediate impact on the stability of the region and the stability of the world.

I take this opportunity to thank the International community for their support to the people of Pakistan in calling upon General Musharaf to lift the curbs of the media, release political prisoners, retire as Army Chief on schedule, and hold elections on schedule.

Democracy is morally right, and even more important to this forum, democracy is the only viable way to contain the growth of extremism, militancy and fanaticism that now threatens the world.

My Party, Nation and I have spent our lives fighting for democracy and for democratic governance. We are fighting now for democracy to safeguard people's rights and also to safeguard the unity of Pakistan. Our goal is to ensure that through empowerment, employment and education, regions of my country cease being the Petri dish for international and national terrorist plots that threaten us all.

Pakistan under dictatorship is a pressure cooker. Without a place to vent, the passion of our people for liberty threatens to explode. The current military dictatorship that rules my country with an iron fist is opposing the inevitable forces of history. There is not enough barbed wire, or bullets, or bayonets to defeat my people's unquestionable desire for democracy, for control over their own lives, for human rights, gender equality, labour and minority rights and for a chance to build a better life for their children. These are indeed the dreams of the Pakistani people and of all people. These are universal values.

It is eleven years since the destabilization of the democratic government I led but it has failed to crush the will and support of the people for a representative government that addresses the bread and butter issues of our people, 74% of whom live in poverty. Poverty has increased as has joblessness since the PPP government, with a 6% growth rate, without 10 billions of aid, was overthrown.

Yesterday my home was surrounded and I was unofficially put under house arrest because this military regime did not want the world to see that the people of Pakistan want change, want freedom, and want liberty. The garrison town of Pindi was cut off and surrounded with blockades, barbed wire and the motorway closed. For 8 hours people were gassed and beaten. The brutal images of police beating innocent women were on every television screen on Earth. Tomorrow people of Sindh will hold protest meetings in every district to show solidarity with the people of Rawalpindi.

In Pakistan we say that there are two tests for the success of a public meeting. When the government does not use coercive methods to stop a meeting, the success is judged by the number of people present.

On October 18 the people of Pakistan held the most historic rally in the history of Pakistan when three million turned out in Karachi, stretching all the way back on the National Highway to receive me to express their hope that my return would be a catalyst of a change from dictatorship to democracy.

The second test of a popular public meeting is the amount of police force and measures used to block the people's participation. The amount of force and restrictive measures is proportionate to the number of people expected to turn out and attend a public meeting.

Force was used to bring closure of Northern Pakistan on November 9, 2007. There was tear-gassing, baton charging, arrest of 5000 activists, including women parliamentarians from Peshwar to Rawalpindi, Karachi to Pindi, Lahore to Pindi and Islamabad to Pindi. The amount of force used was, according to our calculation, to stop a gathering of a million people. By both tests, coercive and non-coercive, the people rose to the occasion.

Now the ball is in the regime's court. We have called for a Long March from Lahore from November 13, 2007. This is the March for Freedom, freedom from dictatorship, from militancy, from poverty, and unemployment.

The current regime has convinced some Nations that General Musharaf alone stands in the way of a nuclear armed, fundamentalist Pakistan. This is a misperception. The religious parties in Pakistan have never received more than 11% of the vote in any election and they would receive less today. It is dictatorship that fuels extremism. The dictatorship of 80's created the Afghan Mujahideen which morphed into Taliban and Al-Qaeda. The political partners of that dictatorship hold key positions in the political, administrative and security institutions of this dictatorship. They cannot, and have not contained extremism nor reduced poverty. They have exacerbated the situation to an extent where nuclear armed Pakistan is threatened with implosion.

Only a popularly elected democratic government with the mandate of the people has the political base to undermine the militants and bring peace to the people of Pakistan.

Dictatorship does not contain fanaticism. Dictatorship causes fanaticism.

Ladies and gentlemen, Pakistan plays a central role in the direction of one billion Muslims on this planet. A Pakistan which is moderate, enlightened, modern can be a model to the people in the Muslim world who have to choose between the forces of the past and the forces of the future. Pakistan can be critical to democratic development and the containment of extremism all over Asia and Africa, and can spare our global community further senseless attacks from the forces of hatred.

The enemies of reconciliation amongst our peoples and our nations are trying desperately to provoke a clash of civilizations. These fanatics thrive on chaos. They thrive on the desperation that comes from dictatorship. Democracy suffocates them by giving people choices and giving people hope.

The PPP and I negotiated with General Musharaf's regime for a peaceful transition to democracy. We worked on a road map for fair elections and transition to democracy.

Tragically, suddenly General Musharaf suspended the constitution and imposed Martial Law. This Martial law has been called an Emergency for International consumption.

The PPP has called upon General Musharaf to:

- Revive the Constitution and with it the judges.
- Retire as Chief of Army Staff on November 15 as scheduled.
- Hold general elections called on November 15 on schedule for January 15, 2008.
- Re-constitute the Election Commission and implement election reforms including an interim government of national consensus to oversee the elections, suspension of the Mayors for the elections period, appointment of impartial officials to important government positions, a fair voter account, no improvised polling stations and other such measures.
- End of political victimization
- Lifting of curbs on the media.

Pakistan is in a crisis. Our country, armed forces, police, women, judges, lawyers, minorities, labour, peasants, students, intellectuals and youth are under assault, some by the militants, others by the military regime.

The choice must not be between the military or the militants. The choice must be for the will of the people, for democracy.

If for no other reason than your own national self-interest, stand with us in our demand for free and fair elections with robust international monitoring.

The Taliban are coming nearer and nearer. I do not want to be melodramatic. But it is the harsh reality that they came from the mountains of Tora Bora to the tribal areas. First one agency fell, then another as the government struck ceasefires and peace treaties with them. Bajaur fell, Khyber fell, Waziristan fell. Now towns in Swat are falling. Madyan fell, Kalam fell. Today they knock on the doors of Shangla Hills.

The Freedom March, the caravan of democracy is not about Benazir Bhutto. This is not about the Pakistan Peoples Party. This is about saving Pakistan from disintegration at the hands of militants who have grown in strength under a military dictatorship. They threaten us all. The people of Pakistan cannot be allowed to fall from one dictatorship to another, from military dictatorship to religious dictatorship. I have returned to help my people. My supporters, often from working families, often young, are risking their lives to save Pakistan by saving democracy. My Father gave his life for the democratic rights of the people.

We don't accept tyranny. Our cause is just, our path is right for it is the path of truth, the path of the people. We appeal to all the people of our country to walk with us on our common destination towards freedom. We ask the international community to give us moral support.

We believe that victory and defeat are in the hands of God, as is life and death, but we must do what is right, what is just. We must raise our voices and begin the journey in the great walk from tyranny to freedom.

Thank you.

SOURCE: Pakistan Peoples Party. "Mohtarma Benazir Bhutto's Address to Diplomats at PPP Foreign Liaison Committee Reception." November 10, 2007. www.ppp.org.pk/mbb/speeches/speeche86.html (accessed March 17, 2008).

OTHER HISTORIC DOCUMENTS OF INTEREST

FROM THIS VOLUME

- U.S. National Intelligence Estimate on terrorist threats, p. 335
- Afghanistan-Pakistan "peace jirga" communiqué, p. 442
- UN secretary-general on the situation in Afghanistan, p. 547

FROM PREVIOUS *HISTORIC DOCUMENTS*

- UN Secretary-General Annan on the situation in Afghanistan, *2006*, p. 525
- Pakistan's president Musharraf on Islamic extremism, *2005*, p. 472

DOCUMENT IN CONTEXT

Intergovernmental Panel of Scientists on Climate Change

NOVEMBER 17, 2007

New scientific reports during 2007 added more urgency to calls for the world to curb emissions of the greenhouse gases, such as carbon dioxide, that were contributing to global climate change. A series of studies by a United Nations panel of scientists, economists, and other experts warned that time was running out for effective actions to limit the consequences of climate change. Even if the world acted promptly, the panel said, "abrupt and irreversible" changes, some of them potentially damaging to large portions of the Earth, already were taking place.

World leaders did act, or at least promised to act. At a conference in Bali, Indonesia, in December, they agreed to a "road map" to negotiate a treaty on climate change policies by 2009. Such a treaty would take the place of one negotiated in 1997 in Kyoto, Japan, which had committed industrialized countries to curb their greenhouse gas emissions. The United States, the world's largest emitter of greenhouse gases, withdrew from the Kyoto treaty in 2001, and developing countries, including rapidly growing China and India, were exempt from its provisions.

As in previous years, much of the world's attention on climate change issues focused on U.S. president George W. Bush, who early in his administration had been skeptical about the causes and extent of climate change and had staunchly opposed mandatory measures to curb greenhouse gas emissions. Bush had gradually moderated his tone, and in January 2007 he made one of his most emphatic statements yet acknowledging climate change as a problem. In his State of the Union address on January 23, Bush called for research toward new technologies that "will help us be better stewards of the environment, and they will help us to confront the serious challenge of global climate change."

Scientific Reports

The UN in 1988 established the Intergovernmental Panel on Climate Change (IPCC) as a group of scientific experts from around the world who would review meteorological and other data to determine whether climate change—including global warming—really was under way and to assess the impact of any changes that were occurring. The panel consisted of about 600 scientists of varying viewpoints from 154 countries who met periodically and drafted reports that were reviewed by their peers and then edited by officials from the countries represented on the panel. The panel issued major reports in 1990, 1995, and 2001. The 1990 report suggested that observed changes in the climate might be due to "natural variability," but the 1995 report was more emphatic in noting a "discernable human influence" on the climate. The 2001 report cited "new and stronger evidence" of human causes for rising temperatures. This report also suggested that by 2100 the world's average temperature would rise 2.5–10.4 degrees Fahrenheit and the sea level would rise 4–35 inches.

The IPCC panel was scheduled to issue new reports in 2007 in time for the UN conference in Bali. As in the past, the panel's deliberations were fraught with controversy. An apparent minority of scientists, including some panel members, insisted the IPCC was being too alarmist with the upper ends of its ranges for temperature increases and sea level rises. Other scientists insisted the panel was too cautious and was hedging its assertions in response to pressure from governments, notably the Bush administration, that had taken a skeptical position on climate change. Most panel members insisted they weighed only the evidence before them and were alarmed by what they saw.

Panel members gathered in Paris in late January to work on the draft of the first of several technical reports to be issued during the year. The panel's leaders said they had been inundated with e-mails and other messages from scientists around the world seeking to influence the outcome. The result, released on February 2, was the panel's most emphatic assertion so far: it said its findings were "unequivocal" that climate change was occurring and that human activity was largely responsible. The panel stated with a 90 percent certainty that concentrations of carbon dioxide (CO_2) and other greenhouse gases (notably methane) would reach twice the preindustrial level by the year 2100. In such a case, the panel forecast, global temperatures would rise by 3.5–8 degrees Fahrenheit (a slightly narrower range than forecast in 2001) and sea levels would rise by 7–23 inches (also a narrower range than in 2001). The panel's estimates for sea levels did not factor in the melting of glaciers and polar ice sheets; both were occurring, but the panel said the effects were difficult to estimate.

In describing the effect of the predicted changes, the panel said it was "very likely that hot extremes, heat waves and heavy precipitation events will continue to become more frequent." Robert Watson, chief scientist for the World Bank and the chair of the negotiations that produced the report, noted that "every government in the world signed off on this document, including the U.S."

The IPCC's second report of the year was issued, in summary form, in Brussels on April 6 and in final form at several regional news conferences around the world on April 10. It focused on the effects of global warming and other aspects of climate change. As had previous reports, this one said that some parts of the world would see some benefits, such as increased rainfall and longer growing seasons in the higher latitudes. Other regions, however, could see calamitous effects; notably, low-lying coastal regions could be inundated by rising seas, much of southern Europe would face extremely hot and dry summers, and many parts of the world (Africa, in particular) would face severe water shortages. Further, about 20–30 percent of the world's animal and plant species would be at risk of extinction if global temperatures rose to the higher end of the range suggested in the February 2 report. Panel scientists said it was clear that the worst effects of climate change would be in poor countries (notably in Africa and South Asia), but even rich countries would face negative consequences.

The report was the product of heated negotiations, during which governmental representatives from China, India, Saudi Arabia, and the United States sought, successfully in some cases, to reduce the level of certainty expressed in the findings. In comments after release of the report, U.S. officials sought to focus attention on what they called the positive aspects of climate change, such as longer growing seasons in some temperate climates.

The main message of a third IPCC report, issued in Bangkok on May 4, was that the world needed to take steps immediately—not at some future time—to mitigate the negative consequences of climate change outlined in the previous reports. Among the changes advocated were increased taxes on the consumption of coal and petroleum products (known as a "carbon tax"), shifts to alternative fuels that produce fewer greenhouse gas emissions, and research into new energy sources that could be used on a large scale. Although this report addressed controversial topics, the negotiations leading to it proved to be less contentious than previous IPCC sessions, according to participants.

Summary for Policymakers

The IPCC's fourth—and in many ways most significant—report of 2007 was issued on November 17 after meetings in Valencia, Spain. Titled "Summary for Policymakers," the report presented, in remarkably clear language for a scientific document, an overview of the panel's previous reports on the causes and likely effects of climate change, along with various proposals for preventing the most harmful aspects of that change. The report was released just more than a month after the Norwegian Nobel Committee had awarded the year's coveted Nobel Peace Prize to the IPCC (as an institution) along with former U.S. vice president Al Gore, perhaps the world's most prominent advocate of forceful action to address climate change.

The report's two bottom lines were that action to curb greenhouse gas emissions was required immediately and that, even if the world did act, changes in climate already under way would be "abrupt and irreversible." The report laid out the main conclusions of the panel's three previous reports during 2007, including that the fact of global warming was "unequivocal," that human activities were largely responsible, that temperatures and sea levels already were rising and would continue to do so, that many animal and plant species faced potential extinction, that extreme weather conditions would become more frequent in many parts of the world, that some areas of the world would see some benefits, and that prompt action to reduce greenhouse gas emissions would mitigate some of these consequences but would not prevent damage that could linger for many centuries.

Releasing the report, UN secretary-general Ban Ki-moon described climate change as "the defining challenge of our age" and said the countries that emitted the greatest amounts of greenhouse gases—citing China and the United States, in particular—needed to play "a more constructive role" in dealing with the problem. IPCC chairman Rajendra Pachauri stressed the urgent nature of the report's message: "If there's no action before 2012, that's too late," he said. That year was the end date, under the Kyoto Protocol, for mandatory reductions in carbon emissions by the industrialized countries. "What we do in the next two to three years will determine our future. This is the defining moment."

Bali Conference

For many scientists and policymakers most concerned about climate change, the "defining moment" described by the IPCC chair was to come in December during a UN-sponsored session on the resort island of Bali, where participants began to develop a follow-up to the Kyoto Protocol. In the months leading up to the conference, political leaders across the world were busy staking out their positions. Probably the most dramatic action came from European Union leaders who, on March 9, agreed on a plan requiring member countries to reduce greenhouse gas emissions by 20 percent from 1990 levels (the benchmark under UN treaties) by 2020. Some EU countries wanted an even deeper cut, while others argued that the 20 percent figure was too much for their economies to absorb. The EU plan called on member countries to derive 20 percent of their energy by 2020 from renewable sources, such as solar and wind power. This would be nearly three times the current level, officials said.

German chancellor Angela Merkel was a driving force behind the EU plan, and she had hoped to use it to win approval of a far-reaching commitment by her fellow leaders of the Group of Eight countries during their meeting in Germany in June. Merkel advocated a plan under which the G-8 nations would pledge to cut their greenhouse gas emissions in half by 2050. President Bush told Merkel he opposed any mandatory goals, however. Merkel withdrew her proposal on June 6, and the G-8 leaders issued a scaled-back statement on the subject.

With these and other national and regional actions on climate change as a background, the negotiations in Bali began early in December but quickly became deadlocked on the central question of whether countries should commit to specific cuts in greenhouse gas emissions. Negotiators from the EU and some developing countries argued for mandatory cuts of about 20 percent, below 1990 levels, by 2020. This stance was opposed as a matter of principle by the United States, which also argued (as it had for many years) that the fast-growing countries of China and India had an obligation to bear their share of the burden in curbing emissions; both those countries, along with all other developing countries, had been spared mandatory emissions cuts under the Kyoto Protocol.

The Bali conference was supposed to end with an agreement by December 14, but as that date approached and negotiators were still deadlocked, acrimony began to mount. European negotiators threatened to boycott a climate change conference Bush had called for the following January unless the United States agreed to a proposed "road map" for a new treaty. And Al Gore, fresh from receiving the Nobel Peace Prize three days earlier, arrived in Bali on December 13 to charge: "My own country, the United States, is principally responsible for obstructing progress here in Bali." Gore, a Democrat, told the delegates to look beyond the Bush administration, which would leave office in January 2009. "Over the next two years the United States is going to be somewhere it is not now," he said, to loud applause from delegates.

Despite the European threat and Gore's tough language, it emerged that negotiators already were breaking the deadlock, in large part because of concessions by the United States. The conference hammered out a compromise agreement, announced on December 15, that met key objections by the United States and several other countries but that committed all countries to negotiate a new climate change treaty by 2009. The compromise contained none of the specific targets the Europeans had wanted. However, the deadline for the new treaty was a significant concession by the Bush administration, which in previous years had blocked even the setting of agendas for the post-Kyoto period. Among other things, the schedule meant that the most important decisions about a new treaty would be made when the United States had a new president; all leading Democratic candidates and one of the leading Republican candidates (Sen. John McCain) had taken positions on climate change sharply different from those of Bush.

Even with the emphasis on the U.S. position, many delegates noted that China was becoming just as significant a factor in the climate change debate. Depending on which figures were used, China either overtook the United States as the world's major emitter of greenhouse gases in 2007 or would do so in 2008 or 2009, at the latest.

Following are excerpts from "Summary for Policymakers," a report prepared by the Intergovernmental Panel on Climate Change and released on November 17, 2007.

DOCUMENT

Summary for Policymakers by Panel on Climate Change

November 17, 2007

Introduction

This Synthesis Report is based on the assessment carried out by the three Working Groups of the Intergovernmental Panel on Climate Change (IPCC). It provides an integrated view of climate change as the final part of the IPCC's Fourth Assessment Report (AR4).

A complete elaboration of the Topics covered in this summary can be found in this Synthesis Report and in the underlying reports of the three Working Groups.

1. Observed changes in climate and their effects

Warming of the climate system is unequivocal, as is now evident from observations of increases in global average air and ocean temperatures, widespread melting of snow and ice and rising global average sea level.

Eleven of the last twelve years (1995–2006) rank among the twelve warmest years in the instrumental record of global surface temperature (since 1850). The 100-year linear trend (1906–2005) of 0.74 [0.56 to 0.92]°C is larger than the corresponding trend of 0.6 [0.4 to 0.8]°C (1901–2000) given in the Third Assessment Report (TAR). The temperature increase is widespread over the globe and is greater at higher northern latitudes. Land regions have warmed faster than the oceans.

[*Note:* Numbers in square brackets indicate a 90% uncertainty interval around a best estimate, i.e. there is an estimated 5% likelihood that the value could be above the range given in square brackets and 5% likelihood that the value could be below that range. Uncertainty intervals are not necessarily symmetric around the corresponding best estimate. Words in italics represent calibrated expressions of uncertainty and confidence.]

Rising sea level is consistent with warming. Global average sea level has risen since 1961 at an average rate of 1.8 [1.3 to 2.3] mm/yr and since 1993 at 3.1[2.4 to 3.8] mm/yr, with contributions from thermal expansion, melting glaciers and ice caps, and the polar ice sheets. Whether the faster rate for 1993 to 2003 reflects decadal variation or an increase in the longer-term trend is unclear.

Observed decreases in snow and ice extent are also consistent with warming. Satellite data since 1978 show that annual average Arctic sea ice extent has shrunk by 2.7 [2.1 to 3.3]% per decade, with larger decreases in summer of 7.4 [5.0 to 9.8]% per decade. Mountain glaciers and snow cover on average have declined in both hemispheres.

From 1900 to 2005, precipitation increased significantly in eastern parts of North and South America, northern Europe and northern and central Asia but declined in the

Sahel, the Mediterranean, southern Africa and parts of southern Asia. Globally, the area affected by drought has *likely* increased since the 1970s.

It is *very likely* that over the past 50 years: cold days, cold nights and frosts have become less frequent over most land areas, and hot days and hot nights have become more frequent. It is *likely* that: heat waves have become more frequent over most land areas, the frequency of heavy precipitation events has increased over most areas, and since 1975 the incidence of extreme high sea level [excluding tsunamis, which are not due to climate change. Extreme high sea level depends on average sea level and on regional weather systems. It is defined here as the highest 1% of hourly values of observed sea level at a station for a given reference period] has increased worldwide.

There is observational evidence of an increase in intense tropical cyclone activity in the North Atlantic since about 1970, with limited evidence of increases elsewhere. There is no clear trend in the annual numbers of tropical cyclones. It is difficult to ascertain longer-term trends in cyclone activity, particularly prior to 1970.

Average Northern Hemisphere temperatures during the second half of the 20th century were *very likely* higher than during any other 50-year period in the last 500 years and *likely* the highest in at least the past 1300 years.

Observational evidence [Based largely on data sets that cover the period since 1970] from all continents and most oceans shows that many natural systems are being affected by regional climate changes, particularly temperature increases.

Changes in snow, ice and frozen ground have with *high confidence* increased the number and size of glacial lakes, increased ground instability in mountain and other permafrost regions and led to changes in some Arctic and Antarctic ecosystems.

There is *high confidence* that some hydrological systems have also been affected through increased runoff and earlier spring peak discharge in many glacier- and snow-fed rivers and through effects on thermal structure and water quality of warming rivers and lakes.

In terrestrial ecosystems, earlier timing of spring events and poleward and upward shifts in plant and animal ranges are with *very high confidence* linked to recent warming. In some marine and freshwater systems, shifts in ranges and changes in algal, plankton and fish abundance are with *high confidence* associated with rising water temperatures, as well as related changes in ice cover, salinity, oxygen levels and circulation.

Of the more than 29,000 observational data series, from 75 studies, that show significant change in many physical and biological systems, more than 89% are consistent with the direction of change expected as a response to warming. However, there is a notable lack of geographic balance in data and literature on observed changes, with marked scarcity in developing countries.

There is* **medium confidence** *that other effects of regional climate change on natural and human environments are emerging, although many are difficult to discern due to adaptation and non-climatic drivers.

They include effects of temperature increases on:

- agricultural and forestry management at Northern Hemisphere higher latitudes, such as earlier spring planting of crops, and alterations in disturbance regimes of forests due to fires and pests

- some aspects of human health, such as heat-related mortality in Europe, changes in infectious disease vectors in some areas, and allergenic pollen in Northern Hemisphere high and mid-latitudes
- some human activities in the Arctic (e.g. hunting and travel over snow and ice) and in lower-elevation alpine areas (such as mountain sports).

2. Causes of change

Changes in atmospheric concentrations of greenhouse gases (GHGs) and aerosols, land cover and solar radiation alter the energy balance of the climate system.

Global GHG emissions due to human activities have grown since pre-industrial times, with an increase of 70% between 1970 and 2004.

Carbon dioxide (CO_2) is the most important anthropogenic GHG. Its annual emissions grew by about 80% between 1970 and 2004. The long-term trend of declining CO_2 emissions per unit of energy supplied reversed after 2000.

Global atmospheric concentrations of CO_2, methane (CH_4) and nitrous oxide (N_2O) have increased markedly as a result of human activities since 1750 and now far exceed pre-industrial values determined from ice cores spanning many thousands of years.

Atmospheric concentrations of CO_2 (379ppm) and CH_4 (1774ppb) in 2005 exceed by far the natural range over the last 650,000 years. Global increases in CO_2 concentrations are due primarily to fossil fuel use, with land-use change providing another significant but smaller contribution. It is *very likely* that the observed increase in CH_4 concentration is predominantly due to agriculture and fossil fuel use. CH_4 growth rates have declined since the early 1990s, consistent with total emissions (sum of anthropogenic and natural sources) being nearly constant during this period. The increase in N_2O concentration is primarily due to agriculture.

There is *very high confidence* that the net effect of human activities since 1750 has been one of warming.

Most of the observed increase in global average temperatures since the mid-20th century is* very likely *due to the observed increase in anthropogenic GHG concentrations. It is* likely *that there has been significant anthropogenic warming over the past 50 years averaged over each continent (except Antarctica).

During the past 50 years, the sum of solar and volcanic forcings would *likely* have produced cooling. Observed patterns of warming and their changes are simulated only by models that include anthropogenic forcings. Difficulties remain in simulating and attributing observed temperature changes at smaller than continental scales.

Advances since the TAR show that discernible human influences extend beyond average temperature to other aspects of climate.

Human influences have:

- *very likely* contributed to sea level rise during the latter half of the 20th century
- *likely* contributed to changes in wind patterns, affecting extra-tropical storm tracks and temperature patterns
- *likely* increased temperatures of extreme hot nights, cold nights and cold days
- *more likely than not* increased risk of heat waves, area affected by drought since the 1970s and frequency of heavy precipitation events.

Anthropogenic warming over the last three decades has* likely *had a discernible influence at the global scale on observed changes in many physical and biological systems.

Spatial agreement between regions of significant warming across the globe and locations of significant observed changes in many systems consistent with warming is *very unlikely* to be due solely to natural variability. Several modelling studies have linked some specific responses in physical and biological systems to anthropogenic warming.

More complete attribution of observed natural system responses to anthropogenic warming is currently prevented by the short time scales of many impact studies, greater natural climate variability at regional scales, contributions of nonclimate factors and limited spatial coverage of studies.

3. Projected climate change and its impacts

There is* high agreement *and* much evidence *that with current climate change mitigation policies and related sustainable development practices, global GHG emissions will continue to grow over the next few decades.

The IPCC Special Report on Emissions Scenarios projects an increase of global GHG emissions by 25 to 90% (CO_2-eq) between 2000 and 2030, with fossil fuels maintaining their dominant position in the global energy mix to 2030 and beyond. More recent scenarios without additional emissions mitigation are comparable in range.

Continued GHG emissions at or above current rates would cause further warming and induce many changes in the global climate system during the 21st century that would* very likely *be larger than those observed during the 20th century.

For the next two decades a warming of about 0.2°C per decade is projected for a range of SRES emissions scenarios. Even if the concentrations of all GHGs and aerosols had been kept constant at year 2000 levels, a further warming of about 0.1°C per decade would be expected. Afterwards, temperature projections increasingly depend on specific emissions scenarios.

The range of projections is broadly consistent with the TAR, but uncertainties and upper ranges for temperature are larger mainly because the broader range of available models suggests stronger climate-carbon cycle feedbacks. Warming reduces terrestrial and ocean uptake of atmospheric CO_2, increasing the fraction of anthropogenic emissions remaining in the atmosphere. The strength of this feedback effect varies markedly among models.

Because understanding of some important effects driving sea level rise is too limited, this report does not assess the likelihood, nor provide a best estimate or an upper bound for sea level rise. [Omitted table] shows model-based projections of global average sea level rise for 2090–2099. The projections do not include uncertainties in climate-carbon cycle feedbacks nor the full effects of changes in ice sheet flow, therefore the upper values of the ranges are not to be considered upper bounds for sea level rise. They include a contribution from increased Greenland and Antarctic ice flow at the rates observed for 1993–2003, but this could increase or decrease in the future.

There is now higher confidence than in the TAR in projected patterns of warming and other regional-scale features, including changes in wind patterns, precipitation and some aspects of extremes and sea ice.

Regional-scale changes include:

- warming greatest over land and at most high northern latitudes and least over Southern Ocean and parts of the North Atlantic Ocean, continuing recent observed trends.
- contraction of snow cover area, increases in thaw depth over most permafrost regions and decrease in sea ice extent; in some projections using SRES scenarios, Arctic late-summer sea ice disappears almost entirely by the latter part of the 21st century
- *very likely* increase in frequency of hot extremes, heat waves and heavy precipitation
- *likely* increase in tropical cyclone intensity; less confidence in global decrease of tropical cyclone numbers
- poleward shift of extra-tropical storm tracks with consequent changes in wind, precipitation and temperature patterns
- *very likely* precipitation increases in high latitudes and *likely* decreases in most subtropical land regions, continuing observed recent trends.

There is *high confidence* that by mid-century, annual river runoff and water availability are projected to increase at high latitudes (and in some tropical wet areas) and decrease in some dry regions in the mid-latitudes and tropics. There is also *high confidence* that many semi-arid areas (e.g. Mediterranean Basin, western United States, southern Africa and north-eastern Brazil) will suffer a decrease in water resources due to climate change.

Studies since the TAR have enabled more systematic understanding of the timing and magnitude of impacts related to differing amounts and rates of climate change.

Some systems, sectors and regions are likely to be especially affected by climate change.

[Identified on the basis of expert judgment of the assessed literature and considering the magnitude, timing and projected rate of climate change, sensitivity, and adaptive capacity.]

Systems and sectors:

- particular ecosystems:
  - terrestrial: tundra, boreal forest and mountain regions because of sensitivity to warming; mediterranean-type ecosystems because of reduction in rainfall; and tropical rainforests where precipitation declines
  - coastal: mangroves and salt marshes, due to multiple stresses
  - marine: coral reefs due to multiple stresses; the sea ice biome because of sensitivity to warming
- water resources in some dry regions at mid-latitudes [including arid and semi-arid regions] and in the dry tropics, due to changes in rainfall and evapotranspiration, and in areas dependent on snow and ice melt
- agriculture in low latitudes, due to reduced water availability
- low-lying coastal systems, due to threat of sea level rise and increased risk from extreme weather events
- human health in populations with low adaptive capacity.

Regions:

- the Arctic, because of the impacts of high rates of projected warming on natural systems and human communities
- Africa, because of low adaptive capacity and projected climate change impacts
- small islands, where there is high exposure of population and infrastructure to projected climate change impacts
- Asian and African megadeltas, due to large populations and high exposure to sea level rise, storm surges and river flooding.

Within other areas, even those with high incomes, some people (such as the poor, young children and the elderly) can be particularly at risk, and also some areas and some activities.

Ocean acidification

The uptake of anthropogenic carbon since 1750 has led to the ocean becoming more acidic with an average decrease in pH of 0.1 units. Increasing atmospheric CO_2 concentrations lead to further acidification. Projections based on SRES scenarios give a reduction in average global surface ocean pH of between 0.14 and 0.35 units over the 21st century. While the effects of observed ocean acidification on the marine biosphere are as yet undocumented, the progressive acidification of oceans is expected to have negative impacts on marine shell-forming organisms (e.g. corals) and their dependent species.

Altered frequencies and intensities of extreme weather, together with sea level rise, are expected to have mostly adverse effects on natural and human systems.

Anthropogenic warming and sea level rise would continue for centuries due to the time scales associated with climate processes and feedbacks, even if GHG concentrations were to be stabilised.

Contraction of the Greenland ice sheet is projected to continue to contribute to sea level rise after 2100. Current models suggest virtually complete elimination of the Greenland ice sheet and a resulting contribution to sea level rise of about 7m if global average warming were sustained for millennia in excess of 1.9 to 4.6°C relative to pre-industrial values. The corresponding future temperatures in Greenland are comparable to those inferred for the last interglacial period 125,000 years ago, when palaeoclimatic information suggests reductions of polar land ice extent and 4 to 6m of sea level rise.

Current global model studies project that the Antarctic ice sheet will remain too cold for widespread surface melting and gain mass due to increased snowfall. However, net loss of ice mass could occur if dynamical ice discharge dominates the ice sheet mass balance.

Anthropogenic warming could lead to some impacts that are abrupt or irreversible, depending upon the rate and magnitude of the climate change.

Partial loss of ice sheets on polar land could imply metres of sea level rise, major changes in coastlines and inundation of low-lying areas, with greatest effects in river deltas and low-lying islands. Such changes are projected to occur over millennial time scales, but more rapid sea level rise on century time scales cannot be excluded.

Climate change is *likely* to lead to some irreversible impacts. There is *medium confidence* that approximately 20 to 30% of species assessed so far are *likely* to be at increased risk of extinction if increases in global average warming exceed 1.5 to 2.5°C (relative to 1980–1999). As global average temperature increase exceeds about 3.5°C, model projections suggest significant extinctions (40 to 70% of species assessed) around the globe.

Based on current model simulations, the meridional overturning circulation (MOC) of the Atlantic Ocean will *very likely* slow down during the 21st century; nevertheless temperatures over the Atlantic and Europe are projected to increase. The MOC is *very unlikely* to undergo a large abrupt transition during the 21st century. Longer-term MOC changes cannot be assessed with confidence. Impacts of large-scale and persistent changes in the MOC are *likely* to include changes in marine ecosystem productivity, fisheries, ocean CO_2 uptake, oceanic oxygen concentrations and terrestrial vegetation. Changes in terrestrial and ocean CO_2 uptake may feed back on the climate system.

4. Adaptation and mitigation options

A wide array of adaptation options is available, but more extensive adaptation than is currently occurring is required to reduce vulnerability to climate change. There are barriers, limits and costs, which are not fully understood.

Societies have a long record of managing the impacts of weather- and climate-related events. Nevertheless, additional adaptation measures will be required to reduce the adverse impacts of projected climate change and variability, regardless of the scale of mitigation undertaken over the next two to three decades. Moreover, vulnerability to climate change can be exacerbated by other stresses. These arise from, for example, current climate hazards, poverty and unequal access to resources, food insecurity, trends in economic globalisation, conflict and incidence of diseases such as HIV/AIDS.

Some planned adaptation to climate change is already occurring on a limited basis. Adaptation can reduce vulnerability, especially when it is embedded within broader sectoral initiatives. There is *high confidence* that there are viable adaptation options that can be implemented in some sectors at low cost, and/or with high benefit-cost ratios. However, comprehensive estimates of global costs and benefits of adaptation are limited.

Adaptive capacity is intimately connected to social and economic development but is unevenly distributed across and within societies.

A range of barriers limits both the implementation and effectiveness of adaptation measures. The capacity to adapt is dynamic and is influenced by a society's productive base, including natural and man-made capital assets, social networks and entitlements, human capital and institutions, governance, national income, health and technology. Even societies with high adaptive capacity remain vulnerable to climate change, variability and extremes.

Both bottom-up and top-down studies indicate that there is* high agreement *and* much evidence *of substantial economic potential for the mitigation of global GHG emissions over the coming decades that could offset the projected growth of global emissions or reduce emissions below current levels. While top-down and bottom-up studies are in line at the global level there are considerable differences at the sectoral level.

No single technology can provide all of the mitigation potential in any sector. The economic mitigation potential, which is generally greater than the market mitigation potential, can only be achieved when adequate policies are in place and barriers removed.

Bottom-up studies suggest that mitigation opportunities with net negative costs have the potential to reduce emissions by around 6 $GtCO_2$-eq/yr in 2030, realising which requires dealing with implementation barriers.

Future energy infrastructure investment decisions, expected to exceed US$20 trillion between 2005 and 2030, will have long-term impacts on GHG emissions, because of the long lifetimes of energy plants and other infrastructure capital stock. The widespread diffusion of low-carbon technologies may take many decades, even if early investments in these technologies are made attractive. Initial estimates show that returning global energy-related CO_2 emissions to 2005 levels by 2030 would require a large shift in investment patterns, although the net additional investment required ranges from negligible to 5 to 10%.

A wide variety of policies and instruments are available to governments to create the incentives for mitigation action. Their applicability depends on national circumstances and sectoral context.

They include integrating climate policies in wider development policies, regulations and standards, taxes and charges, tradable permits, financial incentives, voluntary agreements, information instruments, and research, development and demonstration (RD&D).

An effective carbon-price signal could realise significant mitigation potential in all sectors. Modelling studies show that global carbon prices rising to US$20–80/t$CO_2$-eq by 2030 are consistent with stabilisation at around 550ppm CO_2-eq by 2100. For the same stabilisation level, induced technological change may lower these price ranges to US$5–65/t$CO_2$-eq in 2030.

There is *high agreement* and *much evidence* that mitigation actions can result in near-term co-benefits (e.g. improved health due to reduced air pollution) that may offset a substantial fraction of mitigation costs.

There is *high agreement* and *medium evidence* that Annex I countries' actions may affect the global economy and global emissions, although the scale of carbon leakage remains uncertain.

Fossil fuel exporting nations (in both Annex I and non-Annex I countries) may expect, as indicated in the TAR, lower demand and prices and lower GDP growth due to mitigation policies. The extent of this spillover depends strongly on assumptions related to policy decisions and oil market conditions.

There is also *high agreement* and *medium evidence* that changes in lifestyle, behaviour patterns and management practices can contribute to climate change mitigation across all sectors.

Many options for reducing global GHG emissions through international cooperation exist. There is high agreement and much evidence that notable achievements of the UNFCCC and its Kyoto Protocol are the establishment of a global response to climate change, stimulation of an array of national policies, and the creation of an international carbon market and new institutional mechanisms that may provide the foundation for future mitigation efforts. Progress has also been made in addressing adaptation within the UNFCCC and additional international initiatives have been suggested.

Greater cooperative efforts and expansion of market mechanisms will help to reduce global costs for achieving a given level of mitigation, or will improve environmental effectiveness. Efforts can include diverse elements such as emissions targets; sectoral, local, sub-national and regional actions; RD&D programmes; adopting common policies; implementing development-oriented actions; or expanding financing instruments.

In several sectors, climate response options can be implemented to realise synergies and avoid conflicts with other dimensions of sustainable development. Decisions about macroeconomic and other non-climate policies can significantly affect emissions, adaptive capacity and vulnerability.

Making development more sustainable can enhance mitigative and adaptive capacities, reduce emissions and reduce vulnerability, but there may be barriers to implementation. On the other hand, it is *very likely* that climate change can slow the pace of progress towards sustainable development. Over the next half-century, climate change could impede achievement of the Millennium Development Goals.

5. The long-term perspective

Determining what constitutes "dangerous anthropogenic interference with the climate system" in relation to Article 2 of the UNFCCC involves value judgements. Science can support informed decisions on this issue, including by providing criteria for judging which vulnerabilities might be labelled 'key'.

Key vulnerabilities may be associated with many climate-sensitive systems, including food supply, infrastructure, health, water resources, coastal systems, ecosystems, global biogeochemical cycles, ice sheets and modes of oceanic and atmospheric circulation.

The five 'reasons for concern' identified in the TAR remain a viable framework to consider key vulnerabilities. These 'reasons' are assessed here to be stronger than in the TAR. Many risks are identified with higher confidence. Some risks are projected to be larger or to occur at lower increases in temperature. Understanding about the relationship between impacts (the basis for 'reasons for concern' in the TAR) and vulnerability (that includes the ability to adapt to impacts) has improved.

This is due to more precise identification of the circumstances that make systems, sectors and regions especially vulnerable and growing evidence of the risks of very large impacts on multiple-century time scales.

- ***Risks to unique and threatened systems.*** There is new and stronger evidence of observed impacts of climate change on unique and vulnerable systems (such as polar and high mountain communities and ecosystems), with increasing levels of adverse impacts as temperatures increase further. An increasing risk of species extinction and coral reef damage is projected with higher confidence than in the TAR as warming proceeds. There is *medium confidence* that approximately 20 to 30% of plant and animal species assessed so far are *likely* to be at increased risk of extinction if increases in global average temperature exceed 1.5 to 2.5°C over 1980–1999 levels. Confidence has increased that a 1 to 2°C increase in global mean temperature above 1990 levels (about 1.5 to 2.5°C above preindustrial) poses significant risks to many unique and threatened systems including many biodiversity hotspots. Corals are vulnerable to thermal stress and have low adaptive capacity. Increases in sea surface temperature of about 1 to 3°C are projected to result in more frequent coral bleaching events and widespread mortality, unless there is thermal adaptation or acclimatisation by corals. Increasing vulnerability of indigenous communities in the Arctic and small island communities to warming is projected.
- ***Risks of extreme weather events.*** Responses to some recent extreme events reveal higher levels of vulnerability than the TAR. There is now higher confidence in the projected increases in droughts, heat waves and floods, as well as their adverse impacts.
- ***Distribution of impacts and vulnerabilities.*** There are sharp differences across regions and those in the weakest economic position are often the most vulnerable to climate change. There is increasing evidence of greater vulnerability of

specific groups such as the poor and elderly not only in developing but also in developed countries. Moreover, there is increased evidence that low-latitude and less developed areas generally face greater risk, for example in dry areas and megadeltas.

- ***Aggregate impacts.*** Compared to the TAR, initial net market-based benefits from climate change are projected to peak at a lower magnitude of warming, while damages would be higher for larger magnitudes of warming. The net costs of impacts of increased warming are projected to increase over time.
- ***Risks of large-scale singularities.*** There is *high confidence* that global warming over many centuries would lead to a sea level rise contribution from thermal expansion alone that is projected to be much larger than observed over the 20th century, with loss of coastal area and associated impacts. There is better understanding than in the TAR that the risk of additional contributions to sea level rise from both the Greenland and possibly Antarctic ice sheets may be larger than projected by ice sheet models and could occur on century time scales. This is because ice dynamical processes seen in recent observations but not fully included in ice sheet models assessed in the AR4 could increase the rate of ice loss.

There is* high confidence *that neither adaptation nor mitigation alone can avoid all climate change impacts; however, they can complement each other and together can significantly reduce the risks of climate change.

Adaptation is necessary in the short and longer term to address impacts resulting from the warming that would occur even for the lowest stabilisation scenarios assessed. There are barriers, limits and costs, but these are not fully understood. Unmitigated climate change would, in the long term, be *likely* to exceed the capacity of natural, managed and human systems to adapt. The time at which such limits could be reached will vary between sectors and regions. Early mitigation actions would avoid further locking in carbon intensive infrastructure and reduce climate change and associated adaptation needs.

Many impacts can be reduced, delayed or avoided by mitigation. Mitigation efforts and investments over the next two to three decades will have a large impact on opportunities to achieve lower stabilisation levels. Delayed emission reductions significantly constrain the opportunities to achieve lower stabilisation levels and increase the risk of more severe climate change impacts.

In order to stabilise the concentration of GHGs in the atmosphere, emissions would need to peak and decline thereafter. The lower the stabilisation level, the more quickly this peak and decline would need to occur. . . .

Sea level rise under warming is inevitable. Thermal expansion would continue for many centuries after GHG concentrations have stabilised, for any of the stabilisation levels assessed, causing an eventual sea level rise much larger than projected for the 21st century. The eventual contributions from Greenland ice sheet loss could be several metres, and larger than from thermal expansion, should warming in excess of 1.9 to 4.6°C above pre-industrial be sustained over many centuries. The long time scales of thermal expansion and ice sheet response to warming imply that stabilisation of GHG concentrations at or above present levels would not stabilise sea level for many centuries.

There is* high agreement *and* much evidence *that all stabilisation levels assessed can be achieved by deployment of a portfolio of technologies that are either currently available or expected to be commercialised in coming decades, assuming appropriate and effective incentives are in place for their development, acquisition, deployment and diffusion and addressing related barriers.

All assessed stabilisation scenarios indicate that 60 to 80% of the reductions would come from energy supply and use and industrial processes, with energy efficiency playing a key role in many scenarios. Including non-CO_2 and CO_2 land-use and forestry mitigation options provides greater flexibility and cost-effectiveness. Low stabilisation levels require early investments and substantially more rapid diffusion and commercialisation of advanced low-emissions technologies.

Without substantial investment flows and effective technology transfer, it may be difficult to achieve emission reduction at a significant scale. Mobilising financing of incremental costs of low-carbon technologies is important.

The macro-economic costs of mitigation generally rise with the stringency of the stabilisation target. For specific countries and sectors, costs vary considerably from the global average.

In 2050, global average macro-economic costs for mitigation towards stabilisation between 710 and 445ppm CO_2-eq are between a 1% gain and 5.5% decrease of global GDP. This corresponds to slowing average annual global GDP growth by less than 0.12 percentage points.

Responding to climate change involves an iterative risk management process that includes both adaptation and mitigation and takes into account climate change damages, co-benefits, sustainability, equity and attitudes to risk.

Impacts of climate change are *very likely* to impose net annual costs, which will increase over time as global temperatures increase. Peer-reviewed estimates of the social cost of carbon in 2005 average US$12 per tonne of CO_2, but the range from 100 estimates is large (–$3 to $95/t$CO_2$). This is due in large part to differences in assumptions regarding climate sensitivity, response lags, the treatment of risk and equity, economic and non-economic impacts, the inclusion of potentially catastrophic losses and discount rates. Aggregate estimates of costs mask significant differences in impacts across sectors, regions and populations and *very likely* underestimate damage costs because they cannot include many nonquantifiable impacts.

Limited and early analytical results from integrated analyses of the costs and benefits of mitigation indicate that they are broadly comparable in magnitude, but do not as yet permit an unambiguous determination of an emissions pathway or stabilisation level where benefits exceed costs.

Climate sensitivity is a key uncertainty for mitigation scenarios for specific temperature levels.

Choices about the scale and timing of GHG mitigation involve balancing the economic costs of more rapid emission reductions now against the corresponding medium-term and long-term climate risks of delay.

SOURCE: United Nations. World Meteorological Organization and United Nations Environment Program. Intergovernmental Panel on Climate Change. "Climate Change 2007: Synthesis Report. Summary for Policymakers." November 17, 2007. www.ipcc.ch/pdf/assessment-report/ar4/syr/ar4_syr_spm.pdf (accessed March 25, 2008).

OTHER HISTORIC DOCUMENTS OF INTEREST

FROM THIS VOLUME

- U.S. Supreme Court on EPA regulation of greenhouse gases, p. 139
- Nobel Peace Prize speech by Al Gore, p. 691

FROM PREVIOUS *HISTORIC DOCUMENTS*

- British Treasury report on the economics of climate change, *2006*, p. 615
- Canadian prime minister Martin on climate change, *2005*, p. 919
- Negotiation of the Kyoto Protocol, *1997*, p. 859

DOCUMENT IN CONTEXT

UN Refugee Agency on the Status of Iraqi Refugees

NOVEMBER 23, 2007

New attention was focused during 2007 on the plight of more than 4 million Iraqis who had fled their homes as a result of war and sectarian violence. Thousands of displaced Iraqis attempted to return home, particularly late in the year once sectarian violence in Iraq had eased. The actual number of returnees was unclear, however, and United Nations officials reported that most of the returnees from neighboring Syria had headed home because they had run out of money or no longer had permission to stay in Syria, not because they believed Iraq was becoming safe.

The Iraqi and U.S. governments at first heralded the return of displaced people as evidence that the "surge" of some 30,000 additional U.S. troops into Iraq was working to calm the sectarian violence. Late in the year, Iraq joined the UN in appealing to most Iraqi refugees and displaced people to stay put until security conditions improved and proper arrangements could be made for their return.

The Displaced and Refugees

At the end of 2007, the UN and Iraqi aid agencies officially listed about 2.2 million Iraqi citizens as refugees (meaning they had fled to other countries) and about 2.4 million as internally displaced persons (meaning they had fled their homes but remained within Iraq). This total of 4.6 million meant that about one in six Iraqis had fled their homes; Iraq's total population was about 27.5 million, according to the UN.

The UN High Commissioner for Refugees (UNHCR), the main international agency aiding refugees, estimated that about 1.4 million of the Iraqis who had fled their country were in neighboring Syria. Another 500,000–700,000 had gone to Jordan, an unknown number were in Iran, about 50,000 went to Lebanon, about 20,000 were in Europe, and the rest were spread across other countries.

The vast majority of the refugees had fled Iraq since the U.S. invasion and overthrow of Iraqi leader Saddam Hussein in 2003. According to UN figures, the number of Iraqi refugees in other countries reached 366,000 by the end of 2004, rose to 899,000 by the end of 2005, then climbed dramatically to 1.8 million during 2006, following the explosion of sectarian violence during that year. Iraq experienced a net outflow of about 400,000 refugees during 2007, mostly in the first eight months when violence remained high during the early stages of the U.S. military surge.

Of the 2.4 million Iraqis listed by agencies as displaced persons inside the country, an estimated 1 million had been displaced from their homes even before the U.S. invasion in March 2003. Most were members of the ethnic groups long suppressed by Hussein's government, notably Shiite Muslims and the Kurds. These more or less permanently displaced Iraqis were still living in Iraq at the time of the invasion.

According to displaced persons figures from the Iraqi Red Crescent Society and other agencies, about 100,000 Iraqis fled their homes in the months after the U.S. invasion but remained within Iraq, as did a total of another 150,000 during 2004 and 2005. The explosion of sectarian violence during 2006 sent another 400,000 Iraqis into flight. The Red Crescent Society estimated that about 725,000 more Iraqis were displaced during 2007—most of them during the first eight months. This estimate meant that about 1.4 million Iraqis had been displaced since the U.S. invasion, in addition to the 1 million displaced earlier. The figures were subject to considerable disagreement, however, because the Iraqi Red Crescent Society based its numbers largely on change-of-address requests by families seeking government food subsidies—a measure some Iraqi government officials insisted did not accurately reflect actual displacement.

The UNHCR said in an August 28 statement that Iraqis had given many reasons—in addition to the continuing violence—for leaving their homes in 2007 at a rate of 60,000 or more each month, a much higher rate than in earlier years. Many reported difficulties obtaining social services in their home areas, and many "are choosing to leave ethnically mixed areas before they are forced to do so," the UN said, referring to the fact that sectarian militias were continuing to force either Shiites or Sunnis out of formerly mixed areas, particularly in Baghdad. Some Iraqis also left their homes at the end of the school year, the agency added.

The Iraqi government, the governments of neighboring countries, the UN, and international aid agencies were struggling to cope with the needs of millions of Iraqi refugees and displaced people, many of whom had little in the way of personal resources. A survey of Iraqi refugees living in Syria, conducted in June 2007 by the International Committee of the Red Cross, found that about 25 percent needed a broad range of humanitarian aid, including food, housing, and medical supplies. The UNHCR and its sister agency, the World Food Programme, fed about 51,000 Iraqi refugees in Syria during November and December and planned to increase that number gradually to more than 260,000 by the end of 2008. Although the Syrian Red Crescent Society had overall responsibility for the care for Iraqi refugees in the country, it was "overwhelmed by the needs and lacks the necessary capacity" to deal with so many refugees, according to a November 14 report by Refugees International, a U.S.-based advocacy group.

Nearly half of the internally displaced people in Iraq did not have access to the country's subsidized food system, according to UN figures. The UNHCR said it provided aid, of varying kinds and amounts, to 166,000 displaced people between September and November.

Returning Home

The overall reduction of violence in Iraq during the last four months of 2007 led to expectations by Iraqi government and UN officials that some of the refugees and displaced families might try to return to their homes. The Iraqi government heightened such expectations with an announcement on November 3 that some 3,000 families who had fled the violence in Baghdad had returned since August. Prime Minister Nouri al-Maliki appeared intent on reinforcing the message of safety in Baghdad two days later when he went to a popular gathering spot along the Tigris River in the capital and said the sectarian violence was ending.

U.S. and Iraqi officials went a step further on November 7 with their statements praising the return of refugees as a sign that the U.S. military surge had contained the violence. An Iraqi government spokesman said 46,030 displaced people had returned to their homes in the capital in October; he did not reconcile this figure with the estimate four days earlier than 3,000 families—which would translate into roughly 18,000 individuals—had returned over the previous three months. The commander of U.S. forces in Baghdad, Maj. Gen. Joseph Fil, also said his troops had observed a return of Iraqis to their homes. "There's no question about it," he told reporters.

These and similar statements from the Iraqi and U.S. governments were followed by several news reports in November suggesting that thousands of Iraqis were streaming back to their homes because they believed it was now safe to do so. Television networks interviewed Iraqi families as they headed east into Iraq from Syria. Officials at the Iraqi embassy in Damascus said refugees were being offered free trips home by the Baghdad government, and Iraq's minister for displacement and migration was quoted in mid-November as saying that 1,600 people were returning every day.

Some, but not all, of the news reports noted that most of the Iraqis leaving Syria told officials they were doing so because they had no choice: they had run out of money or their visas had expired. For most of the period since 2003, Syria had allowed Iraqi refugees to stay for three months, after which they had to leave for a short period before returning to Syria for another three months. Starting in October 2007, however, Syria limited entry to a small number of Iraqis who could obtain visas for business, educational, or other reasons—but not as refugees. Syria reportedly took this position at the request of the Iraqi government, which at that point was encouraging refugees to return home.

The UNHCR on December 7 said that between August and November 97,000 Iraqis had entered Syria while 128,000 had left Syria to return to Iraq through the main border crossing at Al Waleed. These figures included all Iraqis transiting the border, not just refugees. The agency reported that Iraqis leaving Syria continued to express concerns about long-term security in Iraq, and many were headed to areas where they felt safe, not necessarily to their homes.

Refugees returning to Baghdad found a city that appeared to be safer than when they left. However, some of them were unable to return to their own homes, which had since been occupied by members of the opposite Muslim sect, either Shiite or Sunni. Entire neighborhoods in the capital that had once housed both Shiites and Sunnis were now inhabited by members of only one sect as the result of ethnic cleansing by sectarian militias. One Shiite family, the Zubaidis, told a *Washington Post* reporter in November that they were too afraid to ask the Sunni family occupying their former home to leave, so they rented an apartment in a different neighborhood, inhabited mostly by Shiites. "Security is better," Melal al-Zubaidi was quoted as saying. "But we still have fear inside ourselves."

Too Soon for Returns

Responding to news reports about the returning refugees, the UNHCR said on November 23 that it was still too soon for a mass return of refugees to Iraq. "UNHCR does not believe that the time has come to promote, organize, or encourage returns," agency spokesperson Jennifer Pagonis told reporters in Geneva. "That would be possible only when proper return conditions are in place—including material and legal support and physical safety." Pagonis added that her agency had seen "no sign of any large-scale return" of refugees to Iraq, despite the news reports and statements by Iraqi and U.S. officials. Specifically, Pagonis said the UN agency could not confirm the Baghdad government's statement that 46,000 Iraqi refugees had returned from Syria in October. There had been a "fluctuating average" each day of about 1,500 Iraqis leaving Syria and 500 Iraqis entering Syria—figures that would produce a net influx into Iraq of about 30,000 a month. However, most of those returning said they were doing so involuntarily, not because they believed conditions were safe in Iraq, she said.

The Iraqi government reversed course on the refugee return matter on December 4, acknowledging that it was too soon for most of the refugees to head back home. "The reality is that we cannot handle a huge influx of people," Abdul Samad al-Sultan, the minister of displacement and migration, told reporters. "The refugees in some countries, we ask them to wait." Sultan's statement came as the Iraqi government and the UN mission in Iraq announced plans to provide aid for only about 5,000 returning families, or a total of about 30,000 people, a fraction of those who might be returning over the subsequent months.

At about the same time, U.S. military officials expressed their concern about the security impact of refugee returns, particularly in cases when families tried to return to homes that had since been occupied by others of different sects. Iraq's government lacked policies for dealing with such situations, U.S. officials said, meaning that American troops, who were still largely responsible for security in much of Iraq, could be put in the position of mediating contentious property disputes. "We have been asking, pleading with the government of Iraq to come up with a policy" so U.S. commanders would not have to make ad hoc decisions about who belonged in which houses, Col. William Rapp, a senior aide to overall U.S. commander Gen. David H. Petraeus, told reporters in late November. Rapp said the ultimate solution probably would be the construction of new housing "as opposed to wholesale evictions and resettlement."

Following is a statement attributed to United Nations High Commissioner for Refugees spokesperson Jennifer Pagonis at a November 23, 2007, news briefing in Geneva, Switzerland, responding to reports that Iraqi refugees were returning home.

DOCUMENT

UN High Commissioner for Refugees on the Status of Iraqi Refugees

November 23, 2007

There have been several public reports in recent days about limited returns to Iraq. We welcome improvements to the security conditions and stand ready to assist people who have decided or will decide to return voluntarily. However, UNHCR does not believe that the time has come to promote, organize or encourage returns. That would be possible only when proper return conditions are in place—including material and legal support and physical safety. Presently, there is no sign of any large-scale return to Iraq as the security situation in many parts of the country remains volatile and unpredictable.

According to a survey done by our staff in Syria, there are many reasons for returns to Iraq other than considerations of improved security. Of some 110 Iraqi families

UNHCR spoke with in Syria the majority said they are returning because they are running out of money and/or resources, face difficult living conditions, or because their visas have expired. As a result of recent visa restrictions, a number of Iraqis have been unable to shuttle back and forth between Iraq and Syria to get additional resources, make some money or collect food distributions or pensions. The incentives offered by the Iraqi government of some $700–$800 to return home, as well as free bus and plane rides, have also played a role in returns. The survey also highlighted that this is the first time in recent years that Iraqi refugees are actually discussing return, which was not the case a few months ago.

UNHCR staff also did quick interviews with returnees in Baghdad, who cited economic difficulties caused by their long displacement as a major reason for going home. Many had run out of or nearly depleted their savings. Some returned as it was the last chance to get their children back into Iraqi schools before the end of the first term. Some were indeed encouraged by the reports regarding improvement of security, but many expressed concern about longer-term security, citing the fact that militias are still around and many areas remain insecure. People have mainly been returning to areas where they feel that local security forces are working properly and are maintaining control.

Although we are not in a position to monitor borders on a 24-hour basis, we have noted more returns to Iraq than arrivals in Syria—with a fluctuating average of 1,500 departures to Iraq and 500 arrivals in Syria per day. We cannot confirm reports that 46,000 Iraqis returned from Syria in October.

Inside Iraq, the number of internally displaced people (IDPs) increased slightly over the last few months. According to the latest figures received by UNHCR (with data from the Iraqi Red Crescent Society, Ministry of Displacement and Migration, Kurdistan Regional Government, local and international NGOs [nongovernmental organizations], UNHCR and the International Organisation of Migration) as of 21 November, it is estimated that over 2.4 million Iraqis are displaced inside Iraq. The breakdown is: 1,021,962 displaced prior to 2003; 190,146 displaced between 2003–2005; and 1,199,491 displaced after the first Samarra bombing in February 2006 (28,017 during October).

Reasons for the increase include better registration of the displaced, but also recent visa restrictions in Syria, which meant more people moved inside Iraq rather than seeking refuge outside. We have also seen more secondary displacement, as governorates close their doors to the newly displaced from elsewhere (11 out of 18 governorates have limited access to new arrivals).

Still, there have been some returns among displaced people. Various families received financial incentives to return. We have some reports of families who returned to very difficult conditions (destroyed or damaged property) while others did not return to their original homes, but settled elsewhere (secondary displacement). Displaced Iraqis say access to shelter, food, work, water/sanitation and legal aid remain the most common needs.

According to government estimates, some 2.2 million Iraqis live outside Iraq—with some 500,000–700,000 in Jordan and up to 1.5 million in Syria.

SOURCE: United Nations. Office of the High Commissioner for Refugees. "Iraq: UNHCR Cautious About Returns." November 23, 2007. www.unhcr.org/cgi-bin/texis/vtx/iraq?page=briefing&id=4746da102 (accessed April 2, 2008).

OTHER HISTORIC DOCUMENTS OF INTEREST

FROM THIS VOLUME

- President Bush on the "surge" of U.S. troops to Iraq, p. 9
- U.S. Ambassador Crocker on U.S.-Iran diplomacy, p. 257
- National Intelligence Estimate and General Petraeus's testimony on the security situation in Iraq, p. 407
- Government Accountability Office on Iraqi government actions to meet "benchmarks," p. 477

FROM PREVIOUS *HISTORIC DOCUMENTS*

- Report of the special inspector general on Iraq reconstruction, *2006,* p. 161
- President Bush and Prime Minister Maliki on the government of Iraq, *2006,* p. 171
- UN Secretary-General Annan on the situation in Iraq, *2006,* p. 702
- Iraq Study Group report on Iraq, *2006,* p. 725
- President Bush and international observers on the Iraqi elections, *2005,* p. 941

DOCUMENT IN CONTEXT

President Bush on Middle East Peace Talks

NOVEMBER 27, 2007

President George W. Bush chose in the second half of 2007 to push for renewed peace talks between the Israelis and the Palestinians. Under intense U.S. pressure, the two sides agreed in late November to begin negotiations intended to reach a significant peace agreement before Bush left office in January 2009. The first talks in December got off to a rocky start, suggesting that the deadline would not be an easy one to meet.

Events on the ground conspired to make 2007 seem an especially unfruitful time for another attempt to negotiate the long-sought peace agreement to end the Israeli-Palestinian conflict. Hamas, the radical Islamist faction of the Palestinians, seized control of the Gaza Strip in mid-June, leaving the Palestinian Authority government, headed by the more moderate Fatah faction, in control of just the West Bank. Both Israeli prime minister Ehud Olmert and Palestinian president Mahmoud Abbas, the head of Fatah, were so weak politically that neither could guarantee that his constituency would support whatever agreement they reached. And the Bush administration—the key outside actor pushing for peace talks—had a recent history of turning its attention elsewhere when the diplomacy became difficult.

The year's diplomatic drive for Middle East peace was the second major one during Bush's presidency. The first came in 2003, after the U.S. invasion of Iraq, when Bush endorsed an international plan called the "Road Map" that laid out specific steps for a peace agreement between the Israelis and Palestinians. Bush met with the key leaders in June 2003 and promised to "ride herd" on them until they reached their goal. A Palestinian suicide bombing two months later threw the negotiations off track. With only a few modest exceptions, the Bush administration abandoned its active search for peace for most of the next four years.

Abbas-Olmert Talks

The one positive sign for the peace process in 2007, and the one most often cited by officials in Washington, was that the Israelis and the Palestinians—or at least one faction of the latter—now had leaders who seemed determined to seek peace and had political interests in doing so. Olmert had come to office in January 2006, upon the incapacitation of his mentor Ariel Sharon, pledging to push for an agreement that would extract Israel from most of the West Bank, just as Sharon had pulled Israel out of the Gaza Strip in 2005. Olmert's moderate Kadima party won an election in March 2006 after campaigning on such a policy. Olmert lacked Sharon's popularity and personal authority, however, and he was widely seen in Israel as a politically weak leader who needed a major achievement, such as a peace agreement, to bolster his authority.

Abbas also had succeeded a popular, if controversial mentor: long-time Palestinian leader Yasir Arafat. Despite winning presidential elections in January 2005, Abbas was neither a dynamic nor a particularly popular figure; he, too, was widely seen as needing a peace agreement to add to his political stature. Abbas had suffered a major embarrassment in January 2006 when Hamas defeated Fatah to take control of the Palestinian Authority government. This left Abbas as a president with power over diplomacy and some security services but little else, and the elections showed how unpopular he and his fellow Fatah leaders had become.

From all reports, Abbas and Olmert had a reasonably positive personal relationship, which they nurtured during numerous meetings, starting in 2006. This relationship helped them get through difficult moments that almost certainly would have torpedoed negotiations among two less compatible leaders.

Arabs Reaffirm Peace Plan

The Bush administration first signaled its renewed interest in working on a peace agreement in late January 2007, when Bush asked Secretary of State Condoleezza Rice to moderate "informal talks" between Abbas and Olmert. Rice deliberately lowered expectations for the talks, saying the goal was to establish a "political horizon" for a final agreement leading to a Palestinian state and not to negotiate the specific details of how such an agreement would be reached. This diplomatic push was greeted with great skepticism in the region because of the inauspicious circumstances under which it was taking place. One positive development—which turned out to be temporary—came on February 7, when Saudi Arabia brokered an agreement between the feuding Fatah and Hamas factions that led one month later to a unity Palestinian government, which collapsed in mid-June.

The first set of talks mediated by Rice, on February 19, demonstrated the difficulties that lay ahead. Abbas and Olmert had what both sides described as an amicable meeting, but their only agreement was that they would continue to meet. Another session on March 11 produced a modest concession by Olmert to expand the operating schedule of the major crossing for cargo shipments from Israel into the Gaza Strip. Israel had opened the crossing only rarely, citing security threats from Palestinian extremists in Gaza. Three days later, Abbas reached agreement with Hamas on the details of the Palestinian unity government negotiated a month earlier. This agreement put new stress on the Abbas-Olmert dialogue because Hamas refused to recognize Israel. Olmert had vowed not to have even indirect dealings with Hamas until it recognized Israel and halted all violence against Israel, notably the steady rain of rockets from Gaza into southern Israel. The Israeli cabinet voted on March 18 to stop the peace talks because of Hamas's participation in the new Palestinian government.

Middle East diplomacy rarely stays static for long, and positive developments quickly followed the freeze that resulted from the Palestinian unity government. In another mission to the region, Rice on March 27 convinced Olmert to resume his discussions with Abbas, although only on interim matters, not on the so-called final status issues involved in the creation of a Palestinian state.

Two days later, the leaders of the Arab League, meeting in Riyadh, Saudi Arabia, reaffirmed a peace initiative they had launched in 2002: it offered Arab recognition of, and peace with, Israel in exchange for establishment of a Palestinian state and an Israeli withdrawal from the territories captured in the June 1967 Arab-Israeli war. The initiative, when first proposed, had been overshadowed in March 2002 by a Palestinian suicide bombing, and Israel had rejected it at the time as inadequate. Five years later, however, Olmert's government officially welcomed the initiative, even while expressing concerns about some of its details. The United States, which also had been ambivalent about the initiative in 2002, now described it as an important opening.

Abbas and Olmert resumed their talks on April 15, saying they would meet at least twice each month, as Rice had requested. These talks continued until early June, when mounting violence between the Fatah and Hamas factions in Gaza overshadowed diplomacy for the moment.

After Gaza, Bush Seizes the Initiative

After a violent round of internecine conflict, Hamas took control of the Gaza Strip in mid-June. This was a defining moment for the Middle East, one that nearly all sides described as offering the potential either for another descent into anarchy and possibly war or a new opportunity for moderates to seize the initiative toward compromise. Abbas took an important step down the latter road by ending the Palestinian unity government that had been in place only since March, appointing a new government with authority over the West Bank, and declaring an end to his dealings with Hamas. Olmert's government responded by agreeing to release, to the new Palestinian government, tens of millions of dollars in tax revenue that Israel had collected on behalf of the Palestinian Authority but had withheld since the Hamas electoral victory in January 2006. Olmert also met on June 25 with Abbas and the leaders of the two Arab countries with which Israel had peace treaties, Egypt and Jordan.

On June 27, the diplomatic consortium known as the Quartet (the European Union, Russia, the United Nations, and the United States) appointed recently resigned British prime minister Tony Blair as its new Middle East envoy. Blair's job was defined as helping the Palestinians improve their economy and develop the institutions necessary for an eventual state—he was not expected to negotiate a peace agreement.

On July 16, Bush announced the second major Middle East peace drive of his presidency. He said he would call a regional conference in the fall to launch a new round of peace negotiations. He specifically cited the Hamas takeover of Gaza one month earlier as putting the Palestinians at the "moment of choice" between rejection of and peace with Israel. The president offered few details of the conference —not even the date or location —except to say that Rice would chair it.

Bush's announcement generally was welcomed in the Middle East, except by Hamas, which rejected it, and several Arab governments, which asked for more details about the purpose of the conference. Numerous Middle East experts in the United States—particularly those who had been associated with diplomacy during the presidency of Bill Clinton (1993–2001)—said Bush should have undertaken this initiative two years earlier, when Abbas won election as Palestinian president and was in a better political position to negotiate with Israel.

Rice spent much of the subsequent four months shuttling among Middle East capitals. She had two goals: encouraging Arab leaders to attend the conference as a sign of regional support for peace, and getting the Israelis and Palestinians to agree ahead of time on a written document outlining the main points of what would be discussed in any substantive negotiations after the conference. By mid-October the administration had settled on the location of the conference—the U.S. Naval Academy in Annapolis, Maryland—and scheduled it for late November.

A "Joint Understanding"

As the November 27 date of the Annapolis conference approached, two important questions remained unanswered. One was whether Syria would join most of its fellow Arab countries in sending a representative. The Bush administration for more than four years had refused to have any dealings with Syria on the grounds that it was fomenting radicalism in the region and aiding anti-U.S. insurgents in Iraq. But now the administration wanted Syria's presence at Annapolis as a sign of support for a new peace process. Officials in Damascus sent word on November 25 that Syria would attend, but only if the conference addressed Syria's main grievance with Israel: the continued Israeli occupation of the Golan Heights, a strategic border area captured in the 1967 Arab-Israeli war.

The other question was whether Rice could persuade Israeli and Palestinian negotiators to agree on a joint statement before the conference. This proved to be an unexpectedly difficult endeavor. By all accounts, Palestinians pushed for a document with specific commitments to what would be negotiated, while the Israelis wanted a vague document that launched peace talks but did not even mention the core issues between the sides. Negotiations over the wording of the document went down to the wire. Bush met with Abbas and Olmert at the White House on November 26, the day before the conference, and urged them to reach agreement but promised that the United States would not impose one. Rice and her aides continued meeting with Israeli and Palestinian negotiators until shortly before Bush was scheduled to convene the conference on the morning of November 27 at the ornate Memorial Hall on the grounds of the Naval Academy.

The resulting agreement—called a "joint understanding"—was so fresh that Bush had to read it to delegates at the start of his remarks to the conference because there had not been enough time to print copies. The understanding had none of the specifics the Palestinians had sought, but both sides did commit "to engage in vigorous, ongoing and continuous negotiations, and [to] make every effort to conclude an agreement before the end of 2008." Bush, in turn, pledged "to devote my effort during my time as president to do all I can do help you achieve this ambitious goal." The president acknowledged that reaching a final peace agreement would not be easy. "If it were easy, it would have happened a long time ago," he said. Bush did not mention, nor did he need to, that President Clinton had pushed very hard for such an agreement in 2000, only to fail in the concluding weeks of his presidency.

Abbas and Olmert, in their speeches to the conference, each promised to set aside past grievances and work honestly toward peace. For Olmert, the conference represented a rare opportunity for an Israeli leader to speak directly to Arab diplomats, and he used it to stress that Israel sought peace with all its neighbors. "We are prepared to make a painful compromise" to reach peace, he said. Bush met again at the White House with Abbas and Olmert the next day and again assured them that "the United States will be actively engaged in the process."

The first round of post-Annapolis talks got off to a difficult start, on December 13, when Palestinian negotiators accused Israel of undermining the prospects for peace by issuing plans to expand a controversial Jewish settlement, Har Homa, in East Jerusalem. This meeting, which was to have laid the groundwork for subsequent negotiating sessions, lasted just ninety minutes and accomplished nothing other than setting the date for another meeting on December 24—a meeting that also appeared to accomplish little. Abbas and Olmert held their last get-together of the year on December 27. Again, no significant progress appeared to result from this session, but Olmert did attempt to calm Palestinian anger over the Har Homa settlement, saying he would freeze further expansion of it. At year's end, the peace process was in a state of suspension pending a planned trip by Bush to the region early in January 2008—the first time in his presidency that he would visit Israel and the West Bank.

Following is the text of remarks by President George W. Bush at the Annapolis Conference on the Middle East, on November 27, 2007, in which he announced a joint understanding between Israeli prime minister Ehud Olmert and Palestinian president Mahmoud Abbas on the convening of negotiations toward a peace treaty.

DOCUMENT

President Bush on an Israeli-Palestinian "Joint Understanding"

November 27, 2007

Thank you for coming. Prime Minister Olmert, President Abbas, Secretary-General Ban, former Prime Minister Blair, distinguished guests: Welcome to one of the finest institutes we have in America, the United States Naval Academy. We appreciate you joining us in what I believe is an historic opportunity to encourage the expansion of freedom and peace in the Holy Land.

We meet to lay the foundation for the establishment of a new nation, a democratic Palestinian state that will live side by side with Israel in peace and security. We meet to help bring an end to the violence that has been the true enemy of the aspirations of both the Israelis and Palestinians.

We're off to a strong start. I'm about to read a statement that was agreed upon by our distinguished guests.

"The representatives of the Government of the State of Israel and the Palestinian Liberation Organization, represented respective by Prime Minister Ehud Olmert and President Mahmoud Abbas in his capacity as Chairman of the PLO Executive Committee and President of the Palestinian Authority, have convened in Annapolis, Maryland, under the auspices of President George W. Bush of the United States of America, and with the support of the participants of this international conference, having concluded the following joint understanding.

"We express our determination to bring an end to bloodshed, suffering, and decades of conflict between our peoples; to usher in a new era of peace, based on freedom, security, justice, dignity, respect, and mutual recognition; to propagate a culture of peace and nonviolence; to confront terrorism and incitement, whether committed by Palestinians or Israelis. In furtherance of the goal of two states, Israel and Palestine living side by side in peace and security, we agree to immediately launch good-faith bilateral negotiations in order to conclude a peace treaty, resolving all outstanding issues, including all core issues, without exception, as specified in previous agreements.

"We agree to engage in vigorous, ongoing, and continuous negotiations and shall make every effort to conclude an agreement before the end of 2008. For this purpose, a steering committee, led jointly by the head of the delegation of each party, will meet continuously, as agreed. The steering committee will develop a joint work plan and establish and oversee the work of negotiations teams to address all issues, to be headed by one lead representative from each party. The first session of the steering committee will be held on 12 December 2007.

"President Abbas and Prime Minister Olmert will continue to meet on a biweekly basis to follow up the negotiations in order to offer all necessary assistance for their advancement.

"The parties also commit to immediately implement their respective obligations under the performance-based roadmap to a permanent two-state solution to the Israeli-Palestinian conflict, issued by the Quartet on 30 April 2003—this is called the roadmap—and agree to form an American, Palestinian, and Israeli mechanism, led by the United States, to follow up on the implementation of the roadmap.

"The parties further commit to continue the implementation of the ongoing obligations of the roadmap until they reach a peace treaty. The United States will monitor and judge the fulfillment of the commitment of both sides of the roadmap. Unless otherwise agreed by the parties, implementation of the future peace treaty will be subject to the implementation of the roadmap as judged by the United States."

Congratulations for your strong leadership.

The Palestinian people are blessed with many gifts and talents. They want the opportunity to use those gifts to better their own lives and build a future for their children. They want the dignity that comes with sovereignty and independence. They want justice and equality under the rule of law. They want freedom from violence and fear.

The people of Israel have just aspirations as well. They want their children to be able to ride a bus or to go to school without fear of suicide bombers. They want an end to rocket attacks and constant threats of assault. They want their nation to be recognized and welcomed in the region where they live.

Today, Palestinians and Israelis each understand that helping the other to realize their aspirations is key to realizing their own aspirations—both require an independent, democratic, viable Palestinian state. Such a state will provide Palestinians with the chance to lead lives of freedom and purpose and dignity. Such a state will help provide the Israelis with something they have been seeking for generations: to live in peace with their neighbors.

Achieving this goal is not going to be easy; if it were easy, it would have happened a long time ago. To achieve freedom and peace, both Israelis and Palestinians will have to make tough choices. Both sides are sober about the work ahead, but having spent time with their leaders, they are ready to take on the tough issues. As Prime Minister Olmert recently put it, "We will avoid none of the historic questions; we will not run from discussing any of them." As President Abbas has said, "I believe that there is an opportunity not only for us but for the Israelis too. We have a historic and important opportunity that we must benefit from." It is with that spirit that we concluded—that they concluded this statement I just read.

Our purpose here in Annapolis is not to conclude an agreement. Rather, it is to launch negotiations between the Israelis and the Palestinians. For the rest of us, our job is to encourage the parties in this effort and to give them the support they need to succeed.

In light of recent developments, some have suggested that now is not the right time to pursue peace. I disagree. I believe now is precisely the right time to begin these negotiations for a number of reasons.

First, the time is right because Palestinians and Israelis have leaders who are determined to achieve peace. President Abbas seeks to fulfill his people's aspirations for statehood, dignity, and security. President Abbas understands that a Palestinian state will not be born of terror and that terrorism is the enemy standing in the way of a state. He and Prime Minister Fayyad have both declared, without hesitation, that they are opposed to terrorism and committed to peace. They're committed to turning these declarations into actions on the ground to combat terror.

The emergence of responsible Palestinian leaders has given Israeli leaders the confidence they need to reach out to the Palestinians in true partnership. Prime Minister Olmert has expressed his understanding of the suffering and indignities felt by the Palestinian people. He's made clear that the security of Israel will be enhanced by the establishment of a responsible, democratic Palestinian state. With leaders of courage and conviction on both sides, now is the time to come together and seek the peace that both sides desire.

Second, the time is right because a battle is underway for the future of the Middle East, and we must not cede victory to the extremists. With their violent actions and contempt for human life, the extremists are seeking to impose a dark vision on the Palestinian people, a vision that feeds on hopelessness and despair to sow chaos in the Holy Land. If this vision prevails, the future of the region will be endless terror, endless war, and endless suffering.

Standing against this dark vision are President Abbas and his Government. They are offering the Palestinian people an alternative vision for the future—a vision of peace, a homeland of their own, and a better life. If responsible Palestinian leaders can deliver on this vision, they will deal the forces of extremism a devastating blow. And when liberty takes root in the rocky soil of the West Bank and Gaza, it will inspire millions across the Middle East who want their societies built on freedom and peace and hope.

By contrast, if Palestinian reformers cannot deliver on this hopeful vision, then the forces of extremism and terror will be strengthened, a generation of Palestinians could be lost to the extremists, and the Middle East will grow in despair. We cannot allow this to happen. Now is the time to show Palestinians that their dream of a free and independent state can be achieved at the table of peace and that the terror and violence preached by Palestinian extremists is the greatest obstacle to a Palestinian state.

Third, the time is right because the world understands the urgency of supporting these negotiations. We appreciate that representatives from so many governments and international institutions have come to join us here in Annapolis, especially the Arab world. We're here because we recognize what is at stake. We are here because we each have a vital role to play in helping Palestinians forge the institutions of a free society. We're here because we understand that the success of these efforts to achieve peace between Israelis and Palestinians will have an impact far beyond the Holy Land.

These are the reasons we've gathered here in Annapolis, and now we begin the difficult work of freedom and peace. The United States is proud to host this meeting, and we reaffirm the path to peace set out in the roadmap. Yet in the end, the outcome of the negotiations they launch here depends on the Israelis and Palestinians themselves. America will do everything in our power to support their quest for peace, but we cannot achieve it for them. The success of these efforts will require that all parties show patience and flexibility and meet their responsibilities.

For these negotiations to succeed, the Palestinians must do their part. They must show the world they understand that while the borders of a Palestinian state are important, the nature of a Palestinian state is just as important. They must demonstrate that a Palestinian state will create opportunity for all its citizens and govern justly and dismantle the infrastructure of terror. They must show that a Palestinian state will accept its responsibility and have the capability to be a source of stability and peace for its own citizens, for the people of Israel, and for the whole region.

The Israelis must do their part. They must show the world that they are ready to begin—to bring an end to the occupation that began in 1967 through a negotiated settlement. This settlement will establish Palestine as a Palestinian homeland, just as Israel is a homeland for the Jewish people. Israel must demonstrate its support for the creation of a prosperous and successful Palestinian state by removing unauthorized outposts, ending settlement expansion, and finding other ways for the Palestinian Authority to exercise its responsibilities without compromising Israel's security.

Arab States also have a vital role to play. Relaunching the Arab League initiative and the Arab League's support for today's conference are positive steps. All Arab States should show their strong support for the Government of President Abbas and provide needed assistance to the Palestinian Authority. Arab States should also reach out to Israel, work toward the normalization of relations, and demonstrate in both word and deed that they believe that Israel and its people have a permanent home in the Middle East. These are vital steps toward the comprehensive peace that we all seek.

Finally, the international community has important responsibilities. Prime Minister Fayyad is finalizing a plan to increase openness and transparency and accountability throughout Palestinian society, and he needs the resources and support from the international community. With strong backing from those gathered here, the Palestinian Government can build the free institutions that will support a free Palestinian state.

The United States will help Palestinian leaders build these free institutions, and the United States will keep its commitment to the security of Israel as a Jewish state and homeland for the Jewish people.

The United States strongly feels that these efforts will yield the peace that we want, and that is why we will continue to support the Lebanese people. We believe democracy brings peace. And democracy in Lebanon is vital, as well, for the peace in the Middle East. Lebanese people are in the process of electing a President. That decision is for the Lebanese people to make, and they must be able to do so free from outside interference and intimidation. As they embark on this process, the people of Lebanon can know that the American people stand with them, and we look forward to the day when the people of Lebanon can enjoy the blessings of liberty without fear of violence or coercion.

The task begun here at Annapolis will be difficult. This is the beginning of the process, not the end of it, and no doubt a lot of work remains to be done. Yet the parties can approach this work with confidence. The time is right. The cause is just. And with hard effort, I know they can succeed.

President Abbas and Prime Minister Olmert, I pledge to devote my effort during my time as President to do all I can to help you achieve this ambitious goal. I give you my personal commitment to support your work with the resources and resolve of the American Government. I believe a day is coming when freedom will yield the peace we desire. And the land that is holy to so many will see the light of peace.

The day is coming when Palestinians will enjoy the blessings that freedom brings and all Israelis will enjoy the security they deserve. That day is coming. The day is coming when the terrorists and extremists who threaten the Israeli and Palestinian people will be marginalized and eventually defeated. And when that day comes, future generations will look to the work we began here at Annapolis. They will give thanks to the leaders who gathered on the banks of the Chesapeake for their vision, their wisdom, and courage to choose a future of freedom and peace.

Thanks for coming. May God bless their work.

SOURCE: U.S. Executive Office of the President. "Remarks at the Annapolis Conference in Annapolis, Maryland." November 27, 2007. *Weekly Compilation of Presidential Documents* 43, no. 48 (December 3, 2007): 1534–1537. Washington, D.C.: National Archives and Records Administration. www.gpoaccess.gov/wcomp/v43no48.html (accessed March 27, 2008).

OTHER HISTORIC DOCUMENTS OF INTEREST

FROM THIS VOLUME

- Palestinian political developments, p. 39
- Israeli political developments, p. 200
- British prime minister Tony Blair announces his resignation, p. 221

FROM PREVIOUS *HISTORIC DOCUMENTS*

- Middle East Quartet and Hamas on the new Palestinian government, *2006*, p. 13
- Interim prime minister Olmert on the new Israeli government, *2006*, p. 193

December

DOCUMENT IN CONTEXT

Gore on Accepting the Nobel Peace Prize for Work on Climate Change

DECEMBER 10, 2007

The Norwegian Nobel Committee awarded the 2007 Nobel Peace Prize to former vice president Al Gore and the Intergovernmental Panel on Climate Change (IPCC) for their work in calling the world's attention to the threat of climate change. The committee characterized Gore as "the single individual who has done most to create worldwide understanding" of climate change and the measures needed to counteract it. The IPCC, a joint panel of the United Nations and the World Meteorological Organization, was cited for creating "an ever-broader informed consensus about the connection between human activities and global warming."

In brief comments upon hearing of the award, Gore said he was honored. "I will accept this award on behalf of all the people that have been working so long and so hard to try to get the message out about this planetary emergency," he said on October 12. Climate change "is the most dangerous challenge we've ever faced, but it is also the greatest opportunity we have ever had to make changes that we should be making for other reasons anyway." Gore and the IPCC split the award of 10 million Swedish kronor (about $1.5 million). Gore said he would donate his share to the Alliance for Climate Protection, a nonprofit that he founded.

It was the fourth time in five years that the Nobel Peace Prize had been awarded to people and organizations for work not directly related to conflict resolution. In 2006 the award was given to Muhammad Yunus, a Bangladeshi economist, and the Grameen Bank, which he founded, for their pioneering work in making credit available to the very poor. Other recent winners were Wangari Maathai of Kenya, recognized in 2004 for her contributions to sustainable development, and Shirin Ebadi, in 2003, for her work as an advocate of human rights and political reform in Iran.

"An Inconvenient Truth"

For most of his public career—as a U.S. representative and then a senator from Tennessee, and as vice president from 1993 to 2001—Al Gore made conservation of the environment in general, and climate change in particular, a key focus of his efforts. He first brought the potential dangers of global warming to popular attention with his best-selling 1992 book, *Earth in the Balance*. As vice president, he held regular seminars on the issue and helped negotiate the Kyoto Protocol in 1997, which committed developed countries, but not developing countries, to specific reductions in greenhouse gas emissions. Although President Bill Clinton signed the protocol, he never submitted it to the Senate for ratification because of opposition from Republicans and many Democrats, who believed the protocol should cover all polluting countries regardless of their development status.

Gore's views on climate change were described by some as exaggerated and alarmist. "When he first started really working on the climate change issue, I remember he was ridiculed in the press and certainly by political opponents as some kind of kook out there in la-la land," Tom Peterson, a climate scientist who worked with the IPCC, told the *New York Times* after the Nobel Prize was announced. One of the politicians who had ridiculed Gore was George W. Bush, who during the 2000 presidential campaign referred to his opponent as "Ozone Man," or sometimes just "Ozone." Bush's stance on climate change was almost diametrically opposed to Gore's. Bush for many years said that the science supporting claims that humans were creating global warming was unsound, and throughout his presidency he adamantly opposed placing any mandatory controls on greenhouse gas emissions, particularly if developing countries such as China were not also required to reduce their emissions.

After conceding the presidential race to Bush in 2000—Gore won the popular vote but lost in the Electoral College after a divided Supreme Court ruled against a recount in Florida—Gore largely put aside politics to travel the world warning of the dangers of failing to act to curb global warming. In 2006, he turned his slide show on climate change into a film entitled *An Inconvenient Truth,* which brought his message to an even wider audience and was honored with an Academy Award for best documentary in 2007. His book of the same title, like his earlier volume, reached the best-seller lists.

Gore received widespread praise when he won the Oscar and even more when he won the Nobel Peace Prize. "It's difficult for Americans to comprehend how Gore is one of the most influential global leaders of our time," Ólafur Ragnar Grímsson, the president of Iceland, told the *Washington Post.* "He is influential not only for his views, but for how he is mobilizing action and awareness in all countries, on all continents." Gore won accolades from several leading Democrats seeking their party's 2008 presidential nomination as well as from Republican John McCain, who called the award "well-deserved." President Bush did not comment publicly, but a White House spokesman said "of course, he's happy" for the former vice president.

As they had in the past, a small number of scientists, including some who remained skeptical that climate change was a serious problem caused largely by human activity, said that Gore was alarmist, because he focused on the most extreme projections of the potential damages resulting from climate change, and that he gave inaccurate representations of the science. In the same week that the Peace Prize was announced, a British judge ruled that there were nine scientific "errors" or exaggerations in *An Inconvenient Truth* and ruled that schoolchildren could be shown the film only if they were apprised of these errors. Another critic, Myron Ebell, of the libertarian Competitive Enterprise Institute, told the *Washington Post* that Gore deserved a prize for "most travel in a corporate jet in a year. Just having good intentions and saying you're out to save the plant and save humanity should not be enough to merit this sort of recognition." Ebell was concerned that public opinion was beginning to line up behind mandatory caps on greenhouse gas emissions, which he opposed.

Gore's Acceptance Speech

The Nobel award sparked a boomlet of supporters urging Gore to enter the Democratic presidential race. He showed no signs of interest in the race—and no evidence that he would back away from his predictions about the effects of climate change. "We, the human species, are confronting a planetary emergency—a threat to the survival of our civilization that is gathering ominous and destructive potential even as we gather here," he said in his Nobel Lecture in Oslo on December 10. He called on heads of state to take personal responsibility for addressing climate change and singled out two governments, the United States and China, for "failing to do enough." He described them as the "two emitters—most of all, my own country—that will need to make the boldest moves, or stand accountable before history for their failure to act." Three days later, Gore arrived in Bali, Indonesia, where he made similar tough remarks to a UN-sponsored conference of world leaders who were meeting to work out a road map for negotiating a follow-up to the Kyoto Protocol.

The IPCC

Created in 1988 by the UN, the Intergovernmental Panel on Climate Change consisted of about 600 scientific experts from 154 countries who studied meteorological and other data to determine whether climate change—including global warming—was under way, its causes, and its potential effects on Earth and its inhabitants. The panel issued major reports outlining its findings in 1990, 1995, 2001, and 2007. In its initial, 1990 report, the panel said that observed changes in climate might be due to "natural variability." By the time it issued its "Summary for Policymakers" in November 2007, the panel had concluded that the fact of global warming was "unequivocal," that human activities were largely responsible, and that immediate action was required to curb greenhouse gas emissions, which trapped heat inside the Earth's atmosphere. Even if the world did act quickly, the panel said, some changes in climate were still likely to be "abrupt and irreversible."

Spencer A. Weart, a science historian at the American Institute of Physics and the author of a recent book on global warming, told the *New York Times* on October 13, after the award was announced, that the panel had been "set up to be the lowest common denominator, to weed out anything anyone could disagree with. It was deliberately created, largely under the influence of the Reagan administration, because governments didn't want a bunch of self-appointed scientists [running around] out there. It's no accident that it's the 'intergovernmental' panel. Even the Saudi government has to agree. That means that when the IPCC says you're in trouble, you're really in trouble."

In accepting the shared Nobel Peace Prize, IPCC chairman Rajendra K. Pachauri, the founder and director of India's leading environmental think tank, said the Norwegian Nobel Committee's selection of the panel was a "signal honor" for its "acknowledgement of three important realities." The first reality, Pachauri, said, was the "power and promise of collective scientific endeavor" to "reach across national boundaries and political differences" for the "larger good of human society." The second was "the importance of the role of knowledge in shaping public policy and guiding global affairs." And the third, he said, was the acknowledgement "of the threats to stability and human security inherent in the impacts of a changing climate" and the need to take "timely and adequate action to avoid such threats in the future."

Following are excerpts from the address delivered by former vice president Al Gore as he accepted the 2007 Nobel Peace Prize in Oslo, Norway, on December 10, 2007.

DOCUMENT

The Nobel Lecture by Nobel Peace Prize Laureate 2007, Al Gore

December 10, 2007

Your Majesties, Your Royal Highnesses, Honorable members of the Norwegian Nobel Committee, Excellencies, Ladies and gentlemen.

I have a purpose here today. It is a purpose I have tried to serve for many years. I have prayed that God would show me a way to accomplish it.

Sometimes, without warning, the future knocks on our door with a precious and painful vision of what might be. . . .

Seven years ago tomorrow, I read my own political obituary in a judgment that seemed to me harsh and mistaken—if not premature. But that unwelcome verdict also brought a precious if painful gift: an opportunity to search for fresh new ways to serve my purpose.

Unexpectedly, that quest has brought me here. Even though I fear my words cannot match this moment, I pray what I am feeling in my heart will be communicated clearly enough that those who hear me will say, "We must act."

The distinguished scientists with whom it is the greatest honor of my life to share this award have laid before us a choice between two different futures—a choice that to my ears echoes the words of an ancient prophet: "Life or death, blessings or curses. Therefore, choose life, that both thou and thy seed may live."

We, the human species, are confronting a planetary emergency—a threat to the survival of our civilization that is gathering ominous and destructive potential even as we gather here. But there is hopeful news as well: we have the ability to solve this crisis and avoid the worst—though not all—of its consequences, if we act boldly, decisively and quickly.

However, despite a growing number of honorable exceptions, too many of the world's leaders are still best described in the words Winston Churchill applied to those who ignored Adolf Hitler's threat: "They go on in strange paradox, decided only to be undecided, resolved to be irresolute, adamant for drift, solid for fluidity, all powerful to be impotent."

So today, we dumped another 70 million tons of global-warming pollution into the thin shell of atmosphere surrounding our planet, as if it were an open sewer. And tomorrow, we will dump a slightly larger amount, with the cumulative concentrations now trapping more and more heat from the sun.

As a result, the earth has a fever. And the fever is rising. The experts have told us it is not a passing affliction that will heal by itself. We asked for a second opinion. And a third. And a fourth. And the consistent conclusion, restated with increasing alarm, is that something basic is wrong.

We are what is wrong, and we must make it right.

Last September 21, as the Northern Hemisphere tilted away from the sun, scientists reported with unprecedented distress that the North Polar ice cap is "falling off a cliff." One study estimated that it could be completely gone during summer in less than 22 years. Another new study, to be presented by U.S. Navy researchers later this week, warns it could happen in as little as 7 years.

Seven years from now.

In the last few months, it has been harder and harder to misinterpret the signs that our world is spinning out of kilter. Major cities in North and South America, Asia and Australia are nearly out of water due to massive droughts and melting glaciers. Desperate farmers are losing their livelihoods. Peoples in the frozen Arctic and on low-lying Pacific islands are planning evacuations of places they have long called home. Unprecedented wildfires have forced a half million people from their homes in one country and caused a national emergency that almost brought down the government in another. Climate refugees have migrated into areas already inhabited by people with different cultures, religions, and traditions, increasing the potential for conflict. Stronger storms in the Pacific and Atlantic have threatened whole cities. Millions have been displaced by massive flooding in South Asia, Mexico, and 18 countries in Africa. As temperature extremes have increased, tens of thousands have lost their lives. We are recklessly burning and clearing our forests and driving more and more species into extinction. The very web of life on which we depend is being ripped and frayed.

We never intended to cause all this destruction, just as Alfred Nobel never intended that dynamite be used for waging war. He had hoped his invention would promote human progress. We shared that same worthy goal when we began burning massive quantities of coal, then oil and methane. . . .

We also find it hard to imagine making the massive changes that are now necessary to solve the crisis. And when large truths are genuinely inconvenient, whole societies can, at least for a time, ignore them. Yet as George Orwell reminds us: "Sooner or later a false belief bumps up against solid reality, usually on a battlefield."

In the years since this prize was first awarded, the entire relationship between humankind and the earth has been radically transformed. And still, we have remained largely oblivious to the impact of our cumulative actions.

Indeed, without realizing it, we have begun to wage war on the earth itself. Now, we and the earth's climate are locked in a relationship familiar to war planners: "Mutually assured destruction."

More than two decades ago, scientists calculated that nuclear war could throw so much debris and smoke into the air that it would block life-giving sunlight from our atmosphere, causing a "nuclear winter." Their eloquent warnings here in Oslo helped galvanize the world's resolve to halt the nuclear arms race.

Now science is warning us that if we do not quickly reduce the global warming pollution that is trapping so much of the heat our planet normally radiates back out of the atmosphere, we are in danger of creating a permanent "carbon summer."

As the American poet Robert Frost wrote, "Some say the world will end in fire; some say in ice." Either, he notes, "would suffice."

But neither need be our fate. It is time to make peace with the planet.

We must quickly mobilize our civilization with the urgency and resolve that has previously been seen only when nations mobilized for war. These prior struggles for survival were won when leaders found words at the 11th hour that released a mighty surge of courage, hope and readiness to sacrifice for a protracted and mortal challenge.

These were not comforting and misleading assurances that the threat was not real or imminent; that it would affect others but not ourselves; that ordinary life might be lived even in the presence of extraordinary threat; that Providence could be trusted to do for us what we would not do for ourselves.

No, these were calls to come to the defense of the common future. They were calls upon the courage, generosity and strength of entire peoples, citizens of every class and condition who were ready to stand against the threat once asked to do so. Our enemies in those times calculated that free people would not rise to the challenge; they were, of course, catastrophically wrong.

Now comes the threat of climate crisis—a threat that is real, rising, imminent, and universal. Once again, it is the 11th hour. The penalties for ignoring this challenge are immense and growing, and at some near point would be unsustainable and unrecoverable. For now we still have the power to choose our fate, and the remaining question is only this: Have we the will to act vigorously and in time, or will we remain imprisoned by a dangerous illusion? . . .

We must abandon the conceit that individual, isolated, private actions are the answer. They can and do help. But they will not take us far enough without collective action. At the same time, we must ensure that in mobilizing globally, we do not invite the establishment of ideological conformity and a new lock-step "ism."

That means adopting principles, values, laws, and treaties that release creativity and initiative at every level of society in multifold responses originating concurrently and spontaneously.

This new consciousness requires expanding the possibilities inherent in all humanity. The innovators who will devise a new way to harness the sun's energy for pennies or invent an engine that's carbon negative may live in Lagos or Mumbai or Montevideo. We must ensure that entrepreneurs and inventors everywhere on the globe have the chance to change the world.

When we unite for a moral purpose that is manifestly good and true, the spiritual energy unleashed can transform us. The generation that defeated fascism throughout the world in the 1940s found, in rising to meet their awesome challenge, that they had gained the moral authority and long-term vision to launch the Marshall Plan, the United Nations, and a new level of global cooperation and foresight that unified Europe and facilitated the emergence of democracy and prosperity in Germany, Japan, Italy and much of the world. One of their visionary leaders said, "It is time we steered by the stars and not by the lights of every passing ship."

In the last year of that war, you gave the Peace Prize to a man from my hometown of 2,000 people, Carthage, Tennessee. Cordell Hull was described by Franklin Roosevelt as the "Father of the United Nations." He was an inspiration and hero to my own father, who followed Hull in the Congress and the U.S. Senate and in his commitment to world peace and global cooperation.

My parents spoke often of Hull, always in tones of reverence and admiration. Eight weeks ago, when you announced this prize, the deepest emotion I felt was when I saw the headline in my hometown paper that simply noted I had won the same prize that Cordell Hull had won. In that moment, I knew what my father and mother would have felt were they alive.

Just as Hull's generation found moral authority in rising to solve the world crisis caused by fascism, so too can we find our greatest opportunity in rising to solve the climate crisis. In the Kanji characters used in both Chinese and Japanese, "crisis" is written with two symbols, the first meaning "danger," the second "opportunity." By facing and removing the danger of the climate crisis, we have the opportunity to gain the moral authority and vision to vastly increase our own capacity to solve other crises that have been too long ignored.

We must understand the connections between the climate crisis and the afflictions of poverty, hunger, HIV-AIDS and other pandemics. As these problems are linked, so too must be their solutions. We must begin by making the common rescue of the global environment the central organizing principle of the world community.

Fifteen years ago, I made that case at the "Earth Summit" in Rio de Janeiro. Ten years ago, I presented it in Kyoto. This week, I will urge the delegates in Bali to adopt a bold mandate for a treaty that establishes a universal global cap on emissions and uses the market in emissions trading to efficiently allocate resources to the most effective opportunities for speedy reductions.

This treaty should be ratified and brought into effect everywhere in the world by the beginning of 2010—two years sooner than presently contemplated. The pace of our response must be accelerated to match the accelerating pace of the crisis itself.

Heads of state should meet early next year to review what was accomplished in Bali and take personal responsibility for addressing this crisis. It is not unreasonable to ask, given the gravity of our circumstances, that these heads of state meet every three months until the treaty is completed.

We also need a moratorium on the construction of any new generating facility that burns coal without the capacity to safely trap and store carbon dioxide.

And most important of all, we need to put a *price* on carbon—with a CO_2 tax that is then rebated back to the people, progressively, according to the laws of each nation, in ways that shift the burden of taxation from employment to pollution. This is by far the most effective and simplest way to accelerate solutions to this crisis.

The world needs an alliance—especially of those nations that weigh heaviest in the scales where earth is in the balance. I salute Europe and Japan for the steps they've taken in recent years to meet the challenge, and the new government in Australia, which has made solving the climate crisis its first priority.

But the outcome will be decisively influenced by two nations that are now failing to do enough: the United States and China. While India is also growing fast in importance, it should be absolutely clear that it is the two largest CO_2 emitters—most of all, my own country—that will need to make the boldest moves, or stand accountable before history for their failure to act.

Both countries should stop using the other's behavior as an excuse for stalemate and instead develop an agenda for mutual survival in a shared global environment.

These are the last few years of decision, but they can be the first years of a bright and hopeful future if we do what we must. No one should believe a solution will be found without effort, without cost, without change. Let us acknowledge that if we wish to redeem squandered time and speak again with moral authority, then these are the hard truths:

The way ahead is difficult. The outer boundary of what we currently believe is feasible is still far short of what we actually must do. Moreover, between here and there, across the unknown, falls the shadow. . . .

We are standing at the most fateful fork in that path. So I want to end as I began, with a vision of two futures—each a palpable possibility—and with a prayer that we will see with vivid clarity the necessity of choosing between those two futures, and the urgency of making the right choice now.

The great Norwegian playwright, Henrik Ibsen, wrote, "One of these days, the younger generation will come knocking at my door."

The future is knocking at our door right now. Make no mistake, the next generation *will* ask us one of two questions. Either they will ask: "What were you thinking; why didn't you act?"

Or they will ask instead: "How did you find the moral courage to rise and successfully resolve a crisis that so many said was impossible to solve?"

We have everything we need to get started, save perhaps political will, but political will is a renewable resource.

So let us renew it, and say together: "We have a purpose. We are many. For this purpose we will rise, and we will act."

SOURCE: The Nobel Foundation. "Nobel Lecture by Al Gore." Oslo, Norway. December 10, 2007. http://nobelprize.org/nobel_prizes/peace/laureates/2007/gore-lecture_en.html (accessed March 24, 2008).

OTHER HISTORIC DOCUMENTS OF INTEREST

FROM THIS VOLUME

- Supreme Court on EPA regulation of greenhouse gases, p. 139
- International panel of scientists on climate change, p. 657

FROM PREVIOUS *HISTORIC DOCUMENTS*

- British Treasury report on the economics of climate change, *2006,* p. 615
- Microcredit pioneer Yunus on accepting the Nobel Peace Prize, *2006,* p. 770
- Supreme Court on Florida election vote recount, *2000,* p. 999
- Gore concession speech, Bush victory speech, *2000,* p. 1025
- UN treaty on global warming (Kyoto), *1997,* p. 859
- Earth Summit in Rio de Janeiro, *1992,* p. 499

DOCUMENT IN CONTEXT

International "Troika" on Independence Options for Kosovo

DECEMBER 10, 2007

Two high-profile diplomatic missions tried but failed during 2007 to bridge differences between local leaders in Kosovo, who were demanding independence for the Serbian province, and the government of Serbia, which insisted on retaining Kosovo as a province but was willing to grant it greater autonomy. The lack of a diplomatic solution by year's end seemed to guarantee that Kosovo would declare independence early in 2008 and would thereby win recognition from the United States and some European countries but anger Serbia and its patron, Russia.

The status of Kosovo was one of two major items left on the international agenda following the wars that had raged in the former Republic of Yugoslavia during the 1990s. The other question concerned Bosnia, which, under a U.S.-brokered peace agreement in 1995, was a multiethnic federation that often seemed on the point of breakdown. Many diplomats and observers, including some of those who drafted the 1995 Dayton Accord, said the political arrangement for Bosnia needed to be redrawn; however, a U.S.-sponsored effort to reach a compromise among Bosnia's factions had collapsed in 2006.

Kosovo had been in a limbo status ever since the United States and its NATO allies waged a bombing war in 1999 to force Serbian security forces to withdraw from the province. The United Nations Security Council adopted a resolution (1244) that left Kosovo technically a part of Serbia but placed a newly formed ethnic Albanian majority-led government there under tight supervision by a UN envoy. Kosovo was protected by a security umbrella provided by NATO troops and a UN-sponsored police force. Elections and opinion polls consistently showed that the vast majority of Kosovo's 900,000-some ethnic Albanians (known as Kosovars) wanted independence, while the dwindling minority of about 100,000 ethnic Serbs feared independence and clung to the hope that the province would remain within Serbia.

The United States and most (but certainly not all) European countries had concluded that the only workable option was a form of independence that was greater than the "autonomy" Serbia was offering but less than the full and absolute independence enjoyed by nearly 200 other countries. Russia, a fellow Slavic state that had long served as Serbia's protector, was the main obstacle to Kosovo's independence because of Moscow's veto power in the UN Security Council, which had the legal power to settle the matter. Several European countries also were reluctant to accept independence for Kosovo because their leaders feared setting a precedent that would heighten separatist pressures within their own borders. A prime example was Spain, which faced a guerrilla movement fighting for independence in the Basque region and strong political pressures for greater autonomy or even independence in several other regions.

Many Rounds of Diplomatic Negotiations

UN secretary-general Kofi Annan early in 2006 appointed former Finnish president Martti Ahtisaari as a special envoy to bridge the differences between Kosovo's Albanian leaders and the Serbian government in Belgrade, the capital of Serbia. Ahtisaari supervised several stages of negotiations all through 2006 but was unable to overcome a resistance to compromise by both sides. He announced in November 2006 that he would propose his own solution early in 2007.

Ahtisaari presented a draft proposal on February 2, calling for a modified form of independence. Serbia rejected the plan. One week later, violent clashes in Pristina, the capital of Kosovo, resulted in the shooting deaths of two ethnic Albanians by UN police and raised concerns that a failure of diplomacy could lead to broader violence. Kosovo generally had been peaceful since rioting in March 2004 that killed nineteen people. However, extremists on both sides of Kosovo's ethnic divide were becoming more vocal, with some Albanians even demanding the departure of the UN so the province could declare independence on its own.

Ahtisaari then held more negotiations that again failed to secure an agreement between the parties. Declaring that his diplomacy was at an end, Ahtisaari submitted a slightly revised version of his earlier proposal to the new UN secretary-general, Ban Ki-moon, on March 15. This plan advocated what Ahtisaari called "supervised independence," allowing Kosovo to gain freedom from Serbia but with the European Union watching closely and providing security to ensure the safety of the minority Serb community. The Kosovo parliament embraced this proposal on April 5, but Serb leaders in Kosovo (who had boycotted the local government) rejected it, as did the Belgrade government and Russia. The full Security Council then traveled to the region in late April for meetings with political leaders in Belgrade and in Pristina, where they gained a firsthand view of the deadlock but failed to break it.

An attempt within the Security Council to broker an agreement got under way in late May, starting with a proposal by the United States and other Western nations that supported, but did not explicitly endorse, Ahtisaari's most recent proposal. Faced with continuing Russian objections, the diplomats drafted alternatives, none of which bridged the divide. The UN's chief representative in Kosovo, Joachim Rücker, told the Security Council on July 9 there was "an undercurrent of anxiety" in Kosovo that the process of diplomacy was unraveling, and he appealed to the council to provide a "road map" for the future there. "The people deserve clarity on status. The people need clarity on status," he said. Four days later the United States circulated yet another proposal. Zalmay Khalilzad, the U.S. ambassador to the UN, warned that Washington, and some of its European allies, might "move forward" to recognize Kosovo's independence if Russia continued blocking action in the Security Council.

"Troika" Negotiations

The major international players on the Kosovo question decided in late July to make one more, and probably final, attempt to negotiate an end to the deadlock. This move came from the so-called Contact Group on Kosovo, consisting of France, Germany, Britain,

Italy, Russia, and the United States. Over the years, the Contact Group had been a diplomatic mechanism for bringing together the powers most interested in the subject. The Contact Group on July 27 appointed a "Troika" of three senior diplomats, who were authorized to negotiate with all of the interested parties: Wolfgang Ischinger, the German ambassador to Britain; Aleksandr Botsan-Kharchenko, head of the Russian Foreign Ministry's Balkans department; and Frank Wisner, a retired U.S. ambassador. The diplomats were given until December 10 to come up with a compromise solution.

The Troika began its work in London on August 9 and held ten meetings with the Kosovo Albanians and the Serbs, with final sessions taking place in Belgrade and Pristina on December 3. Five of the sessions were face-to-face meetings between the opposing parties; other sessions featured separate meetings by the Troika with each side. These negotiations generally took place behind the scenes, but that did not prevent the two disputing parties from offering their own proposals and making threats from the sidelines. Kosovo Albanian officials continued to say they would accept nothing short of full independence, and Serbian president Boris Tadić told the UN General Assembly on September 27 that Belgrade was willing to go no further than accepting "autonomous development of their community [the Kosovars] within the Republic of Serbia."

Early in the process, the Troika diplomats submitted a list of fourteen "principal conclusions" they said should guide any solution. The conclusions sought to steer a middle ground between the two opposing positions but made clear that Serbia fundamentally had lost control of Kosovo. There would be no return to the pre-1999 status (of Kosovo being an integral part of Serbia), and Belgrade would neither govern Kosovo nor reestablish a "physical presence" there, the Troika said.

While the negotiations were in progress, Kosovo on November 17 held regularly scheduled elections for its parliament. These elections resulted in strong gains by the opposition Democratic Party of Kosovo, which had pledged to declare independence if the mediation efforts failed. This led to formation of a coalition government headed by that party's leader, Hashim Thaçi; the coalition partner was the Democratic League of Fatmir Sejdiu, the president. Thaçi had been a leader of the Kosovo Liberation Army, the guerrilla group that fought for independence in the 1990s. As in previous elections, most ethnic Serbs boycotted the polls. Thaçi's victory brought warnings from European leaders against precipitous action by the Kosovars.

The Troika's diplomacy appeared to come to an end after face-to-face sessions in Vienna on November 29 again failed to break the deadlock. Troika diplomats held one more round of separate talks with the parties on December 3, then gave up their task as a failure. In a report submitted to Secretary-General Ban on December 7 and published officially on December 10—the UN's deadline for a solution—the three diplomats needed just one sentence to summarize why their efforts failed: "Neither party was willing to cede its position on the fundamental question of sovereignty over Kosovo." Nevertheless, the Troika report added, in diplomatic language, the outcome was "regrettable, as a negotiated settlement is in the best interests of both parties."

Reaction to the failure of the Troika's mission fell along predictable lines. The United States and some of its Western allies said that they were preparing for the likelihood of a unilateral declaration of independence by Kosovo, probably in February 2008 following a presidential election in Serbia. NATO foreign ministers, meeting in Brussels on December 7, agreed to keep the alliance's military presence in Kosovo even after such a declaration to maintain stability. Russian foreign minister Sergei Lavrov insisted a diplomatic solution—one avoiding an independence declaration—was still possible. In Kosovo, the sense of anticipation among ethnic Albanians grew stronger by the day, and on December 10, thousands of pro-independence demonstrators gathered in Pristina to cheer what they saw as the inevitable next step. Minority ethnic Serbs, however, were becoming increasingly fearful of the consequences of losing all official contact with Belgrade.

Reflecting their own divisions over the possible consequences of independence for Kosovo, European leaders staked out a cautious position during their annual year-end summit, in Brussels on December 14–15. The leaders warned Kosovo against an immediate declaration of independence but also decided not to speed up talks with Serbia on potential EU membership. Cyprus, Romania, and Slovakia—all of which had joined the EU only in recent years and had their own minority communities—adamantly opposed anything that would encourage Kosovo independence. As a compromise, EU leaders did agree to send a mission of as many as 1,800 civil administrators, judges, and police officers to Kosovo early in 2008 to help promote stability in Kosovo no matter what happened regarding independence.

Final Push for Compromise

The last effort of the year—and likely the last one ever—to find a compromise solution took place at the UN Security Council on December 19. Council members listened as Serbian prime minister Vojislav Koštunica and Kosovo president Fatmir Sejdiu staked out their opposing views. Koštunica told the council that forcing Serbia to give up Kosovo would be "to abolish its sovereignty." Sejdiu argued Kosovo needed full independence to end the "lack of clarity about our status" that, he said, had undermined the province's economic prospects. These two positions, British ambassador to the UN John Sawers said, "underlined just how enormous the gulf is between the two parties."

The Security Council drew no formal conclusions from its meeting, but members said it was clear that diplomacy had come to an end. Italy's foreign minister Massimo D'Alema, who presided over the meeting, said much of the blame lay with Russia and the United States, each of which had pushed an extremist position: Serbia's opposition to independence versus Kosovo's demand for independence.

At year's end, the two principal parties again set out their positions. The Serbian parliament on December 26 overwhelmingly adopted a resolution opposing independence for Kosovo and warning that Belgrade might "reconsider" diplomatic and other relations with countries that recognized an independence declaration. In Pristina, the two main ethnic Albanian parties agreed on December 28 to form a unity government focused on the overriding goal of achieving independence.

"Genocide," But No Prosecutions

A major piece of legal business left over from the Balkans wars of the 1990s was wrapped up on February 26, when the International Court of Justice (the World Court) declared that the massacre of about 7,000 Muslim men and boys in the Bosnian town of Srebrenica in July 1995 had been an act of genocide under the provisions of the post–World War II Genocide Convention. Even so, the court ruled that the Serbian government in Belgrade (which at the time of the killings was in the form of the Republic of Yugoslavia) could not be held legally responsible.

The judges said the Belgrade government, then headed by Slobodan Milošević, should have tried to intervene to stop the killings. But Belgrade did not have effective control over the Bosnian Serb militia that carried out the Srebrenica massacre, the court ruled. The ruling was a blow to the government of Bosnia, which had demanded reparations from Serbia. The Srebrenica massacre was the single bloodiest event of all the Balkans wars, and it led the United States to sponsor peace negotiations that ended the war in Bosnia in November 1995.

Following are excerpts from a report of the diplomatic group known as the "Troika" named by the European Union, Russia, and the United States to attempt to negotiate an agreement between the Republic of Serbia and the government of Kosovo. These excerpts include the main report and the text of Annex VI, which lists the Troika's "principal conclusions" on which any settlement would have to be based. Members of the Troika were Wolfgang Ischinger, the German ambassador to Britain; Aleksandr Botsan-Kharchenko, head of the Russian Foreign Ministry's Balkans department; and Frank Wisner, a retired U.S. ambassador. The report, which is dated December 4, 2007, was submitted on December 7 to UN secretary-general Ban Ki-moon and published officially on December 10, 2007.

DOCUMENT

"Troika" Report on Negotiations over the Future of Kosovo

December 10, 2007

Summary

1. We, a Troika of representatives from the European Union, the United States and the Russian Federation, have spent the last four months conducting negotiations between Belgrade and Pristina on the future status of Kosovo. Our objective was to facilitate an agreement between the parties. The negotiations were conducted within the framework of Security Council resolution 1244 (1999) and the guiding principles of the Contact Group (see S/2005/709). In the course of our work, the parties discussed a wide range of options, such as full independence, supervised independence, territorial partition, substantial autonomy, confederal arrangements and even a status silent "agreement to disagree".

2. The Troika was able to facilitate high-level, intense and substantive discussions between Belgrade and Pristina. Nonetheless, the parties were unable to reach an agreement on the final status of Kosovo. Neither party was willing to cede its position on the fundamental question of sovereignty over Kosovo. This is regrettable, as a negotiated settlement is in the best interests of both parties.

Background

3. A political process to determine the future status of Kosovo, the last major issue related to Yugoslavia's collapse, has been under way for over two years. The United

Nations Secretary-General appointed Martti Ahtisaari as his Special Envoy in November 2005 to undertake the future status process envisioned in Security Council resolution 1244 (1999). After 15 months of United Nations-sponsored negotiations, President Ahtisaari prepared a *Comprehensive Proposal for the Kosovo Status Settlement,* which included measures to protect Kosovo's non-Albanian communities, and a recommendation that Kosovo should become independent subject to a period of international supervision. Pristina accepted the Ahtisaari Settlement in its entirety; Belgrade rejected it.

4. After a period of discussions in the Security Council, the Contact Group (France, Germany, Italy, Russia, the United Kingdom and the United States) proposed that a "Troika" of officials from the European Union, the United States and Russia undertake yet another period of negotiations with the goal of achieving a negotiated agreement. On 1 August 2007, the Secretary-General welcomed this initiative, restated his belief that the status quo was unsustainable and requested a report from the Contact Group on these efforts by 10 December 2007. The United Nations Office of the Special Envoy for the Kosovo Future Status Process (UNOSEK) would be associated with the process by standing ready to provide information and clarification on request. . . .

The Troika's mission

5. Upon our appointment as Troika representatives, we vowed to "leave no stone unturned" in the search for a mutually acceptable outcome. In pursuit of this goal, we explained to the parties the principles that would guide our work. First, we reaffirmed that Security Council resolution 1244 (1999) and the November 2005 guiding principles of the Contact Group would continue to be our operating framework. Second, we noted that while the Ahtisaari Settlement was still on the table, we would be prepared to endorse any agreement the parties might be able to reach. Both sides were repeatedly reminded of their responsibility for success or failure of the process.

6. We also explained that the Troika had no intention of imposing a solution. Instead, the burden was on each party to convince the other side of the merits of its position. Although our role would be primarily to facilitate direct dialogue, we also intended to take an active role in identifying areas of possible compromise.

Working schedule

7. During the four months of our mandate, we undertook an intense schedule of meetings with the parties. . . . This schedule comprised 10 sessions, six of which consisted of face-to-face dialogue, including a final intensive three-day conference in Baden, Austria, as well as two trips to the region. During the process, Belgrade was represented by President Boris Tadić, Prime Minister Vojislav Koštunica, Foreign Minister Vuk Jeremić and Minister for Kosovo Slobodan Samardzić. Pristina was represented by the "Team of Unity" composed of President Fatmir Sejdiu, Prime Minister Agim Ĉeku, President of the Assembly Kolë Berisha, Hashim Thaĉi and Veton Surroi. The Troika appreciated the fact that both delegations were represented at the highest possible level, underlining the importance they attached to the process. In addition to the joint sessions we arranged separate meetings with the parties in order to consult with them individually. Our sessions were long and often difficult, as we confronted a legacy of mutual mistrust and sense of historical grievance about the conflicts of the 1990s. The Contact Group supported our work, and its foreign ministers urged the parties to approach the

negotiations with "creativity, boldness and in a spirit of compromise" . . . . We also sought, and received, pledges from the parties that neither would engage in provocative acts or statements during the process . . . .

8. As we began our work, we first explored the well-established positions of each side. Pristina restated its preference for Kosovo's supervised independence and reconfirmed its acceptance of the Ahtisaari proposal. Belgrade rejected the Ahtisaari proposal and restated its preference that Kosovo be autonomous within Serbia. As a result, there was no discussion of the Ahtisaari proposal nor any discussion that it should be modified. Both sides employed historical, functional, legal and practical arguments to support their preferred outcome. Belgrade elaborated its model of substantial autonomy to enhance the powers of an autonomous Kosovo and reduce those that it would reserve. It asserted that there would be no return to the pre-March 1999 situation. Pristina presented a draft "Treaty of Friendship and Cooperation", which describes how Kosovo and Serbia, as independent states, could cooperate on issues of mutual concern, establish common bodies, enhance their commitment to multi-ethnicity and support each other's Euro-Atlantic aspirations.

9. Despite our repeated call for fresh ideas and a spirit of compromise, neither side was able to convince the other to accept its preferred outcome. Encouraged by the Contact Group's Ministerial Statement of 27 September . . . , we undertook a more active approach. We developed our assessment in the form of the "Fourteen Points" of possible overlap in the parties' positions (see annex VI). The parties responded to these points, without accepting them fully.

10. Under our guidance, the parties reviewed outcomes ranging from independence to autonomy, as well as alternate models such as confederal arrangements, and even a model based on an "agreement to disagree" in which neither party would be expected to renounce its position but would nonetheless pursue practical arrangements designed to facilitate cooperation and consultation between them. Other international models, such as Hong Kong, the Åland Islands and the Commonwealth of Independent States, were discussed. While it was broached, we did not dwell on the option of territorial partition, which was deemed unacceptable by both the parties and the Contact Group. None of these models proved to be an adequate basis for compromise. We concluded face-to-face negotiations between the parties at a high-level conference in Baden, Austria, from 26 to 28 November, where we again encouraged both sides to find a way out of the deadlock.

Conclusions

11. Throughout the negotiations both parties were fully engaged. After 120 days of intensive negotiations, however, the parties were unable to reach an agreement on Kosovo's status. Neither side was willing to yield on the basic question of sovereignty.

12. Nevertheless, despite this fundamental difference on status, which the Troika was unable to bridge, we believe this process served a useful purpose. We gave the parties an opportunity to find a solution to their differences. Under our auspices, the parties engaged in the most sustained and intense high-level direct dialogue since hostilities ended in Kosovo in 1999. Through this process, the parties discovered areas where their interests aligned. The parties also agreed on the need to promote and protect multi-ethnic societies and address difficult issues holding back reconciliation, particularly the fate of missing persons and the return of displaced persons. Perhaps most important, Belgrade and Pristina reaffirmed the centrality of their European perspective to their

future relations, with both sides restating their desire to seek a future under the common roof of the European Union.

13. While differences between the parties remain unchanged, the Troika has nevertheless been able to extract important commitments from the parties. In particular, both parties have pledged to refrain from actions that might jeopardize the security situation in Kosovo or elsewhere and not use violence, threats or intimidation (see Annex VII). They made these commitments without prejudice to their positions on status. Both parties must be reminded that their failure to live up to these commitments will affect the achievement of the European future that they both seek.

14. We note that Kosovo and Serbia will continue to be tied together due to the special nature of their relationship, especially in its historical, human, geographical, economical and cultural dimensions. As noted by Contact Group Ministers at their meeting in New York on 27 September, the resolution of Kosovo's status is crucial to the stability and security of the Western Balkans and Europe as a whole. We believe the maintenance of peace in the region and the avoidance of violence is of paramount importance and therefore look to the parties to stand by their commitments. We, furthermore, strongly believe that the settlement of Kosovo's status would contribute to the fulfilment of the European aspirations of both parties. . . .

[Annexes I through V omitted.]

Annex VI

Troika assessment of negotiations: principal conclusions

The Troika has reviewed the positions of the two parties. Without prejudice to the positions of both parties on status, the following principles can open a path to a solution:

1. Belgrade and Pristina will focus on developing the special nature of the relations existing between them especially in their historical, economic, cultural and human dimensions.

2. Belgrade and Pristina will solve future problems between them in a peaceful manner and not engage in actions or dispositions that would be regarded as threatening to the other side.

3. Kosovo will be fully integrated into regional structures, particularly those involving economic cooperation.

4. There will be no return to the pre-1999 status.

5. Belgrade will not govern Kosovo.

6. Belgrade will not re-establish a physical presence in Kosovo.

7. Belgrade and Pristina are determined to make progress towards association and eventually membership of the European Union as well as to move progressively towards Euro-Atlantic structures.

8. Pristina will implement broad measures to enhance the welfare of Kosovo-Serbs as well as other non-Albanian communities, particularly through decentralization of local government, constitutional guarantees and protection of cultural and religious heritage.

9. Belgrade and Pristina will cooperate on issues of mutual concern, including:
   a. Fate of missing persons and return of displaced persons
   b. Protection of minorities
   c. Protection of cultural heritage
   d. Their European perspectives and regional initiatives
   e. Economic issues, including fiscal policy and energy, trade and harmonization with EU standards and development of a joint economic growth and development strategy in line with regional economic initiatives
   f. Free movement of people, goods, capital and services
   g. Banking sector
   h. Infrastructure, transportation and communications
   i. Environmental protection
   j. Public health and social welfare
   k. Fight against crime, particularly in the areas of terrorism, human-, weapon- and drug-trafficking and organized crime
   l. Cooperation between municipalities and the Government of one of the two sides
   m. Education.

10. Belgrade and Pristina will establish common bodies to implement cooperation.

11. Belgrade will not interfere in Pristina's relationship with international financial institutions.

12. Pristina will have full authority over its finances (taxation, public revenues, etc.).

13. Kosovo's EU Stabilization and Association Process (Tracking Mechanism) will continue unhindered by Belgrade.

14. The international community will retain civilian and military presences in Kosovo after status is determined. . . .

SOURCE: United Nations. Security Council. "Letter Dated 10 December 2007 from the Secretary-General to the President of the Security Council. Enclosure: Report of the European Union/United States/Russian Federation Troika on Kosovo." S/2007/723. December 10, 2007. Released December 13, 2007. www.un.org/Docs/journal/asp/ws.asp?m=s/2007/723 (accessed March 26, 2008).

Other Historic Documents of Interest

From this volume

- European leaders on new treaty for the European Union, p. 721

From previous *Historic Documents*

- Delay in Kosovo status, *2006,* p. 367
- UN envoy on violence in Kosovo, *2004,* p. 949
- Kosovo War, *1999,* p. 119
- Srebrenica massacre, *1999,* p. 735

DOCUMENT IN CONTEXT

New York Attorney General on the Student Loan Settlement

DECEMBER 11, 2007

The private student loan industry experienced a major shake-up in 2007 after a far-reaching investigation disclosed that many private lenders were using questionable marketing practices to increase their student loan volume. Among the practices were payments made to colleges and universities that put the student loan companies on a "preferred lender" list and prizes or other inducements offered to students who signed up for the loans. By the end of the year, several college aid officials had lost their jobs, some schools and loan companies had agreed to pay millions of dollars into a fund to educate students and their parents about obtaining student loans, and many lenders had agreed to abide by a new code of conduct for marketing the loans.

The investigation, launched by New York attorney general Andrew M. Cuomo, also bolstered efforts by congressional Democrats to make higher education more affordable to students from low- and middle-income families. A major component of the legislation, which was signed into law in September, shifted billions of dollars in federal subsidies from private lenders to grant programs for needy students.

In November and December, Williams College, Harvard University, and the University of Pennsylvania joined the small but growing list of colleges with sizable endowments that had decided to eliminate or reduce need-based student loans for eligible middle-income students, who would receive grants instead. Princeton was the first college to adopt a no-loan policy, in 2001; since then more than thirty schools had adopted similar programs. These moves came as college costs continued to rise, making higher education increasingly unaffordable for low- and middle-income students unless they saddled themselves with large amounts of loans that they would then have to pay off as they were starting their careers. The College Board reported that the average cost for tuition, room, and board increased nearly 6 percent in 2007 to $13,589 at a four-year public college and $32,307 at a private college. Tuition alone at several top private schools was around $45,000 per year.

By most estimates, college students took out more than $85 billion in federal and private loans in 2006. About three-quarters were federally subsidized and guaranteed loans, borrowed either directly from the government or through private lenders such as Sallie Mae, the nation's largest student lender. Private lenders accounted for about two-thirds of that total. The fastest-growing element of the student loan business, however, was private loans that were not guaranteed by the government. These loans, which had become increasingly popular as tuitions soared across the country, totaled $17 billion in 2006, according to the College Board.

Cuomo Investigation

Concern about the often undisclosed relationships between student loan companies and college aid offices had been growing for some time. At the beginning of 2007, some Democratic legislators and the Education Department were considering action to regulate so-called preferred-lender lists—lists recommending specific loan companies that college financial aid offices gave students. Because students tended to rely on these lists, getting on them was important for loan companies, and many of them offered financial incentives to win the favor of school aid officers. Lenders also used a variety of direct marketing tactics to entice students to take out loans.

It was an aggressive investigation by Andrew Cuomo, New York's attorney general, however, that focused widespread attention on the issue. "We have found that these school-lender relationships are often highly tainted with conflicts of interest," Cuomo said on March 15, announcing his initial findings. "These school-lender relationships are often for the benefit of the schools at the expense of the student." Among the more widespread tactics Cuomo uncovered was a practice known as "revenue-sharing," in which the loan company gave cash back to the college based on the amount students there borrowed. Some schools contracted with lenders operating call centers to field students' questions about financial aid; in some cases, the call center staffers identified themselves as affiliated with the university or college, according to papers obtained by Cuomo's office. Cuomo's office also detailed cases where loan companies gave financial aid officials stock options on preferred terms, hired them as consultants, and paid their travel expenses to student aid conferences, where they were also lavished with gifts like video iPods and DVD players.

Cuomo's investigation also looked into a variety of tactics for marketing loans directly to students that could be considered misleading or deceptive; these included paying students for filling out loan applications or for steering other students to the loan company, sending "checks" to prospective customers that could be cashed only if the customers took out loans with the company, and using brands and logos associated with the school or its sports teams. One company used an image of an eagle that was very similar to the logo of the federal government. "The practices we have found in the direct marketing of loans to students are surprisingly blatant and even involve some companies who portray themselves as arms of the federal government," Cuomo said in a statement issued October 10.

In June Cuomo announced that he would also examine the criteria lenders used in making loans, including whether they might be discriminatory and in violation of civil rights laws. Testifying before the Senate Banking Committee on June 6, Cuomo suggested that some lenders might be setting interest rates for private loans based on factors such as race.

To avoid legal action from Cuomo's office as well as from other state attorneys general, dozens of lenders and schools agreed to cash settlements. In a press release issued on December 11, Cuomo said that ten universities had agreed to reimburse nearly $3.4 million to students who had received loans from companies that made payments to the universities based on loan volume. Twelve student loan companies, including the eight largest, had agreed to contribute $13.7 million to a National Education Fund set up by Cuomo to be used to educate high school students and their families about the college financial aid process. The lenders also agreed to abide by a code of conduct prohibiting them from making payments or giving gifts to colleges and universities and requiring full disclosure of loan terms. A second code of conduct was aimed at preventing lenders from engaging in false and misleading advertising practices.

The investigations also resulted in the suspension or dismissal of several college aid and loan company officials. A senior official in the Education Department's Office of Federal Student Aid was put on leave after it was revealed that he held stock in one of the loan companies he was responsible for monitoring. A financial aid officer at Johns Hopkins University was dropped as a member of an agency task force drafting new federal student loan regulations after she was found to have received $65,000 in consulting fees and tuition from a loan company.

Secretary of Education Margaret Spellings staunchly defended her department against allegations that it was too cozy with the loan companies it was meant to regulate. Nonetheless, the department issued new regulations on November 1, to take effect in July 2008, that applied to lenders participating in the federal student loan program; among other things, the rules prohibited lenders from offering schools gifts or payments to steer student borrowers to them and required that at least three lenders be on preferred-lender lists. Congressional leaders who had been pushing the department to take action generally applauded the new rules, but some said the department should have gone further. "The new regulations . . . should help curtail inappropriate relationships between colleges and lenders that make federal student loans," Sen. Edward M. Kennedy, D-Mass., said in a statement released October 30. "But there's more to be done—lenders who offer private student loans should also be covered by these rules."

Federal Student Aid Legislation

Cuomo's findings helped boost the prospects of Democratic legislation to overhaul the federal student aid system. During their 2006 bid to retake Congress, Democrats had campaigned on college affordability, a winning issue among middle-class swing voters. Their plan centered on slashing subsidies paid to private lenders that offered federally guaranteed student loans and redirecting the money to students through various federal grant programs.

The initial Democratic proposal, which would have cut subsidies by $7.1 billion to pay for halving the interest rates on subsidized student loans, faced a shaky future in the Senate, where it was opposed by several Republican senators who thought the interest rate cut was expensive and poorly targeted; President George W. Bush also announced his opposition to cutting the interest rates. But Cuomo's investigation stirred such a public outcry that legislators found it difficult to justify leaving the existing interest rates in place. Bush helped the Democrats' cause in February, when he outlined a plan in his fiscal 2008 budget to cut nearly $20 billion from lenders to pay for higher Pell grant awards. (The backbone of the federal student aid program, Pell grants were given to low-income students based on need.) Bush's proposal made it nearly impossible for legislators to object to cuts in the subsidies.

The House and Senate passed differing versions of the student loan bill in July. Kennedy, chairman of the Senate education committee, and Rep. George Miller, D-Calif., his counterpart in the House, then spent weeks behind the scenes trying to reach agreement among recalcitrant colleagues, mostly Republicans, and with the White House, which had threatened to veto the House version and expressed "grave concern" about the Senate version. By early September, Miller and Kennedy had built veto-proof majorities in both chambers of Congress and won widespread public support for their compromise version (HR 2669). Bush lifted his veto threat, and the measure was cleared September 7 by overwhelming majorities. Bush signed HR 2669 into law (PL 110-84) on September 27.

As cleared, the bill cut about $20 billion from federal subsidies to private lenders and redirected the money to student aid. It increased the maximum Pell grant award, capped loan repayments at 15 percent of discretionary income, and created a debt-forgiveness program for many public-sector employees, such as teachers, law enforcement professionals, and public health doctors and nurses, who had taken direct loans from the federal government. It also included several of the provisions that had prompted the veto threat, including cutting the interest rate on student loans in half over four years, from 6.8 percent to 3.4 percent. The final version dropped several other proposed entitlement programs that the White House opposed, but it kept a grant program for students committed to teaching subjects in which the demand for teachers was high as well as a grant program for schools that served minorities.

The measure did not contain any provisions prohibiting the questionable marketing practices by lenders uncovered by Cuomo's investigation. Those were expected to be included in a separate bill reauthorizing the entire federal Higher Education Act, which was scheduled to expire in March 2008.

Toward the end of the year, several student lenders warned that the lower subsidies and interest rates, coupled with rising defaults on student loans, might cut back on the availability of student loans in the coming months. They also said tightened credit conditions could affect the amount of money available for lending. Sallie Mae, the country's largest student loan company, was facing particular trouble. A planned sale to a private equity firm had collapsed, its share price had fallen to its lowest level since 2001, and its credit rating was deteriorating. In addition, the Education Department was investigating the company's billing practices for possible fraud, and it was facing a class-action lawsuit, filed in a federal district court in Connecticut, alleging that it had discriminated against minorities by steering them to higher-priced loans.

Following are excerpts from a press release issued on December 11, 2007, by New York attorney general Andrew M. Cuomo, announcing a settlement with a student loan company over deceptive advertising practices.

DOCUMENT

New York Attorney General Cuomo on Settlement of Student Loan Cases

December 11, 2007

New York Attorney General Andrew M. Cuomo today announced a settlement with a student loan consolidation company specializing in the direct marketing of student loans, the first such settlement in this growing segment of the student lending industry.

A four-month investigation found that Clearwater, Florida-based Student Financial Services Inc. (SFS), which also operates under the banner of University Financial Services (UFS), had agreed to pay some of the nation's top universities, school athletic departments, and sports marketing firms for generating loan applications, in a kickback scheme euphemistically known as revenue sharing. The company had contracts at 63 colleges nationwide, 57 of which are National Collegiate Athletic Association (NCAA) Division I schools.

Under these agreements, the company also paid for the rights to use school names, team names, colors, mascots, and logos to advertise their loans directly to students. This practice, known as "co-branding," was intended to imply that the company was the official lender of the school, or that it was actually a part of the school. Schools, athletic departments, and sports marketing firms made these agreements without evaluating the quality of the loans.

"When lenders use deceptive techniques to advertise their loans, they are playing a dangerous game with a student's future," said Cuomo. "Student loan companies incorporate school insignia and colors into advertisements because they know students are more likely to trust a lender if its loan appears to be approved by their college. We cannot allow lenders to exploit this trust with deceptive, co-branded marketing. A student loan is a very serious financial commitment, and choosing the wrong loan can lead to devastating consequences."

Under the settlement, which was joined by Florida Attorney General Bill McCollum, SFS has agreed to:

- **End all lending-related agreements at a total of sixty-three schools** including Georgetown University, Wake Forest University, University of Kansas, Central Michigan University, St. John's University, University of Washington, University of Oregon, University of Texas El Paso, Rutgers University, Georgia Tech, Florida State University, Florida Atlantic University, the University of Central Florida, and the University of Pittsburgh. SFS has until December 31, 2007 to comply;
- **End all lending-related agreements with five sports marketing companies** that, in some instances, were sold the right to market the school's insignia, colors, and mascot, and in turn signed an agreement with SFS. These companies are ESPN Regional Television, Inc., International Sports Properties, Inc., Host Communications, Nelligan Sports Marketing, Inc., and Learfield Communications, Inc. SFS has until December 31, 2007 to comply;
- **Launch a print advertising campaign at 63 schools** alerting students through their top-circulating newspapers that they must protect themselves when shopping for a loan;
- **End the practice of cash-based inducements,** including paying students up to $50 to refer their peers to the company and encouraging students to apply for SFS loans by creating contests where they could win up to $1,000.

Also under the settlement, SFS has agreed to adopt a new Code of Conduct . . . developed by Attorney General Cuomo that prevents false and misleading direct loan marketing to students. The Code expressly prohibits lenders and marketers from buying rights to a college or university's name, team name, colors, logo, and mascot for loan marketing purposes. It also requires lenders and marketers to provide important disclosures to students in connection with loan transactions and prohibits a variety of misleading and deceptive practices identified by the Attorney General's investigation of the industry.

"Deceptive direct marketing schemes are illegal in New York, and will not be tolerated," said Cuomo. "That's why I have developed a new Code of Conduct to protect students from these predatory practices and ensure that any lending information they receive is honest, accurate, and consistent."

"College students and their parents deserve transparency and full disclosure when they are making important financial decisions about student loans and this settlement will extend that protection beyond the universities," said Florida Attorney General Bill McCollum. "The Florida Code of Conduct and the New York Code of Conduct are ground-breaking decisions to put the best interests of our students and parents first."...

Today's settlement is a significant milestone in Cuomo's nationwide investigation into the student loan industry, which has already resulted in agreements with twelve student loan companies, including the eight largest lenders in America—Citibank, Sallie Mae, Nelnet, JP Morgan Chase, Bank of America, Wells Fargo, Wachovia, and College Loan Corporation—as well as Education Finance Partners (EFP), CIT, National City Bank, and Regions Financial Corporation. Citibank, Sallie Mae, Nelnet, CLC, EFP, CIT, Johns Hopkins University, Columbia University, Mercy College, and Career Education Corporation have all agreed to contribute a total of $13.7 million to a National Education Fund established by Attorney General Cuomo. This fund is dedicated to educating the country's high school students and their families about the financial aid process.

In addition, the University of Pennsylvania, New York University, Syracuse University, Texas Christian University, St. John's University, Fordham University, Salve Regina, Long Island University, Drexel University, and DeVry University agreed to reimburse students a total of $3,368,443.

Cuomo's original Code of Conduct has become New York State law as the Student Lending Accountability, Transparency, and Enforcement (SLATE) Act of 2007. Proposed federal legislation regarding the student loan industry also incorporates Cuomo's original Code of Conduct; the Student Loan Sunshine Act has been passed by the U.S. House of Representatives and the U.S. Senate....

SOURCE: State of New York. Office of the Attorney General. "Cuomo Announces Settlement With Student Loan Company Tied to NCAA Division I Schools: Lender to End Kickbacks and Co-Branding Agreements." Press release. December 11, 2007. www.oag.state.ny.us/press/2007/dec/dec11b_07.html (accessed April 3, 2008).

OTHER HISTORIC DOCUMENTS OF INTEREST

FROM PREVIOUS *HISTORIC DOCUMENTS*

- President Bush on Pell Grant increases, *2005*, p. 144; *2001*, p. 59.

DOCUMENT IN CONTEXT

Mitchell Report on Investigation into Steroid Use in Baseball

DECEMBER 13, 2007

Two of the biggest stars in major league baseball history, slugger Barry Bonds and pitcher Roger Clemens, were caught up in a long-brewing scandal over the use of steroids by ball players. Bonds was indicted on November 15 by a federal grand jury on charges that he lied during an investigation into the activities of a San Francisco–area company that allegedly supplied steroids to him and other players. Clemens's was the biggest new name to emerge from an investigation into steroid use conducted for the baseball commissioner's office by former Senate majority leader George J. Mitchell. In his report, made public on December 13, Mitchell listed the names of ninety-two players—Bonds and Clemens most prominent among them—about whom there were credible allegations of steroid use in recent years.

For more than twenty years, just about every big-time sport had endured allegations that athletes had used steroids and other performance-enhancing drugs, some of which were illegal without a prescription in the United States and many other countries. The International Olympic Committee and some specific sports federations began cracking down on steroid use in the early 1980s, but Major League Baseball lagged well behind.

Although baseball banned steroid use in 1991, it did not begin enforcing the ban until after the 2002 season, and it launched a sustained testing program only in 2004. In the meantime, some of the sport's biggest stars were named in books, news reports, and even grand jury testimony as having used steroids during their careers. Among them was Mark McGwire, who set a single-season home run record in 1998 only to lose it three years later to Bonds. McGwire suffered the consequences for his reported use of a legal body-building drug by losing his initial bid for membership in the Baseball Hall of Fame. In the annual voting in January, McGwire's name appeared on only 23.5 percent of the ballots, less than one-third of the 75 percent needed for induction. The rejection of McGwire, despite his home run achievement, was widely seen as an early indication of how he and other players caught up in the steroids scandal would be judged by those who wrote baseball's history books.

THE BARRY BONDS CASE

A central figure in the investigations into the use of steroids by baseball players and other athletes was a San Francisco–area company, the Bay Area Laboratory Cooperative (BALCO). A federal investigation begun in 2003 found that the firm specialized in so-called designer steroids—supplements that allowed athletes to train longer and harder than they might otherwise, and that were difficult to detect through common urine tests. In 2004 a San Francisco newspaper disclosed leaked grand jury testimony naming Bonds, who played for the San Francisco Giants, and former Giants player Jason Giambi as being among BALCO's customers for steroids. The company's founder, Victor Conte, served a brief prison term (late 2005 to early 2006) after he pleaded guilty to felony charges stemming from the investigation.

Bonds was by far the most important figure caught up in the BALCO case, but he consistently denied using steroids or performance-enhancing drugs of any kind. One of baseball's premier hitters for more than fifteen years, Bonds entered 2007 with the expectation that he would become the all-time leader in home runs. That crowning achievement came on August 7, in a home game against the Washington Nationals, when he hit his 756th home run, breaking Hank Aaron's lifetime record, set in 1976. Giants fans cheered lustily—a welcome change for the slugger, who since the steroids allegations surfaced often had faced hostile crowds in ballparks of opposing teams. Even after his record home run, fans in other parks occasionally held up signs saying: 756*. The asterisk signified a questionable achievement, similar to the asterisk that baseball put in the record books after New York Yankees slugger Roger Maris broke Babe Ruth's single-season record of 60 home runs in 1960; in Maris's case, the season had been extended by eight games since Ruth's time, giving him what many fans considered an unfair advantage.

What should have been a year of triumph for Bonds ended with two sad chapters. On September 21, as baseball's regular season drew to a close, Bonds announced that the Giants had told him he would not be invited back for the 2008 season. Speculation began immediately about whether any other team would want the home run champion, despite the allegations about his steroid use.

That speculation came to an abrupt end on November 15, when Bonds was indicted on four counts of perjury and one count of obstruction of justice; the charges resulted from his testimony on December 4, 2003, to the federal grand jury that was investigating BALCO. His lawyer and several legal experts raised questions about why it took federal prosecutors nearly four years to bring the charges, and Bonds pleaded not guilty to all four charges on December 7.

Mitchell's Investigation

Baseball commissioner Allan H. "Bud" Selig asked Mitchell in March 2006 to investigate the extent of steroid use in the sport and make recommendations for dealing with the situation. Working through his New York law firm, Mitchell quietly collected information during 2006, and early in 2007 he began putting pressure on key players in the sport to cooperate with his probe. He told the owners of baseball clubs on January 18 that they needed to cooperate with his investigation because baseball "has a cloud over its head, and that cloud will not just go away."

An important step for Mitchell's probe came on April 27, when Kirk Radomski, a former New York Mets clubhouse attendant, pleaded guilty to felony charges of distributing steroids and laundering money on behalf of players. Radomski admitted distributing anabolic steroids, human growth hormone, and other drugs to several dozen players after he left the Mets in 1995 and became a personal trainer. He agreed to cooperate with Mitchell, and his testimony figured prominently in the final report.

Mitchell also revealed in May that he had asked an unspecified number of active players to testify, even though their union, the Players Association, strongly discouraged such testimony. Jason Giambi of the New York Yankees on August 17 became the first active player to talk with Mitchell. Giambi, who had also been named in the BALCO case, had told *USA Today* in May that he apologized for taking steroids in the past.

Mitchell made his report public at a news conference in New York City on December 13. It was one of the year's most-anticipated news events, and coverage of it dominated all American news media for days afterwards. In a 409-page report, Mitchell listed both active and retired players who he said had been linked to the use of steroids and other performance-enhancing drugs.

By far the most prominent, aside from Bonds, was Roger Clemens, who had never before been mentioned publicly as a possible user of such drugs. The report named Clemens eighty-two times. All the allegations against him came from a former Yankees coach, Brian McNamee, who said he had repeatedly injected Clemens with steroids, starting in 1999.

Clemens for years had been one of the most dominant pitchers in baseball; he won seven Cy Young Awards (given to the year's best pitcher in each league), stood eighth on the all-time list of victories with 354, and was widely considered a shoo-in for the Hall of Fame. His 2007 season with the New York Yankees had been disappointing, but even though he was forty-five years old, there had been suggestions that he wanted to return to baseball in 2008.

Most of the other players named by Mitchell had been the subject of previous reports about steroid use, including Bonds, Giambi, and José Canseco, who had written a book describing steroid use in baseball. Two of the most prominent players listed in the report, but who had not been previously mentioned in reports about steroid use, were Miguel Tejada of the Baltimore Orioles and Andy Pettitte of the Yankees.

In a statement at the December 13 news conference, Mitchell emphasized that the problem of steroid use in baseball was long standing and that responsibility was shared widely. "Everyone involved in baseball over the past two decades—commissioners, club officials, the players' association—shares to some extent the responsibility for the steroids era," he said. "There was a collective failure to recognize the problem as it emerged and to deal with it early on." The use of such drugs is not trivial, his report added, and "poses a serious threat to the integrity of the game" because it "unfairly disadvantages the honest athletes who refuse to use them and raises questions about the validity of baseball records."

Mitchell recommended sweeping changes in the drug-testing system baseball had adopted just two years earlier, including putting an independent authority in charge of testing and making the timing of tests more unpredictable. Since testing began in 2005, there had been numerous reports that some players had learned of tests in advance.

Although his report named some of the sport's top players, and chastised nearly everyone involved in baseball, Mitchell said he believed baseball officials should concentrate on fixing the system to prevent and detect steroids use rather than penalize players for past conduct. Mitchell suggested that the baseball commissioner impose penalties only in "those cases where he determines that the conduct is so serious that discipline is necessary to maintain the integrity of the game."

In a news conference immediately following Mitchell's release of the report, Commissioner Selig said he supported Mitchell's twenty recommendations and would implement "immediately" all those that he had unilateral authority to carry out. Most of the recommendations would require negotiations with the Players Association, he said. Selig further announced that he would deal with the active players named in Mitchell's report "on a case-by-case basis." He also acknowledged a key finding of Mitchell's report: that some players had stopped taking steroids, which could be detected through urine tests, and had begun taking human growth hormones, which could not be detected with any current testing procedures.

Donald M. Fehr, executive director of the Players Association, appeared not to share Mitchell's and Selig's enthusiasm for changes to the current testing program. "The program in place today is a strong and effective one, and has been improved even in the last two years," he commented. Fehr said he was willing to discuss possible changes to the program, but he made it clear that the union would support the players named in Mitchell's report if Selig chose to punish them.

Some of the players mentioned in Mitchell's report denied the allegations against them, notably Clemens; through his attorney, Rusty Hardin, he said he had never used any performance-enhancing drugs. Hardin characterized Mitchell's report as unfairly targeting Clemens, who "is left with no meaningful way to combat what he strongly contends are totally false allegations." Hardin put his objections in more graphic terms in another comment to reporters: "He [Mitchell] has thrown a skunk into the jury box, and we will never be able to remove that smell." In a video posted on the Internet on December 24, Clemens spoke for himself: "I did not use steroids or human growth hormone, and I've never done so."

Although enacting Mitchell's recommendations would toughen baseball's drug-testing procedures, some critics said the suggested changes did not go far enough. Chief among the critics was Dick Pound, chairman of the World Anti-Doping Agency (which supervised testing for the Olympics), who said neither baseball's current procedures nor the changes recommended by Mitchell were adequate to detect steroid abuse. "Baseball's going to set up a special baseball FBI with no power to investigate, no power to seize evidence, no power to compel witnesses to come forward?" he asked. "That's going to be a kind of a farce."

A more positive comment came from President George W. Bush, who had been directly involved in baseball during the 1990s when he was an owner of the Texas Rangers. "The players and the owners must take the Mitchell report seriously," he said in a statement. "I'm confident they will."

Steroids in Other Sports

Baseball's wrestling with steroids generated the most attention in 2007, but other sports, at various levels, also dealt with similar issues. Among them:

- The Texas Legislature on May 28 gave final approval to legislation mandating state-funded drug tests for public high school athletes in all sports. Signed into law by Governor Rick Perry, the bill also provided for suspensions of student athletes who tested positive or refused to be tested. Florida in June set up a one-year pilot program of tests for athletes in baseball, football, and weightlifting. New Jersey in 2006 became the first state to require random tests in high school sports.

- For the second year in a row, the world's premier cycling race, the Tour de France, was rocked by doping scandals. In 2006, winner Floyd Landis, an American, was stripped of his title after he tested positive for a banned substance. Landis fought the action but lost an appeal to an arbitration panel, which upheld the action against him on September 21, 2007. Three top riders were ejected from the 2007 race because they failed tests or avoided taking the tests; among them were a pre-race favorite from Kazakhstan and a Danish rider who had led the race for nine of its twenty-thee days.

- The major professional golf associations agreed on September 20 to begin testing of players in 2008. The PGA Tour and other major tours banned ten classes of drugs and set procedures for mandatory testing. Golf officials insisted that steroid use was not a problem in the sport, but some controversy erupted in July when former champion Gary Player said he knew of some golfers who used illegal drugs.
- Marion Jones, an American track star who won five medals at the 2000 Summer Olympics in Sydney, Australia, pleaded guilty on October 5 to two counts of lying to federal agents about her use of steroids and involvement in a counterfeit check scheme. Jones, who had long battled allegations that she used steroids, took three gold medals and two bronze medals home from Sydney. She was the first female athlete ever to win that many medals at a single Olympics. Jones relinquished her medals on October 8, and the International Olympic Committee on December 12 formally expunged her name from the record books and banned her from participating in, or attending in an official capacity, any future games.

Following are excerpts from the conclusions of the "Summary and Recommendations" section of the "Report to the Commissioner of Baseball of an Independent Investigation into the Illegal Use of Steroids and Other Performance-Enhancing Substances by Players in Major League Baseball," made public on December 13, 2007, by former U.S. senator George J. Mitchell.

DOCUMENT

Conclusions of the Mitchell Report on Steroids in Baseball

December 13, 2007

J. Conclusions

There has been a great deal of speculation about this report. Much of it has focused on players' names: how many and which ones. After considering that issue very carefully I concluded that it is appropriate and necessary to include them in this report. Otherwise I would not have done what I was asked to do: to try to find out what happened and to report what I learned accurately, fairly, and thoroughly.

While the interest in names is understandable, I hope the media and the public will keep that part of the report in context and will look beyond the individuals to the central conclusions and recommendations of this report. In closing, I want to emphasize them:

1. The use of steroids in Major League Baseball was widespread. The response by baseball was slow to develop and was initially ineffective. For many years, citing concerns

for the privacy rights of the players, the Players Association opposed mandatory random drug testing of its members for steroids and other substances. But in 2002, the effort gained momentum after the clubs and the Players Association agreed to and adopted a mandatory random drug testing program. The current program has been effective in that detectable steroid use appears to have declined. However, that does not mean that players have stopped using performance enhancing substances. Many players have shifted to human growth hormone, which is not detectable in any currently available urine test.

2. The minority of players who used such substances were wrong. They violated federal law and baseball policy, and they distorted the fairness of competition by trying to gain an unfair advantage over the majority of players who followed the law and the rules. They—the players who follow the law and the rules—are faced with the painful choice of either being placed at a competitive disadvantage or becoming illegal users themselves. No one should shave to make that choice.

3. Obviously, the players who illegally used performance enhancing substances are responsible for their actions. But they did not act in a vacuum. Everyone involved in baseball over the past two decades—Commissioners, club officials, the Players Association, and players—shares to some extent in the responsibility for the steroids era. There was a collective failure to recognize the problem as it emerged and to deal with it early on. As a result, an environment developed in which illegal use became widespread.

4. Knowledge and understanding of the past are essential if the problem is to be dealt with effectively in the future. But being chained to the past is not helpful. Baseball does not need and cannot afford to engage in a never-ending search for the name of every player who ever used performance enhancing substances. The Commissioner was right to ask for this investigation and report. It would have been impossible to get closure on this issue without it, or something like it.

5. But it is now time to look to the future, to get on with the important and difficult task that lies ahead. Everyone involved in Major League Baseball should join in a well-planned, well-executed, and sustained effort to bring the era of steroids and human growth hormone to an end and to prevent its recurrence in some other form in the future. That is the only way this cloud will be removed from the game. The adoption of the recommendations set forth in this report will be a first step in that direction.

SOURCE: Major League Baseball. "Report to the Commissioner of Baseball of an Independent Investigation into the Illegal Use of Steroids and Other Performance Enhancing Substances by Players in Major League Baseball." Summary and Recommendations. December 13, 2007. mlb.mlb.com/mlb/news/mitchell/index.jsp (accessed January 16, 2008).

OTHER HISTORIC DOCUMENTS OF INTEREST

FROM PREVIOUS *HISTORIC DOCUMENTS*

- Congressional hearings on the use of steroids in baseball, *2005,* p. 212
- Mark McGwire's home run record, *1998,* p. 626

DOCUMENT IN CONTEXT

European Union Leaders on the Treaty of Lisbon

DECEMBER 13, 2007

European leaders in December adopted a new treaty intended to streamline the governing apparatus of the European Union, which had expanded to twenty-seven members and often had trouble making major decisions. The new Treaty of Lisbon still needed to be ratified by EU member nations by 2009. It took the place of a massive 800-page constitution that had died in 2005 after being rejected by voters in France and the Netherlands.

Much of the Lisbon treaty dealt with the arcane details of how the EU would make its decisions, but leaders and commentators said it had broader significance as a symbol of the EU's determination to meet twenty-first-century challenges on a broad range of economic, foreign policy, law enforcement, and other matters. The most noticeable changes would be the appointment of a semi-permanent European president and a foreign policy representative with expanded clout on the world stage.

Two new members joined the EU at the beginning of the year—Bulgaria and Romania—which brought total membership to twenty-seven nations and completed the latest round of expansion, following the addition of ten new members in 2004. Seven other countries were still on the waiting list: Albania, Bosnia, Croatia, Macedonia, Montenegro, Serbia, and Turkey. Each of these countries faced objections of one kind or another from current member states. Turkey was by far the most controversial because it was so big (with about 70 million people), its population was almost entirely Muslim, it was much poorer than other EU countries, and most of its landmass was in Asia, not in Europe. The new president of France, Nicholas Sarkozy, in June forced a slowdown of EU negotiations with Turkey on key elements of its potential membership, thus raising questions about whether Turkey could join by a projected date of 2015. Turkey did win plaudits from the EU during 2007 when a political intervention by the military was rebuffed by voters, an action seen as reinforcing democracy there.

Another landmark for the EU took place on December 21, when nine of the countries that had joined in 2004 became part of the so-called Schengen area (named after a 1985 treaty signed in Schengen, Luxembourg), which meant their citizens could travel freely to and from fifteen other EU member nations without having to go through border controls. The new countries were the Czech Republic, Estonia, Hungary, Latvia, Lithuania, Malta, Poland, Slovakia, and Slovenia. Except for Malta, all had been under communist rule until the collapse of communism, starting in 1989. EU officials portrayed expansion of the Schengen area as a major achievement for continent-wide unity. Even so, it was controversial in many Western European countries, where some citizens worried about a deluge of low-income Eastern Europeans competing for jobs and demanding social services.

The only EU member nations not part of the Schengen area as of 2007 were Cyprus and the brand-new members, Bulgaria and Romania.

Writing a New Treaty

Following the defeat of the constitution in 2005, European leaders agreed on the need for what they variously described as a "pause" or a "period of reflection" before trying again to revise how the EU functioned. German chancellor Angela Merkel, whose country held the EU's presidency for the first six months of 2007, announced on January 17 that the time for action had come. "The reflection pause is over," she said as she set out her priorities, which included negotiating a new EU governance agreement. A similar call to action came on January 26 from foreign ministers of the eighteen states that had ratified the constitution, in some cases even after Dutch and French voters had killed it. During a meeting in Madrid of "friends of the constitution," Spanish foreign minister Miguel Ángel Moratinos called the constitution a "magnificent document" that should be saved and perhaps changed but not cut "into little pieces."

Despite this praise for the constitution, the carving up of that document began almost immediately. At an EU summit on March 8, Merkel unveiled plans for a slimmed-down document, not even called a constitution, to be written in a matter of a few months. Merkel then used her influence and well-known persuasive skills to win her colleagues' approval of a document called the "Berlin Declaration," which was issued at a special summit on March 25 to commemorate the 1957 treaty that led eventually to the European Community and then the EU itself. The declaration praised the goal of a united Europe and spoke of a unanimous aim by the leaders to place the union "on a renewed common basis" before the June 2009 elections for the European Parliament. Not to meet this goal, Merkel said, "would be an historic failure."

An important new actor came onto the European stage two months later. Sarkozy took office as the new president of France on May 16 and wasted no time in making clear that he wanted action on the question of revamping the EU, but not in the constitutional form that had been rejected by voters in his country. "We need to move forward, and a simplified treaty is the way forward," he said during his first visit, as president, to EU headquarters in Brussels, Belgium, on May 23.

Merkel and other leaders moved quickly to get agreement on specifics of a new EU governing plan, but they faced strong opposition from Poland, which feared it would lose voting power under any new arrangement. Poland's objections almost prevented EU leaders from reaching agreement during a summit meeting in Brussels on June 21–23. However, Poland backed down during an all-night negotiating session under intense pressure, notably by Sarkozy, who stepped in to defuse tensions between the German and Polish leaders—tensions that reflected historic animosities between their countries. In a key compromise, the leaders agreed to delay a change in the voting system used by EU leaders until 2014. Sarkozy praised the agreement as renewing the EU's forward momentum: "A Europe that had come to a standstill has begun to move again."

The June summit settled most of the major issues regarding the EU's new governing structure, but other details remained to be hammered out. A first draft of the proposed new treaty was made public early in August. Very few ordinary citizens could understand it, however, because it consisted of hundreds of amendments, written in dense legal language, to the EU's previous treaties. EU leaders approved a near-final version during a summit meeting in Lisbon on October 19, thus branding the new document as the Treaty of Lisbon. During this meeting, leaders were forced to make concessions to Britain's new prime minister, Gordon Brown, who wanted to avoid having to call a referendum on the treaty—a referendum that pollsters and pundits in Britain said he almost certainly would lose. Brown's predecessor, Tony Blair, had promised to submit the previous constitution to a skeptical British electorate but was spared the necessity of doing so once Dutch and French voters killed it. At the Lisbon summit, fellow EU leaders accepted Brown's demands that Britain would retain authority over matters, such as its justice system and foreign policies, that fell within what he called the "red lines" of what British voters and legislators would accept. Brown's decision not to call a referendum proved controversial back at home, with the opposition Conservatives demanding such a public vote.

European leaders formally signed the new treaty at an elaborate ceremony at a monastery in Lisbon on December 13—except Brown, who stayed in London for a parliamentary committee hearing and signed it later, in private. Leaders heralded the occasion as a major step forward in EU history. "Europe was blocked, not knowing how to move forward, and we found the solution," Portuguese prime minister José Sócrates, host of the event, said.

EU leaders said they hoped to complete the process of ratifying the treaty during 2008, or by early 2009 at the latest. As of the end of 2007, only Ireland planned to hold a referendum. This vote was widely seen as the single greatest hurdle for the treaty because approval by Irish voters was far from certain, and by its terms the treaty would take effect only if all twenty-seven EU member nations approved it. Governments in the other twenty-six countries planned to ratify the treaty through action by their parliaments or other procedures short of a public vote.

A New Decision-Making System

Two changes were at the heart of the Lisbon treaty: a complex new mechanism for decision-making at the EU's highest levels and the designation of two senior figures to represent the union, both among member countries and on the broader world stage. Similar changes had been included in the proposed constitution that failed in 2005.

The changes in the decision-making process were intended to assure a broad range of Europeans, including those in small countries, that their voices would be heard and to ensure that having so many participants at the table would not block action when it was required. To enhance the democratic aspect of EU decision making, the treaty gave expanded powers to the European Parliament, which was directly elected by European citizens but long had been derided by critics (and even by some of its members) as little more than a debating society. Under the treaty, the parliament for the first time was given equal standing with the European Council on most legislation; the council was composed of the heads of governments of the member nations and in the past had been considered the EU's highest-ranking decision-making body.

The treaty also included a provision, called "subsidiarity," that allowed the parliaments of individual countries to object when they believed the EU was intruding into national, rather than union-wide, affairs. This was intended to address many countries' complaints that the EU had drafted laws and regulations that unnecessarily diminished the individuality of the countries; among the most controversial of these had been health regulations that essentially barred foods that had been popular for centuries but were found to violate the latest safety standards.

In terms of streamlining the decision-making process, the most important change concerned voting powers at the European Council. Since 2000, the council had used a complex weighted voting system that many countries considered unfair; for example, the system gave Poland as much clout as its much larger neighbor, Germany, thus explaining Poland's resistance to change. The new system, called "qualified majority voting," was intended to provide for majority rule, but with significant protections for minorities, particularly the smaller countries. Under this system, to take effect in 2014, any decision would require a 55 percent majority of member states (in other words, at least fifteen of the twenty-seven current EU nations), representing at least 65 percent of the total EU population. To prevent two or three of the biggest countries (notably Britain, France, and Germany) from blocking action all by themselves under the population requirement, the treaty required that at least four nations had to be part of any blocking minority.

The treaty also revamped two of the most visible EU leadership posts: the president and the foreign affairs representative. In the past, the presidency of the European Council rotated every six months, meaning that a leader had only a short time to achieve anything of significance. The new treaty changed this to a term of two-and-a-half years, renewable for one subsequent term. The treaty also consolidated the two existing positions that represented the EU to the world at large (a commissioner for foreign affairs and a commissioner for external relations) into one position with broader powers and the imposing title of High Representative of the Union for Foreign Affairs and Security Policy.

Following is the text of "Treaty at a Glance," a document produced by the European Union as a summary of the Treaty of Lisbon, which was signed by EU leaders in Lisbon on December 13, 2007.

DOCUMENT

The Treaty of Lisbon at a Glance

December 13, 2007

On 13 December 2007, EU leaders signed the Treaty of Lisbon, thus bringing to an end several years of negotiation about institutional issues.

The Treaty of Lisbon amends the current EU and EC [European Community, the predecessor to the EU] treaties, without replacing them. It will provide the Union with the legal framework and tools necessary to meet future challenges and to respond to citizens' demands.

1. **A more democratic and transparent Europe,** with a strengthened role for the European Parliament and national parliaments, more opportunities for citizens to have their voices heard and a clearer sense of who does what at European and national level.

- A strengthened role for the European Parliament: the European Parliament, directly elected by EU citizens, will see important new powers emerge over the EU legislation, the EU budget and international agreements. In particular, the increase of co-decision procedure in policy-making will ensure the European

Parliament is placed on an equal footing with the Council [the Council of Europe, comprised of heads of governments], representing Member States, for the vast bulk of EU legislation.

- A greater involvement of national parliaments: national parliaments will have greater opportunities to be involved in the work of the EU, in particular thanks to a new mechanism to monitor that the Union only acts where results can be better attained at EU level (subsidiarity). Together with the strengthened role for the European Parliament, it will enhance democracy and increase legitimacy in the functioning of the Union.
- A stronger voice for citizens: thanks to the Citizens' Initiative, one million citizens from a number of Member States will have the possibility to call on the Commission to bring forward new policy proposals.
- Who does what: the relationship between the Member States and the European Union will become clearer with the categorisation of competences.
- Withdrawal from the Union: the Treaty of Lisbon explicitly recognises for the first time the possibility for a Member State to withdraw from the Union.

2. **A more efficient Europe,** with simplified working methods and voting rules, streamlined and modern institutions for a EU of 27 members and an improved ability to act in areas of major priority for today's Union.

- Effective and efficient decision-making: qualified majority voting in the Council will be extended to new policy areas to make decision-making faster and more efficient. From 2014 on, the calculation of qualified majority will be based on the double majority of Member States and people, thus representing the dual legitimacy of the Union. A double majority will be achieved when a decision is taken by 55% of the Member States representing at least 65% of the Union's population.
- A more stable and streamlined institutional framework: the Treaty of Lisbon creates the function of President of the European Council elected for two and a half years, introduces a direct link between the election of the Commission President and the results of the European elections, provides for new arrangements for the future composition of the European Parliament and for a smaller Commission, and includes clearer rules on enhanced cooperation and financial provisions.

  Improving the life of Europeans: the Treaty of Lisbon improves the EU's ability to act in several policy areas of major priority for today's Union and its citizens. This is the case in particular for the policy areas of freedom, security and justice, such as combating terrorism or tackling crime. It also concerns to some extent other areas including energy policy, public health, civil protection, climate change, services of general interest, research, space, territorial cohesion, commercial policy, humanitarian aid, sport, tourism and administrative cooperation.

3. **A Europe of rights and values, freedom, solidarity and security,** promoting the Union's values, introducing the Charter of Fundamental Rights into European primary law, providing for new solidarity mechanisms and ensuring better protection of European citizens.

- Democratic values: the Treaty of Lisbon details and reinforces the values and objectives on which the Union is built. These values aim to serve as a reference point for European citizens and to demonstrate what Europe has to offer its partners worldwide.

- Citizens' rights and Charter of Fundamental Rights: the Treaty of Lisbon preserves existing rights while introducing new ones. In particular, it guarantees the freedoms and principles set out in the Charter of Fundamental Rights and gives its provisions a binding legal force. It concerns civil, political, economic and social rights.

- Freedom of European citizens: the Treaty of Lisbon preserves and reinforces the "four freedoms" and the political, economic and social freedom of European citizens.

- Solidarity between Member States: the Treaty of Lisbon provides that the Union and its Member States act jointly in a spirit of solidarity if a Member State is the subject of a terrorist attack or the victim of a natural or man-made disaster. Solidarity in the area of energy is also emphasised.

- Increased security for all: the Union will get an extended capacity to act on freedom, security and justice, which will bring direct benefits in terms of the Union's ability to fight crime and terrorism. New provisions on civil protection, humanitarian aid and public health also aim at boosting the Union's ability to respond to threats to the security of European citizens.

4. **Europe as an actor on the global stage** will be achieved by bringing together Europe's external policy tools, both when developing and deciding new policies. The Treaty of Lisbon will give Europe a clear voice in relations with its partners worldwide. It will harness Europe's economic, humanitarian, political and diplomatic strengths to promote European interests and values worldwide, while respecting the particular interests of the Member States in Foreign Affairs.

- A new High Representative for the Union in Foreign Affairs and Security Policy, also Vice-President of the Commission, will increase the impact, the coherence and the visibility of the EU's external action.

- A new European External Action Service will provide back-up and support to the High Representative.

- A single legal personality for the Union will strengthen the Union's negotiating power, making it more effective on the world stage and a more visible partner for third countries and international organisations.

- Progress in European Security and Defence Policy will preserve special decision-making arrangements but also pave the way towards reinforced cooperation amongst a smaller group of Member States.

SOURCE: European Union. "The Treaty at a Glance." December 13, 2007. http://europa.eu/lisbon_treaty/glance/index_en.htm (accessed March 31, 2008).

Other Historic Documents of Interest

From this volume

- Political developments in Turkey, p. 190
- Tony Blair on his resignation as British prime minister, p. 221
- Nicolas Sarkozy on his inauguration as president of France, p. 240
- Council of Europe investigation into CIA "secret prisons" in Europe, p. 275
- International "troika" on independence options for Kosovo, p. 699

From previous *Historic Documents*

- French and British leaders on the failure of the EU constitution, *2005*, p. 338

DOCUMENT IN CONTEXT

U.S. Comptroller General on the Nation's Long-Term Deficit

DECEMBER 17, 2007

The federal budget deficit dropped to its lowest level in five years, falling to $163 billion in 2007, down from $248 billion recorded in 2006 and $319 billion in 2005. But even as the short-term deficit was declining, government budget experts were stepping up warnings that long-term debt was likely to bankrupt the nation if left unaddressed. "Our government has made a whole lot of promises that, in the long run, it cannot possibly keep" without slashing federal benefit programs or imposing huge tax increases, Comptroller General David M. Walker said December 17 in a speech to the National Press Club.

Walker, in conjunction with several Washington think tanks and policy groups, had been traveling the country since 2005 on what he called a "Fiscal Wake-up Tour" to call the public's attention to the difficult political decisions that he and others said needed to be made soon to keep the U.S. economy from sinking into a mass of debt. Federal Reserve chairman Ben S. Bernanke and Congressional Budget Office (CBO) director Peter R. Orszag were among the other government officials expressing similar warnings throughout the year in testimony before Congress and in speeches across the country.

Rising health care costs and an aging population nearing or in retirement represented the two biggest threats to the government's long-term fiscal stability. By Walker's reckoning, future spending necessary to make good on promised Social Security and Medicare benefits, as well as on all other federal spending commitments, totaled $53 trillion at the end of 2007, up from $20 trillion since 2000. That amounted to about $450,000 for every household in America. By virtually everyone's reckoning, the U.S. economy would be unable keep up with the federal debt such spending would incur, resulting in a lowering of the standard of living for future generations, unless action were taken to put the deficit on a more sustainable path.

Federal Deficit, National Debt

There are three ways to look at the federal government's annual budget deficit (or surplus), each of which gives a slightly different picture of the current situation. The commonly cited method is the unified budget, which counts any surplus in the Social Security trust fund as revenue, thus making the deficit between total revenues and total outlays appear smaller than it might otherwise be. For example, under the unified budget, the 2007 deficit was $163 billion, but under an accounting system that does not rely on borrowing from the surplus funds in the Social Security trust fund, the 2007 budget deficit was $344 billion. The Treasury Department also did an accounting using the accrual method to show the government's finances in the same way that private corporations are required to show theirs. Under the accrual method, expenses are recorded when they are incurred rather than when they are paid, effectively taking into account the costs of long-term liabilities such as pensions and health insurance. Under accrual accounting, the fiscal 2007 budget deficit totaled $276 billion.

However one accounted for the budget deficit, it was still a deficit and added to the national debt, which stood at $9.1 trillion in December, up from $5.7 billion when President George W. Bush took office in 2001. Since 2002, Congress had boosted the ceiling on the federal debt five times, to $9.815 trillion. The last increase, of $850 billion, was approved in September 2007.

Of the $9.1 trillion national debt, about $4 trillion represented amounts the government had borrowed from the surplus in the Social Security trust fund and other government accounts to finance its current operations. The remaining $5.1 trillion was the public debt—debt held by the public in the form of Treasury bonds and other securities. Nearly half the public debt, $2.23 trillion, was held by foreign governments and foreign investors. According to the Treasury Department, the biggest foreign holder was Japan ($586 billion), followed by China ($400 billion), Britain ($244 billion), and Saudi Arabia and other oil-exporting countries ($123 billion). That level of indebtedness to foreign countries was a concern to some. "Borrowing hundreds of billions of dollars from China and OPEC [Organization of the Petroleum Exporting Countries] puts not only our future economy, but also our national security at risk," said Sen. George V. Voinovich, R-Ohio. "It is critical that we ensure that countries that control our debt do not control our future."

Long-Term Deficits

The rising national debt implied rising interest payments to carry it. Those payments, along with the federal government's commitment to help an aging population pay for its health care and retirement costs, were at the crux of the long-term deficit problem. Both the Office of Management and Budget in the White House and the CBO made long-range projections of federal spending and revenues, which differed somewhat because of variations in assumptions about levels and timing of revenue and spending. Under either set of projections, however, the nation was in trouble. Economists often look at the size of the federal deficit by measuring it as a share of gross domestic product (GDP), the value of all the goods and services produced in the United States. Under CBO's alternative fiscal scenario, which incorporated widely expected changes in policy, federal spending, not counting interest payments, was projected to increase from 18.2 percent of GDP in 2007 to 24.2 percent of GDP by 2030 and to 35.3 percent of GDP by 2082 (the end date in CBO's seventy-five year forecast). Interest payments on the national debt were projected to rise from 1.7 percent of GDP in 2007 to 4.8 percent in 2030 and then to 40.1 percent in 2082. Meanwhile, revenues were expected to remain relatively stable as a percentage of GDP, rising from 18.8 in 2007 to 20.9 in 2082. As a result, the total deficit (including interest payments) was projected to increase from 1.2 percent of GDP in 2007 to 10.1 percent in 2030 and to 54.5 percent in 2082.

Virtually everyone who addressed the long-term deficit problem said something had to be done quickly to control Social Security and Medicare spending. "The longer we wait, the more severe, the more draconian, the more difficult the adjustment is going to be," Fed chairman Bernanke told the Senate Budget Committee on January 18. "I think the right time to start is about ten years ago."

As CBO director Orszag pointed out in his testimony before the House Budget Committee on December 13, delay in taking action would have three consequences: It would cause government debt to rise, squeezing private domestic capital and increasing borrowing from foreign investors. It would exacerbate uncertainty by not giving people time to plan for whatever adjustments would be made in their taxes and benefits. And it would raise the cost of interest payments on the federal debt, requiring Congress to make ever-larger changes in policy—tax hikes, spending cuts, and additional borrowing—to finance ever-increasing interest payments. That, in turn, would make it more difficult for Congress to finance other national priorities, reduce the government's flexibility to deal with unexpected developments such as war or recession, and make the economy more vulnerable to a crisis.

Orszag said the most significant driver of long-term deficits was not the aging population but the soaring growth of health care costs, which were pushing up the costs of the federal Medicare and Medicaid programs. Under current law, Orszag said, federal spending on those two programs was projected to grow from 4 percent of GDP in 2007 to 12 percent in 2050 and to 19 percent in 2082. Both Orzsag and Walker were particularly critical of the additional indebtedness created by the Medicare prescription drug benefit that Congress approved in 2003, suggesting that in their eagerness to woo voters by passing the popular bill, legislators did not fully take into account its future costs, now estimated at $8 trillion through 2082.

In Orszag's view, health care costs were also likely to be the most complicated to address. Because per beneficiary costs of Medicare and Medicaid tended to track health care costs in the private sector, he said, many analysts thought it would be possible to slow Medicare and Medicaid cost growth only in conjunction with slowing health care costs overall. "A variety of evidence suggests that opportunities exist to constrain costs without incurring adverse consequences for health outcomes—and even perhaps to simultaneously reduce cost growth and improve health," Orszag said. He added that moving the country in that direction was "essential" for stabilizing the long-run fiscal situation.

Walker, like many others, urged comprehensive reform of entitlement programs, starting with legislation that would create a bipartisan commission or task force to make recommendations for overhauling benefit programs such as Social Security and Medicare and then submit those recommendations to Congress for an up-or-down vote. "Such an entity could help set the table for action while providing needed political cover for elected officials to act," Walker said. As President Bush discovered in 2005 when his plan to partially privatize Social Security fell flat, any proposal for lowering benefits or raising taxes was likely to stir intense opposition.

Walker's proposal was one of thirteen recommendations his agency, the Government Accountability Office, offered as a blueprint for weeding out expensive and ineffective government programs and ensuring that the remaining programs were oriented toward achieving "real, desirable, and sustainable results." The report, "A Call for Stewardship," urged Congress, the president, and the American public to undertake a "top-to-bottom" review of the federal programs and policies "to decide which federal activities remain priorities, which should be overhauled, and which have simply outlived their usefulness." To begin that process, the report recommended that the government develop a set of "key national indicators" on a range of economic, environmental, safety and security, social, and cultural issues. These indicators could be used to develop programs and to determine whether they were achieving their intended results. The report also called for the president to develop a governmentwide strategic plan that would integrate a range of federal activities and provide a framework for considering organizational changes, allocating resources, and holding key personnel accountable for achieving measurable results.

Difficulty of Getting Control of the Budget

The political difficulty of tightening spending on federal programs or raising taxes to control mounting long-range deficits was amply illustrated by congressional actions in 2007. After campaigning on a platform of fiscal responsibility in 2006, Democrats reestablished "pay-as-you-go" rules soon after taking control of Congress at the beginning of 2007. Also known as PAYGO, the rules required that any new mandatory spending or tax cuts be offset by spending cuts or tax increases. Previous PAYGO rules had helped President Bill Clinton and a Republican-controlled Congress bring the budget into balance in 1997. But by 2001, the rules had been largely abandoned in the GOP's rush to enact tax cut legislation that cost hundreds of millions of dollars but included no offsets. The Medicare prescription drug benefit enacted in 2003, the largest entitlement program approved by Congress in decades, also did not contain any offsets.

The new PAYGO rules sparked frequent and difficult searches for offsets throughout 2007. Democrats maneuvered to avoid breaking the rule, and they survived most of the year without formally abandoning the restriction. But the run ended in December, when the Senate, faced with a tight deadline and the need for an expensive bill to prevent the alternative minimum tax from reaching further into the middle class, ignored the rule and passed the tax cut without any revenue-raising offsets. House members initially balked, but eventually accepted the legislation and sent it on to the president.

Even before Democrats abandoned the rule, Republicans charged that Congress was in many instances following the letter but not the spirit of PAYGO. In at least one case, for example, Democrats found offsets by shifting the dates when certain payments would be made, a change that did not save money in the long run. Some Republicans were also concerned that an extension of the massive tax cuts of 2001, which were scheduled to expire in 2010, would be made more difficult if PAYGO rules requiring offsets were in effect.

Following are excerpts from the text of a speech delivered by Comptroller General David M. Walker at the National Press Club in Washington on December 17, 2007, discussing the government's long-term deficit problems and outlining a plan to help bring the deficits under control.

DOCUMENT

U.S. Comptroller General Walker on Long-Term Financial Problems

December 17, 2007

Good afternoon ladies and gentlemen. Thank you for coming today. I know some of you were in this same room back in 2003 when I gave a speech entitled "Truth and Transparency." In that speech, I urged the federal government to provide more complete and reliable information on where our nation stands financially and where it was headed fiscally. Those who were here will recall that during my speech, I expressed particular concern about the cost of the then pending proposal to add a Medicare prescription drug benefit. A little over four years have passed since I gave that speech, and I felt the time was right to provide an update on the state of America's finances. . . .

When it comes to America's financial situation, are we on the right course? The answer is "yes" and "no." It just depends on your time horizon.

From a short-term perspective, it's true that our federal deficits have declined for three straight years and declining deficits are better than rising deficits. However, we are still running large deficits on an operating basis. What do I mean by an operating deficit? I mean the results of the federal government's operations excluding the Social Security surplus. After all, the federal government spends the entire Social Security surplus on various government operating expenses and replaces the cash with government bonds held in so-called government "trust funds." Given their structure, in my view, they really should be called "trust the government funds." For example, in fiscal 2007, the federal government's cash based operating deficit was about $344 billion, much higher than the widely publicized unified deficit of $163 billion. The accrual based net operating deficit was $276 billion. Of these amounts, about $120 billion related to Iraq and global war on terrorism expenditures.

Candidly, our current deficit and debt levels are not unduly troubling as a percentage of our national economy. However, these deficit levels and related debt burdens are set to escalate dramatically in the near future due to the retirement of the "baby boomers" and rising health care costs. The fact is, absent meaningful reforms, America faces escalating deficit levels and debt burdens that could swamp our ship of state!

This brings me to the longer-range picture. Believe it or not, the federal government's total liabilities and unfunded commitments for future benefits payments promised under the current Social Security and Medicare programs are now estimated at $53 trillion, in current dollar terms, up from about $20 trillion in 2000. This translates into a de facto mortgage of about $455,000 for every American household and there's no house to back this mortgage! In other words, our government has made a whole lot of promises that, in the long run, it cannot possibly keep without huge tax increases.

The Medicare program alone represents about $34 trillion of our current $53 trillion fiscal gap. If there is one thing in particular that could bankrupt America, it's runaway health care costs. And don't forget, the first "baby boomers" will begin to draw their early retirement benefits under Social Security in a couple of weeks! And, just three years later, they will be eligible for Medicare. When "baby boomers" begin to retire in big numbers, it will bring a tsunami of spending that, unlike most tsunamis, will never recede.

The prescription drug benefit alone represents about $8 trillion of Medicare's $34 trillion gap. Incredibly, this number was not disclosed or discussed until after the Congress had voted on the bill and the President had signed it into law. Generations of Americans will be paying the price—with compound interest—for this new entitlement benefit. In many ways, the 2003 Medicare prescription drug episode arguably represents government "truth" and "transparency" at its worst. Unfortunately, based on adding the prescription drug benefit and other spending and tax actions, the federal government seems to be ignoring the first rule of holes in connection with its fiscal affairs. Namely, when you're in a hole, stop digging!

If trillions of dollars aren't big enough to get your attention, believe it or not, in fiscal 2007 over 62 percent of the federal budget was on "auto-pilot" and this percentage is on the rise! Shockingly, the major functions expressly envisioned by our Founding Fathers as a proper role for the federal government—things like national defense, homeland security, foreign policy, the treasury function, the federal judiciary, the Congress and the Executive Office of the President—are in the remaining 38 percent of the federal budget! And this portion of the budget is set to get squeezed.

Unfortunately, many Americans are in denial about the seriousness of our situation. Relatively low interest rates and modest inflation rates are partly to blame for this false sense of security. The truth is, too many American families are following the poor financial practices employed by the federal government. They're spending more money than they make, taking on more debt and incurring compounding interest costs. Both America and many Americans have become addicted to debt both in good times and bad.

One important obstacle to public enlightenment is that key government financial reports are very thick. I'm a CPA and the head of GAO, and I can tell you it's a struggle to get through some of this material. I'm sorry to say the consolidated financial report of the U.S. government, which the Treasury Department released earlier today, falls into this category. As a result, it will not be read by many.

Given that fact, GAO has been working with the Treasury Department and the Office of Management and Budget (OMB) to produce the first ever Summary Annual Report for the federal government. This document will be much more concise and user friendly than the voluminous annual report Treasury issued today. Hopefully, it will be both useful and used.

The first Summary Annual Report is scheduled to be issued in mid January 2008. Keep an eye out for this report. It will be version 1.0 and enhancements can be expected in future years.

I'm hopeful policymakers, the press, and the public will spend a few minutes to read the newly formatted document. I'm confident that if they do, they'll have a much better grasp of the growing fiscal challenge facing our nation. Special recognition and thanks goes to OMB and the Treasury Department, especially Under Secretary of the Treasury Bob Steele who spearheaded the Treasury Department's efforts on this project. Thank you Bob.

The Summary Annual Report is a positive step to help us understand our fiscal challenge, but far more dramatic action is needed to help us solve the problem. GAO is working with members of Congress from both parties as well as OMB, CBO, and others to draft a proposed Transparency and Accountability Act. Among other things, the proposal GAO is working on will require greater transparency on the longer-term cost of major legislation before it is enacted into law. The President would have to include at least a 10-year projection, along with an overall statement of fiscal philosophy, in his or her annual budget submission. The draft bill would also require the U.S. government to periodically issue a comprehensive Fiscal Sustainability Report, similar to the ones now issued by New Zealand, the United Kingdom, and other countries. Notably, this years' annual report on the federal government's consolidated financial statements includes a long-term outlook by the Administration. Their analysis serves to reference that the U.S. Government is on an imprudent and unsustainable long range fiscal path.

But we must do much more. After more than nine years in my position as Comptroller General of the United States and head of the GAO, I have become increasingly frustrated by the wide-spread myopia, tunnel vision, and self-centeredness in Washington. President Reagan had it right when he said that "Washington is an island surrounded by a sea of reality." This gap needs to be closed through an increased public education and civic engagement process. Only through such efforts can the three most powerful words in the Constitution, "We the People," come alive.

Our current state of public debate is not helping in this regard. Just turn on the TV or the radio and you have your choice of shows often billed as public affairs programming. You've seen these programs, where political pundits from the left and the right express their personal views that sometimes are not supported by facts or hard evidence. Many of these shows are essentially "fact free zones." Their focus on ideological debates and partisan attacks are designed to entertain rather than inform the public. I realize that in America, everyone is entitled to his or her own opinion. However, people are not entitled to their own facts!

To counter political spin and help educate the public about the real state of America's finances, I joined with the Concord Coalition, the Brookings Institution, the Heritage Foundation and others to embark on the "Fiscal Wake-up Tour" in September 2005. So far, we've gone to over 30 cities in 25 states. We have also spoken to numerous business and civic groups, editorial boards and local media outlets in these cities. Many more stops are planned in 2008 with an emphasis on critical swing states for the Presidential election.

Media reaction to "The Wake-Up Tour" has been overwhelmingly positive. A segment on the CBS news program "60 Minutes" was shown twice earlier this year and was nominated for an Emmy. I've conducted numerous other interviews, including on such diverse programs as NPR's "Diane Rehm Show" and Comedy Central's "The Colbert Report." There have been numerous op-eds, editorials, and articles in newspapers and periodicals across the country. I appreciate the important role the press has played and continues to play in getting the truth out, but more needs to be done.

Importantly, a commercial documentary based on my four national deficits message, namely our budget, savings, balance of payments and leadership deficits, is nearing completion. The film features Bob Bixby of [the] Concord [Coalition], myself and others. The tentative title is "I.O.U.S.A.," and the documentary is set for general release next spring—in time for the 2008 general election campaign. I.O.U.S.A. was one of 16 out of over 900 films accepted for the 2008 Sundance Film Festival. Who knows, maybe the documentary will be nominated for an Oscar? I don't know about Bob Bixby, but personally, I'm not planning to quit my day job for a Hollywood career!

What have we learned on the Tour? First, the American people have little trust or confidence in the federal government's ability to address serious issues in a timely and constructive manner. Second, most Americans don't have a high opinion of the executive or the legislative branch, or of either major political party. Third, they are starved for two things, truth and leadership. It's time they got more of both.

On a more positive note, the American people are smarter than many elected officials and other individuals give them credit for. We've seen that once citizens are given the facts, most of them get it. In addition, most are willing to make some sacrifices for the future of their country and their families.

Knowledgeable Americans understand and acknowledge that they cannot run their households and businesses the way the federal government operates. They realize that tough choices will be required to put our nation on a more prudent and sustainable path. They also understand that, due to the power of compounding, it is prudent to make these choices sooner rather than later. As Albert Einstein is reported to have said, "The most powerful force on earth is not nuclear energy, it is the power of compounding!"

Very importantly, most Americans, including me, care about their children and grandchildren. We don't want to leave our descendants an indirect burden that could exceed any direct bequest we may be able to give them.

So where do we go from here? First, we need to re-impose tough budget controls, tougher than the ones that expired in 2002.

It's also urgent that we engage in comprehensive Social Security reform, as well as round one of comprehensive health care and tax reform. This may require a capable, credible, and bi-partisan commission or task force like the SAFE Commission proposed by Congressmen [Jim] Cooper [Tennessee] and [Frank R.] Wolf [Virginia], or the task force proposed by Senators [Kent] Conrad [North Dakota] and [Judd] Gregg [New Hampshire]. Such an entity could help set the table for action while providing needed political cover for elected officials to act. We can't afford to wait until a crisis is at hand. At that point, our options will be more limited and far worse. In my view, it's in everyone's best interests to establish such a commission or task force as soon as possible, because time is working against us.

There is little question that it will take committed, courageous, capable, inspired, and sustained leadership to help us see the way forward, reach consensus on meaningful reforms, and discharge our fiduciary and stewardship responsibilities. This is why I believe that the next President of the United States must make fiscal responsibility and inter-generational equity one of his or her top three priorities. If that is the case and the next President is willing to work on a constructive and bi-partisan basis to achieve needed reforms, we can turn things around. If not, I think it's only a matter of time before we do face a major crisis.

I've focused so far on our government's growing fiscal imbalance, but our nation's sustainability challenges go beyond that. For example, it's pretty clear that much of the federal government isn't well aligned with the realities of the 21st century. Too few agencies are well positioned to meet new challenges or capitalize on emerging opportunities.

The truth is, much of government today is on autopilot and reflects social conditions and spending priorities that date back to the 1950s and 1960s. The Cold War is over, our population is aging, the dollar is no longer the only major reserve currency, and globalization is affecting everything from international trade to public health.

Unfortunately, once federal programs or agencies are created, the tendency is to fund them in perpetuity. Washington rarely seems to question the wisdom of its existing commitments. Instead, it simply adds new programs and initiatives on top of the old ones. Again, President Reagan had it right when he said, "The closest thing to eternal life on this earth is a federal program." This continual layering is a key reason our government has grown so large, so expensive, so inefficient, and in some cases, so ineffective.

As Clay Johnson and I know first hand, the federal government wastes huge sums of money each year, largely through not being "results oriented," and through failing to properly target its actions. For example, every year, the U.S. government spends nearly $3 trillion, foregoes tax revenues of over $800 billion as a result of various tax preferences, and issues thousands of pages of regulations. Unfortunately, our government does all this without in many cases knowing which federal activities are making a real and meaningful outcome-based difference and which are not! On this basis alone, I'd venture that "waste" across the federal government could involve hundreds of billions of dollars each year.

From education to infrastructure, policymakers face competing demands in a range of vital areas. The question is how best to target finite resources and get the greatest value for money spent, whether it's through direct spending, government guarantees or tax preferences. In my view, it's critically important that the United States adopt a set of key national indicators to inform strategic planning; enhance government performance and accountability reporting; and facilitate a much needed and long overdue re-examination, re-prioritization, and re-engineering of the base of the federal government.

Gross domestic product, unemployment figures, violent crime statistics, infant mortality rates, math and science proficiency, and air quality indexes are all examples of commonly used indicators. A key national indicator system pulls together these various measures to tell a more complete story on how a country is doing and how we compare to others.

It matters how a nation keeps score. Indicator systems use fact-based information. And with more comprehensive and fact-based information, policymakers are more likely to ask good questions. They are also more likely to propose sound solutions and make wise decisions on spending, legislation, and oversight matters. Given these outcomes, public confidence in government should increase. After all, it can't get much lower!

One possible way to develop a set of key national indicators is through a public/private partnership. This approach is being pursued by the not-for-profit group State of the USA and other organizations. Their efforts deserve more attention and support.

The realty is in addition to our overall fiscal challenge, the U.S. faces several other key sustainability challenges. They include things like our health care system, education, energy, environmental protection, immigration, Iraq and critical infrastructure policies—just to name a few.

How can Congress and the President begin to sort out all these challenges and take steps now before the challenges of today become the crises of tomorrow? At GAO, we are trying to make sure the nation's leaders have the tools to meet this challenge. As I mentioned earlier today, GAO is issuing a new report that lays out a possible path for change. The report is entitled, "A Call for Stewardship: Enhancing the Federal Government's Ability to Address Key Fiscal and Other 21st Century Challenges."

This report provides 13 potential tools for Congress and the Administration to use to begin to confront our long-term fiscal and other challenges. This report is our latest addition to a portfolio of GAO products designed to address major 21st century challenges facing the nation.

As our latest report shows, the United States is not alone in facing a range of major challenges. In fact, representatives from GAO and about 150 other countries' Supreme Audit Institutions (SAIs) gathered in Mexico City last month to address two issues of interest around the globe.

The first issue relates to the need to take steps to ensure sustainable debt levels for nations in the future. The second issue relates to the need to develop a set of key national indicators to help improve government performance, ensure accountability, and enhance citizen engagement. Do these topics sound familiar? In the interest of full and fair disclosure, GAO did have some input on their selection!

Given these themes and the subject of my talk today, I ask that you consider the following potential future. Imagine a day in the future when government leaders are focused on their fiduciary responsibility to generate positive results for the people based on a set of key and results-based indicators. A day when government leaders provide timely and reliable performance information to their citizens in order to assess a nation's position, progress, and standing relative to other nations.

Imagine a day when government leaders understand and exercise their stewardship responsibility with regard to fiscal, environmental, and other key issues of national and global concern. A day when government leaders don't just focus on today but also take steps to create a better tomorrow.

When that day comes, we will have made a real difference—not just for our nations, and our fellow citizens, but for all mankind.

As the head of one of the leading supreme audit institutions in the world, the GAO, I can assure you that GAO will do its part to help address these and other key stewardship issues. All that I ask is that elected officials, political appointees, career civil servants, the press, and other caring citizens do their part as well. If all of us do our part, and if we start making tough choices sooner rather than later, we can keep America great, ensure that our future is better than the past, and ensure that our great nation is the first republic to stand the test of time. To me, that is a cause worth fighting for.

In closing, as my favorite modern day President, Theodore Roosevelt said, "fighting for the right (cause) is the noblest sport the world affords." Please join the fight to keep America great. Your children, your grandchildren, and generations yet unborn, will be glad that you did.

Thank you for your time and attention.

SOURCE: U.S. Congress. Government Accountability Office. "A Call for Stewardship." Speech by Comptroller General David M. Walker at the National Press Club, Washington, D.C. December 17, 2007. GAO-08-371CG. www.gao.gov/cghome/d08371cg.pdf (accessed January 30, 2008).

OTHER HISTORIC DOCUMENTS OF INTEREST

FROM THIS VOLUME

- The U.S. economic situation, p. 449
- The U.S. housing crisis, p. 514
- The international economy, p. 603

FROM PREVIOUS *HISTORIC DOCUMENTS*

- Budget issues facing the U.S. government, *2005,* p. 121
- Withholding Medicare cost estimates from Congress, *2004,* p. 577

DOCUMENT IN CONTEXT

Kenyan President Kibaki on His Reelection to a Second Term

DECEMBER 30, 2007

Apparently rigged election results in Kenya at the end of December led to a period of intense violence—lasting into the early weeks of 2008—that exposed deep tribal rivalries and severely undermined Kenya's reputation as one of the most stable and successful countries in sub-Saharan Africa. Mediation efforts were under way early in 2008, but even optimists in Kenya worried that the societal fissures opened up by the election crisis would be difficult and painful to repair.

The chief candidates in the election were two former political allies who had become bitter enemies following the previous presidential election in 2002: Mwai Kibaki, age seventy-six, who had won the presidency in 2002 by running as a reformer against the entrenched political establishment, and Raila Odinga, age sixty-two, who headed an opposition coalition and was attempting to emulate Kibaki's earlier feat. The campaign was largely peaceful, but the vote counting brought numerous complaints that both the government and the opposition had rigged results. After three days of uncertainty and mounting tension, the election commission declared Kibaki the winner late on December 30, and he promptly had himself sworn into office for a second term. Nairobi's slums, a core constituency for Odinga, immediately exploded in violent protest that quickly swelled into violence around the country, much of it based on tribal rivalries. Several hundred people were killed, and tens of thousands were displaced from their homes, in just the first week of violence.

Background

Like most sub-Saharan African countries, Kenya never had been a model of democracy, but for much of the time since independence in 1963, it enjoyed stability and relative harmony among its numerous ethnic groups. The country was ruled by two strongmen for all of its first four decades: first by Jomo Kenyatta (a leader of the largest tribal population, the Kikuyu), then after his death in 1978, by Daniel Arap Moi, a member of the smaller but influential Kalenjin tribe. Moi banned opposition parties in 1982 and ruled with an autocratic hand all through the 1980s. Widespread violence in 1991 exposed underlying tensions among the country's forty-some ethnic communities and led to strong Western pressure on Moi to open up the political system. He allowed multiparty elections in 1992, which he won by exploiting a split among opposition forces. Moi also benefited from a divided opposition in the next elections held in January 1998; Kibaki and Odinga were among the three candidates who split the opposition vote, enabling Moi to win a fifth term.

Again under strong pressure, both domestic and international, Moi stepped aside for the 2002 elections. He chose as his successor Uhuru Kenyatta, the son of the first president, pushing aside both his vice president, George Saitoti, and Odinga, who had merged his own party with Moi's ruling party in hopes of succeeding Moi. Opposition leaders, for the first time, agreed on a unity candidate: Kibaki, who had been a cabinet minister under both Kenyatta and Moi but had joined the opposition during the 1990s. He campaigned on a platform of rooting out corruption, which he said was ingrained in the nation's business and political affairs. In the December 2002 election, Kibaki won an overwhelming victory, which was widely heralded as ushering in a new era of democratic and unified governance in Kenya.

A key component of Kibaki's victory had been a preelection agreement, called a memorandum of understanding, that he had secretly negotiated with Odinga, a wealthy businessman who was the son of one of Kenya's founders, Oginga Odinga. Under the agreement, Kibaki pledged to push for constitutional amendments creating the post of prime minister, which would then be given to Raila Odinga. After he took office as president, however, Kibaki refused to carry out the understanding, thus embittering Odinga and creating the conditions for their rivalry five years later.

Kibaki had a mixed record during his five years in office. Lacking Moi's autocratic tendencies, he gave the political system more of an open feel than it had had since independence. Some aspects of the economy, notably agriculture and tourism, boomed during Kibaki's first term, but the president failed to attract the industrial development that he had promised would provide jobs for several million unemployed, or underemployed, Kenyans, primarily among the young. Critics said Kibaki's big failure was a lack of follow-through on his key campaign pledge to deal with corruption. Kibaki did appoint a tough anticorruption official, John Githongo, who uncovered large-scale fraud within the government but fled into exile in Britain in 2005 because, he said, his work was not supported by the president.

Kibaki failed to keep another 2002 campaign promise to stem violent crime that had plagued much of the country, particularly Nairobi. Crime soared in the capital all through Kibaki's term, and starting in late 2006, and continuing all through 2007, Kenya experienced a series of violent clashes resulting from disputes over land. Many of the worst were in the western region around Mount Elgon, where about 300 people had been killed by mid-December. Human rights groups and other nongovernmental organizations expressed fear, as early as April, that these clashes would be intensified by the elections—a concern that turned out to be well-founded.

The 2007 Election Campaign

Odinga campaigned in 2007 as a modernizer, arguing that Kenya's rulers had run the country for the benefit of its political and business elites (which included his family) but had not invested in infrastructure necessary for the industrialization he said would pull the mass of Kenyans out of poverty. Odinga cited in particular the country's failure to expand or modernize the railway system built by Britain before independence in 1963. Odinga also said he had a solution for the regional tensions that had emerged periodically in Kenya, largely due to disputes over land among the major tribal factions: "devolution" of some government powers from the central government to the eight provinces. He based this idea on Britain's transfer in the late 1990s of some authority from the London government to regional assemblies in Scotland and Wales. Odinga called his plan *majimbo,* which meant devolution in Swahili, but the term also was controversial in Kenya because of its use in previous battles over regional control of land and natural resources.

As had been the case in the previous three elections since Moi allowed opposition parties in 1992, it was widely assumed that the building of coalitions among the regions and ethnic groups would be the deciding factor in 2007. Opinion surveys and election results showed that more Kenyans voted on the basis of ethnic loyalties than any other factor. The leading candidates were particularly interested in winning over two constituencies with the voting power to swing a close election: the populous Luhya tribe that was dominant in Western province and the Muslim electorate in Coast and North Eastern provinces. Kibaki and Odinga each chose a prominent Luhya politician as a vice presidential running mate, and Odinga had success in courting Muslim voters who believed the government had slighted them. Odinga also explicitly targeted the large population of unemployed young people in urban areas with a pledge that he would create the jobs Kibaki had promised but failed to deliver.

Odinga, a member of the Luo tribe, built a broadly based coalition, called the Orange Democratic Movement (ODM), with prominent leaders from all of the provinces and ethnic groups except for the Kikuyu, the largest single tribal group. Kibaki's Party of National Unity (PNU) had its base in the Kikuyu and other ethnic communities in Central and Eastern provinces. Kibaki also had support from his predecessor, Moi, and leaders of the historically powerful Kalenjin community.

Although the presidential election attracted most of the attention, all 210 seats in Kenya's parliament were at stake in the election. More than 2,600 candidates, representing 108 political parties, contested those 210 seats.

Opinion polls early in the campaign showed Odinga with a wide lead over Kibaki and a third candidate, Kalonzo Musyoka, a former foreign minister who had split from Odinga's party in August and was widely believed to be running to gain name recognition for a later try at the presidency. The race tightened significantly as the campaign proceeded, however, and the last polls before the voting showed that Odinga held a narrow 2 percentage-point lead over Kibaki. Odinga's party accused Kibaki of using government resources—notably state-owned television stations—to skew the campaign in his favor. Even so, opposition politicians said they believed Odinga's broad popular appeal would overcome any bias in the system.

Kibaki Declared the Winner

Election day, December 27, saw a large turnout, with more than 14 million Kenyans voting, many of them after standing in line for hours. With just a few exceptions, voting was peaceful and the process of counting the votes locally appeared to go smoothly, even though about one-third of polling stations opened later and closed later than the specified times. Several outside groups, including the European Union, stationed monitors at polling places around the country and at the national election commission office in Nairobi.

The national tabulation of votes in Kenya typically took a day or two and started with the release of results in the presidential contest followed by parliamentary results. By December 29, however, there was no sign of national results in the presidential election, but results were posted for some parliamentary seats. The election commission chairman, Samuel Kivuitu, said he had lost contact with some commission officials, and he questioned whether the results were being "cooked" in some areas. Results were turned in from dozens of polling places without the required paperwork or the signatures of election officials and agents for the competing parties.

Results from the parliamentary elections showed Odinga's coalition with a strong lead, suggesting Odinga himself had won a landslide victory. His ODM coalition won 99 seats, 7 short of an absolute majority, whereas Kibaki's PNU coalition captured only 43 seats. Just as damaging for the government, more than half of Kibaki's cabinet ministers, including his vice president, were defeated in their constituencies.

Kivuitu finally released presidential results on the afternoon of December 30, even though he later expressed little confidence in them. The official results showed Kibaki with 4.58 million votes (or just under 46 percent of the total), and Odinga with 4.35 million votes (or 43.6 percent of the total). The third candidate, Musyoka, had about 880,000 votes, or nearly 9 percent of the total.

Shortly after Kivuitu's declaration, Kibaki had himself sworn into office for a second term. In contrast to his previous inauguration—a boisterous outdoor event attended by tens of thousands of cheering supporters—this ceremony reportedly was a perfunctory one at the presidential office, which is called the State House; it was observed by only a few dozen political supporters and government officials. In a nationally televised address, Kibaki insisted the election had been "free and fair" and appealed to his citizens to "set aside the passions that were excited by the election process." He also called specifically on Odinga and Musyoka to "set aside their differences" and "work together to build consensus on issues of national importance including a new constitution, and any matter that benefits and improves the welfare of our people."

Odinga immediately rejected the election results and said he would refuse to recognize Kibaki as the legitimate president. "There is a clique of people around Kibaki trying to rob Kenyans of the election," he said on December 30. As Odinga spoke at a news conference, the government shut down all live media broadcasts.

Over the next several days, independent election monitors published reports detailing serious problems with the counting of votes, both at the local level and by the national election commission. The most detailed report, by the European Union's observer mission, was published on January 1, 2008. It said the presidential elections "have fallen short of key international and regional standards for democratic elections. Most significantly, they were marred by a lack of transparency in the processing and tallying of presidential results, which raises concerns about the accuracy of the final result of this election." The parliamentary results, by contrast, appeared "to command greater confidence by election stakeholders," the EU mission said.

An Outbreak of Violence

Several towns in western Kenya and the slums of Nairobi experienced unrest on December 29, when it became clear that election results were being delayed. The unrest exploded into widespread violence on December 30 within minutes of the announcement of official results. The worst violence at this stage took place in the large Kibera slum in Nairobi, which had been a hotbed of support for Odinga. According to all reports, gangs of Luo youth armed with machetes and other weapons attacked Kikuyu neighborhoods; they set houses and shops on fire, killed men and boys who got in their way, and raped dozens of women and girls.

The violence spread quickly to other areas, both in Nairobi and in outlying towns, with members of the Luo and other tribes attacking Kikuyu, and vice versa. Some of the most savage violence took place in the Rift Valley of western Kenya, where members of the Luo and Kikuyu tribes long had disputed ownership of prime agricultural land. One incident in the Rift Valley grabbed the world's attention above all others: In the town of Eldoret, several hundred people, most of them said to be Kikuyus, attempted to seek refuge from the violence in an Assemblies of God church. According to witnesses, a band of youths overpowered guards at the church early on the afternoon of December 31 and set the building on fire. Terrified people, some of them with their clothes in flames, fled the building. At least thirty people, most of them children, were not so lucky; their burned bodies were later recovered from the ruins of the church.

Odinga and Kibaki were both slow to call for an end to the violence, which raged well into January 2008. By most accounts, the death toll had soared past 150 by year's end and climbed rapidly in subsequent weeks. Kenyan civil society groups said tens of thousands of people were forced from their homes. The African Union launched an effort to broker a compromise between Kibaki and Odinga, but most observers said it would take more skilled leadership than Kenya had seen so far, and the passage of time, to heal the wounds laid bare by Kenya's postelection violence.

Following is the text of a speech delivered by Mwai Kibaki on December 30, 2007, shortly after he had been sworn in for a second term as president of Kenya, following disputed elections.

DOCUMENT

President Kibaki on His Acceptance of a Second Term as President

December 30, 2007

Fellow Kenyans,

Following the announcement of the presidential election results by the Electoral Commission of Kenya, I stand before you humbled and grateful for the opportunity you have given me to serve you again as your President for a second five-year term.

I thank all Kenyans who voted in large numbers in these elections. I thank all of you for the trust you have bestowed upon me in renewing my mandate, which I accept with sincere gratitude and humility. I am confident that together, we will succeed in changing our country into a better home for all Kenyans.

The elections were very closely contested. I thank those of our brothers and sisters who voted for me and other presidential candidates for expressing their democratic right and choice. As a democrat, I acknowledge and respect the right of every Kenyan to choose candidates of their choice.

With the elections behind us now, I assure them that as President of Kenya, I will serve everyone equally, irrespective of the person they may have voted for.

I urge all of us to set aside the passions that were excited by the election process, and work together as one people with the single purpose of building a strong, united, prosperous and equitable country.

Fellow Kenyans,

I want to commend my opponents, specifically Honorable Kalonzo Musyoka and Honorable Raila Odinga. They campaigned strongly, and garnered support from across the country. I call on all the political leaders to set aside their differences. Let us all work together to build consensus on issues of national importance including a new Constitution, and any matter that benefits and improves the welfare of our people.

Fellow Kenyans,

We have done our nation proud, and set a good example for the rest of the continent, through the conduct of free and fair elections.

The freedom of choice, the openness and integrity of the electoral process, and the peaceful manner in which we conducted ourselves as people has raised Kenya's democratic profile throughout the world.

I am particularly pleased that millions of Kenyans in the largest voter turnout seen in our country chose to exercise their democratic right to elect a new team of political leaders. Most notable is the fact that so many young people have participated in the democratic process for the first time, either as candidates or as voters. These are good developments that testify to the strength of our nation's democratic culture.

Indeed, I am proud to note that in holding free and fair elections in our country both in the just concluded general elections, and in the referendum two years ago, we have demonstrated to the world that we are politically mature and capable of nurturing and upholding democracy.

I thank the Electoral Commission of Kenya, our security agents, observers and all other stakeholders, for remaining committed to the conduct of honest, orderly and credible elections that have enabled the true verdict of the people to prevail. I call upon all candidates and Kenyans in general to accept the verdict of the people.

With the general election now behind us, it is now the time for healing and reconciliation among all Kenyans. We need to heal the differences that have been created among us, between different communities, regions, and religions.

I urge all of us to set aside the divisive views and opinions we held during the campaign period, and instead embrace one another as brothers and sisters. After all, we belong to one family called Kenya.

On my part as your President, I will personally lead our country in promoting unity, tolerance, peace and harmony among all Kenyans. I appeal to all political and religious leaders to do the same.

As I pledged to Kenyans during the campaigns, I will shortly form a clean hands Government that represents the face of Kenya. The new PNU [Party of National Unity, Kibaki's political party] Government will incorporate the affiliate parties as well as other friendly parties. I also pledge to ensure that our young people and women are fully represented in all public appointments.

I will remain committed to the development of all parts of our country and to ensuring justice and equal treatment of all Kenyans. My Government will be committed to the task of creating and sustaining a prosperous, secure, and equitable future for all Kenyans.

In conclusion, fellow Kenyans, you have given us a vote of confidence in the values and principles of freedom, equality, and development that we began five years ago.

You have chosen the leaders you wish to serve you during the next five years. You have given us the agenda for change you wish to see implemented for the next five years. I humbly respect your choices and your agenda.

In return, I ask all of us, and particularly all leaders to embrace a renewed spirit of national unity, respect the peoples' choice, and maintain peace, law and order. Let us choose to live together in the true democratic spirit of tolerance and mutual respect.

Let us all endeavor to build a society that is in harmony and at peace with itself.

In the last few days, many Kenyans have had a hectic time in preparation for this event. To give all of us deserved rest and to prepare for the New Year celebrations, I declare tomorrow, December 31, 2007, a Public Holiday.

Thank You. God Bless You All. God Bless Kenya.

SOURCE: Republic of Kenya. State House. "Acceptance Speech by His Excellency Hon. Mwai Kibaki, C.G.H., M.P., President and Commander-in-Chief of the Armed Forces of the Republic of Kenya Following His Re-Election to Serve a Second Term." Nairobi, Kenya. December 30, 2007. www.statehousekenya.go.ke/speeches/kibaki/dec07/2007301201.htm (accessed March 30, 2008).

OTHER HISTORIC DOCUMENTS OF INTEREST

FROM THIS VOLUME

- Catholic bishops on human rights abuses in Zimbabwe, p. 149
- UN reports on the Democratic Republic of the Congo, p. 374
- UN secretary-general on the postwar situation in Liberia, p. 435
- World Bank on its agriculture programs in Africa, p. 592

FROM PREVIOUS *HISTORIC DOCUMENTS*

- Kibaki on his inauguration as president of Kenya, *2002,* p. 1044

Credits

Credits are a continuation from page iv.

“Speeches Delivered by First Minister and Deputy First Minister Today in Parliament Buildings, Belfast” and “Statement at Downing Street” Crown copyright © 2007. Reproduced under the terms of the Click-Use License.

“Turkey 2007 Progress Report” and “The Treaty at a Glance” copyright © 2007 European Communities.

“UNAMID Deployment on the Brink: The Road to Security in Darfur Blocked by Government Obstructions” copyright © 2007 Human Rights Watch.

“World Bank Assistance to Agriculture in Sub-Saharan Africa” copyright © 2007 The World Bank Group.

“World Economic Outlook: Globalization and Inequality” copyright © 2007 International Monetary Fund.

Cumulative Index, 2003–2007

*The years in **boldface** type in the entries indicate which volume is being cited.*

D

G

J

O

P

T

U

V

X

Y